area A **width** w **base** b

perimeter P **surface area** S **circumference** C

length l **altitude (height)** h **radius** r slant height s

Rectangle

$$A = lw \qquad P = 2l + 2w$$

Triangle

$$A = \frac{1}{2}bh$$

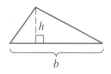

Square

$$A = s^2 \qquad P = 4s$$

Parallelogram

$$A = bh$$

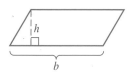

Trapezoid

$$A = \frac{1}{2}h(b_1 + b_2)$$

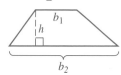

Circle

$$A = \pi r^2 \qquad C = 2\pi r$$

30°–60° Right Triangle

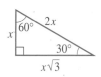

Right Triangle

$$a^2 + b^2 = c^2$$

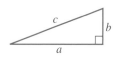

Isosceles Right Triangle

Right Circular Cylinder

$$V = \pi r^2 h \qquad S = 2\pi r^2 + 2\pi rh$$

Sphere

$$S = 4\pi r^2 \qquad V = \frac{4}{3}\pi r^3$$

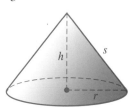

Right Circular Cone

$$V = \frac{1}{3}\pi r^2 h \qquad S = \pi r^2 + \pi rs$$

Pyramid

$$V = \frac{1}{3}Bh$$

Prism

$$V = Bh$$

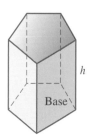

College Algebra

SEVENTH EDITION

College Algebra

Jerome E. Kaufmann

Karen L. Schwitters

Seminole Community College

THOMSON

BROOKS/COLE

Australia • Brazil • Canada • Mexico • Singapore • Spain
United Kingdom • United States

THOMSON

BROOKS/COLE

College Algebra, Seventh Edition

Jerome E. Kaufmann, Karen L. Schwitters

Mathematics Editor: Gary Whalen
Development Editor: Kristin Marrs
Assistant Editor: Natasha Coats
Editorial Assistant: Rebecca Dashiell
Technology Project Manager: Lynh Pham
Marketing Manager: Joe Rogove
Marketing Assistant: Ashley Pickering
Marketing Communications Manager: Darlene
Amidon-Brent
Project Manager, Editorial Production: Hal Humphrey
Art Director: Vernon Boes
Print Buyer: Rebecca Cross

Permissions Editor: Timothy Sisler
Production Service: Susan Graham
Text Designer: John Edeen
Photo Researcher: Sarah Evertson
Copy Editor: Susan Graham
Illustrator: Network Graphics and G&S Typesetters
Cover Designer: Lisa Henry
Cover Image: Doug Smock/Getty Images
Cover Printer: Transcontinental Printing/Interglobe
Compositor: Newgen@Austin
Printer: Transcontinental Printing/Interglobe

Printed in Canada

1 2 3 4 5 6 7 11 10 09 08

Library of Congress Control Number: 2007939902

Student Edition: ISBN-13: 978-0-495-55403-5
ISBN-10: 0-495-55403-0

Thomson Higher Education
10 Davis Drive
Belmont, CA 94002-3098
USA

For more information about our products, contact us at:
Thomson Learning Academic Resource Center
1-800-423-0563

For permission to use material from this text or product, submit a request online at **http://www.thomsonrights.com.** Any additional questions about permissions can be submitted by e-mail to **thomsonrights@thomson.com.**

Contents

Chapter 9   Sequences and Mathematical Induction   642

Preface

College Algebra, Seventh Edition was written for students who need a college algebra course that serves as a prerequisite for the calculus sequence, for the finite math/calculus sequence, or that satisfies a liberal arts requirement. Sample outlines for these three types of courses are included at the end of this preface.

Four major ideas unify this text: solving equations and inequalities, solving problems, developing graphing techniques, and developing and using the concept of a function.

College Algebra, Seventh Edition presents basic concepts of algebra in a simple, straightforward way. Examples motivate students and reinforce algebraic concepts by the application to real-world situations that students can identify with. These examples also guide students to organize their work in a logical fashion and to use meaningful shortcuts whenever appropriate.

In the preparation of this seventh edition, we made a special effort to incorporate improvements suggested by reviewers and users of the earlier editions without sacrificing the book's many successful features.

■ New in This Edition

New features in this edition focus on aiding the student and the instructor.

- *Objectives*. Each section now has a list of objectives so that instructors can tell at a glance what material is covered in that section. Students can use the objectives as a quick check on the type of skills they should have acquired after studying the section.
- *Concept Quiz*. Problems Sets are now preceded by a Concept Quiz. The concept quizzes are predominantly true/false questions that help measure the student's understanding of the material. Answers to the Concept Quiz are located at the end of the Problem Set for quick reference. Concept quizzes can be assigned or can be used as classroom discussion.
- *Example Link*. At the end of almost all examples there is now a "Now Work This Problem" statement that references a parallel type of problem in the following Problem Set. This can aid the instructor by having a readily available problem to present to the class, which is similar to the example. The referenced problem can aid the student by working a problem that has an example readily available as a guide.
- *Enhanced WebAssign Problems*. In the annotated instructor's edition, problems that are available in electronic form in Enhanced WebAssign are identified by

a highlighted problem number. If the instructor wants to create an assessment — whether it is a quiz, homework assignment, or a test — he or she can select the problems by problem number from the identified problems in the text.

Based on suggestions offered by reviewers of the previous edition, a few changes have been made to the text.

- Section 0.1 was reorganized to have real number topics together that then flow into a review of the Cartesian coordinate system.
- A *Further Investigations* problem set was added to Section 0.6 to introduce some problems on simplifying radicals with variables (to disavow the assumption that all variables represent positive numbers).
- Recognizing that graphing piece-wise functions is a difficult skill for students, more problems on graphing piece-wise functions were added to Problem Set 3.3.
- Throughout the text, and particularly in Section 5.2, interest rates and prices have been updated.
- Chapter 10 of the previous edition was removed from the text because most of the topics in Chapter 10 are typically covered in a liberal arts math course and not in College Algebra. The material on the binomial theorem was retained and placed in Appendix A.

■ Other Special Features

- Photos and applications are used in the chapter openers to introduce some concepts presented in the chapter.
- A **Chapter Test** appears at the end of each chapter. Along with the Chapter Review Problem Sets, these practice tests should provide the students with ample opportunity to prepare for the "real" examinations.
- **Cumulative Review Problem Sets** appear at the ends of Chapters 2, 3, 4, 5, 6, and 8. *All* answers for Chapter Review Problem Sets, Chapter Tests, and Cumulative Review Problem Sets are included in the back of the text.
- Problems called **Thoughts into Words** are included in every problem set except the review exercises. These problems are designed to encourage students to express, in written form, their thoughts about various mathematical ideas. For example, see Problem Sets 0.5, 1.2, 1.3, and 4.6.
- Many problem sets contain a special group of problems called **Further Investigations,** which lend themselves to small-group work. These problems encompass a variety of ideas: some are proofs, some exhibit different approaches to topics covered in the text, some bring in supplementary topics and relationships, and some are more challenging problems. Although these problems add variety and flexibility to the problem sets, they can also be omitted entirely without disrupting the continuity of the text. For examples, see Problem Sets 0.6, 1.1, 1.2, 2.1, and 2.3.
- As recommended in the standards produced by NCTM and AMATYC, **problem solving** is an integral part of this text. With problem solving as its focus, Chapter 1 pulls together and expands on a variety of approaches to solving equations and inequalities. Polya's four-phase plans is used as a basis for developing various problem solving strategies. Applications of radical equations are a part of

Section 1.5, and applications of slope are in Section 2.3. Functions are introduced in Chapter 3 and are immediately used to solve problems. Exponential and logarithmic functions become problem solving tools in Chapter 5. Systems of equations provide more problem solving power in Chapter 6. Problem solving is the unifying theme of Chapter 9.

Problems have been chosen so that a variety of problem-solving strategies can be introduced. Sometimes alternate solutions are shown for the same problem (see Problem 3 of Section 1.4), while at other times different problems of the same type are used to illustrate different approaches (see Problems 6, 7, and 8 of Section 1.4). *No attempt is made to dictate a specific problem technique; instead our goal is to introduce the students to a large variety of techniques.*

- Chapter 0, a review of intermediate algebra concepts, was written so that students can work through this material with a minimum of assistance from the instructor. The concepts are reviewed in enough detail to provide a good basis for understanding the many worked-out examples. The problem sets also provide ample practice material. The chapter test and/or the chapter review problem set could be used to determine problem areas.
- Chapter 9, if not covered as part of the regular course, could be used as enrichment material. The easy-to-read explanations along with carefully chosen examples make this chapter accessible to students with minimal help from the instructor.
- The Cartesian coordinate system and the use of graphing utilities are briefly introduced in Section 0.1. This allows instructors (and the authors) *to use the graphing calculator as a teaching tool early in the text.* For example, visual support can be given for the manipulation of algebraic expressions. Students do not need a graphing calculator to benefit from the graphs.
- A graphical analysis of approximating solution sets is introduced in Chapter 1. Then a graphical approach is used to both lend visual support to an algebraic approach and sometimes to *predict approximate solutions before an algebraic approach is shown.*
- Beginning with Problem Set 0.1, a group of problems called **Graphing Calculator Activities** is included in many of the problem sets. These activities, which are good for either individual or small-group work, have been designed to reinforce concepts (see, for example, Problem Set 5.5) as well as to lay groundwork for concepts about to be discussed (see, for example, Problem Set 2.2). Some of these activities ask students to predict shapes and locations of graphs based on previous graphing experiences, and then to use a graphing utility to check their predictions (see, for example, Problem Set 3.5). The graphing calculator is also used as a problem solving tool (see, for example, Problem Set 4.5); when students do these activities, they should become familiar with the capabilities and limitations of a graphing utility.
- Specific graphing ideas (intercepts, symmetry, restrictions, asymptotes and transformations) are introduced and used throughout Chapters 2, 3, 4, 5, and 8. In Section 3.5 the extensive work with graphing parabolas from Section 3.3. is used to motivate definitions for translations, reflections, stretchings, and shrinkings.

These transformations are then applied to the graphs of $f(x) = x^3$, $f(x) = x^4$, $f(x) = \sqrt{x}$, and $f(x) = |x|$. Furthermore, in later chapters the transformations are applied to graphs of exponential, logarithmic, polynomial, and rational functions.

- Reviewers suggested several places where a page, a paragraph, a sentence, an example, or a solution to a problem could be rewritten to further clarify the intended meaning. Sometimes we included a *Remark* to add a little flavor to the discussion.

- Please note the exceptionally pleasing design features of this text, including the functional use of color. The open format makes for a continuous and easy flow of material instead of working through a maze of flags, caution symbols, reminder symbols,etc.

- The following are sample outlines for a college algebra course that (a) serves a as a prerequisite for the standard calculus sequences, (b) serves as a prerequisite for a finite mathematics or business calculus course, or (c) satisfies a liberal arts requirement.

Precalculus	Business	Liberal Arts
0.1–0.8	0.1–0.7	0.1–0.6
Some Basic Concepts of Algebra	Some Basic Concepts of Algebra	Some Basic Concepts of Algebra
1.1–1.7	1.1–1.7	1.1–1.4 and 1.6
Equations, Inequalities, and Problem Solving	Equations, Inequalities, and Problem Solving	Equations, Inequalities, and Problem Solving
2.1–2.5	2.1–2.4	2.1–2.4
Coordinate Geometry and Graphing Techniques	Coordinate Geometry and Graphing Techniques	Coordinate Geometry and Graphing Techniques
3.1–3.7	3.1–3.7	3.1–3.7
Functions	Functions	Functions
4.1–4.7	4.1–4.5	5.1–5.5
Polynomial and Rational Functions	Polynomial and Rational Functions	Exponential and Logarithmic Functions
5.1–5.5	5.1–5.5	6.1–6.3
Exponential and Logarithmic Functions	Exponential and Logarithmic Functions	Systems of Equations
8.1–8.3	6.1–6.5	7.1–7.2
Conic Sections	Systems of Equations	Matrices
9.1–9.3	7.4	9.1–9.3
Sequences	Linear Programming	Sequences
Appendix A:	9.1–9.3	Appendix A:
Binomial Theorem	Sequences	Binomial Theorem

■ Ancillaries

For the Instructor

Annotated Instructor's Edition. Answers are printed next to all respective exercises. Graphs, tables, and other answers appear in a special answer section in the back of the text.

Test Bank. The Test Bank includes multiple tests per chapter as well as final exams. The tests are made up of a combination of multiple-choice, free-response, true/false, and fill in-the-blank questions.

ExamView®. Create, deliver, and customize tests (both print and online) in minutes with this easy-to-use assessment system.

Complete Solutions Manual. The Complete Solutions Manual provides worked-out solutions to all of the problems in the text.

Enhanced WebAssign. WebAssign, the most widely used homework system in higher education, allows you to assign, collect, grade, and record homework assignments via the web. Through a partnership between WebAssign and Thomson Brooks/Cole, this proven homework system has been enhanced to include links to textbook sections, video examples, and problem-specific tutorials.

JoinIn™ Student Response System Featuring TurningPoint®. Thomson Brooks/Cole is pleased to offer you book-specific JoinIn™ content for student response systems tailored to College Algebra, Seventh Edition. You can transform your classroom and assess your students' progress with instant in-class quizzes and polls. JoinIn lets you pose book-specific questions and display students' answers seamlessly within the Microsoft® PowerPoint® slides of your own lecture, in conjunction with the "clicker" hardware of your choice. Enhance how your students interact with you, your lecture, and each other. Contact your local Thomson Brooks/Cole representative to learn more.

Text-Specific DVDs. These text-specific DVDs, available at no charge to qualified adopters of the text, feature 10- to 20-minute problem-solving lessons that cover each section of every chapter.

Website. www.thomsonedu.com/mathematics When you adopt a Thomson Brooks/Cole mathematics text, you and your students will have access to a variety of teaching and learning resources. This website features everything from book-specific resources to newsgroups. It's a great way to make teaching and learning an interactive and intriguing experience.

Enhanced WebAssign problems, In the annotated instructor's edition, problems that are available in electronic form in Enhanced WebAssign are highlighted. If the instructor wants to create an assessment — whether it is a quiz, homework assignment, or a test — he or she can easily select the problems by problem number from the identified problems in the text.

For the Student

Student Solutions Manual. The Student Solutions Manual provides worked-out solutions to the odd-numbered problems in the text.

Website. www.thomsonedu.com/mathematics When you adopt a Thomson Brooks/Cole mathematics text, you and your students will have access to everything from book-specific resources to newsgroups. It's a great way to make teaching and learning an interactive and intriguing experience.

■ Acknowledgments

We would like to take this opportunity to thank the following people who served as reviewers for the seventh edition of *College Algebra*:

Kevin Bolan
Everet Community College

Ramendra Bose
University of Texas–Pan American

John Drake
Cochise College

Rahim Faradineh
East Los Angeles College

Glenn Hunt
Riverside City College

Nam Nguyen
University of Texas – Pan American

Mari Peddycoart
Kingwood College

Ken Reeves
San Antonio College

Lynn Salyer
McCook Community College

Ron Sperber
Keuka College

Alain Togbe
Purdue University

Lynn White
Jones County Junior College

Loris Zucca
Kingwood College

We would like to express our sincere gratitude to the staff of Brook/Cole, especially Gary Whalen, Kristin Marrs, Natasha Coats, and Lynh Pham, for their continuous cooperation and assistance throughout this project; and to Susan Graham and Hal Humphrey, who carry out the many details of production. And finally, very special thanks are due to Arlene Kaufmann, who spends numerous hours reading page proofs.

Jerome E. Kaufmann
Karen L. Schwitters

Some Basic Concepts of Algebra: A Review

Rodin's sculpture, "The Thinker," is an icon of intellectual thought. Your study of algebra should expand your thinking to a deeper level.

© Index Stock Imagery/Jupiter Images

The temperature in Big Lake, Alaska at 3 P.M. was $-4°$F. By 11 P.M. the temperature had dropped another $20°$. We can use the *numerical expression* $-4 - 20$ to determine the temperature at 11 P.M.

Megan has p pennies, n nickels, d dimes, and q quarters. The *algebraic expression* $p + 5n + 10d + 25q$ can be used to represent the total amount of money in cents.

Algebra is often described as a generalized arithmetic. That description does not tell the whole story, but it does convey an important idea: A good understanding of arithmetic provides a sound basis for the study of algebra. In this chapter we will often use arithmetic examples to lead into a review of basic algebraic concepts. Then we will use the algebraic concepts in a wide variety of problem-solving situations. Your study of algebra should make you a better problem solver. Be sure that you can work effectively with the algebraic concepts reviewed in this first chapter.

0.1 Some Basic Ideas

Objectives

- ■ Recognize the vocabulary and symbolism associated with sets.
- ■ Know the various subset classifications of the real number system.
- ■ Determine the absolute value of a number.
- ■ Find distance on a number line.
- ■ Know the real number properties.
- ■ Evaluate algebraic expressions.
- ■ Review the Cartesian coordinate system.

Let's begin by pulling together the basic tools we need for the study of algebra. In arithmetic, symbols such as 6, $\frac{2}{3}$, 0.27, and π are used to represent numbers. The operations of addition, subtraction, multiplication, and division are commonly indicated by the symbols $+$, $-$, $\times$, and $\div$, respectively. These symbols enable us to form specific **numerical expressions.** For example, the indicated sum of 6 and 8 can be written $6 + 8$.

In algebra, we use variables to generalize arithmetic ideas. For example, by using x and y to represent *any* two numbers, we can use the expression $x + y$ to represent the indicated sum of *any* two numbers. The x and y in such an expression are called **variables,** and the phrase $x + y$ is called an **algebraic expression.**

Many of the notational agreements we make in arithmetic can be extended to algebra, with a few modifications. The following chart summarizes those notational agreements regarding the four basic operations.

Operation	Arithmetic	Algebra	Vocabulary
Addition	$4 + 6$	$x + y$	The sum of x and y
Subtraction	$14 - 10$	$a - b$	The difference of a and b
Multiplication	7×5 or $7 \cdot 5$	$a \cdot b$, $a(b)$, $(a)b$, $(a)(b)$, or ab	The product of a and b
Division	$8 \div 4$, $\frac{8}{4}$, $8/4$ or $4\overline{)8}$	$x \div y$, $\frac{x}{y}$, x/y, or $y\overline{)x}$ $(y \neq 0)$	The quotient of x divided by y

Note the different ways of indicating a product, including the use of parentheses. The ab form is the simplest and probably the most widely used form. Expressions such as abc, $6xy$, and $14xyz$ all indicate multiplication. Notice the various forms used to indicate division. In algebra, the fraction forms $\frac{x}{y}$ and x/y are generally used, although the other forms do serve a purpose at times.

■ The Use of Sets

Some of the vocabulary and symbolism associated with the concept of sets can be effectively used in the study of algebra. A **set** is a collection of objects; the objects are called **elements** or **members of the set.** The use of capital letters to name sets and the use of set braces, { }, to enclose the elements or a description of the elements provide a convenient way to communicate about sets. For example, a set A that consists of the vowels of the English alphabet can be represented as follows:

$$A = \{\text{vowels of the English alphabet}\} \qquad \text{Word description}$$

or $\qquad A = \{a, e, i, o, u\}$ List or roster description

or $\qquad A = \{x \mid x \text{ is a vowel}\}$ Set-builder notation

A set consisting of no elements is called the **null set** or **empty set** and is written $\varnothing$.

Set-builder notation combines the use of braces and the concept of a variable. For example, $\{x \mid x \text{ is a vowel}\}$ is read "the set of all x such that x is a vowel." Note that the vertical line is read "such that."

Two sets are said to be **equal** if they contain exactly the same elements. For example, $\{1, 2, 3\} = \{2, 1, 3\}$ because both sets contain exactly the same elements; the order in which the elements are listed does not matter. A slash mark through an equality symbol denotes *not equal to.* Thus if $A = \{1, 2, 3\}$ and $B = \{3, 6\}$, we can write $A \neq B$, which is read "set A is not equal to set B."

■ Real Numbers

The following terminology is commonly used to classify different types of numbers:

$\{1, 2, 3, 4, \ldots\}$ Natural numbers, counting numbers, positive integers

$\{0, 1, 2, 3, \ldots\}$ Whole numbers, nonnegative integers

$\{\ldots, -3, -2, -1\}$ Negative integers

$\{\ldots, -3, -2, -1, 0\}$ Nonpositive integers

$\{\ldots, -2, -1, 0, 1, 2, \ldots\}$ Integers

A **rational number** is defined as any number that can be expressed in the form a/b, where a and b are integers and b is not zero. The following are examples of rational numbers:

$$\frac{2}{3} \qquad\qquad -\frac{3}{4} \qquad\qquad \frac{-1}{7} \qquad\qquad \frac{9}{2}$$

$$6\frac{1}{2} \quad \text{because } 6\frac{1}{2} = \frac{13}{2} \qquad\qquad -4 \quad \text{because } -4 = \frac{-4}{1} = \frac{4}{-1}$$

$$0 \quad \text{because } 0 = \frac{0}{1} = \frac{0}{2} = \frac{0}{3}, \text{etc.} \qquad 0.3 \quad \text{because } 0.3 = \frac{3}{10}$$

A rational number can also be defined in terms of a decimal representation. Before doing so, let's briefly review the different possibilities for decimal representations. Decimals can be classified as **terminating, repeating,** or **non-repeating.** Here are some examples of each:

$$\begin{bmatrix} 0.3 \\ 0.46 \\ 0.789 \\ 0.2143 \end{bmatrix} \qquad \text{Terminating decimals}$$

$$\begin{bmatrix} 0.333\ldots \\ 0.1414\ldots \\ 0.7127127\ldots \\ 0.241717\ldots \end{bmatrix} \qquad \text{Repeating decimals}$$

$$\begin{bmatrix} 0.472195631\ldots \\ 0.21411711191111\ldots \\ 3.141592654\ldots \\ 1.414213562\ldots \end{bmatrix} \qquad \text{Nonrepeating decimals}$$

A **repeating decimal** has a block of digits that repeats indefinitely. This repeating block of digits may be of any size and may or may not begin immediately after the decimal point. A small horizontal bar is commonly used to indicate the repeating block. Thus $0.3333\ldots$ can be expressed as $0.\overline{3}$ and $0.24171717\ldots$ as $0.24\overline{17}$.

In terms of decimals, a rational number is defined as a number with either a terminating or a repeating decimal representation. The following examples illustrate some rational numbers written in $\dfrac{a}{b}$ form and in the equivalent decimal form:

$$\frac{3}{4} = 0.75 \qquad \frac{3}{11} = 0.\overline{27} \qquad \frac{1}{8} = 0.125 \qquad \frac{1}{7} = 0.\overline{142857} \qquad \frac{1}{3} = 0.\overline{3}$$

We define an **irrational number** as a number that cannot be expressed in $\dfrac{a}{b}$ form, where a and b are integers and b is not zero. Furthermore, an irrational number has a nonrepeating, nonterminating decimal representation. Following are some examples of irrational numbers and a partial decimal representation for each number. Note that the decimals do not terminate and do not repeat.

$$\sqrt{2} = 1.414213562373095\ldots$$

$$\sqrt{3} = 1.73205080756887\ldots$$

$$\pi = 3.14159265358979\ldots$$

The entire set of **real numbers** is composed of the rational numbers along with the irrationals. The following tree diagram can be used to summarize the various classifications of the real number system.

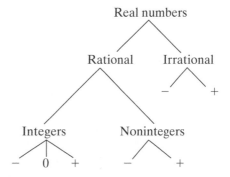

Any real number can be traced down through the tree. Here are some examples:

7 is real, rational, an integer, and positive.

$-\dfrac{2}{3}$ is real, rational, a noninteger, and negative.

$\sqrt{7}$ is real, irrational, and positive.

0.59 is real, rational, a noninteger, and positive.

The concept of a subset is convenient to use at this time. A set A is a **subset** of another set B if and only if every element of A is also an element of B. For example, if $A = \{1, 2\}$ and $B = \{1, 2, 3\}$, then A is a subset of B. This is written $A \subseteq B$ and is read "A is a subset of B." The slash mark can also be used here to denote negation. If $A = \{1, 2, 4, 6\}$ and $B = \{2, 3, 7\}$, we can say A *is not a subset of B* by writing $A \not\subseteq B$. The following statements use the subset vocabulary and symbolism; they are represented in Figure 0.1.

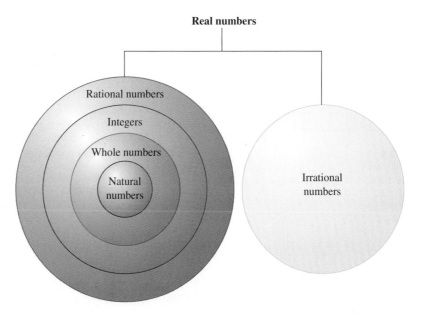

Figure 0.1

1. The set of whole numbers is a subset of the set of integers:

$$\{0, 1, 2, 3, \ldots\} \subseteq \{\ldots, -2, -1, 0, 1, 2, \ldots\}$$

2. The set of integers is a subset of the set of rational numbers:

$$\{\ldots, -2, -1, 0, 1, 2, \ldots\} \subseteq \{x \mid x \text{ is a rational number}\}$$

3. The set of rational numbers is a subset of the set of real numbers:

$$\{x \mid x \text{ is a rational number}\} \subseteq \{y \mid y \text{ is a real number}\}$$

■ Real Number Line and Absolute Value

It is often helpful to have a geometric representation of the set of real numbers in front of us, as indicated in Figure 0.2. Such a representation, called the **real number line,** indicates a one-to-one correspondence between the set of real numbers and the points on a line. In other words, to each real number there corresponds one and only one point on the line, and to each point on the line there corresponds one and only one real number. The number that corresponds to a particular point on the line is called the **coordinate** of that point.

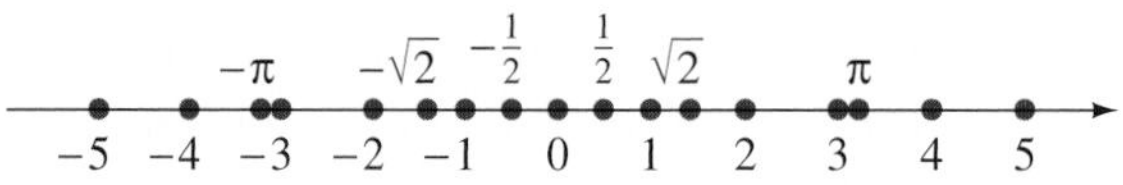

Figure 0.2

Many operations, relations, properties, and concepts pertaining to real numbers can be given a geometric interpretation on the number line. For example, the addition problem $(-1) + (-2)$ can be interpreted on the number line as shown in Figure 0.3.

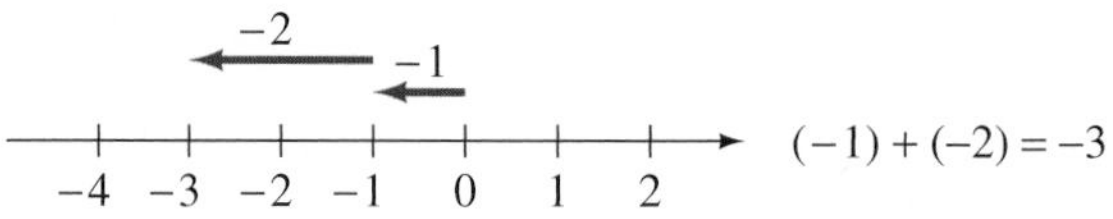

$(-1) + (-2) = -3$

Figure 0.3

The inequality relations also have a geometric interpretation. The statement $a > b$ (read "a is greater than b") means that a is to the right of b, and the statement $c < d$ (read "c is less than d") means that c is to the left of d (see Figure 0.4). The property $-(-x) = x$ can be pictured on the number line in a sequence of steps. See Figure 0.5.

Figure 0.4

1. Choose a point that has a coordinate of x.

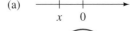
(a)

2. Locate its opposite (written as $-x$) on the other side of zero.

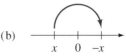
(b)

3. Locate the opposite of $-x$ [written as $-(-x)$] on the other side of zero.

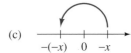
(c)

Figure 0.5

Therefore, we conclude that **the opposite of the opposite of any real number is the number itself,** and we express this symbolically by $-(-x) = x$.

Remark: The symbol -1 can be read "negative one," the "negative of one," the "opposite of one," or the "additive inverse of one." The opposite-of and additive-inverse-of terminology is especially meaningful when working with variables. For example, the symbol $-x$, read "the opposite of x or the additive inverse of x," emphasizes an important issue. Because x can be any real number, $-x$ (opposite of x) can be zero, positive, or negative. If x is positive, then $-x$ is negative. If x is negative, then $-x$ is positive. If x is zero, then $-x$ is zero. For example,

If $x = 4$, then $-x = -(4) = -4$.

If $x = -2$, then $-x = -(-2) = 2$.

If $x = 0$, then $-x = -(0) = 0$.

The concept of absolute value can be interpreted on the number line. Geometrically, the **absolute value** of any real number is the distance between that number and zero on the number line. For example, the absolute value of 2 is 2, the absolute value of -3 is 3, and the absolute value of zero is zero (see Figure 0.6).

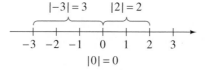

Figure 0.6

Symbolically, absolute value is denoted with vertical bars. Thus we write $|2| = 2$, $|-3| = 3$, and $|0| = 0$. More formally, the concept of absolute value is defined as follows.

Definition 0.1

For all real numbers a,

    1. If $a \geq 0$, then $|a| = a$.

    2. If $a < 0$, then $|a| = -a$.

According to Definition 0.1, we obtain

$|6| = 6$ by applying part 1

$|0| = 0$ by applying part 1

$|-7| = -(-7) = 7$ by applying part 2

Notice that the absolute value of a positive number is the number itself, but the absolute value of a negative number is its opposite. Thus the absolute value of any number except zero is positive, and the absolute value of zero is zero. Together, these facts indicate that the absolute value of any real number is equal to the absolute value of its opposite. All of these ideas are summarized in the following properties.

Properties of Absolute Value

The variables a and b represent any real number.

1. $|a| \geq 0$ The absolute value of a real number is positive or zero.

2. $|a| = |-a|$ The absolute value of a real number is equal to the absolute value of its opposite

3. $|a - b| = |b - a|$ The expressions $a - b$ and $b - a$ are opposites of each other, hence their absolute values are equal.

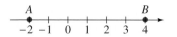

Figure 0.7

In Figure 0.7 the points A and B are located at -2 and 4, respectively. The distance *between* A and B is 6 units and can be calculated by using either $|-2 - 4|$ or $|4 - (-2)|$. In general, if two points on a number line have coordinates x_1 and x_2, then the distance between the two points is determined by using either

$$|x_2 - x_1| \qquad \text{or} \qquad |x_1 - x_2|$$

because, by property 3 above, they are the same quantity.

■ Properties of Real Numbers

As you work with the set of real numbers, the basic operations, and the relations of equality and inequality, the following properties will guide your study. Be sure that you understand these properties, because they not only facilitate manipulations with real numbers but also serve as a basis for many algebraic computations. The variables a, b, and c represent real numbers.

Properties of Real Numbers

Closure properties	$a + b$ is a unique real number.
	ab is a unique real number.
Commutative properties	$a + b = b + a$
	$ab = ba$
Associative properties	$(a + b) + c = a + (b + c)$
	$(ab)c = a(bc)$
Identity properties	There exists a real number 0 such that
	$a + 0 = 0 + a = a.$
	There exists a real number 1 such that
	$a(1) = 1(a) = a.$

Inverse properties	For every real number a, there exists a unique real number $-a$ such that $a + (-a) = (-a) + a = 0$.
	For every nonzero real number a, there exists a unique real number $\dfrac{1}{a}$ such that $$a\left(\frac{1}{a}\right) = \frac{1}{a}(a) = 1.$$
Multiplication property of zero	$a(0) = (0)(a) = 0$
Multiplication property of negative one	$a(-1) = -1(a) = -a$
Distributive property	$a(b + c) = ab + ac$

Let's make a few comments about the properties of real numbers. The set of real numbers is said to be **closed** with respect to addition and multiplication. That is, the sum of two real numbers is a real number, and the product of two real numbers is a real number. **Closure** plays an important role when we are proving additional properties that pertain to real numbers.

Addition and multiplication are said to be **commutative operations.** This means that the order in which you add or multiply two real numbers does not affect the result. For example, $6 + (-8) = -8 + 6$ and $(-4)(-3) = (-3)(-4)$. It is important to realize that subtraction and division are *not* commutative operations; order does make a difference. For example, $3 - 4 = -1$, but $4 - 3 = 1$. Likewise, $2 \div 1 = 2$, but $1 \div 2 = \dfrac{1}{2}$.

Addition and multiplication are **associative operations.** The associative properties are grouping properties. For example, $(-8 + 9) + 6 = -8 + (9 + 6)$; changing the grouping of the numbers does not affect the final sum. Likewise, for multiplication, $[(-4)(-3)](2) = (-4)[(-3)(2)]$. Subtraction and division are *not* associative operations. For example, $(8 - 6) - 10 = -8$, but $8 - (6 - 10) = 12$. An example showing that division is not associative is $(8 \div 4) \div 2 = 1$, but $8 \div (4 \div 2) = 4$.

Zero is the **identity element for addition.** This means that the sum of any real number and zero is identically the same real number. For example, $-87 + 0 = 0 + (-87) = -87$. One is the **identity element for multiplication.** The product of any real number and 1 is identically the same real number. For example, $(-119)(1) = (1)(-119) = -119$.

The real number $-a$ is called the **additive inverse of a** or the **opposite of a.** The sum of a number and its additive inverse is the identity element for addition. For example, 16 and -16 are additive inverses, and their sum is zero. The additive inverse of zero is zero.

The real number $1/a$ is called the **multiplicative inverse** or **reciprocal of a.** The product of a number and its multiplicative inverse is the identity element for multiplication. For example, the reciprocal of 2 is $\dfrac{1}{2}$, and $2\left(\dfrac{1}{2}\right) = \dfrac{1}{2}(2) = 1$.

The product of any real number and zero is zero. For example, $(-17)(0) = (0)(-17) = 0$. The product of any real number and -1 is the opposite of the real number. For example, $(-1)(52) = (52)(-1) = -52$.

The **distributive property** ties together the operations of addition and multiplication. We say that *multiplication distributes over addition.* For example, $7(3 + 8) = 7(3) + 7(8)$. Furthermore, because $b - c = b + (-c)$, it follows that *multiplication also distributes over subtraction.* This can be expressed symbolically as $a(b - c) = ab - ac$. For example, $6(8 - 10) = 6(8) - 6(10)$.

■ Cartesian Coordinate System

Just as real numbers can be associated with points on a line, pairs of real numbers can be associated with points in a plane. To do this, we set up two number lines, one vertical and one horizontal, perpendicular to each other at the point associated with zero on both lines, as shown in Figure 0.8. We refer to these number lines as the **horizontal axis** and the **vertical axis** or together as the **coordinate axes.** They partition a plane into four regions called **quadrants.** The quadrants are numbered counterclockwise from I through IV as indicated in Figure 0.8. The point of intersection of the two axes is called the **origin.**

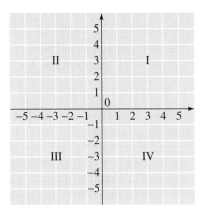

Figure 0.8

The positive direction on the horizontal axis is to the right, and the positive direction on the vertical axis is up. It is now possible to set up a one-to-one correspondence between **ordered pairs** of real numbers and the points in a plane. To each ordered pair of real numbers there corresponds a unique point in the plane, and to each point in the plane there corresponds a unique ordered pair of real numbers. A part of this correspondence is illustrated in Figure 0.9. For example, the ordered pair (3, 2) means that the point A is located 3 units to the right of and 2 units up from the origin. Likewise, the ordered pair $(-3, -5)$ means that the point D is located 3 units to the left of and 5 units down from the origin. The ordered pair (0, 0) is associated with the origin O.

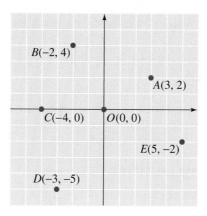

Figure 0.9

In general we refer to the real numbers a and b in an ordered pair (a, b) associated with a point as the **coordinates of the point.** The first number, a, called the **abscissa,** is the directed distance of the point from the vertical axis measured parallel to the horizontal axis. The second number, b, called the **ordinate,** is the directed distance of the point from the horizontal axis measured parallel to the vertical axis (Figure 0.10). Thus in the first quadrant, all points have a positive abscissa and a positive ordinate. In the second quadrant all points have a negative abscissa and a positive ordinate. We have indicated the sign situations for all four quadrants in Figure 0.11. This system of associating points in a plane with pairs of real numbers is called the **rectangular coordinate system** or the **Cartesian coordinate system.**

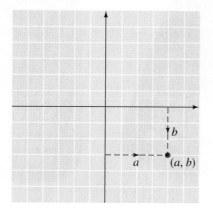

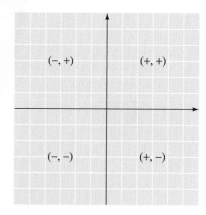

Figure 0.10   **Figure 0.11**

Historically, the rectangular coordinate system provided the basis for the development of the branch of mathematics called **analytic geometry,** or what we presently refer to as **coordinate geometry.** In this discipline, René Descartes, a French 17th-century mathematician, was able to transform geometric problems into an algebraic setting and then use the tools of algebra to solve the problems.

Basically, there are two kinds of problems to solve in coordinate geometry:

1. Given an algebraic equation, find its geometric graph.
2. Given a set of conditions pertaining to a geometric figure, find its algebraic equation.

Throughout this text we will consider a wide variety of situations dealing with both kinds of problems.

For most purposes in coordinate geometry, it is customary to label the horizontal axis the *x* **axis** and the vertical axis the *y* **axis.** Then ordered pairs of real numbers associated with points in the *xy* plane are of the form (x, y); that is, *x* is the first coordinate and *y* is the second coordinate.

■ Graphing Utilities

The term **graphing utility** is used in current literature to refer to either a graphing calculator (see Figure 0.12) or a computer with a graphing software package. (We will frequently use the phrase "use a graphing calculator" to mean either a graphing calculator or a computer with an appropriate software package.) We will introduce various features of graphing calculators as we need them in the text. Because so many different types of graphing utilities are available, we will use mostly generic terminology and let you consult a user's manual for specific key-punching instructions. We urge you to study the graphing calculator examples in this text even if you do not have access to a graphing utility. The examples are chosen to reinforce concepts under discussion. Furthermore, for those who do have access to a graphing utility, we provide "Graphing Calculator Activities" in many of the problem sets.

Courtesy of Texas Instruments

Figure 0.12

Graphing calculators have display windows large enough to show graphs. This window feature is also helpful when you're using a graphing calculator for computational purposes, because it allows you to see the entries of the problem. Figure 0.13 shows a display window for an example of the distributive property. Note that we can check to see that the correct numbers and operational symbols have been entered. Also note that the answer is given below and to the right of the problem.

```
486(324+379)
            341658
486*324+486*379
            341658
```

Figure 0.13

■ Algebraic Expressions

Algebraic expressions such as

$$2x \qquad 8xy \qquad -3xy \qquad -4abc \qquad z$$

are called "terms." A **term** is an indicated product and may have any number of factors. The variables of a term are called "literal factors," and the numerical factor is called the "numerical coefficient." Thus in $8xy$, the x and y are **literal factors,** and 8 is the **numerical coefficient.** Because $1(z) = z$, the numerical coefficient of the term z is understood to be 1. Terms that have the same literal factors are called "similar terms" or "like terms." The distributive property in the form $ba + ca = (b + c)a$ provides the basis for simplifying algebraic expressions by *combining similar terms,* as illustrated in the following examples:

$$3x + 5x = (3 + 5)x = 8x$$

$$-6xy + 4xy = (-6 + 4)xy = -2xy$$

$$4x - x = 4x - 1x = (4 - 1)x = 3x$$

Sometimes we can simplify an algebraic expression by applying the distributive property to remove parentheses and combine similar terms, as the next examples illustrate:

$$4(x + 2) + 3(x + 6) = 4(x) + 4(2) + 3(x) + 3(6)$$

$$= 4x + 8 + 3x + 18$$

$$= 7x + 26$$

$$-5(y + 3) - 2(y - 8) = -5(y) - 5(3) - 2(y) - 2(-8)$$
$$= -5y - 15 - 2y + 16$$
$$= -7y + 1$$

An algebraic expression takes on a numerical value whenever each variable in the expression is replaced by a real number. For example, when x is replaced by 5 and y by 9, the algebraic expression $x + y$ becomes the numerical expression $5 + 9$, which is equal to 14. We say that $x + y$ has a value of 14 when $x = 5$ and $y = 9$.

Consider the following examples, which illustrate the process of finding a value of an algebraic expression. The process is commonly referred to as **evaluating an algebraic expression.**

EXAMPLE 1   Find the value of $3xy - 4z$ when $x = 2$, $y = -4$, and $z = -5$.

Solution

$$3xy - 4z = 3(2)(-4) - 4(-5) \qquad \text{when } x = 2, y = -4, \text{ and } z = -5$$
$$= -24 + 20$$
$$= -4$$

▶ **Now work Problem 59.**   ■

Remark: Notice, at the end of most examples, we give you a problem to work out, which is similar in substance to the example. This problem is selected from the problem set at the end of the section.

EXAMPLE 2   Find the value of $a - [4b - (2c + 1)]$ when $a = -8$, $b = -7$, and $c = 14$.

Solution

$$a - [4b - (2c + 1)] = -8 - [4(-7) - (2(14) + 1)]$$
$$= -8 - [-28 - 29]$$
$$= -8 - [-57]$$
$$= 49$$

▶ **Now work Problem 63.**   ■

EXAMPLE 3   Evaluate $\dfrac{a - 2b}{3c + 5d}$ when $a = 14$, $b = -12$, $c = -3$, and $d = -2$.

Solution

$$\frac{a - 2b}{3c + 5d} = \frac{14 - 2(-12)}{3(-3) + 5(-2)}$$
$$= \frac{14 + 24}{-9 - 10}$$
$$= \frac{38}{-19} = -2$$

▶ **Now work Problem 65.**   ■

Look back at Examples 1–3, and note that we use the following **order of operations** when simplifying numerical expressions.

1. Perform the operations inside the symbols of inclusion (parentheses, brackets, and braces) and above and below each fraction bar. Start with the innermost inclusion symbol.
2. Perform all multiplications and divisions in the order in which they appear, from left to right.
3. Perform all additions and subtractions in the order in which they appear, from left to right.

You should also realize that first simplifying by combining similar terms can sometimes aid in the process of evaluating algebraic expressions. The last example of this section illustrates this idea.

EXAMPLE 4

Evaluate $2(3x + 1) - 3(4x - 3)$ when $x = -5$.

Solution

$$2(3x + 1) - 3(4x - 3) = 2(3x) + 2(1) - 3(4x) - 3(-3)$$
$$= 6x + 2 - 12x + 9$$
$$= -6x + 11$$

Now substituting -5 for x, we obtain

$$-6x + 11 = -6(-5) + 11$$
$$= 30 + 11$$
$$= 41$$

Now work Problem 77.   ■

One calculator method for evaluating an algebraic expression such as $3xy - 4z$ for $x = 2$, $y = -4$, and $z = -5$ (see Example 1) is to replace x with 2, y with -4, and z with -5, and then calculate the resulting numerical expression. Another method is shown in Figure 0.14, in which the values for x, y, and z are stored and then the algebraic expression $3xy - 4z$ is evaluated.

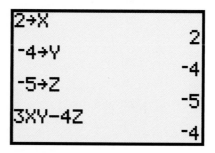

Figure 0.14

CONCEPT QUIZ For Problems 1–8, answer true or false.

1. The null set is written as $\{\varnothing\}$.

2. The sets {a, b, c, d} and {a, d, c, b} are equal sets.

3. Decimal numbers that are classified as repeating or terminating decimals represent rational numbers.

4. The absolute value of x is equal to x.

5. The axes of the rectangular coordinate system intersect in a point called the center.

6. Subtraction is a commutative operation.

7. Every real number has a multiplicative inverse.

8. The associative properties are grouping properties.

Remark: You can find answers to the Concept Quiz questions at the end of the next Problem Set.

Problem Set 0.1

For Problems 1–10, identify each statement as *true* or *false*.

1. Every rational number is a real number.

2. Every irrational number is a real number.

3. Every real number is a rational number.

4. If a number is real, then it is irrational.

5. Some irrational numbers are also rational numbers.

6. All integers are rational numbers.

7. The number zero is a rational number.

8. Zero is a positive integer.

9. Zero is a negative number.

10. All whole numbers are integers.

For Problems 11–18, list those elements of the set of numbers

$$\left\{0, \sqrt{5}, -\sqrt{2}, \frac{7}{8}, -\frac{10}{13}, 7\frac{1}{8}, 0.279, 0.4\overline{67}, -\pi, -14, 46, 6.75\right\}$$

that belong to each of the following sets.

11. The natural numbers

12. The whole numbers

13. The integers

14. The rational numbers

15. The irrational numbers

16. The nonnegative integers

17. The nonpositive integers

18. The real numbers

For Problems 19–32, use the following set designations:

$N = \{x \mid x \text{ is a natural number}\}$
$W = \{x \mid x \text{ is a whole number}\}$
$I = \{x \mid x \text{ is an integer}\}$
$Q = \{x \mid x \text{ is a rational number}\}$
$H = \{x \mid x \text{ is an irrational number}\}$
$R = \{x \mid x \text{ is a real number}\}$

Place $\subseteq$ or $\not\subseteq$ in each blank to make a true statement.

19. N _____ R

20. R _____ N

21. N _____ I

22. I _____ Q

23. H _____ Q

24. Q _____ H

25. W _____ I

26. N _____ W

27. I _____ W

28. I _____ N

29. $\{0, 2, 4, \ldots\}$ _____ W

30. $\{1, 3, 5, 7, \ldots\}$ _____ I

31. $\{-2, -1, 0, 1, 2\}$ _____ W

32. $\{0, 3, 6, 9, \ldots\}$ _____ N

For Problems 33–42, list the elements of each set. For example, the elements of $\{x \mid x$ is a natural number less than 4$\}$ can be listed $\{1, 2, 3\}$.

33. $\{x \mid x$ is a natural number less than 2$\}$

34. $\{x \mid x$ is a natural number greater than 5$\}$

35. $\{n \mid n$ is a whole number less than 4$\}$

36. $\{y \mid y$ is an integer greater than $-3\}$

37. $\{y \mid y$ is an integer less than 2$\}$

38. $\{n \mid n$ is a positive integer greater than $-4\}$

39. $\{x \mid x$ is a whole number less than 0$\}$

40. $\{x \mid x$ is a negative integer greater than $-5\}$

41. $\{n \mid n$ is a nonnegative integer less than 3$\}$

42. $\{n \mid n$ is a nonpositive integer greater than 1$\}$

43. Find the distance on the real number line between two points whose coordinates are the following:

    a. 17 and 35   **b.** -14 and 12

    c. 18 and -21 **d.** -17 and -42

    e. -56 and -21 **f.** 0 and -37

44. Evaluate each of the following if x is a nonzero real number.

    a. $\dfrac{|x|}{x}$ **b.** $\dfrac{x}{|x|}$

    c. $\dfrac{|-x|}{-x}$ **d.** $|x| - |-x|$

In Problems 45–58, state the property that justifies each of the statements. For example, $3 + (-4) = (-4) + 3$ because of the commutative property of addition.

45. $x(2) = 2(x)$

46. $(7 + 4) + 6 = 7 + (4 + 6)$

47. $1(x) = x$

48. $43 + (-18) = (-18) + 43$

49. $(-1)(93) = -93$

50. $109 + (-109) = 0$

51. $5(4 + 7) = 5(4) + 5(7)$

52. $-1(x + y) = -(x + y)$

53. $7yx = 7xy$

54. $(x + 2) + (-2) = x + [2 + (-2)]$

55. $6(4) + 7(4) = (6 + 7)(4)$

56. $\left(\dfrac{2}{3}\right)\left(\dfrac{3}{2}\right) = 1$

57. $4(5x) = (4 \cdot 5)x$

58. $[(17)(8)](25) = (17)[(8)(25)]$

For Problems 59–79, evaluate each of the algebraic expressions for the given values of the variables.

59. $5x + 3y$; $x = -2$ and $y = -4$

60. $7x - 4y$; $x = -1$ and $y = 6$

61. $-3ab - 2c$; $a = -4, b = 7$, and $c = -8$

62. $x - (2y + 3z)$; $x = -3, y = -4$, and $z = 9$

63. $(a - 2b) + (3c - 4)$; $a = 6, b = -5$, and $c = -11$

64. $3a - [2b - (4c + 1)]$; $a = 4, b = 6$, and $c = -8$

65. $\dfrac{-2x + 7y}{x - y}$; $x = -3$ and $y = -2$

66. $\dfrac{x - 3y + 2z}{2x - y}$; $x = 4, y = 9, z = -12$

67. $(5x - 2y)(-3x + 4y)$; $x = -3$ and $y = -7$

68. $(2a - 7b)(4a + 3b)$; $a = 6$ and $b = -3$

69. $5x + 4y - 9y - 2y$; $x = 2$ and $y = -8$

70. $5a + 7b - 9a - 6b$; $a = -7$ and $b = 8$

71. $-5x + 8y + 7y + 8x$; $x = 5$ and $y = -6$

72. $|x - y| - |x + y|$; $x = -4$ and $y = -7$

73. $|3x + y| + |2x - 4y|$; $x = 5$ and $y = -3$

74. $\left|\dfrac{x - y}{y - x}\right|$; $x = -6$ and $y = 13$

75. $\left|\dfrac{2a - 3b}{3b - 2a}\right|$; $a = -4$ and $b = -8$

76. $5(x - 1) + 7(x + 4)$; $x = 3$

77. $2(3x + 4) - 3(2x - 1)$; $x = -2$

78. $-4(2x - 1) - 5(3x + 7)$; $x = -1$

79. $5(a - 3) - 4(2a + 1) - 2(a - 4)$; $a = -3$

■ ■ ■ THOUGHTS INTO WORDS

80. Do you think $3\sqrt{2}$ is a rational or an irrational number? Defend your answer.

81. Explain why $\dfrac{0}{8} = 0$ but $\dfrac{8}{0}$ is undefined.

82. The solution of the following simplification problem is incorrect. The answer should be -11. Find and correct the error.

$$8 \div (-4)(2) - 3(4) \div 2 + (-1) = (-2)(2) - 12 \div 1$$
$$= -4 - 12$$
$$= -16$$

83. Explain the difference between "simplifying a numerical expression" and "evaluating an algebraic expression."

 GRAPHING CALCULATOR ACTIVITIES

84. Different graphing calculators use different sequences of key strokes to evaluate algebraic expressions.

Be sure that you can do Problems 59–79 with your calculator.

Answers to the Concept Quiz

1. False   **2.** True   **3.** True   **4.** False   **5.** False   **6.** False   **7.** False   **8.** True

---

0.2   Exponents

Objectives

■ Evaluate numerical expressions that have integer exponents.

■ Apply the properties of exponents to simplify algebraic expressions.

■ Write numbers in scientific notation.

■ Change numbers from scientific notation to ordinary decimal notation.

■ Perform calculations with numbers in scientific form.

Positive integers are used as *exponents* to indicate repeated multiplication. For example, $4 \cdot 4 \cdot 4$ can be written 4^3, where the raised 3 indicates that 4 is to be used as a factor three times. The following general definition is helpful.

Definition 0.2

If n is a positive integer, and b is any real number, then

$$b^n = \underbrace{bbb \cdots b}_{n \text{ factors of } b}$$

The number b is referred to as the **base,** and n is called the **exponent.** The expression b^n can be read "b to the nth power." The terms **squared** and **cubed** are commonly associated with exponents of 2 and 3, respectively. For example, b^2 is read "b squared" and b^3 as "b cubed." An exponent of 1 is usually not written, so b^1 is simply written b. The following examples illustrate Definition 0.2:

$$2^3 = 2 \cdot 2 \cdot 2 = 8 \qquad \left(\frac{1}{2}\right)^5 = \frac{1}{2} \cdot \frac{1}{2} \cdot \frac{1}{2} \cdot \frac{1}{2} \cdot \frac{1}{2} = \frac{1}{32}$$

$$3^4 = 3 \cdot 3 \cdot 3 \cdot 3 = 81 \qquad (0.7)^2 = (0.7)(0.7) = 0.49$$

$$(-5)^2 = (-5)(-5) = 25 \qquad -5^2 = -(5 \cdot 5) = -25$$

We especially want to call your attention to the last example in each column. Note that $(-5)^2$ means that -5 is the base used as a factor twice. However, -5^2 means that 5 is the base, and after it is squared, we take the opposite of the result.

■ Properties of Exponents

In a previous algebra course, you may have seen some properties pertaining to the use of positive integers as exponents. Those properties can be summarized as follows.

Property 0.1 *Properties of Exponents*

If a and b are real numbers, and m and n are positive integers, then

1. $b^n \cdot b^m = b^{n+m}$

2. $(b^n)^m = b^{mn}$

3. $(ab)^n = a^n b^n$

4. $\left(\dfrac{a}{b}\right)^n = \dfrac{a^n}{b^n}, \qquad b \neq 0$

5. $\dfrac{b^n}{b^m} = b^{n-m} \qquad$ when $n > m, b \neq 0$

$\dfrac{b^n}{b^m} = 1 \qquad$ when $n = m, b \neq 0$

$\dfrac{b^n}{b^m} = \dfrac{1}{b^{m-n}} \qquad$ when $n < m, b \neq 0$

Each part of Property 0.1 can be justified by using Definition 0.2. For example, to justify part 1, we can reason as follows:

$$b^n \cdot b^m = \underbrace{(bbb \cdots b)}_{\substack{n \text{ factors} \\ \text{of } b}} \cdot \underbrace{(bbb \cdots b)}_{\substack{m \text{ factors} \\ \text{of } b}}$$

$$= \underbrace{bbb \cdots b}_{(n + m) \text{ factors of } b}$$

$$= b^{n+m}$$

Similar reasoning can be used to verify the other parts of Property 0.1. The following examples illustrate the use of Property 0.1 along with the commutative and associative properties of the real numbers. We have chosen to show all of the steps; however many of the steps can be performed mentally.

E X A M P L E   1

$$\begin{aligned} (3x^2y)(4x^3y^2) &= 3 \cdot 4 \cdot x^2 \cdot x^3 \cdot y \cdot y^2 \\ &= 12x^{2+3}y^{1+2} \qquad\qquad b^n \cdot b^m = b^{n+m} \\ &= 12x^5y^3 \end{aligned}$$

▶ **Now work Problem 63.**   ■

E X A M P L E   2

$$\begin{aligned} (-2y^3)^5 &= (-2)^5(y^3)^5 \qquad (ab)^n = a^n b^n \\ &= -32y^{15} \qquad\qquad (b^n)^m = b^{mn} \end{aligned}$$

▶ **Now work Problem 65.**   ■

E X A M P L E   3

$$\begin{aligned} \left(\frac{a^2}{b^4}\right)^7 &= \frac{(a^2)^7}{(b^4)^7} \qquad \left(\frac{a}{b}\right)^n = \frac{a^n}{b^n} \\ &= \frac{a^{14}}{b^{28}} \qquad\quad (b^n)^m = b^{mn} \end{aligned}$$

▶ **Now work Problem 67.**   ■

E X A M P L E   4

$$\begin{aligned} \frac{-56x^9}{7x^4} &= -8x^{9-4} \quad \frac{b^n}{b^m} = b^{n-m} \quad \text{when } n > m \\ &= -8x^5 \end{aligned}$$

▶ **Now work Problem 69.**   ■

■ Zero and Negative Integers As Exponents

Now we can extend the concept of an exponent to include the use of zero and negative integers. First, let's consider the use of zero as an exponent. We want to use zero in a way that Property 0.1 will continue to hold. For example, if $b^n \cdot b^m = b^{n+m}$ is to hold, then $x^4 \cdot x^0$ should equal x^{4+0}, which equals x^4. In other words, x^0 *acts like* 1 because $x^4 \cdot x^0 = x^4$. Look at the following definition.

Definition 0.3

If b is a nonzero real number, then
$$b^0 = 1$$

Therefore, according to Definition 0.3, the following statements are all true:

$$5^0 = 1 \qquad\qquad (-413)^0 = 1$$

$$\left(\frac{3}{11}\right)^0 = 1 \qquad (x^3 y^4)^0 = 1 \quad \text{if } x \neq 0 \text{ and } y \neq 0$$

A similar line of reasoning can be used to motivate a definition for the use of negative integers as exponents. Consider the example $x^4 \cdot x^{-4}$. If $b^n \cdot b^m = b^{n+m}$ is to hold, then $x^4 \cdot x^{-4}$ should equal $x^{4+(-4)}$, which equals $x^0 = 1$. Therefore, x^{-4} must be the reciprocal of x^4 because their product is 1. That is, $x^{-4} = 1/x^4$. This suggests the following definition.

Definition 0.4

If n is a positive integer, and b is a nonzero real number, then
$$b^{-n} = \frac{1}{b^n}$$

According to Definition 0.4, the following statements are true:

$$x^{-5} = \frac{1}{x^5} \qquad\qquad 2^{-4} = \frac{1}{2^4} = \frac{1}{16}$$

$$\left(\frac{3}{4}\right)^{-2} = \frac{1}{\left(\frac{3}{4}\right)^2} = \frac{1}{\frac{9}{16}} = \frac{16}{9} \qquad \frac{2}{x^{-3}} = \frac{2}{\frac{1}{x^3}} = 2x^3$$

The first four parts of Property 0.1 hold true *for all integers*. Furthermore, we do not need all three equations in part 5 of Property 0.1. The first equation,

$$\frac{b^n}{b^m} = b^{n-m}$$

can be used for *all integral exponents*. Let's restate Property 0.1 as it pertains to integers. We will include name tags for easy reference.

Property 0.2

If m and n are integers, and a and b are real numbers, with $b \neq 0$ whenever it appears in a denominator, then

1. $b^n \cdot b^m = b^{n+m}$ Product of two powers

2. $(b^n)^m = b^{mn}$ Power of a power

3. $(ab)^n = a^n b^n$ Power of a product

4. $\left(\dfrac{a}{b}\right)^n = \dfrac{a^n}{b^n}$ Power of a quotient

5. $\dfrac{b^n}{b^m} = b^{n-m}$ Quotient of two powers

Having the use of all integers as exponents allows us to work with a large variety of numerical and algebraic expressions. Let's consider some examples that illustrate the various parts of Property 0.2.

Evaluate each of the following numerical expressions.

a. $(2^{-1} \cdot 3^2)^{-1}$ **b.** $\left(\dfrac{2^{-3}}{3^{-2}}\right)^{-2}$

Solution

a. $(2^{-1} \cdot 3^2)^{-1} = (2^{-1})^{-1}(3^2)^{-1}$ Power of a product

$= (2^1)(3^{-2})$ Power of a power

$= (2)\left(\dfrac{1}{3^2}\right)$

$= 2\left(\dfrac{1}{9}\right) = \dfrac{2}{9}$

b. $\left(\dfrac{2^{-3}}{3^{-2}}\right)^{-2} = \dfrac{(2^{-3})^{-2}}{(3^{-2})^{-2}}$ Power of a quotient

$= \dfrac{2^6}{3^4}$ Power of a power

$= \dfrac{64}{81}$

Now work Problems 27 and 29.   ∎

E X A M P L E   6

Find the indicated products and quotients, and express the final results with positive integral exponents only.

a. $(3x^2y^{-4})(4x^{-3}y)$

b. $\dfrac{12a^3b^2}{-3a^{-1}b^5}$

c. $\left(\dfrac{15x^{-1}y^2}{5xy^{-4}}\right)^{-1}$

Solution

a. $(3x^2y^{-4})(4x^{-3}y) = 12x^{2+(-3)}y^{-4+1}$ Product of powers

$= 12x^{-1}y^{-3}$

$= \dfrac{12}{xy^3}$

b. $\dfrac{12a^3b^2}{-3a^{-1}b^5} = -4a^{3-(-1)}b^{2-5}$ Quotient of powers

$= -4a^4b^{-3}$

$= -\dfrac{4a^4}{b^3}$

c. $\left(\dfrac{15x^{-1}y^2}{5xy^{-4}}\right)^{-1} = (3x^{-1-1}y^{2-(-4)})^{-1}$ First simplify inside parentheses

$= (3x^{-2}y^6)^{-1}$

$= 3^{-1}x^2y^{-6}$ Power of a product

$= \dfrac{x^2}{3y^6}$

▶ **Now work Problems 71, 75, and 79.**   ■

The next two examples illustrate the simplification of numerical and algebraic expressions involving sums and differences. In such cases, Definition 0.4 can be used to change from negative to positive exponents so that we can proceed in the usual ways.

E X A M P L E   7

Simplify $2^{-3} + 3^{-1}$.

Solution

$2^{-3} + 3^{-1} = \dfrac{1}{2^3} + \dfrac{1}{3^1}$

$= \dfrac{1}{8} + \dfrac{1}{3}$

$$= \frac{3}{24} + \frac{8}{24}$$

$$= \frac{11}{24}$$

▷ **Now work Problem 37.** ∎

E X A M P L E 8

Simplify $(4^{-1} - 3^{-2})^{-1}$.

Solution

$$(4^{-1} - 3^{-2})^{-1} = \left(\frac{1}{4^1} - \frac{1}{3^2}\right)^{-1}$$

$$= \left(\frac{1}{4} - \frac{1}{9}\right)^{-1}$$

$$= \left(\frac{9}{36} - \frac{4}{36}\right)^{-1}$$

$$= \left(\frac{5}{36}\right)^{-1}$$

$$= \frac{1}{\left(\frac{5}{36}\right)^1} = \frac{36}{5}$$

▷ **Now work Problem 41.** ∎

Figure 0.15 shows calculator windows for Examples 7 and 8. Note that the answers are given in decimal form. If your calculator also handles common fractions, then the display window may appear as in Figure 0.16.

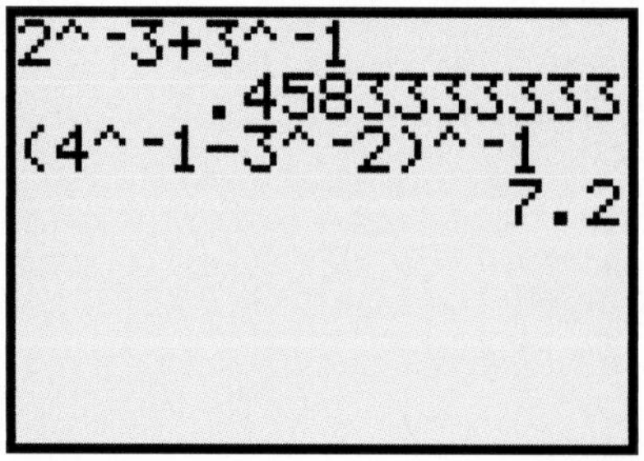

```
2^-3+3^-1
          .4583333333
(4^-1-3^-2)^-1
                  7.2
```

Figure 0.15

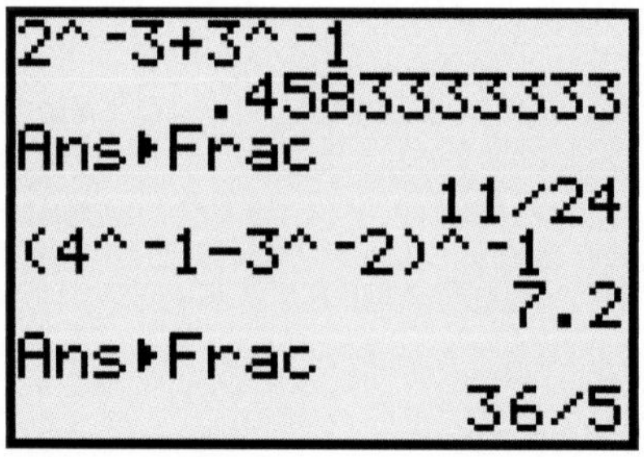

```
2^-3+3^-1
          .4583333333
Ans▶Frac
                11/24
(4^-1-3^-2)^-1
                  7.2
Ans▶Frac
                 36/5
```

Figure 0.16

EXAMPLE 9

Express $a^{-1} + b^{-2}$ as a single fraction involving positive exponents only.

Solution

$$a^{-1} + b^{-2} = \frac{1}{a^1} + \frac{1}{b^2}$$

$$= \left(\frac{1}{a}\right)\left(\frac{b^2}{b^2}\right) + \left(\frac{1}{b^2}\right)\left(\frac{a}{a}\right)$$

$$= \frac{b^2}{ab^2} + \frac{a}{ab^2}$$

$$= \frac{b^2 + a}{ab^2}$$

▷ **Now work Problem 81.**   ■

■ Scientific Notation

The expression $(n)(10)^k$ (where n is a number greater than or equal to 1 and less than 10, written in decimal form, and k is any integer) is commonly called **scientific notation** or the **scientific form** of a number. The following are examples of numbers expressed in scientific form:

$$(4.23)(10)^4 \qquad (8.176)(10)^{12} \qquad (5.02)(10)^{-3} \qquad (1)(10)^{-5}$$

Very large and very small numbers can be conveniently expressed in scientific notation. For example, a light year (the distance that a ray of light travels in one year) is approximately 5,900,000,000,000 miles, and this can be written as $(5.9)(10)^{12}$. The weight of an oxygen molecule is approximately 0.000000000000000000000053 of a gram, and this can be expressed as $(5.3)(10)^{-23}$.

To change from ordinary decimal notation to scientific notation, the following procedure can be used.

Write the given number as the product of a number greater than or equal to 1 and less than 10, and a power of 10. The exponent of 10 is determined by counting the number of places that the decimal point was moved when going from the original number to the number greater than or equal to 1 and less than 10. This exponent is (a) negative if the original number is less than 1, (b) positive if the original number is greater than 10, and (c) 0 if the original number itself is between 1 and 10.

Thus we can write

$$0.00092 = (9.2)(10)^{-4}$$

$$872,000,000 = (8.72)(10)^8$$

$$5.1217 = (5.1217)(10)^0$$

To change from scientific notation to ordinary decimal notation, the following procedure can be used.

> Move the decimal point the number of places indicated by the exponent of 10. Move the decimal point to the right if the exponent is positive. Move it to the left if the exponent is negative.

Thus we can write

$$(3.14)(10)^7 = 31,400,000$$

$$(7.8)(10)^{-6} = 0.0000078$$

Scientific notation can be used to simplify numerical calculations. We merely change the numbers to scientific notation and use the appropriate properties of exponents. Consider the following examples.

E X A M P L E   1 0   Perform the indicated operations.

a. $\dfrac{(0.00063)(960,000)}{(3200)(0.0000021)}$ **b.** $\sqrt{90,000}$

Solution

a. $\dfrac{(0.00063)(960,000)}{(3200)(0.0000021)} = \dfrac{(6.3)(10)^{-4}(9.6)(10)^5}{(3.2)(10)^3(2.1)(10)^{-6}}$

$$= \dfrac{(6.3)(9.6)(10)^1}{(3.2)(2.1)(10^{-3})}$$

$$= (9)(10)^4$$

$$= 90,000$$

b. $\sqrt{90,000} = \sqrt{(9)(10)^4}$

$$= \sqrt{9}\sqrt{10^4}$$

$$= (3)(10)^2$$

$$= 3(100)$$

$$= 300$$

▶ **Now work Problem 109.**   ■

Many calculators are equipped to display numbers in scientific notation. The display panel shows the number between 1 and 10 and the appropriate exponent of 10. For example, evaluating $(3,800,000)^2$ yields

| 1.444E13 |

Thus $(3,800,000)^2 = (1.444)(10)^{13} = 14,440,000,000,000$. Similarly, the answer for $(0.000168)^2$ is displayed as

$$\boxed{2.8224\text{E}-8}$$

Thus $(0.000168)^2 = (2.8224)(10)^{-8} = 0.000000028224$.

Calculators vary in the number of digits they display between 1 and 10 when they represent a number in scientific notation. For example, we used two different calculators to estimate $(6729)^6$ and obtained the following results:

$$\boxed{9.283316768\text{E}22}$$
$$\boxed{9.28331676776\text{E}22}$$

Obviously, you need to know the capabilities of your calculator when working with problems in scientific notation.

Many calculators also allow you to enter a number in scientific notation. Such calculators are equipped with an enter-the-exponent key often labeled $\boxed{\text{EE}}$. Thus a number such as $(3.14)(10)^8$ might be entered as follows.

Enter	Press	Display
3.14	$\boxed{\text{EE}}$	3.14E
8		3.14E8

A $\boxed{\text{MODE}}$ key is often used on calculators to let you choose normal decimal notation, scientific notation, or engineering notation. (The abbreviations Norm, Sci, and Eng are commonly used.) If the calculator is in scientific mode, then a number can be entered and changed to scientific form with the $\boxed{\text{ENTER}}$ key. For example, when we enter 589 and press the $\boxed{\text{ENTER}}$ key, the display will show 5.89E2. Likewise, when the calculator is in scientific mode, the answers to computational problems are given in scientific form. For example, the answer for $(76)(533)$ is given as 4.0508E4.

It should be evident from this brief discussion that even when you are using a calculator, you need to have a thorough understanding of scientific notation.

CONCEPT QUIZ

For Problems 1–8, answer true or false.

1. Exponents are used to indicate repeated multiplications.
2. An exponent cannot be zero.
3. $2^{-2} = -4$
4. $(-1)^{-2} = 2$
5. In the expression 6^3, the number 6 is referred to as the baseline number.
6. $(2 + 5)^2 = 4 + 25$
7. When writing a number in scientific notation, $(n)(10)^k$, the number n must be greater than 1 and less than or equal to 10.
8. Single-digit numbers can be expressed in scientific notation.

Problem Set 0.2

For Problems 1–42, evaluate each numerical expression.

1. 2^{-3}

2. 3^{-2}

3. -10^{-3}

4. 10^{-4}

5. $\dfrac{1}{3^{-3}}$

6. $\dfrac{1}{2^{-5}}$

7. $\left(\dfrac{1}{2}\right)^{-2}$

8. $-\left(\dfrac{1}{3}\right)^{-2}$

9. $\left(-\dfrac{2}{3}\right)^{-3}$

10. $\left(\dfrac{5}{6}\right)^{-2}$

11. $\left(-\dfrac{1}{5}\right)^{0}$

12. $\dfrac{1}{\left(\dfrac{3}{5}\right)^{-2}}$

13. $\dfrac{1}{\left(\dfrac{4}{5}\right)^{-2}}$

14. $\left(\dfrac{4}{5}\right)^{0}$

15. $2^5 \cdot 2^{-3}$

16. $3^{-2} \cdot 3^5$

17. $10^{-6} \cdot 10^4$

18. $10^6 \cdot 10^{-9}$

19. $10^{-2} \cdot 10^{-3}$

20. $10^{-1} \cdot 10^{-5}$

21. $(3^{-2})^{-2}$

22. $((-2)^{-1})^{-3}$

23. $(4^2)^{-1}$

24. $(3^{-1})^{3}$

25. $(3^{-1} \cdot 2^2)^{-1}$

26. $(2^3 \cdot 3^{-2})^{-2}$

27. $(4^2 \cdot 5^{-1})^{2}$

28. $(2^{-2} \cdot 4^{-1})^{3}$

29. $\left(\dfrac{2^{-2}}{5^{-1}}\right)^{-2}$

30. $\left(\dfrac{3^{-1}}{2^{-3}}\right)^{-2}$

31. $\left(\dfrac{3^{-2}}{8^{-1}}\right)^{2}$

32. $\left(\dfrac{4^2}{5^{-1}}\right)^{-1}$

33. $\dfrac{2^3}{2^{-3}}$

34. $\dfrac{2^{-3}}{2^3}$

35. $\dfrac{10^{-1}}{10^4}$

36. $\dfrac{10^{-3}}{10^{-7}}$

37. $3^{-2} + 2^{-3}$

38. $2^{-3} + 5^{-1}$

39. $\left(\dfrac{2}{3}\right)^{-1} - \left(\dfrac{3}{4}\right)^{-1}$

40. $3^{-2} - 2^3$

41. $(2^{-4} + 3^{-1})^{-1}$

42. $(3^{-2} - 5^{-1})^{-1}$

Simplify Problems 43–62; express final results without using zero or negative integers as exponents.

43. $x^3 \cdot x^{-7}$

44. $x^{-2} \cdot x^{-3}$

45. $a^2 \cdot a^{-3} \cdot a^{-1}$

46. $b^{-3} \cdot b^5 \cdot b^{-4}$

47. $(a^{-3})^{2}$

48. $(b^5)^{-2}$

49. $(x^3 y^{-4})^{-1}$

50. $(x^4 y^{-2})^{-2}$

51. $(ab^2 c^{-1})^{-3}$

52. $(a^2 b^{-1} c^{-2})^{-4}$

53. $(2x^2 y^{-1})^{-2}$

54. $(3x^4 y^{-2})^{-1}$

55. $\left(\dfrac{x^{-2}}{y^{-3}}\right)^{-2}$

56. $\left(\dfrac{y^4}{x^{-1}}\right)^{-3}$

57. $\left(\dfrac{2a^{-1}}{3b^{-2}}\right)^{-2}$

58. $\left(\dfrac{3x^2 y}{4a^{-1} b^{-3}}\right)^{-1}$

59. $\dfrac{x^{-5}}{x^{-2}}$

60. $\dfrac{a^{-3}}{a^5}$

61. $\dfrac{a^2 b^{-3}}{a^{-1} b^{-2}}$

62. $\dfrac{x^{-1} y^{-2}}{x^3 y^{-1}}$

For Problems 63–70, find the indicated products, quotients, and powers; express answers without using zero or negative integers as exponents.

63. $(4x^3 y^2)(-5xy^3)$

64. $(-6xy)(3x^2 y^4)$

65. $(-3xy^3)^3$

66. $(-2x^2 y^4)^4$

67. $\left(\dfrac{2x^2}{3y^3}\right)^{3}$

68. $\left(\dfrac{4x}{5y^2}\right)^{3}$

69. $\dfrac{72x^8}{-9x^2}$

70. $\dfrac{108x^6}{-12x^2}$

For Problems 71–80, find the indicated products and quotients; express results using positive integral exponents only.

71. $(2x^{-1} y^2)(3x^{-2} y^{-3})$

72. $(4x^{-2} y^3)(-5x^3 y^{-4})$

73. $(-6a^5 y^{-4})(-a^{-7} y)$

74. $(-8a^{-4} b^{-5})(-6a^{-1} b^8)$

75. $\dfrac{24x^{-1} y^{-2}}{6x^{-4} y^3}$

76. $\dfrac{56xy^{-3}}{8x^2 y^2}$

77. $\dfrac{-35a^3b^{-2}}{7a^5b^{-1}}$

78. $\dfrac{27a^{-4}b^{-5}}{-3a^{-2}b^{-4}}$

79. $\left(\dfrac{14x^{-2}y^{-4}}{7x^{-3}y^{-6}}\right)^{-2}$

80. $\left(\dfrac{24x^5y^{-3}}{-8x^6y^{-1}}\right)^{-3}$

95. $\dfrac{-24y^{5b+1}}{6y^{-b-1}}$

96. $(x^a)^{2b}(x^b)^a$

97. $\dfrac{(xy)^b}{y^b}$

98. $\dfrac{(2x^{2b})(-4x^{b+1})}{8x^{-b+2}}$

For Problems 81–88, express each as a single fraction involving positive exponents only.

81. $x^{-1}+x^{-2}$

82. $x^{-2}+x^{-4}$

83. $x^{-2}-y^{-1}$

84. $2x^{-1}-3y^{-3}$

85. $3a^{-2}+2b^{-3}$

86. $a^{-2}+a^{-1}b^{-2}$

87. $x^{-1}y-xy^{-1}$

88. $x^2y^{-1}-x^{-3}y^2$

For Problems 89–98, find the following products and quotients. Assume that all variables appearing as exponents represent integers. For example,

$$(x^{2b})(x^{-b+1})=x^{2b+(-b+1)}=x^{b+1}$$

89. $(3x^a)(4x^{2a+1})$

90. $(5x^{-a})(-6x^{3a-1})$

91. $(x^a)(x^{-a})$

92. $(-2y^{3b})(-4y^{b+1})$

93. $\dfrac{x^{3a}}{x^a}$

94. $\dfrac{4x^{2a+1}}{2x^{a-2}}$

For Problems 99–102, express each number in scientific notation.

99. 62,000,000

100. 17,000,000,000

101. 0.000412

102. 0.000000078

For Problems 103–106, change each number from scientific notation to ordinary decimal form.

103. $(1.8)(10)^5$

104. $(5.41)(10)^7$

105. $(2.3)(10)^{-6}$

106. $(4.13)(10)^{-9}$

For Problems 107–112, use scientific notation and the properties of exponents to help perform the indicated operations.

107. $\dfrac{0.00052}{0.013}$

108. $\dfrac{(0.000075)(4,800,000)}{(15,000)(0.0012)}$

109. $\sqrt{900,000,000}$

110. $\sqrt{0.000004}$

111. $\sqrt{0.0009}$

112. $\dfrac{(0.00069)(0.0034)}{(0.0000017)(0.023)}$

■ ■ ■ **THOUGHTS INTO WORDS**

113. Explain how you would simplify $(3^{-1}\cdot 2^{-2})^{-1}$ and also how you would simplify $(3^{-1}+2^{-2})^{-1}$.

114. How would you explain why the product of x^2 and x^4 is x^6 and not x^8?

■ **GRAPHING CALCULATOR ACTIVITIES**

115. Use your calculator to check your answers for Problems 107–112.

116. Use your calculator to evaluate each of the following. Express final answers in ordinary notation.

    a. $(27,000)^2$

    b. $(450,000)^2$

c. $(14,800)^2$

d. $(1700)^3$

e. $(900)^4$

f. $(60)^5$

g. $(0.0213)^2$

h. $(0.000213)^2$

i. $(0.000198)^2$

j. $(0.000009)^3$

117. Use your calculator to estimate each of the following. Express final answers in scientific notation with the number between 1 and 10 rounded to the nearest one-thousandth.

a. $(4576)^4$ **b.** $(719)^{10}$

c. $(28)^{12}$ **d.** $(8619)^6$

e. $(314)^5$ **f.** $(145{,}723)^2$

118. Use your calculator to estimate each of the following. Express final answers in ordinary notation rounded to the nearest one-thousandth.

a. $(1.09)^5$ **b.** $(1.08)^{10}$

c. $(1.14)^7$ **d.** $(1.12)^{20}$

e. $(0.785)^4$ **f.** $(0.492)^5$

Answers to the Concept Quiz

1. True   **2.** False   **3.** False   **4.** False   **5.** False   **6.** False   **7.** False   **8.** True

0.3   Polynomials

Objectives

■ Add and subtract polynomials

■ Multiply polynomials.

■ Perform binomial expansions.

■ Divide a polynomial by a monomial.

Recall that algebraic expressions such as $5x$, $-6y^2$, $2x^{-1}y^{-2}$, $14a^2b$, $5x^{-4}$, and $-17ab^2c^3$ are called **terms.** Terms that contain variables with only nonnegative integers as exponents are called **monomials.** Of the previously listed terms, $5x$, $-6y^2$, $14a^2b$, and $-17ab^2c^3$ are monomials. The **degree** of a monomial is the sum of the exponents of the literal factors. For example, $7xy$ is of degree 2, whereas $14a^2b$ is of degree 3, and $-17ab^2c^3$ is of degree 6. If the monomial contains only one variable, then the exponent of that variable is the degree of the monomial. For example, $5x^3$ is of degree 3, and $-8y^4$ is of degree 4. Any nonzero constant term, such as 8, is of degree zero.

A **polynomial** is a monomial or a finite sum of monomials. Thus all of the following are examples of polynomials:

$$4x^2 \qquad\qquad 3x^2 - 2x - 4 \qquad 7x^4 - 6x^3 + 5x^2 - 2x - 1$$

$$3x^2y + 2y \qquad \frac{1}{5}a^2 - \frac{2}{3}b^2 \qquad 14$$

In addition to calling a polynomial with one term a monomial, we classify polynomials with two terms as **binomials** and those with three terms as **trinomials.** The **degree of a polynomial** is the degree of the term with the highest degree in the polynomial. The following examples illustrate some of this terminology:

The polynomial $4x^3y^4$ is a monomial in two variables of degree 7.

The polynomial $4x^2y - 2xy$ is a binomial in two variables of degree 3.

The polynomial $9x^2 - 7x - 1$ is a trinomial in one variable of degree 2.

■ Addition and Subtraction of Polynomials

Both adding polynomials and subtracting them rely on the same basic ideas. The commutative, associative, and distributive properties provide the basis for rearranging, regrouping, and combining similar terms. Consider the following addition problems:

$$(4x^2 + 5x + 1) + (7x^2 - 9x + 4) = (4x^2 + 7x^2) + (5x - 9x) + (1 + 4)$$
$$= 11x^2 - 4x + 5$$

$$(5x - 3) + (3x + 2) + (8x + 6) = (5x + 3x + 8x) + (-3 + 2 + 6)$$
$$= 16x + 5$$

The definition of subtraction as *adding the opposite* $[a - b = a + (-b)]$ extends to polynomials in general. The opposite of a polynomial can be formed by taking the opposite of each term. For example, the opposite of $3x^2 - 7x + 1$ is $-3x^2 + 7x - 1$. Symbolically, this is expressed as

$$-(3x^2 - 7x + 1) = -3x^2 + 7x - 1$$

You can also think in terms of the property $-x = -1(x)$ and the distributive property. Therefore,

$$-(3x^2 - 7x + 1) = -1(3x^2 - 7x + 1) = -3x^2 + 7x - 1$$

Now consider the following subtraction problems.

$$(7x^2 - 2x - 4) - (3x^2 + 7x - 1) = (7x^2 - 2x - 4) + (-3x^2 - 7x + 1)$$
$$= (7x^2 - 3x^2) + (-2x - 7x) + (-4 + 1)$$
$$= 4x^2 - 9x - 3$$

$$(4y^2 + 7) - (-3y^2 + y - 2) = (4y^2 + 7) + (3y^2 - y + 2)$$
$$= (4y^2 + 3y^2) + (-y) + (7 + 2)$$
$$= 7y^2 - y + 9$$

■ Multiplying Polynomials

The distributive property is usually stated as $a(b+c)=ab+ac$, but it can be extended as follows:

$$a(b + c + d) = ab + ac + ad$$
$$a(b + c + d + e) = ab + ac + ad + ae \qquad \text{etc.}$$

The commutative and associative properties, the properties of exponents, and the distributive property work together to form the basis for finding the product of a monomial and a polynomial with more than one term. The following example illustrates this idea:

$$3x^2(2x^2 + 5x + 3) = 3x^2(2x^2) + 3x^2(5x) + 3x^2(3)$$
$$= 6x^4 + 15x^3 + 9x^2$$

Extending the method of finding the product of a monomial and a polynomial to finding the product of two polynomials, each of which has more than one term, is again based on the distributive property:

$$(x + 2)(y + 5) = x(y + 5) + 2(y + 5)$$
$$= x(y) + x(5) + 2(y) + 2(5)$$
$$= xy + 5x + 2y + 10$$

In the next example, notice that each term of the first polynomial multiplies each term of the second polynomial:

$$(x - 3)(y + z + 3) = x(y + z + 3) - 3(y + z + 3)$$
$$= xy + xz + 3x - 3y - 3z - 9$$

Frequently, multiplying polynomials produces similar terms that can be combined, which simplifies the resulting polynomial:

$$(x + 5)(x + 7) = x(x + 7) + 5(x + 7)$$
$$= x^2 + 7x + 5x + 35$$
$$= x^2 + 12x + 35$$

In a previous algebra course, you may have developed a shortcut for multiplying binomials, as illustrated by Figure 0.17.

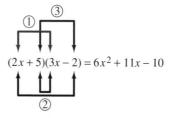

Figure 0.17

STEP 1   Multiply $(2x)(3x)$.

STEP 2   Multiply $(5)(3x)$ and $(2x)(-2)$ and combine.

STEP 3   Multiply $(5)(-2)$.

Remark: Shortcuts can be very helpful for certain manipulations in mathematics. But a word of caution: Do not lose the understanding of what you are doing. Make sure that you are able to do the manipulation without the shortcut.

Keep in mind that the shortcut illustrated in Figure 0.17 applies only to multiplying two binomials. The next example applies the distributive property to find the product of a binomial and a trinomial:

$$(x - 2)(x^2 - 3x + 4) = x(x^2 - 3x + 4) - 2(x^2 - 3x + 4)$$
$$= x^3 - 3x^2 + 4x - 2x^2 + 6x - 8$$
$$= x^3 - 5x^2 + 10x - 8$$

In this example we are claiming that

$$(x - 2)(x^2 - 3x + 4) = x^3 - 5x^2 + 10x - 8$$

for all real numbers. In addition to going back over our work, how can we verify such a claim? Obviously, we cannot try all real numbers, but trying at least one number gives us a partial check. Let's try the number 4:

$$(x - 2)(x^2 - 3x + 4) = (4 - 2)(4^2 - 3(4) + 4)$$
$$= 2(16 - 12 + 4)$$
$$= 2(8)$$
$$= 16$$

$$x^3 - 5x^2 + 10x - 8 = 4^3 - 5(4)^2 + 10(4) - 8$$
$$= 64 - 80 + 40 - 8$$
$$= 16$$

We can also use a graphical approach as a partial check for such a problem. In Figure 0.18, we let $Y_1 = (x - 2)(x^2 - 3x + 4)$ and $Y_2 = x^3 - 5x^2 + 10x - 8$ and graphed them on the same set of axes. Note that the graphs appear to be identical.

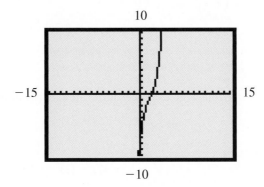

Figure 0.18

Remark: Graphing on the Cartesian coordinate system is not formally reviewed in this text until Chapter 2. However, we feel confident that your knowledge of this topic from previous mathematics courses is sufficient for what we are doing at this time.

Exponents can also be used to indicate repeated multiplication of polynomials. For example, $(3x - 4y)^2$ means $(3x - 4y)(3x - 4y)$, and $(x + 4)^3$ means $(x + 4)(x + 4)(x + 4)$. Therefore, raising a polynomial to a power is merely another multiplication problem.

$$(3x - 4y)^2 = (3x - 4y)(3x - 4y)$$
$$= 9x^2 - 24xy + 16y^2$$

[*Hint*: When squaring a binomial, be careful not to forget the middle term. That is, $(x + 5)^2 \neq x^2 + 25$; instead, $(x + 5)^2 = x^2 + 10x + 25$.]

$$(x + 4)^3 = (x + 4)(x + 4)(x + 4)$$
$$= (x + 4)(x^2 + 8x + 16)$$
$$= x(x^2 + 8x + 16) + 4(x^2 + 8x + 16)$$
$$= x^3 + 8x^2 + 16x + 4x^2 + 32x + 64$$
$$= x^3 + 12x^2 + 48x + 64$$

■ Special Patterns

In multiplying binomials, you should learn to recognize some special patterns. These patterns can be used to find products, and some of them will be helpful later when you are factoring polynomials.

$$(a + b)^2 = a^2 + 2ab + b^2$$
$$(a - b)^2 = a^2 - 2ab + b^2$$
$$(a + b)(a - b) = a^2 - b^2$$
$$(a + b)^3 = a^3 + 3a^2b + 3ab^2 + b^3$$
$$(a - b)^3 = a^3 - 3a^2b + 3ab^2 - b^3$$

The three following examples illustrate the first three patterns, respectively:

$$(2x + 3)^2 = (2x)^2 + 2(2x)(3) + (3)^2$$
$$= 4x^2 + 12x + 9$$
$$(5x - 2)^2 = (5x)^2 - 2(5x)(2) + (2)^2$$
$$= 25x^2 - 20x + 4$$
$$(3x + 2y)(3x - 2y) = (3x)^2 - (2y)^2 = 9x^2 - 4y^2$$

In the first two examples, the resulting trinomial is called a **perfect-square trinomial;** it is the result of squaring a binomial. In the third example, the resulting binomial is called the **difference of two squares.** Later, we will use both of these patterns extensively when factoring polynomials.

The cubing-of-a-binomial patterns are helpful primarily when you are multiplying. These patterns can shorten the work of cubing a binomial, as the next two examples illustrate:

$$(3x + 2)^3 = (3x)^3 + 3(3x)^2(2) + 3(3x)(2)^2 + (2)^3$$
$$= 27x^3 + 54x^2 + 36x + 8$$
$$(5x - 2y)^3 = (5x)^3 - 3(5x)^2(2y) + 3(5x)(2y)^2 - (2y)^3$$
$$= 125x^3 - 150x^2y + 60xy^2 - 8y^3$$

Keep in mind that these multiplying patterns are useful shortcuts, but if you forget them, simply revert to applying the distributive property.

■ Binomial Expansion Pattern

It is possible to write the expansion of $(a + b)^n$, where n is *any* positive integer, without showing all of the intermediate steps of multiplying and combining similar terms. To do this, let's observe some patterns in the following examples; each one can be verified by direct multiplication:

$$(a + b)^1 = a + b$$
$$(a + b)^2 = a^2 + 2ab + b^2$$
$$(a + b)^3 = a^3 + 3a^2b + 3ab^2 + b^3$$
$$(a + b)^4 = a^4 + 4a^3b + 6a^2b^2 + 4ab^3 + b^4$$
$$(a + b)^5 = a^5 + 5a^4b + 10a^3b^2 + 10a^2b^3 + 5ab^4 + b^5$$

First, note the patterns of the exponents for a and b on a term-by-term basis. The exponents of a begin with the exponent of the binomial and decrease by 1, term by term, until the last term, which has $a^0 = 1$. The exponents of b begin with zero ($b^0 = 1$) and increase by 1, term by term, until the last term, which contains b to the power of the original binomial. In other words, the variables in the expansion of $(a + b)^n$ have the pattern

$$a^n, \quad a^{n-1}b, \quad a^{n-2}b^2, \quad \ldots, \quad ab^{n-1}, \quad b^n$$

where, for each term, the *sum* of the exponents of a and b is n.

Next, let's arrange the *coefficients* in a triangular formation; this yields an easy-to-remember pattern.

```
          1       1
       1     2     1
    1     3     3     1
  1     4     6     4     1
1     5    10    10     5     1
```

Row number n in the formation contains the coefficients of the expansion of $(a + b)^n$. For example, the fifth row contains 1 5 10 10 5 1, and these numbers are the coefficients of the terms in the expansion of $(a + b)^5$. Furthermore, each can be formed from the previous row as follows.

1. Start and end each row with 1.

2. All other entries result from adding the two numbers in the row immediately above, one number to the left and one number to the right.

Thus from row 5, we can form row 6.

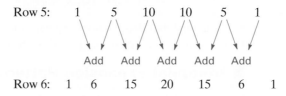

Now we can use these seven coefficients and our discussion about the exponents to write out the expansion for $(a + b)^6$.

$$(a + b)^6 = a^6 + 6a^5b + 15a^4b^2 + 20a^3b^3 + 15a^2b^4 + 6ab^5 + b^6$$

Remark: The triangular formation of numbers that we have been discussing is often referred to as *Pascal's triangle*. This is in honor of Blaise Pascal, a 17th century mathematician, to whom the discovery of this pattern is attributed.

Let's consider two more examples using Pascal's triangle and the exponent relationships.

E X A M P L E   1   Expand $(a - b)^4$.

Solution

We can treat $a - b$ as $a + (-b)$ and use the fourth row of Pascal's triangle $(1, 4, 6, 4, 1)$ to obtain the coefficients:

$$[a + (-b)]^4 = a^4 + 4a^3(-b) + 6a^2(-b)^2 + 4a(-b)^3 + (-b)^4$$
$$= a^4 - 4a^3b + 6a^2b^2 - 4ab^3 + b^4$$

▶ **Now work Problem 57.**   ■

E X A M P L E   2   Expand $(2x + 3y)^5$.

Solution

Let $2x = a$ and $3y = b$. The coefficients $(1, 5, 10, 10, 5, 1)$ come from the fifth row of Pascal's triangle:

$$(2x + 3y)^5 = (2x)^5 + 5(2x)^4(3y) + 10(2x)^3(3y)^2 + 10(2x)^2(3y)^3 + 5(2x)(3y)^4 + (3y)^5$$
$$= 32x^5 + 240x^4y + 720x^3y^2 + 1080x^2y^3 + 810xy^4 + 243y^5$$

▶ **Now work Problem 65.**   ■

■ Dividing Polynomials by Monomials

In Section 0.5 we will review the addition and subtraction of rational expressions using the properties

$$\frac{a}{b} + \frac{c}{b} = \frac{a+c}{b} \quad \text{and} \quad \frac{a}{b} - \frac{c}{b} = \frac{a-c}{b}$$

These properties can also be viewed as

$$\frac{a+c}{b} = \frac{a}{b} + \frac{c}{b} \quad \text{and} \quad \frac{a-c}{b} = \frac{a}{b} - \frac{c}{b}$$

Together with our knowledge of dividing monomials, these properties provide the basis for dividing polynomials by monomials. Consider the following examples:

$$\frac{18x^3 + 24x^2}{6x} = \frac{18x^3}{6x} + \frac{24x^2}{6x} = 3x^2 + 4x$$

$$\frac{35x^2y^3 - 55x^3y^4}{5xy^2} = \frac{35x^2y^3}{5xy^2} - \frac{55x^3y^4}{5xy^2} = 7xy - 11x^2y^2$$

Therefore, to divide a polynomial by a monomial, we divide each term of the polynomial by the monomial. As with many skills, once you feel comfortable with the process, you may then choose to perform some of the steps mentally. Your work could take the following format:

$$\frac{40x^4y^5 + 72x^5y^7}{8x^2y} = 5x^2y^4 + 9x^3y^6$$

$$\frac{36a^3b^4 - 48a^3b^3 + 64a^2b^5}{-4a^2b^2} = -9ab^2 + 12ab - 16b^3$$

CONCEPT QUIZ

For Problems 1–8, answer true or false.

1. The variables of a monomial term have exponents that are either positive integers or zero.

2. The term, 3^2xy^2, is of degree 5.

3. Any nonzero constant term is of degree zero.

4. A polynomial is a monomial or a finite sum of monomials.

5. A polynomial with three terms is classified as a binomial.

6. $(x - 6)^2 = x^2 + 36$

7. A perfect-square trinomial is the result when a trinomial is squared.

8. Row number 4 in Pascal's triangle contains the coefficients of the expansion of $(a + b)^3$.

Problem Set 0.3

For Problems 1–10, perform the indicated operations.

1. $(5x^2 - 7x - 2) + (9x^2 + 8x - 4)$

2. $(-9x^2 + 8x + 4) + (7x^2 - 5x - 3)$

3. $(14x^2 - x - 1) - (15x^2 + 3x + 8)$

4. $(-3x^2 + 2x + 4) - (4x^2 + 6x - 5)$

5. $(3x - 4) - (6x + 3) + (9x - 4)$

6. $(7a - 2) - (8a - 1) - (10a - 2)$

7. $(8x^2 - 6x - 2) + (x^2 - x - 1) - (3x^2 - 2x + 4)$

8. $(12x^2 + 7x - 2) - (3x^2 + 4x + 5) + (-4x^2 - 7x - 2)$

9. $5(x - 2) - 4(x + 3) - 2(x + 6)$

10. $3(2x - 1) - 2(3x + 4) - 4(5x - 1)$

For Problems 11–54, find the indicated products. Remember the special patterns that we discussed in this section.

11. $3xy(4x^2y + 5xy^2)$

12. $-2ab^2(3a^2b - 4ab^3)$

13. $6a^3b^2(5ab - 4a^2b + 3ab^2)$

14. $-xy^4(5x^2y - 4xy^2 + 3x^2y^2)$

15. $(x + 8)(x + 12)$

16. $(x - 9)(x + 6)$

17. $(n - 4)(n - 12)$

18. $(n + 6)(n - 10)$

19. $(s - t)(x + y)$

20. $(a + b)(c + d)$

21. $(3x - 1)(2x + 3)$

22. $(5x + 2)(3x + 4)$

23. $(4x - 3)(3x - 7)$

24. $(4n + 3)(6n - 1)$

25. $(x + 4)^2$

26. $(x - 6)^2$

27. $(2n + 3)^2$

28. $(3n - 5)^2$

29. $(x + 2)(x - 4)(x + 3)$

30. $(x - 1)(x + 6)(x - 5)$

31. $(x - 1)(2x + 3)(3x - 2)$

32. $(2x + 5)(x - 4)(3x + 1)$

33. $(x - 1)(x^2 + 3x - 4)$

34. $(t + 1)(t^2 - 2t - 4)$

35. $(t - 1)(t^2 + t + 1)$

36. $(2x - 1)(x^2 + 4x + 3)$

37. $(3x + 2)(2x^2 - x - 1)$

38. $(3x - 2)(2x^2 + 3x + 4)$

39. $(x^2 + 2x - 1)(x^2 + 6x + 4)$

40. $(x^2 - x + 4)(2x^2 - 3x - 1)$

41. $(5x - 2)(5x + 2)$ **42.** $(3x - 4)(3x + 4)$

43. $(x^2 - 5x - 2)^2$ **44.** $(-x^2 + x - 1)^2$

45. $(2x + 3y)(2x - 3y)$ **46.** $(9x + y)(9x - y)$

47. $(x + 5)^3$ **48.** $(x - 6)^3$

49. $(2x + 1)^3$ **50.** $(3x + 4)^3$

51. $(4x - 3)^3$ **52.** $(2x - 5)^3$

53. $(5x - 2y)^3$ **54.** $(x + 3y)^3$

For Problems 55–66, use Pascal's triangle to help expand each expression.

55. $(a + b)^7$

56. $(a + b)^8$

57. $(x - y)^5$

58. $(x - y)^6$

59. $(x + 2y)^4$

60. $(2x + y)^5$

61. $(2a - b)^6$

62. $(3a - b)^4$

63. $(x^2 + y)^7$

64. $(x + 2y^2)^7$

65. $(2a - 3b)^5$

66. $(4a - 3b)^3$

For Problems 67–72, perform the indicated divisions.

67. $\dfrac{15x^4 - 25x^3}{5x^2}$

68. $\dfrac{-48x^8 - 72x^6}{-8x^4}$

69. $\dfrac{30a^5 - 24a^3 + 54a^2}{-6a}$

70. $\dfrac{18x^3y^2 + 27x^2y^3}{3xy}$

71. $\dfrac{-20a^3b^2 - 44a^4b^5}{-4a^2b}$

72. $\dfrac{21x^5y^6 + 28x^4y^3 - 35x^5y^4}{7x^2y^3}$

For Problems 73–82, find the indicated products. Assume all variables that appear as exponents represent integers.

73. $(x^a + y^b)(x^a - y^b)$

74. $(x^{2a} + 1)(x^{2a} - 3)$

75. $(x^b + 4)(x^b - 7)$

76. $(3x^a - 2)(x^a + 5)$

77. $(2x^b - 1)(3x^b + 2)$

78. $(2x^a - 3)(2x^a + 3)$

79. $(x^{2a} - 1)^2$

80. $(x^{3b} + 2)^2$

81. $(x^a - 2)^3$

82. $(x^b + 3)^3$

■ ■ ■ THOUGHTS INTO WORDS

83. Describe how to multiply two binomials.

84. Describe how to multiply a binomial and a trinomial.

85. Determine the number of terms in the product of $(x + y)$ and $(a + b + c + d)$ without doing the multiplication. Explain how you arrived at your answer.

GRAPHING CALCULATOR ACTIVITIES

86. Use the computing feature of your graphing calculator to check at least one real number for your answers for Problems 29–40.

87. Use the graphing feature of your graphing calculator to give visual support for your answers for Problems 47–52.

88. Some of the product patterns can be used to do arithmetic computations mentally. For example, let's use the pattern $(a + b)^2 = a^2 + 2ab + b^2$ to compute 31^2 mentally. Your thought process should be "$31^2 = (30 + 1)^2 = 30^2 + 2(30)(1) + 1^2 = 961$." Compute each of the following numbers mentally, and then check your answers with your calculator.

   a. 21^2 **b.** 41^2

   c. 71^2 **d.** 32^2

   e. 52^2 **f.** 82^2

89. Use the pattern $(a - b)^2 = a^2 - 2ab + b^2$ to compute each of the following numbers mentally, and then check your answers with your calculator.

   a. 19^2 **b.** 29^2

   c. 49^2 **d.** 79^2

   e. 38^2 **f.** 58^2

90. Every whole number with a units digit of 5 can be represented by the expression $10x + 5$, where x is a whole number. For example, $35 = 10(3) + 5$ and $145 = 10(14) + 5$. Now let's observe the following pattern when squaring such a number:

$$(10x + 5)^2 = 100x^2 + 100x + 25$$
$$= 100x(x + 1) + 25$$

The pattern inside the dashed box can be stated as "add 25 to the product of x, $x + 1$, and 100." Thus to compute 35^2 mentally, we can think "$35^2 = 3(4)(100) + 25 = 1225$." Compute each of the following numbers mentally, and then check your answers with your calculator.

   a. 15^2 **b.** 25^2

   c. 45^2 **d.** 55^2

   e. 65^2 **f.** 75^2

   g. 85^2 **h.** 95^2

   i. 105^2

0.4 Factoring Polynomials

Objectives

■ Factor out a common factor.

■ Factor the difference of two squares.

■ Factor trinomials.

■ Factor the sum or difference of two cubes.

If a polynomial is equal to the product of other polynomials, then each polynomial in the product is called a **factor** of the original polynomial. For example, because $x^2 - 4$ can be expressed as $(x + 2)(x - 2)$, we say that $x + 2$ and $x - 2$ are factors of $x^2 - 4$. The process of expressing a polynomial as a product of polynomials is called **factoring.** In this section we will consider methods of factoring polynomials with integer coefficients.

In general, factoring is the reverse of multiplication, so we can use our knowledge of multiplication to help develop factoring techniques. For example, we previously used the distributive property to find the product of a monomial and a polynomial, as the next examples illustrate.

$$3(x + 2) = 3(x) + 3(2) = 3x + 6$$

$$3x(x + 4) = 3x(x) + 3x(4) = 3x^2 + 12x$$

For factoring purposes, the distributive property [now in the form $ab + ac = a(b + c)$] can be used to reverse the process.

$$3x + 6 = 3(x) + 3(2) = 3(x + 2)$$

$$3x^2 + 12x = 3x(x) + 3x(4) = 3x(x + 4)$$

Polynomials can be factored in a variety of ways. Consider some factorizations of $3x^2 + 12x$:

$$3x^2 + 12x = 3x(x + 4) \quad \text{or} \quad 3x^2 + 12x = 3(x^2 + 4x) \quad \text{or}$$

$$3x^2 + 12x = x(3x + 12) \quad \text{or} \quad 3x^2 + 12x = \frac{1}{2}(6x^2 + 24x)$$

We are, however, primarily interested in the first of these factorization forms; we refer to it as the **completely factored form.** A polynomial with integral coefficients is in completely factored form if:

1. it is expressed as a product of polynomials with *integral coefficients,* and

2. no polynomial, other than a monomial, within the factored form can be further factored into polynomials with integral coefficients.

Do you see why only the first of the factored forms of $3x^2 + 12x$ is said to be in completely factored form? In each of the other three forms, the polynomial inside the parentheses can be factored further. Moreover, in the last form, $\frac{1}{2}(6x^2 + 24x)$, the condition of using only integers is violated.

This application of the distributive property is often referred to as **factoring out the highest common monomial factor.** The following examples illustrate the process:

$$12x^3 + 16x^2 = 4x^2(3x + 4)$$

$$8ab - 18b = 2b(4a - 9)$$

$$6x^2y^3 + 27xy^4 = 3xy^3(2x + 9y)$$

$$30x^3 + 42x^4 - 24x^5 = 6x^3(5 + 7x - 4x^2)$$

Sometimes there may be a common *binomial* factor rather than a common monomial factor. For example, each of the two terms in the expression $x(y + 2) + z(y + 2)$ has a binomial factor of $y + 2$. Thus we can factor $y + 2$ from each term and obtain the following result:

$$x(y + 2) + z(y + 2) = (y + 2)(x + z)$$

Consider a few more examples involving a common binomial factor:

$$a^2(b + 1) + 2(b + 1) = (b + 1)(a^2 + 2)$$

$$x(2y - 1) - y(2y - 1) = (2y - 1)(x - y)$$

$$x(x + 2) + 3(x + 2) = (x + 2)(x + 3)$$

It may seem that a given polynomial exhibits no apparent common monomial or binomial factor. Such is the case with $ab + 3c + bc + 3a$. However, by using the commutative property to rearrange the terms, we can factor it as follows.

$$ab + 3c + bc + 3a = ab + 3a + bc + 3c$$

$$= a(b + 3) + c(b + 3) \qquad \text{Factor } a \text{ from the first two terms and } c \text{ from the last two terms}$$

$$= (b + 3)(a + c) \qquad \text{Factor } b + 3 \text{ from both terms}$$

This factoring process is referred to as **factoring by grouping.** Let's consider another example of this type.

$$ab^2 - 4b^2 + 3a - 12 = b^2(a - 4) + 3(a - 4) \qquad \text{Factor } b^2 \text{ from the first two terms, 3 from the last two}$$

$$= (a - 4)(b^2 + 3) \qquad \text{Factor the common binomial from both terms}$$

■ Difference of Two Squares

In Section 0.3 we called your attention to some special multiplication patterns. One of these patterns was

$$(a + b)(a - b) = a^2 - b^2$$

This same pattern, viewed as a factoring pattern,

$$a^2 - b^2 = (a + b)(a - b)$$

is referred to as the **difference of two squares.** Applying the pattern is a fairly simple process, as these next examples illustrate.

$$x^2 - 16 = (x)^2 - (4)^2 = (x + 4)(x - 4)$$
$$4x^2 - 25 = (2x)^2 - (5)^2 = (2x + 5)(2x - 5)$$

Because multiplication is commutative, the order in which we write the factors is not important. For example, $(x + 4)(x - 4)$ can also be written $(x - 4)(x + 4)$.

You must be careful not to assume an analogous factoring pattern for the *sum* of two squares; *it does not exist.* For example, $x^2 + 4 \neq (x + 2)(x + 2)$ because $(x + 2)(x + 2) = x^2 + 4x + 4$. We say that a polynomial such as $x^2 + 4$ is **not factorable using integers.**

Sometimes the difference-of-two-squares pattern can be applied more than once, as the next example illustrates:

$$16x^4 - 81y^4 = (4x^2 + 9y^2)(4x^2 - 9y^2) = (4x^2 + 9y^2)(2x + 3y)(2x - 3y)$$

It may also happen that the squares are not just simple monomial squares. These next three examples illustrate such polynomials.

$$(x + 3)^2 - y^2 = [(x + 3) + y][(x + 3) - y] = (x + 3 + y)(x + 3 - y)$$
$$4x^2 - (2y + 1)^2 = [2x + (2y + 1)][2x - (2y + 1)]$$
$$= (2x + 2y + 1)(2x - 2y - 1)$$
$$(x - 1)^2 - (x + 4)^2 = [(x - 1) + (x + 4)][(x - 1) - (x + 4)]$$
$$= (x - 1 + x + 4)(x - 1 - x - 4)$$
$$= (2x + 3)(-5)$$

It is possible that both the technique of factoring out a common monomial factor and the pattern of the difference of two squares can be applied to the same problem. *In general, it is best to look first for a common monomial factor.* Consider the following examples.

$$2x^2 - 50 = 2(x^2 - 25)$$
$$= 2(x + 5)(x - 5)$$
$$48y^3 - 27y = 3y(16y^2 - 9)$$
$$= 3y(4y + 3)(4y - 3)$$

$$9x^2 - 36 = 9(x^2 - 4)$$
$$= 9(x + 2)(x - 2)$$

■ Factoring Trinomials

Expressing a trinomial as the product of two binomials is one of the most common factoring techniques used in algebra. As before, to develop a factoring technique we first look at some multiplication ideas. Let's consider the product $(x + a)(x + b)$, using the distributive property to show how each term of the resulting trinomial is formed:

$$(x + a)(x + b) = x(x + b) + a(x + b)$$
$$= x(x) + x(b) + a(x) + a(b)$$
$$= x^2 + (a + b)x + ab$$

Notice that the coefficient of the middle term is the *sum* of a and b and that the last term is the *product* of a and b. These two relationships can be used to factor trinomials. Let's consider some examples.

E X A M P L E   1

Factor $x^2 + 12x + 20$.

Solution

We need two integers whose sum is 12 and whose product is 20. The numbers are 2 and 10, and we can complete the factoring as follows:

$$x^2 + 12x + 20 = (x + 2)(x + 10)$$ ■

E X A M P L E   2

Factor $x^2 - 3x - 54$.

Solution

We need two integers whose sum is -3 and whose product is -54. The integers are -9 and 6, and we can factor as follows:

$$x^2 - 3x - 54 = (x - 9)(x + 6)$$

▶ **Now work Problem 17.** ■

E X A M P L E   3

Factor $x^2 + 7x + 16$.

Solution

We need two integers whose sum is 7 and whose product is 16. The only possible pairs of factors of 16 are $1 \cdot 16$, $2 \cdot 8$, and $4 \cdot 4$. A sum of 7 is not produced by any of these pairs, so the polynomial $x^2 + 7x + 16$ is *not factorable using integers*. ■

■ Trinomials of the Form $ax^2 + bx + c$

Now let's consider factoring trinomials where the coefficient of the squared term is not one. First, let's illustrate an informal trial-and-error technique that works well for certain types of trinomials. This technique is based on our knowledge of multiplication of binomials.

E X A M P L E   4

Factor $3x^2 + 5x + 2$.

Solution

By looking at the first term, $3x^2$, and the positive signs of the other two terms, we know that the binomials are of the form

$$(x + \underline{\hspace{0.3cm}})(3x + \underline{\hspace{0.3cm}})$$

Because the factors of the last term, 2, are 1 and 2, we have only the following two possibilities to try.

$$(x + 2)(3x + 1) \qquad \text{or} \qquad (x + 1)(3x + 2)$$

By checking the middle term formed in each of these products, we find that the second possibility yields the desired middle term of $5x$. Therefore

$$3x^2 + 5x + 2 = (x + 1)(3x + 2)$$

▶ **Now work Problem 25.**   ■

E X A M P L E   5

Factor $8x^2 - 30xy + 7y^2$.

Solution

First, observe that the first term, $8x^2$, can be written as $2x \cdot 4x$ or $x \cdot 8x$. Second, because the middle term is negative and the last term is positive, we know that the binomials are of the form

$$(2x - \underline{\hspace{0.3cm}})(4x - \underline{\hspace{0.3cm}}) \qquad \text{or} \qquad (x - \underline{\hspace{0.3cm}})(8x - \underline{\hspace{0.3cm}})$$

Third, because the factors of the last term, $7y^2$, are $1y$ and $7y$, the following possibilities exist.

$$(2x - 1y)(4x - 7y) \qquad (2x - 7y)(4x - 1y)$$
$$(x - 1y)(8x - 7y) \qquad (x - 7y)(8x - 1y)$$

By checking the middle term formed in each of these products, we find that $(2x - 7y)(4x - 1y)$ produces the desired middle term of $-30xy$. Therefore

$$8x^2 - 30xy + 7y^2 = (2x - 7y)(4x - y)$$

▶ **Now work Problem 43.**   ■

E X A M P L E   6

Factor $10x^2 - 36x - 16$.

Solution

First, note that there is a common factor of 2. By using the distributive property we obtain $10x^2 - 36x - 16 = 2(5x^2 - 18x - 8)$. Now, let's determine if $5x^2 - 18x - 8$ can be factored. The first term, $5x^2$, can be written as $x \cdot 5x$. The last term, -8, can be written as $(-2)(4), (2)(-4), (-1)(8)$, or $(1)(-8)$. Therefore we have the following possibilities to try.

$$(x - 2)(5x + 4) \quad (x + 4)(5x - 2) \quad (x - 1)(5x + 8) \quad (x + 8)(5x - 1)$$

$$(x + 2)(5x - 4) \quad (x - 4)(5x + 2) \quad (x + 1)(5x - 8) \quad (x - 8)(5x + 1)$$

By checking the middle terms, we find that $(x - 4)(5x + 2)$ yields the desired middle term of $-18x$. Thus

$$10x^2 - 36x - 16 = 2(5x^2 - 18x - 8) = 2(x - 4)(5x + 2)$$

▶ **Now work Problem 63.**   ■

E X A M P L E   7

Factor $4x^2 + 6x + 9$.

Solution

The first term, $4x^2$, and the positive signs of the middle and last terms indicate that the binomials are of the form

$$(x + \underline{\ \ })(4x + \underline{\ \ }) \quad \text{or} \quad (2x + \underline{\ \ })(2x + \underline{\ \ })$$

Because the factors of the last term, 9, are 1 and 9 or 3 and 3, we have the following possibilities to try:

$$(x + 1)(4x + 9)$$

$$(x + 9)(4x + 1)$$

$$(x + 3)(4x + 3)$$

$$(2x + 1)(2x + 9)$$

$$(2x + 3)(2x + 3)$$

None of these possibilities yields a middle term of $6x$. Therefore $4x^2 + 6x + 9$ is *not factorable using integers.*

▶ **Now work Problem 45.**   ■

Certainly, as the number of possibilities increases, this trial-and-error technique for factoring becomes more tedious. The key idea is to organize your work so that all possibilities are considered. We have suggested one possible format in the previous examples. However, as you practice such problems, you may devise a format that works better for you. Whatever works best for you is the right approach.

There is another, more systematic technique that you may wish to use with some trinomials. It is an extension of the technique we used earlier with trinomials where the coefficient of the squared term was one. To see the basis of this technique, consider the following general product:

$$(px + r)(qx + s) = px(qx) + px(s) + r(qx) + r(s)$$
$$= (pq)x^2 + ps(x) + rq(x) + rs$$
$$= (pq)x^2 + (ps + rq)x + rs$$

Notice that the product of the coefficient of x^2 and the constant term is $pqrs$. Likewise, the product of the two coefficients of x (ps and rq) is also $pqrs$. Therefore, the coefficient of x must be a sum of the form $ps + rq$, such that the product of the coefficient of x^2 and the constant term is $pqrs$. Now let's see how this works in some specific examples.

EXAMPLE 8   Factor $6x^2 + 17x + 5$.

Solution

$$6x^2 + 17x + 5 \qquad \text{Sum of 17}$$

Product of $6 \cdot 5 = 30$

We need two integers whose sum is 17 and whose product is 30. The integers 2 and 15 satisfy these conditions. Therefore the middle term, $17x$, of the given trinomial can be expressed as $2x + 15x$, and we can proceed as follows:

$$6x^2 + 17x + 5 = 6x^2 + 2x + 15x + 5$$
$$= 2x(3x + 1) + 5(3x + 1) \qquad \text{Factor by grouping}$$
$$= (3x + 1)(2x + 5)$$

▶ **Now work Problem 23.**   ■

EXAMPLE 9   Factor $5x^2 - 18x - 8$.

Solution

$$5x^2 - 18x - 8 \qquad \text{Sum of } -18$$

Product of $5(-8) = -40$

We need two integers whose sum is -18 and whose product is -40. The integers -20 and 2 satisfy these conditions. Therefore the middle term, $-18x$, of the trinomial can be written $-20x + 2x$, and we can factor as follows:

$$5x^2 - 18x - 8 = 5x^2 - 20x + 2x - 8$$
$$= 5x(x - 4) + 2(x - 4)$$
$$= (x - 4)(5x + 2)$$

Now work Problem 39.

EXAMPLE 10

Factor $24x^2 + 2x - 15$.

Solution

$24x^2 + 2x - 15$ Sum of 2

Product of $24(-15) = -360$

We need two integers whose sum is 2 and whose product is -360. To help find these integers, let's factor 360 into primes:

$$360 = 2 \cdot 2 \cdot 2 \cdot 3 \cdot 3 \cdot 5$$

Now by grouping these factors in various ways, we find that $2 \cdot 2 \cdot 5 = 20$ and $2 \cdot 3 \cdot 3 = 18$, so we can use the integers 20 and -18 to produce a sum of 2 and a product of -360. Therefore, the middle term, $2x$, of the trinomial can be expressed as $20x - 18x$, and we can proceed as follows:

$$24x^2 + 2x - 15 = 24x^2 + 20x - 18x - 15$$
$$= 4x(6x + 5) - 3(6x + 5)$$
$$= (6x + 5)(4x - 3)$$

Now work Problem 61.

Probably the best way to check a factoring problem is to make sure the conditions for a polynomial to be completely factored are satisfied, and the product of the factors equals the given polynomial. We can also give some visual support to a factoring problem by graphing the given polynomial and its completely factored form on the same set of axes, as shown for Example 10 in Figure 0.19. Note that the graphs for $Y_1 = 24x^2 + 2x - 15$ and $Y_2 = (6x + 5)(4x - 3)$ appear to be identical.

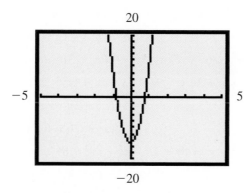

Figure 0.19

■ Sum and Difference of Two Cubes

Earlier in this section we discussed the difference-of-squares factoring pattern. We pointed out that no analogous sum-of-squares pattern exists; that is, a polynomial such as $x^2 + 9$ is not factorable using integers. However, there do exist patterns for both the *sum* and the *difference of two cubes*. These patterns come from the following special products:

$$(x + y)(x^2 - xy + y^2) = x(x^2 - xy + y^2) + y(x^2 - xy + y^2)$$
$$= x^3 - x^2y + xy^2 + x^2y - xy^2 + y^3$$
$$= x^3 + y^3$$
$$(x - y)(x^2 + xy + y^2) = x(x^2 + xy + y^2) - y(x^2 + xy + y^2)$$
$$= x^3 + x^2y + xy^2 - x^2y - xy^2 - y^3$$
$$= x^3 - y^3$$

Thus we can state the following factoring patterns:

$$x^3 + y^3 = (x + y)(x^2 - xy + y^2)$$
$$x^3 - y^3 = (x - y)(x^2 + xy + y^2)$$

Note how these patterns are used in the next three examples:

$$x^3 + 8 = x^3 + 2^3 = (x + 2)(x^2 - 2x + 4)$$
$$8x^3 - 27y^3 = (2x)^3 - (3y)^3 = (2x - 3y)(4x^2 + 6xy + 9y^2)$$
$$8a^6 + 125b^3 = (2a^2)^3 + (5b)^3 = (2a^2 + 5b)(4a^4 - 10a^2b + 25b^2)$$

We do want to leave you with one final word of caution. **Be sure to factor completely.** Sometimes more than one technique needs to be applied, or perhaps the same technique can be applied more than once. Study the following examples very carefully:

$$2x^2 - 8 = 2(x^2 - 4) = 2(x + 2)(x - 2)$$
$$3x^2 + 18x + 24 = 3(x^2 + 6x + 8) = 3(x + 4)(x + 2)$$
$$3x^3 - 3y^3 = 3(x^3 - y^3) = 3(x - y)(x^2 + xy + y^2)$$
$$a^4 - b^4 = (a^2 + b^2)(a^2 - b^2) = (a^2 + b^2)(a + b)(a - b)$$
$$x^4 - 6x^2 - 27 = (x^2 - 9)(x^2 + 3) = (x + 3)(x - 3)(x^2 + 3)$$
$$3x^4y + 9x^2y - 84y = 3y(x^4 + 3x^2 - 28)$$
$$= 3y(x^2 + 7)(x^2 - 4)$$
$$= 3y(x^2 + 7)(x + 2)(x - 2)$$
$$x^2 - y^2 + 8y - 16 = x^2 - (y^2 - 8y + 16)$$
$$= x^2 - (y - 4)^2$$
$$= (x - (y - 4))(x + (y - 4))$$
$$= (x - y + 4)(x + y - 4)$$

CONCEPT QUIZ   For Problems 1–8, answer true or false.

1. The process of expressing a polynomial as a product of polynomials is called factoring.

2. $x^2(5x - 10)$ is the completely factored form of $5x^2 - 10x^2$.

3. The polynomial, $3a^3b - 4c^2d + 5bd$, does not have a common factor.

4. The sum of two squares is not factorable using integers.

5. The sum of two cubes is not factorable using integers.

6. A factoring problem can be partially checked by making sure the product of the factors equals the polynomial.

7. All trinomials are factorable using integers.

8. All common factors are monomial factors.

Problem Set 0.4

Factor each polynomial completely. Indicate any that are not factorable using integers.

1. $6xy - 8xy^2$

2. $4a^2b^2 + 12ab^3$

3. $x(z + 3) + y(z + 3)$

4. $5(x + y) + a(x + y)$

5. $3x + 3y + ax + ay$

6. $ac + bc + a + b$

7. $ax - ay - bx + by$

8. $2a^2 - 3bc - 2ab + 3ac$

9. $9x^2 - 25$

10. $4x^2 + 9$

11. $1 - 81n^2$

12. $9x^2y^2 - 64$

13. $(x + 4)^2 - y^2$

14. $x^2 - (y - 1)^2$

15. $9s^2 - (2t - 1)^2$

16. $4a^2 - (3b + 1)^2$

17. $x^2 - 5x - 14$

18. $a^2 + 5a - 24$

19. $15 - 2x - x^2$

20. $40 - 6x - x^2$

21. $x^2 + 7x - 36$

22. $x^2 - 4xy - 5y^2$

23. $3x^2 - 11x + 10$

24. $2x^2 - 7x - 30$

25. $10x^2 + 17x + 7$

26. $8y^2 + 22y - 21$

27. $x^3 - 8$

28. $x^3 + 64$

29. $64x^3 + 27y^3$

30. $27x^3 - 8y^3$

31. $4x^2 + 16$

32. $n^3 - 49n$

33. $x^3 - 9x$

34. $12n^2 + 59n + 72$

35. $9a^2 - 42a + 49$

36. $1 - 16x^4$

37. $2n^3 + 6n^2 + 10n$

38. $x^2 - (y - 7)^2$

39. $10x^2 + 39x - 27$

40. $3x^2 + x - 5$

41. $36a^2 - 12a + 1$

42. $18n^3 + 39n^2 - 15n$

43. $8x^2 + 2xy - y^2$

44. $12x^2 + 7xy - 10y^2$

45. $2n^2 - n - 5$

46. $25t^2 - 100$

47. $2n^3 + 14n^2 - 20n$

48. $25n^2 + 64$

49. $4x^3 + 32$

50. $2x^3 - 54$

51. $x^4 - 4x^2 - 45$

52. $x^4 - x^2 - 12$

53. $2x^4y - 26x^2y - 96y$

54. $3x^4y - 15x^2y - 108y$

55. $(a + b)^2 - (c + d)^2$

56. $(a - b)^2 - (c - d)^2$

57. $x^2 + 8x + 16 - y^2$

58. $4x^2 + 12x + 9 - y^2$

59. $x^2 - y^2 - 10y - 25z$

60. $y^2 - x^2 + 16x - 64$

61. $60x^2 - 32x - 15$

62. $40x^2 + 37x - 63$

63. $84x^3 + 57x^2 - 60x$

64. $210x^3 - 102x^2 - 180x$

For Problems 65–74, factor each of the following, and assume that all variables appearing as exponents represent integers.

65. $x^{2a} - 16$

66. $x^{4n} - 9$

67. $x^{3n} - y^{3n}$

68. $x^{3a} + y^{6a}$

69. $x^{2a} - 3x^a - 28$

70. $x^{2a} + 10x^a + 21$

71. $2x^{2n} + 7x^n - 30$

72. $3x^{2n} - 16x^n - 12$

73. $x^{4n} - y^{4n}$

74. $16x^{2a} + 24x^a + 9$

75. Suppose that we want to factor $x^2 + 34x + 288$. We need to complete the following with two numbers whose sum is 34 and whose product is 288.

$$x^2 + 34x + 288 = (x + \_\_)(x + \_\_)$$

These numbers can be found as follows: Because we need a product of 288, let's consider the prime factorization of 288.

$$288 = 2^5 \cdot 3^2$$

Now we need to use five 2s and two 3s in the statement

$$(\ \ ) + (\ \ ) = 34$$

Because 34 is divisible by 2 but not by 4, four factors of 2 must be in one number and one factor of 2 in the other number. Also, because 34 is not divisible by 3, both factors of 3 must be in the same number. These facts aid us in determining that

$$(2 \cdot 2 \cdot 2 \cdot 2) + (2 \cdot 3 \cdot 3) = 34$$

or

$$16 \ + \ 18 \ = 34$$

Thus we can complete the original factoring problem:

$$x^2 + 34x + 288 = (x + 16)(x + 18)$$

Use this approach to factor each of the following expressions.

a. $x^2 + 35x + 96$

b. $x^2 + 27x + 176$

c. $x^2 - 45x + 504$

d. $x^2 - 26x + 168$

e. $x^2 + 60x + 896$

f. $x^2 - 84x + 1728$

■ ■ ■ **THOUGHTS INTO WORDS**

76. Describe, in words, the pattern for factoring the sum of two cubes.

77. What does it mean to say that the polynomial $x^2 + 5x + 7$ is not factorable using integers?

78. What role does the distributive property play in the factoring of polynomials?

79. Explain your thought process when factoring $30x^2 + 13x - 56$.

80. Consider the following approach to factoring $12x^2 + 54x + 60$:

$$12x^2 + 54x + 60 = (3x + 6)(4x + 10)$$
$$= 3(x + 2)(2)(2x + 5)$$
$$= 6(x + 2)(2x + 5)$$

Is this factoring process correct? What can you suggest to the person who used this approach?

Answers to the Concept Quiz

1. True   **2.** False   **3.** True   **4.** True   **5.** False   **6.** True   **7.** False   **8.** False

0.5 Rational Expressions

Objectives

- Simplify rational expressions.

- Multiply and divide rational expressions.

- Add and subtract rational expressions.

- Simplify complex fractions.

Indicated quotients of algebraic expressions are called **algebraic fractions** or **fractional expressions.** The indicated quotient of two polynomials is called a **rational expression.** (This is analogous to defining a rational number as the indicated quotient of two integers.) The following are examples of rational expressions:

$$\frac{3x^2}{5} \qquad \frac{x-2}{x+3} \qquad \frac{x^2+5x-1}{x^2-9} \qquad \frac{xy^2+x^2y}{xy} \qquad \frac{a^3-3a^2-5a-1}{a^4+a^3+6}$$

Because division by zero must be avoided, no values can be assigned to variables that will create a denominator of zero. Thus the rational expression $\frac{x-2}{x+3}$ is meaningful for all real number values of x except $x = -3$. Rather than making restrictions for each individual expression, we will merely assume that **all denominators represent nonzero real numbers.**

The basic properties of the real numbers can be used for working with rational expressions. For example, the property

$$\frac{a \cdot k}{b \cdot k} = \frac{a}{b}$$

which is used to reduce rational numbers, is also used to *simplify* rational expressions. Consider the following examples:

$$\frac{15xy}{25y} = \frac{3 \cdot \cancel{5} \cdot x \cdot \cancel{y}}{\cancel{5} \cdot 5 \cdot \cancel{y}} = \frac{3x}{5}$$

$$\frac{-9}{18x^2y} = -\frac{\overset{1}{\cancel{9}}}{\underset{2}{\cancel{18}}x^2y} = -\frac{1}{2x^2y}$$

Note that slightly different formats were used in these two examples. In the first one, we factored the coefficients into primes and then proceeded to simplify; however, in the second problem we simply divided a common factor of 9 out of both the numerator and denominator. This is basically a format issue and depends on your personal preference. Also notice that in the second example, we applied the property $\frac{-a}{b} = -\frac{a}{b}$. This is part of the general property that states

$$\frac{-a}{b} = \frac{a}{-b} = -\frac{a}{b}$$

The properties $(b^n)^m = b^{mn}$ and $(ab)^n = a^n b^n$ may also play a role when simplifying a rational expression, as the next example demonstrates.

$$\frac{(4x^3y)^2}{6x(y^2)^2} = \frac{4^2 \cdot (x^3)^2 \cdot y^2}{6 \cdot x \cdot y^4} = \frac{\overset{8}{\cancel{16}}\overset{x^5}{\cancel{x^6}}y^2}{\underset{3}{\cancel{6}}x\underset{y^2}{\cancel{y^4}}} = \frac{8x^5}{3y^2}$$

The factoring techniques discussed in the previous section can be used to factor numerators and denominators so that the property $(a \cdot k)/(b \cdot k) = a/b$ can be applied. Consider the following examples:

$$\frac{x^2 + 4x}{x^2 - 16} = \frac{x(\cancel{x + 4})}{(x - 4)(\cancel{x + 4})} = \frac{x}{x - 4}$$

$$\frac{5n^2 + 6n - 8}{10n^2 - 3n - 4} = \frac{(\cancel{5n - 4})(n + 2)}{(\cancel{5n - 4})(2n + 1)} = \frac{n + 2}{2n + 1}$$

$$\frac{x^3 + y^3}{x^2 + xy + 2x + 2y} = \frac{(x + y)(x^2 - xy + y^2)}{x(x + y) + 2(x + y)}$$

$$= \frac{(\cancel{x + y})(x^2 - xy + y^2)}{(\cancel{x + y})(x + 2)} = \frac{x^2 - xy + y^2}{x + 2}$$

$$\frac{6x^3y - 6xy}{x^3 + 5x^2 + 4x} = \frac{6xy(x^2 - 1)}{x(x^2 + 5x + 4)} = \frac{6\cancel{x}y(\cancel{x + 1})(x - 1)}{\cancel{x}(\cancel{x + 1})(x + 4)} = \frac{6y(x - 1)}{x + 4}$$

Note that in the last example we left the numerator of the final fraction in factored form. This is often done if expressions other than monomials are involved. Either

$$\frac{6y(x - 1)}{x + 4} \qquad \text{or} \qquad \frac{6xy - 6y}{x + 4}$$

is an acceptable answer.

Remember that the quotient of any nonzero real number and its opposite is -1. For example, $6/-6 = -1$ and $-8/8 = -1$. Likewise, the indicated quotient of any polynomial and its opposite is equal to -1. For example,

$$\frac{a}{-a} = -1 \qquad \text{because } a \text{ and } -a \text{ are opposites}$$

$$\frac{a - b}{b - a} = -1 \qquad \text{because } a - b \text{ and } b - a \text{ are opposites}$$

$$\frac{x^2 - 4}{4 - x^2} = -1 \qquad \text{because } x^2 - 4 \text{ and } 4 - x^2 \text{ are opposites}$$

The next example illustrates how we use this idea when simplifying rational expressions.

$$\frac{4 - x^2}{x^2 + x - 6} = \frac{(2 + x)\boxed{(2 - x)}}{(x + 3)\boxed{(x - 2)}}$$

$$= (-1)\left(\frac{x + 2}{x + 3}\right) \qquad\qquad \frac{2 - x}{x - 2} = -1$$

$$= -\frac{x + 2}{x + 3} \qquad \text{or} \qquad \frac{-x - 2}{x + 3}$$

■ Multiplying and Dividing Rational Expressions

Multiplication of rational expressions is based on the following property:

$$\frac{a}{b} \cdot \frac{c}{d} = \frac{ac}{bd}$$

In other words, we multiply numerators and we multiply denominators and express the final product in simplified form. Study the following examples carefully and pay special attention to the formats used to organize the computational work.

$$\frac{3x}{4y} \cdot \frac{8y^2}{9x} = \frac{3 \cdot \overset{2}{8} \cdot x \cdot \overset{y}{y^2}}{\underset{3}{4 \cdot 9} \cdot x \cdot y} = \frac{2y}{3}$$

$$\frac{12x^2y}{-18xy} \cdot \frac{-24xy^2}{56y^3} = \frac{\overset{2}{12} \cdot \overset{8}{24} \cdot \overset{x^2}{x^3} \cdot y^3}{\underset{3}{18} \cdot \underset{7}{56} \cdot x \cdot \underset{y}{y^4}} = \frac{2x^2}{7y} \qquad \frac{12x^2y}{-18xy} = -\frac{12x^2y}{18xy} \quad \text{and} \quad \frac{-24xy^2}{56y^3} = -\frac{24xy^2}{56y^3}$$
$$\text{so the product is positive.}$$

$$\frac{y}{x^2 - 4} \cdot \frac{x + 2}{y^2} = \frac{y(x + 2)}{\underset{y}{y^2}(x + 2)(x - 2)} = \frac{1}{y(x - 2)}$$

$$\frac{x^2 - x}{x + 5} \cdot \frac{x^2 + 5x + 4}{x^4 - x^2} = \frac{x(x - 1)(x + 1)(x + 4)}{(x + 5)(\underset{x}{x^2})(x + 1)(x - 1)} = \frac{x + 4}{x(x + 5)}$$

To divide rational expressions, we merely apply the following property:

$$\frac{a}{b} \div \frac{c}{d} = \frac{a}{b} \cdot \frac{d}{c} = \frac{ad}{bc}$$

That is, the quotient of two rational expressions is the product of the first expression times the reciprocal of the second. Consider the following examples:

$$\frac{16x^2y}{24xy^3} \div \frac{9xy}{8x^2y^2} = \frac{16x^2y}{24xy^3} \cdot \frac{8x^2y^2}{9xy} = \frac{16 \cdot 8 \cdot \overset{x^2}{x^4} \cdot y^3}{\underset{3}{24} \cdot 9 \cdot x^2 \cdot \underset{y}{y^4}} = \frac{16x^2}{27y}$$

$$\frac{3a^2 + 12}{3a^2 - 15a} \div \frac{a^4 - 16}{a^2 - 3a - 10} = \frac{3a^2 + 12}{3a^2 - 15a} \cdot \frac{a^2 - 3a - 10}{a^4 - 16}$$

$$= \frac{3(a^2 + 4)(a - 5)(a + 2)}{3a(a - 5)(a^2 + 4)(a + 2)(a - 2)}$$

$$= \frac{1}{a(a - 2)}$$

■ Adding and Subtracting Rational Expressions

The following two properties provide the basis for adding and subtracting rational expressions:

$$\frac{a}{b} + \frac{c}{b} = \frac{a + c}{b}$$

$$\frac{a}{b} - \frac{c}{b} = \frac{a - c}{b}$$

These properties state that rational expressions with a common denominator can be added (or subtracted) by adding (or subtracting) the numerators and placing the result over the common denominator. Let's illustrate this idea.

$$\frac{8}{x - 2} + \frac{3}{x - 2} = \frac{8 + 3}{x - 2} = \frac{11}{x - 2}$$

$$\frac{9}{4y} - \frac{7}{4y} = \frac{9 - 7}{4y} = \frac{2}{4y} = \frac{1}{2y}$$

Don't forget to simplify the final result.

$$\frac{n^2}{n - 1} - \frac{1}{n - 1} = \frac{n^2 - 1}{n - 1} = \frac{(n + 1)(n - 1)}{n - 1} = n + 1$$

If we need to add or subtract rational expressions that do not have a common denominator, then we apply the property $a/b = (a \cdot k)/(b \cdot k)$ to obtain equivalent fractions with a common denominator. Study the next examples and again pay special attention to the format we used to organize our work.

Remark: Remember that the **least common multiple** of a set of whole numbers is the smallest nonzero whole number divisible by each of the numbers in the set. When we add or subtract rational numbers, the least common multiple of the denominators of those numbers is the **least common denominator (LCD).** This concept of a least common denominator can be extended to include polynomials.

E X A M P L E   1

Add $\dfrac{x + 2}{4} + \dfrac{3x + 1}{3}$.

Solution

By inspection we see that the LCD is 12.

$$\frac{x + 2}{4} + \frac{3x + 1}{3} = \left(\frac{x + 2}{4}\right) + \left(\frac{3x + 1}{3}\right)$$

$$= \frac{3(x + 2)}{12} + \frac{4(3x + 1)}{12}$$

$$= \frac{3x + 6 + 12x + 4}{12}$$

$$= \frac{15x + 10}{12}$$

▶ **Now work Problem 35.** ■

EXAMPLE 2 Perform the indicated operations.

$$\frac{x + 3}{10} + \frac{2x + 1}{15} - \frac{x - 2}{18}$$

Solution

If you cannot determine the LCD by inspection, then use the prime-factored forms of the denominators:

$$10 = 2 \cdot 5 \qquad 15 = 3 \cdot 5 \qquad 18 = 2 \cdot 3 \cdot 3$$

The LCD must contain one factor of 2, two factors of 3, and one factor of 5. Thus the LCD is $2 \cdot 3 \cdot 3 \cdot 5 = 90$.

$$\frac{x + 3}{10} + \frac{2x + 1}{15} - \frac{x - 2}{18} = \left(\frac{x + 3}{10}\right)\left(\frac{9}{9}\right) + \left(\frac{2x + 1}{15}\right)\left(\frac{6}{6}\right) - \left(\frac{x - 2}{18}\right)\left(\frac{5}{5}\right)$$

$$= \frac{9(x + 3)}{90} + \frac{6(2x + 1)}{90} - \frac{5(x - 2)}{90}$$

$$= \frac{9x + 27 + 12x + 6 - 5x + 10}{90}$$

$$= \frac{16x + 43}{90}$$

▶ **Now work Problem 37.** ■

The presence of variables in the denominators does not create any serious difficulty; our approach remains the same. Study the following examples very carefully. For each problem we use the same basic procedure: (1) Find the LCD. (2) Change each fraction to an equivalent fraction having the LCD as its denominator. (3) Add or subtract numerators and place this result over the LCD. (4) Look for possibilities to simplify the resulting fraction.

EXAMPLE 3

Add $\dfrac{3}{2x} + \dfrac{5}{3y}$.

Solution

Using an LCD of $6xy$, we can proceed as follows:

$$\frac{3}{2x} + \frac{5}{3y} = \left(\frac{3}{2x}\right)\left(\frac{3y}{3y}\right) + \left(\frac{5}{3y}\right)\left(\frac{2x}{2x}\right)$$

$$= \frac{9y}{6xy} + \frac{10x}{6xy}$$

$$= \frac{9y + 10x}{6xy}$$

▶ **Now work Problem 39.** ■

EXAMPLE 4

Subtract $\dfrac{7}{12ab} - \dfrac{11}{15a^2}$.

Solution

We can factor the numerical coefficients of the denominators into primes to help find the LCD.

$$\left.\begin{array}{l} 12ab = 2\cdot 2\cdot 3\cdot a\cdot b \\ 15a^2 = 3\cdot 5\cdot a^2 \end{array}\right\} \qquad \text{LCD} = 2\cdot 2\cdot 3\cdot 5\cdot a^2\cdot b = 60a^2b$$

$$\frac{7}{12ab} - \frac{11}{15a^2} = \left(\frac{7}{12ab}\right)\left(\frac{5a}{5a}\right) - \left(\frac{11}{15a^2}\right)\left(\frac{4b}{4b}\right)$$

$$= \frac{35a}{60a^2b} - \frac{44b}{60a^2b}$$

$$= \frac{35a - 44b}{60a^2b}$$

▶ **Now work Problem 40.** ■

EXAMPLE 5

Add $\dfrac{8}{x^2 - 4x} + \dfrac{2}{x}$.

Solution

$$\left.\begin{array}{l} x^2 - 4x = x(x - 4) \\ x = x \end{array}\right\} \qquad \text{LCD} = x(x - 4)$$

$$\frac{8}{x(x-4)} + \frac{2}{x} = \frac{8}{x(x-4)} + \left(\frac{2}{x}\right)\left(\frac{x-4}{x-4}\right)$$

$$= \frac{8}{x(x-4)} + \frac{2(x-4)}{x(x-4)}$$

$$= \frac{8 + 2x - 8}{x(x-4)}$$

$$= \frac{2\cancel{x}}{\cancel{x}(x-4)}$$

$$= \frac{2}{x-4}$$

▶ **Now work Problem 47.** ■

In Figure 0.20 we give some visual support for our answer in Example 5 by graphing $Y_1 = \frac{8}{x^2 - 4x} + \frac{2}{x}$ and $Y_2 = \frac{2}{x-4}$. Certainly their graphs appear to be identical, but a word of caution is needed here. Actually, the graph of $Y_1 = \frac{8}{x^2 - 4x} + \frac{2}{x}$ has a hole at $\left(0, -\frac{1}{2}\right)$ because x cannot equal zero. When you use a graphing calculator, this hole may not be detected. Except for the hole, the graphs are identical, and we are claiming that $\frac{8}{x^2 - 4x} + \frac{2}{x} = \frac{2}{x-4}$ for all values of x except 0 and 4.

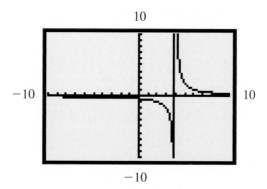

Figure 0.20

E X A M P L E   6

Add $\dfrac{3n}{n^2 + 6n + 5} + \dfrac{4}{n^2 - 7n - 8}$.

Solution

$$\left.\begin{array}{l} n^2 + 6n + 5 = (n+5)(n+1) \\ n^2 - 7n - 8 = (n-8)(n+1) \end{array}\right\} \quad \text{LCD} = (n+1)(n+5)(n-8)$$

$$\frac{3n}{n^2 + 6n + 5} + \frac{4}{n^2 - 7n - 8} = \left[\frac{3n}{(n+5)(n+1)}\right]\left(\frac{n-8}{n-8}\right) + \left[\frac{4}{(n-8)(n+1)}\right]\left(\frac{n+5}{n+5}\right)$$

$$= \frac{3n(n-8)}{(n+5)(n+1)(n-8)} + \frac{4(n+5)}{(n+5)(n+1)(n-8)}$$

$$= \frac{3n^2 - 24n + 4n + 20}{(n+5)(n+1)(n-8)}$$

$$= \frac{3n^2 - 20n + 20}{(n+5)(n+1)(n-8)}$$

▶ **Now work Problem 57.** ∎

■ Simplifying Complex Fractions

Fractional forms that contain rational expressions in the numerator and/or the denominator are called **complex fractions.** The following examples illustrate some approaches to simplifying complex fractions.

EXAMPLE 7 Simplify $\dfrac{\dfrac{3}{x} + \dfrac{2}{y}}{\dfrac{5}{x} - \dfrac{6}{y^2}}$.

Solution A

Treating the numerator as the sum of two rational expressions and the denominator as the difference of two rational expressions, we can proceed as follows.

$$\frac{\dfrac{3}{x} + \dfrac{2}{y}}{\dfrac{5}{x} - \dfrac{6}{y^2}} = \frac{\left(\dfrac{3}{x}\right)\left(\dfrac{y}{y}\right) + \left(\dfrac{2}{y}\right)\left(\dfrac{x}{x}\right)}{\left(\dfrac{5}{x}\right)\left(\dfrac{y^2}{y^2}\right) - \left(\dfrac{6}{y^2}\right)\left(\dfrac{x}{x}\right)}$$

$$= \frac{\dfrac{3y}{xy} + \dfrac{2x}{xy}}{\dfrac{5y^2}{xy^2} - \dfrac{6x}{xy^2}} = \frac{\dfrac{3y + 2x}{xy}}{\dfrac{5y^2 - 6x}{xy^2}}$$

$$= \frac{3y + 2x}{\cancel{xy}} \cdot \frac{\cancel{x}\cancel{y^2}\,^y}{5y^2 - 6x}$$

$$= \frac{y(3y + 2x)}{5y^2 - 6x}$$

Solution B

The LCD of all four denominators (x, y, x, and y^2) is xy^2. Let's multiply the entire complex fraction by a form of 1 — namely, $(xy^2)/(xy^2)$:

$$\dfrac{\dfrac{3}{x} + \dfrac{2}{y}}{\dfrac{5}{x} - \dfrac{6}{y^2}} = \left(\dfrac{\dfrac{3}{x} + \dfrac{2}{y}}{\dfrac{5}{x} - \dfrac{6}{y^2}}\right)\left(\dfrac{xy^2}{xy^2}\right)$$

$$= \dfrac{(xy^2)\left(\dfrac{3}{x}\right) + (xy^2)\left(\dfrac{2}{y}\right)}{(xy^2)\left(\dfrac{5}{x}\right) - (xy^2)\left(\dfrac{6}{y^2}\right)}$$

$$= \dfrac{3y^2 + 2xy}{5y^2 - 6x} \quad \text{or} \quad \dfrac{y(3y + 2x)}{5y^2 - 6x}$$

▶ **Now work Problem 71.** ■

Certainly either approach (Solution A or Solution B) will work with a problem such as Example 7. We suggest that you study Solution B very carefully. This approach works effectively with complex fractions when the LCD of all the denominators is easy to find. Let's look at a type of complex fraction used in certain calculus problems.

E X A M P L E   8

Simplify $\dfrac{\dfrac{1}{x + h} - \dfrac{1}{x}}{h}$.

Solution

$$\dfrac{\dfrac{1}{x + h} - \dfrac{1}{x}}{\dfrac{h}{1}} = \left[\dfrac{x(x + h)}{x(x + h)}\right]\left[\dfrac{\dfrac{1}{x + h} - \dfrac{1}{x}}{\dfrac{h}{1}}\right]$$

$$= \dfrac{x(x + h)\left(\dfrac{1}{x + h}\right) - x(x + h)\left(\dfrac{1}{x}\right)}{x(x + h)(h)}$$

$$= \dfrac{x - (x + h)}{hx(x + h)} = \dfrac{x - x - h}{hx(x + h)}$$

$$= \dfrac{-h}{hx(x + h)} = -\dfrac{1}{x(x + h)}$$

▶ **Now work Problem 83.** ■

Example 9 illustrates another way to simplify complex fractions.

E X A M P L E   9

Simplify $1 - \dfrac{n}{1 - \dfrac{1}{n}}$.

Solution

We first simplify the complex fraction by multiplying by n/n:

$$\left(\dfrac{n}{1 - \dfrac{1}{n}}\right)\left(\dfrac{n}{n}\right) = \dfrac{n^2}{n - 1}$$

Now we can perform the subtraction:

$$1 - \dfrac{n^2}{n - 1} = \left(\dfrac{n - 1}{n - 1}\right)\left(\dfrac{1}{1}\right) - \dfrac{n^2}{n - 1}$$

$$= \dfrac{n - 1}{n - 1} - \dfrac{n^2}{n - 1}$$

$$= \dfrac{n - 1 - n^2}{n - 1} \quad \text{or} \quad \dfrac{-n^2 + n - 1}{n - 1}$$

▶ **Now work Problem 79.**   ∎

Finally, we need to recognize that complex fractions are sometimes the result of applying the definition $b^{-n} = \dfrac{1}{b^n}$. Our final example illustrates this idea.

E X A M P L E   1 0

Simplify $\dfrac{2x^{-1} + y^{-1}}{x - 3y^{-2}}$.

Solution

First, let's apply $b^{-n} = \dfrac{1}{b^n}$.

$$\dfrac{2x^{-1} + y^{-1}}{x - 3y^{-2}} = \dfrac{\dfrac{2}{x} + \dfrac{1}{y}}{x - \dfrac{3}{y^2}}$$

Now we can proceed as in the previous examples:

$$\left(\dfrac{\dfrac{2}{x} + \dfrac{1}{y}}{x - \dfrac{3}{y^2}}\right)\left(\dfrac{xy^2}{xy^2}\right) = \dfrac{\dfrac{2}{x}(xy^2) + \dfrac{1}{y}(xy^2)}{x(xy^2) - \dfrac{3}{y^2}(xy^2)}$$

$$= \dfrac{2y^2 + xy}{x^2y^2 - 3x}$$

▶ **Now work Problem 91.**   ∎

CONCEPT QUIZ For Problems 1–7, answer true or false.

1. The indicated quotient of two polynomials is called a rational expression.

2. The rational expressions $\dfrac{3x - 4}{x + 2}$ is defined for all values of x.

3. The rational expressions $\dfrac{a^2 - 4}{b - 2}$ and $-\dfrac{4 - a^2}{2 - b}$ are equivalent.

4. The quotient of any nonzero polynomial and its opposite is -1.

5. To multiply rational expressions that do not have a common denominator we need to obtain equivalent fractions with a common denominator.

6. Complex fractions are fractional forms that contain rational expressions in the numerator and/or the denominator.

7. The difference of $\dfrac{3x - 4}{7x + 8}$ and $\dfrac{5x - 1}{7x + 8}$ would equal zero if $3x - 4 = 5x - 1$.

8. Under what conditions would the product of $\dfrac{x + 2}{x}$ and $\dfrac{x - 2}{x}$ be equal to zero?

Problem Set 0.5

For Problems 1–18, simplify each rational expression.

1. $\dfrac{14x^2y}{21xy}$

2. $\dfrac{-26xy^2}{65y}$

3. $\dfrac{-63xy^4}{-81x^2y}$

4. $\dfrac{x^2 - y^2}{x^2 + xy}$

5. $\dfrac{(2x^2y^2)^3}{(3xy)^2}$

6. $\dfrac{(3a^3b)^2}{6a^2(b^2)^2}$

7. $\dfrac{a^2 + 7a + 12}{a^2 - 6a - 27}$

8. $\dfrac{6x^2 + x - 15}{8x^2 - 10x - 3}$

9. $\dfrac{2x^3 + 3x^2 - 14x}{x^2y + 7xy - 18y}$

10. $\dfrac{3x - x^2}{x^2 - 9}$

11. $\dfrac{x^3 - y^3}{x^2 + xy - 2y^2}$

12. $\dfrac{ax - 3x + 2ay - 6y}{2ax - 6x + ay - 3y}$

13. $\dfrac{2y - 2xy}{x^2y - y}$

14. $\dfrac{16x^3y + 24x^2y^2 - 16xy^3}{24x^2y + 12xy^2 - 12y^3}$

15. $\dfrac{8x^2 + 4xy - 2x - y}{4x^2 - 4xy - x + y}$

16. $\dfrac{2x^3 + 2y^3}{2x^2 + 6x + 2xy + 6y}$

17. $\dfrac{27x^3 + 8y^3}{3x^2 - 15x + 2xy - 10y}$

18. $\dfrac{x^3 + 64}{3x^2 + 11x - 4}$

For Problems 19–68, perform the indicated operations involving rational expressions. Express final answers in simplest form.

19. $\dfrac{4x^2}{5y^2} \cdot \dfrac{15xy}{24x^2y^2}$

20. $\dfrac{5xy}{8y^2} \cdot \dfrac{18x^2y}{15}$

21. $\dfrac{-14xy^4}{18y^2} \cdot \dfrac{24x^2y^3}{35y^2}$

22. $\dfrac{6xy}{9y^4} \cdot \dfrac{30x^3y}{-48x}$

23. $\dfrac{7a^2b}{9ab^3} \div \dfrac{3a^4}{2a^2b^2}$

24. $\dfrac{9a^2c}{12bc^2} \div \dfrac{21ab}{14c^3}$

25. $\dfrac{5xy}{x + 6} \cdot \dfrac{x^2 - 36}{x^2 - 6x}$

26. $\dfrac{2a^2 + 6}{a^2 - a} \cdot \dfrac{a^3 - a^2}{8a - 4}$

27. $\dfrac{5a^2 + 20a}{a^3 - 2a^2} \cdot \dfrac{a^2 - a - 12}{a^2 - 16}$

28. $\dfrac{t^4 - 81}{t^2 - 6t + 9} \cdot \dfrac{6t^2 - 11t - 21}{5t^2 + 8t - 21}$

29. $\dfrac{x^2 + 5xy - 6y^2}{xy^2 - y^3} \cdot \dfrac{2x^2 + 15xy + 18y^2}{xy + 4y^2}$

30. $\dfrac{10n^2 + 21n - 10}{5n^2 + 33n - 14} \cdot \dfrac{2n^2 + 6n - 56}{2n^2 - 3n - 20}$

31. $\dfrac{9y^2}{x^2 + 12x + 36} \div \dfrac{12y}{x^2 + 6x}$

32. $\dfrac{x^2 - 4xy + 4y^2}{7xy^2} \div \dfrac{4x^2 - 3xy - 10y^2}{20x^2y + 25xy^2}$

33. $\dfrac{2x^2 + 3x}{2x^3 - 10x^2} \cdot \dfrac{x^2 - 8x + 15}{3x^3 - 27x} \div \dfrac{14x + 21}{x^2 - 6x - 27}$

34. $\dfrac{a^2 - 4ab + 4b^2}{6a^2 - 4ab} \cdot \dfrac{3a^2 + 5ab - 2b^2}{6a^2 + ab - b^2} \div \dfrac{a^2 - 4b^2}{8a + 4b}$

35. $\dfrac{x + 4}{6} + \dfrac{2x - 1}{4}$

36. $\dfrac{3n - 1}{9} - \dfrac{n + 2}{12}$

37. $\dfrac{x + 1}{4} + \dfrac{x - 3}{6} - \dfrac{x - 2}{8}$

38. $\dfrac{x - 2}{5} - \dfrac{x + 3}{6} + \dfrac{x + 1}{15}$

39. $\dfrac{7}{16a^2b} + \dfrac{3a}{20b^2}$

40. $\dfrac{5b}{24a^2} - \dfrac{11a}{32b}$

41. $\dfrac{1}{n^2} + \dfrac{3}{4n} - \dfrac{5}{6}$

42. $\dfrac{3}{n^2} - \dfrac{2}{5n} + \dfrac{4}{3}$

43. $\dfrac{3}{4x} + \dfrac{2}{3y} - 1$

44. $\dfrac{5}{6x} - \dfrac{3}{4y} + 2$

45. $\dfrac{3}{2x + 1} + \dfrac{2}{3x + 4}$

46. $\dfrac{5}{x - 1} - \dfrac{3}{2x - 3}$

47. $\dfrac{4x}{x^2 + 7x} + \dfrac{3}{x}$

48. $\dfrac{6}{x^2 + 8x} - \dfrac{3}{x}$

49. $\dfrac{4a - 4}{a^2 - 4} - \dfrac{3}{a + 2}$

50. $\dfrac{6a + 4}{a^2 - 1} - \dfrac{5}{a - 1}$

51. $\dfrac{3}{x - 1} - \dfrac{2}{4x - 4}$

52. $\dfrac{3x + 2}{4x - 12} + \dfrac{2x}{6x - 18}$

53. $\dfrac{4}{n^2 - 1} + \dfrac{2}{3n + 3}$

54. $\dfrac{5}{n^2 - 4} - \dfrac{7}{3n - 6}$

55. $\dfrac{3}{x + 1} + \dfrac{x + 5}{x^2 - 1} - \dfrac{3}{x - 1}$

56. $\dfrac{5}{x} - \dfrac{5x - 30}{x^2 + 6x} + \dfrac{x}{x + 6}$

57. $\dfrac{5}{x^2 + 10x + 21} + \dfrac{4}{x^2 + 12x + 27}$

58. $\dfrac{8}{a^2 - 3a - 18} - \dfrac{10}{a^2 - 7a - 30}$

59. $\dfrac{5}{x^2 - 1} - \dfrac{2}{x^2 + 6x - 16}$

60. $\dfrac{4}{x^2 + 2} - \dfrac{7}{x^2 + x - 12}$

61. $\dfrac{3x}{x^2 - 6x + 9} - \dfrac{2}{x - 3}$

62. $\dfrac{6}{x^2 - 9} - \dfrac{9}{x^2 - 6x + 9}$

63. $x - \dfrac{x^2}{x - 1} + \dfrac{1}{x^2 - 1}$

64. $x - \dfrac{x^2}{x + 7} - \dfrac{x}{x^2 - 49}$

65. $\dfrac{2n^2}{n^4 - 16} - \dfrac{n}{n^2 - 4} + \dfrac{1}{n + 2}$

66. $\dfrac{n}{n^2 + 1} + \dfrac{n^2 + 3n}{n^4 - 1} - \dfrac{1}{n - 1}$

67. $\dfrac{2x + 1}{x^2 - 3x - 4} + \dfrac{3x - 2}{x^2 + 3x - 28}$

68. $\dfrac{3x - 4}{2x^2 - 9x - 5} - \dfrac{2x - 1}{3x^2 - 11x - 20}$

69. Consider the addition problem $\dfrac{8}{x - 2} + \dfrac{5}{2 - x}$. Note that the denominators are opposites of each other. If the property $\dfrac{a}{-b} = -\dfrac{a}{b}$ is applied to the second fraction, we obtain $\dfrac{5}{2 - x} = -\dfrac{5}{x - 2}$. Thus we can proceed as follows:

$$\dfrac{8}{x - 2} + \dfrac{5}{2 - x} = \dfrac{8}{x - 2} - \dfrac{5}{x - 2}$$

$$= \dfrac{8 - 5}{x - 2} = \dfrac{3}{x - 2}$$

Use this approach to do the following problems.

a. $\dfrac{7}{x - 1} + \dfrac{2}{1 - x}$

b. $\dfrac{5}{2x - 1} + \dfrac{8}{1 - 2x}$

c. $\dfrac{4}{a-3} - \dfrac{1}{3-a}$

d. $\dfrac{10}{a-9} - \dfrac{5}{9-a}$

e. $\dfrac{x^2}{x-1} - \dfrac{2x-3}{1-x}$

f. $\dfrac{x^2}{x-4} - \dfrac{3x-28}{4-x}$

For Problems 70–92, simplify each complex fraction.

70. $\dfrac{\dfrac{2}{x}+\dfrac{7}{y}}{\dfrac{3}{x}-\dfrac{10}{y}}$

71. $\dfrac{\dfrac{5}{x^2}-\dfrac{3}{x}}{\dfrac{1}{y}+\dfrac{2}{y^2}}$

72. $\dfrac{\dfrac{1}{x}+3}{\dfrac{2}{y}+4}$

73. $\dfrac{1+\dfrac{1}{x}}{1-\dfrac{1}{x}}$

74. $\dfrac{3-\dfrac{2}{n-4}}{5+\dfrac{4}{n-4}}$

75. $\dfrac{1-\dfrac{1}{n+1}}{1+\dfrac{1}{n-1}}$

76. $\dfrac{\dfrac{2}{x-3}-\dfrac{3}{x+3}}{\dfrac{5}{x^2-9}-\dfrac{2}{x-3}}$

77. $\dfrac{\dfrac{-2}{x}-\dfrac{4}{x+2}}{\dfrac{3}{x^2+2x}+\dfrac{3}{x}}$

78. $\dfrac{\dfrac{-1}{y-2}+\dfrac{5}{x}}{\dfrac{3}{x}-\dfrac{4}{xy-2x}}$

79. $1+\dfrac{x}{1+\dfrac{1}{x}}$

80. $2-\dfrac{x}{3-\dfrac{2}{x}}$

81. $\dfrac{a}{\dfrac{1}{a}+4}+1$

82. $\dfrac{3a}{2-\dfrac{1}{a}}-1$

83. $\dfrac{\dfrac{1}{(x+h)^2}-\dfrac{1}{x^2}}{h}$

84. $\dfrac{\dfrac{1}{(x+h)^3}-\dfrac{1}{x^3}}{h}$

85. $\dfrac{\dfrac{1}{x+h+1}-\dfrac{1}{x+1}}{h}$

86. $\dfrac{\dfrac{3}{x+h}-\dfrac{3}{x}}{h}$

87. $\dfrac{\dfrac{2}{2x+2h-1}-\dfrac{2}{2x-1}}{h}$

88. $\dfrac{\dfrac{3}{4x+4h+5}-\dfrac{3}{4x+5}}{h}$

89. $\dfrac{x^{-1}+y^{-1}}{x-y}$

90. $\dfrac{x+y}{x^{-1}+y^{-1}}$

91. $\dfrac{x+2x^{-1}y^{-2}}{4x^{-1}-3y^{-2}}$

92. $\dfrac{x^{-2}-2y^{-1}}{3x^{-1}+y^{-2}}$

THOUGHTS INTO WORDS

93. What role does factoring play in the simplifying of rational expressions?

94. Explain in your own words how to multiply two rational expressions.

95. Give a step-by-step description of how to add $\dfrac{2x-1}{4}+\dfrac{3x+5}{14}$.

96. Look back at the two approaches shown in Example 7. Which approach would you use to simplify $\dfrac{\dfrac{1}{4}+\dfrac{1}{6}}{\dfrac{1}{2}-\dfrac{3}{4}}$? Which approach would you use to simplify $\dfrac{\dfrac{5}{8}+\dfrac{4}{9}}{\dfrac{5}{14}-\dfrac{2}{21}}$? Explain the reason for your choice of approach for each problem.

GRAPHING CALCULATOR ACTIVITIES

97. Use the graphing feature of your graphing calculator to give visual support for your answers for Problems 60–68.

98. For each of the following, use your graphing calculator to help you decide whether the two given expressions are equivalent for all defined values of x.

a. $\dfrac{6x^2 - 7x + 2}{8x^2 + 6x - 5}$ and $\dfrac{3x - 2}{4x + 5}$

b. $\dfrac{4x^2 - 15x - 54}{4x^2 + 13x + 9}$ and $\dfrac{x - 6}{x + 1}$

c. $\dfrac{2x^2 + 3x - 2}{12x^2 + 19x + 5}$ and $\dfrac{2x - 1}{4x + 5}$

d. $\dfrac{x^3 + 2x^2 - 3x}{x^3 + 6x^2 + 5x - 12}$ and $\dfrac{x}{x + 4}$

e. $\dfrac{-5x^2 - 11x + 2}{3x^2 - 13x + 14}$ and $\dfrac{-5x - 1}{3x - 7}$

Answers to the Concept Quiz

1. True   **2.** False   **3.** False   **4.** True   **5.** False   **6.** True   **7.** True   **8.** If $x = 2$ or $x = -2$

0.6   Radicals

Objectives

■ Evaluate roots of numbers.

■ Write radical expressions in simplest radical form.

■ Simplify an indicated sum of radical expressions.

■ Rationalize the denominator.

Recall from our work with exponents that to **square a number** means to raise it to the second power — that is, to use the number as a factor twice. For example, $4^2 = 4 \cdot 4 = 16$, and $(-4)^2 = (-4)(-4) = 16$. A **square root of a number** is one of its two equal factors. Thus 4 and -4 are both square roots of 16. In general, a is a square root of b if $a^2 = b$. The following statements generalize these ideas.

1. Every positive real number has two square roots; one is positive and the other is negative. They are opposites of each other.

2. Negative real numbers have no real-number square roots because the square of any nonzero real number is positive.

3. The square root of zero is zero.

The symbol $\sqrt{}$, called a **radical sign,** is used to designate the *nonnegative* square root, which is called the **principal square root.** The number under the radical sign is called the **radicand,** and the entire expression, such as $\sqrt{16}$, is referred to as a **radical.**

The following examples demonstrate the use of the square root notation.

$\sqrt{16} = 4$ $\sqrt{16}$ indicates the *nonnegative* or *principal square root* of 16.

$-\sqrt{16} = -4$ $-\sqrt{16}$ indicates the negative square root of 16.

$\sqrt{0} = 0$ Zero has only one square root. Technically, we could also write $-\sqrt{0} = -0 = 0$.

$\sqrt{-4}$ Not a real number

$-\sqrt{-4}$ Not a real number

To **cube a number** means to raise it to the third power — that is, to use the number as a factor three times. For example, $2^3 = 2 \cdot 2 \cdot 2 = 8$ and $(-2)^3 = (-2)(-2)(-2) = -8$. A **cube root of a number** is one of its three equal factors. Thus 2 is a cube root of 8, and as we will discuss later, it is the only real number that is a cube root of 8. Furthermore, -2 is the only real number that is a cube root of -8. In general, a is a cube root of b if $a^3 = b$. The following statements generalize these ideas.

1. Every positive real number has one positive real-number cube root.

2. Every negative real number has one negative real-number cube root.

3. The cube root of zero is zero.

Remark: Every nonzero real number has three cube roots, but only one of them is a real number. The other roots are complex numbers, which we will discuss in Section 0.8.

The symbol $\sqrt[3]{}$ is used to designate the cube root of a number. Thus we can write

$$\sqrt[3]{8} = 2 \qquad \sqrt[3]{-8} = -2 \qquad \sqrt[3]{\frac{1}{27}} = \frac{1}{3} \qquad \sqrt[3]{-\frac{1}{27}} = -\frac{1}{3}$$

The concept of root can be extended to fourth roots, fifth roots, sixth roots, and, in general, nth roots. If n is an *even positive integer,* then the following statements are true.

1. Every positive real number has exactly two real nth roots, one positive and one negative. For example, the real fourth roots of 16 are 2 and -2.

2. Negative real numbers do not have real nth roots. For example, there are no real fourth roots of -16.

If n is an *odd positive integer* greater than 1, then the following statements are true.

1. Every real number has exactly one real nth root.

2. The real nth root of a positive number is positive. For example, the fifth root of 32 is 2.

3. The real nth root of a negative number is negative. For example, the fifth root of -32 is -2.

In general, the following definition is useful.

Definition 0.5

$\sqrt[n]{b} = a$ if and only if $a^n = b$

In Definition 0.5, if n is an even positive integer, then a and b are both nonnegative. If n is an odd positive integer greater than 1, then a and b are both nonnegative or both negative. The symbol $\sqrt[n]{}$ designates the principal root.

The following examples are applications of Definition 0.5.

$$\sqrt[4]{81} = 3 \qquad \text{because } 3^4 = 81$$

$$\sqrt[5]{32} = 2 \qquad \text{because } 2^5 = 32$$

$$\sqrt[5]{-32} = -2 \qquad \text{because } (-2)^5 = -32$$

To complete our terminology, the n in the radical $\sqrt[n]{b}$ is called the **index** of the radical. If $n = 2$, we commonly write $\sqrt{b}$ instead of $\sqrt[2]{b}$. In this text, when we use symbols such as $\sqrt[n]{b}$, $\sqrt[m]{y}$, and $\sqrt[r]{x}$, we will assume the previous agreements relative to the existence of real roots without listing the various restrictions, unless a special restriction is needed.

From Definition 0.5 we see that if n is any positive integer greater than 1 and $\sqrt[n]{b}$ exists, then

$$(\sqrt[n]{b})^n = b$$

For example, $(\sqrt{4})^2 = 4$, $(\sqrt[3]{-8})^3 = -8$, and $(\sqrt[4]{81})^4 = 81$. Furthermore, if $b \geq 0$ and n is any positive integer greater than 1, or if $b < 0$ and n is an odd positive integer greater than 1, then

$$\sqrt[n]{b^n} = b$$

For example, $\sqrt{4^2} = 4$, $\sqrt[3]{(-2)^3} = -2$, and $\sqrt[5]{6^5} = 6$. But we must be careful because

$$\sqrt{(-2)^2} \neq -2 \qquad \text{and} \qquad \sqrt[4]{(-2)^4} \neq -2$$

■ Simplest Radical Form

Let's use some examples to motivate another useful property of radicals.

$$\sqrt{16 \cdot 25} = \sqrt{400} = 20 \qquad \text{and} \qquad \sqrt{16} \cdot \sqrt{25} = 4 \cdot 5 = 20$$

$$\sqrt[3]{8 \cdot 27} = \sqrt[3]{216} = 6 \qquad \text{and} \qquad \sqrt[3]{8} \cdot \sqrt[3]{27} = 2 \cdot 3 = 6$$

$$\sqrt[3]{-8 \cdot 64} = \sqrt[3]{-512} = -8 \qquad \text{and} \qquad \sqrt[3]{-8} \cdot \sqrt[3]{64} = -2 \cdot 4 = -8$$

In general, the following property can be stated.

Property 0.3

$$\sqrt[n]{bc} = \sqrt[n]{b}\,\sqrt[n]{c} \quad \text{if } \sqrt[n]{b} \text{ and } \sqrt[n]{c} \text{ are real numbers.}$$

Property 0.3 states that **the nth root of a product is equal to the product of the nth roots.**

The definition of nth root, along with Property 0.3, provides the basis for changing radicals to simplest radical form. The concept of **simplest radical form** takes on additional meaning as we encounter more complicated expressions, but for now it simply means that the radicand does not contain any perfect powers of the index. Consider the following examples of reductions to simplest radical form:

$$\sqrt{45} = \sqrt{9 \cdot 5} = \sqrt{9}\sqrt{5} = 3\sqrt{5}$$
$$\sqrt{52} = \sqrt{4 \cdot 13} = \sqrt{4}\sqrt{13} = 2\sqrt{13}$$
$$\sqrt[3]{24} = \sqrt[3]{8 \cdot 3} = \sqrt[3]{8}\sqrt[3]{3} = 2\sqrt[3]{3}$$

A variation of the technique for changing radicals with index n to simplest form is to factor the radicand into primes and then to look for the perfect nth powers in exponential form, as in the following examples:

$$\sqrt{80} = \sqrt{2^4 \cdot 5} = \sqrt{2^4}\sqrt{5} = 2^2\sqrt{5} = 4\sqrt{5}$$
$$\sqrt[3]{108} = \sqrt[3]{2^2 \cdot 3^3} = \sqrt[3]{3^3}\sqrt[3]{2^2} = 3\sqrt[3]{4}$$

The distributive property can be used to combine radicals that have the same index and the same radicand:

$$3\sqrt{2} + 5\sqrt{2} = (3 + 5)\sqrt{2} = 8\sqrt{2}$$
$$7\sqrt[3]{5} - 3\sqrt[3]{5} = (7 - 3)\sqrt[3]{5} = 4\sqrt[3]{5}$$

Sometimes it is necessary to simplify the radicals first and then to combine them by applying the distributive property:

$$3\sqrt{8} + 2\sqrt{18} - 4\sqrt{2} = 3\sqrt{4}\sqrt{2} + 2\sqrt{9}\sqrt{2} - 4\sqrt{2}$$
$$= 6\sqrt{2} + 6\sqrt{2} - 4\sqrt{2}$$
$$= (6 + 6 - 4)\sqrt{2}$$
$$= 8\sqrt{2}$$

Property 0.3 can also be viewed as $\sqrt[n]{b}\sqrt[n]{c} = \sqrt[n]{bc}$. Then, along with the commutative and associative properties of the real numbers, it provides the basis for multiplying radicals that have the same index. Consider the following two examples:

$$(7\sqrt{6})(3\sqrt{8}) = 7 \cdot 3 \cdot \sqrt{6} \cdot \sqrt{8}$$
$$= 21\sqrt{48}$$
$$= 21\sqrt{16}\sqrt{3}$$
$$= 21 \cdot 4 \cdot \sqrt{3}$$
$$= 84\sqrt{3}$$

$$(2\sqrt[3]{6})(5\sqrt[3]{4}) = 2 \cdot 5 \cdot \sqrt[3]{6} \cdot \sqrt[3]{4}$$
$$= 10\sqrt[3]{24}$$
$$= 10\sqrt[3]{8}\sqrt[3]{3}$$
$$= 10 \cdot 2 \cdot \sqrt[3]{3}$$
$$= 20\sqrt[3]{3}$$

The distributive property, along with Property 0.3, provides a way of handling special products involving radicals, as the next examples illustrate:

$$2\sqrt{2}\,(4\sqrt{3} - 5\sqrt{6}) = (2\sqrt{2})(4\sqrt{3}) - (2\sqrt{2})(5\sqrt{6})$$
$$= 8\sqrt{6} - 10\sqrt{12}$$
$$= 8\sqrt{6} - 10\sqrt{4}\sqrt{3}$$
$$= 8\sqrt{6} - 20\sqrt{3}$$

$$(2\sqrt{2} - \sqrt{7})(3\sqrt{2} + 5\sqrt{7}) = 2\sqrt{2}(3\sqrt{2} + 5\sqrt{7}) - \sqrt{7}(3\sqrt{2} + 5\sqrt{7})$$
$$= (2\sqrt{2})(3\sqrt{2}) + (2\sqrt{2})(5\sqrt{7}) - (\sqrt{7})(3\sqrt{2}) - (\sqrt{7})(5\sqrt{7})$$
$$= 6 \cdot 2 + 10\sqrt{14} - 3\sqrt{14} - 5 \cdot 7$$
$$= -23 + 7\sqrt{14}$$

$$(\sqrt{5} + \sqrt{2})(\sqrt{5} - \sqrt{2}) = \sqrt{5}(\sqrt{5} - \sqrt{2}) + \sqrt{2}(\sqrt{5} - \sqrt{2})$$
$$= (\sqrt{5})(\sqrt{5}) - (\sqrt{5})(\sqrt{2}) + (\sqrt{2})(\sqrt{5}) - (\sqrt{2})(\sqrt{2})$$
$$= 5 - \sqrt{10} + \sqrt{10} - 2$$
$$= 3$$

Pay special attention to the last example. It fits the special-product pattern $(a + b)(a - b) = a^2 - b^2$. We will use that idea in a moment.

■ More About Simplest Radical Form

Another property of nth roots is motivated by the following examples:

$$\sqrt{\frac{36}{9}} = \sqrt{4} = 2 \qquad \text{and} \qquad \frac{\sqrt{36}}{\sqrt{9}} = \frac{6}{3} = 2$$

$$\sqrt[3]{\frac{64}{8}} = \sqrt[3]{8} = 2 \qquad \text{and} \qquad \frac{\sqrt[3]{64}}{\sqrt[3]{8}} = \frac{4}{2} = 2$$

In general, the following property can be stated.

Property 0.4

$$\sqrt[n]{\frac{b}{c}} = \frac{\sqrt[n]{b}}{\sqrt[n]{c}} \quad \text{if } \sqrt[n]{b} \text{ and } \sqrt[n]{c} \text{ are real numbers, and } c \neq 0$$

Property 0.4 states that **the *n*th root of a quotient is equal to the quotient of the *n*th roots.**

To evaluate radicals such as $\sqrt{\dfrac{4}{25}}$ and $\sqrt[3]{\dfrac{27}{8}}$, where the numerator and the denominator of the fractional radicands are perfect *n*th powers, we can either use Property 0.4 or rely on the definition of *n*th root.

$$\sqrt{\frac{4}{25}} = \frac{\sqrt{4}}{\sqrt{25}} = \frac{2}{5} \qquad \text{or} \qquad \sqrt{\frac{4}{25}} = \frac{2}{5} \qquad \text{because } \frac{2}{5} \cdot \frac{2}{5} = \frac{4}{25}$$

$$\sqrt[3]{\frac{27}{8}} = \frac{\sqrt[3]{27}}{\sqrt[3]{8}} = \frac{3}{2} \qquad \text{or} \qquad \sqrt[3]{\frac{27}{8}} = \frac{3}{2} \qquad \text{because } \frac{3}{2} \cdot \frac{3}{2} \cdot \frac{3}{2} = \frac{27}{8}$$

Radicals such as $\sqrt{\dfrac{28}{9}}$ and $\sqrt[3]{\dfrac{24}{27}}$, where only the denominators of the radicand are perfect *n*th powers, can be simplified as follows:

$$\sqrt{\frac{28}{9}} = \frac{\sqrt{28}}{\sqrt{9}} = \frac{\sqrt{4}\sqrt{7}}{3} = \frac{2\sqrt{7}}{3}$$

$$\sqrt[3]{\frac{24}{27}} = \frac{\sqrt[3]{24}}{\sqrt[3]{27}} = \frac{\sqrt[3]{8}\sqrt[3]{3}}{3} = \frac{2\sqrt[3]{3}}{3}$$

Before we consider more examples, let's summarize some ideas about simplifying radicals. A radical is said to be in **simplest radical form** if the following conditions are satisfied.

1. No fraction appears within a radical sign.　　Thus $\sqrt{\dfrac{3}{4}}$ violates this condition.

2. No radical appears in the denominator.　　Thus $\dfrac{\sqrt{2}}{\sqrt{3}}$ violates this condition.

3. No radicand contains a perfect power of the index.　　Thus $\sqrt{7^2 \cdot 5}$ violates this condition.

Now let's consider an example in which neither the numerator nor the denominator of the radicand is a perfect *n*th power:

$$\sqrt{\frac{2}{3}} = \frac{\sqrt{2}}{\sqrt{3}} = \frac{\sqrt{2}}{\sqrt{3}} \cdot \frac{\sqrt{3}}{\sqrt{3}} = \frac{\sqrt{6}}{3}$$

Form of 1

The process used to simplify the radical in this example is referred to as **rationalizing the denominator.** There is more than one way to rationalize the denominator, as illustrated by the next example.

EXAMPLE 1   Simplify $\dfrac{\sqrt{5}}{\sqrt{8}}$.

Solution A

$$\frac{\sqrt{5}}{\sqrt{8}} = \frac{\sqrt{5}}{\sqrt{8}} \cdot \frac{\sqrt{8}}{\sqrt{8}} = \frac{\sqrt{40}}{8} = \frac{\sqrt{4}\sqrt{10}}{8} = \frac{2\sqrt{10}}{8} = \frac{\sqrt{10}}{4}$$

Solution B

$$\frac{\sqrt{5}}{\sqrt{8}} = \frac{\sqrt{5}}{\sqrt{8}} \cdot \frac{\sqrt{2}}{\sqrt{2}} = \frac{\sqrt{10}}{\sqrt{16}} = \frac{\sqrt{10}}{4}$$

Solution C

$$\frac{\sqrt{5}}{\sqrt{8}} = \frac{\sqrt{5}}{\sqrt{4}\sqrt{2}} = \frac{\sqrt{5}}{2\sqrt{2}} = \frac{\sqrt{5}}{2\sqrt{2}} \cdot \frac{\sqrt{2}}{\sqrt{2}} = \frac{\sqrt{10}}{4}$$

▶ **Now work Problem 31.** ■

The three approaches in Example 1 again illustrate the need to think first and then push the pencil. You may find one approach easier than another.

EXAMPLE 2   Simplify $\dfrac{\sqrt{6}}{\sqrt{8}}$.

Solution

$$\frac{\sqrt{6}}{\sqrt{8}} = \sqrt{\frac{6}{8}} \qquad \text{Remember that } \frac{\sqrt{a}}{\sqrt{b}} = \sqrt{\frac{a}{b}}$$

$$= \sqrt{\frac{3}{4}} \qquad \text{Reduce the fraction}$$

$$= \frac{\sqrt{3}}{\sqrt{4}}$$

$$= \frac{\sqrt{3}}{2}$$

▶ **Now work Problem 33.** ■

EXAMPLE 3   Simplify $\dfrac{\sqrt[3]{5}}{\sqrt[3]{9}}$.

Solution

$$\frac{\sqrt[3]{5}}{\sqrt[3]{9}} = \frac{\sqrt[3]{5}}{\sqrt[3]{9}} \cdot \frac{\sqrt[3]{3}}{\sqrt[3]{3}}$$

$$= \frac{\sqrt[3]{15}}{\sqrt[3]{27}}$$

$$= \frac{\sqrt[3]{15}}{3}$$

▶ **Now work Problem 41.** ∎

Now let's consider an example in which the denominator is of binomial form.

EXAMPLE 4

Simplify $\dfrac{4}{\sqrt{5} + \sqrt{2}}$ by rationalizing the denominator.

Solution

Remember that a moment ago we found that $(\sqrt{5} + \sqrt{2})(\sqrt{5} - \sqrt{2}) = 3$. Let's use that idea here:

$$\frac{4}{\sqrt{5} + \sqrt{2}} = \left(\frac{4}{\sqrt{5} + \sqrt{2}}\right)\left(\frac{\sqrt{5} - \sqrt{2}}{\sqrt{5} - \sqrt{2}}\right)$$

$$= \frac{4(\sqrt{5} - \sqrt{2})}{(\sqrt{5} + \sqrt{2})(\sqrt{5} - \sqrt{2})} = \frac{4(\sqrt{5} - \sqrt{2})}{3}$$

▶ **Now work Problem 71.** ∎

The process of rationalizing the denominator does agree with the previously listed conditions. However, for certain problems in calculus, it is necessary to **rationalize the numerator.** Again, the fact that $(\sqrt{a} + \sqrt{b})(\sqrt{a} - \sqrt{b}) = a - b$ can be used.

EXAMPLE 5

Change the form of $\dfrac{\sqrt{x + h} - \sqrt{x}}{h}$ by rationalizing the *numerator.*

Solution

$$\frac{\sqrt{x + h} - \sqrt{x}}{h} = \left(\frac{\sqrt{x + h} - \sqrt{x}}{h}\right)\left(\frac{\sqrt{x + h} + \sqrt{x}}{\sqrt{x + h} + \sqrt{x}}\right)$$

$$= \frac{(x + h) - x}{h(\sqrt{x + h} + \sqrt{x})}$$

$$= \frac{\cancel{h}}{\cancel{h}(\sqrt{x + h} + \sqrt{x})}$$

$$= \frac{1}{\sqrt{x + h} + \sqrt{x}}$$

▶ **Now work Problem 81.** ∎

■ Radicals Containing Variables

Before we illustrate how to simplify radicals that contain variables, there is one important point we should call to your attention. Let's look at some examples to illustrate the idea.

Consider the radical $\sqrt{x^2}$ for different values of x.

Let $x = 3$; then $\sqrt{x^2} = \sqrt{3^2} = \sqrt{9} = 3$.

Let $x = -3$; then $\sqrt{x^2} = \sqrt{(-3)^2} = \sqrt{9} = 3$.

Thus if $x \geq 0$, then $\sqrt{x^2} = x$, but if $x < 0$, then $\sqrt{x^2} = -x$. Using the concept of absolute value, we can state that **for all real numbers, $\sqrt{x^2} = |x|$.**

Now consider the radical $\sqrt{x^3}$. Because x^3 is negative when x is negative, we need to restrict x to the nonnegative real numbers when working with $\sqrt{x^3}$. Thus we can write

$$\text{if } x \geq 0, \quad \text{then } \sqrt{x^3} = \sqrt{x^2}\sqrt{x} = x\sqrt{x}$$

and no absolute value sign is needed.

Finally, let's consider the radical $\sqrt[3]{x^3}$.

Let $x = 2$; then $\sqrt[3]{x^3} = \sqrt[3]{2^3} = \sqrt[3]{8} = 2$.

Let $x = -2$; then $\sqrt[3]{x^3} = \sqrt[3]{(-2)^2} = \sqrt[3]{-8} = -2$.

Thus it is correct to write

$$\sqrt[3]{x^3} = x \quad \text{for all real numbers}$$

and again, no absolute value sign is needed.

The previous discussion indicates that, technically, every radical expression with variables in the radicand needs to be analyzed individually to determine the necessary restrictions on the variables. However, to avoid having to do this on a problem-by-problem basis, we shall merely **assume that all variables represent positive real numbers.**

Let's conclude this section by simplifying some radical expressions that contain variables.

$$\sqrt{72x^2y^7} = \sqrt{36x^2y^6}\sqrt{2xy} = 6xy^3\sqrt{2xy}$$

$$\sqrt[3]{40x^4y^8} = \sqrt[3]{8x^3y^6}\,\sqrt[3]{5xy^2} = 2xy^2\,\sqrt[3]{5xy^2}$$

$$\frac{\sqrt{5}}{\sqrt{12a^3}} = \frac{\sqrt{5}}{\sqrt{12a^3}} \cdot \frac{\sqrt{3a}}{\sqrt{3a}} = \frac{\sqrt{15a}}{\sqrt{36a^4}} = \frac{\sqrt{15a}}{6a^2}$$

$$\frac{3}{\sqrt[3]{4x}} = \frac{3}{\sqrt[3]{4x}} \cdot \frac{\sqrt[3]{2x^2}}{\sqrt[3]{2x^2}} = \frac{3\sqrt[3]{2x^2}}{\sqrt[3]{8x^3}} = \frac{3\sqrt[3]{2x^2}}{2x}$$

CONCEPT QUIZ For Problems 1–8, answer true or false.

1. The symbol $\sqrt{}$ is used to designate the principal square root.

2. Every positive real number has two principal square roots.

3. The square root of zero does not exist in the real number system.

4. Every real number has one real number cube root.

5. The $\sqrt{25}$ could be 5 or -5.

6. $\sqrt{(-3)^2} = -3$

7. If $x < 0$, then $\sqrt{x^2} = -x$.

8. For real numbers, the process of rationalizing the denominator changes the denominator from an irrational number to a rational number.

Problem Set 0.6

For Problems 1–8, evaluate.

1. $\sqrt{81}$

2. $-\sqrt{49}$

3. $\sqrt[3]{125}$

4. $\sqrt[4]{81}$

5. $\sqrt{\dfrac{36}{49}}$

6. $\sqrt{\dfrac{256}{64}}$

7. $\sqrt[3]{-\dfrac{27}{8}}$

8. $\sqrt[3]{\dfrac{64}{27}}$

For Problems 9–44, express each in simplest radical form. All variables represent positive real numbers.

9. $\sqrt{24}$

10. $\sqrt{54}$

11. $\sqrt{112}$

12. $6\sqrt{28}$

13. $-3\sqrt{44}$

14. $-5\sqrt{68}$

15. $\dfrac{3}{4}\sqrt{20}$

16. $\dfrac{3}{8}\sqrt{72}$

17. $\sqrt{12x^2}$

18. $\sqrt{45xy^2}$

19. $\sqrt{64x^4y^7}$

20. $3\sqrt{32a^3}$

21. $\dfrac{3}{7}\sqrt{45xy^6}$

22. $\sqrt[3]{32}$

23. $\sqrt[3]{128}$

24. $\sqrt[3]{54x^3}$

25. $\sqrt[3]{16x^4}$

26. $\sqrt[3]{81x^5y^6}$

27. $\sqrt[4]{48x^5}$

28. $\sqrt[4]{162x^6y^7}$

29. $\sqrt{\dfrac{12}{25}}$

30. $\sqrt{\dfrac{75}{81}}$

31. $\sqrt{\dfrac{7}{8}}$

32. $\dfrac{\sqrt{35}}{\sqrt{7}}$

33. $\dfrac{4\sqrt{6}}{\sqrt{10}}$

34. $\dfrac{\sqrt{27}}{\sqrt{18}}$

35. $\dfrac{6\sqrt{3}}{7\sqrt{6}}$

36. $\sqrt{\dfrac{3x}{2y}}$

37. $\dfrac{\sqrt{5}}{\sqrt{12x^4}}$

38. $\dfrac{\sqrt{5y}}{\sqrt{18x^3}}$

39. $\dfrac{\sqrt{12a^2b}}{\sqrt{5a^3b^3}}$

40. $\dfrac{5}{\sqrt[3]{3}}$

41. $\dfrac{\sqrt[3]{27}}{\sqrt[3]{4}}$

42. $\sqrt[3]{\dfrac{5}{2x}}$

43. $\dfrac{\sqrt[3]{2y}}{\sqrt[3]{3x}}$

44. $\dfrac{\sqrt[3]{12xy}}{\sqrt[3]{3x^2y^5}}$

For Problems 45–52, use the distributive property to help simplify each. For example,

$$3\sqrt{8} + 5\sqrt{2} = 3\sqrt{4}\sqrt{2} + 5\sqrt{2}$$
$$= 6\sqrt{2} + 5\sqrt{2}$$
$$= (6 + 5)\sqrt{2}$$
$$= 11\sqrt{2}$$

45. $5\sqrt{12} + 2\sqrt{3}$

46. $4\sqrt{50} - 9\sqrt{32}$

47. $2\sqrt{28} - 3\sqrt{63} + 8\sqrt{7}$

48. $4\sqrt[3]{2} + 2\sqrt[3]{16} - \sqrt[3]{54}$

49. $\dfrac{5}{6}\sqrt{48} - \dfrac{3}{4}\sqrt{12}$

50. $\dfrac{2}{5}\sqrt{40} + \dfrac{1}{6}\sqrt{90}$

51. $\dfrac{2\sqrt{8}}{3} - \dfrac{3\sqrt{18}}{5} - \dfrac{\sqrt{50}}{2}$

52. $\dfrac{3\sqrt[3]{54}}{2} + \dfrac{5\sqrt[3]{16}}{3}$

For Problems 53–68, multiply and express the results in simplest radical form. All variables represent nonnegative real numbers.

53. $(4\sqrt{3})(6\sqrt{8})$

54. $(5\sqrt{8})(3\sqrt{7})$

55. $2\sqrt{3}(5\sqrt{2} + 4\sqrt{10})$

56. $3\sqrt{6}(2\sqrt{8} - 3\sqrt{12})$

57. $3\sqrt{x}(\sqrt{6xy} - \sqrt{8y})$

58. $\sqrt{6y}(\sqrt{8x} + \sqrt{10y^2})$

59. $(\sqrt{3} + 2)(\sqrt{3} + 5)$

60. $(\sqrt{2} - 3)(\sqrt{2} + 4)$

61. $(4\sqrt{2} + \sqrt{3})(3\sqrt{2} + 2\sqrt{3})$

62. $(2\sqrt{6} + 3\sqrt{5})(3\sqrt{6} + 4\sqrt{5})$

63. $(6 + 2\sqrt{5})(6 - 2\sqrt{5})$

64. $(7 - 3\sqrt{2})(7 + 3\sqrt{2})$

65. $(\sqrt{x} + \sqrt{y})^2$

66. $(2\sqrt{x} - 3\sqrt{y})^2$

67. $(\sqrt{a} + \sqrt{b})(\sqrt{a} - \sqrt{b})$

68. $(3\sqrt{x} + 5\sqrt{y})(3\sqrt{x} - 5\sqrt{y})$

For Problems 69–80, rationalize the denominator and simplify. All variables represent positive real numbers.

69. $\dfrac{3}{\sqrt{5} + 2}$

70. $\dfrac{7}{\sqrt{10} - 3}$

71. $\dfrac{4}{\sqrt{7} - \sqrt{3}}$

72. $\dfrac{2}{\sqrt{5} + \sqrt{3}}$

73. $\dfrac{\sqrt{2}}{2\sqrt{5} + 3\sqrt{7}}$

74. $\dfrac{5}{5\sqrt{2} - 3\sqrt{5}}$

75. $\dfrac{\sqrt{x}}{\sqrt{x} - 1}$

76. $\dfrac{\sqrt{x}}{\sqrt{x} + 2}$

77. $\dfrac{\sqrt{x}}{\sqrt{x} + \sqrt{y}}$

78. $\dfrac{2\sqrt{x}}{\sqrt{x} - \sqrt{y}}$

79. $\dfrac{2\sqrt{x} + \sqrt{y}}{3\sqrt{x} - 2\sqrt{y}}$

80. $\dfrac{3\sqrt{x} - 2\sqrt{y}}{2\sqrt{x} + 5\sqrt{y}}$

For Problems 81–84, *rationalize the numerator*. All variables represent positive real numbers.

81. $\dfrac{\sqrt{2x + 2h} - \sqrt{2x}}{h}$

82. $\dfrac{\sqrt{x + h + 1} - \sqrt{x + 1}}{h}$

83. $\dfrac{\sqrt{x + h - 3} - \sqrt{x - 3}}{h}$

84. $\dfrac{2\sqrt{x + h} - 2\sqrt{x}}{h}$

■ ■ ■ THOUGHTS INTO WORDS

85. Is the equation $\sqrt{x^2 y} = x\sqrt{y}$ true for all real-number values for x and y? Defend your answer.

86. Is the equation $\sqrt{x^2 y^2} = xy$ true for all real-number values for x and y? Defend your answer.

87. Give a step-by-step description of how you would change $\sqrt{252}$ to simplest radical form.

88. Why is $\sqrt{-9}$ not a real number?

89. How could you find a whole number approximation for $\sqrt{2750}$ if you did not have a calculator or table available?

■ ■ ■ FURTHER INVESTIGATIONS

Do the following problems, where the variable could be any real number as long as the radical represents a real number. Use absolute-value signs in the answers as necessary.

90. $\sqrt{125x^2}$

91. $\sqrt{16x^4}$

92. $\sqrt{8b^3}$

93. $\sqrt{3y^5}$

94. $\sqrt{288x^6}$

95. $\sqrt{28m^8}$

96. $\sqrt{128c^{10}}$

97. $\sqrt{18d^7}$

98. $\sqrt{49x^2}$

99. $\sqrt{80n^{20}}$

100. $\sqrt{81h^3}$

GRAPHING CALCULATOR ACTIVITIES

101. Sometimes it is more convenient to express a large or very small number as a product of a power of 10 and a number that is not between 1 and 10. For example, suppose that we want to calculate $\sqrt{640,000}$. We can proceed as follows:

$$\sqrt{640,000} = \sqrt{(64)(10)^4}$$
$$= ((64)(10)^4)^{1/2}$$
$$= (64)^{1/2}(10^4)^{1/2}$$
$$= (8)(10)^2$$
$$= 8(100) = 800$$

Compute each of the following without a calculator, and then use a calculator to check your answers.

a. $\sqrt{49,000,000}$ b. $\sqrt{0.0025}$

c. $\sqrt{14,400}$ d. $\sqrt{0.000121}$

e. $\sqrt[3]{27,000}$ f. $\sqrt[3]{0.000064}$

102. There are several methods of approximating square roots without using a calculator. One such method works on a "clamping between values" principle. For example, to find a whole-number approximation for $\sqrt{128}$, we can proceed as follows: $11^2 = 121$ and $12^2 = 144$. Therefore $11 < \sqrt{128} < 12$. Because

128 is closer to 121 than to 144, we say that 11 is a whole-number approximation for $\sqrt{128}$. If a more precise approximation is needed, we can do more clamping. We would find that $(11.3)^2 = 127.69$ and $(11.4)^2 = 129.96$. Because 128 is closer to 127.69 than to 129.96, we conclude that $\sqrt{128} = 11.3$, to the nearest tenth.

For each of the following, use the clamping idea to find a whole-number approximation. Then check your answers using a calculator and the square root key.

a. $\sqrt{52}$ b. $\sqrt{93}$ c. $\sqrt{174}$

d. $\sqrt{200}$ e. $\sqrt{275}$ f. $\sqrt{350}$

103. The clamping process discussed in Problem 102 works for any whole-number root greater than or equal to 2. For example, a whole-number approximation for $\sqrt[3]{80}$ is 4 because $4^3 = 64$ and $5^3 = 125$, and 80 is closer to 64 than to 125.

For each of the following, use the clamping idea to find a whole-number approximation. Then use your calculator and the appropriate root keys to check your answers.

a. $\sqrt[3]{24}$ b. $\sqrt[3]{32}$ c. $\sqrt[3]{150}$

d. $\sqrt[3]{200}$ e. $\sqrt[4]{50}$ f. $\sqrt[4]{250}$

Answers to the Concept Quiz

1. True   **2.** False   **3.** False   **4.** True   **5.** False   **6.** False   **7.** True   **8.** True

0.7   Relationship Between Exponents and Roots

Objectives

■ Use rational exponents to express the root of a number.

■ Simplify expressions with rational exponents.

■ Apply rational exponents to simplify radical expressions.

Recall that we used the basic properties of positive integral exponents to motivate a definition of negative integers as exponents. In this section, we shall use the properties of integral exponents to motivate definitions for rational numbers as exponents. These definitions will tie together the concepts of *exponent* and *root*. Let's consider the following comparisons.

From our study of radicals we know that	If $(b^m)^n = b^{nm}$ is to hold when m is a rational number of the form $1/p$, where p is a positive integer greater than 1 and $n = p$, then
$(\sqrt{5})^2 = 5$	$(5^{1/2})^2 = 5^{2(1/2)} = 5^1 = 5$
$(\sqrt[3]{8})^3 = 8$	$(8^{1/3})^3 = 8^{3(1/3)} = 8^1 = 8$
$(\sqrt[4]{21})^4 = 21$	$(21^{1/4})^4 = 21^{4(1/4)} = 21^1 = 21$

Such examples motivate the following definition.

Definition 0.6

If b is a real number, n is a positive integer greater than 1, and $\sqrt[n]{b}$ exists, then
$$b^{1/n} = \sqrt[n]{b}$$

Definition 0.6 states that $b^{1/n}$ means the nth root of b. We shall assume that b and n are chosen so that $\sqrt[n]{b}$ exists in the real number system. For example, $(-25)^{1/2}$ is not meaningful at this time because $\sqrt{-25}$ is not a real number. The following examples illustrate the use of Definition 0.6.

$$25^{1/2} = \sqrt{25} = 5 \qquad 16^{1/4} = \sqrt[4]{16} = 2$$
$$8^{1/3} = \sqrt[3]{8} = 2 \qquad (-27)^{1/3} = \sqrt[3]{-27} = -3$$

Now the following definition provides the basis for the use of *all* rational numbers as exponents.

Definition 0.7

If m/n is a rational number expressed in lowest terms, where n is a positive integer greater than 1, and m is any integer, and if b is a real number such that $\sqrt[n]{b}$ exists, then
$$b^{m/n} = \sqrt[n]{b^m} = (\sqrt[n]{b})^m$$

In Definition 0.7, whether we use the form $\sqrt[n]{b^m}$ or $(\sqrt[n]{b})^m$ for computational purposes depends somewhat on the magnitude of the problem. Let's use both forms on the following two problems:

$$8^{2/3} = \sqrt[3]{8^2} = \sqrt[3]{64} = 4 \quad \text{or} \quad 8^{2/3} = (\sqrt[3]{8})^2 = (2)^2 = 4$$

$$27^{2/3} = \sqrt[3]{27^2} = \sqrt[3]{729} = 9 \quad \text{or} \quad 27^{2/3} = (\sqrt[3]{27})^2 = (3)^2 = 9$$

To compute $8^{2/3}$, both forms work equally well. However, to compute $27^{2/3}$, the form $(\sqrt[3]{27})^2$ is much easier to handle. The following examples further illustrate Definition 0.7:

$$25^{3/2} = (\sqrt{25})^3 = 5^3 = 125$$

$$(32)^{-2/5} = \frac{1}{(32)^{2/5}} = \frac{1}{(\sqrt[5]{32})^2} = \frac{1}{2^2} = \frac{1}{4}$$

$$(-64)^{2/3} = (\sqrt[3]{-64})^2 = (-4)^2 = 16$$

$$-8^{4/3} = -(\sqrt[3]{8})^4 = -(2)^4 = -16$$

It can be shown that all of the results pertaining to integral exponents listed in Property 0.2 (on page 22) also hold for all rational exponents. Let's consider some examples to illustrate each of those results.

$$x^{1/2} \cdot x^{2/3} = x^{1/2+2/3} \qquad\qquad b^n \cdot b^m = b^{n+m}$$

$$= x^{3/6+4/6}$$

$$= x^{7/6}$$

$$(a^{2/3})^{3/2} = a^{(3/2)(2/3)} \qquad\qquad (b^n)^m = b^{nm}$$

$$= a^1 = a$$

$$(16y^{2/3})^{1/2} = (16)^{1/2}(y^{2/3})^{1/2} \qquad (ab)^n = a^n b^n$$

$$= 4y^{1/3}$$

$$\frac{y^{3/4}}{y^{1/2}} = y^{3/4-1/2} \qquad\qquad \frac{b^n}{b^m} = b^{n-m}$$

$$= y^{3/4-2/4}$$

$$= y^{1/4}$$

$$\left(\frac{x^{1/2}}{y^{1/3}}\right)^6 = \frac{(x^{1/2})^6}{(y^{1/3})^6} \qquad\qquad \left(\frac{a}{b}\right)^n = \frac{a^n}{b^n}$$

$$= \frac{x^3}{y^2}$$

The link between exponents and roots provides a basis for multiplying and dividing some radicals even if they have different indexes. The general procedure is to change from radical to exponential form, apply the properties of exponents, and then change back to radical form. Let's apply these procedures in the next three examples:

$$\sqrt{2}\sqrt[3]{2} = 2^{1/2} \cdot 2^{1/3} = 2^{1/2+1/3} = 2^{5/6} = \sqrt[6]{2^5} = \sqrt[6]{32}$$

$$\sqrt{xy}\,\sqrt[5]{x^2y} = (xy)^{1/2}(x^2y)^{1/5}$$

$$= x^{1/2}y^{1/2}x^{2/5}y^{1/5}$$

$$= x^{1/2+2/5}y^{1/2+1/5}$$

$$= x^{9/10}y^{7/10}$$

$$= (x^9y^7)^{1/10} = \sqrt[10]{x^9y^7}$$

$$\frac{\sqrt{5}}{\sqrt[3]{5}} = \frac{5^{1/2}}{5^{1/3}} = 5^{1/2-1/3} = 5^{1/6} = \sqrt[6]{5}$$

Earlier we agreed that a radical such as $\sqrt[3]{x^4}$ is not in simplest form because the radicand contains a perfect power of the index. Thus we simplified $\sqrt[3]{x^4}$ by expressing it as $\sqrt[3]{x^3}\sqrt[3]{x}$, which in turn can be written $x\sqrt[3]{x}$. Such simplification can also be done in exponential form, as follows:

$$\sqrt[3]{x^4} = x^{4/3} = x^{3/3} \cdot x^{1/3} = x \cdot x^{1/3} = x\sqrt[3]{x}$$

Note the use of this type of simplification in the following examples.

EXAMPLE 1

Perform the indicated operations and express the answers in simplest radical form.

a. $\sqrt[3]{x^2}\sqrt[4]{x^3}$ **b.** $\sqrt{2}\sqrt[3]{4}$ **c.** $\dfrac{\sqrt{27}}{\sqrt[3]{3}}$

Solutions

a. $\sqrt[3]{x^2}\sqrt[4]{x^3} = x^{2/3} \cdot x^{3/4} = x^{2/3+3/4} = x^{17/12} = x^{12/12} \cdot x^{5/12} = x\sqrt[12]{x^5}$

b. $\sqrt{2}\sqrt[3]{4} = 2^{1/2} \cdot 4^{1/3} = 2^{1/2}(2^2)^{1/3} = 2^{1/2} \cdot 2^{2/3}$

$$= 2^{1/2+2/3} = 2^{7/6} = 2^{6/6} \cdot 2^{1/6} = 2\sqrt[6]{2}$$

c. $\dfrac{\sqrt{27}}{\sqrt[3]{3}} = \dfrac{27^{1/2}}{3^{1/3}} = \dfrac{(3^3)^{1/2}}{3^{1/3}} = \dfrac{3^{3/2}}{3^{1/3}} = 3^{3/2-1/3} = 3^{7/6}$

$$= 3^{6/6} \cdot 3^{1/6} = 3\sqrt[6]{3}$$

Now work Problems 33, 35, and 43.   ∎

The process of rationalizing the denominator can sometimes be handled more easily in exponential form. Consider the following examples, which illustrate this procedure.

EXAMPLE 2

Rationalize the denominator and express the answer in simplest radical form.

a. $\dfrac{2}{\sqrt[3]{x}}$ **b.** $\dfrac{\sqrt[3]{x}}{\sqrt{y}}$

Solutions

a. $\dfrac{2}{\sqrt[3]{x}} = \dfrac{2}{x^{1/3}} = \dfrac{2}{x^{1/3}} \cdot \dfrac{x^{2/3}}{x^{2/3}} = \dfrac{2x^{2/3}}{x} = \dfrac{2\sqrt[3]{x^2}}{x}$

b. $\dfrac{\sqrt[3]{x}}{\sqrt{y}} = \dfrac{x^{1/3}}{y^{1/2}} = \dfrac{x^{1/3}}{y^{1/2}} \cdot \dfrac{y^{1/2}}{y^{1/2}} = \dfrac{x^{1/3} \cdot y^{1/2}}{y} = \dfrac{x^{2/6} \cdot y^{3/6}}{y} = \dfrac{\sqrt[6]{x^2 y^3}}{y}$

▷ **Now work Problems 49 and 51.** ∎

Note in part b that if we had changed back to radical form at the step $\dfrac{x^{1/3} y^{1/2}}{y}$, we would have obtained the product of two radicals, $\sqrt[3]{x}\sqrt{y}$, in the numerator. Instead we used the exponential form to find this product and express the final result with a single radical in the numerator. Finally, let's consider an example involving *the root of a root*.

EXAMPLE 3

Simplify $\sqrt[3]{\sqrt{2}}$.

Solution

$$\sqrt[3]{\sqrt{2}} = (2^{1/2})^{1/3} = 2^{1/6} = \sqrt[6]{2}$$

▷ **Now work Problem 57a.** ∎

CONCEPT QUIZ

For Problems 1–4, select the equivalent radical form.

1. $x^{\frac{3}{5}}$ a. $\sqrt[3]{x^5}$ b. $x\sqrt[3]{x^2}$ c. $\sqrt[5]{x^3}$

2. $y^{-\frac{1}{3}}$ a. $\dfrac{1}{\sqrt[3]{y}}$ b. $\dfrac{\sqrt[3]{y^2}}{y}$ c. $-\sqrt[3]{y}$

3. $-w^{-\frac{1}{2}}$ a. $-\dfrac{\sqrt{w}}{w}$ b. $\sqrt{w}$ c. $-\sqrt{w}$

4. $\sqrt[n]{x}\sqrt[m]{x}$ a. $\sqrt[mn]{x}$ b. $x^{\frac{m+n}{mn}}$ c. $x^{\frac{1}{m+n}}$

For Problems 5–8, answer true or false.

5. Assuming the nth root of x exists, $\sqrt[n]{x}$ can be expressed as $x^{\frac{1}{n}}$.

6. The expression $\sqrt[n]{x^m}$ is $(\sqrt[n]{x})^m$.

7. The process of rationalizing the denominator can be done with rational exponents.

8. An exponent of $\dfrac{1}{3}$ indicates the cube root.

Problem Set 0.7

For Problems 1–16, evaluate.

1. $49^{1/2}$

2. $64^{1/3}$

3. $32^{3/5}$

4. $(-8)^{1/3}$

5. $-8^{2/3}$

6. $64^{-1/2}$

7. $\left(\dfrac{1}{4}\right)^{-1/2}$

8. $\left(-\dfrac{27}{8}\right)^{-1/3}$

9. $16^{3/2}$

10. $(0.008)^{1/3}$

11. $(0.01)^{3/2}$

12. $\left(\dfrac{1}{27}\right)^{-2/3}$

13. $64^{-5/6}$

14. $-16^{5/4}$

15. $\left(\dfrac{1}{8}\right)^{-1/3}$

16. $\left(-\dfrac{1}{8}\right)^{2/3}$

For Problems 17–32, perform the indicated operations and simplify. Express final answers using positive exponents only.

17. $(3x^{1/4})(5x^{1/3})$

18. $(2x^{2/5})(6x^{1/4})$

19. $(y^{2/3})(y^{-1/4})$

20. $(2x^{1/3})(x^{-1/2})$

21. $(4x^{1/4}y^{1/2})^3$

22. $(5x^{1/2}y)^2$

23. $\dfrac{24x^{3/5}}{6x^{1/3}}$

24. $\dfrac{18x^{1/2}}{9x^{1/3}}$

25. $\dfrac{56a^{1/6}}{8a^{1/4}}$

26. $\dfrac{48b^{1/3}}{12b^{3/4}}$

27. $\left(\dfrac{2x^{1/3}}{3y^{1/4}}\right)^4$

28. $\left(\dfrac{6x^{2/5}}{7y^{2/3}}\right)^2$

29. $\left(\dfrac{x^2}{y^3}\right)^{-1/2}$

30. $\left(\dfrac{a^3}{b^{-2}}\right)^{-1/3}$

31. $\left(\dfrac{4a^2x}{2a^{1/2}x^{1/3}}\right)^3$

32. $\left(\dfrac{3ax^{-1}}{a^{1/2}x^{-2}}\right)^2$

For Problems 33–48, perform the indicated operations and express the answer in simplest radical form.

33. $\sqrt{2}\sqrt[4]{2}$

34. $\sqrt[3]{3}\sqrt{3}$

35. $\sqrt[3]{x}\sqrt[4]{x}$

36. $\sqrt[3]{x^2}\sqrt[5]{x^3}$

37. $\sqrt{xy}\sqrt[4]{x^3y^5}$

38. $\sqrt[3]{x^2y^4}\sqrt[4]{x^3y}$

39. $\sqrt[3]{a^2b^2}\sqrt[4]{a^3b}$

40. $\sqrt{ab}\sqrt[3]{a^4b^5}$

41. $\sqrt[3]{4}\sqrt{8}$

42. $\sqrt[3]{9}\sqrt{27}$

43. $\dfrac{\sqrt{2}}{\sqrt[3]{2}}$

44. $\dfrac{\sqrt{9}}{\sqrt[3]{3}}$

45. $\dfrac{\sqrt[3]{8}}{\sqrt[4]{4}}$

46. $\dfrac{\sqrt[3]{16}}{\sqrt[6]{4}}$

47. $\dfrac{\sqrt[4]{x^9}}{\sqrt[3]{x^2}}$

48. $\dfrac{\sqrt[5]{x^7}}{\sqrt[3]{x}}$

For Problems 49–56, rationalize the denominator and express the final answer in simplest radical form.

49. $\dfrac{5}{\sqrt[3]{x}}$

50. $\dfrac{3}{\sqrt[3]{x^2}}$

51. $\dfrac{\sqrt{x}}{\sqrt[3]{y}}$

52. $\dfrac{\sqrt[4]{x}}{\sqrt{y}}$

53. $\dfrac{\sqrt[4]{x^3}}{\sqrt[5]{y^3}}$

54. $\dfrac{2\sqrt{x}}{3\sqrt[3]{y}}$

55. $\dfrac{5\sqrt[3]{y^2}}{4\sqrt[4]{x}}$

56. $\dfrac{\sqrt{xy}}{\sqrt[3]{a^2b}}$

57. Simplify each of the following, expressing the final result as one radical. For example,

$$\sqrt{\sqrt{3}} = (3^{1/2})^{1/2} = 3^{1/4} = \sqrt[4]{3}$$

a. $\sqrt[3]{\sqrt{2}}$

b. $\sqrt[3]{\sqrt[4]{3}}$

c. $\sqrt[3]{\sqrt{x^3}}$

d. $\sqrt{\sqrt[3]{x^4}}$

■ ■ ■ THOUGHTS INTO WORDS

58. Your friend keeps getting an error message when evaluating $-4^{5/2}$ on his calculator. What error is he probably making?

59. Explain how you would evaluate $27^{2/3}$ without a calculator.

■ ■ ■ FURTHER INVESTIGATIONS

Sometimes we meet the following type of simplification problem in calculus:

$$\frac{(x - 1)^{1/2} - x(x - 1)^{-(1/2)}}{[(x - 1)^{1/2}]^2}$$

$$= \left(\frac{(x - 1)^{1/2} - x(x - 1)^{-(1/2)}}{(x - 1)^{2/2}}\right) \cdot \left(\frac{(x - 1)^{1/2}}{(x - 1)^{1/2}}\right)$$

$$= \frac{x - 1 - x(x - 1)^0}{(x - 1)^{3/2}}$$

$$= \frac{x - 1 - x}{(x - 1)^{3/2}}$$

$$= \frac{-1}{(x - 1)^{3/2}} \quad \text{or} \quad -\frac{1}{(x - 1)^{3/2}}$$

For Problems 60–65, simplify each expression as we did in the previous example.

60. $\dfrac{2(x + 1)^{1/2} - x(x + 1)^{-(1/2)}}{[(x + 1)^{1/2}]^2}$

61. $\dfrac{2(2x - 1)^{1/2} - 2x(2x - 1)^{-(1/2)}}{[(2x - 1)^{1/2}]^2}$

62. $\dfrac{2x(4x + 1)^{1/2} - 2x^2(4x + 1)^{-(1/2)}}{[(4x + 1)^{1/2}]^2}$

63. $\dfrac{(x^2 + 2x)^{1/2} - x(x + 1)(x^2 + 2x)^{-(1/2)}}{[(x^2 + 2x)^{1/2}]^2}$

64. $\dfrac{(3x)^{1/3} - x(3x)^{-(2/3)}}{[(3x)^{1/3}]^2}$

65. $\dfrac{3(2x)^{1/3} - 2x(2x)^{-(2/3)}}{[(2x)^{1/3}]^2}$

GRAPHING CALCULATOR ACTIVITIES

66. Use your calculator to evaluate each of the following.

   a. $\sqrt[3]{1728}$ **b.** $\sqrt[3]{5832}$

   c. $\sqrt[4]{2401}$ **d.** $\sqrt[4]{65,536}$

   e. $\sqrt[5]{161,051}$ **f.** $\sqrt[5]{6,436,343}$

67. In Definition 0.7 we stated that $b^{m/n} = \sqrt[n]{b^m} = (\sqrt[n]{b})^m$. Use your calculator to verify each of the following.

   a. $\sqrt[3]{27^2} = (\sqrt[3]{27})^2$ **b.** $\sqrt[3]{8^5} = (\sqrt[3]{8})^5$

   c. $\sqrt[4]{16^3} = (\sqrt[4]{16})^3$ **d.** $\sqrt[3]{16^2} = (\sqrt[3]{16})^2$

   e. $\sqrt[5]{9^4} = (\sqrt[5]{9})^4$ **f.** $\sqrt[3]{12^4} = (\sqrt[3]{12})^4$

68. Use your calculator to evaluate each of the following.

   a. $16^{5/2}$ **b.** $25^{7/2}$

   c. $16^{9/4}$ **d.** $27^{5/3}$

   e. $343^{2/3}$ **f.** $512^{4/3}$

69. Use your calculator to estimate each of the following to the nearest thousandth.

   a. $7^{4/3}$ **b.** $10^{4/5}$

   c. $12^{2/5}$ **d.** $19^{2/5}$

   e. $7^{3/4}$ **f.** $10^{5/4}$

0.8   Complex Numbers

Objectives

■ Express the square root of a negative number in terms of i.

■ Add and subtract complex numbers.

■ Multiply and divide complex numbers.

So far we have dealt only with real numbers. However, as we get ready to solve equations in the next chapter, there is a need for *more numbers*. There are some very simple equations that do not have solutions if we restrict ourselves to the set of real numbers. For example, the equation $x^2 + 1 = 0$ has no solutions among the real numbers. To solve such equations, we need to extend the real number system. In this section we will introduce a set of numbers that contains some numbers with squares that are negative real numbers. Then in the next chapter and in Chapter 4 we will see that this set of numbers, called the **complex numbers,** provides solutions not only for equations such as $x^2 + 1 = 0$ but also for *any* polynomial equation in general.

Let's begin by defining a number i such that

$$i^2 = -1$$

The number i is not a real number and is often called the **imaginary unit,** but the number i^2 is the real number -1. The imaginary unit i is used to define a complex number as follows.

Definition 0.8

A **complex number** is any number that can be expressed in the form

   $a + bi$

where a and b are real numbers, and i is the imaginary unit.

The form $a + bi$ is called the **standard form** of a complex number. The real number a is called the **real part** of the complex number, and b is called the **imaginary part.** (Note that b is a real number even though it is called the imaginary part.) Each of the following represents a complex number:

$6 + 2i$	is already expressed in the form $a + bi$. Traditionally, complex numbers for which $a \neq 0$ and $b \neq 0$ have been called imaginary numbers.
$5 - 3i$	can be written $5 + (-3i)$ even though the form $5 - 3i$ is often used.
$-8 + i\sqrt{2}$	can be written $-8 + \sqrt{2}i$. It is easy to mistake $\sqrt{2}i$ for $\sqrt{2i}$. Thus we commonly write $i\sqrt{2}$ instead of $\sqrt{2}i$ to avoid any difficulties with the radical sign.
$-9i$	can be written $0 + (-9i)$. Complex numbers such as $-9i$, for which $a = 0$ and $b \neq 0$, traditionally have been called **pure imaginary numbers**.
5	can be written $5 + 0i$.

The set of real numbers is a subset of the set of complex numbers. The following diagram indicates the organizational format of the complex number system:

Complex numbers

$a + bi$, where a and b
are real numbers

Real numbers

$a + bi$, where $b = 0$

Imaginary numbers

$a + bi$, where $b \neq 0$

Pure imaginary numbers

$a + bi$, where $a = 0$ and $b \neq 0$

Two complex numbers $a + bi$ and $c + di$ are said to be *equal* if and only if $a = c$ and $b = d$. In other words, two complex numbers are equal if and only if their real parts are equal and their imaginary parts are equal.

■ Adding and Subtracting Complex Numbers

The following definition provides the basis for adding complex numbers:

$$(a + bi) + (c + di) = (a + c) + (b + d)i$$

We can use this definition to find the sum of two complex numbers.

$$(4 + 3i) + (5 + 9i) = (4 + 5) + (3 + 9)i = 9 + 12i$$

$$(-6 + 4i) + (8 - 7i) = (-6 + 8) + (4 - 7)i = 2 - 3i$$

$$\left(\frac{1}{2} + \frac{3}{4}i\right) + \left(\frac{2}{3} + \frac{1}{5}i\right) = \left(\frac{1}{2} + \frac{2}{3}\right) + \left(\frac{3}{4} + \frac{1}{5}\right)i$$

$$= \left(\frac{3}{6} + \frac{4}{6}\right) + \left(\frac{15}{20} + \frac{4}{20}\right)i = \frac{7}{6} + \frac{19}{20}i$$

$$(3 + i\sqrt{2}) + (-4 + i\sqrt{2}) = [3 + (-4)] + (\sqrt{2} + \sqrt{2})i = -1 + 2i\sqrt{2}$$

Note the form for writing $2\sqrt{2}i$

The set of complex numbers is **closed with respect to addition;** that is, the sum of two complex numbers is a complex number. Furthermore, the commutative and associative properties of addition hold for all complex numbers. The additive identity element is $0 + 0i$, or simply the real number 0. The additive inverse of $a + bi$ is $-a - bi$ because

$$(a + bi) + (-a - bi) = [a + (-a)] + [b + (-b)]i = 0$$

Therefore, to *subtract* $c + di$ from $a + bi$, we add the additive inverse of $c + di$:

$$(a + bi) - (c + di) = (a + bi) + (-c - di)$$
$$= (a - c) + (b - d)i$$

The following examples illustrate the subtraction of complex numbers:

$$(9 + 8i) - (5 + 3i) = (9 - 5) + (8 - 3)i = 4 + 5i$$

$$(3 - 2i) - (4 - 10i) = (3 - 4) + [-2 - (-10)]i = -1 + 8i$$

$$\left(-\frac{1}{2} + \frac{1}{3}i\right) - \left(\frac{3}{4} + \frac{1}{2}i\right) = \left(-\frac{1}{2} - \frac{3}{4}\right) + \left(\frac{1}{3} - \frac{1}{2}\right)i = -\frac{5}{4} - \frac{1}{6}i$$

■ Multiplying and Dividing Complex Numbers

Because $i^2 = -1$, the number i is a square root of -1, so we write $i = \sqrt{-1}$. It should also be evident that $-i$ is a square root of -1 because

$$(-i)^2 = (-i)(-i) = i^2 = -1$$

Therefore, in the set of complex numbers, -1 has two square roots — namely, i and $-i$. This is expressed symbolically as

$$i = \sqrt{-1} \qquad \text{and} \qquad -i = -\sqrt{-1}$$

Let's extend the definition so that in the set of complex numbers, every negative real number has two square roots. For any positive real number b,

$$(i\sqrt{b})^2 = i^2(b) = -1(b) = -b$$

Therefore, let's denote the **principal square root of** $-b$ by $\sqrt{-b}$ and define it to be

$$\sqrt{-b} = i\sqrt{b}$$

where b is any positive real number. In other words, the principal square root of any negative real number can be represented as the product of a real number and the imaginary unit i. Consider the following examples:

$$\sqrt{-4} = i\sqrt{4} = 2i$$

$$\sqrt{-17} = i\sqrt{17}$$

$$\sqrt{-24} = i\sqrt{24} = i\sqrt{4}\sqrt{6} = 2i\sqrt{6} \qquad \text{Note that we simplified the radical } \sqrt{24} \text{ to } 2\sqrt{6}$$

We should also observe that $-\sqrt{-b}$, where $b > 0$, is a square root of $-b$ because

$$(-\sqrt{-b})^2 = (-i\sqrt{b})^2 = i^2(b) = (-1)b = -b$$

Thus in the set of complex numbers, $-b$ (where $b > 0$) has two square roots: $i\sqrt{b}$ and $-i\sqrt{b}$. These are expressed as

$$\sqrt{-b} = i\sqrt{b} \quad \text{and} \quad -\sqrt{-b} = -i\sqrt{b}$$

We must be careful with the use of the symbol $\sqrt{-b}$, where $b > 0$. Some properties that are true in the set of real numbers involving the square root symbol do not hold if the square root symbol does not represent a real number. For example, $\sqrt{a}\sqrt{b} = \sqrt{ab}$ *does not hold if a and b are both negative numbers.*

Correct $\sqrt{-4}\sqrt{-9} = (2i)(3i) = 6i^2 = 6(-1) = -6$

Incorrect $\sqrt{-4}\sqrt{-9} = \sqrt{(-4)(-9)} = \sqrt{36} = 6$

To avoid difficulty with this idea, you should rewrite all expressions of the form $\sqrt{-b}$, where $b > 0$, in the form $i\sqrt{b}$ *before* doing any computations. The following examples further illustrate this point:

$$\sqrt{-5}\sqrt{-7} = (i\sqrt{5})(i\sqrt{7}) = i^2\sqrt{35} = (-1)\sqrt{35} = -\sqrt{35}$$

$$\sqrt{-2}\sqrt{-8} = (i\sqrt{2})(i\sqrt{8}) = i^2\sqrt{16} = (-1)(4) = -4$$

$$\sqrt{-2}\sqrt{8} = (i\sqrt{2})(\sqrt{8}) = i\sqrt{16} = 4i$$

$$\sqrt{-6}\sqrt{-8} - (i\sqrt{6})(i\sqrt{8}) = i^2\sqrt{48} = i^2\sqrt{16}\sqrt{3} = 4i^2\sqrt{3} = -4\sqrt{3}$$

$$\frac{\sqrt{-2}}{\sqrt{3}} = \frac{i\sqrt{2}}{\sqrt{3}} = \frac{i\sqrt{2}}{\sqrt{3}} \cdot \frac{\sqrt{3}}{\sqrt{3}} = \frac{i\sqrt{6}}{3}$$

$$\frac{\sqrt{-48}}{\sqrt{12}} = \frac{i\sqrt{48}}{\sqrt{12}} = i\sqrt{\frac{48}{12}} = i\sqrt{4} = 2i$$

Because complex numbers have a *binomial form,* we can find the product of two complex numbers in the same way that we find the product of two binomials. Then, by replacing i^2 with -1 we can simplify and express the final product in the standard form of a complex number. Consider the following examples:

$$\begin{aligned}
(2 + 3i)(4 + 5i) &= 2(4 + 5i) + 3i(4 + 5i) \\
&= 8 + 10i + 12i + 15i^2 \\
&= 8 + 22i + 15(-1) \\
&= 8 + 22i - 15 \\
&= -7 + 22i
\end{aligned}$$

$$\begin{aligned}
(1 - 7i)^2 &= (1 - 7i)(1 - 7i) \\
&= 1(1 - 7i) - 7i(1 - 7i) \\
&= 1 - 7i - 7i + 49i^2 \\
&= 1 - 14i + 49(-1) \\
&= 1 - 14i - 49 \\
&= -48 - 14i
\end{aligned}$$

$$(2 + 3i)(2 - 3i) = 2(2 - 3i) + 3i(2 - 3i)$$
$$= 4 - 6i + 6i - 9i^2$$
$$= 4 - 9(-1)$$
$$= 4 + 9$$
$$= 13$$

Remark: Don't forget that when multiplying complex numbers, we can also use the multiplication patterns

$$(a + b)^2 = a^2 + 2ab + b^2$$
$$(a - b)^2 = a^2 - 2ab + b^2$$
$$(a + b)(a - b) = a^2 - b^2$$

The last example illustrates an important idea. The complex numbers $2 + 3i$ and $2 - 3i$ are called *conjugates* of each other. In general, the two complex numbers $a + bi$ and $a - bi$ are called **conjugates** of each other, and **the product of a complex number and its conjugate is a real number.** This can be shown as follows:

$$(a + bi)(a - bi) = a(a - bi) + bi(a - bi)$$
$$= a^2 - abi + abi - b^2i^2$$
$$= a^2 - b^2(-1)$$
$$= a^2 + b^2$$

Conjugates are used to simplify an expression such as $3i/(5 + 2i)$, which *indicates the quotient of two complex numbers*. To eliminate i in the denominator and to change the indicated quotient to the standard form of a complex number, we can multiply both the numerator and denominator by the conjugate of the denominator.

$$\frac{3i}{5 + 2i} = \frac{3i}{5 + 2i} \cdot \frac{5 - 2i}{5 - 2i}$$
$$= \frac{3i(5 - 2i)}{(5 + 2i)(5 - 2i)}$$
$$= \frac{15i - 6i^2}{25 - 4i^2}$$
$$= \frac{15i - 6(-1)}{25 - 4(-1)}$$
$$= \frac{6 + 15i}{29}$$
$$= \frac{6}{29} + \frac{15}{29}i$$

The following examples further illustrate the process of dividing complex numbers.

$$\frac{2 - 3i}{4 - 7i} = \frac{2 - 3i}{4 - 7i} \cdot \frac{4 + 7i}{4 + 7i}$$

$$= \frac{(2 - 3i)(4 + 7i)}{(4 - 7i)(4 + 7i)}$$

$$= \frac{8 + 14i - 12i - 21i^2}{16 - 49i^2}$$

$$= \frac{8 + 2i - 21(-1)}{16 - 49(-1)}$$

$$= \frac{29 + 2i}{65} = \frac{29}{65} + \frac{2}{65}i$$

$$\frac{4 - 5i}{2i} = \frac{4 - 5i}{2i} \cdot \frac{-2i}{-2i}$$

$$= \frac{(4 - 5i)(-2i)}{(2i)(-2i)}$$

$$= \frac{-8i + 10i^2}{-4i^2}$$

$$= \frac{-8i + 10(-1)}{-4(-1)}$$

$$= \frac{-10 - 8i}{4} = -\frac{5}{2} - 2i$$

For a problem such as the last one, in which the denominator is a pure imaginary number, we can change to standard form by choosing a multiplier other than the conjugate of the denominator. Consider the following alternative approach:

$$\frac{4 - 5i}{2i} = \frac{4 - 5i}{2i} \cdot \frac{i}{i}$$

$$= \frac{(4 - 5i)(i)}{(2i)(i)}$$

$$= \frac{4i - 5i^2}{2i^2}$$

$$= \frac{4i - 5(-1)}{2(-1)}$$

$$= \frac{5 + 4i}{-2}$$

$$= -\frac{5}{2} - 2i$$

CONCEPT QUIZ   For Problems 1–8, answer true or false.

1. The number i is not a real number.

2. The number i^2 is a real number.

3. The form $ai + b$ is called the standard form of a complex number.

4. Every real number is a member of the set of complex numbers.

5. The principal square root of any negative real number can be represented as the product of a real number and the imaginary unit i.

6. $6 - 4i$ and $-6 + 4i$ are additive inverses.

7. The conjugate of the number $-2 - 3i$ is $2 + 3i$.

8. The product of a complex number and its conjugate is a real number.

Problem Set 0.8

For Problems 1–14, add or subtract as indicated.

1. $(5 + 2i) + (8 + 6i)$ **2.** $(-9 + 3i) + (4 + 5i)$

3. $(8 + 6i) - (5 + 2i)$ **4.** $(-6 + 4i) - (4 + 6i)$

5. $(-7 - 3i) + (-4 + 4i)$ **6.** $(6 - 7i) - (7 - 6i)$

7. $(-2 - 3i) - (-1 - i)$

8. $\left(\dfrac{1}{3} + \dfrac{2}{5}i\right) + \left(\dfrac{1}{2} + \dfrac{1}{4}i\right)$

9. $\left(-\dfrac{3}{4} - \dfrac{1}{4}i\right) + \left(\dfrac{3}{5} + \dfrac{2}{3}i\right)$

10. $\left(\dfrac{5}{8} + \dfrac{1}{2}i\right) - \left(\dfrac{7}{8} + \dfrac{1}{5}i\right)$

11. $\left(\dfrac{3}{10} - \dfrac{3}{4}i\right) - \left(-\dfrac{2}{5} + \dfrac{1}{6}i\right)$

12. $(4 + i\sqrt{3}) + (-6 - 2i\sqrt{3})$

13. $(5 + 3i) + (7 - 2i) + (-8 - i)$

14. $(5 - 7i) - (6 - 2i) - (-1 - 2i)$

For Problems 15–30, write each in terms of i and simplify. For example,

$$\sqrt{-20} = i\sqrt{20} = i\sqrt{4}\sqrt{5} = 2i\sqrt{5}$$

15. $\sqrt{-9}$ **16.** $\sqrt{-49}$ **17.** $\sqrt{-19}$

18. $\sqrt{-31}$ **19.** $\sqrt{-\dfrac{4}{9}}$ **20.** $\sqrt{-\dfrac{25}{36}}$

21. $\sqrt{-8}$ **22.** $\sqrt{-18}$ **23.** $\sqrt{-27}$

24. $\sqrt{-32}$ **25.** $\sqrt{-54}$ **26.** $\sqrt{-40}$

27. $3\sqrt{-36}$ **28.** $5\sqrt{-64}$ **29.** $4\sqrt{-18}$

30. $6\sqrt{-8}$

For Problems 31–44, write each in terms of i, perform the indicated operations, and simplify. For example,

$$\sqrt{-9}\sqrt{-16} = (i\sqrt{9})(i\sqrt{16}) = (3i)(4i)$$

$$= 12i^2 = 12(-1) = -12$$

31. $\sqrt{-4}\sqrt{-16}$ **32.** $\sqrt{-25}\sqrt{-9}$

33. $\sqrt{-2}\sqrt{-3}$ **34.** $\sqrt{-3}\sqrt{-7}$

35. $\sqrt{-5}\sqrt{-4}$ **36.** $\sqrt{-7}\sqrt{-9}$

37. $\sqrt{-6}\sqrt{-10}$ **38.** $\sqrt{-2}\sqrt{-12}$

39. $\sqrt{-8}\sqrt{-7}$ **40.** $\sqrt{-12}\sqrt{-5}$

41. $\dfrac{\sqrt{-36}}{\sqrt{-4}}$ **42.** $\dfrac{\sqrt{-64}}{\sqrt{-16}}$

43. $\dfrac{\sqrt{-54}}{\sqrt{-9}}$ **44.** $\dfrac{\sqrt{-18}}{\sqrt{-3}}$

For Problems 45–64, find each product and express the answers in standard form.

45. $(3i)(7i)$ **46.** $(-5i)(8i)$

47. $(4i)(3 - 2i)$ **48.** $(5i)(2 + 6i)$

49. $(3 + 2i)(4 + 6i)$

50. $(7 + 3i)(8 + 4i)$

51. $(4 + 5i)(2 - 9i)$

52. $(1 + i)(2 - i)$

53. $(-2 - 3i)(4 + 6i)$

54. $(-3 - 7i)(2 + 10i)$

55. $(6 - 4i)(-1 - 2i)$

56. $(7 - 3i)(-2 - 8i)$

57. $(3 + 4i)^2$

58. $(4 - 2i)^2$

59. $(-1 - 2i)^2$

60. $(-2 + 5i)^2$

61. $(8 - 7i)(8 + 7i)$

62. $(5 + 3i)(5 - 3i)$

63. $(-2 + 3i)(-2 - 3i)$

64. $(-6 - 7i)(-6 + 7i)$

For Problems 65–78, find each quotient and express the answers in standard form.

65. $\dfrac{4i}{3 - 2i}$

66. $\dfrac{3i}{6 + 2i}$

67. $\dfrac{2 + 3i}{3i}$

68. $\dfrac{3 - 5i}{4i}$

69. $\dfrac{3}{2i}$

70. $\dfrac{7}{4i}$

71. $\dfrac{3 + 2i}{4 + 5i}$

72. $\dfrac{2 + 5i}{3 + 7i}$

73. $\dfrac{4 + 7i}{2 - 3i}$

74. $\dfrac{3 + 9i}{4 - i}$

75. $\dfrac{3 - 7i}{-2 + 4i}$

76. $\dfrac{4 - 10i}{-3 + 7i}$

77. $\dfrac{-1 - i}{-2 - 3i}$

78. $\dfrac{-4 + 9i}{-3 - 6i}$

79. Using $a + bi$ and $c + di$ to represent two complex numbers, verify the following properties.

   a. The conjugate of the sum of two complex numbers is equal to the sum of the conjugates of the two numbers.

   b. The conjugate of the product of two complex numbers is equal to the product of the conjugates of the numbers.

■ ■ ■ THOUGHTS INTO WORDS

80. Is every real number also a complex number? Explain your answer.

81. Can the product of two nonreal complex numbers be a real number? Explain your answer.

■ ■ ■ FURTHER INVESTIGATIONS

82. Observe the following powers of i:

$$i = \sqrt{-1}$$
$$i^2 = -1$$
$$i^3 = i^2 \cdot i = -1(i) = -i$$
$$i^4 = i^2 \cdot i^2 = (-1)(-1) = 1$$

Any power of i greater than 4 can be simplified to i, -1, $-i$, or 1 as follows:

$$i^9 = (i^4)^2(i) = (1)(i) = i$$
$$i^{14} = (i^4)^3(i^2) = (1)(-1) = -1$$
$$i^{19} = (i^4)^4(i^3) = (1)(-i) = -i$$
$$i^{28} = (i^4)^7 = (1)^7 = 1$$

Express each of the following as $i, -1, -i$, or 1.

   a. i^5 **b.** i^6 **c.** i^{11}

   d. i^{12} **e.** i^{16} **f.** i^{22}

   g. i^{33} **h.** i^{63}

83. We can use the information from Problem 82 and the binomial expansion patterns to find powers of complex numbers as follows:

$$(3 + 2i)^3 = (3)^3 + 3(3)^2(2i) + 3(3)(2i)^2 + (2i)^3$$
$$= 27 + 54i + 36i^2 + 8i^3$$
$$= 27 + 54i + 36(-1) + 8(-i)$$
$$= -9 + 46i$$

Find the indicated power of each expression:

a. $(2 + i)^3$ **b.** $(1 - i)^3$

c. $(1 - 2i)^3$ **d.** $(1 + i)^4$

e. $(2 - i)^4$ **f.** $(-1 + i)^5$

84. Some of the solution sets for quadratic equations in the next chapter will contain complex numbers such as $(-4 + \sqrt{-12})/2$ and $(-4 - \sqrt{-12})/2$. We can simplify the first number as follows.

$$\frac{-4 + \sqrt{-12}}{2} = \frac{-4 + i\sqrt{12}}{2} =$$

$$\frac{-4 + 2i\sqrt{3}}{2} = \frac{\cancel{2}(-2 + i\sqrt{3})}{\cancel{2}} = -2 + i\sqrt{3}$$

Simplify each of the following complex numbers.

a. $\dfrac{-4 - \sqrt{-12}}{2}$ **b.** $\dfrac{6 + \sqrt{-24}}{4}$

c. $\dfrac{-1 - \sqrt{-18}}{2}$ **d.** $\dfrac{-6 + \sqrt{-27}}{3}$

e. $\dfrac{10 + \sqrt{-45}}{4}$ **f.** $\dfrac{4 - \sqrt{-48}}{2}$

Answers to the Concept Quiz

1. True   **2.** True   **3.** False   **4.** True   **5.** True   **6.** True   **7.** False   **8.** True

Chapter 0 Summary

Be sure of the following key concepts from this chapter: set, null set, equal sets, subset, natural numbers, whole numbers, integers, rational numbers, irrational numbers, real numbers, complex numbers, absolute value, similar terms, exponent, monomial, binomial, polynomial, degree of a polynomial, perfect-square trinomial, factoring polynomials, rational expression, least common denominator, radical, simplest radical form, root, and conjugate of a complex number.

The following properties of the real numbers provide a basis for arithmetic and algebraic computation: closure for addition and multiplication, commutativity for addition and multiplication, associativity for addition and multiplication, identity properties for addition and multiplication, inverse properties for addition and multiplication, multiplication property of zero, multiplication property of negative one, and distributive property.

The following properties of absolute value are useful:

1. $|a| \geq 0$

2. $|a| = |-a|$ a and b are real numbers

3. $|a - b| = |b - a|$

The following properties of exponents provide the basis for much of our computational work with polynomials:

1. $b^n \cdot b^m = b^{n+m}$

2. $(b^n)^m = b^{mn}$ m and n are rational numbers and

3. $(ab)^n = a^n b^n$ a and b are real numbers, except $b \neq 0$ whenever it appears in the denominator

4. $\left(\dfrac{a}{b}\right)^n = \dfrac{a^n}{b^n}$

5. $\dfrac{b^n}{b^m} = b^{n-m}$

The following product patterns are helpful to recognize when multiplying polynomials:

1. $(a + b)^2 = a^2 + 2ab + b^2$

2. $(a - b)^2 = a^2 - 2ab + b^2$

3. $(a + b)(a - b) = a^2 - b^2$

4. $(a + b)^3 = a^3 + 3a^2b + 3ab^2 + b^3$

5. $(a - b)^3 = a^3 - 3a^2b + 3ab^2 - b^3$

Be sure you know how to do the following:

1. Factor out the highest common monomial factor.

2. Factor by grouping.

3. Factor a trinomial into the product of two binomials.

4. Recognize some basic factoring patterns:
$$a^2 + 2ab + b^2 = (a + b)^2$$
$$a^2 - 2ab + b^2 = (a - b)^2$$
$$a^2 - b^2 = (a + b)(a - b)$$
$$a^3 + b^3 = (a + b)(a^2 - ab + b^2)$$
$$a^3 - b^3 = (a - b)(a^2 + ab + b^2)$$

Be sure that you can simplify, add, subtract, multiply, and divide rational expressions using the following properties and definitions:

1. $\dfrac{a \cdot k}{b \cdot k} = \dfrac{a}{b}$

2. $\dfrac{-a}{b} = \dfrac{a}{-b} = -\dfrac{a}{b}$

3. $\dfrac{a}{b} \cdot \dfrac{c}{d} = \dfrac{ac}{bd}$

4. $\dfrac{a}{b} \div \dfrac{c}{d} = \dfrac{a}{b} \cdot \dfrac{d}{c} = \dfrac{ad}{bc}$

5. $\dfrac{a}{c} + \dfrac{b}{c} = \dfrac{a + b}{c}$

6. $\dfrac{a}{c} - \dfrac{b}{c} = \dfrac{a - b}{c}$

Be sure that you can simplify, add, subtract, multiply, and divide radicals using the following definitions and properties:

1. $\sqrt[n]{b} = a$ if and only if $a^n = b$

2. $\sqrt[n]{bc} = \sqrt[n]{b}\sqrt[n]{c}$

3. $\sqrt[n]{\dfrac{b}{c}} = \dfrac{\sqrt[n]{b}}{\sqrt[n]{c}}$

The following definition provides the link between exponents and roots:
$$b^{m/n} = \sqrt[n]{b^m} = (\sqrt[n]{b})^m$$

This link, along with the properties of exponents, allows us (1) to multiply and divide some radicals with different indexes, (2) to change to simplest radical form while in exponential form, and (3) to simplify expressions that are roots of roots.

A complex number is any number that can be expressed in the form $a + bi$, where a and b are real numbers and i is the imaginary unit such that $i^2 = -1$.

Addition and subtraction of complex numbers are defined as follows:

$$(a + bi) + (c + di) = (a + c) + (b + d)i$$
$$(a + bi) - (c + di) = (a - c) + (b - d)i$$

Because complex numbers have a binomial form, we can multiply two complex numbers in the same way that we multiply two binomials. Thus i^2 can be replaced with -1, and the final result can be expressed in the standard form of a complex number. For example,

$$(3 + 2i)(4 - 3i) = 12 - i - 6i^2$$
$$= 12 - i - 6(-1)$$
$$= 18 - i$$

The two complex numbers $a + bi$ and $a - bi$ are called conjugates of each other. The product $(a - bi)(a + bi)$ equals the real number $a^2 + b^2$, and this property is used to help with dividing complex numbers.

Chapter 0   Review Problem Set

For Problems 1–10, evaluate.

1. 5^{-3}

2. -3^{-4}

3. $\left(\dfrac{3}{4}\right)^{-2}$

4. $\dfrac{1}{\left(\dfrac{1}{3}\right)^{-2}}$

5. $-\sqrt{64}$

6. $\sqrt[3]{\dfrac{27}{8}}$

7. $\sqrt[5]{-\dfrac{1}{32}}$

8. $36^{-1/2}$

9. $\left(\dfrac{1}{8}\right)^{-2/3}$

10. $-32^{3/5}$

For Problems 11–18, perform the indicated operations and simplify. Express the final answers using positive exponents only.

11. $(3x^{-2}y^{-1})(4x^4y^2)$

12. $(5x^{2/3})(-6x^{1/2})$

13. $(-8a^{-1/2})(-6a^{1/3})$

14. $(3x^{-2/3}y^{1/5})^3$

15. $\dfrac{64x^{-2}y^3}{16x^3y^{-2}}$

16. $\dfrac{56x^{-1/3}y^{2/5}}{7x^{1/4}y^{-3/5}}$

17. $\left(\dfrac{-8x^2y^{-1}}{2x^{-1}y^2}\right)^2$

18. $\left(\dfrac{36a^{-1}b^4}{-12a^2b^5}\right)^{-1}$

For Problems 19–34, perform the indicated operations.

19. $(-7x - 3) + (5x - 2) + (6x + 4)$

20. $(12x + 5) - (7x - 4) - (8x + 1)$

21. $3(a - 2) - 2(3a + 5) + 3(5a - 1)$

22. $(4x - 7)(5x + 6)$

23. $(-3x + 2)(4x - 3)$

24. $(7x - 3)(-5x + 1)$

25. $(x + 4)(x^2 - 3x - 7)$

26. $(2x + 1)(3x^2 - 2x + 6)$

27. $(5x - 3)^2$

28. $(3x + 7)^2$

29. $(2x - 1)^3$

30. $(3x + 5)^3$

31. $(x^2 - 2x - 3)(x^2 + 4x + 5)$

32. $(2x^2 - x - 2)(x^2 + 6x - 4)$

33. $\dfrac{24x^3y^4 - 48x^2y^3}{-6xy}$

34. $\dfrac{-56x^2y + 72x^3y^2}{8x^2}$

For Problems 35–46, factor each polynomial completely. Indicate any that are not factorable using integers.

35. $9x^2 - 4y^2$

36. $3x^3 - 9x^2 - 120x$

37. $4x^2 + 20x + 25$

38. $(x - y)^2 - 9$

39. $x^2 - 2x - xy + 2y$

40. $64x^3 - 27y^3$

41. $15x^2 - 14x - 8$

42. $3x^3 + 36$

43. $2x^2 - x - 8$

44. $3x^3 + 24$

45. $x^4 - 13x^2 + 36$

46. $4x^2 - 4x + 1 - y^2$

For Problems 47–56, perform the indicated operations involving rational expressions. Express final answers in simplest form.

47. $\dfrac{8xy}{18x^2y} \cdot \dfrac{24xy^2}{16y^3}$

48. $\dfrac{-14a^2b^2}{6b^3} \div \dfrac{21a}{15ab}$

49. $\dfrac{x^2 + 3x - 4}{x^2 - 1} \cdot \dfrac{3x^2 + 8x + 5}{x^2 + 4x}$

50. $\dfrac{9x^2 - 6x + 1}{2x^2 + 8} \cdot \dfrac{8x + 20}{6x^2 + 13x - 5}$

51. $\dfrac{3x - 2}{4} + \dfrac{5x - 1}{3}$

52. $\dfrac{2x - 6}{5} - \dfrac{x + 4}{3}$

53. $\dfrac{3}{n^2} + \dfrac{4}{5n} - \dfrac{2}{n}$

54. $\dfrac{5}{x^2 + 7x} - \dfrac{3}{x}$

55. $\dfrac{3x}{x^2 - 6x - 40} + \dfrac{4}{x^2 - 16}$

56. $\dfrac{2}{x - 2} - \dfrac{2}{x + 2} - \dfrac{4}{x^3 - 4x}$

For Problems 57– 59, simplify each complex fraction.

57. $\dfrac{\dfrac{3}{x} - \dfrac{2}{y}}{\dfrac{5}{x^2} + \dfrac{7}{y}}$

58. $\dfrac{3 - \dfrac{2}{x}}{4 + \dfrac{3}{x}}$

59. $\dfrac{\dfrac{3}{(x + h)^2} - \dfrac{3}{x^2}}{h}$

60. Simplify the expression

$$\dfrac{6(x^2 + 2)^{1/2} - 6x^2(x^2 + 2)^{-1/2}}{[(x^2 + 2)^{1/2}]^2}$$

For Problems 61–68, express each in simplest radical form. All variables represent positive real numbers.

61. $5\sqrt{48}$

62. $3\sqrt{24x^3}$

63. $\sqrt[3]{32x^4y^5}$

64. $\dfrac{3\sqrt{8}}{2\sqrt{6}}$

65. $\sqrt{\dfrac{5x}{2y^2}}$

66. $\dfrac{3}{\sqrt{2} + 5}$

67. $\dfrac{4\sqrt{2}}{3\sqrt{2} + \sqrt{3}}$

68. $\dfrac{3\sqrt{x}}{\sqrt{x} - 2\sqrt{y}}$

For Problems 69–74, perform the indicated operations and express the answers in simplest radical form.

69. $\sqrt{5}\sqrt[3]{5}$

70. $\sqrt[3]{x^2}\sqrt[4]{x}$

71. $\sqrt{x^3}\sqrt[3]{x^4}$

72. $\sqrt{xy}\sqrt[5]{x^3y^2}$

73. $\dfrac{\sqrt{5}}{\sqrt[3]{5}}$

74. $\dfrac{\sqrt[3]{x^2}}{\sqrt[4]{x^3}}$

For Problems 75–86, perform the indicated operations and express the resulting complex number in standard form.

75. $(-7 + 3i) + (-4 - 9i)$

76. $(2 - 10i) - (3 - 8i)$

77. $(-1 + 4i) - (-2 + 6i)$

78. $(3i)(-7i)$

79. $(2 - 5i)(3 + 4i)$

80. $(-3 - i)(6 - 7i)$

81. $(4 + 2i)(-4 - i)$

82. $(5 - 2i)(5 + 2i)$

83. $\dfrac{5}{3i}$

84. $\dfrac{2 + 3i}{3 - 4i}$

85. $\dfrac{-1 - 2i}{-2 + i}$

86. $\dfrac{-6i}{5 + 2i}$

For Problems 87–92, write each in terms of i and simplify.

87. $\sqrt{-100}$

88. $\sqrt{-40}$

89. $4\sqrt{-80}$

90. $(\sqrt{-9})(\sqrt{-16})$

91. $(\sqrt{-6})(\sqrt{-8})$

92. $\dfrac{\sqrt{-24}}{\sqrt{-3}}$

For Problems 93 and 94, use scientific notation and the properties of exponents to help with the computations.

93. $\dfrac{(0.0064)(420,000)}{(0.00014)(0.032)}$

94. $\dfrac{(8600)(0.0000064)}{(0.0016)(0.000043)}$

1. Evaluate each of the following.

    a. -7^{-2} **b.** $\left(\dfrac{3}{2}\right)^{-3}$

    c. $\left(\dfrac{4}{9}\right)^{3/2}$ **d.** $\sqrt[3]{\dfrac{27}{64}}$

2. Find the product $(-3x^{-1}y^{2})(5x^{-3}y^{-4})$ and express the result using positive exponents only.

For Problems 3–7, perform the indicated operations.

3. $(-3x - 4) - (7x - 5) + (-2x - 9)$

4. $(5x - 2)(-6x + 4)$

5. $(x + 2)(3x^{2} - 2x - 7)$

6. $(4x - 1)^{3}$

7. $\dfrac{-18x^{4}y^{3} - 24x^{5}y^{4}}{-2xy^{2}}$

For Problems 8–11, factor each polynomial completely.

8. $18x^{3} - 15x^{2} - 12x$

9. $30x^{2} - 13x - 10$

10. $8x^{3} + 64$

11. $x^{2} + xy - 2y - 2x$

For Problems 12–16, perform the indicated operations involving rational expressions. Express final answers in simplest form.

12. $\dfrac{6x^{3}y^{2}}{5xy} \div \dfrac{8y}{7x^{3}}$

13. $\dfrac{x^{2} - 4}{2x^{2} + 5x + 2} \cdot \dfrac{2x^{2} + 7x + 3}{x^{3} - 8}$

14. $\dfrac{3n - 2}{4} - \dfrac{4n + 1}{6}$

15. $\dfrac{5}{2x^{2} - 6x} + \dfrac{4}{3x^{2} + 6x}$

16. $\dfrac{4}{n^{2}} - \dfrac{3}{2n} - \dfrac{5}{n}$

17. Simplify the complex fraction $\dfrac{\dfrac{2}{x} - \dfrac{5}{y}}{\dfrac{3}{x} + \dfrac{4}{y^{2}}}$.

For Problems 18–21, express each radical expression in simplest radical form. All variables represent positive real numbers.

18. $6\sqrt{28x^{5}}$

19. $\dfrac{5\sqrt{6}}{3\sqrt{12}}$

20. $\dfrac{\sqrt{6}}{2\sqrt{2} - \sqrt{3}}$

21. $\sqrt[3]{48x^{4}y^{5}}$

For Problems 22–25, perform the indicated operations and express the resulting complex numbers in standard form.

22. $(-2 - 4i) - (-1 + 6i) + (-3 + 7i)$

23. $(5 - 7i)(4 + 2i)$

24. $(7 - 6i)(7 + 6i)$

25. $\dfrac{1 + 2i}{3 - i}$

Equations, Inequalities, and Problem Solving

Chapter Outline

We can use the equation $0.2(8) - 0.2x + x = 0.9(8)$ to determine the change of the mixture in a radiator from 20% antifreeze to 90% antifreeze. We would drain 7 liters of coolant from a radiator that holds 8 liters and replace them with pure antifreeze, which will change the protection against temperature from 12°F to −20°F.

© Jim Craigmyle/Getty Images/First Light

A precinct reported that 316 people voted in an election. The number of Republican voters was 6 more than two-thirds the number of Democrat voters. The equation $d + \left(\frac{2}{3}d + 6\right) = 316$ can be used to determine the number of voters from each of the two parties.

If a ring costs a jeweler $250, at what price should it be sold to make a profit of 60% based on the selling price? The equation $s = 250 + 0.6s$ can be used to determine the selling price.

How much water must be evaporated from 20 gallons of a 10% salt solution in order to obtain a 20% salt solution? The guideline *water in original solution minus water evaporated equals water in final solution* can be used to set up the equation $18 - x = 0.8(20 - x)$. Solving this equation produces the amount of water, x, to be evaporated.

A common sequence in precalculus algebra courses is to develop algebraic skills, then use the skills to solve equations and inequalities, and finally use equations and inequalities to solve applied problems. In this chapter we shall review and extend a variety of concepts related to that process.

1.1 Linear Equations and Problem Solving

Objectives

- Know the properties of equality.
- Solve linear equations.
- Solve application problems involving linear equations.

An algebraic equation such as $5x + 2 = 12$ is neither true nor false as it stands; it is sometimes referred to as an **open sentence.** Each time that a number is substituted for x, the algebraic equation $5x + 2 = 12$ becomes a **numerical statement,** which is either true or false. For example, if $x = 5$, then $5x + 2 = 12$ becomes $5(5) + 2 = 12$, which is a false statement. If $x = 2$, then $5x + 2 = 12$ becomes $5(2) + 2 = 12$, which is a true statement. "Solving an equation" refers to the process of finding the number (or numbers) that make(s) an algebraic equation a true numerical statement. Such numbers are called the **solutions** or **roots** of the equation and are said to *satisfy the equation.* The set of all solutions of an equation is called its **solution set.** Thus {2} is the solution set of $5x + 2 = 12$.

An equation that is satisfied by all numbers that can meaningfully replace the variable is called an **identity.** For example,

$$3(x + 2) = 3x + 6 \qquad x^2 - 4 = (x + 2)(x - 2) \qquad \text{and}$$

$$\frac{1}{x} + \frac{1}{2} = \frac{2 + x}{2x}$$

are all identities. In the last identity, x cannot equal zero; thus the statement

$$\frac{1}{x} + \frac{1}{2} = \frac{2 + x}{2x}$$

is true for all real numbers except zero. An equation that is true for some but not all permissible values of the variable is called a **conditional equation.** Thus the equation $5x + 2 = 12$ is a conditional equation.

Equivalent equations are equations that have the same solution set. For example,

$$7x - 1 = 20 \qquad 7x = 21 \qquad \text{and} \qquad x = 3$$

are all equivalent equations because {3} is the solution set of each. The general procedure for solving an equation is to continue replacing the given equation with equivalent but simpler equations until an equation of the form *variable = constant* or *constant = variable* is obtained. Thus in the example above, $7x - 1 = 20$ was simplified to $7x = 21$, which was further simplified to $x = 3$, which gives us the solution set, {3}.

Techniques for solving equations revolve around properties of equality. The following list summarizes some basic properties of equality.

Property 1.1 Properties of Equality

For all real numbers, a, b, and c,

1. $a = a$ Reflexive property

2. If $a = b$, then $b = a$ Symmetric property

3. If $a = b$ and $b = c$, then $a = c$ Transitive property

4. If $a = b$, then a may be replaced by b, or b may be replaced by a, in any statement without changing the meaning of the statement.
Substitution property

5. $a = b$ if and only if $a + c = b + c$ Addition property

6. $a = b$ if and only if $ac = bc$, where $c \neq 0$ Multiplication property

---

The addition property of equality states that any number can be added to both sides of an equation to produce an equivalent equation. The multiplication property of equality states that an equivalent equation is produced whenever both sides of an equation are multiplied by the same nonzero real number.

■ Linear Equations

Now let's consider how these properties of equality can be used to solve a variety of linear equations. A **linear equation** in the variable x is one that can be written in the form

$$ax + b = 0$$

where a and b are real numbers and $a \neq 0$.

EXAMPLE 1

Solve the equation $-4x - 3 = 2x + 9$.

Solution

$$-4x - 3 = 2x + 9$$
$$-4x - 3 + (-2x) = 2x + 9 + (-2x) \qquad \text{Add } -2x \text{ to both sides}$$
$$-6x - 3 = 9$$
$$-6x - 3 + 3 = 9 + 3 \qquad \text{Add 3 to both sides}$$
$$-6x = 12$$
$$-\frac{1}{6}(-6x) = -\frac{1}{6}(12) \qquad \text{Multiply both sides by } -\frac{1}{6}$$
$$x = -2$$

✔ **Check**

To check an apparent solution, we can substitute it into the original equation to see whether we obtain a true numerical statement.

$$-4x - 3 = 2x + 9$$
$$-4(-2) - 3 \stackrel{?}{=} 2(-2) + 9$$
$$8 - 3 \stackrel{?}{=} -4 + 9$$
$$5 = 5$$

Now we know that the solution set is $\{-2\}$.

▶ **Now work Problem 7.**   ■

E X A M P L E   2   Solve $4(n - 2) - 3(n - 1) = 2(n + 6)$.

Solution

First let's use the distributive property to remove parentheses and combine similar terms:

$$4(n - 2) - 3(n - 1) = 2(n + 6)$$
$$4n - 8 - 3n + 3 = 2n + 12$$
$$n - 5 = 2n + 12$$

Now we can apply the addition property of equality:

$$n - 5 + (-n) = 2n + 12 + (-n)$$
$$-5 = n + 12$$
$$-5 + (-12) = n + 12 + (-12)$$
$$-17 = n$$

✔ **Check**

$$4(n - 2) - 3(n - 1) = 2(n + 6)$$
$$4(-17 - 2) - 3(-17 - 1) \stackrel{?}{=} 2(-17 + 6)$$
$$4(-19) - 3(-18) \stackrel{?}{=} 2(-11)$$
$$-76 + 54 \stackrel{?}{=} -22$$
$$-22 = -22$$

The solution set is $\{-17\}$.

▶ **Now work Problem 13.**   ■

As you study these examples, pay special attention to the steps shown in the solutions. Certainly there are no rules about which steps should be performed mentally; this is an individual decision. We suggest that you show enough steps so that the flow of the process is understood and so that the chances of making careless computational errors are minimized. We shall discontinue showing the check for each problem, but remember that checking an answer is the only way to be sure of your result.

E X A M P L E   3

Solve $\dfrac{1}{4}x - \dfrac{2}{3}x = \dfrac{5}{6}$.

Solution

$$\dfrac{1}{4}x - \dfrac{2}{3}x = \dfrac{5}{6}$$

$$12\left(\dfrac{1}{4}x - \dfrac{2}{3}x\right) = 12\left(\dfrac{5}{6}\right) \qquad \text{Multiply both sides by 12, the LCD}$$

$$12\left(\dfrac{1}{4}x\right) - 12\left(\dfrac{2}{3}x\right) = 12\left(\dfrac{5}{6}\right) \qquad \begin{array}{l}\text{Apply the distributive property on}\\ \text{the left side}\end{array}$$

$$3x - 8x = 10$$

$$-5x = 10$$

$$x = -2$$

The solution set is $\{-2\}$.

⊙ **Now work Problem 21.**   ∎

E X A M P L E   4

Solve $\dfrac{2y - 3}{3} + \dfrac{y + 1}{2} = 3$.

Solution

$$\dfrac{2y - 3}{3} + \dfrac{y + 1}{2} = 3$$

$$6\left(\dfrac{2y - 3}{3} + \dfrac{y + 1}{2}\right) = 6(3) \qquad \text{Multiply both sides by 6, the LCD}$$

$$6\left(\dfrac{2y - 3}{3}\right) + 6\left(\dfrac{y + 1}{2}\right) = 6(3) \qquad \begin{array}{l}\text{Apply the distributive property on}\\ \text{the left side}\end{array}$$

$$2(2y - 3) + 3(y + 1) = 18$$

$$4y - 6 + 3y + 3 = 18$$

$$7y - 3 = 18$$

$$7y = 21$$

$$y = 3$$

The solution set is $\{3\}$. (Check it!).

⊙ **Now work Problem 27.**   ∎

E X A M P L E   5

Solve $\dfrac{4x - 1}{10} - \dfrac{5x + 2}{4} + 3 = 0$.

Solution

$$\dfrac{4x - 1}{10} - \dfrac{5x + 2}{4} + 3 = 0$$

$$20\left(\dfrac{4x - 1}{10} - \dfrac{5x + 2}{4} + 3\right) = 20(0)$$

$$20\left(\frac{4x-1}{10}\right) - 20\left(\frac{5x+2}{4}\right) + 20(3) = 20(0)$$

$$2(4x-1) - 5(5x+2) + 60 = 0$$

$$8x - 2 - 25x - 10 + 60 = 0$$

$$-17x + 48 = 0$$

$$-17x = -48$$

$$x = \frac{48}{17}$$

The solution set is $\left\{\dfrac{48}{17}\right\}$.

▶ **Now work Problem 33.**   ■

In Example 5 checking $\dfrac{48}{17}$ in the original equation is a bit messy. So let's give ourselves a partial check by looking at a picture of this situation. Figure 1.1 shows a graph of the equation $y = \dfrac{4x-1}{10} - \dfrac{5x+2}{4} + 3$. If we let $y = 0$, then the equation $y = \dfrac{4x-1}{10} - \dfrac{5x+2}{4} + 3$ is the given equation in Example 5. Graphically speaking, we see that y equals zero at the point where the line crosses the x axis, which is between 2 and 3. Our solution of $\dfrac{48}{17}$ is also between 2 and 3, so at least we have a partial check.

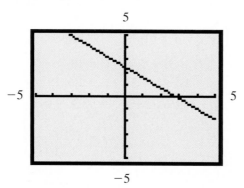

Figure 1.1

We need to emphasize two points pertaining to the previous discussion. First, it is possible to use some features of a graphing calculator to obtain a much better approximation than simply *between 2 and 3*. We will use some of these features later in the text, but for now our focus is on the relationship between the solutions of an algebraic equation and the x intercepts of a geometric graph. Second, we give the following precise definition of an x intercept: The x coordinates of the points that a graph has in common with the x axis are called the ***x* intercepts** of the graph. (To compute the x intercepts, let $y = 0$ and solve for x.)

■ Problem Solving

The ability to use the tools of algebra to solve problems requires that we be able to translate the English language into the language of algebra. More specifically, at this time we need to translate *English sentences* into *algebraic equations* so that we can use our equation-solving skills. Let's work through an example and then comment on some of the problem-solving aspects of it.

P R O B L E M   1

During an online promotion to sell two models of MP3 players, a company sold 1003 players in five hours. The number of 2-gigabyte models sold was 25 less than three times the number of 1-gigabyte models sold. Find the number of 1-gigabyte models sold.

Solution

Let n represent the number of 1-gigabyte models sold. The statement "The number of 2-gigabyte models sold was 25 less than three times the number of 1-gigabyte models sold" tells us that the $3n - 25$ represents the number of 2-gigabyte models sold. Then we can form an equation knowing that the sum of the number of 1-gigabyte models sold plus the number of 2-gigabyte models sold equals 1003. Solving this equation, we obtain

$$n + 3n - 25 = 1003$$
$$4n - 25 = 1003$$
$$4n = 1028$$
$$n = 257$$

There were 257 of the 1-gigabyte models sold.

▶ **Now work Problem 43.** ■

Now let's make a few comments about our approach to Problem 1. Making a statement such as *Let n represent the number to be found* is often referred to as **declaring the variable.** It amounts to choosing a letter to use as a variable and indicating what the variable represents for a specific problem. This may seem like an insignificant idea, but as the problems become more complex, the process of declaring the variable becomes more important. It is also a good idea to choose a *meaningful* variable. For example, if the problem involves finding the width of a rectangle, then a choice of w for the variable is reasonable. Furthermore, it is true that some people can solve a problem such as Problem 1 without setting up an algebraic equation. However, as problems increase in difficulty, the translation from English to algebra becomes a key issue. Therefore, even with these relatively easy problems, we suggest that you concentrate on the translation process.

To check our answer for Problem 1, we must determine whether it satisfies the conditions stated in the original problem. Because the number of 2-gigabyte models is $3n - 25 = 3(257) - 25 = 746$, and $257 + 746 = 1003$, we know our answer of 257 is correct. Remember, when you are checking a potential answer for a word problem, it is *not* sufficient to check the result in the equation used to solve the problem, because the equation itself may be in error.

Sometimes it is necessary not only to declare the variable but also to represent other unknown quantities in terms of that variable. Let's consider a problem that illustrates this idea.

PROBLEM 2

Find three consecutive integers whose sum is -45.

Solution

Let n represent the smallest integer; then $n + 1$ is the next integer and $n + 2$ is the largest of the three integers. Because the sum of the three consecutive integers is to be -45, we have the following equation.

$$n + (n + 1) + (n + 2) = -45$$
$$3n + 3 = -45$$
$$3n = -48$$
$$n = -16$$

If $n = -16$, then $n + 1$ is -15 and $n + 2$ is -14. Thus the three consecutive integers are -16, -15, and -14.

⊳ **Now work Problem 47.**   ■

Frequently, the translation from English to algebra can be made easier by recognizing a guideline that can be used to set up an appropriate equation. Pay special attention to the guidelines used in the solutions of the next two problems.

PROBLEM 3

Tina is paid time-and-a-half for each hour worked over 40 hours in a week. Last week she worked 45 hours and earned $380. What is her normal hourly rate?

Solution

Let r represent Tina's normal hourly rate. Then $\frac{3}{2}r$ represents $1\frac{1}{2}$ times her normal hourly rate (time-and-a-half). The following guideline can be used to help set up the equation:

Regular wages for first 40 hours	+	Wages for 5 hours of overtime	=	Total wages
$40r$	+	$5\left(\frac{3}{2}r\right)$	=	$380

Solving this equation, we obtain

$$2\left[40r + 5\left(\frac{3}{2}r\right)\right] = 2(380)$$
$$2(40r) + 2\left[5\left(\frac{3}{2}r\right)\right] = 760$$

$$80r + 15r = 760$$
$$95r = 760$$
$$r = 8$$

Her normal hourly rate is thus $8 per hour. (Check the answer in the original statement of the problem!)

(▶)   **Now work Problem 53.**   ■

PROBLEM 4

Rafael's present age is 12 years older Rosa's present age. Ten years ago Rafael's age was three times Rosa's age at that time. Find the present ages of Rafael and Rosa.

Solution

Let x represent Rosa's present age; then $x + 12$ represents Rafael's present age. Ten years ago Rosa and Rafael were 10 years younger. To represent their ages 10 years ago, subtract 10 from their present age. Therefore, $x - 10$ represent Rosa's age 10 years ago, and $x + 12 - 10 = x + 2$ represents Rafael's age 10 years ago. The statement "ten years ago Rafael's age was three times Rosa's age at that time" translates into the equation $x + 2 = 3(x - 10)$. Solving the equation, we obtain.

$$x + 2 = 3(x - 10)$$
$$x + 2 = 3x - 30$$
$$-2x = -32$$
$$x = 16$$

So Rosa's present age is 16 years and Rafael's present age is $x + 12 = 16 + 12 = 28$ years.

(▶)   **Now work Problem 61.**   ■

CONCEPT QUIZ

For Problems 1–7, answer true or false.

1. An algebraic equation is called an open sentence because it is neither a true nor a false statement.
2. Equivalent equations have the same solution set.
3. An equation that is true for all meaningful values of x is called an identity.
4. The addition property of equality states that if the same number is added to both sides of an equation, the result is an equivalent equation.
5. The equation $8x = 0$ does not have a solution.
6. Changing an equation from $3 = x$ to $x = 3$ is the application of the symmetric property.
7. If $x = b$ and $b = 6$, then by the reflexive property we know $x = 6$.
8. If the equation $2(x + 4) = 2x + 5$ is simplified, does the equivalent equation fit the definition of a linear equation?

Problem Set 1.1

For Problems 1–42, solve each equation.

1. $9x - 3 = -21$

2. $-5x + 4 = -11$

3. $13 - 2x = 14$

4. $17 = 6a + 5$

5. $3n - 2 = 2n + 5$

6. $4n + 3 = 5n - 9$

7. $-5a + 3 = -3a + 6$

8. $4x - 3 + 2x = 8x - 3 - x$

9. $-3(x + 1) = 7$

10. $5(2x - 1) = 13$

11. $4(2x - 1) = 3(3x + 2)$

12. $5x - 4(x - 6) = -11$

13. $3(n - 1) = -2(n + 4) + 6(n - 3)$

14. $-3(2t - 5) = 2(4t + 7)$

15. $3(2t - 1) - 2(5t + 1) = 4(3t + 4)$

16. $-(3x - 1) + (2x + 3) = -4 + 3(x - 1)$

17. $-2(y - 4) - (3y - 1) = -2 + 5(y + 1)$

18. $\dfrac{-3x}{4} = \dfrac{9}{2}$

19. $-\dfrac{6x}{7} = 12$

20. $\dfrac{n}{2} - \dfrac{1}{3} = \dfrac{13}{6}$

21. $\dfrac{3}{4}n - \dfrac{1}{12}n = 6$

22. $\dfrac{2}{3}x - \dfrac{1}{5}x = 7$

23. $\dfrac{h}{2} + \dfrac{h}{5} = 1$

24. $\dfrac{4y}{5} - 7 = \dfrac{y}{10}$

25. $\dfrac{y}{5} - 2 = \dfrac{y}{2} + 1$

26. $\dfrac{x + 2}{3} + \dfrac{x - 1}{4} = \dfrac{9}{2}$

27. $\dfrac{c + 5}{7} + \dfrac{c - 3}{4} = \dfrac{5}{14}$

28. $\dfrac{2x - 5}{6} - \dfrac{3x - 4}{8} = 0$

29. $\dfrac{n - 3}{2} - \dfrac{4n - 1}{6} = \dfrac{2}{3}$

30. $\dfrac{3x - 1}{2} + \dfrac{x - 3}{4} = \dfrac{1}{2}$

31. $\dfrac{2t + 3}{6} - \dfrac{t - 9}{4} = 5$

32. $\dfrac{2x + 7}{9} - 4 = \dfrac{x - 7}{12}$

33. $\dfrac{3n - 1}{8} - 2 = \dfrac{2n + 5}{7}$

34. $\dfrac{x + 2}{3} + \dfrac{3x + 1}{4} + \dfrac{2x - 1}{6} = 2$

35. $\dfrac{2t - 3}{6} + \dfrac{3t - 2}{4} + \dfrac{5t + 6}{12} = 4$

36. $\dfrac{3y - 1}{8} + y - 2 = \dfrac{y + 4}{4}$

37. $\dfrac{2x + 1}{14} - \dfrac{3x + 4}{7} = \dfrac{x - 1}{2}$

38. $n + \dfrac{2n - 3}{9} - 2 = \dfrac{2n + 1}{3}$

39. $(x - 3)(x - 1) - x(x + 2) = 7$

40. $(3n + 4)(n - 2) - 3n(n + 3) = 3$

41. $(2y + 1)(3y - 2) - (6y - 1)(y + 4) = -20y$

42. $(4t - 3)(t + 2) - (2t + 3)^2 = -1$

Solve each of Problems 43–62 by setting up and solving an algebraic equation.

43. A meal of a chicken sandwich and some pasta salad had 100 grams of carbohydrates. The pasta salad had 10 grams more than twice the grams of carbohydrates in the chicken sandwich. Find the number of grams of carbohydrates for both.

44. The sum of three consecutive integers is 21 larger than twice the smallest integer. Find the integers.

45. Find three consecutive even integers such that if the largest integer is subtracted from four times the smallest, the result is 6 more than twice the middle integer.

46. Find three consecutive odd integers such that three times the largest is 23 less than twice the sum of the two smallest integers.

47. Find two consecutive positive integers such that the difference of their squares is 37.

48. Find three consecutive integers such that the product of the two largest is 20 more than the square of the smallest integer.

49. Find four consecutive integers such that the product of the two largest is 46 more than the product of the two smallest integers.

50. Over the weekend, Mario bicycled 69 miles. On Sunday he rode 9 miles more than two-thirds of his distance on Saturday. Find the number of miles he rode each day.

51. For a given triangle, the measure of angle A is $10°$ less than three times the measure of angle B. The measure of angle C is one-fifth the sum of the measures of angles A and B. Knowing that the sum of the measures of the angles of a triangle equals $180°$, find the measure of each angle.

52. Jennifer went on a shopping spree, spending a total of $124 on a skirt, a sweater, and a pair of shoes. The cost of the sweater was $\frac{8}{7}$ of the cost of the skirt. The shoes cost $8 less than the skirt. Find the cost of each item.

53. Barry is paid double-time for each hour worked over 40 hours in a week. Last week he worked 47 hours and earned $486. What is his normal hourly rate?

54. The average of the salaries of Kelly, Renee, and Nina is $20,000 a year. If Kelly earns $4000 less than Renee, and Nina's salary is two-thirds of Renee's salary, find the salary of each person.

55. Greg had 80 coins consisting of pennies, nickels, and dimes. The number of nickels was 5 more than one-third the number of pennies, and the number of dimes was 1 less than one-fourth the number of pennies. How many coins of each kind did he have?

56. Rita has a collection of 105 coins consisting of nickels, dimes, and quarters. The number of dimes is 5 more than one-third the number of nickels, and the number of quarters is twice the number of dimes. How many coins of each kind does she have?

57. In a class of 43 students, the number of males is 8 less than twice the number of females. How many females and how many males are there in the class?

58. A precinct reported that 316 people had voted in an election. The number of Republican voters was 6 more than two-thirds the number of Democrats. How many Republicans and how many Democrats voted in that precinct?

59. Two years ago Janie was half as old as she will be 9 years from now. How old is she now?

60. The sum of the present ages of Eric and his father is 58 years. In 10 years, his father will be twice as old as Eric will be at that time. Find their present ages.

61. Brad is 6 years older than Pedro. Five years ago Pedro's age was three-fourths of Brad's age at that time. Find the present ages of Brad and Pedro.

62. Tina is 4 years older than Sherry. In 5 years the sum of their ages will be 48. Find their present ages.

■ ■ ■ THOUGHTS INTO WORDS

63. Explain the difference between a numerical statement and an algebraic equation.

64. Are the equations $9 = 3x - 2$ and $3x - 2 = 9$ equivalent equations? Defend your answer.

65. How do you defend the statement that the equation $x + 3 = x + 2$ has no real number solutions?

66. How do you defend the statement that the solution set of the equation $3(x - 4) = 3x - 12$ is the entire set of real numbers?

■ ■ ■ FURTHER INVESTIGATIONS

67. Verify that for any three consecutive integers, the sum of the smallest and the largest is equal to twice the middle integer.

68. Verify that no four consecutive integers can be found such that the product of the smallest and the largest is equal to the product of the other two integers.

69. Some algebraic identities provide a basis for shortcuts to do mental arithmetic. For example, the identity $(x + y)(x - y) = x^2 - y^2$ indicates that a multiplication problem such as $(31)(29)$ can be treated as $(30 + 1)(30 - 1) = 30^2 - 1^2 = 900 - 1 = 899$.

For each of the following, use the given identity to provide a way of mentally performing the indicated computations. Check your answers with a calculator.

a. $(x + y)(x - y) = x^2 - y^2$: $(21)(19)$; $(39)(41)$; $(22)(18)$; $(42)(38)$; $(47)(53)$

b. $(x + y)^2 = x^2 + 2xy + y^2$: $(21)^2$; $(32)^2$; $(51)^2$; $(62)^2$; $(43)^2$

c. $(x - y)^2 = x^2 - 2xy + y^2$: $(29)^2$; $(49)^2$; $(18)^2$; $(38)^2$; $(67)^2$

d. $(10t + 5)^2 = 100t^2 + 100t + 25 = 100t\,(t + 1) + 25$: $(15)^2$; $(35)^2$; $(45)^2$; $(65)^2$; $(85)^2$

GRAPHING CALCULATOR ACTIVITIES

70. Graph the appropriate equations to give visual support for your solutions for Problems 38–42.

71. For each of the following, first graph the appropriate equation and use your graph to predict an approximate solution for the given equation. Then solve the equation algebraically to see how close your prediction came.

a. $5(x - 2) - (2x + 3) = 4$

b. $\dfrac{2x - 1}{4} - \dfrac{3x + 1}{5} = 0$

c. $\dfrac{4x + 1}{3} = \dfrac{x - 3}{2}$

d. $(x - 2)(x + 3) - (x - 2)(x + 1) = 0$

e. $(x + 1)(x - 1) - (x + 6)(x - 2) = 0$

f. $(x + 3)(x + 2) - (x + 4)(x + 1) = 0$

Answers to the Concept Quiz

1. True   **2.** True   **3.** True   **4.** True   **5.** False   **6.** True   **7.** False   **8.** No

1.2   More Equations and Applications

Objectives

- Find restricted values of the variable when solving equations with variables in the denominator.

- Solve proportions.

- Solve absolute value equations.

- Solve formulas for a specified variable.

- Solve application problems.

In the previous section we considered linear equations, such as

$$\frac{x-1}{3} + \frac{x+2}{4} = \frac{1}{6}$$

that have fractional coefficients with constants as denominators. Now let's consider equations that contain the variable in one or more of the denominators. Our approach to solving such equations remains essentially the same except **we must avoid any values of the variable that make a denominator zero.** Consider the following examples.

E X A M P L E   1

Solve $\dfrac{5}{3x} - \dfrac{1}{9} = \dfrac{1}{x}$.

Solution

First we need to realize that *x cannot equal zero.* Let's indicate this restriction so that it is not forgotten; then we can proceed as follows.

$$\frac{5}{3x} - \frac{1}{9} = \frac{1}{x}, \qquad x \neq 0$$

$$9x\left(\frac{5}{3x} - \frac{1}{9}\right) = 9x\left(\frac{1}{x}\right) \qquad \text{Multiply both sides by the LCD}$$

$$9x\left(\frac{5}{3x}\right) - 9x\left(\frac{1}{9}\right) = 9x\left(\frac{1}{x}\right)$$

$$15 - x = 9$$

$$-x = -6$$

$$x = 6$$

The solution set is {6}. (Check it!)

▶ **Now work Problem 5.**   ■

E X A M P L E   2

Solve $\dfrac{65-n}{n} = 4 + \dfrac{5}{n}$.

Solution

$$\frac{65-n}{n} = 4 + \frac{5}{n}, \qquad n \neq 0$$

$$n\left(\frac{65-n}{n}\right) = n\left(4 + \frac{5}{n}\right)$$

$$65 - n = 4n + 5$$

$$60 = 5n$$

$$12 = n$$

The solution set is {12}.

▶ **Now work Problem 9.**   ■

EXAMPLE 3

Solve $\dfrac{a}{a-2} + \dfrac{2}{3} = \dfrac{2}{a-2}$.

Solution

$$\frac{a}{a-2} + \frac{2}{3} = \frac{2}{a-2}, \qquad a \neq 2$$

$$3(a-2)\left(\frac{a}{a-2} + \frac{2}{3}\right) = 3(a-2)\left(\frac{2}{a-2}\right)$$

$$3a + 2(a-2) = 6$$

$$3a + 2a - 4 = 6$$

$$5a = 10$$

$$a = 2$$

Because our initial restriction was $a \neq 2$, we conclude that this equation *has no solution*. The solution set is $\varnothing$.

▶ **Now work Problem 21.**   ■

    Example 3 illustrates the importance of recognizing the restrictions that must be made to exclude division by zero. By the way, what should happen if we graph $y = \dfrac{x}{x-2} + \dfrac{2}{3} - \dfrac{2}{x-2}$? (We had to change the variable a to x for graphing purposes.) Do you agree that the graph must not have any x intercept? Figure 1.2 shows the graph; we should feel good that our answer was an empty solution set.

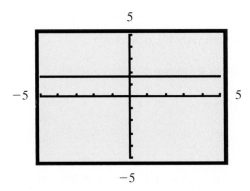

Figure 1.2

■ Ratio and Proportion

A **ratio** is the comparison of two numbers by division. The fractional form is frequently used to express ratios. For example, the ratio of a to b can be written a/b. A statement of equality between two ratios is called a **proportion.** Thus if a/b and c/d are equal ratios, then the proportion $a/b = c/d$ ($b \neq 0$ and $d \neq 0$) can be formed. There is a useful property of proportions:

$$\text{If } \frac{a}{b} = \frac{c}{d}, \text{ then } ad = bc.$$

This property can be deduced as follows:

$$\frac{a}{b} = \frac{c}{d}, \qquad b \neq 0 \text{ and } d \neq 0$$

$$bd\left(\frac{a}{b}\right) = bd\left(\frac{c}{d}\right) \qquad \text{Multiply both sides by } bd$$

$$ad = bc$$

This is sometimes referred to as the **cross-multiplication property of proportions.**

Some equations can be treated as proportions and solved by using the cross-multiplication idea, as the next example illustrates.

EXAMPLE 4

Solve $\dfrac{3}{3x - 2} = \dfrac{4}{2x + 1}$.

Solution

$$\frac{3}{3x - 2} = \frac{4}{2x + 1}, \qquad x \neq \frac{2}{3}, x \neq -\frac{1}{2}$$

$$3(2x + 1) = 4(3x - 2) \qquad \text{Apply the cross-multiplication property}$$

$$6x + 3 = 12x - 8$$

$$11 = 6x$$

$$\frac{11}{6} = x$$

The solution set is $\left\{\dfrac{11}{6}\right\}$.

▶ **Now work Problem 11.**　　　　　　　　　　　　　　　　　　　■

■ Linear Equations Involving Decimals

To solve an equation such as $x + 2.4 = 0.36$, we can add -2.4 to both sides. However, as equations containing decimals become more complex, it is often easier to begin by *clearing the equation of all decimals,* which we accomplish by multiplying both sides by an appropriate power of 10. Let's consider two examples.

EXAMPLE 5

Solve $0.12t - 2.1 = 0.07t - 0.2$.

Solution

$$0.12t - 2.1 = 0.07t - 0.2$$

$$100(0.12t - 2.1) = 100(0.07t - 0.2) \qquad \text{Multiply both sides by 100}$$

$$12t - 210 = 7t - 20$$

$$5t = 190$$
$$t = 38$$

The solution set is {38}.

▶ **Now work Problem 23.** ■

E X A M P L E   6

Solve $0.8x + 0.9(850 - x) = 715$.

Solution

$$0.8x + 0.9(850 - x) = 715$$
$$10[0.8x + 0.9(850 - x)] = 10(715) \qquad \text{Multiply both sides by 10}$$
$$10(0.8x) + 10[0.9(850 - x)] = 10(715)$$
$$8x + 9(850 - x) = 7150$$
$$8x + 7650 - 9x = 7150$$
$$-x = -500$$
$$x = 500$$

The solution set is {500}.

▶ **Now work Problem 25.** ■

■ Changing Forms of Formulas

Many practical applications of mathematics involve the use of formulas. For example, to find the distance traveled in 4 hours at a rate of 55 miles per hour, we multiply the rate times the time; thus the distance is $55(4) = 220$ miles. The rule *distance equals rate times time* is commonly stated as a formula: $d = rt$. When we use a formula, it is sometimes convenient first to change its form. For example, multiplying both sides of $d = rt$ by $1/t$ produces the equivalent form $r = d/t$. Multiplying both sides of $d = rt$ by $1/r$ produces another equivalent form, $t = d/r$. The following two examples further illustrate the process of obtaining equivalent forms of certain formulas.

E X A M P L E   7

If P dollars are invested at a simple rate of r percent, then the amount, A, accumulated after t years is given by the formula $A = P + Prt$. Solve this formula for P.

Solution

$$A = P + Prt$$
$$A = P(1 + rt) \qquad \text{Apply the distributive property to factor the right side}$$
$$\frac{A}{1 + rt} = P \qquad \text{Multiply both sides by } \frac{1}{1 + rt}$$
$$P = \frac{A}{1 + rt} \qquad \text{Apply the symmetric property of equality}$$

▶ **Now work Problem 35.** ■

E X A M P L E   8

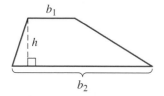

Figure 1.3

The area, A, of a trapezoid (see Figure 1.3) is given by the formula $A = \frac{1}{2}h(b_1 + b_2)$. Solve this equation for b_1.

Solution

$$A = \frac{1}{2}h(b_1 + b_2)$$

$2A = h(b_1 + b_2)$ Multiply both sides by 2

$2A = hb_1 + hb_2$ Apply the distributive property to the right side

$2A - hb_2 = hb_1$ Add $-hb_2$ to both sides

$\dfrac{2A - hb_2}{h} = b_1$ Multiply both sides by $\dfrac{1}{h}$

▶ **Now work Problem 41.** ■

Notice that in Example 7 the distributive property was used to change from the form $P + Prt$ to $P(1 + rt)$. However, in Example 8 the distributive property was used to change $h(b_1 + b_2)$ to $hb_1 + hb_2$. In both examples the goal is to *isolate the term* containing the variable being solved for so that an appropriate application of the multiplication property will produce the desired result. Also note the use of *subscripts* to identify the two bases of the trapezoid. Subscripts allow us to use the same letter b to identify the bases, but b_1 represents one base and b_2 the other.

■ More on Problem Solving

Volumes have been written on the topic of problem solving, but certainly one of the best-known sources is George Polya's book *How to Solve It.** In this book, Polya suggests the following four-phase plan for solving problems.

1. *Understand the problem.*

2. *Devise a plan* to solve the problem.

3. *Carry out the plan* to solve the problem.

4. *Look back* at the completed solution to review and discuss it.

We will comment briefly on each of the phases and offer some suggestions for using an algebraic approach to solve problems.

Understand the Problem   Read the problem carefully, making certain that you understand the meanings of all the words. Be especially alert for any technical terms used in the statement of the problem. Often it is helpful to sketch a figure, diagram, or chart to visualize and organize the conditions of the problem. Determine the known and unknown facts, and if one of the previously mentioned

---

*Polya, George. 1945. *How to Solve It.* Princeton, NJ: Princeton University Press.

pictorial devices is used, record these facts in the appropriate places on the diagram or chart.

Devise a Plan This is the key part of the four-phase plan. It is sometimes referred to as the *analysis* of the problem. There are numerous strategies and techniques used to solve problems. We shall discuss some of these strategies at various places throughout this text; however, at this time we offer the following general suggestions.

1. Choose a meaningful *variable* to represent an unknown quantity in the problem (perhaps t if time is an unknown quantity), and represent any other unknowns in terms of that variable.

2. Look for a *guideline* that can be used to set up an equation. A guideline might be a formula, such as $A = P + Prt$ from Example 7, or a statement of a relationship, such as *the sum of the two numbers is 28*. Sometimes a relationship suggested by a pictorial device can be used as a guideline for setting up the equation. Also, be alert to the possibility that this *new* problem might really be an *old* problem in a new setting, perhaps even stated with different vocabulary.

3. Form an *equation* containing the variable so that the conditions of the guideline are translated from English into algebra.

Carry out the Plan This phase is sometimes referred to as the *synthesis* of the plan. If phase 2 has been successfully completed, then carrying out the plan may simply be a matter of solving the equation and doing any further computations to answer all of the questions in the problem. Confidence in your plan creates a better working atmosphere for carrying it out. It is also in this phase that the calculator may become a valuable tool. The type of data and the amount of complexity involved in the computations are two factors that can influence your decision whether to use one.

Look Back This is an important but often overlooked part of problem solving. The following list of questions suggests some things for you to consider in this phase.

1. Is your answer to the problem a *reasonable* answer?

2. Have you *checked* your answer by substituting it back into the conditions stated in the problem?

3. Looking back over your solution, do you now see another plan that could be used to solve the problem?

4. Do you see a way of generalizing your procedure for this problem that could be used to solve other problems of this type?

5. Do you now see that this problem is closely related to another problem that you have previously solved?

6. Have you tucked away for future reference the technique used to solve this problem?

Looking back over the solution of a newly solved problem can lay important groundwork for solving problems in the future.

Remark: If you are interested in finding out more about George Polya and his insights into problem solving, check the Internet. For example, Google has some interesting information about his problem-solving techniques.

Keep Polya's suggestions in mind as we tackle some more word problems. Perhaps it would also be helpful for you to attempt to solve these problems on your own before looking at our approach.

Sometimes we can use the concepts of ratio and proportion to set up an equation and solve a problem, as the next problem illustrates.

P R O B L E M   1

At a community college, the ratio of transfer students to career education students is 12 to 5. If there is a total of 18,360 students enrolled in the two areas, find the number of transfer students and the number of career students.

Solution

Let x represent the number of transfer students; then $18{,}360 - x$ represents the number of career students. The following proportion can be set up and solved:

$$\frac{x}{18{,}360 - x} = \frac{12}{5}$$
$$5x = 12(18{,}360 - x)$$
$$5x = 220{,}320 - 12x$$
$$17x = 220{,}320$$
$$x = 12{,}960$$

Therefore, there are 12,960 transfer students and $18{,}360 - 12{,}960 = 5{,}400$ career education students.

Now work Problem 49. ∎

The next problem has a geometric setting. In such cases, the use of figures is very helpful.

P R O B L E M   2

The length of a rectangular 4:3 aspect ratio television screen is 7 inches more than the width. If the width is increased by 4.5 inches, and the length is increased by 18.5 inches, the rectangular screen would then be in the widescreen aspect ratio of 16:9, and the area of the screen would be 632.25 square inches more than the 4:3 aspect ratio screen. Find the dimensions of both televisions screens.

Solution

Let x represent the width of the 4:3 aspect ratio television screen. Then $x + 7$ represents the length of the 4:3 aspect ratio television screen. Then for the 16:9 aspect ratio television screen, $x + 4.5$ represents the width, and $x + 7 + 18.5 = x + 25.5$ represents the length. Because the area of the 16:9 aspect ratio television

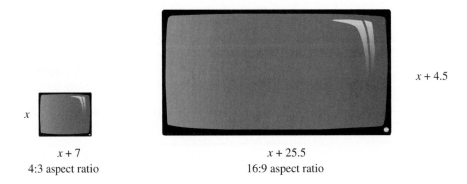

x

$x + 7$
4:3 aspect ratio

$x + 4.5$

$x + 25.5$
16:9 aspect ratio

Figure 1.4

screen is 632.25 square inches more than the 4:3 aspect ratio television screen (see Figure 1.4), the following equation can be set up and solved:

$$\begin{array}{ccc} \text{Area of 4:3} & \text{Increase} & \text{Area of 16:9} \\ \text{television} & + \text{in area} & = \text{television} \end{array}$$

$$x(x + 7) + 632.25 = (x + 4.5)(x + 25.5)$$
$$x^2 + 7x + 632.25 = x^2 + 25.5x + 4.5x + 114.75$$
$$7x + 632.25 = 30x + 114.75$$
$$517.5 = 23x$$
$$x = 22.5$$

For the 4:3 aspect ratio
television screen:

Width $= x = 22.5$ inches

Length $= x + 7 = 22.5 + 7$
$= 29.5$ inches

For the 16:9 aspect ratio
television screen:

Width $= x + 4.5 = 22.5 + 4.5$
$= 27$ inches
Length $= x + 25.5 = 22.5 + 25.5$
$= 48$ inches

Thus the dimensions for the 4:3 aspect ratio television screen are 22.5 inches by 29.5 inches, and the dimensions for the 16:9 aspect ratio television screen are 27 inches by 48 inches.

▶ **Now work Problem 71.**   ■

Many consumer problems can be solved by using an algebraic approach. For example, let's consider a discount sale problem involving the relationship *original selling price minus discount equals discount sale price.*

PROBLEM   3

Jim bought a pair of jeans at a 30% discount sale for $28. What was the original price of the jeans?

Solution

Let p represent the original price of the jeans.

Original price   −   Discount   =   Discount sale price

$$(100\%)(p) \quad - \quad (30\%)(p) = \qquad \$28$$

We switch this equation to decimal form to solve it.

$$p - 0.3p = 28$$
$$0.7p = 28$$
$$p = 40$$

The original price of the jeans was $40.

▶ **Now work Problem 54.**   ∎

Another basic relationship pertaining to consumer problems is *selling price equals cost plus profit*. Profit (also called markup, markon, and margin of profit) may be stated in different ways. It can be expressed as a percent of the cost, as a percent of the selling price, or simply in terms of dollars and cents. Let's consider a problem where the profit is stated as a percent of the selling price.

P R O B L E M   4

A retailer of sporting goods bought a putter for $25. He wants to price the putter to make a profit of 20% of the selling price. What price should he mark on the putter?

Solution

Let s represent the selling price.

Selling price   =   Cost   +   Profit

$$s \qquad = \qquad \$25 \quad + \quad (20\%)(s)$$

Solving this equation involves using the methods we developed earlier for working with decimals:

$$s = 25 + (20\%)(s)$$
$$s = 25 + 0.2s$$
$$10s = 250 + 2s$$
$$8s = 250$$
$$s = 31.25$$

The selling price should be $31.25.

▶ **Now work Problem 57.**   ∎

Certain types of investment problems can be solved by using an algebraic approach. As our final example of this section, let's consider one such problem.

PROBLEM 5

Cindy invested a certain amount of money at 6% interest and $1500 more than that amount at 8% interest. Her total yearly interest was $540. How much did she invest at each rate?

Solution

Let d represent the amount invested at 6%; then $d + 1500$ represents the amount invested at 8%. The following guideline can be used to set up an equation:

Interest earned at 6% + Interest earned at 8% = Total interest

$$(6\%)(d) \quad + \quad (8\%)(d + 1500) \quad = \quad \$540$$

We can solve this equation by multiplying both sides by 100:

$$0.06d + 0.08(d + 1500) = 540$$
$$6d + 8(d + 1500) = 54,000$$
$$6d + 8d + 12,000 = 54,000$$
$$14d = 42,000$$
$$d = 3000$$

Cindy invested $3000 at 6% and $3000 + $1500 = $4500 at 8%.

▶ **Now work Problem 66.**   ∎

Don't forget phase 4 of Polya's problem-solving plan. We have not taken the space to look back over and discuss each of our examples. However, it would be beneficial for you to do so, keeping in mind the questions posed earlier regarding this phase.

CONCEPT QUIZ   For Problems 1–6, answer true or false.

1. When solving equations, restricted values are any values of the variable that make the numerator or the denominator zero.

2. A statement of equality between two ratios is called a proportion.

3. The equation $4 + \dfrac{3}{x - 2} = \dfrac{5}{3x + 1}$ is an example of a proportion.

4. Profit is always expressed as a percent of the cost.

5. The goal in solving a formula for a specified variable is to isolate that variable in a single term and then apply the multiplication property to produce the result.

6. The formulas $d = rt$, $r = \dfrac{d}{t}$, and $t = \dfrac{d}{r}$ are equivalent.

Problem Set 1.2

For Problems 1–32, solve each equation.

1. $\dfrac{x-2}{3} + \dfrac{x+1}{4} = \dfrac{1}{6}$

2. $\dfrac{5n-1}{4} - \dfrac{2n-3}{10} = \dfrac{3}{5}$

3. $\dfrac{5}{x} + \dfrac{1}{3} = \dfrac{8}{x}$

4. $\dfrac{5}{3n} - \dfrac{1}{9} = \dfrac{1}{n}$

5. $\dfrac{1}{3n} + \dfrac{1}{2n} = \dfrac{1}{4}$

6. $\dfrac{1}{x} - \dfrac{3}{2x} = \dfrac{1}{5}$

7. $\dfrac{35-x}{x} = 7 + \dfrac{3}{x}$

8. $\dfrac{n}{46-n} = 5 + \dfrac{4}{46-n}$

9. $\dfrac{n+67}{n} = 5 + \dfrac{11}{n}$

10. $\dfrac{n+52}{n} = 4 + \dfrac{1}{n}$

11. $\dfrac{5}{3x-2} = \dfrac{1}{x-4}$

12. $\dfrac{-2}{5x-3} = \dfrac{4}{4x-1}$

13. $\dfrac{4}{2y-3} - \dfrac{7}{3y-5} = 0$

14. $\dfrac{3}{2n+1} + \dfrac{5}{3n-4} = 0$

15. $\dfrac{n}{n+1} + 3 = \dfrac{4}{n+1}$

16. $\dfrac{a}{a+5} - 2 = \dfrac{3a}{a+5}$

17. $\dfrac{3x}{2x-1} - 4 = \dfrac{x}{2x-1}$

18. $\dfrac{x}{x-8} - 4 = \dfrac{8}{x-8}$

19. $\dfrac{3}{x+3} - \dfrac{1}{x-2} = \dfrac{5}{2x+6}$

20. $\dfrac{6}{x+3} + \dfrac{20}{x^2+x-6} = \dfrac{5}{x-2}$

21. $\dfrac{n}{n-3} - \dfrac{3}{2} = \dfrac{3}{n-3}$

22. $\dfrac{4}{x-2} + \dfrac{x}{x+1} = \dfrac{x^2-2}{x^2-x-2}$

23. $s = 9 + 0.25s$

24. $s = 1.95 + 0.35s$

25. $0.09x + 0.1(700 - x) = 67$

26. $0.08x + 0.09(950 - x) = 81$

27. $0.09x + 0.11(x + 125) = 68.75$

28. $0.08(x + 200) = 0.07x + 20$

29. $0.8(t - 2) = 0.5(9t + 10)$

30. $0.3(2n - 5) = 11 - 0.65n$

31. $0.92 + 0.9(x - 0.3) = 2x - 5.95$

32. $0.5(3x + 0.7) = 20.6$

For Problems 33–46, solve each formula for the indicated variable.

33. $P = 2l + 2w$ for w (Perimeter of a rectangle)

34. $V = \dfrac{1}{3}Bh$ for B (Volume of a pyramid)

35. $A = 2lw + 2lh + 2wh$ for h (Surface area of rectangular box)

36. $z = \dfrac{x - \mu}{\sigma}$ for x (z-score in statistics)

37. $A = 2\pi r^2 + 2\pi rh$ for h (Surface area of a right circular cylinder)

38. $A = \dfrac{1}{2}h(b_1 + b_2)$ for h (Area of a trapezoid)

39. $C = \dfrac{5}{9}(F - 32)$ for F (Fahrenheit to Celsius)

40. $F = \dfrac{9}{5}C + 32$ for C (Celsius to Fahrenheit)

41. $V = C\left(1 - \dfrac{T}{N}\right)$ for T (Linear depreciation)

42. $V = C\left(1 - \dfrac{T}{N}\right)$ for N (Linear depreciation)

43. $I = kl(T - t)$ for T (Expansion allowance in highway construction)

44. $S = \dfrac{CRD}{12d}$ for d (Cutting speed of a circular saw)

45. $\dfrac{1}{R_n} = \dfrac{1}{R_1} + \dfrac{1}{R_2}$ for R_n (Resistance in parallel circuit design)

46. $f = \dfrac{1}{\dfrac{1}{a} + \dfrac{1}{b}}$ for b (Focal length of a camera lens)

For Problems 47–73, set up an equation and solve each problem.

47. Working as a waiter, Tom made $157.50 in tips. Assuming that every customer tipped 15% of the cost of the meal, find the cost of all the meals Tom served.

48. A realtor who is paid 7% of the selling price in commission recently received $10,794 in commission on the sale of a property. What was the selling price of the property?

49. A total of $2250 for a house painting job is to be divided between two painters in the ratio of 2 to 3. How much does each painter receive?

50. One type of motor requires a mixture of oil and gasoline in a ratio of 1 to 15 (that is, 1 part of oil to 15 parts of gasoline). How many liters of each are contained in a 20-liter mixture?

51. The ratio of 2-wheel-drive trucks to 4-wheel-drive trucks at an auto dealership is 8 to 1. If the total number of trucks at the dealership is 189, find the number of each.

52. The ratio of the weight of sodium to that of chlorine in common table salt is 5 to 3. Find the amount of each element in a salt compound weighing 200 pounds.

53. Gary bought an MP3 player at a 20% discount sale for $52. What was the original price of the MP3 player?

54. Roya bought a pair of jeans at a 30% discount sale for $33.60. What was the original price of the jeans?

55. After a 7% increase in salary, Laurie makes $1647.80 per month. How much did she earn per month before the increase?

56. Russ bought a used car for $11,025, including 5% sales tax. What was the selling price of the used car without the tax?

57. A retailer has some shoes that cost $28 per pair. At what price should they be sold to obtain a profit of 15% of the cost?

58. If a head of lettuce costs a retailer $0.80, at what price should it be sold to make a profit of 45% of the cost?

59. Karla sold a bicycle on e-store for $97.50. This selling price represented a 30% profit for her, based on what she had originally paid for the bike. Find Karla's original cost for the bicycle.

60. If a ring costs a jeweler $250, at what price should it be sold to make a profit of 60% of the selling price?

61. A retailer has some skirts that cost $18 each. She wants to sell them at a profit of 40% of the selling price. What price should she charge for the skirts?

62. Suppose that an item costs a retailer $50. How much more profit could be gained by fixing a 50% profit based on selling price rather than a 50% profit based on cost?

63. Derek has some nickels and dimes worth $3.60. The number of dimes is one more than twice the number of nickels. How many nickels and dimes does he have?

64. Robin has a collection of nickels, dimes, and quarters worth $38.50. She has 10 more dimes than nickels and twice as many quarters as dimes. How many coins of each kind does she have?

65. A collection of 70 coins consisting of dimes, quarters, and half-dollars has a value of $17.75. There are three times as many quarters as dimes. Find the number of each kind of coin.

66. A certain amount of money is invested at 8% per year, and $1500 more than that amount is invested at 9% per year. The annual interest from the 9% investment exceeds the annual interest from the 8% investment by $160. How much is invested at each rate?

67. A total of $5500 was invested, part of it at 5% per year and the remainder at 7% per year. If the total yearly interest amounted to $345, how much was invested at each rate?

68. A sum of $3500 is split between two investments, one paying 9% yearly interest and the other 11%. If the return on the 11% investment exceeds that on the 9% investment by $85 the first year, how much is invested at each rate?

69. Celia has invested $2500 at 6% yearly interest. How much must she invest at 5% so that the interest from both investments totals $350 after a year?

70. The length of a rectangle is 2 inches less than three times its width. If the perimeter of the rectangle is 108 inches, find its length and width.

71. The length of a rectangle is 4 centimeters more than its width. If the width is increased by 2 centimeters and the length is increased by 3 centimeters, a new rectangle is formed that has an area of 44 square centimeters more than the area of the original rectangle. Find the dimensions of the original rectangle.

72. The length of a picture without its border is 7 inches less than twice its width. If the border is 1 inch wide,

and its area is 62 square inches, what are the dimensions of the picture alone?

73. If two opposite sides of a square are each increased by 3 centimeters, and the other two sides are each decreased by 2 centimeters, the area is increased by 8 square centimeters. Find the length of a side of the square.

■ ■ ■ THOUGHTS INTO WORDS

74. Give a step-by-step description of how you would solve the formula $F = \dfrac{9}{5}C + 32$ for C.

75. What does the phrase "declare a variable" mean when following the steps to solve a word problem?

76. Why must potential answers to word problems be checked against the original statement of the problem?

77. From a consumer's viewpoint, would you prefer that retailers figure their profit on the basis of the cost or the selling price? Explain your answer.

78. Some people multiply by 2 and add 30 to estimate the change from a Celsius reading to a Fahrenheit reading. Why does this give an estimate? How good is the estimate?

■ ■ ■ FURTHER INVESTIGATIONS

79. Is a 10% discount followed by a 20% discount equal to a 30% discount? Defend your answer.

80. Is a 10% discount followed by a 30% discount the same as a 30% discount followed by a 10% discount? Justify your answer.

81. A retailer buys an item for $90, resells it for $100, and claims that he is making only a 10% profit. Is his claim correct?

82. The following formula can be used to determine the selling price of an item when the profit is based on a percent of the selling price.

$$\text{Selling price} = \frac{\text{Cost}}{100\% - \text{Percent of profit}}$$

Show how this formula is developed.

83. Use the formula from Problem 82 to determine the selling price of each of the following items. The given percent of profit is based on the selling price. Be sure to check each answer.

   a. $0.80 bottle of water;   20% profit
   b. $8.50 music CD;   25% profit
   c. $50 pair of athletic shoes;   40% profit
   d. $200 digital camera;   50% profit
   e. $18,000 car;   15% profit

 GRAPHING CALCULATOR ACTIVITIES

84. Graph the appropriate equations to give visual support for your solutions for Problems 15–21.

Answers to the Concept Quiz

1. False   **2.** True   **3.** False   **4.** False   **5.** True   **6.** True

1.3   Quadratic Equations

Objectives

■ Solve quadratic equations by factoring.

■ Solve quadratic equations by the square root property.

■ Solve quadratic equations by completing the square.

■ Solve quadratic equations by using the quadratic formula.

■ Use the discriminant to determine the nature of the solutions of a quadratic equation.

■ Solve application problems involving quadratic equations.

A **quadratic equation** in the variable x is defined as any equation that can be written in the form

$$ax^2 + bx + c = 0$$

where a, b, and c are real numbers and $a \neq 0$. The form $ax^2 + bx + c = 0$ is called the **standard form** of a quadratic equation. The choice of x for the variable is arbitrary. An equation such as $3t^2 + 5t - 4 = 0$ is a quadratic equation in the variable t.

Quadratic equations such as $x^2 + 2x - 15 = 0$, where the polynomial is factorable, can be solved by applying the following property: **$ab = 0$ if and only if $a = 0$ or $b = 0$.** Our work might take on the following format.

$$x^2 + 2x - 15 = 0$$
$$(x + 5)(x - 3) = 0$$
$$x + 5 = 0 \qquad \text{or} \qquad x - 3 = 0$$
$$x = -5 \qquad \text{or} \qquad x = 3$$

The solution set for this equation is $\{-5, 3\}$.

Let's consider another example of this type.

E X A M P L E   1   Solve the equation $n = -6n^2 + 12$.

Solution

$$n = -6n^2 + 12$$
$$6n^2 + n - 12 = 0$$
$$(3n - 4)(2n + 3) = 0$$
$$3n - 4 = 0 \qquad \text{or} \qquad 2n + 3 = 0$$
$$3n = 4 \qquad \text{or} \qquad 2n = -3$$
$$n = \frac{4}{3} \qquad \text{or} \qquad n = -\frac{3}{2}$$

The solution set is $\left\{ -\frac{3}{2}, \frac{4}{3} \right\}$.

▶   **Now work Problem 1.**   ■

Now suppose that we want to solve $x^2 = k$, where k is any real number. We can proceed as follows:

$$x^2 = k$$
$$x^2 - k = 0$$
$$(x + \sqrt{k})(x - \sqrt{k}) = 0$$
$$x + \sqrt{k} = 0 \qquad \text{or} \qquad x - \sqrt{k} = 0$$
$$x = -\sqrt{k} \qquad \text{or} \qquad x = \sqrt{k}$$

Thus we can state the following property for any real number k.

Property 1.2

The solution set of $x^2 = k$ is $\{-\sqrt{k}, \sqrt{k}\}$, which can also be written $\{\pm\sqrt{k}\}$.

Property 1.2, along with our knowledge of square root, makes it very easy to solve quadratic equations of the form $x^2 = k$.

E X A M P L E   2

Solve each of the following.

a. $x^2 = 72$ **b.** $(3n - 1)^2 = 26$ **c.** $(y + 2)^2 = -24$

Solutions

a. $x^2 = 72$
$$x = \pm\sqrt{72}$$
$$x = \pm 6\sqrt{2}$$

The solution set is $\{\pm 6\sqrt{2}\}$.

▶ **Now work Problem 7.**

b. $(3n - 1)^2 = 26$
$$3n - 1 = \pm\sqrt{26}$$
$$3n - 1 = \sqrt{26} \qquad \text{or} \qquad 3n - 1 = -\sqrt{26}$$
$$3n = 1 + \sqrt{26} \qquad \text{or} \qquad 3n = 1 - \sqrt{26}$$
$$n = \frac{1 + \sqrt{26}}{3} \qquad \text{or} \qquad n = \frac{1 - \sqrt{26}}{3}$$

The solution set is $\left\{\dfrac{1 \pm \sqrt{26}}{3}\right\}$.

▶ **Now work Problem 9.**

c. $(y + 2)^2 = -24$
$$y + 2 = \pm\sqrt{-24}$$

$$y + 2 = \pm 2i\sqrt{6} \qquad \text{Remember that } \sqrt{-24} = i\sqrt{24} = i\sqrt{4}\sqrt{6} = 2i\sqrt{6}$$

$$y + 2 = 2i\sqrt{6} \qquad \text{or} \qquad y + 2 = -2i\sqrt{6}$$

$$y = -2 + 2i\sqrt{6} \qquad \text{or} \qquad y = -2 - 2i\sqrt{6}$$

The solution set is $\{-2 \pm 2i\sqrt{6}\}$.

▶ **Now work Problem 13.**   ■

■ Completing the Square

A factoring technique we reviewed in Chapter 0 relied on recognizing *perfect-square trinomials*. In each of the following examples, the perfect-square trinomial on the right side of the identity is the result of squaring the binomial on the left side.

$$(x + 5)^2 = x^2 + 10x + 25 \qquad (x - 7)^2 = x^2 - 14x + 49$$
$$(x + 9)^2 = x^2 + 18x + 81 \qquad (x - 12)^2 = x^2 - 24x + 144$$

Note that in each of the square trinomials, the constant term is equal to the square of one-half of the coefficient of the x term. This relationship allows us to *form* a perfect-square trinomial by adding a proper constant term. For example, suppose that we want to form a perfect-square trinomial from $x^2 + 8x$. Because $\frac{1}{2}(8) = 4$ and $4^2 = 16$, the perfect-square trinomial is $x^2 + 8x + 16$. Now let's use this idea to solve a quadratic equation.

E X A M P L E   3   Solve $x^2 + 8x - 2 = 0$.

Solution

$$x^2 + 8x - 2 = 0$$
$$x^2 + 8x = 2$$
$$x^2 + 8x + 16 = 2 + 16 \qquad \text{We added 16 to the left side to form}$$
$$\qquad\qquad\qquad\qquad\qquad \text{a perfect-square trinomial; thus 16 has}$$
$$(x + 4)^2 = 18 \qquad\qquad \text{to be added to the right side}$$
$$x + 4 = \pm\sqrt{18}$$
$$x + 4 = \pm 3\sqrt{2}$$
$$x + 4 = 3\sqrt{2} \qquad \text{or} \qquad x + 4 = -3\sqrt{2}$$
$$x = -4 + 3\sqrt{2} \qquad \text{or} \qquad x = -4 - 3\sqrt{2}$$

The solution set is $\{-4 \pm 3\sqrt{2}\}$.

▶ **Now work Problem 19.**   ■

We have been using a relationship for a perfect-square trinomial that states *the constant term is equal to the square of one-half of the coefficient of the x term.* This relationship holds only if the coefficient of x^2 is 1. Thus we need to make a slight adjustment when we are solving quadratic equations that have a coefficient of x^2 other than 1. The next example shows how to make this adjustment.

E X A M P L E   4

Solve $2x^2 + 6x - 3 = 0$.

Solution

$$2x^2 + 6x - 3 = 0$$
$$2x^2 + 6x = 3$$
$$x^2 + 3x = \frac{3}{2} \qquad \text{Multiply both sides by } \frac{1}{2}$$
$$x^2 + 3x + \frac{9}{4} = \frac{3}{2} + \frac{9}{4} \qquad \text{Add } \frac{9}{4} \text{ to both sides}$$
$$\left(x + \frac{3}{2}\right)^2 = \frac{15}{4}$$
$$x + \frac{3}{2} = \pm\frac{\sqrt{15}}{2}$$

$$x + \frac{3}{2} = \frac{\sqrt{15}}{2} \qquad \text{or} \qquad x + \frac{3}{2} = -\frac{\sqrt{15}}{2}$$
$$x = -\frac{3}{2} + \frac{\sqrt{15}}{2} \qquad \text{or} \qquad x = -\frac{3}{2} - \frac{\sqrt{15}}{2}$$
$$x = \frac{-3 + \sqrt{15}}{2} \qquad \text{or} \qquad x = \frac{-3 - \sqrt{15}}{2}$$

The solution set is $\left\{\dfrac{-3 \pm \sqrt{15}}{2}\right\}$.

▶ **Now work Problem 25.**   ■

Again let's pause for a moment and take another look at the relationship between the solutions of an algebraic equation and the x intercepts of a geometric graph. Figure 1.5 shows a graph of $y = 2x^2 + 6x - 3$. Note that one x intercept is between -4 and -3, and the other x intercept is between 0 and 1. The solution $\dfrac{-3 - \sqrt{15}}{2} \approx -3.4$, and the solution $\dfrac{-3 + \sqrt{15}}{2} \approx 0.4$. So our geometric analysis appears to agree with our algebraic solutions.

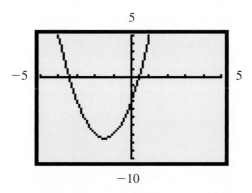

Figure 1.5

■ Quadratic Formula

The process used in Examples 3 and 4 is called **completing the square.** It can be used to solve *any* quadratic equation. If we use this process of completing the square to solve the general quadratic equation $ax^2 + bx + c = 0$, we obtain a formula known as the **quadratic formula.** The details are as follows:

$$ax^2 + bx + c = 0, \qquad a \neq 0$$

$$ax^2 + bx = -c$$

$$x^2 + \frac{b}{a}x = -\frac{c}{a} \qquad \text{Multiply both sides by } \frac{1}{a}$$

$$x^2 + \frac{b}{a}x + \frac{b^2}{4a^2} = -\frac{c}{a} + \frac{b^2}{4a^2} \qquad \text{Complete the square by adding } \frac{b^2}{4a^2} \text{ to both sides}$$

$$\left(x + \frac{b}{2a}\right)^2 = \frac{b^2 - 4ac}{4a^2} \qquad \text{Combine the right side into a single fraction}$$

$$x + \frac{b}{2a} = \pm\sqrt{\frac{b^2 - 4ac}{4a^2}}$$

$$x + \frac{b}{2a} = \pm\frac{\sqrt{b^2 - 4ac}}{\sqrt{4a^2}}$$

$$x + \frac{b}{2a} = \pm\frac{\sqrt{b^2 - 4ac}}{2a} \qquad \sqrt{4a^2} = |2a| \text{ but } 2a \text{ can be used because of the use of } \pm$$

$$x + \frac{b}{2a} = \frac{\sqrt{b^2 - 4ac}}{2a} \qquad \text{or} \qquad x + \frac{b}{2a} = -\frac{\sqrt{b^2 - 4ac}}{2a}$$

$$x = -\frac{b}{2a} + \frac{\sqrt{b^2 - 4ac}}{2a} \qquad \text{or} \qquad x = -\frac{b}{2a} - \frac{\sqrt{b^2 - 4ac}}{2a}$$

$$x = \frac{-b + \sqrt{b^2 - 4ac}}{2a} \qquad \text{or} \qquad x = \frac{-b - \sqrt{b^2 - 4ac}}{2a}$$

The quadratic formula can be stated as follows.

Quadratic Formula

If $a \neq 0$, then the solutions (roots) of the equation $ax^2 + bx + c = 0$ are given by

$$x = \frac{-b \pm \sqrt{b^2 - 4ac}}{2a}$$

We can use the quadratic formula to solve *any* quadratic equation by expressing the equation in the standard form $ax^2 + bx + c = 0$ and then substituting the values for a, b, and c into the formula. Let's consider some examples and use a graphical approach to predict approximate solutions whenever possible.

EXAMPLE 5

Solve each of the following by using the quadratic formula.

a. $3x^2 - x - 5 = 0$ **b.** $25n^2 - 30n = -9$ **c.** $t^2 - 2t + 4 = 0$

Solutions

a. $y = 3x^2 - x - 5$ (See Figure 1.6.)

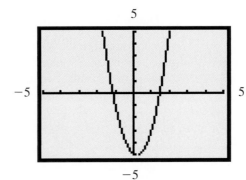

One intercept is between −2 and −1, and the other is between 1 and 2

Figure 1.6

We need to think of $3x^2 - x - 5 = 0$ as $3x^2 + (-x) + (-5) = 0$; thus $a = 3$, $b = -1$, and $c = -5$. We then substitute these values into the quadratic formula and simplify:

$$x = \frac{-b \pm \sqrt{b^2 - 4ac}}{2a}$$

$$x = \frac{-(-1) \pm \sqrt{(-1)^2 - 4(3)(-5)}}{2(3)}$$

$$= \frac{1 \pm \sqrt{61}}{6}$$

The solution set is $\left\{ \dfrac{1 \pm \sqrt{61}}{6} \right\}$. (You should evaluate these solutions to be sure they agree with the intercepts.)

▶ **Now work Problem 35.**

b. $y = 25x^2 - 30x + 9$ (See Figure 1.7.)

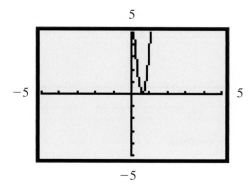

There appears to be one intercept between 0 and 1

Figure 1.7

The quadratic formula is usually stated in terms of the variable x, but again the choice of variable is arbitrary. The given equation, $25n^2 - 30n = -9$, needs to be changed to standard form: $25n^2 - 30n + 9 = 0$. From this we obtain $a = 25$, $b = -30$, and $c = 9$. Now we use the formula:

$$n = \frac{-(-30) \pm \sqrt{(-30)^2 - 4(25)(9)}}{2(25)}$$

$$= \frac{30 \pm \sqrt{0}}{50}$$

$$= \frac{3}{5}$$

The solution set is $\left\{\dfrac{3}{5}\right\}$.

▶ **Now work Problem 43.**

c. $y = x^2 - 2x + 4$ (See Figure 1.8.)

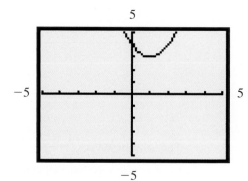

There are no x intercepts

Figure 1.8

We substitute $a = 1$, $b = -2$, and $c = 4$ into the quadratic formula:

$$t = \frac{-(-2) \pm \sqrt{(-2)^2 - 4(1)(4)}}{2(1)}$$

$$= \frac{2 \pm \sqrt{-12}}{2}$$

$$= \frac{2 \pm 2i\sqrt{3}}{2}$$

$$= \frac{2(1 \pm i\sqrt{3})}{2}$$

The solution set is $\{1 \pm i\sqrt{3}\}$.

▶ **Now work Problem 39.** ∎

From Example 5 we see that different kinds of solutions are obtained depending upon the radicand $(b^2 - 4ac)$ inside the radical in the quadratic formula. For this reason, the number $b^2 - 4ac$ is called the **discriminant** of the quadratic equation. It can be used to determine the nature of the solutions as follows.

1. If $b^2 - 4ac > 0$, the equation has two unequal real solutions.

2. If $b^2 - 4ac = 0$, the equation has one real solution.

3. If $b^2 - 4ac < 0$, the equation has two complex but nonreal solutions.

The following examples illustrate each of these situations. (You may want to solve the equations completely to verify our conclusions.)

Equation	Discriminant	Nature of Solutions
$4x^2 - 7x - 1 = 0$	$b^2 - 4ac = (-7)^2 - 4(4)(-1)$ $= 49 + 16$ $= 65$	Two real solutions
$4x^2 + 12x + 9 = 0$	$b^2 - 4ac = (12)^2 - 4(4)(9)$ $= 144 - 144$ $= 0$	One real solution
$5x^2 + 2x + 1 = 0$	$b^2 - 4ac = (2)^2 - 4(5)(1)$ $= 4 - 20$ $= -16$	Two complex solutions

There is another useful relationship involving the solutions of a quadratic equation of the form $ax^2 + bx + c = 0$ and the numbers a, b, and c. Suppose that we let x_1 and x_2 be the two roots of the equation. (If $b^2 - 4ac = 0$, then $x_1 = x_2$ and the one-solution situation can be thought of as two equal solutions.) By the quadratic formula we have

$$x_1 = \frac{-b + \sqrt{b^2 - 4ac}}{2a} \quad \text{and} \quad x_2 = \frac{-b - \sqrt{b^2 - 4ac}}{2a}$$

Now let's consider both the sum and the product of the two roots:

Sum
$$x_1 + x_2 = \frac{-b + \sqrt{b^2 - 4ac}}{2a} + \frac{-b - \sqrt{b^2 - 4ac}}{2a}$$

$$= \frac{-2b}{2a} = -\frac{b}{a}$$

Product
$$(x_1)(x_2) = \left(\frac{-b + \sqrt{b^2 - 4ac}}{2a}\right)\left(\frac{-b - \sqrt{b^2 - 4ac}}{2a}\right)$$

$$= \frac{b^2 - (b^2 - 4ac)}{4a^2}$$

$$= \frac{b^2 - b^2 + 4ac}{4a^2}$$

$$= \frac{4ac}{4a^2} = \frac{c}{a}$$

These relationships provide another way of checking potential solutions when solving quadratic equations. We will illustrate this point in a moment.

■ Solving Quadratic Equations: Which Method?

Which method should you use to solve a particular quadratic equation? There is no definite answer to that question; it depends on the type of equation and perhaps on your personal preference. However, it is to your advantage to be able to use all three techniques and to know the strengths and weaknesses of each. In the next two examples we will give our reasons for choosing a specific technique.

E X A M P L E   6   Solve $x^2 - 4x - 192 = 0$.

Solution

The size of the constant term makes the factoring approach a little cumbersome for this problem. However, because the coefficient of the x^2 term is 1, and the coefficient of the x term is even, the method for completing the square should work effectively.

$$x^2 - 4x - 192 = 0$$
$$x^2 - 4x = 192$$
$$x^2 - 4x + 4 = 192 + 4$$
$$(x - 2)^2 = 196$$
$$x - 2 = \pm\sqrt{196}$$

$$x - 2 = \pm 14$$

$$x - 2 = 14 \quad \text{or} \quad x - 2 = -14$$

$$x = 16 \quad \text{or} \quad x = -12$$

✔ **Check**

Sum of roots $16 + (-12) = 4$ and $-\dfrac{b}{a} = -\left(\dfrac{-4}{1}\right) = 4$

Product of roots $(16)(-12) = -192$ and $\dfrac{c}{a} = \dfrac{-192}{1} = -192$

The solution set is $\{-12, 16\}$.

⊙ **Now work Problem 47.** ∎

E X A M P L E   7

Solve $2x^2 - x + 3 = 0$.

Solution

It would be reasonable first to try factoring the polynomial $2x^2 - x + 3$. Unfortunately, it is not factorable using integers; thus we must solve the equation by completing the square or by using the quadratic formula. The coefficient of the x^2 term is not 1, so let's avoid completing the square and use the formula instead.

$$x = \frac{-b \pm \sqrt{b^2 - 4ac}}{2a}$$

$$= \frac{-(-1) \pm \sqrt{(-1)^2 - 4(2)(3)}}{2(2)}$$

$$= \frac{1 \pm \sqrt{-23}}{4}$$

$$= \frac{1 \pm i\sqrt{23}}{4}$$

✔ **Check**

Sum of roots $\dfrac{1 + i\sqrt{23}}{4} + \dfrac{1 - i\sqrt{23}}{4} = \dfrac{2}{4} = \dfrac{1}{2}$ and

$$-\frac{b}{a} = -\frac{-1}{2} = \frac{1}{2}$$

Product of roots $\left(\dfrac{1 + i\sqrt{23}}{4}\right)\left(\dfrac{1 - i\sqrt{23}}{4}\right) = \dfrac{1 - 23i^2}{16}$

$$= \frac{1 + 23}{16} = \frac{24}{16} = \frac{3}{2} \quad \text{and} \quad \frac{c}{a} = \frac{3}{2}$$

The solution set is $\left\{\dfrac{1 \pm i\sqrt{23}}{4}\right\}$.

⊙ **Now work Problem 49.** ∎

The ability to solve quadratic equations enables us to solve more word problems. Some of these problems involve geometric formulas and relationships. We have included a brief summary of some basic geometric formulas in the back sheets of this text.

PROBLEM   1

One leg of a right triangle is 7 meters longer than the other leg. If the length of the hypotenuse is 17 meters, find the length of each leg.

Solution

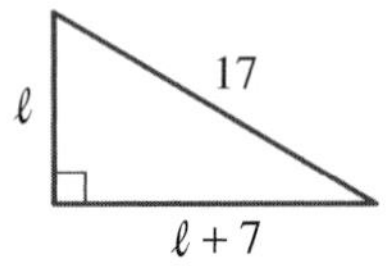

Look at Figure 1.9. Let l represent the length of one leg; then $l + 7$ represents the length of the other leg. Using the Pythagorean theorem as a guideline, we can set up and solve a quadratic equation:

$$l^2 + (l + 7)^2 = 17^2$$
$$l^2 + l^2 + 14l + 49 = 289$$
$$2l^2 + 14l - 240 = 0$$
$$l^2 + 7l - 120 = 0$$
$$(l + 15)(l - 8) = 0$$
$$l + 15 = 0 \qquad \text{or} \qquad l - 8 = 0$$
$$l = -15 \qquad \text{or} \qquad l = 8$$

The negative solution must be disregarded (because l is a length), so the length of one leg is 8 meters. The other leg, represented by $l + 7$, is $8 + 7 = 15$ meters long.

▶ **Now work Problem 66.** ∎

Figure 1.9

CONCEPT QUIZ

For Problems 1–8, answer true or false.

1. The product of two factors is equal to zero if either of the factors equals zero.

2. Solving a quadratic equation by factoring gives the same result as solving the equation by completing the square.

3. For the perfect-square trinomial $x^2 + bx + c$, the constant term is equal to one-half the square of the coefficient of the x term.

4. Any quadratic equation can be solved by completing the square.

5. To use the quadratic formula to solve quadratic equations, the equation needs to be in the standard form, $ax^2 + bx + c = 0$.

6. Every quadratic equation has two real number solutions.

7. The discriminant is $\sqrt{b^2 - 4ac}$.

8. The equation $ax^2 + bc + c = 0$ has two complex nonreal number solutions if $b^2 - 4ac < 0$.

Problem Set 1.3

For Problems 1–16, solve each equation by factoring or by using the property *If* $x^2 = k$, *then* $x = \pm\sqrt{k}$.

1. $x^2 - 3x - 28 = 0$

2. $x^2 - 4x - 12 = 0$

3. $3x^2 + 5x - 12 = 0$

4. $2x^2 - 13x + 6 = 0$

5. $2x^2 - 3x = 0$

6. $3n^2 = 3n$

7. $9y^2 = 12$

8. $(4n - 1)^2 = 16$

9. $(2n + 1)^2 = 20$

10. $3(4x - 1)^2 + 1 = 16$

11. $15n^2 + 19n - 10 = 0$

12. $6t^2 + 23t - 4 = 0$

13. $(x - 2)^2 = -4$

14. $24x^2 + 23x - 12 = 0$

15. $10y^2 + 33y - 7 = 0$

16. $(x - 3)^2 = -9$

For Problems 17–30, use the method of completing the square to solve each equation. Check your solutions by using the sum-and-product-of-roots relationships.

17. $x^2 - 10x + 24 = 0$

18. $x^2 + x - 20 = 0$

19. $n^2 + 10n - 2 = 0$

20. $n^2 + 6n - 1 = 0$

21. $y^2 - 3y = -1$

22. $y^2 + 5y = -2$

23. $x^2 + 4x + 6 = 0$

24. $x^2 - 6x + 21 = 0$

25. $2t^2 + 12t - 5 = 0$

26. $3p^2 + 12p - 2 = 0$

27. $x(x - 2) = 288$

28. $x(x + 4) = 221$

29. $3n^2 + 5n - 1 = 0$

30. $2n^2 + n - 4 = 0$

For Problems 31–44, use the quadratic formula to solve each equation. Check your solutions by using the sum-and-product-of-roots relationships.

31. $n^2 - 3n - 54 = 0$

32. $y^2 + 13y + 22 = 0$

33. $3x^2 + 16x = -5$

34. $10x^2 - 29x - 21 = 0$

35. $y^2 - 2y - 4 = 0$

36. $n^2 - 6n - 3 = 0$

37. $2a(a - 3) = -1$

38. $x(2x + 3) - 1 = 0$

39. $n^2 - 3n = -7$

40. $n^2 - 5n = -8$

41. $x^2 + 4 = 8x$

42. $x^2 + 31 = -14x$

43. $4x^2 - 4x + 1 = 0$

44. $x^2 + 24 = 0$

For Problems 45–60, solve each quadratic equation by using the method that seems most appropriate to you.

45. $8x^2 + 10x - 3 = 0$

46. $18x^2 - 39x + 20 = 0$

47. $x^2 + 2x = 168$

48. $x^2 + 28x = -187$

49. $2t^2 - 3t + 7 = 0$

50. $3n^2 - 2n + 5 = 0$

51. $(3n - 1)^2 + 2 = 18$

52. $20y^2 + 17y - 10 = 0$

53. $4y(y + 1) = 1$

54. $(5n + 2)^2 + 1 = -27$

55. $x^2 - 16x + 14 = 0$

56. $x^2 - 18x + 15 = 0$

57. $t^2 + 20t = 25$

58. $n(n - 18) = 9$

59. $5x^2 - 2x - 1 = 0$

60. $-x^2 + 11x - 18 = 0$

61. Find the discriminant of each of the following quadratic equations, and determine whether the equation has (1) two complex but nonreal solutions, (2) one real solution, or (3) two unequal real solutions.

    a. $4x^2 + 20x + 25 = 0$

    b. $x^2 + 4x + 7 = 0$

    c. $x^2 - 18x + 81 = 0$

    d. $36x^2 - 31x + 3 = 0$

    e. $2x^2 + 5x + 7 = 0$

    f. $16x^2 = 40x - 25$

    g. $6x^2 - 4x - 7 = 0$

    h. $5x^2 - 2x - 4 = 0$

For Problems 62–77, set up a quadratic equation and solve each problem.

62. Find two consecutive positive even integers whose product is 528.

63. Find two consecutive whole numbers such that the sum of their squares is 265.

64. For a remodeling job, an architect suggested increasing the sides of a square patio by 3 feet per side. This made the area of the new patio 49 square feet. What was the area of the original patio?

65. A sailboat has a triangular sail with an area of 30 square feet. The height of the sail is 7 feet more than the length of the base of the sail. Find the height of the sail.

66. One leg of a right triangle is 4 inches longer than the other leg. If the length of the hypotenuse is 20 inches, find the length of each leg.

67. The sum of the lengths of the two legs of a right triangle is 34 meters. If the length of the hypotenuse is 26 meters, find the length of each leg.

68. The lengths of the three sides of a right triangle are consecutive even integers. Find the length of each side.

69. The perimeter of a rectangle is 44 inches and its area is 112 square inches. Find the length and width of the rectangle.

70. A page of a magazine contains 70 square inches of type. The height of the page is twice the width. If the margin around the type is 2 inches uniformly, what are the dimensions of the page?

71. The length of a rectangle is 4 meters more than twice its width. If the area of the rectangle is 126 square meters, find its length and width.

72. The length of one side of a triangle is 3 centimeters less than twice the length of the altitude to that side. If the area of the triangle is 52 square centimeters, find the length of the side and the length of the altitude to that side.

73. A rectangular plot of ground measuring 12 meters by 20 meters is surrounded by a sidewalk of uniform width. The area of the sidewalk is 68 square meters. Find the width of the sidewalk.

74. A piece of wire 60 inches long is cut into two pieces and then each piece is bent into the shape of a square. If the sum of the areas of the two squares is 117 square inches, find the length of each piece of wire.

75. A rectangular piece of cardboard is 4 inches longer than it is wide. From each of its corners, a square piece 2 inches on a side is cut out. The flaps are then turned up to form an open box, which has a volume of 42 cubic inches. Find the length and width of the original piece of cardboard. See Figure 1.10.

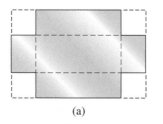
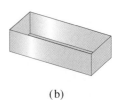

(a)   (b)

Figure 1.10

76. The area of a rectangular region is 52 square feet. If the length of the rectangle is increased by 4 feet and the width by 2 feet, then the area is increased by 50 square feet. Find the length and width of the original rectangular region.

77. The area of a circular region is numerically equal to four times the circumference of the circle. Find the length of a radius of the circle.

■ ■ ■ **THOUGHTS INTO WORDS**

78. Explain how you would solve $(x - 3)(x + 4) = 0$ and also how you would solve $(x - 3)(x + 4) = 8$.

79. Explain the process of completing the square to solve a quadratic equation.

80. Explain how to use the quadratic formula to solve $3x = x^2 - 2$.

81. Your friend states that the equation $-2x^2 + 4x - 1 = 0$ must be changed to $2x^2 - 4x + 1 = 0$ (by multiplying both sides by -1) before the quadratic formula can be applied. Is she right about this, and if not, how would you convince her?

■ ■ ■ **FURTHER INVESTIGATIONS**

82. Solve each of the following equations for x.

  a. $x^2 - 7kx = 0$ **b.** $x^2 = 25kx$

  c. $x^2 - 3kx - 10k^2 = 0$ **d.** $6x^2 + kx - 2k^2 = 0$

  e. $9x^2 - 6kx + k^2 = 0$ **f.** $k^2x^2 - kx - 6 = 0$

  g. $x^2 + \sqrt{2}x - 3 = 0$ **h.** $x^2 - \sqrt{3}x + 5 = 0$

83. Solve each of the following for the indicated variable. (Assume that all letters represent positive numbers.)

  a. $A = \pi r^2$ for r

  b. $E = c^2m - c^2m_0$ for c

  c. $s = \dfrac{1}{2}gt^2$ for t **d.** $\dfrac{x^2}{a^2} + \dfrac{y^2}{b^2} = 1$ for x

  e. $\dfrac{x^2}{a^2} - \dfrac{y^2}{b^2} = 1$ for y **f.** $s = \dfrac{1}{2}gt^2 + V_0t$ for t

For Problems 84–86, use the discriminant to help solve each problem.

84. Determine k so that the solutions of $x^2 - 2x + k = 0$ are complex but nonreal.

85. Determine k so that $4x^2 - kx + 1 = 0$ has two equal real solutions.

86. Determine k so that $3x^2 - kx - 2 = 0$ has real solutions.

■ **GRAPHING CALCULATOR ACTIVITIES**

87. The solution set for $x^2 - 4x - 37 = 0$ is $\{2 \pm \sqrt{41}\}$. With a calculator, we found a rational approximation, to the nearest one-thousandth, for each of these solutions.

$$2 - \sqrt{41} = -4.403$$

$$2 + \sqrt{41} = 8.403$$

Thus the solution set is $\{-4.403, 8.403\}$, with answers rounded to the nearest one-thousandth.

  Solve each of the following equations and express the solutions to the nearest one-thousandth.

  a. $x^2 - 6x - 10 = 0$ **b.** $x^2 - 16x - 24 = 0$

  c. $x^2 + 6x - 44 = 0$ **d.** $x^2 + 10x - 46 = 0$

  e. $x^2 + 8x + 2 = 0$ **f.** $x^2 + 9x + 3 = 0$

  g. $4x^2 - 6x + 1 = 0$ **h.** $5x^2 - 9x + 1 = 0$

  i. $2x^2 - 11x - 5 = 0$ **j.** $3x^2 - 12x - 10 = 0$

88. Graph the appropriate equations to give visual support for your solutions for Problems 45–60.

89. For each of the following, first graph the appropriate equation and use your graph to predict approximate solutions for the given equation. Then solve the equation to see how well you predicted.

  a. $4x^2 - 12x + 9 = 0$ **b.** $2x^2 - x + 2 = 0$
  c. $-3x^2 + 2x - 4 = 0$ **d.** $2x^2 - 23x - 12 = 0$
  e. $x^2 + 2\sqrt{5}x + 5 = 0$ **f.** $-x^2 + 2\sqrt{3}x - 3 = 0$

1.4   Applications of Linear and Quadratic Equations

Objectives

- Solve fractional equations.
- Solve application problems involving rate, time, and distance.
- Solve application problems involving rate of work.
- Solve application problems involving mixture of solutions.

Let's begin this section by considering three fractional equations, one that is equivalent to a linear equation and two that are equivalent to quadratic equations.

EXAMPLE 1   Solve $\dfrac{3}{2x-8} - \dfrac{x-5}{x^2-2x-8} = \dfrac{7}{x+2}$.

Solution

$$\frac{3}{2x-8} - \frac{x-5}{x^2-2x-8} = \frac{7}{x+2}$$

$$\frac{3}{2(x-4)} - \frac{x-5}{(x-4)(x+2)} = \frac{7}{x+2}, \qquad x \neq 4,\ x \neq -2$$

$$2(x-4)(x+2)\left(\frac{3}{2(x-4)} - \frac{x-5}{(x-4)(x+2)}\right) = 2(x-4)(x+2)\left(\frac{7}{x+2}\right)$$

$$3(x+2) - 2(x-5) = 14(x-4)$$

$$3x + 6 - 2x + 10 = 14x - 56$$

$$x + 16 = 14x - 56$$

$$72 = 13x$$

$$\frac{72}{13} = x$$

The solution set is $\left\{\dfrac{72}{13}\right\}$.

▶ **Now work Problem 3.**   ∎

In Example 1, notice that we did not indicate the restrictions until the denominators were expressed in factored form. It is usually easier to determine the necessary restrictions at that step.

EXAMPLE 2   Solve $\dfrac{3n}{n^2+n-6} + \dfrac{2}{n^2+4n+3} = \dfrac{n}{n^2-n-2}$.

Solution

$$\frac{3n}{n^2+n-6} + \frac{2}{n^2+4n+3} = \frac{n}{n^2-n-2}$$

$$\frac{3n}{(n+3)(n-2)} + \frac{2}{(n+3)(n+1)} = \frac{n}{(n-2)(n+1)}, \qquad n \neq -3,\ n \neq 2,\ n \neq -1$$

$$(n + 3)(n - 2)(n + 1)\left(\frac{3n}{(n + 3)(n - 2)} + \frac{2}{(n + 3)(n + 1)}\right) = (n + 3)(n - 2)(n + 1)\left(\frac{n}{(n - 2)(n + 1)}\right)$$

$$3n(n + 1) + 2(n - 2) = n(n + 3)$$
$$3n^2 + 3n + 2n - 4 = n^2 + 3n$$
$$3n^2 + 5n - 4 = n^2 + 3n$$
$$2n^2 + 2n - 4 = 0$$
$$n^2 + n - 2 = 0$$
$$(n + 2)(n - 1) = 0$$

$$n + 2 = 0 \quad \text{or} \quad n - 1 = 0$$
$$n = -2 \quad \text{or} \quad n = 1$$

The solution set is $\{-2, 1\}$.

▶ **Now work Problem 13.**   ■

EXAMPLE 3   Solve $\dfrac{x}{x + 1} - \dfrac{2}{x + 3} = \dfrac{4}{x^2 + 4x + 3}$

Solution

$$\frac{x}{x + 1} - \frac{2}{x + 3} = \frac{4}{x^2 + 4x + 3}$$

$$\frac{x}{x + 1} - \frac{2}{x + 3} = \frac{4}{(x + 1)(x + 3)}, \quad x \neq -1, x \neq -3$$

$$(x + 1)(x + 3)\left(\frac{x}{x + 1} - \frac{2}{x + 3}\right) = \left(\frac{4}{(x + 1)(x + 3)}\right)(x + 1)(x + 3)$$

$$x(x + 3) - 2(x + 1) = 4$$
$$x^2 + 3x - 2x - 2 = 4$$
$$x^2 + x - 6 = 0$$
$$(x + 3)(x - 2) = 0$$

$$x + 3 = 0 \quad \text{or} \quad x - 2 = 0$$
$$x = -3 \quad \text{or} \quad x = 2$$

Because -3 produces a denominator of zero in the original equation, it must be discarded. Thus the solution set is $\{2\}$.

▶ **Now work Problem 15.**   ■

Example 3 reinforces the importance of recognizing the restrictions that must be made to exclude division by zero. Even though the original equation was transformed into a quadratic equation with two roots, one of them had to be discarded because of the restrictions.

■ More on Problem Solving

Before tackling a variety of applications of linear and quadratic equations, let's restate some suggestions made earlier in this chapter for solving word problems.

Suggestions for Solving Word Problems

1. Read the problem carefully, making certain that you understand the meanings of all the words. Be especially alert for any technical terms used in the statement of the problem.

2. Read the problem a second time (perhaps even a third time) to get an overview of the situation being described and to determine the known facts as well as what is to be found.

3. Sketch any figure, diagram, or chart that might be helpful in analyzing the problem.

4. Choose a meaningful variable to represent an unknown quantity in the problem (for example, *l* for the length of a rectangle), and represent any other unknowns in terms of that variable.

5. Look for a *guideline* that can be used in setting up an equation. A guideline might be a formula, such as $A = lw$, or a relationship, such as *the fractional part of the job done by Bill plus the fractional part of the job done by Mary equals the total job*.

6. Form an equation containing the variable to translate the conditions of the guideline from English to algebra.

7. Solve the equation and use the solution to determine all the facts requested in the problem.

8. Check all answers against the *original statement of the problem*.

Suggestion 5 is a key part of the analysis of a problem. A formula to be used as a guideline may or may not be explicitly stated in the problem. Likewise, a relationship to be used as a guideline may not be actually stated in the problem but must be determined from what is stated. Let's consider some examples.

PROBLEM 1

A theater contains 120 chairs. The number of chairs per row is one less than twice the number of rows. Find the number of rows and the number of chairs per row.

Solution

Let r represent the number of rows. Then $2r - 1$ represents the number of chairs per row. The statement of the problem implies a formation of chairs such that the total number of chairs is equal to the number of rows times the number of chairs per row. This gives us an equation:

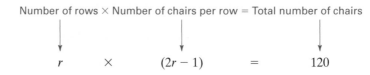

Number of rows × Number of chairs per row = Total number of chairs

$$r \quad \times \quad (2r - 1) \quad = \quad 120$$

We solve this equation by the factorization method:

$$2r^2 - r = 120$$
$$2r^2 - r - 120 = 0$$
$$(2r + 15)(r - 8) = 0$$
$$2r + 15 = 0 \qquad \text{or} \qquad r - 8 = 0$$
$$2r = -15 \qquad \text{or} \qquad r = 8$$
$$r = -\frac{15}{2} \qquad \text{or} \qquad r = 8$$

The solution $-\dfrac{15}{2}$ must be disregarded, so there are 8 rows and $2(8) - 1 = 15$ chairs per row.

▶ **Now work Problem 21.** ∎

The basic relationship *distance equals rate times time* is used to help solve a variety of *uniform-motion problems*. This relationship may be expressed by any one of the following equations.

$$d = rt \qquad r = \frac{d}{t} \qquad t = \frac{d}{r}$$

P R O B L E M   2

Domenica and Javier start from the same location at the same time and ride their bicycles in opposite directions for 4 hours, at which time they are 140 miles apart. If Domenica rides 3 miles per hour faster than Javier, find the rate of each rider.

Solution

Let r represent Javier's rate; then $r + 3$ represents Domenica's rate. A sketch such as Figure 1.11 may help in our analysis. The fact that the total distance is 140 miles can be used as a guideline. We use the $d = rt$ equation.

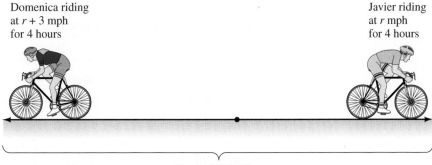

Domenica riding
at $r + 3$ mph
for 4 hours

Javier riding
at r mph
for 4 hours

Total of 140 miles

Figure 1.11

Distance Domenica rides + Distance Javier rides = 140

$$4(r + 3) \qquad + \qquad 4r \qquad = 140$$

Solving this equation yields Javier's speed.

$$4r + 12 + 4r = 140$$
$$8r = 128$$
$$r = 16$$

Thus Javier rides at 16 miles per hour and Domenica at $16 + 3 = 19$ miles per hour.

▶ **Now work Problem 23.**   ■

Remark: An important part of problem solving is the ability to sketch a meaningful figure that can be used to record the given information and help in the analysis of the problem. Our sketches were done by professional artists for aesthetic purposes. Your sketches can be drawn very roughly as long as they depict the situation in a way that helps you analyze the problem.

Note that in the solution of Problem 2 we used a figure and a simple arrow diagram to record and organize the information pertinent to the problem. Some people find it helpful to use a chart for that purpose. We shall use a chart in Problem 3 and show alternative solutions. Keep in mind that we are not trying to dictate a particular approach; you decide what works best for you.

P R O B L E M   3

Riding on a moped, Sue takes 2 hours less time to travel 60 miles than Ann takes to travel 50 miles on a bicycle. Sue travels 10 miles per hour faster than Ann. Find the times and rates of both girls.

Solution A

Let t represent Ann's time; then $t - 2$ represents Sue's time. We can record the information in a table as shown below. The fact that Sue travels 10 miles per hour faster than Ann can be used as a guideline.

	Distance	Time	$r = \dfrac{d}{t}$
Ann	50	t	$\dfrac{50}{t}$
Sue	60	$t - 2$	$\dfrac{60}{t - 2}$

Sue's rate = Ann's rate + 10

$$\frac{60}{t-2} = \frac{50}{t} + 10$$

Solving this equation yields Ann's time.

$$t(t-2)\left(\frac{60}{t-2}\right) = t(t-2)\left(\frac{50}{t} + 10\right), \qquad t \neq 0,\ t \neq 2$$

$$60t = 50(t-2) + 10t(t-2)$$

$$60t = 50t - 100 + 10t^2 - 20t$$

$$0 = 10t^2 - 30t - 100$$

$$0 = t^2 - 3t - 10$$

$$0 = (t-5)(t+2)$$

$$t - 5 = 0 \qquad \text{or} \qquad t + 2 = 0$$

$$t = 5 \qquad \text{or} \qquad t = -2$$

The solution -2 must be disregarded because we're solving for time. Therefore Ann rides for 5 hours at $\frac{50}{5} = 10$ miles per hour, and Sue rides for $5 - 2 = 3$ hours at $\frac{60}{3} = 20$ miles per hour.

Solution B

Let r represent Ann's rate; then $r + 10$ represents Sue's rate. Again, let's record the information in a table.

	Distance	Rate	$t = \dfrac{d}{r}$
Ann	50	r	$\dfrac{50}{r}$
Sue	60	$r + 10$	$\dfrac{60}{r+10}$

This time, let's use as a guideline the fact that Sue's time is 2 hours less than Ann's time.

Sue's time = Ann's time − 2

$$\frac{60}{r + 10} = \frac{50}{r} - 2$$

Solving this equation yields Ann's rate.

$$r(r + 10)\left(\frac{60}{r + 10}\right) = r(r + 10)\left(\frac{50}{r} - 2\right), \qquad r \neq -10, r \neq 0$$

$$60r = (r + 10)(50) - 2r(r + 10)$$

$$60r = 50r + 500 - 2r^2 - 20r$$

$$2r^2 + 30r - 500 = 0$$

$$r^2 + 15r - 250 = 0$$

$$(r + 25)(r - 10) = 0$$

$$r + 25 = 0 \qquad \text{or} \qquad r - 10 = 0$$

$$r = -25 \qquad \text{or} \qquad r = 10$$

The solution −25 must be disregarded because we are solving for a rate. Therefore Ann rides at 10 miles per hour for $\frac{50}{10} = 5$ hours, and Sue rides at $10 + 10 = 20$ miles per hour for $\frac{60}{20} = 3$ hours.

▶ **Now work Problem 27.**   ■

   Take a good look at both Solution A and Solution B for Problem 3. Both are reasonable approaches, but note that the approach in Solution A generates a quadratic equation that is a little easier to solve than the one generated in Solution B. We might have expected this to happen because the *times* in a motion problem are frequently smaller numbers than the *rates*. Thus thinking first before pushing the pencil can make things a bit easier.

   Now let's consider a problem that is often referred to as a *mixture problem*. There is no basic formula that applies to all of these problems, but we suggest that you think in terms of a pure substance, which is often helpful in setting up a guideline. Also keep in mind that the phrase "a 40% solution of some substance" means that the solution contains 40% of that particular substance and 60% of something else mixed with it. For example, a 40% salt solution contains 40% salt, and the other 60% is something else, probably water. Now let's illustrate what we mean by the suggestion *think in terms of a pure substance.*

P R O B L E M   4   How many milliliters of pure acid must be added to 50 milliliters of a 40% acid solution to obtain a 50% acid solution?

Solution

Let a represent the number of milliliters of pure acid to be added. Thinking in terms of pure acid, we know that *the amount of pure acid to start with, plus the amount of pure acid added, equals the amount of pure acid in the final solution.* Let's use that as a guideline and set up an equation.

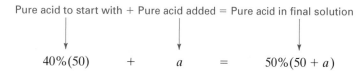

Pure acid to start with + Pure acid added = Pure acid in final solution

$$40\%(50) \quad + \quad a \quad = \quad 50\%(50 + a)$$

Solving this equation, we obtain the amount of acid we must add.

$$0.4(50) + a = 0.5(50 + a)$$
$$4(50) + 10a = 5(50 + a)$$
$$200 + 10a = 250 + 5a$$
$$5a = 50$$
$$a = 10$$

We need to add 10 milliliters of pure acid.

▶ **Now work Problem 29.** ■

There is another class of problems commonly referred to as *work problems*, or sometimes as *rate-time problems*. For example, if a certain machine produces 120 items in 10 minutes, then we say that it is working at a rate of $\frac{120}{10} = 12$ items per minute. Likewise, a person who can do a certain job in 5 hours is working at a rate of $\frac{1}{5}$ of the job per hour. In general, if Q is the quantity of something done in t units of time, then the rate, r, is given by $r = Q/t$. The rate is stated in terms of *so much quantity per unit of time.* The uniform-motion problems discussed earlier are a special kind of rate-time problem where the quantity is distance. Using tables to organize information (as we illustrated with the motion problems) is a convenient aid for rate-time problems in general. Let's consider some problems.

PROBLEM 5

It takes Amy twice as long to deliver newspapers as it does Nancy. How long does it take each girl by herself if they can deliver the papers together in 40 minutes?

Solution

Let m represent the number of minutes that it takes Nancy by herself. Then $2m$ represents Amy's time by herself. Thus the information can be organized as shown below. (Note that the *quantity* is 1; there is one job to be done.)

	Quantity	Time	Rate
Nancy	1	m	$\dfrac{1}{m}$
Amy	1	$2m$	$\dfrac{1}{2m}$

Because their combined rate is $\dfrac{1}{40}$, we can solve the following equation:

$$\frac{1}{m} + \frac{1}{2m} = \frac{1}{40}, \qquad m \neq 0$$

$$40m\left(\frac{1}{m} + \frac{1}{2m}\right) = 40m\left(\frac{1}{40}\right)$$

$$40 + 20 = m$$

$$60 = m$$

Therefore, Nancy can deliver the papers by herself in 60 minutes, and Amy can deliver them by herself in $2(60) = 120$ minutes.

▶ **Now work Problem 41.** ∎

The next problem illustrates another approach that some people find works well for rate-time problems. The basic idea used in this approach involves representing the fractional parts of a job. For example, if a man can do a certain job in 7 hours, then at the end of 3 hours he has finished $\dfrac{3}{7}$ of the job. (Again, a constant rate of work is assumed.) At the end of 5 hours he has finished $\dfrac{5}{7}$ of the job, and, in general, at the end of h hours he has finished $\dfrac{h}{7}$ of the job.

PROBLEM 6

Carlos can mow a lawn in 45 minutes, and Felipe can mow the same lawn in 30 minutes. How long would it take the two of them working together to mow the lawn?

Solution

(Before you read any further, *estimate* an answer for this problem. Remember that Felipe can mow the lawn by himself in 30 minutes.) Let m represent the number of minutes that it takes them working together. Then we can set up the following equation.

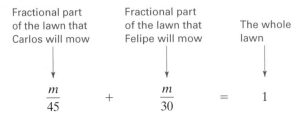

Fractional part of the lawn that Carlos will mow		Fractional part of the lawn that Felipe will mow		The whole lawn
$\dfrac{m}{45}$	$+$	$\dfrac{m}{30}$	$=$	1

Solving this equation yields the time that it will take when they work together.

$$90\left(\frac{m}{45} + \frac{m}{30}\right) = 90(1)$$
$$2m + 3m = 90$$
$$5m = 90$$
$$m = 18$$

It should take them 18 minutes to mow the lawn when they work together.

⊙ **Now work Problem 35.**   ∎

PROBLEM 7

Walt can mow a lawn in 50 minutes, and his son Mike can mow the same lawn in 40 minutes. One day Mike started to mow the lawn by himself and worked for 10 minutes. Then Walt joined him with another mower and they finished the lawn. How long did it take them to finish mowing the lawn after Walt started to help?

Solution

Let m represent the number of minutes that it takes for them to finish the mowing after Walt starts to help. Because Mike has been mowing for 10 minutes, he has done $\frac{10}{40}$, or $\frac{1}{4}$, of the lawn when Walt starts. Thus there is $\frac{3}{4}$ of the lawn yet to mow.

The following guideline can be used to set up an equation:

| Fractional part of the remaining 3/4 of the lawn that Mike will mow in m minutes | + | Fractional part of the remaining 3/4 of the lawn that Walt will mow in m minutes | = | $\frac{3}{4}$ |

$$\frac{m}{40} + \frac{m}{50} = \frac{3}{4}$$

Solving this equation yields the time they mow the lawn together.

$$200\left(\frac{m}{40} + \frac{m}{50}\right) = 200\left(\frac{3}{4}\right)$$
$$5m + 4m = 150$$
$$9m = 150$$
$$m = \frac{150}{9} = \frac{50}{3}$$

They should finish mowing the lawn in $16\frac{2}{3}$ minutes.

⊙ **Now work Problem 39.**   ∎

PROBLEM 8

John is being paid $864 to do a landscape job. It took him 6 hours less than he expected, so he earned $2 per hour more than he originally calculated. How long had he anticipated it would take to do the landscaping?

Solution

Let x represent the number of hours he anticipated the landscaping would take. Then the hourly rate of pay he expected was $\dfrac{484}{x}$. Because he worked 6 hours less than expected, his actual rate of pay was $\dfrac{864}{x-6}$. Now we can form an equation knowing that his actual rate of pay was \$2 more than his expected rate of pay.

$$\frac{864}{x} + 2 = \frac{864}{x-6}$$

Solving this equation gives the number of hours he anticipated for the landscaping.

$$x(x-6)\left(\frac{864}{x} + 2\right) = x(x-6)\left(\frac{864}{x-6}\right)$$

$$864(x-6) + 2x(x-6) = 864x$$
$$864x - 5184 + 2x^2 - 12x = 864x$$
$$2x^2 + 852x - 5184 = 864x$$
$$2x^2 - 12x - 5184 = 0$$
$$2(x^2 - 6x - 2592) = 0$$
$$2(x-54)(x+48) = 0$$
$$x = 54 \qquad \text{or} \qquad x = -48$$

The solution -48 must be disregarded because we are solving for the number of hours. Therefore John anticipated the landscaping would take 54 hours.

▶ **Now work Problem 45.** ∎

   As you tackle word problems throughout this text, keep in mind that our primary objective is to expand your repertoire of problem-solving techniques. We have chosen problems that provide you with the opportunity to use a variety of approaches to solving problems. Don't fall into the trap of thinking "I will never be faced with this kind of problem." That is not the issue; the development of problem-solving techniques is the goal. In the examples we are sharing some of our ideas for solving problems, but don't hesitate to use your own ingenuity. Furthermore, don't become discouraged — all of us have difficulty with some problems. Give each your best shot!

CONCEPT QUIZ   For Problems 1–5, answer true or false.

1. The only restrictions for the equation $\dfrac{5}{x-3} + \dfrac{7}{2x-1} = \dfrac{x}{x^2-3x}$ are that x can not equal $\dfrac{1}{2}$ or 3.

2. In general, if x is the quantity of something done in y units of time, then the rate is given by $r = \dfrac{x}{y}$.

3. Five hundred milliliters of a 30% alcohol solution would contain 150 milliliters of pure alcohol.

4. Kay works at a rate of solving 18 math problems in an hour. She has worked for one hour. She will need to work another three hours to solve 54 problems.

5. Lawton travels 359 miles on 11.5 gallons of gasoline. The proportion $\dfrac{11.5}{359} = \dfrac{x}{35}$ when solved for x would determine how many miles Lawton can travel on 35 gallons of gasoline.

6. John can mow a lawn in 30 minutes, and Marcos can mow the same lawn in 20 minutes. Would a time of 40 minutes for mowing the lawn together be reasonable?

Problem Set 1.4

For Problems 1–20, solve each equation.

1. $\dfrac{x}{2x - 8} + \dfrac{16}{x^2 - 16} = \dfrac{1}{2}$

2. $\dfrac{3}{n - 5} - \dfrac{2}{2n + 1} = \dfrac{n + 3}{2n^2 - 9n - 5}$

3. $\dfrac{5t}{2t + 6} - \dfrac{4}{t^2 - 9} = \dfrac{5}{2}$ **4.** $\dfrac{x}{4x - 4} + \dfrac{5}{x^2 - 1} = \dfrac{1}{4}$

5. $2 + \dfrac{4}{n - 2} = \dfrac{8}{n^2 - 2n}$ **6.** $3 + \dfrac{6}{t - 3} = \dfrac{6}{t^2 - 3t}$

7. $\dfrac{a}{a + 2} + \dfrac{3}{a + 4} = \dfrac{14}{a^2 + 6a + 8}$

8. $\dfrac{3}{x + 1} + \dfrac{2}{x + 3} = 2$

9. $\dfrac{-2}{3x + 2} + \dfrac{x - 1}{9x^2 - 4} = \dfrac{3}{12x - 8}$

10. $\dfrac{-1}{2x - 5} + \dfrac{2x - 4}{4x^2 - 25} = \dfrac{5}{6x + 15}$

11. $\dfrac{n}{2n - 3} + \dfrac{1}{n - 3} = \dfrac{n^2 - n - 3}{2n^2 - 9n + 9}$

12. $\dfrac{3y}{y^2 + y - 6} + \dfrac{2}{y^2 + 4y + 3} = \dfrac{y}{y^2 - y - 2}$

13. $\dfrac{3y + 1}{3y^2 - 4y - 4} + \dfrac{9}{9y^2 - 4} = \dfrac{2y - 2}{3y^2 - 8y + 4}$

14. $\dfrac{4n + 10}{2n^2 - n - 6} - \dfrac{3n + 1}{2n^2 - 5n + 2} = \dfrac{2}{4n^2 + 4n - 3}$

15. $\dfrac{x + 1}{2x^2 + 7x - 4} - \dfrac{x}{2x^2 - 7x + 3} = \dfrac{1}{x^2 + x - 12}$

16. $\dfrac{3}{x - 2} + \dfrac{5}{x + 3} = \dfrac{8x - 1}{x^2 + x - 6}$

17. $\dfrac{7x + 2}{12x^2 + 11x - 15} - \dfrac{1}{3x + 5} = \dfrac{2}{4x - 3}$

18. $\dfrac{2n}{6n^2 + 7n - 3} - \dfrac{n - 3}{3n^2 + 11n - 4} = \dfrac{5}{2n^2 + 11n + 12}$

19. $\dfrac{x}{x + 2} - \dfrac{3}{x + 4} = \dfrac{6}{x^2 + 6x + 8}$

20. $\dfrac{x}{x - 1} + \dfrac{3}{x - 2} = \dfrac{-1}{x^2 - 3x + 2}$

For Problems 21–45, solve each problem.

21. An apple orchard contains 126 trees. The number of trees in each row is 4 less than twice the number of rows. Find the number of rows and the number of trees per row.

22. The sum of a number and its reciprocal is $\dfrac{10}{3}$. Find the number.

23. Jill starts at city A and travels toward city B at 50 miles per hour. At the same time, Russ starts at city B and travels on the same highway toward city A at 52 miles per hour. How long will it take before they meet if the two cities are 459 miles apart?

24. Two cars, which are 510 miles apart and whose speeds differ by 6 miles per hour, are moving toward each other. If they meet in 5 hours, find the speed of each car.

25. Rita rode her bicycle out into the country at a speed of 20 miles per hour and returned along the same route at 15 miles per hour. If the round trip took 5 hours and 50 minutes, how far out did she ride?

26. A jogger who can run an 8-minute mile starts a half-mile ahead of a jogger who can run a 6-minute mile. How long will it take the faster jogger to catch the slower jogger?

27. It takes a freight train 2 hours more to travel 300 miles than it takes an express train to travel 280 miles. The rate of the express train is 20 miles per hour faster than the rate of the freight train. Find the rates of both trains.

28. An airplane travels 2050 miles in the same time that a car travels 260 miles. If the rate of the plane is 358 miles per hour faster than the rate of the car, find the rate of the plane.

29. A container has 6 liters of a 40% alcohol solution in it. How much pure alcohol should be added to raise it to a 60% solution?

30. How many liters of a 60% acid solution must be added to 14 liters of a 10% acid solution to produce a 25% acid solution?

31. One solution contains 50% alcohol and another solution contains 80% alcohol. How many liters of each solution should be mixed to produce 10.5 liters of a 70% alcohol solution?

32. A contractor has a 24-pound mixture that is one-fourth cement and three-fourths sand. How much of a mixture that is half cement and half sand needs to be added to produce a mixture that is one-third cement?

33. A 10-quart radiator contains a 40% antifreeze solution. How much of the solution needs to be drained out and replaced with pure antifreeze in order to raise the solution to 70% antifreeze?

34. How much water must be evaporated from 20 gallons of a 10% salt solution in order to obtain a 20% salt solution?

35. One pipe can fill a tank in 4 hours, and another pipe can fill the tank in 6 hours. How long will it take to fill the tank if both pipes are used?

36. Lolita and Doug working together can paint a shed in 3 hours and 20 minutes. If Doug can paint the shed by himself in 10 hours, how long would it take Lolita to paint the shed by herself?

37. An inlet pipe can fill a tank in 10 minutes. A drain can empty the tank in 12 minutes. If the tank is empty and both the pipe and drain are open, how long will it be before the tank overflows?

38. Pat and Mike working together can assemble a bookcase in 6 minutes. It takes Mike, working by himself, 9 minutes longer than it takes Pat working by himself to assemble the bookcase. How long does it take each, working alone, to do the job?

39. Mark can overhaul an engine in 20 hours, and Phil can do the same job by himself in 30 hours. If they both work together for a time and then Mark finishes the job by himself in 5 hours, how long did they work together?

40. A printing company purchased a new copier that is twice as fast as the old copier. With both copiers working at the same time, it takes 5 hours to do a job. How long would it take the new copier working alone?

41. A professor can grade three tests in the time it takes a student assistant to grade one test. Working together, they can grade the tests for a class in 2 hours. How long would it take the student assistant working alone?

42. A car that averages 16 miles per gallon of gasoline for city driving and 22 miles per gallon for highway driving uses 14 gallons in 296 miles of driving. How much of the driving was city driving?

43. Angie bought some candy bars for $14. If each candy bar had cost $0.25 less, she could have purchased one more bar for the same amount of money. How many candy bars did Angie buy?

44. A new labor contract provides for a wage increase of $1 per hour and a reduction of 5 hours in the workweek. A worker who received $320 per week under the old contract will receive $315 per week under the new contract. How long was the workweek under the old contract?

45. Todd contracted to paint a house for $480. It took him 4 hours longer than he had anticipated, so he earned $0.50 per hour less than he originally calculated. How long had he anticipated it would take him to paint the house?

■ ■ ■ **THOUGHTS INTO WORDS**

46. One of our problem-solving suggestions is to *look for a guideline that can be used to help determine an equation.* What does this suggestion mean to you?

47. Write a paragraph or two summarizing the various problem-solving ideas presented in this chapter.

1.5   Miscellaneous Equations

Objectives

■   Solve polynomial equations by factoring.

■   Solve radical equations.

■   Solve equations that are quadratic in form.

Our previous work with solving linear and quadratic equations provides us with a basis for solving a variety of other types of equations. For example, the technique of factoring and applying the property

$$ab = 0 \quad \text{if and only if } a = 0 \text{ or } b = 0$$

can sometimes be used for solving equations other than quadratic equations.

EXAMPLE 1

Solve $x^3 - 8 = 0$.

Solution

$$x^3 - 8 = 0$$
$$(x - 2)(x^2 + 2x + 4) = 0$$
$$x - 2 = 0 \quad \text{or} \quad x^2 + 2x + 4 = 0$$
$$x = 2 \quad \text{or} \quad x = \frac{-2 \pm \sqrt{4 - 16}}{2}$$
$$= \frac{-2 \pm \sqrt{-12}}{2}$$
$$= \frac{-2 \pm 2i\sqrt{3}}{2}$$
$$= -1 \pm i\sqrt{3}$$

The solution set is $\{2, \, -1 \pm i\sqrt{3}\}$.

▶   **Now work Problem 1.**   ■

EXAMPLE 2

Solve $x^3 + 2x^2 - 9x - 18 = 0$.

Solution

$$x^3 + 2x^2 - 9x - 18 = 0$$
$$x^2(x + 2) - 9(x + 2) = 0$$

$$(x + 2)(x^2 - 9) = 0$$
$$(x + 2)(x + 3)(x - 3) = 0$$

$x + 2 = 0 \qquad$ or $\qquad x + 3 = 0 \qquad$ or $\qquad x - 3 = 0$

$\qquad x = -2 \qquad$ or $\qquad x = -3 \qquad$ or $\qquad x = 3$

The solution set is $\{-3, -2, 3\}$.

▶ **Now work Problem 5.** ■

E X A M P L E   3   Solve $3x^5 + 5x^4 = 3x^3 + 5x^2$.

Solution

$$3x^5 + 5x^4 = 3x^3 + 5x^2$$
$$3x^5 + 5x^4 - 3x^3 - 5x^2 = 0$$
$$x^4(3x + 5) - x^2(3x + 5) = 0$$
$$(3x + 5)(x^4 - x^2) = 0$$
$$(3x + 5)(x^2)(x^2 - 1) = 0$$
$$(3x + 5)(x^2)(x + 1)(x - 1) = 0$$

$3x + 5 = 0 \qquad$ or $\qquad x^2 = 0 \qquad$ or $\qquad x + 1 = 0 \qquad$ or $\qquad x - 1 = 0$

$3x = -5$

$x = -\dfrac{5}{3} \qquad$ or $\qquad x = 0 \qquad$ or $\qquad x = -1 \qquad$ or $\qquad x = 1$

The solution set is $\left\{ -\dfrac{5}{3}, 0, -1, 1 \right\}$.

▶ **Now work Problem 9.** ■

Be careful with an equation like the one in Example 3. Don't be tempted to divide both sides of the equation by x^2. In so doing, you will lose the solution of zero. *In general, don't divide both sides of an equation by an expression that contains the variable.*

Figure 1.12 shows the graph of $y = 3x^5 + 5x^4 - 3x^3 - 5x^2$. Note that the x intercepts appear to agree with our solution set for Example 3. Again it's good to have a visual confirmation of what we did algebraically.

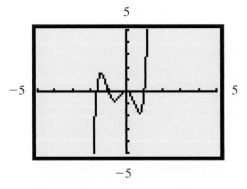

Figure 1.12

■ Radical Equations

An equation such as

$$\sqrt{2x - 4} = x - 2$$

which contains a radical with the variable in the radicand, is often referred to as a **radical equation.** To solve radical equations, we need the following additional property of equality.

Property 1.3

Let a and b be real numbers and n a positive integer.

If $a = b$, then $a^n = b^n$.

Property 1.3 states that *we can raise both sides of an equation to a positive integral power.* However, we must be very careful when applying Property 1.3. Raising both sides of an equation to a positive integral power sometimes produces results that do not satisfy the original equation. Consider the following examples.

EXAMPLE 4   Solve $\sqrt{3x + 1} = 7$.

Solution

$$\sqrt{3x + 1} = 7$$
$$(\sqrt{3x + 1})^2 = 7^2 \qquad \text{Square both sides}$$
$$3x + 1 = 49$$
$$3x = 48$$
$$x = 16$$

Check

$$\sqrt{3x + 1} = 7$$
$$\sqrt{3(16) + 1} \stackrel{?}{=} 7$$
$$\sqrt{49} \stackrel{?}{=} 7$$
$$7 = 7$$

The solution set is $\{16\}$.

Now work Problem 15.   ■

EXAMPLE 5   Solve $\sqrt{2x - 1} = -5$.

Solution

$$\sqrt{2x - 1} = -5$$
$$(\sqrt{2x - 1})^2 = (-5)^2 \qquad \text{Square both sides}$$

$$2x - 1 = 25$$
$$2x = 26$$
$$x = 13$$

✔ **Check**

$$\sqrt{2x - 1} = -5$$
$$\sqrt{2(13) - 1} \overset{?}{=} -5$$
$$\sqrt{25} \overset{?}{=} -5$$
$$5 \neq -5$$

Because 13 does not check, the equation has no solutions; the solution set is $\varnothing$.

▶ **Now work Problem 19.** ■

Remark: It is true that the equation in Example 5 could be solved by inspection because the symbol $\sqrt{}$ refers to nonnegative numbers. However, we did want to demonstrate what happens if Property 1.3 is used.

E X A M P L E   6

Solve $\sqrt{x} + 6 = x$.

Solution

$$\sqrt{x} + 6 = x$$
$$\sqrt{x} = x - 6$$
$$(\sqrt{x})^2 = (x - 6)^2 \qquad \text{Square both sides}$$
$$x = x^2 - 12x + 36$$
$$0 = x^2 - 13x + 36$$
$$0 = (x - 4)(x - 9)$$
$$x - 4 = 0 \qquad \text{or} \qquad x - 9 = 0$$
$$x = 4 \qquad \text{or} \qquad x = 9$$

✔ **Check**

$$\sqrt{x} = x - 6 \qquad \sqrt{x} = x - 6$$
$$\sqrt{4} \overset{?}{=} 4 - 6 \qquad \sqrt{9} \overset{?}{=} 9 - 6$$
$$2 \neq -2 \qquad 3 = 3$$

The only solution is 9, so the solution set is $\{9\}$.

▶ **Now work Problem 27.** ■

Remark: Notice what happens when we square both sides of the original equation. We obtain $x + 12\sqrt{x} + 36 = x^2$, an equation that is more complex than the original one and still contains a radical. Therefore, it is important first to isolate the term that contains the radical on one side of the equation and then to square both sides of the equation.

In general, raising both sides of an equation to a positive integral power produces an equation that has all of the solutions of the original equation, *but* it may also have some extra solutions that will not satisfy the original equation. Such extra solutions are called **extraneous solutions.** Therefore, when using Property 1.3, you *must* check each potential solution in the original equation.

E X A M P L E   7

Solve $\sqrt[3]{2x + 3} = -3$.

Solution

$$\sqrt[3]{2x + 3} = -3$$

$$(\sqrt[3]{2x + 3})^3 = (-3)^3 \qquad \text{Cube both sides}$$

$$2x + 3 = -27$$

$$2x = -30$$

$$x = -15$$

✔ **Check**

$$\sqrt[3]{2x + 3} = -3$$

$$\sqrt[3]{2(-15) + 3} \stackrel{?}{=} -3$$

$$\sqrt[3]{-27} \stackrel{?}{=} -3$$

$$-3 = -3$$

The solution set is $\{-15\}$.

▶ **Now work Problem 25.**   ∎

E X A M P L E   8

Solve $\sqrt{x + 4} = \sqrt{x - 1} + 1$.

Solution

$$\sqrt{x + 4} = \sqrt{x - 1} + 1$$

$$(\sqrt{x + 4})^2 = (\sqrt{x - 1} + 1)^2 \qquad \text{Square both sides}$$

$$x + 4 = x - 1 + 2\sqrt{x - 1} + 1 \qquad \begin{array}{l}\text{Remember the middle term when}\\ \text{squaring the binomial}\end{array}$$

$$4 = 2\sqrt{x - 1}$$

$$2 = \sqrt{x - 1}$$

$$2^2 = (\sqrt{x - 1})^2 \qquad \text{Square both sides}$$

$$4 = x - 1$$

$$5 = x$$

✔ **Check**

$$\sqrt{x + 4} = \sqrt{x - 1} + 1$$

$$\sqrt{5 + 4} \overset{?}{=} \sqrt{5 - 1} + 1$$

$$\sqrt{9} \overset{?}{=} \sqrt{4} + 1$$

$$3 = 3$$

The solution set is {5}.

▶ **Now work Problem 33.**   ∎

■ Equations of Quadratic Form

An equation such as $x^4 + 5x^2 - 36 = 0$ is not a quadratic equation. However, if we let $u = x^2$, then we get $u^2 = x^4$. Substituting u for x^2 and u^2 for x^4 in $x^4 + 5x^2 - 36 = 0$ produces

$$u^2 + 5u - 36 = 0$$

which is a quadratic equation. In general, an equation in the variable x is said to be of **quadratic form** if it can be written in the form

$$au^2 + bu + c = 0$$

where $a \neq 0$ and u is some algebraic expression in x. We have two basic approaches to solving equations of quadratic form, as illustrated by the next two examples.

E X A M P L E   9   Solve $x^{2/3} + x^{1/3} - 6 = 0$.

Solution

Let $u = x^{1/3}$; then $u^2 = x^{2/3}$ and the given equation can be rewritten $u^2 + u - 6 = 0$. Solving this equation yields two solutions.

$$u^2 + u - 6 = 0$$

$$(u + 3)(u - 2) = 0$$

$$u + 3 = 0 \qquad \text{or} \qquad u - 2 = 0$$

$$u = -3 \qquad \text{or} \qquad u = 2$$

Now, substituting $x^{1/3}$ for u, we have

$$x^{1/3} = -3 \qquad \text{or} \qquad x^{1/3} = 2$$

from which we obtain

$$(x^{1/3})^3 = (-3)^3 \qquad \text{or} \qquad (x^{1/3})^3 = 2^3$$

$$x = -27 \qquad \text{or} \qquad x = 8$$

✔ **Check**

$$x^{2/3} + x^{1/3} - 6 = 0 \qquad\qquad x^{2/3} + x^{1/3} - 6 = 0$$

$$(-27)^{2/3} + (-27)^{1/3} - 6 \overset{?}{=} 0 \qquad (8)^{2/3} + (8)^{1/3} - 6 \overset{?}{=} 0$$

$$9 + (-3) - 6 \overset{?}{=} 0 \qquad\qquad 4 + 2 - 6 \overset{?}{=} 0$$

$$0 = 0 \qquad\qquad\qquad 0 = 0$$

The solution set is $\{-27, 8\}$.

▶ **Now work Problem 51.** ■

E X A M P L E   1 0

Solve $x^4 + 5x^2 - 36 = 0$.

Solution

$$x^4 + 5x^2 - 36 = 0$$

$$(x^2 + 9)(x^2 - 4) = 0$$

$$x^2 + 9 = 0 \qquad \text{or} \qquad x^2 - 4 = 0$$

$$x^2 = -9 \qquad \text{or} \qquad x^2 = 4$$

$$x = \pm 3i \qquad \text{or} \qquad x = \pm 2$$

The solution set is $\{\pm 3i,\ \pm 2\}$.

▶ **Now work Problem 47.** ■

Notice in Example 9 that we made a substitution (u for $x^{1/3}$) to change the original equation to a quadratic equation in terms of the variable u. Then, after solving for u, we substituted $x^{1/3}$ for u to obtain the solutions of the original equation. However, in Example 10 we factored the given polynomial and proceeded without changing to a quadratic equation. Which approach you use may depend on the complexity of the given equation and your own personal preference.

E X A M P L E   1 1

Solve $15x^{-2} - 11x^{-1} - 12 = 0$.

Solution

Let $u = x^{-1}$; then $u^2 = x^{-2}$ and the given equation can be written and solved as follows:

$$15u^2 - 11u - 12 = 0$$

$$(5u + 3)(3u - 4) = 0$$

$$5u + 3 = 0 \qquad \text{or} \qquad 3u - 4 = 0$$

$$5u = -3 \qquad \text{or} \qquad 3u = 4$$

$$u = -\frac{3}{5} \qquad \text{or} \qquad u = \frac{4}{3}$$

Now, substituting x^{-1} back for u, we have

$$x^{-1} = -\frac{3}{5} \quad \text{or} \quad x^{-1} = \frac{4}{3}$$

from which we obtain

$$\frac{1}{x} = \frac{-3}{5} \quad \text{or} \quad \frac{1}{x} = \frac{4}{3}$$

$$-3x = 5 \quad \text{or} \quad 4x = 3$$

$$x = -\frac{5}{3} \quad \text{or} \quad x = \frac{3}{4}$$

The solution set is $\left\{ -\dfrac{5}{3}, \dfrac{3}{4} \right\}$.

▶ **Now work Problem 55.**   ∎

CONCEPT QUIZ   For Problems 1–5, answer true or false.

1. To solve the equation $2x^3 - 5x^2 = 4x$, both sides of the equation can be divided by x.

2. The equations $\sqrt{3x + 2} = -4$ and $(\sqrt{3x + 2})^2 = (-4)^2$ are equivalent equations.

3. The equation produced from raising both sides of an equation to a positive integral power may have solutions that the original equation does not have.

4. An equation in the variable x is quadratic in form if the equation can be written in the form $au^2 + bu + c = 0$ where u is an algebraic expression for x and $a \neq 0$.

5. An equation such as $\sqrt{2x - 1} = 9$ is referred to as a rational equation.

Problem Set 1.5

For Problems 1–66, solve each equation. Don't forget that you *must* check potential solutions whenever Property 1.3 is applied.

1. $x^3 + 8 = 0$

2. $x^3 - 27 = 0$

3. $x^3 = 1$

4. $x^4 - 9 = 0$

5. $x^3 + x^2 - 4x - 4 = 0$

6. $x^3 - 5x^2 - x + 5 = 0$

7. $2x^3 - 3x^2 + 2x - 3 = 0$

8. $3x^3 + 5x^2 + 12x + 20 = 0$

9. $8x^5 + 10x^4 = 4x^3 + 5x^2$

10. $10x^5 + 15x^4 = 2x^3 + 3x^2$

11. $x^{3/2} = 4x$

12. $5x^4 = 6x^3$

13. $n^{-2} = n^{-3}$

14. $n^{4/3} = 4n$

15. $\sqrt{3x - 2} = 4$

16. $\sqrt{5x - 1} = -4$

17. $\sqrt{3x - 8} - \sqrt{x - 2} = 0$

18. $\sqrt{2x - 3} = 1$

19. $\sqrt{4x - 3} = -2$

20. $\sqrt{3x - 1} + 1 = 4$

21. $\sqrt{2n + 3} - 2 = -1$

22. $\sqrt{5n + 1} - 6 = -4$

23. $\sqrt{4x - 1} - 3 = 2$

24. $\sqrt{2x - 1} - \sqrt{x + 2} = 0$

25. $\sqrt[3]{2x + 3} + 5 = 2$

26. $\sqrt[3]{n^2 - 1} + 4 = 3$

27. $2\sqrt{n + 3} = n$

28. $\sqrt{3t} - t = -6$

29. $\sqrt{3x - 2} = 3x - 2$

30. $5x - 4 = \sqrt{5x - 4}$

31. $\sqrt{2t - 1} + 2 = t$

32. $p = \sqrt{-4p + 17} + 3$

33. $\sqrt{x + 2} - 1 = \sqrt{x - 3}$

34. $\sqrt{x + 5} - 2 = \sqrt{x - 7}$

35. $\sqrt{7n + 23} - \sqrt{3n + 7} = 2$

36. $\sqrt{5t + 31} - \sqrt{t + 3} = 4$

37. $\sqrt{3x + 1} + \sqrt{2x + 4} = 3$

38. $\sqrt{2x - 1} - \sqrt{x + 3} = 1$

39. $\sqrt{x - 2} - \sqrt{2x - 11} = \sqrt{x - 5}$

40. $\sqrt{-2x - 7} + \sqrt{x + 9} = \sqrt{8 - x}$

41. $\sqrt{1 + 2\sqrt{x}} = \sqrt{x + 1}$

42. $\sqrt{7 + 3\sqrt{x}} = \sqrt{x + 1}$

43. $x^4 - 5x^2 + 4 = 0$

44. $x^4 - 25x^2 + 144 = 0$

45. $2n^4 - 9n^2 + 4 = 0$

46. $3n^4 - 4n^2 + 1 = 0$

47. $x^4 - 2x^2 - 35 = 0$

48. $2x^4 + 5x^2 - 12 = 0$

49. $x^4 - 4x^2 + 1 = 0$

50. $x^4 - 8x^2 + 11 = 0$

51. $x^{2/3} + 3x^{1/3} - 10 = 0$

52. $x^{2/3} + x^{1/3} - 2 = 0$

53. $6x^{2/3} - 5x^{1/3} - 6 = 0$

54. $3x^{2/3} - 11x^{1/3} - 4 = 0$

55. $x^{-2} + 4x^{-1} - 12 = 0$

56. $12t^{-2} - 17t^{-1} - 5 = 0$

57. $x - 11\sqrt{x} + 30 = 0$

58. $2x - 11\sqrt{x} + 12 = 0$

59. $x + 3\sqrt{x} - 10 = 0$

60. $6x - 19\sqrt{x} - 7 = 0$

61. $4x^{-4} - 17x^{-2} + 4 = 0$

62. $x^3 - 7x^{3/2} - 8 = 0$

63. $x^{-4/3} - 5x^{-2/3} + 4 = 0$

64. $3x^{-2} + 2x^{-1} - 8 = 0$

65. $10x^{-2} + 13x^{-1} - 3 = 0$

66. $7x^{-2} - 26x^{-1} - 8 = 0$

For Problems 67–70, solve each problem.

67. The formula for the slant height of a right circular cone is $s = \sqrt{r^2 + h^2}$, where r is the length of a radius of the base, and h is the altitude of the cone. Find the altitude of a cone whose slant height is 13 inches and whose radius is 5 inches.

68. A clockmaker wants to build a grandfather clock with a pendulum whose period will be 1.5 seconds. He knows the formula for the period is $T = 2\pi\sqrt{\dfrac{L}{32.144}}$, where T represents the period in seconds, and L represents the length of the pendulum in feet. What length should the clockmaker use for the pendulum? Express your answer to the nearest hundredth of a foot.

69. Police sometimes use the formula $S = \sqrt{30Df}$ to correlate the speed of a car and the length of skid marks when the brakes have been applied. In this formula, S represents the speed of the car in miles per hour, D represents the length of skid marks measured in feet, and f represents a coefficient of friction. For a particular situation, the coefficient of friction is a constant that depends on the type and condition of the road surface. Using 0.35 as a coefficient of friction, determine, to the nearest foot, how far a car will skid if the brakes are applied when the car is traveling at a speed of 58 miles per hour.

70. Using the formula given in Problem 69 and a coefficient of friction of 0.95, determine, to the nearest foot, how far a car will skid if the brakes are applied when the car is traveling at a speed of 65 miles per hour.

■ ■ ■ THOUGHTS INTO WORDS

71. Explain the concept of extraneous solutions.

72. What does it mean to say that an equation is of quadratic form?

73. Your friend attempts to solve the equation $3 + 2\sqrt{x} = x$ as follows:

$$(3 + 2\sqrt{x})^2 = x^2$$
$$9 + 12\sqrt{x} + 4x = x^2$$

At this step, he stops and doesn't know how to proceed. What help would you give him?

■ ■ ■ FURTHER INVESTIGATIONS

74. Verify that $x = a$ and $x^2 = a^2$ are *not* equivalent equations.

75. Solve the following equations, and express the solutions to the nearest hundredth.

a. $x^4 - 3x^2 + 1 = 0$ **b.** $x^4 - 5x^2 + 2 = 0$

c. $2x^4 - 7x^2 + 2 = 0$ **d.** $3x^4 - 9x^2 + 1 = 0$

e. $x^4 - 100x^2 + 2304 = 0$

f. $4x^4 - 373x^2 + 3969 = 0$

GRAPHING CALCULATOR ACTIVITIES

76. Graph the appropriate equations to give visual support for your solutions for Examples 1, 2, 4, and 6–11.

77. Graph the appropriate equations to give visual support for your solutions for the odd-numbered Problems 1–65.

Answers to the Concept Quiz

1. False   **2.** False   **3.** True   **4.** True   **5.** False

---

1.6   Inequalities

Objectives

■ Solve algebraic inequalities.

■ Write intervals in interval notation.

■ Solve quadratic inequalities.

■ Solve application problems involving inequalities.

Just as we use the symbol = to represent "is equal to," we also use the symbols < and > to represent "is less than" and "is greater than," respectively. Thus various **statements of inequality** can be made:

$a < b$ means a is less than b.

$a \le b$ means a is less than or equal to b.

$a > b$ means a is greater than b.

$a \geq b$ means a is greater than or equal to b.

The following are examples of **numerical statements of inequality:**

$7 + 8 > 10$ $-4 + (-6) \geq -10$

$\qquad -4 > -6$ $7 - 9 \leq -2$

$7 - 1 < 20$ $3 + 4 > 12$

$8(-3) < 5(-3)$ $7 - 1 < 0$

Notice that only $3 + 4 > 12$ and $7 - 1 < 0$ are *false;* the other six are *true* numerical statements.

Algebraic inequalities contain one or more variables. The following are examples of algebraic inequalities:

$\qquad x + 4 > 8$ $3x + 2y \leq 4$

$(x - 2)(x + 4) \geq 0$ $x^2 + y^2 + z^2 \leq 16$

An algebraic inequality such as $x + 4 > 8$ is neither true nor false as it stands and is called an **open sentence.** For each numerical value substituted for x, the algebraic inequality $x + 4 > 8$ becomes a numerical statement of inequality that is true or false. For example, if $x = -3$, then $x + 4 > 8$ becomes $-3 + 4 > 8$, which is false. If $x = 5$, then $x + 4 > 8$ becomes $5 + 4 > 8$, which is true. **Solving an algebraic inequality** refers to the process of finding the numbers that make it a true numerical statement. Such numbers are called the **solutions** of the inequality and are said to **satisfy** it.

The general process for solving inequalities closely parallels that for solving equations. We repeatedly replace the given inequality with equivalent but simpler inequalities until the solution set is obvious. The following property provides the basis for producing equivalent inequalities.

Property 1.4

1. For all real numbers a, b, and c,

$\qquad a > b$ if and only if $a + c > b + c$

2. For all real numbers a, b, and c, **with $c > 0$,**

$\qquad a > b$ if and only if $ac > bc$

3. For all real numbers a, b, and c, **with $c < 0$,**

$\qquad a > b$ if and only if $ac < bc$

Similar properties exist if $>$ is replaced by $<$, $\leq$, or $\geq$. Part 1 of Property 1.4 is commonly called the **addition property of inequality.** Parts 2 and 3 together make up the **multiplication property of inequality.** Pay special attention to part 3. **If both sides of an inequality are multiplied by a negative number, the inequality symbol**

must be reversed. For example, if both sides of $-3 < 5$ are multiplied by -2, the equivalent inequality $6 > -10$ is produced. Now let's consider using the addition and multiplication properties of inequality to help solve some inequalities.

E X A M P L E   1   Solve $3(2x - 1) < 8x - 7$.

Solution

$$3(2x - 1) < 8x - 7$$

$$6x - 3 < 8x - 7 \qquad \text{Apply distributive property to left side}$$

$$-2x - 3 < -7 \qquad \text{Add } -8x \text{ to both sides}$$

$$-2x < -4 \qquad \text{Add 3 to both sides}$$

$$-\frac{1}{2}(-2x) > -\frac{1}{2}(-4) \qquad \text{Multiply both sides by } -\frac{1}{2}, \text{ which reverses the inequality}$$

$$x > 2$$

The solution set is $\{x | x > 2\}$.   ■

A graph of the solution set $\{x | x > 2\}$ in Example 1 is shown in Figure 1.13. The parenthesis indicates that 2 does not belong to the solution set.

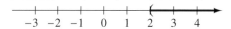

Figure 1.13

Checking the solutions of an inequality presents a problem. Obviously, we cannot check all of the infinitely many solutions for a particular inequality. However, by checking at least one solution, especially when the multiplication property has been used, we might catch a mistake of forgetting to reverse the inequality. In Example 1 we are claiming that all numbers greater than 2 will satisfy the original inequality. Let's check the number 3.

$$3(2x - 1) < 8x - 7$$

$$3[2(3) - 1] \overset{?}{<} 8(3) - 7$$

$$3(5) \overset{?}{<} 17$$

$$15 < 17 \qquad \text{It checks!}$$

We also can get some visual support from a graphical analysis of the inequality. Figure 1.14 shows a graph of the equation $y = 3(2x - 1) - 8x + 7$. To satisfy the inequality $3(2x - 1) - 8x + 7 < 0$, which is equivalent to the given inequality in Example 1, y has to be less than zero. As we see in Figure 1.14, $y < 0$ when $x > 2$. This agrees with our solution set.

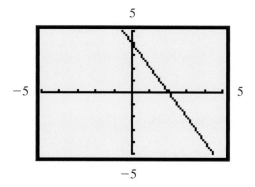

Figure 1.14

■ Interval Notation

It is also convenient to express solution sets of inequalities by using **interval nota-tion.** For example, the symbol $(2, \infty)$ refers to the interval of all real numbers greater than 2. As on the graph in Figure 1.13, the left-hand parenthesis indicates that 2 is not to be included. The infinity symbol, ∞, along with the right-hand parenthesis, indicates that there is no right-hand endpoint. Following is a partial list of interval notations, along with the sets and graphs that they represent. Note the use of square brackets to *include* endpoints.

Set	Graph	Interval Notation
$\{x\lvert x > a\}$		(a, ∞)
$\{x\lvert x \geq a\}$		$[a, \infty)$
$\{x\lvert x < b\}$		$(-\infty, b)$
$\{x\lvert x \leq b\}$		$(-\infty, b]$

EXAMPLE   2   Solve $\dfrac{-3x + 1}{2} > 4.$

Solution

$$\frac{-3x + 1}{2} > 4$$

$$2\left(\frac{-3x + 1}{2}\right) > 2(4) \qquad \text{Multiply both sides by 2}$$

$$-3x + 1 > 8$$

$$-3x > 7$$

$$-\frac{1}{3}(-3x) < -\frac{1}{3}(7) \qquad \text{Multiply both sides by } -\frac{1}{3}, \text{which reverses the inequality}$$

$$x < -\frac{7}{3}$$

The solution set is $\left(-\infty, -\frac{7}{3}\right)$.

▶ **Now work Problem 21.** ∎

E X A M P L E   3

Solve $\dfrac{x-4}{6} - \dfrac{x-2}{9} \le \dfrac{5}{18}$.

Solution

$$\frac{x-4}{6} - \frac{x-2}{9} \le \frac{5}{18}$$

$$18\left(\frac{x-4}{6} - \frac{x-2}{9}\right) \le 18\left(\frac{5}{18}\right) \qquad \text{Multiply both sides by the LCD}$$

$$18\left(\frac{x-4}{6}\right) - 18\left(\frac{x-2}{9}\right) \le 18\left(\frac{5}{18}\right)$$

$$3(x-4) - 2(x-2) \le 5$$

$$3x - 12 - 2x + 4 \le 5$$

$$x - 8 \le 5$$

$$x \le 13$$

The solution set is $(-\infty, 13]$.

▶ **Now work Problem 29.** ∎

■ Compound Statements

We use the words "and" and "or" in mathematics to form **compound statements.** The following are examples of some compound numerical statements that use "and." We call such statements **conjunctions.** We agree to call a conjunction true only if all of its component parts are true. Statements 1 and 2 below are true, but statements 3, 4, and 5 are false.

1. $3 + 4 = 7$ and $-4 < -3$ True

2. $-3 < -2$ and $-6 > -10$ True

3. $6 > 5$ and $-4 < -8$ False

4. $4 < 2$ and $0 < 10$ False

5. $-3 + 2 = 1$ and $5 + 4 = 8$ False

We call compound statements that use "or" **disjunctions.** The following are some examples of disjunctions that involve numerical statements:

6. $0.14 > 0.13$ or $0.235 < 0.237$ True

7. $\dfrac{3}{4} > \dfrac{1}{2}$ or $-4 + (-3) = 10$ True

8. $-\dfrac{2}{3} > \dfrac{1}{3}$ or $(0.4)(0.3) = 0.12$ True

9. $\dfrac{2}{5} < -\dfrac{2}{5}$ or $7 + (-9) = 16$ False

A disjunction is true if at least one of its component parts is true. In other words, disjunctions are false only if all of the component parts are false. In the statements above, 6, 7, and 8 are true, but 9 is false.

Now let's consider finding solutions for some compound statements that involve algebraic inequalities. Keep in mind that our previous agreements for labeling conjunctions and disjunctions true or false form the basis for our reasoning.

E X A M P L E   4

Graph the solution set for the conjunction $x > -1$ *and* $x < 3$.

Solution

The key word is *and,* so we need to satisfy both inequalities. Thus all numbers between -1 and 3 are solutions, and we can indicate this on a number line as in Figure 1.15.

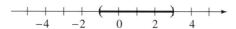

Figure 1.15

Using interval notation, we can represent the interval enclosed in parentheses in Figure 1.15 by $(-1, 3)$. Using set-builder notation, we can express the same interval as $\{x \mid -1 < x < 3\}$, where the statement $-1 < x < 3$ is read "negative one is less than x and x is less than three." In other words, x is between -1 and 3. ∎

Example 4 represents another concept that pertains to sets. The set of all elements common to two sets is called the **intersection** of the two sets. Thus in Example 4 we found the intersection of the two sets $\{x \mid x > -1\}$ and $\{x \mid x < 3\}$ to be the set $\{x \mid -1 < x < 3\}$. In general, we define the intersection of two sets as follows.

Definition 1.1

> The **intersection** of two sets A and B (written $A \cap B$) is the set of all elements that are in both A and B. Using set-builder notation, we can write
>
> $$A \cap B = \{x \mid x \in A \quad and \quad x \in B\}$$

We can solve a conjunction such as $\dfrac{3x+2}{2} > -2$ and $\dfrac{3x+2}{2} < 7$, in which the same algebraic expression is contained in both inequalities, by using the compact form $-2 < \dfrac{3x+2}{2} < 7$ as follows.

EXAMPLE 5

Solve $-2 < \dfrac{3x+2}{2} < 7$.

Solution

$$-2 < \dfrac{3x+2}{2} < 7$$

$$2(-2) < 2\left(\dfrac{3x+2}{2}\right) < 2(7) \qquad \text{Multiply through by 2}$$

$$-4 < 3x+2 < 14$$

$$-6 < 3x < 12 \qquad \text{Add } -2 \text{ to all three quantities}$$

$$-2 < x < 4 \qquad \text{Multiply through by } \dfrac{1}{3}$$

The solution set is the interval $(-2, 4)$.

▶ **Now work Problem 35.** ∎

The word "and" ties the concept of a conjunction to the set concept of intersection. In a like manner, the word "or" links the idea of a disjunction to the set concept of **union.** We define the union of two sets as follows.

Definition 1.2

> The **union** of two sets A and B (written $A \cup B$) is the set of all elements that are in A or in B or in both. Using set-builder notation, we can write
>
> $$A \cup B = \{x | x \in A \quad or \quad x \in B\}$$

EXAMPLE 6

Graph the solution set for the disjunction $x < -1$ *or* $x > 2$, and express it using interval notation.

Solution

The key word is "or," so all numbers that satisfy either inequality (or both) are solutions. Thus all numbers less than -1, along with all numbers greater than 2, are the solutions. The graph of the solution set is shown in Figure 1.16. Using interval notation and the set concept of union, we can express the solution set as $(-\infty, -1)$ $\cup (2, \infty)$.

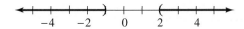

Figure 1.16   ■

Example 6 illustrates that in terms of set vocabulary, the solution set of a disjunction is the union of the solution sets of the component parts of the disjunction. Note that there is *no compact form* for writing $x < -1$ or $x > 2$ *or for any disjunction*.

The following agreements on the use of interval notation should be added to the list on page 159.

Set	Graph	Interval Notation
$\{x \mid a < x < b\}$		(a, b)
$\{x \mid a \leq x < b\}$		$[a, b)$
$\{x \mid a < x \leq b\}$		$(a, b]$
$\{x \mid a \leq x \leq b\}$		$[a, b]$
$\{x \mid x \text{ is a real number}\}$		$(-\infty, \infty)$

■ Quadratic Inequalities

The equation $ax^2 + bx + c = 0$ has been referred to as the standard form of a quadratic equation in one variable. Similarly, the form $ax^2 + bx + c < 0$ is used to represent a **quadratic inequality**. (The symbol $<$ can be replaced by $>$, $\leq$, or $\geq$ to produce other forms of quadratic inequalities.)

The number line can be used to help solve quadratic inequalities where the quadratic polynomial is factorable. Let's consider two examples to illustrate this procedure.

E X A M P L E   7

Solve $x^2 + x - 6 < 0$.

Solution

First, let's factor the polynomial.

$$x^2 + x - 6 < 0$$
$$(x + 3)(x - 2) < 0$$

Second, let's locate the values where the product $(x + 3)(x - 2)$ is equal to zero. The numbers -3 and 2 divide the number line into three intervals (see Figure 1.17):

the numbers less than -3

the numbers between -3 and 2

the numbers greater than 2

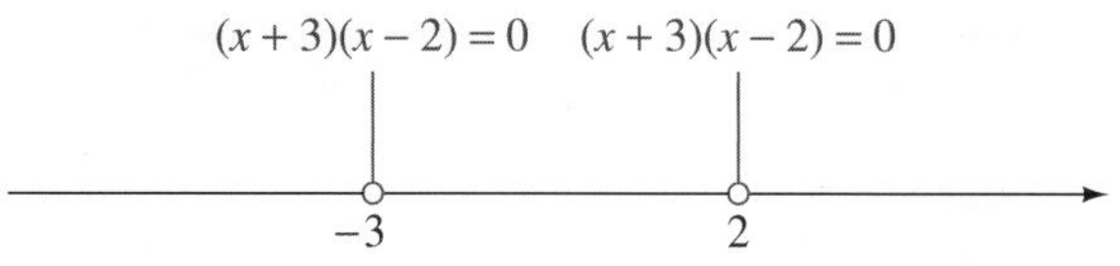

$$(x + 3)(x - 2) = 0 \qquad (x + 3)(x - 2) = 0$$

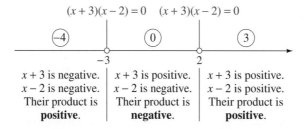

Figure 1.17

We can choose a **test number** from each of these intervals and see how it affects the signs of the factors $x + 3$ and $x - 2$ and, consequently, the sign of the product of these factors. For example, if $x < -3$ (try $x = -4$), then $x + 3$ is negative and $x - 2$ is negative; thus their product is positive. If $-3 < x < 2$ (try $x = 0$), then $x + 3$ is positive and $x - 2$ is negative; thus their product is negative. If $x > 2$ (try $x = 3$), then $x + 3$ is positive and $x - 2$ is positive; thus their product is positive. This information can be conveniently arranged by using a number line, as in Figure 1.18.

$$(x + 3)(x - 2) = 0 \qquad (x + 3)(x - 2) = 0$$

$x + 3$ is negative.	$x + 3$ is positive.	$x + 3$ is positive.
$x - 2$ is negative.	$x - 2$ is negative.	$x - 2$ is positive.
Their product is **positive**.	Their product is **negative**.	Their product is **positive**.

Figure 1.18

Therefore, the given inequality, $x^2 + x - 6 < 0$, is satisfied by the numbers between -3 and 2. That is, the solution set is the open interval $(-3, 2)$.

▶ **Now work Problem 39.** ∎

    A graphical analysis of a quadratic inequality like the one in Example 7 can also be very helpful. In Figure 1.19 we show the graph of $y = x^2 + x - 6$. It certainly appears that $y < 0$ in the x interval $(-3, 2)$, which agrees with our solution set in Example 7.

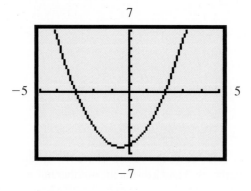

Figure 1.19

    Numbers such as -3 and 2 in the preceding example, where the given polynomial or algebraic expression equals zero or is undefined, are referred to as

critical numbers. Let's consider another example where we make use of critical numbers and test numbers.

E X A M P L E   8

Solve $6x^2 + 17x - 14 \geq 0$.

Solution

First, we factor the polynomial.

$$6x^2 + 17x - 14 \geq 0$$
$$(2x + 7)(3x - 2) \geq 0$$

Second, we locate the values where the product $(2x + 7)(3x - 2)$ equals zero. We suggest putting dots at $-\dfrac{7}{2}$ and $\dfrac{2}{3}$ (see Figure 1.20) to remind ourselves that these

$(2x + 7)(3x - 2) = 0$ $(2x + 7)(3x - 2) = 0$

$-\dfrac{7}{2}$ $\dfrac{2}{3}$

Figure 1.20

two numbers must be included in the solution set because the given statement includes equality. Now let's choose a test number from each of the three intervals and observe the sign behavior of the factors, as in Figure 1.21.

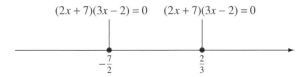

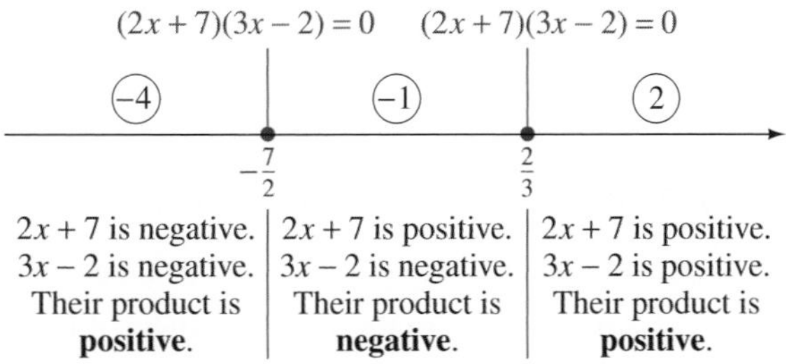

$(2x + 7)(3x - 2) = 0$ $(2x + 7)(3x - 2) = 0$

$-\dfrac{7}{2}$ $\dfrac{2}{3}$

$2x + 7$ is negative.	$2x + 7$ is positive.	$2x + 7$ is positive.
$3x - 2$ is negative.	$3x - 2$ is negative.	$3x - 2$ is positive.
Their product is **positive**.	Their product is **negative**.	Their product is **positive**.

Figure 1.21

Using the concept of set union, we can write the solution set $\left(-\infty, -\dfrac{7}{2}\right]$ $\cup \left[\dfrac{2}{3}, \infty\right)$.

▶ **Now work Problem 45.**   ■

As you work with quadratic inequalities like those in Examples 7 and 8, you may be able to use a more abbreviated format than the one we demonstrated. Basically, it is necessary to keep track of the sign of each factor in each of the intervals. Furthermore, this number line approach is not restricted to just quadratic inequalities. Polynomial inequalities, like $(x - 1)(x + 2)(x - 3) > 0$, can be analyzed nicely on a number line. We will have you do this in the next problem set.

Let's conclude this section by considering a word problem that involves an inequality. All of the problem-solving techniques offered earlier continue to apply except that now we look for a guideline that can be used to generate an inequality rather than an equation.

P R O B L E M   1

Lance has $5000 to invest. If he invests $3000 at 6%, at what rate must he invest the remaining $2000 so that the total yearly interest from the two investments exceeds $330?

Solution

Let r represent the unknown rate of interest. The following guideline can be used to set up an inequality:

$$\text{Interest from 6\% investment} + \text{Interest from } r\% \text{ investment} > \$330$$

$$(6\%)(\$3000) \qquad + \qquad r(\$2000) \qquad > \$330$$

We solve this inequality using methods we have already acquired:

$$(0.06)(3000) + 2000r > 330$$

$$180 + 2000r > 330$$

$$2000r > 150$$

$$r > 0.075$$

The other $2000 must be invested at a rate higher than 7.5%. ∎

C O N C E P T   Q U I Z

For Problems 1–8, answer true or false.

1. The inequality $3x - 4 \geq -2x - 3$ is a true statement if $x = 1$.

2. If both sides of an inequality are divided by a negative number, the inequality symbol must be reversed.

3. The solution set $\{x | x \leq -3\}$ is written as $[-3, -\infty)$.

4. The solution set for the conjunction $x > -3$ and $x < 7$ is the set of all real numbers.

5. The solution set for the disjunction $x < 5$ or $x > 1$ is the set of all real numbers.

6. The quadratic inequality $x^2 + 4x + 5 < 0$ does not have a real number solution.

7. The compound statement $2x + 5 > -3$ and $2x + 5 < 13$ can be rewritten using the compact form $-3 < 2x + 5 < 13$.

8. The compound statement $3x - 4 > 12$ or $3x - 4 < -8$ can be rewritten using the compact form $-8 < 3x - 4 > 12$.

Problem Set 1.6

For Problems 1–12, express each solution set in interval notation and graph each solution set.

1. $x \leq -2$

2. $x > -1$

3. $1 < x < 4$

4. $-1 < x \leq 2$

5. $2 > x > 0$

6. $-3 \geq x$

7. $-2 \leq x \leq -1$

8. $1 \leq x$

9. $x < 1$ or $x > 3$

10. $x > 2$ or $x < -1$

11. $x > -2$ or $x > 2$

12. $x > 2$ or $x < 4$

For Problems 13–20, solve each conjunction by using the compact form and express the solution sets in interval notation.

13. $-17 \leq 3x - 2 \leq 10$

14. $-25 \leq 4x + 3 \leq 19$

15. $2 > 2x - 1 > -3$

16. $4 > 3x + 1 > 1$

17. $-4 < \dfrac{x-1}{3} < 4$

18. $-1 \leq \dfrac{x+2}{4} \leq 1$

19. $-3 < 2 - x < 3$

20. $-4 < 3 - x < 4$

For Problems 21–72, solve each inequality and express the solution sets in interval notation.

21. $-2x + 1 > 5$

22. $6 - 3x < 12$

23. $-3n + 5n - 2 \geq 8n - 7 - 9n$

24. $3n - 5 > 8n + 5$

25. $6(2t - 5) - 2(4t - 1) \geq 0$

26. $3(2x + 1) - 2(2x + 5) < 5(3x - 2)$

27. $\dfrac{2}{3}x - \dfrac{3}{4} \leq \dfrac{1}{4}x + \dfrac{2}{3}$

28. $\dfrac{3}{5} - \dfrac{x}{2} \geq \dfrac{1}{2} + \dfrac{x}{5}$

29. $\dfrac{n+2}{4} + \dfrac{n-3}{8} < 1$

30. $\dfrac{2n+1}{6} + \dfrac{3n-1}{5} > \dfrac{2}{15}$

31. $\dfrac{x}{2} - \dfrac{x-1}{5} \geq \dfrac{x+2}{10} - 4$

32. $\dfrac{4x-3}{6} - \dfrac{2x-1}{12} < -2$

33. $0.09x + 0.1(x + 200) > 77$

34. $0.06x + 0.08(250 - x) \geq 19$

35. $0 < \dfrac{5x-1}{3} < 2$

36. $-3 \leq \dfrac{4x+3}{2} \leq 1$

37. $3 \geq \dfrac{7-x}{2} \geq 1$

38. $-2 \leq \dfrac{5-3x}{4} \leq \dfrac{1}{2}$

39. $x^2 + 3x - 4 < 0$

40. $x^2 - 4 < 0$

41. $x^2 - 2x - 15 > 0$

42. $x^2 - 12x + 32 \geq 0$

43. $n^2 - n \leq 2$

44. $n^2 + 5n \leq 6$

45. $3t^2 + 11t - 4 > 0$

46. $2t^2 - 9t - 5 > 0$

47. $15x^2 - 26x + 8 \leq 0$

48. $6x^2 + 25x + 14 \leq 0$

49. $4x^2 - 4x + 1 > 0$

50. $9x^2 + 6x + 1 \leq 0$

51. $4 - x^2 < 0$

52. $2x^2 - 18 \geq 0$

53. $4(x^2 - 36) < 0$

54. $-4(x^2 - 36) \geq 0$

55. $5x^2 + 20 > 0$

56. $-3x^2 - 27 \geq 0$

57. $x^2 - 2x \geq 0$

58. $2x^2 + 6x < 0$

59. $3x^3 + 12x^2 > 0$

60. $2x^3 + 4x^2 \leq 0$

61. $(x + 1)(x - 3) > (x + 1)(2x - 1)$

62. $(x - 2)(2x + 5) > (x - 2)(x - 3)$

63. $(x + 1)(x - 2) \geq (x - 4)(x + 6)$

64. $(2x - 1)(x + 4) \geq (2x + 1)(x - 3)$

65. $(x - 1)(x - 2)(x + 4) > 0$

66. $(x + 1)(x - 3)(x + 7) \geq 0$

67. $(x + 2)(2x - 1)(x - 5) \leq 0$

68. $(x - 3)(3x + 2)(x + 4) < 0$

69. $x^3 - 2x^2 - 24x \geq 0$ **70.** $x^3 + 2x^2 - 3x > 0$

71. $(x - 2)^2(x + 3) > 0$ **72.** $(x + 4)^2(x + 5) > 0$

For Problems 73–82, use inequalities to help solve each problem.

73. Felix has $10,000 to invest. Suppose he invests $5000 at 6% interest. At what rate must he invest the other $5000 so that the two investments yield more than $800 of yearly interest?

74. Suppose that Annette invests $7000 at 7%. How much must she invest at 11% so that the total yearly interest from the two investments exceeds $974?

75. Rhonda had scores of 94, 84, 86, and 88 on her first four history exams of the semester. What score must she obtain on the fifth exam to have an average of 90 or higher for the five exams?

76. The average height of the two forwards and the center of a basketball team is 6 feet 8 inches. What must the average height of the two guards be so that the team average is at least 6 feet 4 inches?

77. At the food booths at a festival, the area of a pizza must be 150 square inches or more for the pizza to be classified as large. What must the diameter be for a pizza to be classified as large?

78. If the temperature for a 24-hour period ranged between 41°F and 59°F, inclusive, what was the range in Celsius degrees? $\left(F = \dfrac{9}{5}C + 32 \right)$

79. If the temperature for a 24-hour period ranged between $-20°$C and $-5°$C, inclusive, what was the range in Fahrenheit degrees? $\left(C = \dfrac{5}{9}(F - 32) \right)$

80. A person's intelligence quotient (IQ) is found by dividing mental age (M), as indicated by standard tests, by chronological age (C), and then multiplying this ratio by 100. The formula IQ $= 100M/C$ can be used. If the IQ range of a group of 11-year-olds is given by $80 \leq$ IQ ≤ 140, find the mental-age range of this group.

81. A car can be rented from agency A at $75 per day plus $0.10 a mile or from agency B at $50 a day plus $0.20 a mile. If the car is driven m miles, for what values of m does it cost less to rent from agency A?

82. In statistics the formula for a z-score is $z = \dfrac{x - \bar{x}}{s}$, where x is a score, $\bar{x}$ is the mean, and s is the standard deviation. To give credibility to our results in a statistical claim, we want to determine the values of x that will produce a z-score greater than 2.5 when $\bar{x} = 8.7$ and $s = 1.2$. Find such values of x.

■ ■ ■ **THOUGHTS INTO WORDS**

83. Explain the difference between a conjunction and a disjunction. Give an example of each (outside the field of mathematics).

84. How do you know by inspection that the solution set of the inequality $x + 3 > x + 2$ is the entire set of real numbers?

85. Give a step-by-step description of how you would solve the inequality $-4 < 2(x - 1) - 3(x + 2)$.

86. Explain how you would solve the inequality $(x - 1)^2(x + 2)^2 > 0$.

87. Find the solution set for each of the following compound statements and explain your reasoning in each case.

a. $x < 3$ and $5 > 2$
b. $x < 3$ or $5 > 2$
c. $x < 3$ and $6 < 4$
d. $x < 3$ or $6 < 4$

■ ■ ■ FURTHER INVESTIGATIONS

88. The product $(x - 2)(x + 3)$ is positive if both factors are negative *or* if both factors are positive. Therefore, we can solve $(x - 2)(x + 3) > 0$ as follows.

$(x - 2 < 0 \text{ and } x + 3 < 0)$ or $(x - 2 > 0 \text{ and } x + 3 > 0)$

$\qquad (x < 2 \text{ and } x < -3)$ or $\quad (x > 2 \text{ and } x > -3)$

$\qquad\qquad x < -3$ or $\qquad x > 2$

The solution set is $(-\infty, -3) \cup (2, \infty)$. Use this type of analysis to solve each inequality.

a. $(x - 1)(x + 5) > 0$

b. $(x + 2)(x - 4) \geq 0$

c. $(x + 4)(x - 3) < 0$

d. $(2x - 1)(x + 5) \leq 0$

e. $(x + 4)(x + 1)(x - 2) > 0$

f. $(x + 2)(x - 1)(x - 3) < 0$

89. If $a > b > 0$, verify that $1/a < 1/b$.

90. If $a > b$, is it always true that $1/a < 1/b$? Defend your answer.

■ GRAPHING CALCULATOR ACTIVITIES

91. Show a graphical analysis for Examples 2, 3, and 8 of this section.

92. Show a graphical analysis for Problems 39–72.

Answers to the Concept Quiz

1. Ture   **2.** True   **3.** False   **4.** False   **5.** True   **6.** True   **7.** True   **8.** False

---

1.7   Inequalities Involving Quotients, Absolute Value Equations, and Inequalities

Objectives

■ Solve inequalities involving quotients.

■ Solve absolute value equations.

■ Solve absolute value inequalities.

The same type of number-line analysis that we did in the previous section can be used for indicated quotients as well as for indicated products. In other words, inequalities such as

$$\frac{x - 2}{x + 3} > 0$$

can be solved very efficiently using the same basic approach that we used with quadratic inequalities in the previous section. Let's illustrate this procedure.

EXAMPLE 1   Solve $\dfrac{x-2}{x+3} > 0$.

Solution

First we find that at $x = 2$ the quotient $\dfrac{x-2}{x+3}$ equals zero, and that at $x = -3$ the quotient is undefined. The critical numbers -3 and 2 divide the number line into three intervals. Then, using a test number from each interval (such as $-4, 1,$ and 3), we can observe the sign behavior of the quotient, as in Figure 1.22. Therefore, the solution set for $\dfrac{x-2}{x+3} > 0$ is $(-\infty, -3) \cup (2, \infty)$.

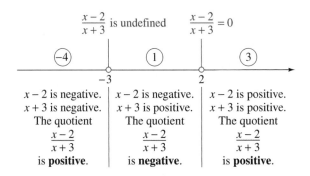

Figure 1.22

▶ **Now work Problem 1.**   ∎

In Figure 1.23 we show the graph of $y = \dfrac{x-2}{x+3}$. Notice that y is greater than zero when x is less than -3 and when x is greater than 2. Again this agrees with our solution set in Example 1.

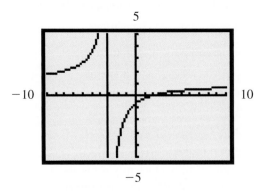

Figure 1.23

EXAMPLE 2   Solve $\dfrac{x+2}{x+4} \le 3$.

Solution

First, let's change the form of the given inequality:

$$\frac{x+2}{x+4} \le 3$$

$$\frac{x+2}{x+4} - 3 \le 0$$

$$\frac{x+2-3(x+4)}{x+4} \le 0$$

$$\frac{x+2-3x-12}{x+4} \le 0$$

$$\frac{-2x-10}{x+4} \le 0$$

Now we can proceed as before. If $x = -5$, then the quotient $\dfrac{-2x-10}{x+4}$ equals zero, and if $x = -4$, the quotient is undefined. Using test numbers such as -6, $-4\frac{1}{2}$, and -3, we are able to study the sign behavior of the quotient, as in Figure 1.24. Therefore, the solution set for $\dfrac{x+2}{x+4} \le 3$ is $(-\infty, -5] \cup (-4, \infty)$.

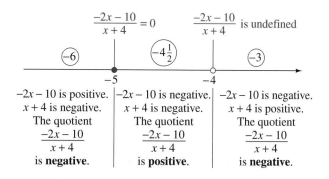

Figure 1.24

▶ **Now work Problem 7.**   ■

■ Absolute Value

In Section 0.1 we defined the **absolute value** of a real number by

$$|a| = \begin{cases} a & \text{if } a \ge 0 \\ -a & \text{if } a < 0 \end{cases}$$

We also interpreted the absolute value of any real number to be the distance between the number and zero on the real number line. For example, $|6| = 6$ because the distance between 6 and 0 is six units. Likewise, $|-8| = 8$ because the distance between -8 and 0 is eight units.

Both the definition and the number-line interpretation of absolute value provide ways of analyzing a variety of equations and inequalities involving absolute value. For example, suppose that we need to solve the equation $|x| = 4$. If we think in terms of distance on the number line, the equation $|x| = 4$ means that we are looking for numbers that are four units from zero. Thus x must be 4 or -4. From the definition viewpoint, we could proceed as follows.

If $x \geq 0$, then $|x| = x$, and the equation $|x| = 4$ becomes $x = 4$.

If $x < 0$, then $|x| = -x$, and the equation $|x| = 4$ becomes $-x = 4$, which is equivalent to $x = -4$.

Using either approach, we see that the solution set for $|x| = 4$ is $\{-4, 4\}$.

The following property should seem reasonable from the distance interpretation and can be verified using the definition of absolute value.

Property 1.5

For any real number $k > 0$,

$$|x| = k \quad \text{if and only if } x = k \text{ or } x = -k$$

E X A M P L E   3

Solve $|3x - 2| = 7$.

Solution

$$|3x - 2| = 7$$

$$3x - 2 = 7 \quad \text{or} \quad 3x - 2 = -7$$

$$3x = 9 \quad \text{or} \quad 3x = -5$$

$$x = 3 \quad \text{or} \quad x = -\frac{5}{3}$$

The solution set is $\left\{ -\frac{5}{3}, 3 \right\}$.

▶ **Now work Problem 15.** ∎

E X A M P L E   4

Solve the equation $|3x - 1| = |x + 4|$.

Solution

We could solve this equation by applying the definition of absolute value to both expressions; however, let's approach it in a less formal way. For the two numbers,

$3x - 1$ and $x + 4$, to have the same absolute value, they must either be equal or be opposites of each other. Therefore the equation $|3x - 1| = |x + 4|$ is equivalent to $3x - 1 = x + 4$ or $3x - 1 = -(x + 4)$, which can be solved as follows:

$$3x - 1 = x + 4 \qquad \text{or} \qquad 3x - 1 = -(x + 4)$$
$$2x = 5 \qquad \text{or} \qquad 3x - 1 = -x - 4$$
$$x = \frac{5}{2} \qquad \text{or} \qquad 4x = -3$$
$$x = \frac{5}{2} \qquad \text{or} \qquad x = -\frac{3}{4}$$

The solution set is $\left\{-\dfrac{3}{4}, \dfrac{5}{2}\right\}$.

▶ **Now work Problem 35.**   ■

The distance interpretation for absolute value also provides a good basis for solving some inequalities. For example, to solve $|x| < 4$, we know that the distance between x and 0 must be less than four units. In other words, x is to be less than four units away from zero. Thus $|x| < 4$ is equivalent to $-4 < x < 4$, and the solution set is the interval $(-4, 4)$. We will have you use the definition of absolute value and verify the following general property in the next set of exercises.

Property 1.6

For any real number $k > 0$,

$$|x| < k \quad \text{if and only if } -k < x < k$$

Example 5 illustrates the use of Property 1.6.

EXAMPLE 5   Solve $|2x + 1| < 5$.

Solution

$$|2x + 1| < 5$$
$$-5 < 2x + 1 < 5$$
$$-6 < 2x < 4$$
$$-3 < x < 2$$

The solution set is the interval $(-3, 2)$.

▶ **Now work Problem 49.**   ■

Property 1.6 can also be expanded to include the $\leq$ situation; that is, $|x| \leq k$ if and only if $-k \leq x \leq k$.

EXAMPLE 6   Solve $|-3x - 2| \le 6$.

Solution

$$|-3x - 2| \le 6$$
$$-6 \le -3x - 2 \le 6$$
$$-4 \le -3x \le 8$$
$$\frac{4}{3} \ge x \ge -\frac{8}{3} \qquad \begin{array}{l} \text{Note that multiplying through by } -\frac{1}{3} \\ \text{reverses the inequalities.} \end{array}$$

The statement $\frac{4}{3} \ge x \ge -\frac{8}{3}$ is equivalent to $-\frac{8}{3} \le x \le \frac{4}{3}$. Therefore, the solution set is $\left[-\frac{8}{3}, \frac{4}{3} \right]$.

▶ **Now work Problem 55.**   ■

Now suppose that we want to solve $|x| > 4$. The distance between x and zero must be more than four units; in other words, x is to be more than four units away from zero. Therefore $|x| > 4$ is equivalent to $x < -4$ or $x > 4$, and the solution set is $(-\infty, -4) \cup (4, \infty)$. The following general property can be verified by using the definition of absolute value.

Property 1.7

For any real number $k > 0$,

$$|x| > k \quad \text{if and only if } x < -k \text{ or } x > k$$

EXAMPLE 7   Solve $|4x - 3| > 9$.

Solution

$$|4x - 3| > 9$$
$$4x - 3 < -9 \quad \text{or} \quad 4x - 3 > 9$$
$$4x < -6 \quad \text{or} \quad 4x > 12$$
$$x < -\frac{6}{4} \quad \text{or} \quad x > 3$$
$$x < -\frac{3}{2} \quad \text{or} \quad x > 3$$

The solution set is $\left(-\infty, -\frac{3}{2} \right) \cup (3, \infty)$.

▶ **Now work Problem 51.**   ■

Property 1.7 can also be expanded to include the $\geq$ situation; that is, $|x| \geq k$ if and only if $x \leq -k$ or $x \geq k$.

EXAMPLE 8

Solve $|-2 - x| \geq 9$.

Solution

$$|-2 - x| \geq 9$$

$-2 - x \leq -9$	or	$-2 - x \geq 9$
$-x \leq -7$	or	$-x \geq 11$
$x \geq 7$	or	$x \leq -11$

The solution set is $(-\infty, -11] \cup [7, \infty)$.

▶ **Now work Problem 59.** ∎

Properties 1.5, 1.6, and 1.7 provide a sound basis for solving many equations and inequalities involving absolute value. However, if at any time you become doubtful about which property applies, don't forget the definition and the distance interpretation for absolute value.

We should also note that in Properties 1.5, 1.6, and 1.7, k is a positive number. This is not a serious restriction because problems where k is nonpositive are easily solved as follows:

$|x - 2| = 0$ The solution set is {2} because $x - 2$ has to equal zero

$|3x - 7| = -4$ The solution set is $\varnothing$; for any real number, the absolute value of $3x - 7$ will always be nonnegative

$|2x - 1| < -3$ The solution set is $\varnothing$; for any real number, the absolute value of $2x - 1$ will always be nonnegative

$|5x + 2| > -4$ The solution set is $(-\infty, \infty)$; the absolute value of $5x + 2$, regardless of which real number is substituted for x, will always be greater than -4

The number-line approach used in Examples 1 and 2 of this section, along with Properties 1.6 and 1.7, provide a systematic way of solving absolute value inequalities that have the variable in the denominator of a fraction. Let's analyze one such problem.

EXAMPLE 9

Solve $\left| \dfrac{x - 2}{x + 3} \right| < 4$.

Solution

By Property 1.6, $\left| \dfrac{x - 2}{x + 3} \right| < 4$ becomes $-4 < \dfrac{x - 2}{x + 3} < 4$, which can be written

$$\frac{x - 2}{x + 3} > -4 \quad \text{and} \quad \frac{x - 2}{x + 3} < 4$$

Each part of this *and* statement can be solved as we handled Example 2 earlier.

(a)   **(b)**

$$\frac{x-2}{x+3} > -4 \qquad \text{and} \qquad \frac{x-2}{x+3} < 4$$

$$\frac{x-2}{x+3} + 4 > 0 \qquad \text{and} \qquad \frac{x-2}{x+3} - 4 < 0$$

$$\frac{x-2+4(x+3)}{x+3} > 0 \qquad \text{and} \qquad \frac{x-2-4(x+3)}{x+3} < 0$$

$$\frac{x-2+4x+12}{x+3} > 0 \qquad \text{and} \qquad \frac{x-2-4x-12}{x+3} < 0$$

$$\frac{5x+10}{x+3} > 0 \qquad \text{and} \qquad \frac{-3x-14}{x+3} < 0$$

This solution set is   This solution set is
shown in Figure 1.25(a).   shown in Figure 1.25(b).

(a)   (b)

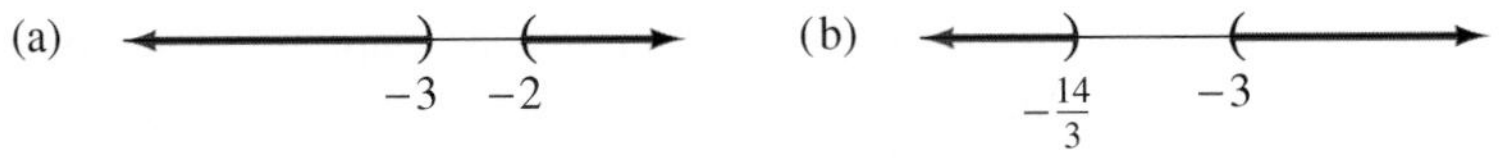

Figure 1.25

The intersection of the two solution sets pictured is the set shown in Figure 1.26.

Therefore, the solution set of $\left|\dfrac{x-2}{x+3}\right| < 4$ is $\left(-\infty, -\dfrac{14}{3}\right) \cup (-2, \infty)$.

Figure 1.26

▶ **Now work Problem 73.**   ■

Yes, Example 9 is a little messy, but it does illustrate the weaving together of previously used techniques to solve a more complicated problem. Don't be in a hurry when doing such problems. First analyze the general approach to be taken and then carry out the details in a neatly organized format to minimize your chances of making careless errors.

Finally, let's show a graphical analysis of Example 9. Figure 1.27 shows the graph of $y = \left|\dfrac{x-2}{x+3}\right| - 4$. Notice that one x intercept is between -5 and -4, and the other intercept appears to be at -2. Furthermore, y is less than zero when x is less than that intercept between -5 and -4, and y is less than zero when x is greater than -2. So our solution set for Example 9 looks good.

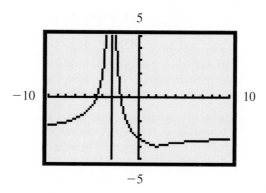

Figure 1.27

CONCEPT QUIZ For Problems 1–3, given $\dfrac{(x + 5)(x - 2)}{x + 1} > 0$, answer true or false.

1. For solving the inequality, three critical values, -5, -1, and 2 would be used to divide the number line into four regions.

2. For all x in the interval $(-5, -1)$, the numerator of the quotient $(x + 5)(x - 2)$ is positive.

3. The quotient $\dfrac{(x + 5)(x - 2)}{x + 1}$ would be undefined if $x = 2$.

For Problems 4–8, answer true or false.

4. The quotient $\dfrac{x + 4}{x - 1}$ will have the same sign (positive or negative) for all values of x for $-4 < x < 1$.

5. The absolute value equation $|2x - 6| = 0$ has two solutions.

6. Every absolute value equation has at least one solution.

7. The solution set for the absolute value inequality $(x - 4)^2 > 0$ is all real numbers.

8. The solution set for the absolute value inequality $(3x - 5)^2 > -2$ is all real numbers.

Problem Set 1.7

For Problems 1–14, solve each inequality and express the solution set in interval notation.

1. $\dfrac{x + 1}{x - 5} > 0$

2. $\dfrac{x + 2}{x + 4} \le 0$

3. $\dfrac{2x - 1}{x + 2} < 0$

4. $\dfrac{3x + 2}{x - 1} > 0$

5. $\dfrac{-x + 3}{3x - 1} \ge 0$

6. $\dfrac{-n - 2}{n + 4} < 0$

7. $\dfrac{n}{n + 2} \ge 3$

8. $\dfrac{x}{x - 1} > 2$

9. $\dfrac{x - 1}{x + 2} < 2$

10. $\dfrac{t - 1}{t - 5} \le 2$

11. $\dfrac{t-3}{t+5} > 1$

12. $\dfrac{x+2}{x+7} < 1$

13. $\dfrac{1}{x-2} < \dfrac{1}{x+3}$

14. $\dfrac{2}{x+1} > \dfrac{3}{x-4}$

For Problems 15–40, solve each equation.

15. $|x-2| = 6$

16. $|x+3| = 4$

17. $\left|x + \dfrac{1}{4}\right| = \dfrac{2}{5}$

18. $\left|x - \dfrac{2}{3}\right| = \dfrac{3}{4}$

19. $|2n-1| = 7$

20. $|2n+1| = 11$

21. $|3x+4| = 5$

22. $|5x-3| = 10$

23. $|7x-1| = -4$

24. $|-2x-1| = 6$

25. $|-3x-2| = 8$

26. $|5x-4| = -3$

27. $|x-3| - 2 = 4$

28. $|2x+1| + 3 = 8$

29. $|5x+1| = 0$

30. $|4x-3| = 0$

31. $|2n+3| - 7 = -2$

32. $|3n-1| - 6 = -4$

33. $\left|\dfrac{3}{k-1}\right| = 4$

34. $\left|\dfrac{-2}{n+3}\right| = 5$

35. $|3x-1| = |2x+3|$

36. $|2x+1| = |4x-3|$

37. $|-2n+1| = |-3n-1|$

38. $|-4n+5| = |-3n-5|$

39. $|x-2| = |x+4|$

40. $|2x-3| = |2x+5|$

For Problems 41–80, solve each inequality and express the solution set in interval notation.

41. $|x| < 6$

42. $|x| \geq 4$

43. $|x| > 8$

44. $|x| \leq 1$

45. $|x| \geq -4$

46. $|x| < -5$

47. $|t-3| > 5$

48. $|n+2| < 1$

49. $|2x-1| \leq 7$

50. $|2x+1| \geq 3$

51. $|3n+2| > 9$

52. $|5n-2| < 2$

53. $|4x-3| < -5$

54. $|2-x| > 1$

55. $|3-2x| < 4$

56. $|4x+5| > -3$

57. $|7x+2| \geq -2$

58. $|-2-x| \leq 5$

59. $|-1-x| \geq 8$

60. $|x-1| + 2 < 4$

61. $|x+3| - 2 < 1$

62. $|x-5| + 4 \leq 2$

63. $|x+4| - 1 > 1$

64. $|x-2| + 3 > 6$

65. $3|x-2| \geq 6$

66. $2|x+1| < 8$

67. $-2|x+1| > -10$

68. $-|x-4| \leq -4$

69. $2|3x-1| - 3 \geq 5$

70. $3|2x+1| + 1 < 7$

71. $-2|x+3| - 1 > 4$

72. $-2|x-1| - 3 > -5$

73. $\left|\dfrac{x+1}{x-2}\right| < 3$

74. $\left|\dfrac{x-1}{x-4}\right| < 2$

75. $\left|\dfrac{x-1}{x+3}\right| > 1$

76. $\left|\dfrac{x+4}{x-5}\right| \geq 3$

77. $\left|\dfrac{n+2}{n}\right| \geq 4$

78. $\left|\dfrac{t+6}{t-2}\right| < 1$

79. $\left|\dfrac{k}{2k-1}\right| \leq 2$

80. $\left|\dfrac{k}{k+2}\right| > 4$

■ ■ ■ **THOUGHTS INTO WORDS**

81. Explain how you would solve the inequality
$\dfrac{x-2}{(x+1)^2} > 0.$

82. Explain how you would solve the inequality
$|3x-7| > -2.$

83. Why is $\left\{\dfrac{3}{2}\right\}$ the solution set for $|2x-3| \leq 0$?

84. Consider the following approach for solving the inequality in Example 2 of this section.

$$\frac{x + 2}{x + 4} \le 3$$

$$(x + 4)\left(\frac{x + 2}{x + 4}\right) \le 3(x + 4)$$

$$x + 2 \le 3x + 12$$

$$-2x \le 10$$

$$x \ge -5$$

Obviously, the solution set that we obtain using this approach differs from what we obtained in the text. What is wrong with this approach? Can we make any adjustments so that this basic approach works?

■ ■ ■ FURTHER INVESTIGATIONS

85. Use the definition of absolute value and prove Property 1.5.

86. Use the definition of absolute value and prove Property 1.6.

87. Use the definition of absolute value and prove Property 1.7.

88. Solve each of the following inequalities by using the definition of absolute value. Do not use Properties 1.6 and 1.7.

   a. $|x + 5| < 11$ **b.** $|x - 4| \le 10$

   c. $|2x - 1| > 7$ **d.** $|3x + 2| \ge 1$

   e. $|2 - x| < 5$ **f.** $|3 - x| > 6$

GRAPHING CALCULATOR ACTIVITIES

89. Show a graphical analysis for Examples 2–8 of this section.

90. Show a graphical analysis for Problems 9–14, 35–40, and 73–80.

Answers to the Concept Quiz

1. True   **2.** False   **3.** False   **4.** True   **5.** False   **6.** False   **7.** False   **8.** True

This chapter covers three large topics: (1) solving equations, (2) solving inequalities, and (3) problem solving.

■ Solving Equations

The following properties are used extensively in the equation-solving process:

1. $a = b$ if and only if $a + c = b + c$.

> Addition property of equality

2. $a = b$ if and only if $ac = bc$, $c \neq 0$.

> Multiplication property of equality

3. If $ab = 0$, then $a = 0$ or $b = 0$.

4. If $a = b$, then $a^n = b^n$, where n is a positive integer.

Remember that applying the fourth property may result in some extraneous solutions, so you *must check* all potential solutions.

The cross-multiplication property of proportions (if $a/b = c/d$, then $ad = bc$) can be used to solve some equations.

Quadratic equations can be solved by (1) factoring, (2) completing the square, or (3) using the quadratic formula, which can be stated as

$$x = \frac{-b \pm \sqrt{b^2 - 4ac}}{2a}$$

The discriminant of a quadratic equation, $b^2 - 4ac$, indicates the nature of the solutions of the equation.

1. If $b^2 - 4ac > 0$, the equation has two unequal real solutions.

2. If $b^2 - 4ac = 0$, the equation has one real solution.

3. If $b^2 - 4ac < 0$, the equation has two complex but nonreal solutions.

If x_1 and x_2 are the solutions of a quadratic equation $ax^2 + bx + c = 0$, then (1) $x_1 + x_2 = -b/a$ and (2) $x_1 x_2 = c/a$. These relationships can be used to check potential solutions.

The property *if* $|x| = k$, *then* $x = k$ *or* $x = -k (k > 0)$ is often helpful for solving equations that involve absolute value.

■ Solving Inequalities

The following properties form a basis for solving inequalities:

1. If $a > b$, then $a + c > b + c$.

2. If $a > b$ and $c > 0$, then $ac > bc$.

3. If $a > b$ and $c < 0$, then $ac < bc$.

To solve compound statements that involve inequalities, we proceed as follows:

1. Solve separately each inequality in the compound statement.

2. If it is a conjunction, the solution set is the intersection of the solution sets of the inequalities.

3. If it is a disjunction, the solution set is the union of the solution sets of the inequalities.

Quadratic inequalities such as $(x + 3)(x - 7) > 0$ can be solved by considering the *sign behavior* of the individual factors.

The following properties play an important role in solving inequalities that involve absolute value.

1. If $|x| < k$, where $k > 0$, then $-k < x < k$.

2. If $|x| > k$, where $k > 0$, then $x > k$ or $x < -k$.

■ Problem Solving

It would be helpful for you to reread the pages of this chapter that pertain to problem solving. Some key problem-solving ideas are illustrated on these pages.

Chapter 1   Review Problem Set

For Problems 1–22, solve each equation.

1. $2(3x - 1) - 3(x - 2) = 2(x - 5)$

2. $\dfrac{n - 1}{4} - \dfrac{2n + 3}{5} = 2$

3. $\dfrac{2}{x + 2} + \dfrac{5}{x - 4} = \dfrac{7}{2x - 8}$

4. $0.07x + 0.12(550 - x) = 56$

5. $(3x - 1)^2 = 16$ **6.** $4x^2 - 29x + 30 = 0$

7. $x^2 - 6x + 10 = 0$ **8.** $n^2 + 4n = 396$

9. $15x^3 + x^2 - 2x = 0$

10. $\dfrac{t + 3}{t - 1} - \dfrac{2t + 3}{t - 5} = \dfrac{3 - t^2}{t^2 - 6t + 5}$

11. $\dfrac{5 - x}{2 - x} - \dfrac{3 - 2x}{2x} = 1$ **12.** $x^4 + 4x^2 - 45 = 0$

13. $2n^{-4} - 11n^{-2} + 5 = 0$

14. $\left(x - \dfrac{2}{x}\right)^2 + 4\left(x - \dfrac{2}{x}\right) = 5$

15. $\sqrt{5 + 2x} = 1 + \sqrt{2x}$

16. $\sqrt{3 + 2n} + \sqrt{2 - 2n} = 3$

17. $\sqrt{3 - t} - \sqrt{3 + t} = \sqrt{t}$

18. $|5x - 1| = 7$ **19.** $|2x + 5| = |3x - 7|$

20. $\left|\dfrac{-3}{n - 1}\right| = 4$ **21.** $x^3 + x^2 - 2x - 2 = 0$

22. $2x^{2/3} + 5x^{1/3} - 12 = 0$

For Problems 23–40, solve each inequality. Express the solution sets using interval notation.

23. $3(2 - x) + 2(x - 4) > -2(x + 5)$

24. $\dfrac{3}{5}x - \dfrac{1}{3} \le \dfrac{2}{3}x + \dfrac{3}{4}$

25. $\dfrac{n - 1}{3} - \dfrac{2n + 1}{4} > \dfrac{1}{6}$

26. $0.08x + 0.09(700 - x) \ge 59$

27. $-16 \le 7x - 2 \le 5$ **28.** $5 > \dfrac{3y + 4}{2} > 1$

29. $x^2 - 3x - 18 < 0$ **30.** $n^2 - 5n \ge 14$

31. $(x - 1)(x - 4)(x + 2) < 0$

32. $\dfrac{x + 4}{2x - 3} \le 0$ **33.** $\dfrac{5n - 1}{n - 2} > 0$

34. $\dfrac{x - 1}{x + 3} \ge 2$ **35.** $\dfrac{t + 5}{t - 4} < 1$

36. $|4x - 3| > 5$ **37.** $|3x + 5| \le 14$

38. $|-3 - 2x| < 6$ **39.** $\left|\dfrac{x - 1}{x}\right| > 2$

40. $\left|\dfrac{n + 1}{n + 2}\right| < 1$

For Problems 41–56, solve each problem.

41. The sum of three consecutive odd integers is 31 less than four times the largest integer. Find the integers.

42. The ratio of men to women on a crew team is 7 to 2. If there is a total of 63 members on the crew team, find the number of men on the team.

43. The perimeter of a rectangle is 38 centimeters and its area is 84 square centimeters. Find the dimensions of the rectangle.

44. A sum of money amounting to $13.55 consists of nickels, dimes, and quarters. There are three times as many dimes as nickels and three fewer quarters than dimes. How many coins of each denomination are there?

45. A retailer has some computer flash drives that cost him $14 each. He wants to sell them to make a profit of 30% of the selling price. What price should he charge for the flash drives?

46. How many gallons of a solution of glycerine and water containing 55% glycerine should be added to 15 gallons of a 20% solution to give a 40% solution?

47. The sum of the present ages of Rosie and her mother is 47 years. In 5 years, Rosie will be one-half as old as her mother at that time. Find the present ages of both Rosie and her mother.

48. Kelly invested $8000, part of it at 7% and the remainder at 8.5%. Her total yearly interest from the two investments was $627.50. How much did she invest at each rate?

49. Regina has scores of 93, 88, 89, and 95 on her first four math exams. What score must she get on the fifth exam to have an average of 92 or higher for the five exams?

50. It takes Angie twice as long as Amy to clean a house. How long does it take each girl by herself if they can clean the house together in 3 hours?

51. Russ started to mow the lawn, a task that usually takes him 40 minutes. After he had been working for 15 minutes, his friend Jay came along with his mower and began to help Russ. Working together, they finished the lawn in 10 minutes. How long would it have taken Jay to mow the lawn by himself?

52. Barry bought a number of shares of stock for $600. A week later the value of the stock increased $3 per share, and he sold all but 10 shares and regained his original investment of $600. How many shares did he sell and at what price per share?

53. Larry drove 156 miles in one hour more than it took Mike to drive 108 miles. Mike drove at an average rate of 2 miles per hour faster than Larry. How fast did each one travel?

54. It takes Bill 2 hours longer to do a certain job than it takes Cindy. They worked together for 2 hours; then Cindy left and Bill finished the job in 1 hour. How long would it take each of them to do the job alone?

55. One leg of a right triangle is 5 centimeters longer than the other leg. The hypotenuse is 25 centimeters long. Find the length of each leg.

56. The area of a rectangle is 35 square inches. If both the length and width are increased by 3 inches, the area is increased by 45 square inches. Find the dimensions of the original rectangle.

Chapter 1   Test

For Problems 1–14, solve each equation.

1. $3(2x - 1) - 4(x + 2) = -7$

2. $10x^2 + 13x - 3 = 0$

3. $(5x + 2)^2 = 25$

4. $\dfrac{3n + 4}{4} - \dfrac{2n - 1}{10} = \dfrac{11}{20}$

5. $2x^2 - x + 4 = 0$

6. $(n - 2)(n + 7) = -18$

7. $0.06x + 0.08(1400 - x) = 100$

8. $|3x - 4| = 7$

9. $3x^2 - 2x - 2 = 0$

10. $3x^3 + 21x^2 - 54x = 0$

11. $\dfrac{x}{2x + 1} - 1 = \dfrac{-4}{7(x - 2)}$

12. $\sqrt{2x} = x - 4$

13. $\sqrt{x + 1} + 2 = \sqrt{x}$

14. $2n^{-2} + 5n^{-1} - 12 = 0$

For Problems 15–21, solve each inequality and express the solution set using interval notation.

15. $2(x - 1) - 3(3x + 1) \geq -6(x - 5)$

16. $\dfrac{x - 2}{6} - \dfrac{x + 3}{9} > -\dfrac{1}{2}$

17. $|6x - 4| < 10$

18. $|4x + 5| \geq 6$

19. $2x^2 - 9x - 5 \leq 0$

20. $\dfrac{3x - 1}{x + 2} > 0$

21. $\dfrac{x - 2}{x + 6} \geq 3$

For Problems 22–25, solve each problem.

22. How many cups of grapefruit juice must be added to 30 cups of a punch that contains 8% grapefruit juice to obtain a punch that is 10% grapefruit juice?

23. Lian can ride her bike 60 miles in one hour less time than it takes Tasya to ride 60 miles. Lian's rate is 3 miles per hour faster than Tasya's rate. Find Lian's rate.

24. Abdul bought a number of shares of stock for a total of $3000. Three months later the stock had increased in value by $5 per share, and he sold all but 50 shares and regained his original investment of $3000. How many shares did he sell?

25. The perimeter of a rectangle is 46 centimeters and its area is 126 square centimeters. Find the dimensions of the rectangle.

2

Coordinate Geometry and Graphing Techniques

René Descartes, a French philosopher and mathematician, developed a system to locate a point on a plane. That system is our current rectangular coordinate grid, called the Cartesian coordinate system, that we use for graphing.

© Erich Lessing/Art Resource, NY

A section of a certain highway has a 2% grade. How many feet does it rise in a horizontal distance of 1 mile? The equation $\dfrac{2}{100} = \dfrac{Y}{5280}$ can be used to determine that the amount of rise in a horizontal distance of 1 mile is 105.6 feet. This equation is based on the concept of slope.

René Descartes, a French mathematician of the 17th century, was able to transform geometric problems into an algebraic setting so that he could use the tools of algebra to solve the problems. This merging of algebraic and geometric ideas is the foundation of a branch of mathematics called **analytic geometry,** now more commonly called **coordinate geometry.** Basically, there are two kinds of problems in coordinate geometry: (1) given an algebraic equation, find its geometric graph, and (2) given a set of conditions pertaining to a geometric graph, find its algebraic equation. We will discuss problems of both types in this chapter.

René Descartes lived only to the age of 54, but in his short life he made some great contributions to both mathematics and philosophy. You may find it interesting to browse through some of his contributions on the Internet.

2.1   Coordinate Geometry

Objectives

■ Find the distance between two points on a number line.

■ On the number line, find the coordinate of a point located a specified distance between two points.

■ Find the distance between two points in the rectangular coordinate system.

■ Find the midpoint of a line segment in the rectangular coordinate system.

■ In the rectangular coordinate system, find the coordinate of a point located a specified distance between two points.

■ Solve application problems involving the length of line segments.

Recall from Section 0.1 that the real number line exhibits a one-to-one correspondence between the set of real numbers and the points on a line (see Figure 2.1). That is, to each real number there corresponds one and only one point on the line, and to each point on the line there corresponds one and only one real number. The number that corresponds to a particular point on the line is called the **coordinate** of that point. Also recall that the distance *between* any two points with coordinates x_1 and x_2 can be found by using either $|x_2 - x_1|$ or $|x_1 - x_2|$.

Figure 2.1

      Suppose that on the number line we want to know the distance *from −2 to 6*. The *from–to* vocabulary implies a **directed distance,** which is $6 - (-2) = 8$ units. In other words, it is 8 units in a *positive direction* from −2 to 6. Likewise, the distance from 9 to −4 is $-4 - 9 = -13$; it is 13 units in a *negative direction*. In general, if x_1 and x_2 are the coordinates of two points on the number line, then the distance **from x_1 to x_2** is given by $x_2 - x_1$, and the distance **from x_2 to x_1** is given by $x_1 - x_2$.

      Sometimes it is necessary to find the coordinate of a point located somewhere between the two given points. For example, in Figure 2.2 suppose that we want to find the coordinate, x, of the point located two-thirds of the distance *from 2 to 8*. Because the total distance from 2 to 8 is $8 - 2 = 6$ units, we can start at 2 and move $\frac{2}{3}(6) = 4$ units toward 8. Thus

$$x = 2 + \frac{2}{3}(6) = 2 + 4 = 6$$

Figure 2.2

The following examples further illustrate the process of finding the coordinate of a point somewhere between two given points (see Figure 2.3).

Problem	**Solution**
a. Three-fourths of the distance from -2 to 10	$x = -2 + \dfrac{3}{4}[10 - (-2)]$
	$= -2 + \dfrac{3}{4}(12)$
	$= 7$

Figure 2.3(a)

b. Two-fifths of the distance from -1 to 7	$x = -1 + \dfrac{2}{5}[7 - (-1)]$
	$= -1 + \dfrac{2}{5}(8)$
	$= \dfrac{11}{5}$

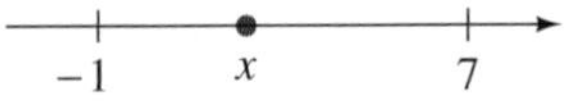

Figure 2.3(b)

c. One-third of the distance from 9 to 1	$x = 9 + \dfrac{1}{3}(1 - 9)$
	$= 9 + \dfrac{1}{3}(-8)$
	$= \dfrac{19}{3}$

Figure 2.3(c)

d. a/b of the distance from x_1 to x_2	$x = x_1 + \dfrac{a}{b}(x_2 - x_1)$

Figure 2.3(d)

Part d indicates that a general formula can be developed for this type of problem. However, it may be easier to remember the basic approach than it is to memorize the formula.

As we saw in Chapter 1, the real number line provides a geometric model for graphing solutions of algebraic equations and inequalities involving *one variable*. For example, the solutions of $x > 2$ or $x \leq -1$ are graphed in Figure 2.4.

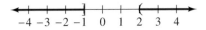

Figure 2.4

■ Rectangular Coordinate System

Recall from Section 0.1 that a pair of real number lines, perpendicular to each other at the point associated with zero on both lines (see Figure 2.5), can be used to exhibit a one-to-one correspondence between pairs of real numbers and points in a plane. Figure 2.6 shows examples of this correspondence.

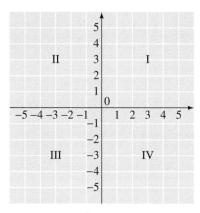

Figure 2.5

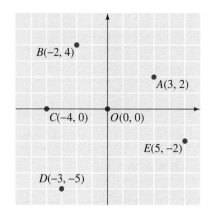

Figure 2.6

Remark: We used the notation $(-2, 4)$ in Chapter 1 to indicate an interval of the real number line. Now we are using the same notation to indicate an ordered pair of real numbers. This double meaning should not be confusing because the context of the material will definitely indicate the meaning at a particular time. Throughout this chapter we will be using the ordered-pair interpretation.

■ Distance Between Two Points

As we work with the rectangular coordinate system, it is sometimes necessary to express the length of certain line segments. In other words, we need to be able to find the *distance* between two points. Let's first consider two specific examples and then develop a general distance formula.

EXAMPLE 1

Find the distance between the points $A(2, 2)$ and $B(5, 2)$ and also between the points $C(-2, 5)$ and $D(-2, -4)$.

Solution

Let's plot the points and draw $\overline{AB}$ and $\overline{CD}$ as in Figure 2.7. (The symbol $\overline{AB}$ denotes the line segment with endpoints A and B.) Because $\overline{AB}$ is parallel to the horizontal axis, its length can be expressed as $|5 - 2|$ or $|2 - 5|$. Thus the length of $\overline{AB}$ (we will use the notation AB to represent the length of $\overline{AB}$) is $AB = 3$ units. Likewise, because $\overline{CD}$ is parallel to the vertical axis, we obtain $CD = |5 - (-4)| = 9$ units.

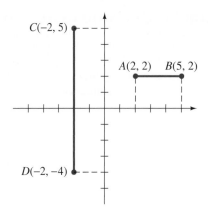

Figure 2.7

▶ **Now work Problem 5.**   ■

EXAMPLE 2   Find the distance between the points $A(2, 3)$ and $B(5, 7)$.

Solution

Let's plot the points and form a right triangle using point D, as indicated in Figure 2.8. Note that the coordinates of point D are $(5, 3)$. Because $\overline{AD}$ is parallel to the horizontal axis, as in Example 1, we have $AD = |5 - 2| = 3$ units. Likewise, $\overline{DB}$ is parallel to the vertical axis, and therefore $DB = |7 - 3| = 4$ units. Applying the Pythagorean theorem, we obtain

$$(AB)^2 = (AD)^2 + (DB)^2$$
$$= 3^2 + 4^2$$
$$= 9 + 16$$
$$= 25$$

Thus

$$AB = \sqrt{25} = 5 \text{ units}$$

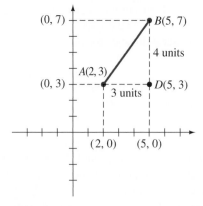

Figure 2.8

▶ **Now work Problem 15.**   ■

Let $P_1(x_1, y_1)$ and $P_2(x_2, y_2)$ represent any two points in the xy plane. We can form a right triangle using point R, as indicated in Figure 2.9. The coordinates of the vertex of the right angle at point R are (x_2, y_1). The length of $\overline{P_1R}$ is $|x_2 - x_1|$, and the length of $\overline{RP_2}$ is $|y_2 - y_1|$. We let d represent the length of $\overline{P_1P_2}$ and apply the Pythagorean theorem to obtain

$$d^2 = |x_2 - x_1|^2 + |y_2 - y_1|^2$$

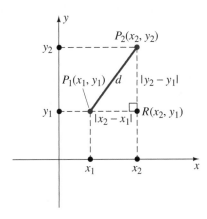

Figure 2.9

Because $|a|^2 = a^2$ for any real number a, the distance formula can be stated

$$d = \sqrt{(x_2 - x_1)^2 + (y_2 - y_1)^2}$$

It makes no difference which point you call P_1 and which you call P_2. Also, remember that if you forget the formula, there is no need to panic: form a right triangle and apply the Pythagorean theorem as we did in Example 2.

Let's consider some examples that illustrate the use of the distance formula.

EXAMPLE 3

Find the distance between $(-2, 5)$ and $(1, -1)$.

Solution

Let $(-2, 5)$ be P_1 and $(1, -1)$ be P_2. Use the distance formula to obtain

$$\begin{aligned}
d &= \sqrt{(x_2 - x_1)^2 + (y_2 - y_1)^2} \\
&= \sqrt{[1 - (-2)]^2 + (-1 - 5)^2} \\
&= \sqrt{3^2 + (-6)^2} \\
&= \sqrt{9 + 36} \\
&= \sqrt{45} = 3\sqrt{5}
\end{aligned}$$

The distance between the two points is $3\sqrt{5}$ units.

▶ **Now work Problem 17.**   ■

In Example 3 note the simplicity of the approach when we use the distance formula. No diagram was needed; we merely plugged in the values and did the computation. However, many times a figure *is* helpful in the analysis of the problem, as we will see in the next example.

E X A M P L E   4

Verify that the points $(-3, 6)$, $(3, 4)$, and $(1, -2)$ are vertices of an isosceles triangle. (An isosceles triangle has two sides of the same length.)

Solution

Let's plot the points and draw the triangle (see Figure 2.10). The lengths d_1, d_2, and d_3 can all be found by using the distance formula.

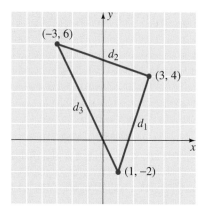

Figure 2.10

$$d_1 = \sqrt{(3-1)^2 + [4-(-2)]^2}$$
$$= \sqrt{4 + 36}$$
$$= \sqrt{40} = 2\sqrt{10}$$
$$d_2 = \sqrt{(-3-3)^2 + (6-4)^2}$$
$$= \sqrt{36 + 4}$$
$$= \sqrt{40} = 2\sqrt{10}$$
$$d_3 = \sqrt{(-3-1)^2 + [6-(-2)]^2}$$
$$= \sqrt{16 + 64}$$
$$= \sqrt{80} = 4\sqrt{5}$$

Because $d_1 = d_2$, it is an isosceles triangle.

▶ **Now work Problem 41.**

■ Points of Division of a Line Segment

Earlier in this section we discussed the process of finding the coordinate of a point on a number line, given that it is located somewhere between two other points on the line. This same type of problem can occur in the xy plane, and the approach we used earlier can be extended to handle it. Let's consider some examples.

EXAMPLE   5

Find the coordinates of the point P, which is two-thirds of the distance from $A(1, 2)$ to $B(7, 5)$.

Solution

In Figure 2.11 we plotted the given points A and B and completed a figure to help us analyze the problem. To find the coordinates of point P, we can proceed as follows. Point D is two-thirds of the distance from A to C because parallel lines cut off

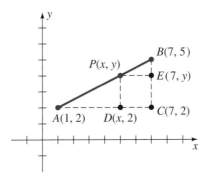

Figure 2.11

proportional segments on every transversal that intersects the lines. Therefore, because $\overline{AC}$ is parallel to the x axis, it can be treated as a segment of the number line (see Figure 2.12). Thus we have

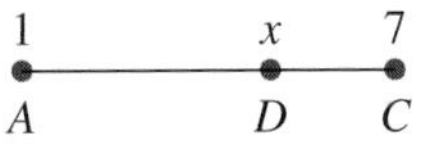

$$x = 1 + \frac{2}{3}(7 - 1) = 1 + \frac{2}{3}(6) = 5$$

Similarly, $\overline{CB}$ is parallel to the y axis, so it can also be treated as a segment of the number line (see Figure 2.13). Thus we obtain

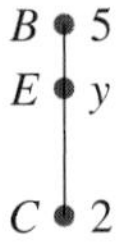

$$y = 2 + \frac{2}{3}(5 - 2)$$

$$= 2 + \frac{2}{3}(3) = 4$$

The point P has the coordinates $(5, 4)$.

Figure 2.12

Figure 2.13

▶   **Now work Problem 27.**   ■

EXAMPLE 6

Find the coordinates of the midpoint of the line segment determined by the points $P_1(x_1, y_1)$ and $P_2(x_2, y_2)$.

Solution

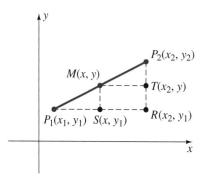

Figure 2.14

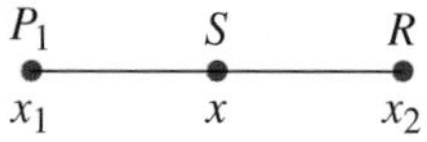

Figure 2.15

Figure 2.14 helps with the analysis of the problem. The line segment $\overline{P_1R}$ is parallel to the x axis, and $S(x, y_1)$ is the midpoint of $\overline{P_1R}$ (see Figure 2.15). Thus we can determine the x coordinate of S:

$$x = x_1 + \frac{1}{2}(x_2 - x_1)$$

$$= x_1 + \frac{1}{2}x_2 - \frac{1}{2}x_1$$

$$= \frac{1}{2}x_1 + \frac{1}{2}x_2 = \frac{x_1 + x_2}{2}$$

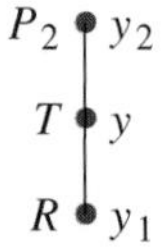

Figure 2.16

Similarly, $\overline{RP_2}$ is parallel to the y axis, and $T(x_2, y)$ is the midpoint of $\overline{RP_2}$ (see Figure 2.16). Therefore we can calculate the y coordinate of T:

$$y = y_1 + \frac{1}{2}(y_2 - y_1)$$

$$= y_1 + \frac{1}{2}y_2 - \frac{1}{2}y_1$$

$$= \frac{1}{2}y_1 + \frac{1}{2}y_2 = \frac{y_1 + y_2}{2}$$

Thus the coordinates of the midpoint of a line segment determined by $P_1(x_1, y_1)$ and $P_2(x_2, y_2)$ are

$$\left(\frac{x_1 + x_2}{2}, \frac{y_1 + y_2}{2} \right)$$

∎

EXAMPLE 7

Find the coordinates of the midpoint of the line segment determined by the points $(-2, 4)$ and $(6, -1)$.

Solution

Using the midpoint formula, we obtain

$$\left(\frac{x_1 + x_2}{2}, \frac{y_1 + y_2}{2}\right) = \left(\frac{-2 + 6}{2}, \frac{4 + (-1)}{2}\right)$$

$$= \left(\frac{4}{2}, \frac{3}{2}\right)$$

$$= \left(2, \frac{3}{2}\right)$$

▶ **Now work Problem 21.**   ■

We want to emphasize two ideas that emerge from Examples 5, 6, and 7. If we want to find a point of division of a line segment, then we use the same approach as in Example 5. However, for the special case of the midpoint, the formula developed in Example 6 is convenient to use.

CONCEPT QUIZ   For Problems 1–8, answer true or false.

1. The statement "the distance from x_1 to x_2" implies a directed distance.

2. For a directed distance, the distance could be negative.

3. The symbol $\overline{EF}$ denotes the line segment with endpoints E and F.

4. When applying the distance formula $\sqrt{(x_2 - x_1)^2 + (y_2 - y_1)^2}$ to find the distance between two points, you can designate either of the points as P_1.

5. The formula $\left(\dfrac{x_1 + x_2}{2}, \dfrac{y_1 + y_2}{2}\right)$ for determining the coordinates of the midpoint between two points can be described as finding the average of the x coordinates and the average of the y coordinates.

6. The distance formula $\sqrt{(x_2 - x_1)^2 + (y_2 - y_1)^2}$ can be derived by applying the Pythagorean theorem to a right triangle formed by the points $P_1(x_1, y_1)$, $P_2(x_2, y_2)$, and $R\ (x_1, y_2)$.

7. The distance between the two points (a, y) and (b, y) is $|b - a|$.

8. An isosceles triangle has three sides of the same length.

Problem Set 2.1

For Problems 1–4, find the indicated distances on a number line.

1. From -4 to 6

2. From 5 to -14

3. From -6 to -11

4. From -7 to 10

For Problems 5–8, find the distance between the points A and B.

5. $A\ (3, -7), B\ (3, -2)$

6. $A\ (-1, 6), B\ (-1, -2)$

7. $A\ (-4, 8), B\ (5, 8)$

8. $A\ (3, -4), B\ (-1, -4)$

For Problems 9–14, find the coordinate of the indicated point on a number line.

9. Two-thirds of the distance from 1 to 10

10. Three-fourths of the distance from -2 to 14

11. One-third of the distance from -3 to 7

12. Two-fifths of the distance from -5 to 6

13. Three-fifths of the distance from -1 to -11

14. Five-sixths of the distance from 3 to -7

For Problems 15–20, find the distance between the points A and B.

15. A (4, 1), B (7, 5)

16. A (-7, 5), B (-2, 17)

17. A (-1, 4), B (3, -2)

18. A (3, -5), B (-4, -2)

19. $A(2, 1)$, $B(10, 7)$

20. $A(-2, -1)$, $B(7, 11)$

For Problems 21–26, find the midpoint of the line segment determined by the given points.

21. $A(1, -1)$, $B(3, -4)$

22. $A(-5, 2)$, $B(-1, 6)$

23. $A(6, -4)$, $B(9, -7)$

24. $A(-3, 3)$, $B(0, -3)$

25. $A\left(\dfrac{1}{2}, \dfrac{1}{3}\right)$, $B\left(-\dfrac{1}{3}, \dfrac{3}{2}\right)$

26. $A\left(-\dfrac{3}{4}, 2\right)$, $B\left(-1, -\dfrac{5}{4}\right)$

For Problems 27–36, find the coordinates of the indicated point in the xy plane.

27. One-third of the distance from (2, 3) to (5, 9)

28. Two-thirds of the distance from (1, 4) to (7, 13)

29. Two-fifths of the distance from (-2, 1) to (8, 11)

30. Three-fifths of the distance from (2, -3) to (-3, 8)

31. Five-eighths of the distance from (-1, -2) to (4, -10)

32. Seven-eighths of the distance from (-2, 3) to (-1, -9)

33. Five-sixths of the distance from (-7, 2) to (-1, -4)

34. Three-fourths of the distance from (-1, -6) to (-5, 2)

35. Three-eighths of the distance from (6, 8) to (-2, 4)

36. One-third of the distance from (4, -1) to (-3, -5)

For Problems 37–52, solve each of the problems.

37. Find the coordinates of the point that is one-fourth of the distance from (2, 4) to (10, 13) by (a) using the midpoint formula twice and (b) using the same approach as for Problems 27–36.

38. If one endpoint of a line segment is (-6, 4), and the midpoint of the segment is (-2, 7), find the other endpoint.

39. Use the distance formula to verify that the points (-2, 7), (2, 1), and (4, -2) lie on a straight line.

40. Use the distance formula to verify that the points (-3, 8), (7, 4), and (5, -1) are vertices of a right triangle.

41. Verify that the points (0, 3), (2, -3), and (-4, -5) are vertices of an isosceles triangle.

42. Verify that the points (7, 12) and (11, 18) divide the line segment joining (3, 6) and (15, 24) into three segments of equal length.

43. Find the perimeter of the triangle with vertices (-6, -4), (0, 8), and (6, 5).

44. Verify that (-4, 9), (8, 4), (3, -8), and (-9, -3) are vertices of a square.

45. Verify that the points (4, -5), (6, 7), and (-8, -3) lie on a circle that has its center at (-1, 2).

46. Suppose that (-2, 5), (6, 3) and (-4, -1) are three vertices of a parallelogram. How many possibilities are there for the fourth vertex? Find the coordinates of each of these points. [*Hint:* The diagonals of a parallelogram bisect each other.]

47. Find x such that the line segment determined by (x, -2) and (-2, -14) is 13 units long.

48. Consider the triangle with vertices (4, -6), (2, 8), and (-4, 2). Verify that the medians of this triangle intersect at a point that is two-thirds of the distance from a vertex to the midpoint of the opposite side. (A **median** of the triangle is the line segment determined by a vertex and the midpoint of the opposite side. Every triangle has three medians.)

49. Consider the line segment determined by $A(-1, 2)$ and $B(5, 11)$. Find the coordinates of a point P such that $\overline{AP}/\overline{PB} = 2/1$.

50. Verify that the midpoint of the hypotenuse of the right triangle formed by the points $A(4, 0)$, $B(0, 0)$, and $C(0, 6)$ is the same distance from all three vertices.

51. Consider the parallelogram determined by the points $A(1, 1)$, $B(5, 1)$, $C(6, 4)$, and $D(2, 4)$. Verify that the diagonals of this parallelogram bisect each other.

52. Consider the quadrilateral determined by the points $A(5, -3)$, $B(3, 4)$, $C(-2, 1)$, and $D(-1, -2)$. Verify that the line segments joining the midpoints of the opposite sides of this quadrilateral bisect each other.

■■■ THOUGHTS INTO WORDS

53. Consider the line segment determined by the two endpoints $A(2, 1)$ and $B(5, 10)$. Describe how you would find the coordinates of the point that is two-thirds of the distance from A to B. Then describe how you would find the point that is two-thirds of the distance from B to A.

54. How would you define the term "coordinate geometry" to a group of elementary algebra students?

■■■ FURTHER INVESTIGATIONS

55. The tools of coordinate geometry can be used to prove various geometric properties. For example, consider the following way of proving that the diagonals of a rectangle are equal in length.

First we draw a rectangle and display it on coordinate axes by using a convenient position for the origin. Now we can use the distance formula to find the lengths of the diagonals $\overline{AC}$ and $\overline{BD}$ (see Figure 2.17):

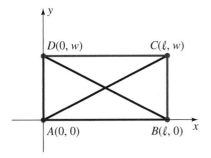

Figure 2.17

$$AC = \sqrt{(l - 0)^2 + (w - 0)^2} = \sqrt{l^2 + w^2}$$

$$BD = \sqrt{(0 - l)^2 + (w - 0)^2} = \sqrt{l^2 + w^2}$$

Thus $AC = BD$, and we have proved that the diagonals are equal in length. Prove each of the following:

a. The diagonals of an isosceles trapezoid are equal in length.

b. The line segment joining the midpoints of two sides of a triangle is equal in length to one-half of the third side.

c. The midpoint of the hypotenuse of a right triangle is equally distant from all three vertices.

d. The diagonals of a parallelogram bisect each other.

e. The line segments joining the midpoints of the opposite sides of a quadrilateral bisect each other.

f. The medians of a triangle intersect at a point that is two-thirds of the distance from a vertex to the midpoint of the opposite side. (See Problem 48.)

Answers to the Concept Quiz

1. True   **2.** True   **3.** True   **4.** True   **5.** True   **6.** True   **7.** True   **8.** False

2.2 Graphing Techniques: Linear Equations and Inequalities

Objectives

■ Graph linear equations.

■ Find the x and y intercepts for the graph of a linear equation.

■ Graph linear inequalities.

We have been showing different graphs throughout the text to emphasize the relationship between geometric graphs and solutions of algebraic equations. You have not been required to actually sketch any graphs. In this chapter we will discuss some specific graphing techniques and get you involved in the graphing process.

First, let's briefly review some basic ideas by considering the solutions for the equation $y = x + 2$. A **solution** of an equation in two variables is an ordered pair of real numbers that satisfy the equation. When the variables are x and y, the ordered pairs are of the form (x, y). We see that $(1, 3)$ is a solution for $y = x + 2$ because replacing x by 1 and y by 3 yields a true numerical statement: $3 = 1 + 2$. Likewise, $(-2, 0)$ is a solution because $0 = -2 + 2$ is a true statement. We can find an infinite number of pairs of real numbers that satisfy $y = x + 2$ by arbitrarily choosing values for x and, for each value of x chosen, determining a corresponding value for y. Let's use a table to record some of the solutions for $y = x + 2$.

Choose x	Determine y from $y = x + 2$	Solutions for $y = x + 2$
0	2	$(0, 2)$
1	3	$(1, 3)$
3	5	$(3, 5)$
5	7	$(5, 7)$
-2	0	$(-2, 0)$
-4	-2	$(-4, -2)$
-6	-4	$(-6, -4)$

Plotting the points associated with the ordered pairs from the table produces Figure 2.18(a). The straight line that contains the points is called the **graph of the equation** $y = x + 2$ [see Figure 2.18(b)].

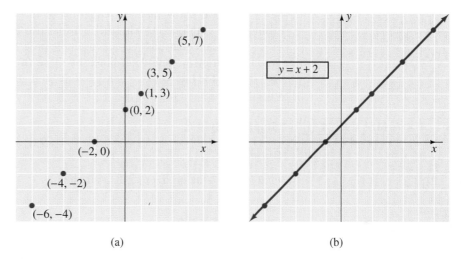

(a)   (b)

Figure 2.18

■ Graphing Linear Equations

Probably the most valuable graphing technique is the ability to recognize the kind of graph that is produced by a particular type of equation. For example, from previous mathematics courses you may remember that any equation of the form $Ax + By = C$, where A, B, and C are constants (A and B not both zero) and x and y are variables, is a **linear equation** and that its graph is a **straight line.** Two comments about this description of a linear equation should be made. First, the choice of x and y as variables is arbitrary; any two letters can be used to represent the variables. For example, an equation such as $3r + 2s = 9$ is also a linear equation in two variables. To avoid constantly changing the labeling of the coordinate axes when graphing equations, we will use the same two variables, x and y, in all equations. Second, the statement "any equation of the form" $Ax + By = C$" technically means any equation of that form or equivalent to that form. For example, the equation $y = 2x - 1$ is equivalent to $-2x + y = -1$ and therefore is linear and produces a straight-line graph.

Before we graph some linear equations, let's define in general the **intercepts** of a graph.

The x coordinates of the points that a graph has in common with the x axis are called the **x intercepts** of the graph. (To compute the x intercepts, let $y = 0$ and solve for x.)

The y coordinates of the points that a graph has in common with the y axis are called the **y intercepts** of the graph. (To compute the y intercepts, let $x = 0$ and solve for y.)

Once we know that any equation of the form $Ax + By = C$ produces a straight-line graph, along with the fact that two points determine a straight line, graphing linear equations becomes a simple process. We can find two points on the graph and draw the line determined by those two points. Usually the two points that involve the intercepts are easy to find, and generally it's a good idea to plot a third point to serve as a check.

EXAMPLE 1

Graph $3x - 2y = 6$.

Solution

First let's find the intercepts. If $x = 0$, then

$$3(0) - 2y = 6$$
$$-2y = 6$$
$$y = -3$$

Therefore the point $(0, -3)$ is on the line. If $y = 0$, then

$$3x - 2(0) = 6$$
$$3x = 6$$
$$x = 2$$

Thus the point $(2, 0)$ is also on the line. Now let's find a check point. If $x = -2$, then

$$3(-2) - 2y = 6$$
$$-6 - 2y = 6$$
$$-2y = 12$$
$$y = -6$$

Thus the point $(-2, -6)$ is also on the line. In Figure 2.19 the three points are plotted, and the graph of $3x - 2y = 6$ is drawn.

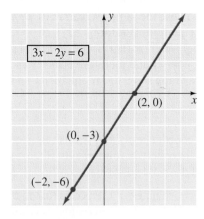

Figure 2.19

▶ **Now work Problem 1.**   ■

Note in Example 1 that we did not solve the given equation for y in terms of x or for x in terms of y. Because we know the graph is a straight line, there is no need for an extensive table of values; thus there is no need to change the form of the original equation. Furthermore, the point $(-2, -6)$ served as a check point. If it had not been on the line determined by the two intercepts, then we would have known that we had made an error in finding the intercepts.

EXAMPLE 2

Graph $y = -2x$.

Solution

If $x = 0$, then $y = -2(0) = 0$, so the origin $(0, 0)$ is on the line. Because both intercepts are determined by the point $(0, 0)$, another point is necessary to determine the line. Then a third point should be found as a check point. The graph of $y = -2x$ is shown in Figure 2.20.

x	y
0	0
1	-2
-1	2

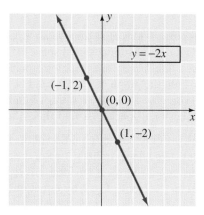

Figure 2.20

 Now work Problem 11. ∎

Example 2 illustrates the general concept that for the form $Ax + By = C$, if $C = 0$, then the line contains the origin. Stated another way, *the graph of any equation of the form $y = kx$, where k is any real number, is a straight line containing the origin.*

EXAMPLE 3

Graph $x = 2$.

Solution

Because we are considering linear equations *in two variables*, the equation $x = 2$ is equivalent to $x + 0(y) = 2$. Any value of y can be used, but the x value must always be 2. Therefore some of the solutions are $(2, 0)$, $(2, 1)$, $(2, 2)$, $(2, -1)$, and $(2, -2)$. The graph of $x = 2$ is the vertical line shown in Figure 2.21.

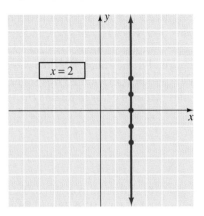

Figure 2.21

 Now work Problem 13.   ∎

Remark: It is important to realize that we are presently graphing **equations in two variables** (graphing in two-dimensional space). Thus as shown in Example 3, the graph of $x = 2$ is a line. If we were graphing **equations in one variable** (graphing on a number line), then the graph of $x = 2$ would be a dot at 2. In subsequent mathematics courses, you may do some graphing of **equations in three variables** (graphing in three-dimensional space). At that time, the graph of $x = 2$ will be a plane.

In general, the graph of any equation of the form $Ax + By = C$, where $A = 0$ or $B = 0$ (not both), is a line parallel to one of the axes. More specifically, **any equation of the form $x = a$,** where a is any nonzero real number, is a *line parallel to the y axis* having an x intercept of a. **Any equation of the form $y = b$,** where b is a nonzero real number, is a *line parallel to the x axis* having a y intercept of b.

∎ Graphing Linear Inequalities

Linear inequalities in two variables are of the form $Ax + By > C$ or $Ax + By < C$, where A, B, and C are real numbers. (*Combined linear equality and inequality statements* are of the form $Ax + By \geq C$ or $Ax + By \leq C$.) Graphing linear inequalities is almost as easy as graphing linear equations. The following discussion will lead us to a simple, step-by-step process.

Let's consider the following equation and related inequalities:

$$x + y = 2 \qquad x + y > 2 \qquad x + y < 2$$

The straight line in Figure 2.22 is the graph of $x + y = 2$. The line divides the plane into two half-planes, one above the line and one below the line. For each point in the half-plane *above* the line, the ordered pair (x, y) associated with the point satisfies the inequality $x + y > 2$. For example, the ordered pair $(3, 4)$ produces the true statement $3 + 4 > 2$. Likewise, for each point in the half-plane *below* the line, the ordered pair (x, y) associated with the point satisfies the inequality $x + y < 2$. For example, $(-3, 1)$ produces the true statement $-3 + 1 < 2$.

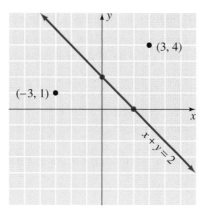

Figure 2.22

Now let's use these ideas to help graph some inequalities.

EXAMPLE 4

Graph $x - 2y > 4$.

Solution

First, we graph $x - 2y = 4$ as a dashed line, because equality is not included in $x - 2y > 4$ (see Figure 2.23). Second, because *all* of the points in a specific half-plane satisfy either $x - 2y > 4$ or $x - 2y < 4$, we must try a *test point*. For example, consider the origin:

$x - 2y > 4$ becomes $0 - 2(0) > 4$, which is a false statement

Because the ordered pairs in the half-plane containing the origin do not satisfy $x - 2y > 4$, the ordered pairs in the other half-plane must satisfy it. Therefore the graph of $x - 2y > 4$ is the half-plane *below* the line, as indicated by the shaded portion in Figure 2.24.

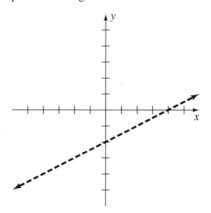

Figure 2.23

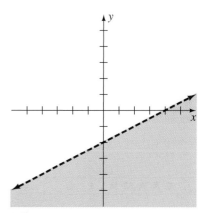

Figure 2.24

▶ **Now work Problem 17.**   ■

To graph a linear inequality, we suggest the following steps:

1. Graph the corresponding equality. Use a solid line if equality is included in the original statement and a dashed line if equality is not included.
2. Choose a *test point* not on the line and substitute its coordinates into the inequality. (The origin is a convenient point if it is not on the line.)
3. The graph of the original inequality is
   a. the half-plane containing the test point if the inequality is satisfied by that point;
   b. the half-plane not containing the test point if the inequality is not satisfied by the point.

EXAMPLE 5

Graph $2x + 3y \geq -6$.

Solution

Step 1   Graph $2x + 3y = -6$ as a solid line (see Figure 2.25).

Step 2   Choose the origin as a test point:

$$2x + 3y \geq -6 \quad \text{becomes} \quad 2(0) + 3(0) \geq -6$$

which is true.

Step 3   The test point satisfies the given inequality, so all points in the same half-plane as the test point satisfy it. The graph of $2x + 3y \geq -6$ is the line and the half-plane above the line (see Figure 2.25).

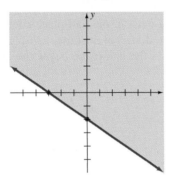

Figure 2.25

▶ **Now work Problem 21.**   ∎

EXAMPLE 6

Use a graphing utility to obtain a graph of the line $2.1x + 5.3y = 7.9$.

Solution

First, we need to solve the equation for y in terms of x. (If you are using a computer for this problem, you may not need to change the form of the given equation. Some software packages will allow you to graph two-variable equations without solving for y.)

$$2.1x + 5.3y = 7.9$$
$$5.3y = 7.9 - 2.1x$$
$$y = \frac{7.9 - 2.1x}{5.3}$$

Now we can enter the expression $\dfrac{7.9 - 2.1x}{5.3}$ for Y_1 and obtain the graph as shown in Figure 2.26.

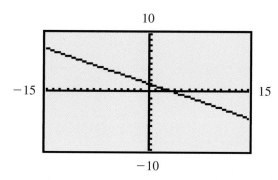

Figure 2.26

For Problems 1–8, answer true or false.

1. A solution of an equation in two variables is an ordered pair of real numbers that satisfies the equation.

2. The equation $y = x + 5$ has a finite number of solutions.

3. The plot of points associated with the solutions of an equation is called a graph of the equation.

4. If an equation is of the form $y = mx$, where m is a real number, its graph will pass through the origin.

5. If an equation is of the form $Ax = B$, then the graph is a line parallel to the x axis.

6. The graph of the solution of $2x - y > 4$ is either all the points in the half plane above the line $2x - y = 4$ or all the points in the half plane below the line $2x - y = 4$.

7. When graphing in two-dimensional space, the graph of the equation $x = 2$ is one point at (2,0).

8. When graphing in the rectangular coordinate system, the graph of $x > 0$ is all the points in quadrant I and quadrant II.

Problem Set 2.2

For Problems 1–16, find the x and y intercepts and graph each linear equation.

1. $x - 2y = 4$

2. $2x + y = -4$

3. $3x + 2y = 6$

4. $2x - 3y = 6$

5. $4x - 5y = 20$

6. $5x + 4y = 20$

7. $x - y = 3$

8. $-x + y = 4$

9. $y = 3x - 1$

10. $y = -2x + 3$

11. $y = -x$

12. $y = 4x$

13. $x = 0$

14. $y = -1$

15. $y = \dfrac{2}{3}x$

16. $y = -\dfrac{1}{2}x$

For Problems 17–30, graph each linear inequality.

17. $x + 2y > 4$

18. $2x - y < -4$

19. $3x - 2y < 6$

20. $2x + 3y < 6$

21. $2x + 5y \le 10$

22. $4x + 5y \le 20$

23. $y > -x - 1$

24. $y < 3x - 2$

25. $y \le -x$

26. $y \ge x$

27. $x + 2y < 0$

28. $3x - y > 0$

29. $x > -1$

30. $y < 3$

■ ■ ■ THOUGHTS INTO WORDS

31. Explain how you would graph the inequality $-x + 2y > -4$.

32. What is the graph of the disjunction $x = 0$ *or* $y = 0$? What is the graph of the conjunction $x = 0$ *and* $y = 0$? Explain your answers.

■ ■ ■ FURTHER INVESTIGATIONS

From our work with absolute value, we know that $|x + y| = 4$ is equivalent to $x + y = 4$ or $x + y = -4$. Therefore the graph of $|x + y| = 4$ is the two lines $x + y = 4$ and $x + y = -4$. For Problems 33–38, graph each equation.

33. $|x - y| = 2$

34. $|2x + y| = 1$

35. $|x - 2y| \le 4$

36. $|3x - 2y| \ge 6$

37. $|2x + 3y| > 6$

38. $|5x + 2y| < 10$

From the definition of absolute value, the equation $y = |x| + 2$ becomes $y = x + 2$ for $x \ge 0$ and $y = -x + 2$ for $x < 0$. Therefore the graph of $y = |x| + 2$ is as shown in Figure 2.27. For Problems 39–44, graph each equation.

39. $y = |x| - 1$

40. $y = |x| + 1$

41. $y = |x + 2|$

42. $y = |x - 1|$

43. $y = 2|x|$

44. $y = \dfrac{1}{3}|x|$

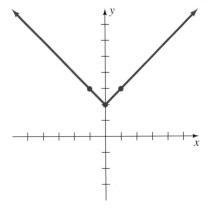

Figure 2.27

GRAPHING CALCULATOR ACTIVITIES

The following problems are designed to lay some ground-work for concepts we will present in the next section. Set your boundaries so that the distance between tic marks is the same on both axes.

45. a. Graph $y = 4x$, $y = 4x - 3$, $y = 4x + 2$, and $y = 4x + 5$ on the same set of axes. Do they appear to be parallel lines?

b. Graph $y = -2x + 1$, $y = -2x + 4$, $y = -2x - 2$, and $y = -2x - 5$ on the same set of axes. Do they appear to be parallel lines?

c. Graph $y = -\dfrac{1}{2}x + 3$, $y = -\dfrac{1}{2}x + 1$, $y = -\dfrac{1}{2}x - 1$, and $y = -\dfrac{1}{2}x - 4$ on the same set of axes. Do they appear to be parallel lines?

d. Graph $2x + 5y = 1$, $2x + 5y = -3$, $2x + 5y = 4$, and $2x + 5y = -5$ on the same set of axes. Do they appear to be parallel lines?

e. Graph $3x - 4y = 7$, $-3x + 4y = 8$, $3x - 4y = -2$, and $4x - 3y = 6$ on the same set of axes. Do they appear to be parallel lines?

f. On the basis of your results in parts a–e, make a statement about how we can recognize parallel lines from their equations.

46. a. Graph $y = 4x$ and $y = -\dfrac{1}{4}x$ on the same set of axes. Do they appear to be perpendicular lines?

b. Graph $y = 3x$ and $y = \dfrac{1}{3}x$ on the same set of axes. Do they appear to be perpendicular lines?

c. Graph $y = \dfrac{2}{5}x - 1$ and $y = -\dfrac{5}{2}x + 2$ on the same set of axes. Do they appear to be perpendicular lines?

d. Graph $y = \dfrac{3}{4}x - 3$, $y = \dfrac{4}{3}x + 2$, and $y = -\dfrac{4}{3}x + 2$ on the same set of axes. Does there appear to be a pair of perpendicular lines?

e. On the basis of your results in parts a–d, make a statement about how we can recognize perpendicular lines from their equations.

47. For each of the following pairs of equations, (1) predict whether they represent parallel lines, perpendicular lines, or lines that intersect but are not perpendicular, and (2) graph each pair of lines to check your prediction.

a. $5.2x + 3.3y = 9.4$ and $5.2x + 3.3y = 12.6$

b. $1.3x - 4.7y = 3.4$ and $1.3x - 4.7y = 11.6$

c. $2.7x + 3.9y = 1.4$ and $2.7x - 3.9y = 8.2$

d. $5x - 7y = 17$ and $7x + 5y = 19$

e. $9x + 2y = 14$ and $2x + 9y = 17$

f. $2.1x + 3.4y = 11.7$ and $3.4x - 2.1y = 17.3$

Answers to the Concept Quiz

1. True   **2.** False   **3.** True   **4.** True   **5.** False   **6.** True   **7.** False   **8.** False

---

2.3   Determining the Equation of a Line

Objectives

- Find the slope of a line determined by two points.
- Write the equation of a line, given a slope and a point contained in the line.
- Write the equation of a line, given two points contained in the line.
- Write the equation of a line parallel or perpendicular to another line and containing a specified point.
- Solve application problems involving the slope of a line.

As we stated earlier, there are basically two types of problems in coordinate geometry: given an algebraic equation, find its geometric graph, and given a set of conditions pertaining to a geometric figure, find its algebraic equation. In the previous section we considered some of the first type of problem; that is, we did some graphing. Now we want to consider some problems of the second type that deal specifically with straight lines; in other words, given certain facts about a line, we need to be able to determine its algebraic equation.

As we work with straight lines, it is often helpful to be able to refer to the *steepness* or *slant* of a particular line. The concept of *slope* is used as a measure of the slant of a line. The **slope** of a line is the ratio of the vertical change of distance to the horizontal change of distance as we move from one point on a line to another. Consider the line in Figure 2.28. From point A to point B there is a vertical change of two units and a horizontal change of three units; therefore the slope of the line is $\frac{2}{3}$.

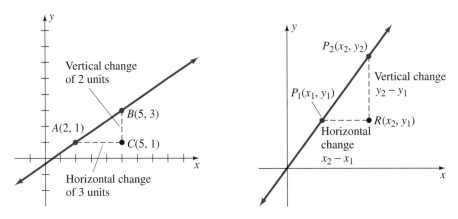

Figure 2.28   **Figure 2.29**

A precise definition for slope can be given by considering the coordinates of the points P_1, P_2, and R in Figure 2.29. The horizontal change of distance as we move from P_1 to P_2 is $x_2 - x_1$, and the vertical change is $y_2 - y_1$. Thus we have the following definition.

Definition 2.1

If P_1 and P_2 are any two different points on a line, P_1 with coordinates (x_1, y_1) and P_2 with coordinates (x_2, y_2), then the **slope** of the line (denoted by m) is

$$m = \frac{y_2 - y_1}{x_2 - x_1}, \qquad x_2 \neq x_1$$

Because

$$\frac{y_2 - y_1}{x_2 - x_1} = \frac{y_1 - y_2}{x_1 - x_2}$$

how we designate P_1 and P_2 is not important. Let's use Definition 2.1 to find the slopes of some lines.

EXAMPLE 1

Find the slope of the line determined by each of the following pairs of points and graph each line.

    a. $(-1, 1)$ and $(3, 2)$ **b.** $(4, -2)$ and $(-1, 5)$ **c.** $(2, -3)$ and $(-3, -3)$

Solutions

    a. Let $(-1, 1)$ be P_1 and $(3, 2)$ be P_2 (see Figure 2.30):

$$m = \frac{y_2 - y_1}{x_2 - x_1} = \frac{2 - 1}{3 - (-1)} = \frac{1}{4}$$

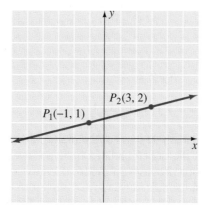

Figure 2.30

    b. Let $(4, -2)$ be P_1 and $(-1, 5)$ be P_2 (see Figure 2.31):

$$m = \frac{5 - (-2)}{-1 - 4} = \frac{7}{-5} = -\frac{7}{5}$$

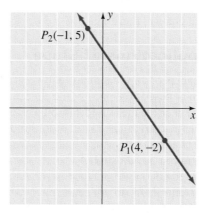

Figure 2.31

c. Let $(2, -3)$ be P_1 and $(-3, -3)$ be P_2 (see Figure 2.32):

$$m = \frac{-3 - (-3)}{-3 - 2} = \frac{0}{-5} = 0$$

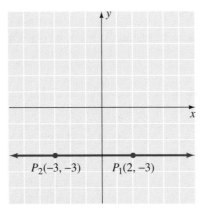

$P_2(-3, -3)$ $P_1(2, -3)$

Figure 2.32

▶ **Now work Problems 1 and 5.** ■

The three parts of Example 1 illustrate the three basic possibilities for slope; that is, the slope of a line can be positive, negative, or zero. A line that has a positive slope rises as we move from left to right, as in Figure 2.30. A line that has a negative slope falls as we move from left to right, as in Figure 2.31. A horizontal line, as in Figure 2.32, has a slope of zero. Finally, we need to realize that **the concept of slope is undefined for vertical lines.** This is because, for any vertical line, the horizontal change is zero as we move from one point on the line to another. Thus the ratio $(y_2 - y_1)/(x_2 - x_1)$ will have a denominator of zero and be undefined. Hence the restriction $x_2 \neq x_1$ is included in Definition 2.1.

Don't forget that **the slope of a line is a ratio,** the ratio of vertical change to horizontal change. For example, a slope of $\frac{2}{3}$ means that for every two units of vertical change, there must be a corresponding three units of horizontal change.

■ Applications of Slope

The concept of slope has many real-world applications even though the word "slope" is often not used. Technically, the concept of slope applies in most situations where the idea of an incline is used. Hospital beds are constructed so that both the head end and the foot end can be raised or lowered; that is, the slope of either end of the bed can be changed. Likewise, treadmills are designed so that the incline (slope) of the platform can be adjusted. A roofer, when making an estimate to replace a roof, is concerned not only about the total area to be covered but also about the pitch of the roof. (Contractors do not define pitch exactly in accordance with the mathematical definition of slope, but both concepts refer to "steepness.") In Figure 2.33, the two roofs might require the same amount of shingles, but the roof on the left will take longer to complete because the pitch is so great that scaffolding will be required.

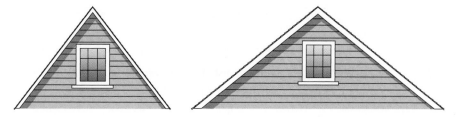

Figure 2.33

The concept of slope is also used in the construction of stairways. The steepness (slope) of stairs can be expressed as the ratio of *rise* to *run.* In Figure 2.34 the stairs on the left, which have a ratio of $\frac{10}{11}$, are steeper than the stairs on the right, which have a ratio of $\frac{7}{11}$.

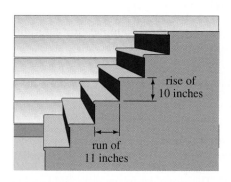
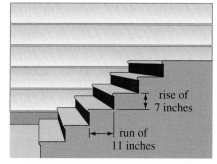

Figure 2.34

In highway construction, the word *grade* is used to describe the slope. For example, the highway in Figure 2.35 is said to have a grade of 17%. This means that for every horizontal distance of 100 feet, the highway rises or drops 17 feet. In other words, the slope of the highway is $\frac{17}{100}$.

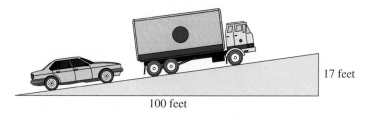

Figure 2.35

EXAMPLE 2

A certain highway has a 3% grade. How many feet does it rise in a horizontal distance of 1 mile?

Solution

A 3% grade means a slope of $\dfrac{3}{100}$. Therefore, if we let y represent the unknown vertical distance and use the fact that 1 mile = 5280 feet, we can set up and solve the following proportion:

$$\frac{3}{100} = \frac{y}{5280}$$

$$100y = 3(5280) = 15{,}840$$

$$y = 158.4$$

The highway rises 158.4 feet in a horizontal distance of 1 mile.

▶ **Now work Problem 87.**   ■

■ Equations of Lines

Now let's consider some techniques for determining the equation of a line when given certain facts about the line.

EXAMPLE 3

Find the equation of the line that has a slope of $\dfrac{2}{5}$ and contains the point $(3, 1)$.

Solution

First, let's draw the line and record the given information, as in Figure 2.36. Then we choose a point (x, y) that represents any point on the line other than the given point $(3, 1)$. The slope determined by $(3, 1)$ and (x, y) is to be $\dfrac{2}{5}$. Thus

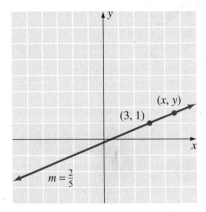

Figure 2.36

$$\frac{y - 1}{x - 3} = \frac{2}{5}$$

$$2(x - 3) = 5(y - 1)$$

$$2x - 6 = 5y - 5$$

$$2x - 5y = 1 \qquad \blacksquare$$

_____

E X A M P L E   4

_____

Find the equation of the line determined by $(1, -2)$ and $(-3, 4)$.

Solution

First, let's draw the line determined by the two given points, as in Figure 2.37. These two points determine the slope of the line.

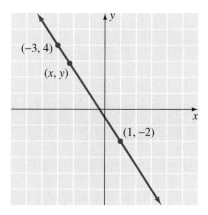

Figure 2.37

$$m = \frac{4 - (-2)}{-3 - 1} = \frac{6}{-4} = -\frac{3}{2}$$

Now we can use the same approach as in Example 3. We form an equation using one of the two given points, a point (x, y), and a slope of $-\frac{3}{2}$:

$$\frac{y + 2}{x - 1} = \frac{3}{-2}$$

$$3(x - 1) = -2(y + 2)$$

$$3x - 3 = -2y - 4$$

$$3x + 2y = -1$$

▶   **Now work Problem 31.** $\qquad \blacksquare$

EXAMPLE 5   Find the equation of the line that has a slope of $\dfrac{1}{4}$ and a y intercept of 2.

Solution

A y intercept of 2 means that the point $(0, 2)$ is on the line (see Figure 2.38). Choosing a point (x, y), we can proceed as in the previous examples:

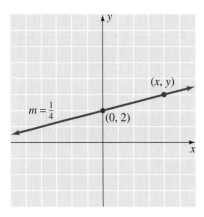

Figure 2.38

$$\frac{y - 2}{x - 0} = \frac{1}{4}$$
$$1(x - 0) = 4(y - 2)$$
$$x = 4y - 8$$
$$x - 4y = -8$$

■

At this point you might pause for a moment and look back over Examples 3, 4, and 5. Note that we used the same basic approach in all three examples: We chose a point (x, y) and used it to determine the equation that satisfies the conditions stated in the problem. We will use this same approach later with figures other than straight lines. Furthermore, you should realize that this approach can be used to develop some general forms of equations of straight lines.

■ Point–Slope Form

EXAMPLE 6   Find the equation of the line that has a slope of m and contains the point (x_1, y_1).

Solution

Choosing (x, y) to represent another point on the line (see Figure 2.39), we can give the slope of the line:

$$m = \frac{y - y_1}{x - x_1}, \qquad x \neq x_1$$

from which we obtain

$$y - y_1 = m(x - x_1)$$

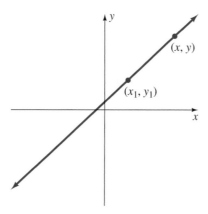

Figure 2.39

We refer to the equation

$$y - y_1 = m(x - x_1)$$

as the **point–slope form** of the equation of a straight line. Therefore, instead of using the approach of Example 3, we can substitute information into the point–slope form to write the equation of a line with a given slope that contains a given point. For example, the equation of the line that has a slope of $\dfrac{3}{5}$ and contains the point $(2, 4)$ can be determined this way. We substitute $(2, 4)$ for (x_1, y_1) and $\dfrac{3}{5}$ for m in the point–slope equation:

$$y - 4 = \frac{3}{5}(x - 2)$$
$$5(y - 4) = 3(x - 2)$$
$$5y - 20 = 3x - 6$$
$$-14 = 3x - 5y$$

■ Slope–Intercept Form

EXAMPLE 7

Find the equation of the line that has a slope of m and a y intercept of b.

Solution

A y intercept of b means that $(0, b)$ is on the line (see Figure 2.40). Therefore, using the point–slope form with $(x_1, y_1) = (0, b)$, we obtain

$$y - y_1 = m(x - x_1)$$
$$y - b = m(x - 0)$$
$$y - b = mx$$
$$y = mx + b$$

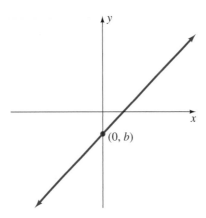

Figure 2.40 ∎

We refer to the equation

$$y = mx + b$$

as the **slope–intercept form** of the equation of a straight line. It can be used for two primary purposes, as the next two examples illustrate.

E X A M P L E   8   Find the equation of the line that has a slope of $\dfrac{1}{4}$ and a y intercept of 2.

Solution

This is a restatement of Example 5, but this time we will use the slope–intercept form ($y = mx + b$) of the equation of a line to write its equation. Because $m = \dfrac{1}{4}$ and $b = 2$, we obtain

$$y = mx + b$$
$$y = \frac{1}{4}x + 2$$
$$4y = x + 8$$
$$-8 = x - 4y \qquad \text{Same result as in Example 5}$$

▶ **Now work Problem 41.** ∎

Remark: Sometimes we leave linear equations in slope–intercept form. We did not do so in Example 8 because we wanted to show that it was the same result as in Example 5.

E X A M P L E   9   Find the slope and y intercept of the line that has an equation $2x - 3y = 7$.

Solution

We can solve the equation for y in terms of x and then compare the result to the general slope–intercept form.

$$2x - 3y = 7$$
$$-3y = -2x + 7$$
$$y = \frac{2}{3}x - \frac{7}{3} \qquad y = mx + b$$

The slope of the line is $\frac{2}{3}$ and the y intercept is $-\frac{7}{3}$.

▶ **Now work Problem 69.**   ■

In general, **if the equation of a nonvertical line is written in slope–intercept form, the coefficient of x is the slope of the line and the constant term is the y intercept.**

■ Parallel and Perpendicular Lines

Because the concept of slope is used to indicate the slant of a line, it seems reasonable to expect slope to be related to the concepts of parallelism and perpendicularity. Such is the case, and the following two properties summarize this link.

Property 2.1

> If two nonvertical lines have slopes of m_1 and m_2, then
>
> **1.** The two lines are parallel if and only if $m_1 = m_2$.
>
> **2.** The two lines are perpendicular if and only if $m_1 m_2 = -1$.

We will test your ingenuity in devising proofs of these properties in the next problem set; here we will illustrate their use.

E X A M P L E   1 0

a. Verify that the graphs of $3x + 2y = 9$ and $6x + 4y = 19$ are parallel lines.

b. Verify that the graphs of $5x - 3y = 12$ and $3x + 5y = 27$ are perpendicular lines.

Solution

a. Let's change each equation to slope–intercept form.

$$3x + 2y = 9 \longrightarrow 2y = -3x + 9$$
$$y = -\frac{3}{2}x + \frac{9}{2}$$

$$6x + 4y = 19 \longrightarrow 4y = -6x + 19$$
$$y = -\frac{6}{4}x + \frac{19}{4}$$
$$y = -\frac{3}{2}x + \frac{19}{4}$$

The two lines have the same slope but different y intercepts. Therefore they are parallel.

b. Change each equation to slope–intercept form.

$$5x - 3y = 12 \longrightarrow -3y = -5x + 12$$

$$y = \frac{5}{3}x - 4$$

$$3x + 5y = 27 \longrightarrow 5y = -3x + 27$$

$$y = -\frac{3}{5}x + \frac{27}{5}$$

Because $\left(\dfrac{5}{3}\right)\left(-\dfrac{3}{5}\right) = -1$, the product of the two slopes is -1, and the lines are perpendicular.

▶ **Now work Problem 63.** ■

Remark: The statement "the product of two slopes is -1" is equivalent to saying that the two slopes are **negative reciprocals** of each other — that is, $m_1 = -1/m_2$.

E X A M P L E   1 1

Find the equation of the line that contains the point $(-1, 2)$ and is parallel to the line with the equation $2x - y = 4$.

Solution

First, we draw Figure 2.41 to help in our analysis of the problem. Because the line through $(-1, 2)$ is to be parallel to the given line, it must have the same slope. Let's find the slope by changing $2x - y = 4$ to slope–intercept form:

$$2x - y = 4$$
$$-y = -2x + 4$$
$$y = 2x - 4$$

The slope of both lines is 2. Now, using the point–slope form with $(x_1, y_1) = (-1, 2)$, we obtain the equation of the line:

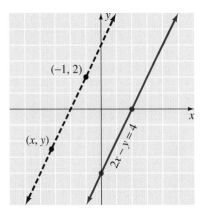

Figure 2.41

$$y - y_1 = m(x - x_1)$$
$$y - 2 = 2[x - (-1)]$$
$$y - 2 = 2(x + 1)$$
$$y - 2 = 2x + 2$$
$$-4 = 2x - y$$

▶ **Now work Problem 53.**   ∎

E X A M P L E   1 2

Find the equation of the line that contains the point $(-1, -3)$ and is perpendicular to the line determined by $3x + 4y = 12$.

Solution

Again let's start by drawing a figure to help with our analysis (see Figure 2.42). Because the line through $(-1, -3)$ is to be perpendicular to the given line, its slope must be the negative reciprocal of the slope of the line with the equation $3x + 4y = 12$. Let's find the slope of $3x + 4y = 12$ by changing to slope–intercept form:

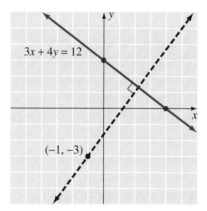

Figure 2.42

$$3x + 4y = 12$$
$$4y = -3x + 12$$
$$y = -\frac{3}{4}x + 3$$

The slope of the desired line is $\frac{4}{3}$ $\left(\text{the negative reciprocal of } -\frac{3}{4}\right)$, and we can proceed as before to obtain its equation:

$$y - y_1 = m(x - x_1)$$
$$y - (-3) = \frac{4}{3}[x - (-1)]$$

$$y + 3 = \frac{4}{3}(x + 1)$$
$$3y + 9 = 4x + 4$$
$$5 = 4x - 3y$$

▶ **Now work Problem 55.**   ■

Two forms of equations of straight lines are used extensively. They are the *standard form* and the *slope–intercept form*.

Standard Form $Ax + By = C$, where B and C are integers and A is a nonnegative integer (A and B are not both zero).

Slope–Intercept Form $y = mx + b$, where m is a real number representing the slope of the line, and b is a real number representing the y intercept.

CONCEPT QUIZ   For Problems 1–8, answer true or false.

1. The slope of a line is the ratio of the horizontal change in distance to the vertical change in distance as we move from one point on the line to another point on the line.

2. The slope of a line can be zero.

3. If P_1 and P_2 are two different points on a nonvertical line, the slope from P_2 to P_1 is the opposite of the slope from P_1 to P_2.

4. The concept of slope is undefined for vertical lines.

5. If the equation of a line is written in the form $y = mx + b$, then m is the slope of the line, and b is the x intercept.

6. If the equation of a line is $Ax + By = C$, where $B \neq 0$, then the slope of the line is $\dfrac{-A}{B}$.

7. The two lines whose equations are $Ax + By = C$ and $Ax + By = D$ are parallel lines.

8. The two lines whose equations are $Ax + By = C$ and $Bx - Ay = C$ are perpendicular lines.

Problem Set 2.3

For Problems 1–8, find the slope of the line determined by each pair of points.

1. $(3, 1)$ and $(7, 4)$

2. $(-1, 2)$ and $(5, -3)$

3. $(-2, -1)$ and $(-1, -6)$

4. $(-2, -4)$ and $(3, 7)$

5. $(-4, 2)$ and $(-2, 2)$

6. $(4, -5)$ and $(-1, -5)$

7. $(a, 0)$ and $(0, b)$

8. (a, b) and (c, d)

9. Find x if the line through $(-2, 4)$ and $(x, 7)$ has a slope of $\dfrac{2}{9}$.

10. Find y if the line through $(1, y)$ and $(4, 2)$ has a slope of $\dfrac{5}{3}$.

11. Find x if the line through $(x, 4)$ and $(2, -6)$ has a slope of $-\dfrac{9}{4}$.

12. Find y if the line through $(5, 2)$ and $(-3, y)$ has a slope of $-\dfrac{7}{8}$.

For each of the lines in Problems 13–20, you are given one point and the slope of the line. Find the coordinates of three other points on the line.

13. $(3, 2)$; $m = \dfrac{2}{3}$ **14.** $(-4, 4)$; $m = \dfrac{5}{6}$

15. $(-1, -4)$; $m = 4$ **16.** $(-5, -3)$; $m = 2$

17. $(2, -1)$; $m = -\dfrac{3}{5}$ **18.** $(5, -1)$; $m = -\dfrac{2}{3}$

19. $(-3, 2)$; $m = -4$ **20.** $(-5, 6)$; $m = -3$

For Problems 21–30, write the equation of the line that has the indicated slope and contains the indicated point. Express final equations in standard form.

21. $m = \dfrac{1}{3}$; $(2, 4)$ **22.** $m = \dfrac{3}{5}$; $(-1, 4)$

23. $m = 2$; $(-1, -2)$ **24.** $m = -3$; $(2, 5)$

25. $m = -\dfrac{2}{3}$; $(4, -3)$ **26.** $m = -\dfrac{1}{5}$; $(-3, 7)$

27. $m = 0$; $(5, -2)$ **28.** $m = \dfrac{4}{3}$; $(-4, -5)$

29. m is undefined; $(3, -4)$

30. $m = 0$; $(-4, -6)$

For Problems 31–40, write the equation of each line that contains the indicated pair of points. Express final equations in standard form.

31. $(2, 3)$ and $(9, 8)$ **32.** $(1, -4)$ and $(4, 4)$

33. $(-1, 7)$ and $(5, 2)$ **34.** $(-3, 1)$ and $(6, -2)$

35. $(4, 2)$ and $(-1, 3)$ **36.** $(2, 7)$ and $(2, 5)$

37. $(4, -3)$ and $(-7, -3)$ **38.** $(-4, 2)$ and $(2, -3)$

39. $(-2, 6)$ and $(-2, -7)$ **40.** $(4, 5)$ and $(-1, 5)$

For Problems 41–48, write the equation of each line that has the indicated slope, m, and y intercept, b. Express final equations in slope–intercept form.

41. $m = \dfrac{1}{2}$, $b = 3$ **42.** $m = \dfrac{5}{3}$, $b = -1$

43. $m = -\dfrac{3}{7}$, $b = 2$ **44.** $m = -3$, $b = -4$

45. $m = 4$, $b = \dfrac{3}{2}$ **46.** $m = \dfrac{2}{3}$, $b = \dfrac{3}{5}$

47. $m = -\dfrac{5}{6}$, $b = \dfrac{1}{4}$ **48.** $m = -\dfrac{4}{5}$, $b = 0$

For Problems 49–60, write the equation of each line that satisfies the given conditions. Express final equations in standard form.

49. The x intercept is 4 and the y intercept is -5.

50. Contains the point $(3, -1)$ and is parallel to the x axis

51. Contains the point $(-4, 3)$ and is parallel to the y axis

52. Contains the point $(1, 2)$ and is parallel to the line $3x - y = 5$

53. Contains the point $(4, -3)$ and is parallel to the line $5x + 2y = 1$

54. Contains the origin and is parallel to the line $5x - 2y = 10$

55. Contains the point $(-2, 6)$ and is perpendicular to the line $x - 4y = 7$

56. Contains the point $(-3, -5)$ and is perpendicular to the line $3x + 7y = 4$

57. Contains the point $(1, -6)$ and is parallel to the line $x = 4$

58. The x intercept is -2 and the y intercept is 5

59. Contains the point $(-3, 5)$ and is perpendicular to the line $y = 3$

60. Contains the point $(-1, -4)$ and is perpendicular to the line $x = 5$

For each pair of lines in Problems 61–68, determine whether they are parallel, perpendicular, or intersecting lines that are not perpendicular.

61. $y = \dfrac{5}{6}x + 2$
$y = \dfrac{5}{6}x - 4$

62. $y = 5x - 1$
$y = -\dfrac{1}{5}x + \dfrac{2}{3}$

63. $5x - 7y = 14$
$7x + 5y = 12$

64. $2x - y = 4$
$4x - 2y = 17$

65. $4x + 9y = 13$
$-4x + y = 11$

66. $y = 5x$
$y = -5x$

67. $x + y = 0$
$x - y = 0$

68. $2x - y = 14$
$3x - y = 17$

For Problems 69–76, find the slope and the y intercept of each line.

69. $2x - 3y = 4$

70. $3x + 4y = 7$

71. $x - 2y = 7$

72. $2x + y = 9$

73. $y = -3x$

74. $x - 5y = 0$

75. $7x - 5y = 12$

76. $-5x + 6y = 13$

77. The slope-intercept form of a line can also be used for graphing purposes. Suppose that we want to graph $y = \dfrac{2}{3}x + 1$. Because the y intercept is 1, the point $(0, 1)$ is on the line. Furthermore, because the slope is $\dfrac{2}{3}$, another point can be found by moving two units *up* and three units to the *right*. Thus the point $(3, 3)$ is also on the line. The two points $(0, 1)$ and $(3, 3)$ determine the line.

Use the slope-intercept form to help graph each of the following lines.

a. $y = \dfrac{3}{4}x + 2$

b. $y = \dfrac{1}{2}x - 4$

c. $y = -\dfrac{4}{5}x + 1$

d. $y = -\dfrac{2}{3}x - 6$

e. $y = -2x + \dfrac{5}{4}$

f. $y = x - \dfrac{3}{2}$

78. Use the concept of slope to verify that $(-4, 6)$, $(6, 10)$, $(10, 0)$, and $(0, -4)$ are the vertices of a square.

79. Use the concept of slope to verify that $(6, 6)$, $(2, -2)$, $(-8, -5)$, and $(-4, 3)$ are vertices of a parallelogram.

80. Use the concept of slope to verify that the triangle determined by $(4, 3)$, $(5, 1)$, and $(3, 0)$ is a right triangle.

81. Use the concept of slope to verify that the quadrilateral with vertices $(0, 7)$, $(-2, -1)$, $(2, -2)$, and $(4, 6)$ is a rectangle.

82. Use the concept of slope to verify that the points $(8, -3)$, $(2, 1)$, and $(-4, 5)$ lie on a straight line.

For Problems 83–90, solve the problem.

83. The midpoints of the sides of a triangle are $(-3, 4)$, $(1, -4)$, and $(7, 2)$. Find the equations of the lines that contain the sides of the triangle.

84. The vertices of a certain triangle are $(2, 6)$, $(5, 1)$, and $(1, -4)$. Find the equations of the lines that contain the three altitudes of the triangle. (An altitude of a triangle is the perpendicular line segment from a vertex to the opposite side.)

85. The vertices of a certain triangle are $(1, -6)$, $(3, 1)$, and $(-2, 2)$. Find the equations of the lines that contain the three medians of the triangle. (A median of a triangle is the line segment from a vertex to the midpoint of the opposite side.)

86. A certain highway has a 2% grade. How many feet does it rise in a horizontal distance of 1 mile? (1 mile = 5280 feet)

87. The grade of a highway up a hill is 30%. How much change in horizontal distance is there if the vertical height of the hill is 75 feet?

88. If the ratio of rise to run is to be $\dfrac{3}{5}$ for some stairs and the rise is 19 centimeters, find the measure of the run to the nearest centimeter.

89. If the ratio of rise to run is to be $\dfrac{2}{3}$ for some stairs and the run is 28 centimeters, find the rise to the nearest centimeter.

90. Suppose that a county ordinance requires a $2\dfrac{1}{4}\%$ fall for a sewage pipe from the house to the main pipe at the street. How much vertical drop must there be for a horizontal distance of 45 feet? Express the answer to the nearest tenth of a foot.

■ ■ ■ **THOUGHTS INTO WORDS**

91. How would you explain the concept of slope to someone who was absent from class the day it was discussed?

92. If one line has a slope of $\dfrac{2}{5}$, and another line has a slope of $\dfrac{3}{7}$, which line is steeper? Explain your answer.

93. What does it mean to say that two points *determine* a line? Do three points *determine* a line? Explain your answers.

94. Explain how you would find the slope of the line $y = 2$.

■ ■ ■ **FURTHER INVESTIGATIONS**

95. The form

$$\frac{y - y_1}{x - x_1} = \frac{y_2 - y_1}{x_2 - x_1}$$

is called the **two-point form** of the equation of a straight line. Using points (x_1, y_1) and (x_2, y_2), develop the two-point form for the equation of a line. Then use the two-point form to write the equation of each of the following lines that contain the indicated pair of points. Express the final equations in standard form.

  a. $(4, 3)$ and $(5, 6)$ **b.** $(-3, 5)$ and $(2, -1)$

  c. $(0, 0)$ and $(-7, 2)$

  d. $(-3, -4)$ and $(5, -1)$

96. The form $(x/a) + (y/b) = 1$ is called the **intercept form** of the equation of a straight line. Using a to represent the x intercept and b to represent the y intercept, develop the intercept form. Then use the intercept form to write the equation of each of the following lines. Express the final equations in standard form.

  a. $a = 2, b = 5$ **b.** $a = -3, b = 1$

  c. $a = 6, b = -4$ **d.** $a = -1, b = -2$

97. Prove each of the following statements.

  a. Two nonvertical parallel lines have the same slope.

  b. Two lines with the same slope are parallel.

  c. If two nonvertical lines are perpendicular, then their slopes are negative reciprocals of each other.

  d. If the slopes of two lines are negative reciprocals of each other, then the lines are perpendicular.

98. Let $Ax + By = C$ and $A'x + B'y = C'$ represent two lines. Verify each of the following properties.

  a. If $(A/A') = (B/B') \neq (C/C')$, then the lines are parallel.

  b. If $AA' = -BB'$, then the lines are perpendicular.

99. The properties in Problem 98 give us another way to write the equation of a line parallel or perpendicular to a given line through a point not on the given line. For example, suppose we want the equation of the line perpendicular to $3x + 4y = 6$ that contains the point $(1, 2)$. The form $4x - 3y = k$, where k is a constant, represents a family of lines perpendicular to $3x + 4y = 6$ because we have satisfied the condition $AA' = -BB'$. Therefore, to find the specific line of the family containing $(1, 2)$, we substitute 1 for x and 2 for y to determine k:

$$4x - 3y = k$$

$$4(1) - 3(2) = k$$

$$-2 = k$$

Thus the equation of the desired line is $4x - 3y = -2$. Use the properties from Problem 98 to help write the equation of each of the following lines.

  a. Contains $(5, 6)$ and is parallel to the line $2x - y = 1$

  b. Contains $(-3, 4)$ and is parallel to the line $3x + 7y = 2$

  c. Contains $(2, -4)$ and is perpendicular to the line $2x - 5y = 9$

  d. Contains $(-3, -5)$ and is perpendicular to the line $4x + 6y = 7$

100. Some real-world situations can be described by the use of linear equations in two variables. If two pairs of values are known, then the equation can be determined by using the approach we used in Example 4 of this section. For each of the following, assume that the relationship can be expressed as a linear equation in two variables, and use the given information to determine the equation. Express the equation in standard form.

  a. A company produces 10 fiberglass shower stalls for $2015 and 15 stalls for $3015. Let y be the cost and x the number of stalls.

  b. A company can produce 6 boxes of candy for $8 and 10 boxes of candy for $13. Let y represent the cost and x the number of boxes of candy.

c. Two banks on opposite corners of a town square have signs displaying the up-to-date temperature. One bank displays the temperature in Celsius degrees and the other in Fahrenheit. A temperature of 10°C was displayed at the same time as a temperature of 50°F. On another day a temperature of −5°C was displayed at the same time as a temperature of 23°F. Let y represent the temperature in Fahrenheit and x the temperature in Celsius.

101. The relationships that tie slope to parallelism and perpendicularity are powerful tools for constructing coordinate geometry proofs. Prove each of the following using a coordinate geometry approach.

a. The diagonals of a square are perpendicular.
b. The line segment joining the midpoints of two sides of a triangle is parallel to the third side.
c. The line segments joining successive midpoints of the sides of a quadrilateral form a parallelogram.
d. The line segments joining successive midpoints of the sides of a rectangle form a rhombus. (A rhombus is a parallelogram with all sides of the same length.)

 GRAPHING CALCULATOR ACTIVITIES

102. Predict whether each of the following pairs of equations represents parallel lines, perpendicular lines, or lines that intersect but are not perpendicular. Then graph each pair of lines to check your predictions. (The properties presented in Problem 98 should be very helpful.)

a. $5.2x + 3.3y = 9.4$ and $5.2x + 3.3y = 12.6$
b. $1.3x − 4.7y = 3.4$ and $1.3x − 4.7y = 11.6$
c. $2.7x + 3.9y = 1.4$ and $2.7x − 3.9y = 8.2$
d. $5x − 7y = 17$ and $7x + 5y = 19$
e. $9x + 2y = 14$ and $2x + 9y = 17$
f. $2.1x + 3.4y = 11.7$ and $3.4x − 2.1y = 17.3$
g. $7.1x − 2.3y = 6.2$ and $2.3x + 7.1y = 9.9$
h. $−3x + 9y = 12$ and $9x − 3y = 14$
i. $2.6x − 5.3y = 3.4$ and $5.2x − 10.6y = 19.2$
j. $4.8x − 5.6y = 3.4$ and $6.1x + 7.6y = 12.3$

Answers to the Concept Quiz

1. False **2.** True **3.** False **4.** True **5.** False **6.** True **7.** True **8.** True

2.4 More on Graphing

Objectives

■ Graph nonlinear equations.

■ Determine if the graph of an equation is symmetric to the x axis, the y axis, or the origin.

As we stated earlier, it is very helpful to recognize that a certain type of equation produces a particular kind of graph. In a later chapter, we will pursue that idea in much more detail. However, we also need to develop some general graphing techniques to use with equations when we do not recognize the graph. Let's begin with the following suggestions and then add to the list throughout the remainder of the text. (You may recognize some of the graphs in this section from previous graphing experiences, but keep in mind that the primary objective at this time is to develop some additional graphing techniques.)

1. Find the intercepts.
2. Solve the equation for y in terms of x or for x in terms of y if it is not already in such a form.
3. Set up a table of ordered pairs that satisfy the equation.
4. Plot the points associated with the ordered pairs and connect them with a smooth curve.

E X A M P L E   1

Graph $y = x^2 - 4$.

Solution

First, let's find the intercepts. If $x = 0$, then

$$y = 0^2 - 4$$
$$y = -4$$

This determines the point $(0, -4)$. If $y = 0$, then

$$0 = x^2 - 4$$
$$4 = x^2$$
$$\pm 2 = x$$

Thus the points $(2, 0)$ and $(-2, 0)$ are determined.

Second, because the given equation expresses y in terms of x, the form is convenient for setting up a table of ordered pairs. Plotting these points, and connecting them with a smooth curve, produces Figure 2.43.

x	y	
0	−4	
2	0	intercepts
−2	0	
1	−3	
−1	−3	other
3	5	points
−3	5	

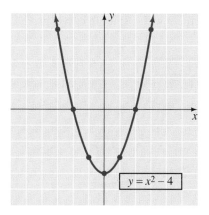

Figure 2.43

▶ **Now work Problem 33.**   ■

The curve in Figure 2.43 is said to be **symmetric with respect to the y axis.** Stated another way, each half of the curve is a mirror image of the other half through the y axis. Note in the table of values that for each ordered pair (x, y), the ordered pair $(-x, y)$ is also a solution. Thus a general test for y axis symmetry can be stated as follows.

y Axis Symmetry

> The graph of an equation is symmetric with respect to the y axis if replacing x with $-x$ results in an equivalent equation.

Thus the equation $y = x^2 - 4$ exhibits y axis symmetry because replacing x with $-x$ produces $y = (-x)^2 - 4 = x^2 - 4$. Likewise, the equations $y = x^2 + 6$, $y = x^4$, and $y = x^4 + 2x^2$ exhibit y axis symmetry.

EXAMPLE 2   Graph $x - 1 = y^2$.

Solution

If $x = 0$, then

$$0 - 1 = y^2$$
$$-1 = y^2$$

The equation $y^2 = -1$ has no real number solutions; therefore, this graph has no points on the y axis. If $y = 0$, then

$$x - 1 = 0$$
$$x = 1$$

Thus the point $(1, 0)$ is determined. Solving the original equation for x produces $x = y^2 + 1$, for which the table of values is easily determined. Plotting these points, and connecting them with a smooth curve, produces Figure 2.44.

x	y	
1	0	intercept
2	1	
2	−1	other
5	2	points
5	−2	

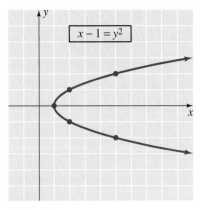

Figure 2.44

▶ **Now work Problem 55.**   ■

The curve in Figure 2.44 is said to be **symmetric with respect to the** x **axis.** That is to say, each half of the curve is a mirror image of the other half through the x axis. Note in the table of values that for each ordered pair (x, y), the ordered pair $(x, -y)$ is also a solution. The following general test of x axis symmetry can be stated.

x Axis Symmetry

> The graph of an equation is symmetric with respect to the x axis if replacing y with $-y$ results in an equivalent equation.

Thus the equation $x - 1 = y^2$ exhibits x axis symmetry because replacing y with $-y$ produces $x - 1 = (-y)^2 = y^2$. Likewise, the equations $x = y^2$, $x = y^4 + 2$, and $x^3 = y^2$ exhibit x axis symmetry.

EXAMPLE 3

Graph $y = x^3$.

Solution

If $x = 0$, then

$$y = 0^3 = 0$$

Thus the origin $(0, 0)$ is on the graph. The table of values is easily determined from the equation. Plotting these points, and connecting them with a smooth curve, produces Figure 2.45.

x	y	
0	0	intercept
1	1	
2	8	other
−1	−1	points
−2	−8	

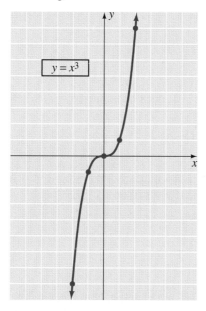

$$y = x^3$$

Figure 2.45

Now work Problem 37.

The curve in Figure 2.45 is said to be **symmetric with respect to the origin.** Each half of the curve is a mirror image of the other half through the origin. In the table of values, we see that for each ordered pair (x, y), the ordered pair $(-x, -y)$ is also a solution. The following general test for origin symmetry can be stated.

Origin Symmetry

The graph of an equation is symmetric with respect to the origin if replacing x with $-x$ and y with $-y$ results in an equivalent equation.

The equation $y = x^3$ exhibits origin symmetry because replacing x with $-x$ and y with $-y$ produces $-y = -x^3$, which is equivalent to $y = x^3$. (Multiplying both sides of $-y = -x^3$ by -1 produces $y = x^3$.) Likewise, the equations $xy = 4$, $x^2 + y^2 = 10$, and $4x^2 - y^2 = 12$ exhibit origin symmetry.

Remark: From the symmetry tests, we observe that if a curve has both x axis and y axis symmetry, then it must have origin symmetry. However, it is possible for a curve to have origin symmetry and not be symmetric to either axis. Figure 2.45 is an example of such a curve.

Another graphing consideration is that of **restricting a variable** to ensure real number solutions. The following example illustrates this point.

E X A M P L E   4

Graph $y = \sqrt{x - 1}$.

Solution

The radicand, $x - 1$, must be nonnegative. Therefore,

$$x - 1 \geq 0$$
$$x \geq 1$$

The restriction $x \geq 1$ indicates that there is no y intercept. The x intercept can be found as follows: If $y = 0$, then

$$0 = \sqrt{x - 1}$$
$$0 = x - 1$$
$$1 = x$$

The point $(1, 0)$ is on the graph. Now, keeping that restriction in mind, we can determine the table of values. Plotting these points, and connecting them with a smooth curve, produces Figure 2.46.

x	y	
1	0	intercept
2	1	other
5	2	points
10	3	

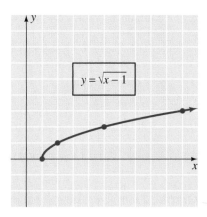

$$y = \sqrt{x-1}$$

Figure 2.46

▶ **Now work Problem 51.**   ∎

Now let's restate and add the concepts of symmetry and restrictions to the list of graphing suggestions. The order of the suggestions also indicates the order in which we usually attack a graphing problem if it is a new graph — that is, one that we do not recognize from its equation.

1. Determine what type of symmetry the equation exhibits.

2. Find the intercepts.

3. Solve the equation for y in terms of x or for x in terms of y, if it is not already in such a form.

4. Determine the restrictions necessary to ensure real number solutions.

5. Set up a table of ordered pairs that satisfy the equation. The type of symmetry and the restrictions will affect your choice of values in the table.

6. Plot the points associated with the ordered pairs and connect them with a smooth curve. Then, if appropriate, reflect this curve according to the symmetry possessed by the graph.

The final two examples of this section should help you pull these ideas together and demonstrate the power of having these techniques at your fingertips.

E X A M P L E   5

Graph $x = -y^2 - 3$.

Solution

Symmetry   The graph is symmetric with respect to the x axis, because replacing y with $-y$ produces $x = -(-y)^2 - 3$, which is equivalent to $x = -y^2 - 3$.

Intercepts   If $x = 0$, then

$$0 = -y^2 - 3$$
$$y^2 = -3$$

Therefore the graph contains no points on the y axis. If $y = 0$, then

$$x = -0^2 - 3$$
$$x = -3$$

Thus the point $(-3, 0)$ is on the graph.

Restrictions   Because $x = -y^2 - 3$, y can take on any real number value, and for every value of y, x will be less than or equal to -3.

Table of Values   Because of the x axis symmetry, let's choose only nonnegative values for y.

Plotting the Graph   Plotting the points determined by the table, and connecting them with a smooth curve, produces Figure 2.47(a). Then reflecting that portion of the curve across the x axis produces the complete curve in Figure 2.47(b).

x	y
-3	0
$-\dfrac{13}{4}$	$\dfrac{1}{2}$
-4	1
$-\dfrac{21}{4}$	$\dfrac{3}{2}$
-7	2

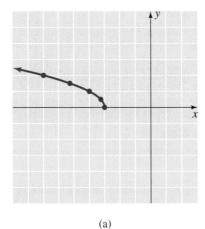

(a)

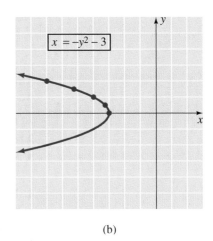

(b)

Figure 2.47

▶ **Now work Problem 57.**   ■

E X A M P L E   6   Graph $x^2 - y^2 = 4$.

Solution

Symmetry   The graph is symmetric with respect to both axes and the origin, because replacing x with $-x$ and y with $-y$ produces $(-x)^2 - (-y)^2 = 4$, which is equivalent to $x^2 - y^2 = 4$.

Intercepts   If $x = 0$, then

$$0^2 - y^2 = 4$$
$$-y^2 = 4$$
$$y^2 = -4$$

Therefore the graph contains no points on the y axis. If $y = 0$, then

$$x^2 - 0^2 = 4$$
$$x^2 = 4$$
$$x = \pm 2$$

Thus the points $(2, 0)$ and $(-2, 0)$ are on the graph.

Restrictions   Solving the given equation for y produces

$$x^2 - y^2 = 4$$
$$-y^2 = 4 - x^2$$
$$y^2 = x^2 - 4$$
$$y = \pm\sqrt{x^2 - 4}$$

Therefore $x^2 - 4 \geq 0$, which is equivalent to $x \geq 2$ or $x \leq -2$.

Table of Values   Because of the restrictions and symmetries, we need only choose values corresponding to $x \geq 2$.

Plotting the Graph   Plotting the points in the table of values, and connecting them with a smooth curve, produces Figure 2.48(a). Because of the symmetry with respect to both axes and the origin, the portion of the curve in Figure 2.48(a) can be reflected across both axes and through the origin to produce the complete curve shown in Figure 2.48(b).

x	y
2	0
3	$\sqrt{5} \approx 2.2$
4	$2\sqrt{3} \approx 3.5$
5	$\sqrt{21} \approx 4.6$
6	$4\sqrt{2} \approx 5.7$

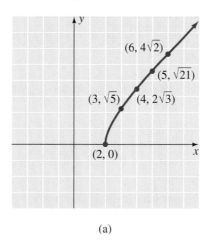

(a)

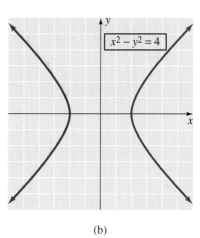

(b)

Figure 2.48

▶ **Now work Problem 41.**   ■

Even when you are using a graphing utility, it is often helpful to determine symmetry, intercepts, and restrictions before graphing the equations. This can serve as a partial check against using the utility incorrectly.

E X A M P L E   7

Use a graphing utility to obtain the graph of $y = \sqrt{x^2 - 49}$.

Solution

Symmetry   The graph is symmetric with respect to the y axis because replacing x with $-x$ produces the same equation.

Intercepts   If $x = 0$, then $y = \sqrt{-49}$; thus the graph has no points on the y axis. If $y = 0$, then $x = \pm 7$; thus the points $(7, 0)$ and $(-7, 0)$ are on the graph.

Restrictions   Because $x^2 - 49$ has to be nonnegative, we know that $x \leq -7$ or $x \geq 7$.

Now let's enter the expression $\sqrt{x^2 - 49}$ for Y_1 and obtain the graph in Figure 2.49. Note that the graph does exhibit the symmetry, intercepts, and restrictions that we determined earlier.

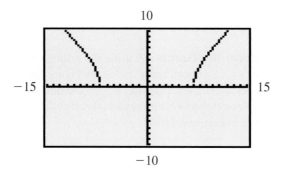

Figure 2.49   ■

C O N C E P T   Q U I Z

1. Give the coordinates of the point that is symmetric to the point (x, y) with respect to the x axis.

2. Give the coordinates of the point that is symmetric to the point (x, y) with respect to the y axis.

3. Give the coordinates of the point that is symmetric to the point (x, y) with respect to the origin.

For Problems 4–8, answer true or false.

4. The graph of the line $y = 3x$ is symmetric to the y axis.

5. If a graph is symmetric to both the x axis and the y axis, the graph is symmetric to the origin.

6. If a graph is symmetric to the origin, then it is symmetric to both the x axis and the y axis.

7. The graph of a straight line is symmetric to the origin only if the graph passes through the origin.

8. Every straight line that passes through the origin is symmetric with respect to the origin.

Problem Set 2.4

For Problems 1–6, determine the points that are symmetric to the given point with respect to the x axis, the y axis, and the origin.

1. $(4, 3)$

2. $(-2, 5)$

3. $(-6, -1)$

4. $(3, -7)$

5. $(0, 4)$

6. $(-5, 0)$

For Problems 7–30, determine the type of symmetry (x axis, y axis, origin) possessed by each graph. Do not sketch the graph.

7. $y = x^2 - 6$

8. $x = y^2 + 1$

9. $x^3 = y^2$

10. $x^2 y^2 = 4$

11. $x^2 + 2y^2 = 6$

12. $3x^2 - y^2 + 4x = 6$

13. $x^2 - 2x + y^2 - 3y - 4 = 0$

14. $xy = 4$

15. $y = x$

16. $2x - 3y = 15$

17. $y = x^3 + 2$

18. $y = x^4 + x^2$

19. $5x^2 - y^2 + 2y - 1 = 0$

20. $x^2 + y^2 - 2y - 4 = 0$

21. $x^2 - xy = y^2$

22. $x^2 + y^2 = 1$

23. $y^2 = x^2 + 2x - 1$

24. $x^2 + y^2 - 2x - y = 4$

25. $y = x^2 + 2x - 1$

26. $y = |x|$

27. $y = \sqrt{x - 2}$

28. $x = 2y^2 + 3y$

29. $y = \dfrac{4}{x}$

30. $2x^2 + 3xy - y^2 = 0$

For Problems 31–58, use symmetry, intercepts, restrictions, and point plotting to help graph each equation.

31. $y = x^2$

32. $y = -x^2$

33. $y = x^2 + 2$

34. $y = -x^2 - 1$

35. $xy = 4$

36. $xy = -2$

37. $y = -x^3$

38. $y = x^3 + 2$

39. $y^2 = x^3$

40. $y^3 = x^2$

41. $y^2 - x^2 = 4$

42. $x^2 - 2y^2 = 8$

43. $y = -\sqrt{x}$

44. $y = \sqrt{x + 1}$

45. $x^2 y = 4$

46. $xy^2 = 4$

47. $x^2 + 2y^2 = 8$

48. $2x^2 + y^2 = 4$

49. $y = \dfrac{4}{x^2 + 1}$

50. $y = \dfrac{-2}{x^2 + 1}$

51. $y = \sqrt{x - 2}$

52. $y = \sqrt{3 - x}$

53. $-xy = 3$

54. $-x^2 y = 4$

55. $x = y^2 + 2$

56. $x = -y^2 + 4$

57. $x = -y^2 - 1$

58. $x = y^2 - 3$

■■■ THOUGHTS INTO WORDS

59. How does the concept of symmetry help when we are graphing equations?

60. Explain how you would go about graphing $x^2 y^2 = 4$.

 GRAPHING CALCULATOR ACTIVITIES

61. Graph $y = \dfrac{4}{x^2}$, $y = \dfrac{4}{(x - 2)^2}$, $y = \dfrac{4}{(x - 4)^2}$, and $y = \dfrac{4}{(x + 2)^2}$ on the same set of axes. Now predict the graph for $y = \dfrac{4}{(x - 6)^2}$. Check your prediction.

62. Graph $y = \sqrt{x}$, $y = \sqrt{x + 1}$, $y = \sqrt{x - 2}$, and $y = \sqrt{x - 4}$ on the same set of axes. Now predict the graph for $y = \sqrt{x + 3}$. Check your prediction.

63. Graph $y = \sqrt{x}$, $y = 2\sqrt{x}$, $y = 4\sqrt{x}$, and $y = 7\sqrt{x}$ on the same set of axes. How does the constant in front of the radical seem to affect the graph?

64. Graph $y = \dfrac{8}{x^2}$ and $y = -\dfrac{8}{x^2}$ on the same set of axes. How does the negative sign seem to affect the graph?

65. Graph $y = \sqrt{x}$ and $y = -\sqrt{x}$ on the same set of axes. How does the negative sign seem to affect the graph?

66. Graph $y = \sqrt{x}$, $y = \sqrt{x} + 2$, $y = \sqrt{x} + 4$, and $y = \sqrt{x} - 3$ on the same set of axes. How does the constant term seem to affect the graph?

67. Graph $y = \sqrt{x}$, $y = \sqrt{x+3}$, $y = \sqrt{x-1}$, and $y = \sqrt{x-5}$ on the same set of axes. How are the graphs related? Predict the location of $y = \sqrt{x+5}$. Check your prediction.

68. To graph $x = y^2$ we need first to solve for y in terms of x. This produces $y = \pm\sqrt{x}$. Now we can let $Y_1 = \sqrt{x}$ and $Y_2 = -\sqrt{x}$ and graph the two equations on the same set of axes. Then graph $x = y^2 + 4$ on this same set of axes. How are the graphs related? Predict the location of the graph of $x = y^2 - 4$. Check your prediction.

69. To graph $x = y^2 + 2y$ we need first to solve for y in terms of x. Let's complete the square to do this:

$$y^2 + 2y = x$$

$$y^2 + 2y + 1 = x + 1$$

$$(y+1)^2 = (\sqrt{x+1})^2$$

$$y + 1 = \sqrt{x+1} \qquad \text{or} \qquad y + 1 = -\sqrt{x+1}$$

$$y = -1 + \sqrt{x+1} \qquad \text{or} \qquad y = -1 - \sqrt{x+1}$$

Thus let's make the assignments $Y_1 = -1 + \sqrt{x+1}$ and $Y_2 = -1 - \sqrt{x+1}$ and graph them on the same set of axes to produce the graph of $x = y^2 + 2y$. Then graph $x = y^2 + 2y - 4$ on this same set of axes. Now predict the location of the graph of $x = y^2 + 2y + 4$. Check your prediction.

Answers to the Concept Quiz

1. $(x, -y)$ **2.** $(-x, y)$ **3.** $(-x, -y)$ **4.** False **5.** True **6.** False **7.** True **8.** True

---

2.5 Circles, Ellipses, and Hyperbolas

Objectives

- Write the equation of a circle.

- Find the center and length of a radius, given the equation of a circle.

- Draw graphs of circles, ellipses, and hyperbolas.

■ Circles

When we apply the distance formula

$$d = \sqrt{(x_2 - x_1)^2 + (y_2 - y_1)^2}$$

(developed in Section 2.1) to the definition of a circle, we get what is known as the **standard form of the equation of a circle.** We start with a precise definition of a circle.

Definition 2.2

A **circle** is the set of all points in a plane equidistant from a given fixed point called the **center.** A line segment determined by the center and any point on the circle is called a **radius.**

Now let's consider a circle that has a radius of length r and a center at (h, k) on a coordinate system (see Figure 2.50). For any point P on the circle with coordinates (x, y), the length of a radius, denoted by r, can be expressed as

$$r = \sqrt{(x - h)^2 + (y - k)^2}$$

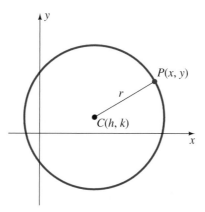

Figure 2.50

Squaring both sides of this equation, we obtain the standard form of the equation of a circle:

$$(x - h)^2 + (y - k)^2 = r^2$$

This form of the equation of a circle can be used to solve the two basic kinds of problems: (1) given the coordinates of the center of a circle and the length of a radius of a circle, find its equation, and (2) given the equation of a circle, determine its graph. Let's illustrate each of these types of problems.

E X A M P L E   1

Find the equation of a circle that has its center at $(-3, 5)$ and has a radius of length four units.

Solution

Substitute -3 for h, 5 for k, and 4 for r in the standard equation and simplify to give the equation of the circle.

$$(x - h)^2 + (y - k)^2 = r^2$$
$$[x - (-3)]^2 + (y - 5)^2 = 4^2$$
$$(x + 3)^2 + (y - 5)^2 = 4^2$$
$$x^2 + 6x + 9 + y^2 - 10y + 25 = 16$$
$$x^2 + y^2 + 6x - 10y + 18 = 0$$

▶ **Now work Problem 1.**   ∎

Note that in Example 1 we simplified the equation to the form $x^2 + y^2 + Dx + Ey + F = 0$, where D, E, and F are constants. This is another form that we commonly use when working with circles.

E X A M P L E   2

Graph $x^2 + y^2 - 6x + 4y + 9 = 0$.

Solution

We can change the given equation into the standard form for a circle by completing the square on x and on y:

$$x^2 + y^2 - 6x + 4y + 9 = 0$$
$$(x^2 - 6x) + (y^2 + 4y) = -9$$
$$(x^2 - 6x + 9) + (y^2 + 4y + 4) = -9 + 9 + 4$$

| Add 9 to complete the square on x | Add 4 to complete the square on y | Add 9 and 4 to compensate for the 4 and 9 added on the left side |

$$(x - 3)^2 + (y + 2)^2 = 2^2$$
$$(x - 3)^2 + [y - (-2)]^2 = 2^2$$

$$\qquad\quad h \qquad\qquad k \qquad r$$

The center is at $(3, -2)$, and the length of a radius is two units. The circle is drawn in Figure 2.51.

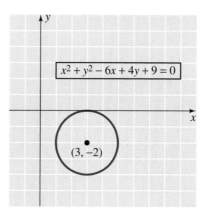

Figure 2.51

▶ **Now work Problem 9.**   ∎

E X A M P L E   3

Find the center and length of a radius of the circle

$$4x^2 + 4x + 4y^2 - 12y - 26 = 0$$

Solution

$$4x^2 + 4x + 4y^2 - 12y - 26 = 0$$

$$4(x^2 + x + \underline{\ \ }) + 4(y^2 - 3y + \underline{\ \ }) = 26$$

$$4\left(x^2 + x + \frac{1}{4}\right) + 4\left(y^2 - 3y + \frac{9}{4}\right) = 26 + 1 + 9$$

$$4\left(x + \frac{1}{2}\right)^2 + 4\left(y - \frac{3}{2}\right)^2 = 36$$

$$\left(x + \frac{1}{2}\right)^2 + \left(y - \frac{3}{2}\right)^2 = 9$$

$$\left[x - \left(-\frac{1}{2}\right)\right]^2 + \left(y - \frac{3}{2}\right)^2 = 3^2$$

$$\underset{h}{\uparrow} \qquad\qquad \underset{k}{\uparrow} \quad \underset{r}{\uparrow}$$

Therefore the center is at $\left(-\dfrac{1}{2}, \dfrac{3}{2}\right)$ and the length of a radius is three units.

▶ **Now work Problem 17.**   ∎

Now suppose that we substitute 0 for h and 0 for k in the standard form of the equation of a circle.

$$(x - h)^2 + (y - k)^2 = r^2$$
$$(x - 0)^2 + (y - 0)^2 = r^2$$
$$x^2 + y^2 = r^2$$

The form $x^2 + y^2 = r^2$ is called the **standard form of the equation of a circle that has its center at the origin.** For example, by inspection we can recognize that $x^2 + y^2 = 9$ is a circle with its center at the origin and a radius of length three units. Likewise, the equation $5x^2 + 5y^2 = 10$ is equivalent to $x^2 + y^2 = 2$; therefore its graph is a circle with its center at the origin and a radius of length $\sqrt{2}$ units. Furthermore, we can easily determine that the equation of the circle with its center at the origin and a radius of eight units is $x^2 + y^2 = 64$.

■ Ellipses

Generally it is true that any equation of the form $Ax^2 + By^2 = F$ (where $A = B$ and A, B, and F are nonzero constants that have the same sign) is a circle with its center at the origin. We can use the general equation $Ax^2 + By^2 = F$ to describe other geometric figures by changing the restrictions on A and B. For example, if A, B, and F are of the same sign but $A \neq B$, then the graph of the equation $Ax^2 + By^2 = F$ is an **ellipse.** Let's consider two examples.

E X A M P L E   4   Graph $4x^2 + 9y^2 = 36$.

Solution

Let's find the intercepts. If $x = 0$, then

$$4(0)^2 + 9y^2 = 36$$
$$9y^2 = 36$$
$$y^2 = 4$$
$$y = \pm 2$$

Thus the points $(0, 2)$ and $(0, -2)$ are on the graph. If $y = 0$, then

$$4x^2 + 9(0)^2 = 36$$
$$4x^2 = 36$$
$$x^2 = 9$$
$$x = \pm 3$$

Thus the points $(3, 0)$ and $(-3, 0)$ are on the graph. Because we know that it is an ellipse, plotting the four points that we have gives us a pretty good sketch of the figure (see Figure 2.52).

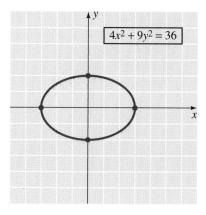

Figure 2.52

▶ **Now work Problem 39.**   ■

    In Figure 2.52, the line segment with endpoints at $(-3, 0)$ and $(3, 0)$ is called the **major axis** of the ellipse. The shorter segment with endpoints at $(0, -2)$ and $(0, 2)$ is called the **minor axis.** Establishing the endpoints of the major and minor axes provides a basis for sketching an ellipse. Also note that the equation $4x^2 + 9y^2 = 36$ exhibits symmetry with respect to both axes and the origin, as we see in Figure 2.52.

EXAMPLE   5

Graph $25x^2 + y^2 = 25$.

Solution

The endpoints of the major and minor axes can be determined by finding the intercepts. If $x = 0$, then

$$25(0)^2 + y^2 = 25$$
$$y^2 = 25$$
$$y = \pm 5$$

The endpoints of the major axis are therefore at $(0, 5)$ and $(0, -5)$. If $y = 0$, then

$$25x^2 + (0)^2 = 25$$
$$25x^2 = 25$$
$$x^2 = 1$$
$$x = \pm 1$$

The endpoints of the minor axis are at $(1, 0)$ and $(-1, 0)$. The ellipse is sketched in Figure 2.53.

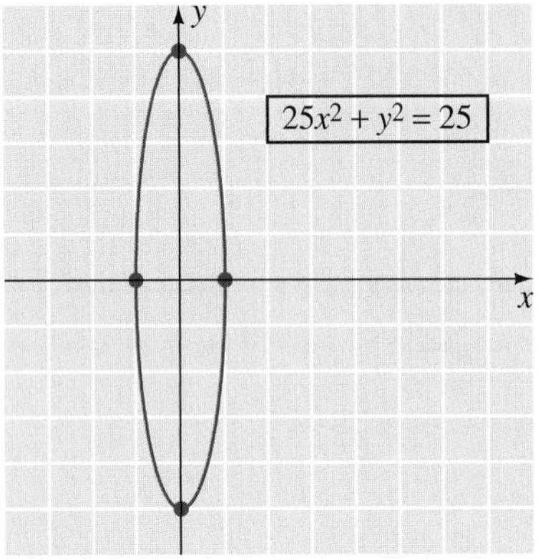

Figure 2.53

▶ **Now work Problem 33.** ■

■ Hyperbolas

The graph of an equation of the form $Ax^2 + By^2 = F$, where A and B are of *unlike* signs, is a **hyperbola.** The next two examples illustrate the graphing of hyperbolas.

EXAMPLE 6

Graph $x^2 - 4y^2 = 4$.

Solution

If we let $y = 0$, then

$$x^2 - 4(0)^2 = 4$$
$$x^2 = 4$$
$$x = \pm 2$$

Thus the points $(2, 0)$ and $(-2, 0)$ are on the graph. If we let $x = 0$, then

$$0^2 - 4y^2 = 4$$
$$-4y^2 = 4$$
$$y^2 = -1$$

Because $y^2 = -1$ has no real number solutions, there are no points of the graph on the y axis.

Note that the equation $x^2 - 4y^2 = 4$ exhibits symmetry with respect to both axes and the origin. Now let's solve the given equation for y to get a more convenient form for finding other solutions.

$$x^2 - 4y^2 = 4$$
$$-4y^2 = 4 - x^2$$

$$4y^2 = x^2 - 4$$

$$y^2 = \frac{x^2 - 4}{4}$$

$$y = \frac{\pm \sqrt{x^2 - 4}}{2}$$

Because the radicand, $x^2 - 4$, must be nonnegative, the values chosen for x must be such that $x \geq 2$ or $x \leq -2$. Symmetry and the points determined by the table provide the basis for sketching Figure 2.54.

x	y	
2	0	intercepts
−2	0	
3	$\pm\frac{\sqrt{5}}{2}$	
4	$\pm\sqrt{3}$	other points
5	$\pm\frac{\sqrt{21}}{2}$	

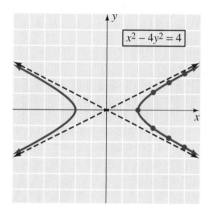

$x^2 - 4y^2 = 4$

Figure 2.54

▶ **Now work Problem 47.**   ■

Note the dashed lines in Figure 2.54; they are called **asymptotes**. Each **branch** of the hyperbola approaches one of these lines but does not intersect it. Therefore, being able to sketch the asymptotes of a hyperbola is very helpful for graphing purposes. Fortunately, the equations of the asymptotes are easy to determine. They can be found by replacing the constant term in the given equation of the hyperbola with zero and then solving for y. (The reason this works will be discussed in a later chapter.) For the hyperbola in Example 6, we obtain

$$x^2 - 4y^2 = 0$$

$$-4y^2 = -x^2$$

$$y^2 = \frac{1}{4}x^2$$

$$y = \pm\frac{1}{2}x$$

Thus the lines $y = \dfrac{1}{2}x$ and $y = -\dfrac{1}{2}x$ are the asymptotes indicated by the dashed lines in Figure 2.54.

EXAMPLE 7   Graph $4y^2 - 9x^2 = 36$.

Solution

If $x = 0$, then

$$4y^2 - 9(0)^2 = 36$$
$$4y^2 = 36$$
$$y^2 = 9$$
$$y = \pm 3$$

The points $(0, 3)$ and $(0, -3)$ are on the graph. If $y = 0$, then

$$4(0)^2 - 9x^2 = 36$$
$$-9x^2 = 36$$
$$x^2 = -4$$

Because $x^2 = -4$ has no real number solutions, we know that this hyperbola does not intersect the x axis. Solving the equation for y yields

$$4y^2 - 9x^2 = 36$$
$$4y^2 = 9x^2 + 36$$
$$y^2 = \frac{9x^2 + 36}{4}$$
$$y = \frac{\pm\sqrt{9x^2 + 36}}{2}$$
$$y = \frac{\pm\sqrt{9(x^2 + 4)}}{2}$$
$$y = \pm\frac{3\sqrt{x^2 + 4}}{2}$$

The table shows some additional solutions. The equations of the asymptotes are determined as follows:

$$4y^2 - 9x^2 = 0$$
$$4y^2 = 9x^2$$
$$y^2 = \frac{9}{4}x^2$$
$$y = \pm\frac{3}{2}x$$

Sketching the asymptotes, plotting the points from the table, and using symmetry, we determine the hyperbola in Figure 2.55.

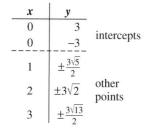

x	y	
0	3	intercepts
0	−3	
1	$\pm\frac{3\sqrt{5}}{2}$	
2	$\pm 3\sqrt{2}$	other points
3	$\pm\frac{3\sqrt{13}}{2}$	

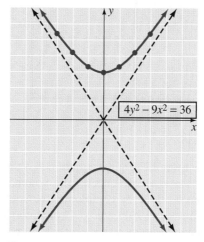

$$4y^2 - 9x^2 = 36$$

Figure 2.55

  Now work Problem 41. ■

When using a graphing utility, we may find it necessary to change the boundaries on x or y (or both) to obtain a complete graph. Consider the following example.

EXAMPLE 8

Use a graphing utility to graph $x^2 - 40x + y^2 + 351 = 0$.

Solution

First we need to solve for y in terms of x:

$$x^2 - 40x + y^2 + 351 = 0$$
$$y^2 = -x^2 + 40x - 351$$
$$y = \pm\sqrt{-x^2 + 40x - 351}$$

Now we can make the following assignments:

$$Y_1 = \sqrt{-x^2 + 40x - 351}$$
$$Y_2 = -Y_1$$

(Note that we assigned Y_2 in terms of Y_1. By doing this, we avoid repetitive key strokes and reduce the chance for errors. You may need to consult your user's manual for instructions on how to key stroke $-Y_1$.) Figure 2.56 shows the graph.

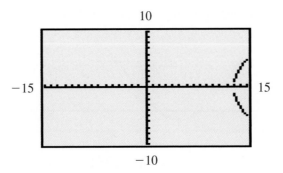

Figure 2.56

We know from the original equation that this graph should be a circle, so we need to make some adjustments on the boundaries in order to get a complete graph. This can be done by completing the square on the original equation to change its form to $(x - 20)^2 + y^2 = 49$ or simply by a trial-and-error process. By changing the boundaries on x such that $-15 \leq x \leq 30$, we obtain Figure 2.57.

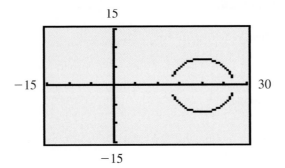

Figure 2.57   ■

In summarizing this section, we do want you to be aware of the continuity pattern used. We started by using the definition of a circle to generate the standard form of the equation of a circle. Then we discussed ellipses and hyperbolas, not from a definition viewpoint, but by considering variations of the general equation of a circle with its center at the origin ($Ax^2 + By^2 = F$, where A, B, and F are of the same sign and $A = B$). In Chapter 8 we will develop parabolas, ellipses, and hyperbolas from a definition viewpoint. In other words, we first define each of the concepts and then use those definitions to generate standard forms for their equations.

CONCEPT QUIZ   For Problems 1–3, match each equation with its name.

1. $x^2 - y^2 = 36$ A. Circle

2. $4x^2 + y^2 = 36$ B. Ellipse

3. $x^2 + y^2 = 36$ C. Hyperbola

For Problems 4–8, answer true or false.

4. A circle is the set of all points in a plane that are equidistant from a fixed point.

5. The circle $(x - 2)^2 + (y + 4)^2 = 20$ has its center at $(2, 4)$.

6. The circle $(x - 2)^2 + (y + 4)^2 = 20$ has a radius of length 20 units.

7. For ellipses, the major axis is always parallel to the x axis.

8. The graph of a hyperbola has two branches.

Problem Set 2.5

For Problems 1–8, write the equation of each circle. Express the final equations in the form $x^2 + y^2 + Dx + Ey + F = 0$.

1. Center at $(2, 3)$ and $r = 5$

2. Center at $(-3, 4)$ and $r = 2$

3. Center at $(-1, -5)$ and $r = 3$

4. Center at $(4, -2)$ and $r = 1$

5. Center at $(3, 0)$ and $r = 3$

6. Center at $(0, -4)$ and $r = 6$

7. Center at the origin and $r = 7$

8. Center at the origin and $r = 1$

For Problems 9–24, find the center and length of a radius of each circle.

9. $x^2 + y^2 - 6x - 10y + 30 = 0$

10. $x^2 + y^2 + 8x - 12y + 43 = 0$

11. $x^2 + y^2 + 10x + 14y + 73 = 0$

12. $x^2 + y^2 + 6y - 7 = 0$

13. $x^2 + y^2 - 10x = 0$

14. $x^2 + y^2 - 4x + 2y = 0$

15. $x^2 + y^2 = 8$ **16.** $4x^2 + 4y^2 = 1$

17. $4x^2 + 4y^2 - 4x - 8y - 11 = 0$

18. $36x^2 + 36y^2 + 48x - 36y - 11 = 0$

19. $x^2 + y^2 - 4y - 2 = 0$ **20.** $x^2 + y^2 + 8x + 4 = 0$

21. $3x^2 + 3y^2 - 6x + 12y - 1 = 0$

22. $2x^2 + 2y^2 - 10x - 2y + 2 = 0$

23. $2x^2 + 2y^2 + 6x + 14y + 4 = 0$

24. $4x^2 + 4y^2 + 8x - 3 = 0$

25. Find the equation of the circle that passes through the origin and has its center at $(6, -8)$.

26. Find the equation of the circle where the line segment determined by the points $(3, -4)$ and $(-3, 2)$ is a diameter.

27. Find the equation of the circle where the line segment determined by $(-4, 9)$ and $(10, -3)$ is a diameter.

28. Find the equation of the circle that passes through the origin and has its center at $(-3, -4)$.

29. Find the equation of the circle that is tangent to both axes, has a radius of length seven units, and has its center in the fourth quadrant.

30. Find the equation of the circle that passes through the origin, has an x intercept of -6, and has a y intercept of 12. (The perpendicular bisector of a chord contains the center of the circle.)

31. Find the equations of the circles that are tangent to the x axis and have a radius of length five units. In each case, the abscissa of the center is -3. (There is more than one circle that satisfies these conditions.)

For Problems 32–48, graph each equation.

32. $4x^2 + 25y^2 = 100$ **33.** $9x^2 + 4y^2 = 36$

34. $x^2 - y^2 = 4$ **35.** $y^2 - x^2 = 9$

36. $x^2 + y^2 - 4x - 2y - 4 = 0$

37. $x^2 + y^2 - 4x = 0$ **38.** $4x^2 + y^2 = 4$

39. $x^2 + 9y^2 = 36$

40. $x^2 + y^2 + 2x - 6y - 6 = 0$

41. $y^2 - 3x^2 = 9$ **42.** $4x^2 - 9y^2 = 16$

43. $x^2 + y^2 + 4x + 6y - 12 = 0$

44. $2x^2 + 5y^2 = 50$ **45.** $4x^2 + 3y^2 = 12$

46. $x^2 + y^2 - 6x + 8y = 0$

47. $3x^2 - 2y^2 = 3$ **48.** $y^2 - 8x^2 = 9$

The graphs of equations of the form $xy = k$, where k is a nonzero constant, are also hyperbolas, sometimes referred to as **rectangular hyperbolas.** For Problems 49–52, graph each rectangular hyperbola.

49. $xy = 2$ **50.** $xy = 4$

51. $xy = -3$ **52.** $xy = -2$

■ ■ ■ THOUGHTS INTO WORDS

53. What is the graph of $xy = 0$? Explain your answer.

54. We have graphed various equations of the form $Ax^2 + By^2 = F$, where F is a nonzero constant. Describe the graph of each of the following and explain your answers.

   a. $x^2 + y^2 = 0$ **b.** $2x^2 + 3y^2 = 0$

   c. $x^2 - y^2 = 0$ **d.** $4x^2 - 9y^2 = 0$

■ ■ ■ FURTHER INVESTIGATIONS

55. By expanding $(x - h)^2 + (y - k)^2 = r^2$, we obtain $x^2 - 2hx + h^2 + y^2 - 2ky + k^2 - r^2 = 0$. Comparing this result to the form $x^2 + y^2 + Dx + Ey + F = 0$, we see that $D = -2h$, $E = -2k$, and $F = h^2 + k^2 - r^2$. Therefore the center and the length of a radius of a circle can be found by using $h = D/-2$, $k = E/-2$, and $r = \sqrt{h^2 - k^2 - F}$. Use these relationships to find the center and the length of a radius of each of the following circles.

   a. $x^2 + y^2 - 2x - 8y + 8 = 0$

   b. $x^2 + y^2 + 4x - 14y + 49 = 0$

   c. $x^2 + y^2 + 12x + 8y - 12 = 0$

   d. $x^2 + y^2 - 16x + 20y + 115 = 0$

   e. $x^2 + y^2 - 12y - 45 = 0$

   f. $x^2 + y^2 + 14x = 0$

56. Use a coordinate geometry approach to prove that an angle inscribed in a semicircle is a right angle (see Figure 2.58).

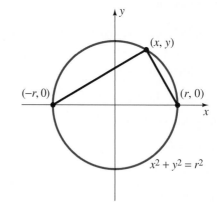

Figure 2.58

57. Use a coordinate geometry approach to prove that a line segment from the center of a circle bisecting a chord is perpendicular to the chord. [*Hint*: Let the ends of the chord be $(r, 0)$ and (a, b).]

GRAPHING CALCULATOR ACTIVITIES

58. For each of the following equations, predict the type and location of the graph, and then use your graphing calculator to check your prediction.

    a. $x^2 + y^2 = 9$ **b.** $2x^2 + y^2 = 4$

    c. $x^2 - y^2 = 9$ **d.** $4x^2 - y^2 = 16$

    e. $x^2 + 2x + y^2 - 4 = 0$

    f. $x^2 + y^2 - 4y - 2 = 0$

    g. $(x - 2)^2 + (y + 1)^2 = 4$

    h. $(x + 3)^2 - (y - 4)^2 = 9$

    i. $9y^2 - 4x^2 = 36$ **j.** $9y^2 + 4x^2 = 36$

Answers to the Concept Quiz

1. C   **2.** B   **3.** A   **4.** True   **5.** False   **6.** False   **7.** False   **8.** True

Chapter 2   Summary

We emphasized throughout this chapter that coordinate geometry deals with two basic kinds of problems:

1. Given an algebraic equation, determine its geometric graph.
2. Given a set of conditions pertaining to a geometric figure, determine its algebraic equation.

Let's review this chapter in terms of those two kinds of problems.

■ Graphing

The following graphing techniques were discussed in this chapter.

1. Recognize the type of graph that a certain kind of equation produces.

   a. $Ax + By = C$ produces a straight line.
   b. $x^2 + y^2 + Dx + Ey + F = 0$ produces a circle. The center and the length of a radius can be found by completing the square and comparing it to the standard form of the equation of a circle:

   $$(x - h)^2 + (y - k)^2 = r^2$$

   c. $Ax^2 + By^2 = F$, where A, B, and F have the same sign and $A = B$, produces a circle with the center at the origin.
   d. $Ax^2 + By^2 = F$, where A, B, and F are of the same sign but $A \neq B$, produces an ellipse.
   e. $Ax^2 + By^2 = F$, where A and B are of unlike signs, produces a hyperbola.

2. Determine the symmetry that a graph possesses.

   a. The graph of an equation is symmetric with respect to the y axis if replacing x with $-x$ results in an equivalent equation.
   b. The graph of an equation is symmetric with respect to the x axis if replacing y with $-y$ results in an equivalent equation.
   c. The graph of an equation is symmetric with respect to the origin if replacing x with $-x$ and y with $-y$ results in an equivalent equation.

3. Find the intercepts. The x intercept is found by letting $y = 0$ and solving for x. The y intercept is found by letting $x = 0$ and solving for y.

4. Determine the restrictions necessary to ensure real number solutions.

5. Set up a table of ordered pairs that satisfy the equation. The type of symmetry and the restrictions will affect your choice of values in the table. Furthermore, it may be convenient to change the form of the original equation by solving for y in terms of x or for x in terms of y.

6. Plot the points associated with the ordered pairs in the table and connect them with a smooth curve. Then, if appropriate, reflect the curve according to any symmetries possessed by the graph.

■ Determining Equations When Given Certain Conditions

You should review Examples 3, 4, and 5 of Section 2.3 to be sure you are thoroughly familiar with the general approach of choosing a point (x, y) and using it to determine the equation that satisfies the conditions stated in the problem.

We developed some special forms that can be used to determine equations:

Point–slope form of a straight line: $y - y_1 = m(x - x_1)$

Slope–intercept form of a straight line: $y = mx + b$

Standard form of a circle: $(x - h)^2 + (y - k)^2 = r^2$

The following formulas were used in different parts of the chapter:

Distance formula: $d = \sqrt{(x_2 - x_1)^2 + (y_2 - y_1)^2}$

Midpoint formula: The coordinates of the midpoint of a line segment determined by (x_1, y_1) and (x_2, y_2) are

$$\left(\frac{x_1 + x_2}{2}, \frac{y_1 + y_2}{2} \right)$$

Slope formula: $m = \dfrac{y_2 - y_1}{x_2 - x_1}, \qquad x_1 \neq x_2$

Chapter 2   **Review Problem Set**

1. On a number line, find the coordinate of the point located three-fifths of the distance from -4 to 11.

2. On a number line, find the coordinate of the point located four-ninths of the distance from 3 to -15.

3. In the xy plane, find the coordinates of the point located five-sixths of the distance from $(-1, -3)$ to $(11, 1)$.

4. If one endpoint of a line segment is at $(8, 14)$, and the midpoint of the segment is $(3, 10)$, find the coordinates of the other endpoint.

5. Verify that the points $(2, 2)$, $(6, 4)$, and $(5, 6)$ are vertices of a right triangle.

6. Verify that the points $(-3, 1)$, $(1, 3)$, and $(9, 7)$ lie in a straight line.

For Problems 7–12, identify any symmetries (x axis, y axis, origin) that the equation exhibits.

7. $x = y^2 + 4$
8. $y = x^2 + 6x - 1$
9. $5x^2 - y^2 = 4$
10. $x^2 + y^2 - 2y - 4 = 0$
11. $y = -x$
12. $y = \dfrac{6}{x^2 + 4}$

For Problems 13–22, graph each of the following.

13. $x^2 + y^2 - 6x + 4y - 3 = 0$
14. $x^2 + 4y^2 = 16$
15. $x^2 - 4y^2 = 16$
16. $-2x + 3y = 6$
17. $2x - y < 4$
18. $x^2 y^2 = 4$
19. $4y^2 - 3x^2 = 8$
20. $x^2 + y^2 + 10y = 0$
21. $9x^2 + 2y^2 = 36$
22. $y \leq -2x - 3$

23. Find the slope of the line determined by $(-3, -4)$ and $(-5, 6)$.

24. Find the slope of the line with equation $5x - 7y = 12$.

For Problems 25–28, write the equation of the line that satisfies the stated conditions. Express final equations in standard form ($Ax + By = C$).

25. Contains the point $(7, 2)$ and has a slope of $-\dfrac{3}{4}$

26. Contains the points $(-3, -2)$ and $(1, 6)$

27. Contains the point $(2, -4)$ and is parallel to $4x + 3y = 17$

28. Contains the point $(-5, 4)$ and is perpendicular to $2x - y = 7$

For Problems 29–32, write the equation of the circle that satisfies the stated conditions. Express final equations in the form $x^2 + y^2 + Dx + Ey + F = 0$.

29. Center at $(5, -6)$ and $r = 1$

30. The endpoints of a diameter are $(-2, 4)$ and $(6, 2)$.

31. Center at $(-5, 12)$ and passes through the origin

32. Tangent to both axes, $r = 4$, and center in the third quadrant

1. On a number line, find the coordinate of the point located two-thirds of the distance from -4 to 14.

2. In the xy plane, find the coordinates of the point located three-fourths of the distance from $(2, -3)$ to $(-6, 9)$.

3. If one endpoint of a line segment is at $(-2, -1)$, and the midpoint of the segment is at $\left(2, -\dfrac{5}{2} \right)$, find the coordinates of the other endpoint.

4. Find the slope of the line determined by $(-4, -2)$ and $(5, -6)$.

5. Find the slope of the line determined by the equation $2x - 7y = -9$.

For Problems 6–10, determine the equation of the line that satisfies the stated conditions. Express final equations in standard form.

6. Has a slope of $-\dfrac{3}{4}$ and a y intercept of -3

7. Contains the points $(1, -4)$ and $(4, 7)$

8. Contains the point $(-1, 4)$ and is parallel to $x - 5y = 5$

9. Contains the point $(3, 5)$ and is perpendicular to $4x + 7y = 3$

10. Contains the point $(-2, -4)$ and is perpendicular to the x axis

For Problems 11–13, determine the equation of the circle that satisfies the stated conditions. Express final equations in the form $x^2 + y^2 + Dx + Ey + F = 0$.

11. Center at $(-3, -6)$ and a radius of length four units

12. The endpoints of a diameter are at $(-1, 3)$ and $(5, 5)$.

13. Center at $(4, -3)$ and passes through the origin

14. Find the center and the length of a radius of the circle $x^2 + 16x + y^2 - 10y + 80 = 0$.

15. Find the lengths of the three sides of the triangle determined by $(3, 2)$, $(5, -2)$, and $(-1, -1)$. Express the lengths in simplest radical form.

16. Find the x intercepts of the graph of the equation $x^2 - 6x + y^2 + 2y + 5 = 0$.

17. Find the y intercepts of the graph of the equation $5x^2 + 12y^2 = 36$.

18. Find the length of the major axis of the ellipse $9x^2 + 2y^2 = 18$.

19. Find the equations of the asymptotes for the hyperbola $9x^2 - 16y^2 = 48$.

20. Identify any symmetries (x axis, y axis, origin) that the equation exhibits.

    a. $x^2 + 2x + y^2 - 6 = 0$

    b. $xy = -4$

    c. $y = \dfrac{4}{x^2 + 1}$

    d. $x^2 y^2 = 5$

21. Graph the inequality $3x - y \le 6$.

For Problems 22–25, graph the equation.

22. $y^2 - 2x^2 = 9$

23. $x = y^2 - 4$

24. $3x^2 + 5y^2 = 45$

25. $x^2 + 4x + y^2 - 12 = 0$

For Problems 1-6, evaluate each expression.

1. 3^{-3}

2. -4^{-2}

3. $\left(\dfrac{2}{3}\right)^{-2}$

4. $-\sqrt[3]{\dfrac{8}{27}}$

5. $\left(\dfrac{1}{27}\right)^{-2/3}$

6. $\dfrac{1}{\left(\dfrac{3}{4}\right)^{-2}}$

For Problems 7-12, perform the indicated operations and simplify. Express final answers using positive exponents only.

7. $(5x^{-3}y^{-2})(4xy^{-1})$

8. $(-7a^{-3}b^2)(8a^4b^{-3})$

9. $\left(\dfrac{1}{2}x^{-2}y^{-1}\right)^{-2}$

10. $\dfrac{80x^{-3}y^{-4}}{16xy^{-6}}$

11. $\left(\dfrac{102x^{2/3}y^{3/4}}{6xy^{-1}}\right)^{-1}$

12. $\left(\dfrac{14a^3b^{-4}}{7a^{-1}b^3}\right)^2$

For Problems 13-20, express each in simplest radical form. All variables represent positive real numbers.

13. $-5\sqrt{72}$

14. $2\sqrt{27x^3y^2}$

15. $\sqrt[3]{56x^4y^7}$

16. $\dfrac{3\sqrt{18}}{5\sqrt{12}}$

17. $\sqrt{\dfrac{3x}{7y}}$

18. $\dfrac{5}{\sqrt{2}-3}$

19. $\dfrac{3\sqrt{7}}{2\sqrt{2}-\sqrt{6}}$

20. $\dfrac{4\sqrt{x}}{\sqrt{x}+3\sqrt{y}}$

For Problems 21-26, perform the indicated operations involving rational expressions. Express final answers in simplest form.

21. $\dfrac{12x^2y}{18x}\cdot\dfrac{9x^3y^3}{16xy^2}$

22. $\dfrac{-15ab^2}{14a^3b}\div\dfrac{20a}{7b^2}$

23. $\dfrac{3x^2+5x-2}{x^2-4}\cdot\dfrac{5x^2-9x-2}{3x^2-x}$

24. $\dfrac{2x-1}{4}+\dfrac{3x+2}{6}-\dfrac{x-1}{8}$

25. $\dfrac{5}{3n^2}-\dfrac{2}{n}+\dfrac{3}{2n}$

26. $\dfrac{5x}{x^2+6x-27}+\dfrac{3}{x^2-9}$

For Problems 27-38, solve each equation.

27. $3(-2x-1)-2(3x+4)=-4(2x-3)$

28. $(2x-1)(3x+4)=(x+2)(6x-5)$

29. $\dfrac{3x-1}{4}-\dfrac{2x-1}{5}=\dfrac{1}{10}$

30. $9x^2-4=0$

31. $5x^3+10x^2-40x=0$

32. $7t^2-31t+12=0$

33. $x^4+15x^2-16=0$

34. $|5x-2|=3$

35. $2x^2-3x-1=0$

36. $(3x-2)(x+4)=(2x-1)(x-1)$

37. $\sqrt{5-t}+1=\sqrt{7+2t}$

38. $(2x-1)^2+4=0$

For Problems 39-48, solve each inequality. Express the solution sets using interval notation.

39. $-2(x-1)+(3-2x)>4(x+1)$

40. $2n+1+\dfrac{3n-1}{4}\geq\dfrac{n-1}{2}$

41. $0.09x+0.12(450-x)\geq46.5$

42. $n^2+5n>24$

43. $6x^2+7x-3<0$

44. $(2x - 1)(x + 3)(x - 4) > 0$

45. $\dfrac{3x - 2}{x + 1} \leq 0$ **46.** $\dfrac{x + 5}{x - 1} \geq 2$

47. $|3x - 1| > 5$ **48.** $|5x - 3| < 12$

For Problems 49–54, graph each equation.

49. $x^2 + 4y^2 = 36$ **50.** $4x^2 - y^2 = 4$

51. $y = -x^3 - 1$ **52.** $y = -x + 3$

53. $y^2 - 5x^2 = 9$ **54.** $y = -\dfrac{3}{4}x - 1$

For Problems 55–58, solve each problem.

55. Find the center and the length of a radius of the circle with equation $x^2 + y^2 + 14x - 8y + 56 = 0$.

56. Write the equation of the line that is parallel to $3x - 4y = 17$ and contains the point $(2, 8)$.

57. Find the coordinates of the point located one-fifth of the distance from $(-3, 4)$ to $(2, 14)$.

58. Write the equation of the perpendicular bisector of the line segment determined by $(-3, 4)$ and $(5, 10)$.

For Problems 59–65, set up an equation and solve the problem.

59. A retailer has some shirts that cost $22 per shirt. At what price should they be sold to obtain a profit of 30% of the cost? At what price should they be sold to obtain a profit of 30% of the selling price?

60. A total of $7500 was invested, part of it at 5% yearly interest and the remainder at 6%. The total yearly interest was $420. How much was invested at each rate?

61. The length of a rectangle is 1 inch less than twice the width. The area of the rectangular region is 36 square inches. Find the length and width of the rectangle.

62. The length of one side of a triangle is 4 centimeters less than three times the length of the altitude to that side. The area of the triangle is 80 square centimeters. Find the length of the side and the length of the altitude to that side.

63. How many milliliters of pure acid must be added to 40 milliliters of a 30% acid solution to obtain a 50% acid solution?

64. Amanda rode her bicycle out into the country at a speed of 15 miles per hour and returned along the same route at 10 miles per hour. The round trip took 5 hours. How far out did she ride?

65. If two inlet pipes are both open, they can fill a pool in 1 hour and 12 minutes. One of the pipes can fill the pool by itself in 2 hours. How long would it take the other pipe to fill the pool by itself?

3

Functions

*The volume of a right circular
cylinder is a function of its
height and the length of
a radius of its base.*

© Alvis Uptis/Getty Images/Riser

Suppose that the cost of burning a 60-watt light bulb is determined by the function $c(h) = 0.0036h$, where h represents the number of hours that the bulb burns. Using this *linear function,* we can determine the cost of burning the bulb for any specific number of hours. For example, burning the bulb for a 24-hour period would cost $c(24) = (0.0036)(24) = 0.0864$, or approximately 9 cents.

A golf pro shop manager finds that she can sell 30 sets of golf clubs at $500 per set in a year. Furthermore, she predicts that for each $25 decrease in price, she could sell an additional 3 sets of clubs. At what price should she sell the clubs to maximize gross income? The *quadratic function* $f(x) = -75x^2 + 750x + 15,000$ can be used to determine that she should sell the clubs at $375 per set.

One of the fundamental concepts of mathematics is that of a function. Functions are used to unify different areas of mathematics, and they also serve as a meaningful way of applying mathematics to many real-world problems. Functions provide a means of studying quantities that vary with one another — that is, quantities such that a change in one produces a corresponding change in another. In this chapter we will (1) introduce the basic ideas pertaining to functions, (2) use the idea of a function to unify some concepts from Chapter 2, and (3) discuss some applications in which functions are used.

3.1   Concept of a Function

Objectives

- Know the definition of a function.
- Apply the vertical line test to determine if a graph represents a function.
- Evaluate a function for a given input value.
- Evaluate a piecewise-defined function for a given input value.
- Find the difference quotient of a function.
- Determine the domain and range of a function.
- Determine if a function is even, odd, or neither even nor odd.
- Solve application problems involving functions.

The notion of correspondence is used in everyday situations and is central to the concept of a function. Consider the following correspondences:

1. To each person in a class, there corresponds an assigned seat.

2. To each day of a year, there corresponds an assigned integer that represents the average temperature for that day in a certain geographic location.

3. To each book in a library, there corresponds a whole number that represents the number of pages in the book.

Such correspondences can be depicted as in Figure 3.1. To each member in set *A* there corresponds *one and only one* member in set *B*. For example, in correspondence 1, set *A* would consist of the students in a class, and set *B* would be the assigned seats. In the second example, set *A* would consist of the days of a year, and set *B* would be a set of integers. Furthermore, the same integer might be assigned to more than one day of the year. (Different days might have the same average temperature.) The key idea is that *one and only one* integer is assigned to *each* day of the year. Likewise, in the third example, more than one book may have the same number of pages, but to each book there is assigned one and only one number of pages.

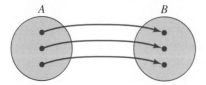

Figure 3.1

Mathematically, the general concept of a function can be defined as follows:

Definition 3.1

A **function** f is a correspondence between two sets X and Y that assigns to each element x of set X one and only one element y of set Y. The element y being assigned is called the **image** of x. The set X is called the **domain** of the function, and the set of all images is called the **range** of the function.

In Definition 3.1 the image y is usually denoted by $f(x)$. Thus the symbol $f(x)$, which is read *f of x* or *the value of f at x*, represents the element in the range associated with the element x from the domain. Figure 3.2 depicts this situation. Again, we emphasize that each member of the domain has precisely one image in the range; however, different members in the domain, such as a and b in Figure 3.2, may have the same image.

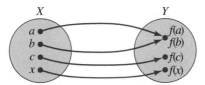

Figure 3.2

In Definition 3.1 we named the function f. It is common to name functions by means of a single letter, and the letters f, g, and h are often used. We suggest you make more meaningful choices when functions are used in real-world situations. For example, if a problem involves a profit function, then naming the function p or even P seems natural. Be careful not to confuse f and $f(x)$. Remember that f is used to name a function, whereas $f(x)$ is an element of the range — namely, the element assigned to x by f.

The assignments made by a function are often expressed as ordered pairs. For example, the assignments in Figure 3.2 could be expressed as $(a, f(a))$, $(b, f(b))$, $(c, f(c))$, and $(x, f(x))$, where the first components are from the domain, and the second components are from the range. Thus a function can also be thought of as **a set of ordered pairs in which no two of the ordered pairs have the same first component.**

Remark: In some texts the concept of a **relation** is introduced first, and then functions are defined as special kinds of relations. A relation is defined as *a set of ordered pairs,* and a function is defined as *a relation in which no two ordered pairs have the same first element.*

The ordered pairs that represent a function can be generated by various means, such as a graph or a chart. However, one of the most common ways of generating ordered pairs is by using equations. For example, the equation $f(x) = 2x + 3$ indicates that to each value of x in the domain, we assign $2x + 3$ from the range. For example,

$$f(1) = 2(1) + 3 = 5 \qquad \text{produces the ordered pair } (1, 5)$$
$$f(4) = 2(4) + 3 = 11 \qquad \text{produces the ordered pair } (4, 11)$$
$$f(-2) = 2(-2) + 3 = -1 \qquad \text{produces the ordered pair } (-2, -1)$$

It may be helpful for you to picture the concept of a function as a *function machine*, as illustrated in Figure 3.3. Each time a value of x is put into the machine, the equation $f(x) = 2x + 3$ is used to generate one and only one value for $f(x)$ to be ejected from the machine.

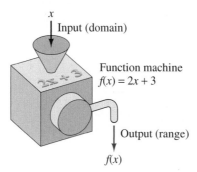

Figure 3.3

Using the ordered-pair interpretation of a function, we can define the **graph** of a function f to be the set of all points in a plane of the form $(x, f(x))$, where x is from the domain of f. In other words, the graph of f is the same as the graph of the equation $y = f(x)$. Furthermore, because $f(x)$, or y, takes on only one value for each value of x, we can easily tell whether a given graph represents a function. For example, in Figure 3.4(a), for any choice of x there is only one value for y. Geometrically, this means that no vertical line intersects the curve in more than one point. On the other hand, Figure 3.4(b) does not represent the graph of a function because certain values of x (all positive values) produce more than one value for y. In other words, some vertical lines intersect the curve in more than one point, as illustrated in Figure 3.4(b). A **vertical line test** for functions can be stated as follows:

Vertical Line Test

If each vertical line intersects a graph in no more than one point, then the graph represents a function.

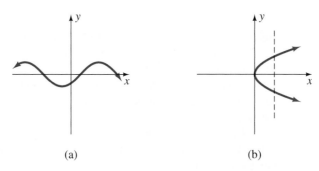

(a)   (b)

Figure 3.4

Let's consider some examples to help pull together some of these ideas about functions.

EXAMPLE 1

If $f(x) = x^2 - x + 4$ and $g(x) = x^3 - x^2$, find $f(3), f(-2), g(4)$, and $g(-3)$.

Solution

$$f(3) = 3^2 - 3 + 4 = 10 \qquad f(-2) = (-2)^2 - (-2) + 4 = 10$$
$$g(4) = 4^3 - 4^2 = 48 \qquad g(-3) = (-3)^3 - (-3)^2 = -36$$

▶ **Now work Problem 1.**   ■

Note that in Example 1 we were working with two different functions in the same problem. That is why we used two different names, f and g. Sometimes the rule of assignment for a function may consist of more than one part. We often refer to such functions as **piecewise-defined functions.** An everyday example of this concept is that the price of admission to a theme park depends on whether you are a child, an adult, or a senior citizen. Let's consider an example of such a function.

PROBLEM 1

A progressive county government collects taxes for schools based on the income of an individual rather than real estate taxes. The county uses the following function to determine the amount of tax due, where x represents an individual's income in dollars.

$$f(x) = \begin{cases} 0.05x & 0 < x \le 20{,}000 \\ 0.06x & 20{,}000 < x \le 50{,}000 \\ 0.08x & 50{,}000 < x \end{cases}$$

If Darren, Martha, Tonieka, and Caleb, respectively, have incomes of $23,000, $18,500, $55,000, and $20,000, find the amount of tax for each person.

Solution

Because Darren has an income of $23,000, the function to be used is $f(x) = 0.06x$. Therefore, the tax is figured as $f(23,000) = 0.06(23,000) = 1380$. Darren will pay $1,380 in tax.

Because Martha has an income of $18,500, the function to be used is $f(x) = 0.05x$. Therefore, the tax is figured as $f(18,500) = 0.05(18,500) = 925$. Martha will pay $925 in tax.

Because Tonieka has an income of $55,000, the function to be used is $f(x) = 0.08x$. Therefore, the tax is figured as $f(55,000) = 0.08(55,000) = 4400$. Tonieka will pay $4,400 in tax.

Because Caleb has an income of $20,000 the function to be used is $f(x) = 0.06x$. Therefore, the tax is figured as $f(20,000) = 0.06(20,000) = 1200$. Caleb will pay $1200 in tax.

▶ **Now work Problem 87.** ■

E X A M P L E   2   If $f(x) = \begin{cases} 2x + 1 & \text{for } x \geq 0 \\ 3x - 1 & \text{for } x < 0 \end{cases}$,   find $f(2), f(4), f(-1)$, and $f(-3)$.

Solution

For $x \geq 0$, we use the assignment $f(x) = 2x + 1$.

$$f(2) = 2(2) + 1 = 5$$
$$f(4) = 2(4) + 1 = 9$$

For $x < 0$, we use the assignment $f(x) = 3x - 1$.

$$f(-1) = 3(-1) - 1 = -4$$
$$f(-3) = 3(-3) - 1 = -10$$

▶ **Now work Problem 9.** ■

The quotient $\dfrac{f(a + h) - f(a)}{h}$ is often called a **difference quotient.** We use it extensively with functions when studying the limit concept in calculus. The next examples illustrate finding the difference quotient for specific functions.

E X A M P L E   3   Find $\dfrac{f(a + h) - f(a)}{h}$ for each of the following functions.

a. $f(x) = 3x + 7$ **b.** $f(x) = 2x^2 + 3x - 4$ **c.** $f(x) = \dfrac{1}{x}$

Solutions

a. $f(a) = 3a + 7$

$$f(a + h) = 3(a + h) + 7 = 3a + 3h + 7$$

Therefore

$$f(a + h) - f(a) = 3a + 3h + 7 - (3a + 7)$$
$$= 3a + 3h + 7 - 3a - 7 = 3h$$

and

$$\frac{f(a + h) - f(a)}{h} = \frac{3\cancel{h}}{\cancel{h}} = 3$$

b. $\quad f(a) = 2a^2 + 3a - 4$

$$f(a + h) = 2(a + h)^2 + 3(a + h) - 4$$
$$= 2(a^2 + 2ha + h^2) + 3a + 3h - 4$$
$$= 2a^2 + 4ha + 2h^2 + 3a + 3h - 4$$

Therefore

$$f(a + h) - f(a) = (2a^2 + 4ha + 2h^2 + 3a + 3h - 4) - (2a^2 + 3a - 4)$$
$$= 2a^2 + 4ha + 2h^2 + 3a + 3h - 4 - 2a^2 - 3a + 4$$
$$= 4ha + 2h^2 + 3h$$

and

$$\frac{f(a + h) - f(a)}{h} = \frac{4ha + 2h^2 + 3h}{h}$$
$$= \frac{\cancel{h}(4a + 2h + 3)}{\cancel{h}}$$
$$= 4a + 2h + 3$$

c. $\quad f(a) = \dfrac{1}{a}$

$$f(a + h) = \frac{1}{a + h}$$

Therefore

$$f(a + h) - f(a) = \frac{1}{a + h} - \frac{1}{a}$$
$$= \frac{a - (a + h)}{a(a + h)}$$
$$= \frac{a - a - h}{a(a + h)}$$
$$= \frac{-h}{a(a + h)} \qquad \text{or} \qquad -\frac{h}{a(a + h)}$$

and

$$\frac{f(a + h) - f(a)}{h} = \frac{-\dfrac{h}{a(a + h)}}{h}$$

$$= -\frac{h}{a(a + h)} \cdot \frac{1}{h}$$

$$= -\frac{1}{a(a + h)}$$

▶ **Now work Problems 15, 23, and 27.** ∎

For our purposes in this text, if the domain of a function is not specifically in-dicated or determined by a real-world application, then we will assume the domain to be *all real number* replacements for the variable, provided they represent ele-ments in the domain and produce *real number* functional values.

E X A M P L E   4

For the function $f(x) = \sqrt{x - 1}$, (a) specify the domain, (b) determine the range, and (c) evaluate $f(5), f(50),$ and $f(25)$.

Solutions

  a. The radicand must be nonnegative, so $x - 1 \geq 0$, and thus $x \geq 1$. There-fore the domain, D, is

  $$D = \{x|x \geq 1\}$$

  b. The symbol $\sqrt{}$ indicates the nonnegative square root; thus the range, R, is

  $$R = \{f(x)|f(x) \geq 0\}$$

  c. $f(5) = \sqrt{4} = 2$

  $f(50) = \sqrt{49} = 7$

  $f(25) = \sqrt{24} = 2\sqrt{6}$ ∎

As we will see later, the range of a function is often easier to determine after we have graphed the function. However, our equation- and inequality-solving pro-cesses are frequently sufficient to determine the domain of a function. Let's con-sider some examples.

E X A M P L E   5

Determine the domain for each of the following functions.

  a. $f(x) = \dfrac{3}{2x - 5}$ **b.** $g(x) = \dfrac{1}{x^2 - 9}$ **c.** $f(x) = \sqrt{x^2 + 4x - 12}$

Solutions

a. We can replace x with any real number except $\frac{5}{2}$ because $\frac{5}{2}$ makes the denominator zero. Thus the domain is

$$D = \left\{ x \,\middle|\, x \neq \frac{5}{2} \right\}$$

b. We need to eliminate any values of x that will make the denominator zero. Therefore let's solve the equation $x^2 - 9 = 0$:

$$x^2 - 9 = 0$$
$$x^2 = 9$$
$$x = \pm 3$$

The domain is thus the set

$$D = \{x \,|\, x \neq 3 \text{ and } x \neq -3\}$$

c. The radicand, $x^2 + 4x - 12$, must be nonnegative. Therefore let's use a number-line approach, as we did in Chapter 2, to solve the inequality $x^2 + 4x - 12 \geq 0$ (see Figure 3.5):

$$x^2 + 4x - 12 \geq 0$$
$$(x + 6)(x - 2) \geq 0$$

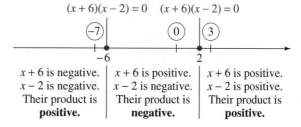

Figure 3.5

The product $(x + 6)(x - 2)$ is nonnegative if $x \leq -6$ or $x \geq 2$. Using interval notation, we can express the domain as $(-\infty, -6] \cup [2, \infty)$.

▶ **Now work Problems 53, 61, and 71.** ■

Functions and function notation provide the basis for describing many real-world relationships. The next example illustrates this point.

E X A M P L E   6

Suppose a factory determines that the overhead for producing a quantity of a certain item is $500, and the cost for each item is $25. Express the total expenses as a function of the number of items produced, and compute the expenses for producing 12, 25, 50, 75, and 100 items.

Solution

Let n represent the number of items produced. Then $25n + 500$ represents the total expenses. Using E to represent the *expense function,* we have

$$E(n) = 25n + 500 \qquad \text{where } n \text{ is a whole number}$$

Therefore we obtain

$$E(12) = 25(12) + 500 = 800$$
$$E(25) = 25(25) + 500 = 1125$$
$$E(50) = 25(50) + 500 = 1750$$
$$E(75) = 25(75) + 500 = 2375$$
$$E(100) = 25(100) + 500 = 3000$$

Thus the total expenses for producing 12, 25, 50, 75, and 100 items are $800, $1125, $1750, $2375, and $3000, respectively.

▶ **Now work Problem 91.** ∎

As we stated before, an equation such as $f(x) = 5x - 7$ that is used to determine a function can also be written $y = 5x - 7$. In either form, we refer to x as the **independent variable** and to y [or $f(x)$] as the **dependent variable.** Many formulas in mathematics and other related areas also determine functions. For example, the area formula for a circular region, $A = \pi r^2$, assigns to each positive real value for r a unique value for A. This formula determines a function f, where $f(r) = \pi r^2$. The variable r is the independent variable, and A [or $f(r)$] is the dependent variable.

Many functions that we will study throughout this text can be classified as even or odd functions. A function f having the property that $f(-x) = f(x)$ for every x in the domain of f is called an **even function.** A function f having the property that $f(-x) = -f(x)$ for every x in the domain of f is called an **odd function.**

EXAMPLE 7

For each of the following, classify the function as even, odd, or neither even nor odd.

a. $f(x) = 2x^3 - 4x$ **b.** $f(x) = x^4 - 7x^2$ **c.** $f(x) = x^2 + 2x - 3$

Solutions

a. The function $f(x) = 2x^3 - 4x$ is an odd function because $f(-x) = 2(-x)^3 - 4(-x) = -2x^3 + 4x$, which equals $-f(x)$.

b. The function $f(x) = x^4 - 7x^2$ is an even function because $f(-x) = (-x)^4 - 7(-x)^2 = x^4 - 7x^2$, which equals $f(x)$.

c. The function $f(x) = x^2 + 2x - 3$ is neither even nor odd because $f(-x) = (-x)^2 + 2(-x) - 3 = x^2 - 2x - 3$, which does not equal either $f(x)$ or $-f(x)$.

▶ **Now work Problem 81.** ∎

CONCEPT QUIZ   For Problems 1–8, answer true or false.

1. For a function, each member of the domain has precisely one image in the range.

2. The set of ordered pairs {(1,2), (2,2), (3,2), (4,2)} is a function.

3. The graph of a function g is the set of all points in a plane of the form $(x, g(x))$, where x is from the domain of g.

4. If a vertical line intersects a graph at more than one point, then the graph does not represent a function.

5. A piecewise-defined function assigns different rules to subsets of the domain.

6. The quotient $\dfrac{f(a) - f(a - h)}{a}$ is called the difference quotient.

7. For the function $f(r) = 2\pi r$, r is the dependent variable.

8. Every function can be classified as either an even or odd function.

Problem Set 3.1

1. If $f(x) = -2x + 5$, find $f(3)$, $f(5)$, and $f(-2)$.

2. If $f(x) = x^2 - 3x - 4$, find $f(2)$, $f(4)$, and $f(-3)$.

3. If $g(x) = -2x^2 + x - 5$, find $g(3)$, $g(-1)$, and $g(-4)$.

4. If $g(x) = -x^2 - 4x + 6$, find $g(0)$, $g(5)$, and $g(-5)$.

5. If $h(x) = \dfrac{2}{3}x - \dfrac{3}{4}$, find $h(3)$, $h(4)$, and $h\left(-\dfrac{1}{2}\right)$.

6. If $h(x) = -\dfrac{1}{2}x + \dfrac{2}{3}$, find $h(-2)$, $h(6)$, and $h\left(-\dfrac{2}{3}\right)$.

7. If $f(x) = \sqrt{2x - 1}$, find $f(5)$, $f\left(\dfrac{1}{2}\right)$, and $f(23)$.

8. If $f(x) = \sqrt{3x + 2}$, find $f\left(\dfrac{14}{3}\right)$, $f(10)$, and $f\left(-\dfrac{1}{3}\right)$.

9. If $f(x) = \begin{cases} x & \text{for } x \ge 0 \\ x^2 & \text{for } x < 0 \end{cases}$, find $f(4)$, $f(10)$, $f(-3)$, and $f(-5)$.

10. If $f(x) = \begin{cases} 3x + 2 & \text{for } x \ge 0 \\ 5x - 1 & \text{for } x < 0 \end{cases}$, find $f(2)$, $f(6)$, $f(-1)$, and $f(-4)$.

11. If $f(x) = \begin{cases} 2x & \text{for } x \ge 0 \\ -2x & \text{for } x < 0 \end{cases}$, find $f(3)$, $f(5)$, $f(-3)$, and $f(-5)$.

12. If $f(x) = \begin{cases} 2 & \text{for } x < 0 \\ x^2 + 1 & \text{for } 0 \le x \le 4 \\ -1 & \text{for } x > 4 \end{cases}$, find $f(3)$, $f(6)$, $f(0)$, and $f(-3)$.

13. If $f(x) = \begin{cases} 1 & \text{for } x > 0 \\ 0 & \text{for } -1 < x \le 0 \\ -1 & \text{for } x \le -1 \end{cases}$, find $f(2)$, $f(0)$, $f\left(-\dfrac{1}{2}\right)$, and $f(-4)$.

For Problems 14–29, find $\dfrac{f(a + h) - f(a)}{h}$.

14. $f(x) = 4x + 5$

15. $f(x) = -7x - 2$

16. $f(x) = -3x + 12$

17. $f(x) = \dfrac{1}{2}x + 4$

18. $f(x) = 4$

19. $f(x) = -3$

20. $f(x) = x^2 - 3x$

21. $f(x) = -x^2 + 4x - 2$

22. $f(x) = 2x^2 + 7x - 4$

23. $f(x) = 3x^2 - x - 4$

24. $f(x) = x^3$

25. $f(x) = x^3 - x^2 + 2x - 1$

26. $f(x) = \dfrac{1}{x + 1}$

27. $f(x) = \dfrac{2}{x - 1}$

28. $f(x) = \dfrac{x}{x + 1}$

29. $f(x) = \dfrac{1}{x^2}$

For Problems 30–37 (Figures 3.6 through 3.13), determine whether the indicated graph represents a function of x.

30.

31.

Figure 3.6   **Figure 3.7**

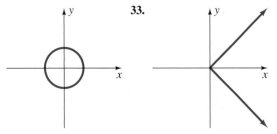

32.

33.

Figure 3.8   **Figure 3.9**

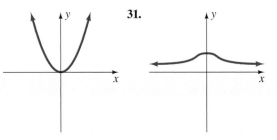

34.

35.

Figure 3.10   **Figure 3.11**

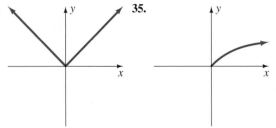

36.   **37.**

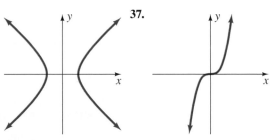

Figure 3.12   **Figure 3.13**

For Problems 38–51, determine the domain and the range of the given function.

38. $f(x) = \sqrt{x}$

39. $f(x) = \sqrt{3x - 4}$

40. $f(x) = x^2 + 1$

41. $f(x) = x^2 - 2$

42. $f(x) = x^3$

43. $f(x) = |x|$

44. $f(x) = x^4$

45. $f(x) = -\sqrt{x}$

46. $f(x) = \sqrt{2x - 5}$

47. $f(x) = \sqrt{x - 2} + 3$

48. $f(x) = \sqrt{x + 4} - 2$

49. $f(x) = |x| + 5$

50. $f(x) = |x - 1| - 3$

51. $f(x) = -|x| - 6$

For Problems 52–65, determine the domain of the given function.

52. $f(x) = \dfrac{3}{x - 4}$

53. $f(x) = \dfrac{-4}{x + 2}$

54. $f(x) = \dfrac{2x}{(x - 2)(x + 3)}$

55. $f(x) = \dfrac{5}{(2x - 1)(x + 4)}$

56. $f(x) = \sqrt{5x + 1}$

57. $f(x) = \dfrac{1}{x^2 - 4}$

58. $g(x) = \dfrac{3}{x^2 + 5x + 6}$

59. $f(x) = \dfrac{4x}{x^2 - x - 12}$

60. $g(x) = \dfrac{5}{x^2 + 4x}$

61. $g(x) = \dfrac{x}{6x^2 + 13x - 5}$

62. $f(x) = \dfrac{x + 2}{x^2 + 1}$

63. $f(x) = \sqrt{-3x - 1}$

64. $f(x) = \sqrt{-2x + 5}$

65. $f(x) = \dfrac{2x - 1}{x^2 + 4}$

For Problems 66–75, express the domain of the given function using interval notation.

66. $f(x) = \sqrt{x^2 - 1}$

67. $f(x) = \sqrt{x^2 - 16}$

68. $f(x) = \sqrt{x^2 + 4}$

69. $f(x) = \sqrt{x^2 + 1} - 4$

70. $f(x) = \sqrt{x^2 - 2x - 24}$

71. $f(x) = \sqrt{x^2 - 3x - 40}$

72. $f(x) = \sqrt{12x^2 + x - 6}$

73. $f(x) = -\sqrt{8x^2 + 6x - 35}$

74. $f(x) = \sqrt{4 - x^2}$

75. $f(x) = \sqrt{16 - x^2}$

For Problems 76–85, determine whether f is even, odd, or neither even nor odd.

76. $f(x) = x^2$

77. $f(x) = x^3$

78. $f(x) = x^2 + 1$

79. $f(x) = 3x - 1$

80. $f(x) = x^2 + x$

81. $f(x) = x^3 + 1$

82. $f(x) = x^5$

83. $f(x) = x^4 + x^2 + 1$

84. $f(x) = -x^3$

85. $f(x) = x^5 + x^3 + x$

For Problems 86–95, solve each problem.

86. An equipment rental agency charges rent in dollars for a small backhoe according to the following function: h represents the number of hours the backhoe is rented. Find the rent charged when the backhoe is rented for 6.5 hours; 3 hours; and 10 hours.

$$f(h) = \begin{cases} 100 + 50h & 0 < h \le 3 \\ 160 + 30h & 3 < h \le 8 \\ 200 + 25h & 8 < h \end{cases}$$

87. A copy center charges for copies depending on the number of copies made. The following functions are used to determine the cost in dollars of color or black and white copies, where n is the number of copies.

Color Copies

$$c(n) = \begin{cases} 0.89n & 0 < n \le 20 \\ 0.79n & 20 < n \le 50 \\ 0.69n & 50 < n \end{cases}$$

Black and White Copies

$$b(n) = \begin{cases} 0.09n & 0 < n \le 50 \\ 0.08n & 50 < n \le 200 \\ 0.06n & 200 < n \end{cases}$$

Isaac is producing a cookbook that requires him to make 20 color copies and 210 black and white copies. What will it cost Isaac to make the copies?

88. Suppose that the profit function for selling n items is given by

$$P(n) = -n^2 + 500n - 61{,}500$$

Evaluate $P(200)$, $P(230)$, $P(250)$, and $P(260)$.

89. The equation $A(r) = \pi r^2$ expresses the area of a circular region as a function of the length of a radius, r. Compute $A(2)$, $A(3)$, $A(12)$, and $A(17)$, and express your answers to the nearest hundredth.

90. In a physics experiment it is found that the equation $V(t) = 1667t - 6940t^2$ expresses the velocity of an object as a function of time, t. Compute $V(0.1)$, $V(0.15)$, and $V(0.2)$.

91. The height of a projectile fired vertically into the air (neglecting air resistance) at an initial velocity of 64 feet per second is a function of the time, t, and is given by the equation $h(t) = 64t - 16t^2$. Compute $h(1)$, $h(2)$, $h(3)$, and $h(4)$.

92. A car rental agency charges $50 per day plus $0.32 a mile. Therefore the daily charge for renting a car is a function of the number of miles traveled, m, and can be expressed as $C(m) = 50 + 0.32m$. Compute $C(75)$, $C(150)$, $C(225)$, and $C(650)$.

93. The equation $I(r) = 500r$ expresses the amount of simple interest earned by an investment of $500 for one year as a function of the rate of interest, r. Compute $I(0.11)$, $I(0.12)$, $I(0.135)$, and $I(0.15)$.

94. Suppose that the height of a semielliptical archway is given by the function $h(x) = \sqrt{64 - 4x^2}$, where x is the distance from the center line of the arch. Compute $h(0)$, $h(2)$, and $h(4)$.

95. The equation $A(r) = 2\pi r^2 + 16\pi r$ expresses the total surface area of a right circular cylinder of height 8 centimeters as a function of the length of a radius, r. Compute $A(2)$, $A(4)$, and $A(8)$ and express your answers to the nearest hundredth.

■ ■ ■ THOUGHTS INTO WORDS

96. Expand Definition 3.1 to include a definition for the concept of a relation.

97. What does it mean to say that the domain of a function may be restricted if the function represents a real-world situation? Give three examples of such functions.

98. Does $f(a + b) = f(a) + f(b)$ for all functions? Defend your answer.

99. Are there any functions for which $f(a + b) = f(a) + f(b)$? Defend your answer.

Answers to the Concept Quiz

1. True **2.** True **3.** True **4.** True **5.** True **6.** False **7.** False **8.** False

3.2 Linear Functions and Applications

Objectives

■ Graph linear functions.

■ Determine a linear function for specified conditions.

■ Solve application problems involving linear functions.

As we use the function concept in our study of mathematics, it is helpful to classify certain types of functions and become familiar with their equations, characteristics, and graphs. This will enhance our problem-solving capabilities.

Any function that can be written in the form

$$f(x) = ax + b$$

where a and b are real numbers, is called a **linear function.** The following are examples of linear functions.

$$f(x) = -2x + 4 \qquad f(x) = 3x - 6 \qquad f(x) = \frac{2}{3}x + \frac{5}{6}$$

The equation $f(x) = ax + b$ can also be written $y = ax + b$. From our work in Section 2.3, we know that $y = ax + b$ is the equation of a straight line that has a slope of a and a y intercept of b. This information can be used to graph linear functions, as illustrated by the following example.

E X A M P L E   1

Graph $f(x) = -2x + 4$.

Solution

Because the y intercept is 4, the point $(0, 4)$ is on the line. Furthermore, because the slope is -2, we can move two units down and one unit to the right of $(0, 4)$ to determine the point $(1, 2)$. The line determined by $(0, 4)$ and $(1, 2)$ is shown in Figure 3.14.

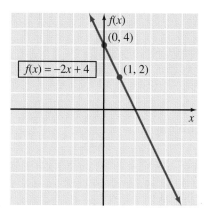

Figure 3.14

▶ **Now work Problem 3.**   ∎

Note that in Figure 3.14 we labeled the vertical axis $f(x)$. It could also be labeled y because $y = f(x)$. We will use the label $f(x)$ for most of our work with functions; however, we will continue to refer to y axis symmetry instead of $f(x)$ axis symmetry.

Recall from Section 2.2 that we can also graph linear equations by finding the two intercepts. This same approach can be used with linear functions, as illustrated by the next two examples.

E X A M P L E   2   Graph $f(x) = 3x - 6$.

Solution

First, we see that $f(0) = -6$; thus the point $(0, -6)$ is on the graph. Second, by setting $3x - 6$ equal to zero and solving for x, we obtain

$$3x - 6 = 0$$
$$3x = 6$$
$$x = 2$$

Therefore $f(2) = 3(2) - 6 = 0$, and the point $(2, 0)$ is on the graph. The line determined by $(0, -6)$ and $(2, 0)$ is drawn in Figure 3.15.

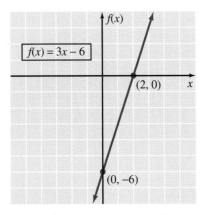

Figure 3.15

▶ **Now work Problem 1.**   ■

E X A M P L E   3   Graph the function $f(x) = \dfrac{2}{3}x + \dfrac{5}{6}$.

Solution

Because $f(0) = \dfrac{5}{6}$ the point $\left(0, \dfrac{5}{6}\right)$ is on the graph. By setting $\dfrac{2}{3}x + \dfrac{5}{6}$ equal to zero
and solving for x, we obtain

$$\frac{2}{3}x + \frac{5}{6} = 0$$
$$\frac{2}{3}x = -\frac{5}{6}$$
$$x = -\frac{5}{4}$$

Therefore $f\left(-\dfrac{5}{4}\right) = 0$ and the point $\left(-\dfrac{5}{4}, 0\right)$ is on the graph. The line determined by the two points $\left(0, \dfrac{5}{6}\right)$ and $\left(-\dfrac{5}{4}, 0\right)$ is shown in Figure 3.16.

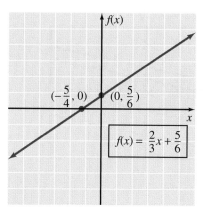

Figure 3.16

▶ **Now work Problem 13.** ∎

   As you graph functions using function notation, it is often helpful to think of the ordinate of every point on the graph as the value of the function at a specific value of x. Geometrically, the functional value is the directed distance of the point from the x axis. This idea is illustrated in Figure 3.17 for the function $f(x) = x$ and in Figure 3.18 for the function $f(x) = 2$. The linear function $f(x) = x$ is often called the **identity function.** Any linear function of the form $f(x) = ax + b$, where $a = 0$ [that is, $f(x) = b$], is called a **constant function,** and its graph is a horizontal line.

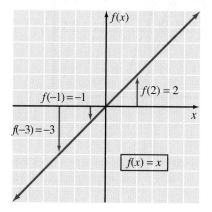

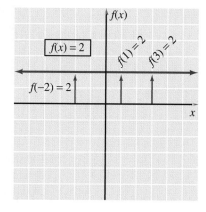

Figure 3.17   **Figure 3.18**

   From our previous work with linear equations, we know that parallel lines have equal slopes and that two perpendicular lines have slopes that are negative reciprocals of each other. Thus when we work with linear functions of the form

$f(x) = ax + b$, it is easy to recognize parallel and perpendicular lines. For example, the lines determined by $f(x) = 0.21x + 4$ and $g(x) = 0.21x - 3$ are parallel lines because both lines have a slope of 0.21. Let's use a graphing calculator to graph these two functions along with $h(x) = 0.21x + 2$ and $p(x) = 0.21x - 7$ (see Figure 3.19).

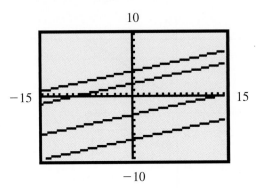

Figure 3.19

The graphs of the functions $f(x) = \dfrac{2}{5}x + 8$ and $g(x) = -\dfrac{5}{2}x - 4$ are perpendicular lines because the slopes $\left(\dfrac{2}{5} \text{ and } -\dfrac{5}{2}\right)$ of the two lines are negative reciprocals of each other. Again using our graphing calculator, let's graph these two functions along with $h(x) = -\dfrac{5}{2}x + 2$ and $p(x) = -\dfrac{5}{2}x - 6$ (see Figure 3.20).

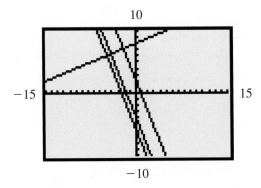

Figure 3.20

Remark: A property of plane geometry states that *If two or more lines are perpendicular to the same line, then they are parallel lines.* Figure 3.20 is a good illustration of that property.

The function notation can also be used to determine linear functions that satisfy certain conditions. Let's see how this works.

E X A M P L E 4

Determine the linear function whose graph is a line with a slope of $\dfrac{1}{4}$ that contains the point (2, 5).

Solution

We can substitute $\frac{1}{4}$ for a in the equation $f(x) = ax + b$ to obtain $f(x) = \frac{1}{4}x + b$.

The fact that the line contains the point $(2, 5)$ means that $f(2) = 5$. Therefore

$$f(2) = \frac{1}{4}(2) + b = 5$$

$$b = \frac{9}{2}$$

and the function is $f(x) = \frac{1}{4}x + \frac{9}{2}$.

▶ **Now work Problem 17.**   ■

■ Applications of Linear Functions

Now let's consider some applications that use the concept of a linear function to connect mathematics to the physical world.

E X A M P L E   5   The cost for burning a 60-watt light bulb is given by the function $c(h) = 0.0036h$, where h represents the number of hours that the bulb is burning.

　　a. How much does it cost to burn a 60-watt bulb for 3 hours per night for a 30-day month?
　　b. Graph the function $c(h) = 0.0036h$.
　　c. Suppose that a 60-watt light bulb is left burning in a closet for a week before it is discovered and turned off. Use the graph from part b to approximate the cost of allowing the bulb to burn for a week. Then use the function to find the exact cost.

Solutions

　　a. $c(90) = 0.0036(90) = 0.324$; the cost, to the nearest cent, is \$0.32.
　　b. Because $c(0) = 0$ and $c(100) = 0.36$, we can use the points $(0, 0)$ and $(100, 0.36)$ to graph the linear function $c(h) = 0.0036h$ (see Figure 3.21).

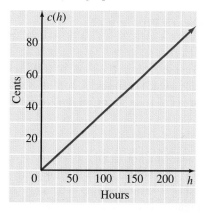

Figure 3.21

c. If the bulb burns for 24 hours per day for a week, it burns for 24(7) = 168 hours. Reading from the graph, we can approximate 168 on the horizontal axis, read up to the line, and then read across to the vertical axis. It looks as though it will cost approximately 60 cents. Using $c(h) = 0.0036h$, we obtain exactly $c(168) = 0.0036(168) = 0.6048$.

▶ **Now work Problem 23.** ■

E X A M P L E 6

The Clear Call Cellular phone company has a fixed monthly charge plus an amount per minute of airtime. In May, Anna used 720 minutes of airtime and had a bill of $54.80. For the month of June, she used 510 minutes of airtime and had a bill of $46.40. Determine the linear function that Clear Call Cellular uses to determine its monthly bills.

Solution

The linear function $f(x) = ax + b$, where x represents the number of airtime minutes, models this situation. Anna's two monthly bills can be represented by the ordered pairs (720, 54.80) and (510, 46.40). From these two ordered pairs, we can determine a, which is the slope of the line:

$$a = \frac{46.40 - 54.80}{510 - 720} = \frac{-8.4}{-210} = 0.04$$

Thus $f(x) = ax + b$ becomes $f(x) = 0.04x + b$. Now either ordered pair can be used to determine the value of b. Using (510, 46.40), we have $f(510) = 46.40$, so

$$f(510) = 0.04(510) + b = 46.40$$

$$b = 26$$

The linear function is $f(x) = 0.04x + 26$. In other words, Clear Call Cellular charges a monthly fee of $26.00 plus $0.04 per minute of airtime.

▶ **Now work Problem 25.** ■

E X A M P L E 7

Suppose that Anna (Example 6) is thinking of switching to Simple Cellular phone company, which charges a monthly fee of $14 plus $0.06 per minute of airtime. Should Anna use Clear Cellular from Example 6 or Simple Cellular?

Solution

The linear function $g(x) = 0.06x + 14$, where x represents the number of airtime minutes, can be used to determine the monthly bill from Simple Cellular. Let's graph this function and $f(x) = 0.04\,x + 26$ from Example 6 on the same set of axes (see Figure 3.22).

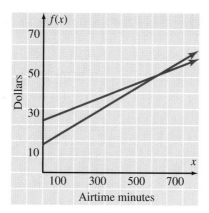

Figure 3.22

Now we see that the two functions have equal values at the point of intersection of the two lines. To find the coordinates of this point, we can set $0.06x + 14$ equal to $0.04x + 26$ and solve for x:

$$0.06x + 14 = 0.04x + 26$$
$$0.02x = 12$$
$$x = 600$$

If $x = 600$, then $0.06(600) + 14 = 50$ and the point of intersection is $(600, 50)$. Again from the lines in Figure 3.22, Anna should switch to Simple Cellular if she uses less than 600 minutes of airtime, but she should stay with Clear Cellular if she plans on using more than 600 minutes of airtime.

▶ **Now work Problem 26.**   ■

CONCEPT QUIZ   For Problems 1–8, answer true or false.

1. Any function of the form of $f(x) = ax^n + b$, where a, b, and n are real numbers, is a linear function.

2. Geometrically, the functional value is the directed distance from the y axis.

3. The graph of a horizontal line represents a function.

4. The linear function $f(x) = 1$ is called the identity function.

5. The graphs of $f(x) = mx + b$ and $f(x) = -mx + b$ are perpendicular lines.

6. Every straight line graph represents a function.

7. If a city has a 7% sales tax on the dollars spent for hotel rooms, then the sales tax is a linear function of the dollars spent on hotel rooms.

8. If the amount of a paycheck varies directly with the number of hours worked, then the amount of the paycheck is a linear function of the hours worked.

Problem Set 3.2

For Problems 1–16, graph each of the linear functions.

1. $f(x) = 2x - 4$

2. $f(x) = 3x + 3$

3. $f(x) = -x + 3$

4. $f(x) = -2x + 6$

5. $f(x) = 3x + 9$

6. $f(x) = 2x - 6$

7. $f(x) = -4x - 4$

8. $f(x) = -x - 5$

9. $f(x) = -3x$

10. $f(x) = -4x$

11. $f(x) = -3$

12. $f(x) = -1$

13. $f(x) = \dfrac{1}{2}x + 3$

14. $f(x) = \dfrac{2}{3}x + 4$

15. $f(x) = -\dfrac{3}{4}x - 6$

16. $f(x) = -\dfrac{1}{2}x - 1$

17. Determine the linear function whose graph is a line with a slope of $\dfrac{2}{3}$ and contains the point $(-1, 3)$.

18. Determine the linear function whose graph is a line with a slope of $-\dfrac{3}{5}$ and contains the point $(4, -5)$.

19. Determine the linear function whose graph is a line that contains the points $(-3, -1)$ and $(2, -6)$.

20. Determine the linear function whose graph is a line that contains the points $(-2, -3)$ and $(4, 3)$.

21. Determine the linear function whose graph is a line that is perpendicular to the line $g(x) = 5x - 2$ and contains the point $(6, 3)$.

22. Determine the linear function whose graph is a line that is parallel to the line $g(x) = -3x - 4$ and contains the point $(2, 7)$.

23. The cost for burning a 75-watt light bulb is given by the function $c(h) = 0.0045h$, where h represents the number of hours that the bulb burns.

   a. How much does it cost to burn a 75-watt bulb for 3 hours per night for a 31-day month? Express your answer to the nearest cent.

   b. Graph the function $c(h) = 0.0045h$.

   c. Use the graph in part b to approximate the cost of burning a 75-watt bulb for 225 hours.

   d. Use $c(h) = 0.0045h$ to find the exact cost, to the nearest cent, of burning a 75-watt bulb for 225 hours.

24. The Rent-Me Car Rental charges $15 per day plus $0.22 per mile to rent a car. Determine a linear function that can be used to calculate daily car rentals. Then use that function to determine the cost of renting a car for a day and driving 175 miles; 220 miles; 300 miles; 460 miles.

25. Suppose ABC Car Rental charges a fixed amount per day plus an amount per mile for renting a car. Heidi rented a car one day and paid $80 for 200 miles. On another day she rented a car from the same agency and paid $117.50 for 350 miles. Determine the linear function that the agency could use to determine its daily rental charges.

26. Suppose that Heidi (of Problem 25) also has access to Speedy Car Rental, which charges a daily fee of $15.00 plus $0.31 per mile. Should Heidi use ABC Car Rental from Problem 25 or Speedy Car Rental?

27. A car rental agency uses the function $f(x) = 26$ for any daily use of a car up to and including 200 miles. For driving more than 200 miles per day, it uses the function $g(x) = 26 + 0.15(x - 200)$ to determine the charges. How much would the company charge for daily driving of 150 miles? 230 miles? 360 miles? 430 miles?

28. A retailer has a number of items that she wants to sell and make a profit of 40% of the cost of each item. The function $s(c) = c + 0.4c = 1.4c$, where c represents the cost of an item, can be used to determine the selling price. Find the selling price of items that cost $1.50, $3.25, $14.80, $21, and $24.20.

29. "All Items 20% Off Marked Price" is a sign at a local golf shop. Create a function and then use it to determine how much one has to pay for each of the following marked items: a $9.50 hat; a $15 umbrella; a $75 pair of golf shoes; a $12.50 golf glove; a $750 set of golf clubs.

30. The linear depreciation method assumes that an item depreciates the same amount each year. Suppose a new piece of machinery costs $32,500, and it depreciates $1950 each year for t years.

a. Set up a linear function that yields the value of the machinery after t years.

b. Find the value of the machinery after 5 years.

c. Find the value of the machinery after 8 years.

d. Graph the function from part a.

e. Use the graph from part d to approximate how many years it takes for the value of the machinery to become zero.

f. Use the function to determine how long it takes for the value of the machinery to become zero.

■ ■ ■ THOUGHTS INTO WORDS

31. Is $f(x) = (3x + 2) - (2x + 1)$ a linear function? Explain your answer.

32. Suppose that Bianca walks at a constant rate of 3 miles per hour. Explain what it means that the distance Bianca walks is a linear function of the time that she walks.

■ ■ ■ FURTHER INVESTIGATIONS

For Problems 33–37, graph each of the functions.

33. $f(x) = |x|$

34. $f(x) = x + |x|$

35. $f(x) = x - |x|$

36. $f(x) = |x| - x$

37. $f(x) = \dfrac{x}{|x|}$

GRAPHING CALCULATOR ACTIVITIES

38. Use a graphing calculator to check your graphs for Problems 1–16.

39. Use a graphing calculator to do parts b and c of Example 5.

40. Use a graphing calculator to check our solution for Example 7.

41. Use a graphing calculator to do parts b and c of Problem 23.

42. Use a graphing calculator to do parts d and e of Problem 30.

43. Use a graphing calculator to check your graphs for Problems 33–37.

44. a. Graph $f(x) = |x|$, $f(x) = 2|x|$, $f(x) = 4|x|$, and $f(x) = \dfrac{1}{2}|x|$ on the same set of axes.

    b. Graph $f(x) = |x|$, $f(x) = -|x|$, $f(x) = -3|x|$, and $f(x) = -\dfrac{1}{2}|x|$ on the same set of axes.

    c. Use your results from parts a and b to make a conjecture about the graphs of $f(x) = a|x|$, where a is a nonzero real number.

    d. Graph $f(x) = |x|$, $f(x) = |x| + 3$, $f(x) = |x| - 4$, and $f(x) = |x| + 1$ on the same set of axes. Make a conjecture about the graphs of $f(x) = |x| + k$, where k is a nonzero real number.

    e. Graph $f(x) = |x|$, $f(x) = |x - 3|$, $f(x) = |x - 1|$, and $f(x) = |x + 4|$ on the same set of axes. Make a conjecture about the graphs of $f(x) = |x - h|$, where h is a nonzero real number.

    f. On the basis of your results from parts a–e, sketch each of the following graphs. Then use a graphing calculator to check your sketches.

        (1) $f(x) = |x - 2| + 3$

        (2) $f(x) = |x + 1| - 4$

        (3) $f(x) = 2|x - 4| - 1$

        (4) $f(x) = -3|x + 2| + 4$

        (5) $f(x) = -\dfrac{1}{2}|x - 3| - 2$

3.3 Quadratic Functions

Objectives

■ Graph quadratic functions.

■ Change the form $f(x) = ax^2 + bx + c$ to the form $f(x) = a(x - h)^2 + k$.

■ Graph piecewise-defined functions.

In Chapter 1 we used graphs of quadratic equations both to predict approximate solutions and to give visual support for solutions that were arrived at algebraically. Those graphs were obtained by using a graphing calculator, but now we want to develop some specific graphing techniques using quadratic functions as examples.

Any function that can be written in the form

$$f(x) = ax^2 + bx + c$$

where a, b, and c are real numbers with $a \neq 0$, is called a **quadratic function.** The graph of any quadratic function is a **parabola.** As we work with parabolas, we will use the vocabulary illustrated in Figure 3.23.

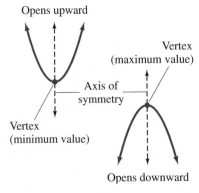

Figure 3.23

■ Graphing Parabolas

The process of graphing parabolas begins with three steps as follows.

1. Finding the vertex.

2. Determining whether parabolas open upward or downward.

3. Locating two points on opposite sides of the axis of symmetry.

It is also helpful to compare the parabolas produced by various types of equations, such as $f(x) = x^2 + k$, $f(x) = ax^2$, $f(x) = (x - h)^2$, and $f(x) = a(x - h)^2 + k$. We are especially interested in how they compare to the **basic parabola** produced by the equation $f(x) = x^2$. The graph of $f(x) = x^2$ is shown in Figure 3.24. Note that the graph of $f(x) = x^2$ is symmetric with respect to the y, or $f(x)$, axis. Remember that an equation exhibits y axis symmetry if replacing x with $-x$ produces an equivalent equation. Therefore, because $f(-x) = (-x)^2 = x^2$, the equation $f(x) = x^2$ exhibits y axis symmetry.

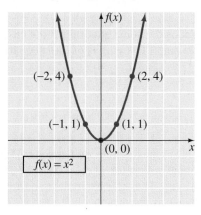

Figure 3.24

Now let's consider an equation of the form $f(x) = x^2 + k$, where k is a constant. (Keep in mind that all such equations exhibit y axis symmetry.)

EXAMPLE 1

Graph $f(x) = x^2 - 2$.

Solution

It should be observed that functional values for $f(x) = x^2 - 2$ are 2 less than corresponding functional values for $f(x) = x^2$. For example, $f(1) = -1$ for $f(x) = x^2 - 2$, but $f(1) = 1$ for $f(x) = x^2$. Thus the graph of $f(x) = x^2 - 2$ is the same as the graph of $f(x) = x^2$ except that it is *moved down two units* (see Figure 3.25).

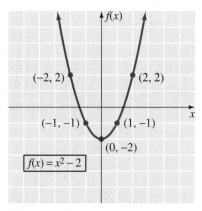

Figure 3.25

▶ **Now work Problem 1.**

> In general, the graph of a quadratic function of the form $f(x) = x^2 + k$ is the same as the graph of $f(x) = x^2$ except that it is moved up or down $|k|$ units, depending on whether k is positive or negative. We say that the graph of $f(x) = x^2 + k$ is a **vertical translation** of the graph of $f(x) = x^2$.

Now let's consider some quadratic functions of the form $f(x) = ax^2$, where a is a nonzero constant. (The graphs of these equations also have y axis symmetry.)

EXAMPLE 2

Graph $f(x) = 2x^2$.

Solution

Let's set up a table to make some comparisons of functional values. Note that in the table the functional values for $f(x) = 2x^2$ are *twice* the corresponding functional values for $f(x) = x^2$. Thus the parabola associated with $f(x) = 2x^2$ has the same vertex (the origin) as the graph of $f(x) = x^2$, but it is *narrower,* as shown in Figure 3.26.

x	$f(x) = x^2$	$f(x) = 2x^2$
0	0	0
1	1	2
2	4	8
−1	1	2
−2	4	8

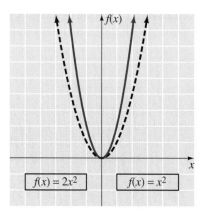

Figure 3.26

▶ **Now work Problem 3.**   ■

EXAMPLE 3

Graph $f(x) = \dfrac{1}{2}x^2$.

Solution

As we see from the table, the functional values for $f(x) = \dfrac{1}{2}x^2$ are *one-half* of the corresponding functional values for $f(x) = x^2$. Therefore the parabola associated with $f(x) = \dfrac{1}{2}x^2$ is *wider* than the basic parabola, as shown in Figure 3.27.

x	$f(x) = x^2$	$f(x) = \frac{1}{2}x^2$
0	0	0
1	1	$\frac{1}{2}$
2	4	2
−1	1	$\frac{1}{2}$
−2	4	2

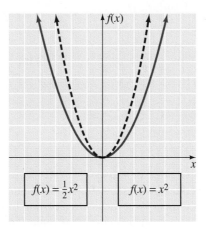

Figure 3.27   ■

E X A M P L E   4

Graph $f(x) = -x^2$.

Solution

It should be evident that the functional values for $f(x) = -x^2$ are the *opposites* of the corresponding functional values for $f(x) = x^2$. Therefore the graph of $f(x) = -x^2$ is a reflection across the x axis of the basic parabola, as shown in Figure 3.28.

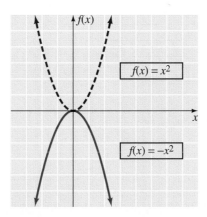

Figure 3.28

▶ **Now work Problem 5.**   ■

In general, the graph of a quadratic function of the form $f(x) = ax^2$ has its vertex at the origin and opens upward if a is positive and downward if a is negative. The parabola is narrower than the basic parabola if $|a| > 1$ and wider if $|a| < 1$.

Let's continue our investigation of quadratic functions by considering those of the form $f(x) = (x - h)^2$, where h is a nonzero constant.

EXAMPLE 5   Graph $f(x) = (x - 3)^2$.

Solution

A fairly extensive table of values illustrates a pattern. Note that $f(x) = (x - 3)^2$ and $f(x) = x^2$ take on the same functional values, but for different values of x. More specifically, if $f(x) = x^2$ achieves a certain functional value at a specific value of x, then $f(x) = (x - 3)^2$ achieves that same functional value at x *plus three*. In other words, the graph of $f(x) = (x - 3)^2$ is the graph of $f(x) = x^2$ *moved three units to the right* (see Figure 3.29).

x	$f(x) = x^2$	$f(x) = (x - 3)^2$
−1	1	16
0	0	9
1	1	4
2	4	1
3	9	0
4	16	1
5	25	4
6	36	9
7	49	16

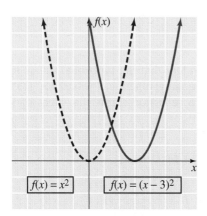

Figure 3.29

▶ **Now work Problem 7.**   ■

In general, the graph of a quadratic function of the form $f(x) = (x - h)^2$ is the same as the graph of $f(x) = x^2$ except that it is moved to the right h units if h is positive or moved to the left $|h|$ units if h is negative. We say that the graph of $f(x) = (x - h)^2$ is a **horizontal translation** of the graph of $f(x) = x^2$.

The following diagram summarizes our work thus far for graphing quadratic functions.

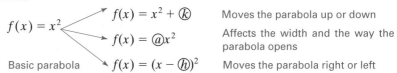

$f(x) = x^2$

$f(x) = x^2 + \textcircled{k}$ Moves the parabola up or down

$f(x) = \textcircled{a}x^2$ Affects the width and the way the parabola opens

Basic parabola $f(x) = (x - \textcircled{h})^2$ Moves the parabola right or left

We have studied, separately, the effects a, h, and k have on the graph of a quadratic function. However, we need to consider the general form of a quadratic function when all of these effects are present.

> In general, the graph of a quadratic function of the form $f(x) = a(x - h)^2 + k$ has its vertex at (h, k) and opens upward if a is positive and downward if a is negative. The parabola is narrower than the basic parabola if $|a| > 1$ and wider if $|a| < 1$.

Now let's consider two examples that combine these ideas.

EXAMPLE   6

Graph $f(x) = 3(x - 2)^2 + 1$.

Solution

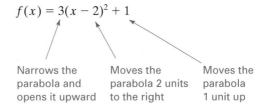

$f(x) = 3(x - 2)^2 + 1$

Narrows the parabola and opens it upward

Moves the parabola 2 units to the right

Moves the parabola 1 unit up

The vertex is $(2, 1)$, and the line $x = 2$ is the axis of symmetry. If $x = 1$, then $f(1) = 3(1 - 2)^2 + 1 = 4$. Thus the point $(1, 4)$ is on the graph, and so is its reflection, $(3, 4)$, across the line of symmetry. The parabola is shown in Figure 3.30.

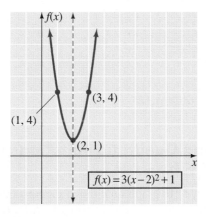

$f(x) = 3(x-2)^2 + 1$

Figure 3.30

EXAMPLE 7   Graph $f(x) = -\dfrac{1}{2}(x + 1)^2 - 3$.

Solution

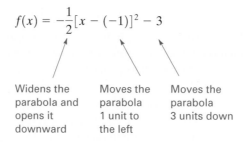

$$f(x) = -\frac{1}{2}[x - (-1)]^2 - 3$$

Widens the parabola and opens it downward

Moves the parabola 1 unit to the left

Moves the parabola 3 units down

The vertex is at $(-1, -3)$, and the line $x = -1$ is the axis of symmetry. If $x = 0$, then $f(0) = -\dfrac{1}{2}(0 + 1)^2 - 3 = -\dfrac{7}{2}$. Thus the point $\left(0, -\dfrac{7}{2}\right)$ is on the graph, and so is its reflection, $\left(-2, -\dfrac{7}{2}\right)$, across the line of symmetry. The parabola is shown in Figure 3.31.

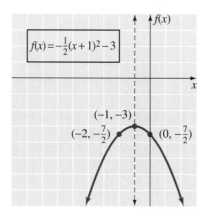

Figure 3.31

▶ **Now work Problem 13.** ∎

■ Graphing Quadratic Functions of the Form $f(x) = ax^2 + bx + c$

We are now ready to graph quadratic functions of the form $f(x) = ax^2 + bx + c$. The general approach is to change from the form $f(x) = ax^2 + bx + c$ to the form $f(x) = a(x - h)^2 + k$ and then proceed as we did in Examples 6 and 7. The process of *completing the square* serves as the basis for making the change in form. Let's consider two examples to illustrate the details.

EXAMPLE 8

Graph $f(x) = x^2 - 4x + 3$.

Solution

$$f(x) = x^2 - 4x + 3$$
$$= (x^2 - 4x) + 3$$
$$= (x^2 - 4x + 4) + 3 - 4$$
$$= (x - 2)^2 - 1$$

Add 4, which is the square of one-half of the coefficient of x

Subtract 4 to compensate for the 4 that was added

The graph of $f(x) = (x - 2)^2 - 1$ is the basic parabola moved two units to the right and one unit down, as in Figure 3.32.

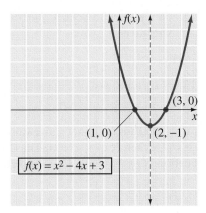

$f(x) = x^2 - 4x + 3$

(3, 0)

(1, 0)   (2, −1)

Figure 3.32

▶ **Now work Problem 15.** ∎

EXAMPLE 9

Graph $f(x) = -2x^2 - 4x + 1$.

Solution

$$f(x) = -2x^2 - 4x + 1$$
$$= -2(x^2 + 2x) + 1$$
$$= -2(x^2 + 2x + 1) + 1 + 2$$
$$= -2(x + 1)^2 + 3$$

Factor −2 from the first two terms

Add 1 inside the parentheses to complete the square

Add 2 to compensate for the 1 inside the parentheses times the factor −2

The graph of $f(x) = -2(x + 1)^2 + 3$ is shown in Figure 3.33.

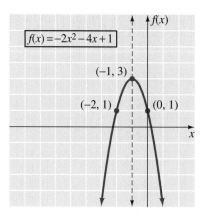

Figure 3.33

▶ **Now work Problem 19.**   ∎

Now let's graph a piecewise-defined function that involves both linear and quadratic rules of assignment.

EXAMPLE 10

Graph $f(x) = \begin{cases} 2x & \text{for } x \ge 0 \\ x^2 + 1 & \text{for } x < 0 \end{cases}$.

Solution

If $x \ge 0$, then $f(x) = 2x$. Thus for nonnegative values of x, we graph the linear function $f(x) = 2x$. If $x < 0$, then $f(x) = x^2 + 1$. Thus for negative values of x, we graph the quadratic function $f(x) = x^2 + 1$. The complete graph is shown in Figure 3.34.

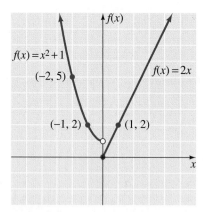

Figure 3.34

▶ **Now work Problem 29.**   ∎

What we know about parabolas and the process of completing the square can be helpful when we are using a graphing utility to graph a quadratic function. Consider the following example:

EXAMPLE 11

Use a graphing utility to obtain the graph of the quadratic function

$$f(x) = -x^2 + 37x - 311$$

Solution

First, we know that the parabola opens downward, and its width is the same as that of the basic parabola $f(x) = x^2$. Then we can start the process of completing the square to determine an approximate location of the vertex.

$$f(x) = -x^2 + 37x - 311$$

$$= -(x^2 - 37x) - 311$$

$$= -\left(x^2 - 37x + \left(\frac{37}{2}\right)^2\right) - 311 + \left(\frac{37}{2}\right)^2$$

$$= -(x^2 - 37x + (18.5)^2) - 311 + 342.25$$

Thus the vertex is near $x = 18$ and $y = 31$. Therefore, setting the boundaries of the viewing rectangle so that $-2 \leq x \leq 25$ and $-10 \leq y \leq 35$, we obtain the graph shown in Figure 3.35.

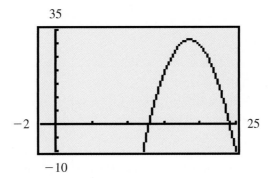

Figure 3.35   ■

Remark: The graph in Figure 3.35 is sufficient for most purposes because it shows the vertex and the x intercepts of the parabola. Certainly, we could use other boundaries that would also give this information.

CONCEPT QUIZ

For Problems 1–5, answer true or false.

1. The graph of any quadratic function is a parabola.

2. For the quadratic function $f(x) = ax^2 + bx + c$, the vertex of the parabola is always the minimum value of the function.

3. The graph of $y = \frac{1}{2}x^2$ is a parabola that opens downward.

4. If the vertex of a parabola that opens upward is located at the point (a, b), then the axis of symmetry is $x = a$.

5. If the point $(1, 4)$ is on the graph of a parabola, then the point $(-1, 4)$ is also on the parabola.

For Problems 6–8, match the quadratic function with its graphs (Figures 3.36–3.38).

6. $y = x^2 + 2$ **7.** $y = (x - 2)^2$ **8.** $y = 2x^2$

A.

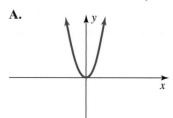

Figure 3.36

B.

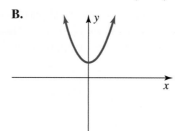

Figure 3.37

C.

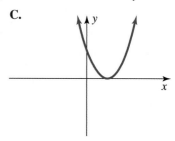

Figure 3.38

Problem Set 3.3

For Problems 1–26, graph each quadratic function.

1. $f(x) = x^2 + 1$

2. $f(x) = x^2 - 3$

3. $f(x) = 3x^2$

4. $f(x) = -2x^2$

5. $f(x) = -x^2 + 2$

6. $f(x) = -3x^2 - 1$

7. $f(x) = (x + 2)^2$

8. $f(x) = (x - 1)^2$

9. $f(x) = -2(x + 1)^2$

10. $f(x) = 3(x - 2)^2$

11. $f(x) = (x - 1)^2 + 2$

12. $f(x) = -(x + 2)^2 + 3$

13. $f(x) = \frac{1}{2}(x - 2)^2 - 3$

14. $f(x) = 2(x - 3)^2 - 1$

15. $f(x) = x^2 + 2x + 4$

16. $f(x) = x^2 - 4x + 2$

17. $f(x) = x^2 - 3x + 1$

18. $f(x) = x^2 + 5x + 5$

19. $f(x) = 2x^2 + 12x + 17$

20. $f(x) = 3x^2 - 6x$

21. $f(x) = -x^2 - 2x + 1$

22. $f(x) = -2x^2 + 12x - 16$

23. $f(x) = 2x^2 - 2x$

24. $f(x) = 2x^2 + 3x - 1$

25. $f(x) = -2x^2 - 5x + 1$

26. $f(x) = -3x^2 + x - 2$

For Problems 27–44, graph each piecewise-defined function.

27. $f(x) = \begin{cases} x^2 & \text{for } x < 0 \\ x^2 + 3 & \text{for } x \geq 0 \end{cases}$

28. $f(x) = \begin{cases} x^2 & \text{for } x < 0 \\ -x^2 + 2 & \text{for } x \geq 0 \end{cases}$

29. $f(x) = \begin{cases} -x^2 & \text{for } x < 0 \\ x^2 + 1 & \text{for } x \geq 0 \end{cases}$

30. $f(x) = \begin{cases} -x^2 & \text{for } x < 0 \\ x^2 - 3 & \text{for } x \geq 0 \end{cases}$

31. $f(x) = \begin{cases} -x^2 + 2 & \text{for } x < 0 \\ 2 & \text{for } 0 \leq x < 3 \\ x^2 - 7 & \text{for } x \geq 3 \end{cases}$

32. $f(x) = \begin{cases} x^2 + 1 & \text{for } x < 0 \\ 1 & \text{for } 0 \leq x < 3 \\ -x^2 + 10 & \text{for } x \geq 3 \end{cases}$

33. $f(x) = \begin{cases} -x^2 & \text{for } x < -1 \\ x & \text{for } -1 \leq x < 1 \\ x^2 & \text{for } x \geq 1 \end{cases}$

34. $f(x) = \begin{cases} x^2 & \text{for } x < -2 \\ \frac{1}{2}x + 5 & \text{for } -2 \leq x < 2 \\ x^2 + 2 & \text{for } x \geq 2 \end{cases}$

35. $f(x) = \begin{cases} -x^2 & \text{for } x < 0 \\ x + 2 & \text{for } 0 \leq x < 4 \\ 2x & \text{for } x \geq 4 \end{cases}$

36. $f(x) = \begin{cases} -2x & \text{for } x < 0 \\ x^2 & \text{for } 0 \leq x < 2 \\ \frac{1}{2}x + 2 & \text{for } x \geq 2 \end{cases}$

37. $f(x) = \begin{cases} 3x & \text{for } x < 0 \\ x & \text{for } x \geq 0 \end{cases}$

38. $f(x) = \begin{cases} 4x & \text{for } x < 0 \\ -x & \text{for } x \geq 0 \end{cases}$

39. $f(x) = \begin{cases} x^2 & \text{for } x < 0 \\ 2x + 1 & \text{for } x \geq 0 \end{cases}$

40. $f(x) = \begin{cases} 2x^2 & \text{for } x < 0 \\ -x^2 & \text{for } x \geq 0 \end{cases}$

41. $f(x) = \begin{cases} -1 & \text{if } x < 0 \\ 2 & \text{if } x \geq 0 \end{cases}$

42. $f(x) = \begin{cases} -1 & \text{if } x \leq 0 \\ 1 & \text{if } 0 < x \leq 2 \\ 2 & \text{if } x > 2 \end{cases}$

43. $f(x) = \begin{cases} 1 & \text{if } 0 \leq x < 1 \\ 2 & \text{if } 1 \leq x < 2 \\ 3 & \text{if } 2 \leq x < 3 \\ 4 & \text{if } 3 \leq x < 4 \end{cases}$

44. $f(x) = \begin{cases} 2x + 3 & \text{if } x < 0 \\ x^2 & \text{if } 0 \leq x < 2 \\ 1 & \text{if } x \geq 2 \end{cases}$

45. The **greatest integer function** is defined by the equation $f(x) = [x]$, where $[x]$ refers to the largest integer less than or equal to x. For example, $[2.6] = 2$, $[\sqrt{2}] = 1$, $[4] = 4$, and $[-1.4] = -2$. Graph $f(x) = [x]$ for $-4 \leq x < 4$.

■ ■ ■ **THOUGHTS INTO WORDS**

46. Explain the concept of a piecewise-defined function.

47. Give a step-by-step description of how you would use the ideas presented in this section to graph $f(x) = 5x^2 + 10x + 4$.

 GRAPHING CALCULATOR ACTIVITIES

48. This problem is designed to reinforce ideas presented in this section. For each part, first predict the shapes and locations of the parabolas, and then use your graphing calculator to graph them on the same set of axes.

a. $f(x) = x^2, f(x) = x^2 - 4, f(x) = x^2 + 1, f(x) = x^2 + 5$

b. $f(x) = x^2, \; f(x) = (x - 5)^2, \; f(x) = (x + 5)^2, f(x) = (x - 3)^2$

c. $f(x) = x^2, f(x) = 5x^2, f(x) = \frac{1}{3}x^2, f(x) = -2x^2$

d. $f(x) = x^2, f(x) = (x - 7)^2 - 3, f(x) = -(x + 8)^2 + 4, f(x) = -3x^2 - 4$

e. $f(x) = x^2 - 4x - 2, \; f(x) = -x^2 + 4x + 2, f(x) = -x^2 - 16x - 58, f(x) = x^2 + 16x + 58$

49. a. Graph both $f(x) = x^2 - 14x + 51$ and $f(x) = x^2 + 14x + 51$ on the same set of axes. What relationship seems to exist between the two graphs?

b. Graph both $f(x) = x^2 + 12x + 34$ and $f(x) = x^2 - 12x + 34$ on the same set of axes. What relationship seems to exist between the two graphs?

c. Graph both $f(x) = -x^2 + 8x - 20$ and $f(x) = -x^2 - 8x - 20$ on the same set of axes. What relationship seems to exist between the two graphs?

d. Make a statement that generalizes your findings in parts a–c.

50. Use your graphing calculator to graph the piecewise-defined functions in Problems 27–44. You may need to consult your user's manual for instructions on graphing these functions.

Answers to the Concept Quiz

1. True   **2.** False   **3.** False   **4.** True   **5.** False   **6.** B   **7.** C   **8.** A

3.4 More Quadratic Functions and Applications

Objectives

■ Graph parabolas using a formula to locate the vertex.

■ Determine the x and y intercepts for a parabola.

■ Solve application problems involving quadratic functions.

In the previous section we used the process of completing the square to change a quadratic function such as $f(x) = x^2 - 4x + 3$ to the form $f(x) = (x - 2)^2 - 1$. From the form $f(x) = (x - 2)^2 - 1$, it is easy to identify the vertex $(2, -1)$ and the axis of symmetry $x = 2$ of the parabola. In general, if we complete the square on

$$f(x) = ax^2 + bx + c$$

we obtain

$$f(x) = a\left(x^2 + \frac{b}{a}x \right) + c$$

$$= a\left(x^2 + \frac{b}{a}x + \frac{b^2}{4a^2} \right) + c - \frac{b^2}{4a}$$

$$= a\left(x + \frac{b}{2a} \right)^2 + \frac{4ac - b^2}{4a}$$

Therefore the parabola associated with the function $f(x) = ax^2 + bx + c$ has its vertex at

$$\left(-\frac{b}{2a}, \frac{4ac - b^2}{4a} \right)$$

and the equation of its axis of symmetry is $x = -b/2a$. These facts are illustrated in Figure 3.39.

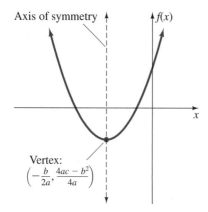

Figure 3.39

By using the information from Figure 3.39, we now have another way of graphing quadratic functions of the form $f(x) = ax^2 + bx + c$, as indicated by the following steps:

1. Determine whether the parabola opens upward (if $a > 0$) or downward (if $a < 0$).
2. Find $-b/2a$, which is the x coordinate of the vertex.
3. Find $f(-b/2a)$, which is the y coordinate of the vertex, or find the y coordinate by evaluating

$$\frac{4ac - b^2}{4a}$$

4. Locate another point on the parabola and also locate its image across the axis of symmetry, which is the line with equation $x = -b/2a$.

The three points found in steps 2, 3, and 4 should determine the general shape of the parabola. Let's illustrate this procedure with two examples.

EXAMPLE 1

Graph $f(x) = 3x^2 - 6x + 5$.

Solution

Step 1   Because $a > 0$, the parabola opens upward.

Step 2 $-\dfrac{b}{2a} = -\dfrac{-6}{6} = 1$

Step 3 $f\left(-\dfrac{b}{2a}\right) = f(1) = 3 - 6 + 5 = 2$. Thus the vertex is at $(1, 2)$.

Step 4   Letting $x = 2$, we obtain $f(2) = 12 - 12 + 5 = 5$. Thus $(2, 5)$ is on the graph, and so is its reflection, $(0, 5)$, across the line of symmetry, $x = 1$.

The three points $(1, 2)$, $(2, 5)$, and $(0, 5)$ are used to graph the parabola in Figure 3.40.

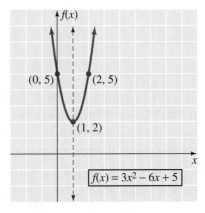

Figure 3.40

▶ **Now work Problem 7.**   ∎

E X A M P L E   2   Graph $f(x) = -x^2 - 4x - 7$.

Solution

Step 1   Because $a < 0$, the parabola opens downward.

Step 2 $-\dfrac{b}{2a} = -\dfrac{-4}{-2} = -2$

Step 3 $f\left(-\dfrac{b}{2a}\right) = f(-2) = -(-2)^2 - 4(-2) - 7 = -3$. Thus the vertex is at $(-2, -3)$.

Step 4   Letting $x = 0$, we obtain $f(0) = -7$. Thus $(0, -7)$ is on the graph, and so is its reflection, $(-4, -7)$, across the line of symmetry, $x = -2$.

The three points $(-2, -3)$, $(0, -7)$, and $(-4, -7)$ are used to draw the parabola in Figure 3.41.

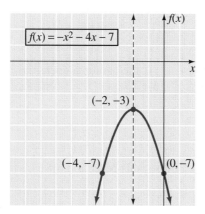

Figure 3.41

▶ **Now work Problem 5.**   ∎

In summary, we have two basic methods to graph a quadratic function.

1. We can express the function in the form $f(x) = a(x - h)^2 + k$ and use the values of a, h, and k to determine the parabola.

2. We can express the function in the form $f(x) = ax^2 + bx + c$ and use the approach demonstrated in Examples 1 and 2.

Parabolas possess various properties that make them very useful. For example, if a parabola is rotated about its axis, then a parabolic surface is formed and such surfaces are used for light and sound reflectors. A projectile fired into the air will follow the curvature of a parabola. The *trend line* of profit and cost functions sometimes follows a parabolic curve. In most applications of the parabola, we are primarily interested in the x intercepts and the vertex. Let's consider some examples of finding the x intercepts and the vertex.

EXAMPLE 3

Find the x intercepts and the vertex for each of the following parabolas.

a. $f(x) = -x^2 + 11x - 18$ **b.** $f(x) = x^2 - 8x - 3$

c. $f(x) = 2x^2 - 12x + 23$

Solutions

a. To find the x intercepts, let $f(x) = 0$ and solve the resulting equation.

$$-x^2 + 11x - 18 = 0$$
$$x^2 - 11x + 18 = 0$$
$$(x - 2)(x - 9) = 0$$
$$x - 2 = 0 \quad \text{or} \quad x - 9 = 0$$
$$x = 2 \quad \text{or} \quad x = 9$$

Therefore the x intercepts are 2 and 9. To find the vertex, let's determine the point $\left(-\dfrac{b}{2a}, f\left(-\dfrac{b}{2a}\right)\right)$:

$$f(x) = -x^2 + 11x - 18$$

$$-\frac{b}{2a} = -\frac{11}{2(-1)} = -\frac{11}{-2} = \frac{11}{2}$$

$$f\left(\frac{11}{2}\right) = -\left(\frac{11}{2}\right)^2 + 11\left(\frac{11}{2}\right) - 18$$

$$= -\frac{121}{4} + \frac{121}{2} - 18$$

$$= \frac{-121 + 242 - 72}{4}$$

$$= \frac{49}{4}$$

Therefore the vertex is at $\left(\dfrac{11}{2}, \dfrac{49}{4}\right)$.

b. To find the x intercepts, let $f(x) = 0$ and solve the resulting equation:

$$x^2 - 8x - 3 = 0$$

$$x = \frac{-(-8) \pm \sqrt{(-8)^2 - 4(1)(-3)}}{2(1)}$$

$$= \frac{8 \pm \sqrt{76}}{2}$$

$$= \frac{8 \pm 2\sqrt{19}}{2}$$

$$= 4 \pm \sqrt{19}$$

Therefore the x intercepts are $4 + \sqrt{19}$ and $4 - \sqrt{19}$. This time, to find the vertex, let's complete the square on x:

$$f(x) = x^2 - 8x - 3$$
$$= x^2 - 8x + 16 - 3 - 16$$
$$= (x - 4)^2 - 19$$

Therefore the vertex is at $(4, -19)$.

c. To find the x intercepts, let $f(x) = 0$ and solve the resulting equation:

$$2x^2 - 12x + 23 = 0$$

$$x = \frac{-(-12) \pm \sqrt{(-12)^2 - 4(2)(23)}}{2(2)}$$

$$= \frac{12 \pm \sqrt{-40}}{4}$$

Because these solutions are nonreal complex numbers, there are no x intercepts. To find the vertex, let's determine the point $\left(-\frac{b}{2a}, f\left(-\frac{b}{2a} \right) \right)$:

$$f(x) = 2x^2 - 12x + 23$$
$$-\frac{b}{2a} = -\frac{-12}{2(2)}$$
$$= 3$$
$$f(3) = 2(3)^2 - 12(3) + 23$$
$$= 18 - 36 + 23$$
$$= 5$$

Therefore the vertex is at $(3, 5)$.

▶ **Now work Problem 35.** ∎

Remark: Note that in parts a and c we used the general point

$$\left(-\frac{b}{2a}, f\left(-\frac{b}{2a} \right) \right)$$

to find the vertices. In part b, however, we completed the square and used that form to determine the vertex. Which approach you use is up to you. We chose to complete the square in part b because the algebra involved was quite easy.

In Problem 60 of Problem Set 1.3 you were asked to solve the quadratic equation $-x^2 + 11x - 18 = 0$. You should have obtained a solution set of $\{2, 9\}$. Here, in part a of Example 3, we solved the same equation to determine that the x intercepts of the graph of the function $f(x) = -x^2 + 11x - 18$ are 2 and 9. The numbers 2 and 9 are also called the **real number zeros** of the function. In part b, the real numbers $4 + \sqrt{19}$ and $4 - \sqrt{19}$ are the x intercepts of the graph of the function $f(x) = x^2 - 8x - 3$ and are the real number zeros of the function. In part c, the nonreal complex numbers $\dfrac{12 \pm \sqrt{-40}}{4}$, which simplify to $\dfrac{6 \pm i\sqrt{10}}{2}$, indicate that the graph of the function $f(x) = 2x^2 - 12x + 23$ has no points on the x axis. The complex numbers are zeros of the function, but they have no physical significance for the graph other than indicating that the graph has no points on the x axis.

Figure 3.42 shows the result we got when we used a graphing calculator to graph the three functions of Example 3 on the same set of axes. This gives us a visual interpretation of the conclusions drawn regarding the x intercepts and vertices.

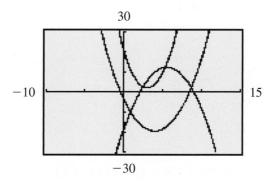

Figure 3.42

■ Back to Problem Solving

As we have seen, the vertex of the graph of a quadratic function is either the lowest or the highest point on the graph. Thus we often speak of the **minimum value** or **maximum value** of a function in applications of the parabola. The x value of the vertex indicates where the minimum or maximum occurs, and $f(x)$ yields the minimum or maximum value of the function. Let's consider some problems that illustrate these ideas.

P R O B L E M   1

A farmer has 120 rods of fencing and wants to enclose a rectangular plot of land that requires fencing on only three sides because it is bounded by a river on one side. Find the length and width of the plot that will maximize the area.

Solution

Let x represent the width; then $120 - 2x$ represents the length, as indicated in Figure 3.43.

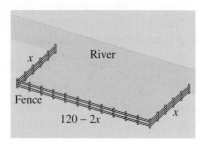

Figure 3.43

The function $A(x) = x(120 - 2x)$ represents the area of the plot in terms of the width x. Because

$$A(x) = x(120 - 2x)$$
$$= 120x - 2x^2$$
$$= -2x^2 + 120x$$

we have a quadratic function with $a = -2$, $b = 120$, and $c = 0$. Therefore the *maximum* value ($a < 0$ so the parabola opens downward) of the function is obtained where the x value is

$$-\frac{b}{2a} = -\frac{120}{2(-2)} = 30$$

If $x = 30$, then $120 - 2x = 120 - 2(30) = 60$. Thus the farmer should make the plot 30 rods wide and 60 rods long to maximize the area at $(30)(60) = 1800$ square rods.

▶ **Now work Problem 49.**   ■

PROBLEM 2

Find two numbers whose sum is 30, such that the sum of their squares is a minimum.

Solution

Let x represent one of the numbers; then $30 - x$ represents the other number. By expressing the sum of their squares as a function of x, we obtain

$$f(x) = x^2 + (30 - x)^2$$

which can be simplified to

$$f(x) = x^2 + 900 - 60x + x^2$$
$$= 2x^2 - 60x + 900$$

This is a quadratic function with $a = 2$, $b = -60$, and $c = 900$. Therefore the *minimum value* ($a > 0$ so the parabola opens upward) of the function is obtained where the x value is

$$-\frac{b}{2a} = -\frac{-60}{4}$$

$$= 15$$

If $x = 15$, then $30 - x = 30 - 15 = 15$. Thus the two numbers are both 15.

▶ **Now work Problem 45.**   ■

P R O B L E M   3

A golf pro shop operator finds that she can sell 30 sets of golf clubs at $500 per set in a year. Furthermore, she predicts that for each $25 decrease in price, she could sell three extra sets of golf clubs. At what price should she sell the clubs to maximize gross income, and what will the gross income be at that price?

Solution

When we analyze such a problem, it sometimes helps to start by setting up a table.

	Number of Sets	Price per Set	Income
Three additional sets can be sold for	30	$500	$15,000
	33	$475	$15,675
a $25 decrease in price.	36	$450	$16,200

Let x represent the number of $25 decreases in price. Then the income can be expressed as a function of x:

$$f(x) = (30 + 3x)(500 - 25x)$$

Number   Price per
of sets   set

Simplifying this, we obtain

$$f(x) = 15{,}000 - 750x + 1500x - 75x^2$$

$$= -75x^2 + 750x + 15{,}000$$

This is a quadratic function with $a = -75$, $b = 750$, and $c = 15{,}000$. Therefore the *maximum value* ($a < 0$ so the parabola opens downward) of the function is obtained where the x value is

$$-\frac{b}{2a} = -\frac{750}{2(-75)} = -\frac{750}{-150} = 5$$

Thus 5 decreases of $25 each — that is, a $125 reduction in price — will give a maximum gross income. The golf clubs should be sold at $500 − $125 = $375 per set. The maximum gross income is the functional value at $x = 5$:

$$f(5) = -75(5)^2 + 750(5) + 15{,}000 = 16{,}875$$

So the maximum gross income is $16,875.

▶ **Now work Problem 51.**   ■

We have determined that the vertex of a parabola associated with $f(x) = ax^2 + bx + c$ is located at $\left(-\dfrac{b}{2a}, f\left(-\dfrac{b}{2a}\right)\right)$ and that the x intercepts of the graph can be found by solving the quadratic equation $ax^2 + bx + c = 0$. Therefore a graphing utility does not provide us with much extra power when we are working with quadratic functions. However, as functions become more complex, a graphing utility becomes more helpful. Let's build our confidence in the use of a graphing utility at this time, while we have a way of checking our results.

P R O B L E M   4

Use a graphing utility to graph $f(x) = x^2 - 8x - 3$ and find the x intercepts of the graph. (This is the parabola from part b of Example 3.)

Solution

A graph of the parabola is shown in Figure 3.44. One x intercept appears to be between 0 and −1 and the other between 8 and 9. Let's zoom in on the x intercept between 8 and 9. This produces a graph like Figure 3.45. Now we can use the trace function to determine that this x intercept is at approximately 8.4. (This agrees with the answer of $4 + \sqrt{19}$ that we got in Example 3.) In a similar fashion, we can determine that the other x intercept is at −0.4.

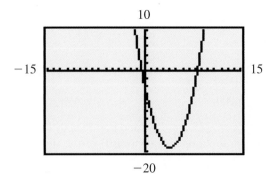

Figure 3.44

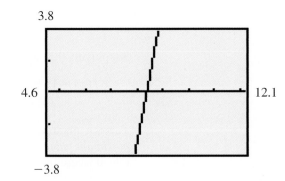

Figure 3.45   ■

CONCEPT QUIZ   For Problems 1–8, answer true or false.

1. For the parabola associated with $f(x) = ax^2 + bx + c$, $f\left(-\dfrac{b}{2a}\right) = \dfrac{4ac - b^2}{4a}$.

2. For the parabola associated with $f(x) = ax^2 + bx + c$, the axis of symmetry is $x = -\dfrac{b}{2a}$.

3. For the parabola associated with $f(x) = ax^2 + bx + c$, the parabola will always open upward if b is positive.

4. For the parabola associated with $f(x) = ax^2 + bx + c$, the x intercepts can be found by using the quadratic formula $x = \dfrac{-b \pm \sqrt{b^2 - 4ac}}{2a}$.

5. Every graph of a quadratic function $f(x) = ax^2 + bx + c$ has x intercepts.

6. Every graph of a quadratic function $f(x) = ax^2 + bx + c$ has a y intercept.

7. For the quadratic function $f(x) = -4x^2 + 3x + 1$, the vertex of its parabola will be the highest point on the graph.

8. For the parabola associated with $f(x) = 2x^2 + 5x + 8$, $f\left(\dfrac{-5}{4}\right)$ is the minimum value of the function.

Problem Set 3.4

For Problems 1–12, use the approach of Examples 1 and 2 of this section to graph each quadratic function.

1. $f(x) = x^2 - 8x + 15$
2. $f(x) = x^2 + 6x + 11$

3. $f(x) = 2x^2 + 20x + 52$
4. $f(x) = 3x^2 - 6x - 1$

5. $f(x) = -x^2 + 4x - 7$
6. $f(x) = -x^2 - 6x - 5$

7. $f(x) = -3x^2 + 6x - 5$
8. $f(x) = -2x^2 - 4x + 2$

9. $f(x) = x^2 + 3x - 1$
10. $f(x) = x^2 + 5x + 2$

11. $f(x) = -2x^2 + 5x + 1$
12. $f(x) = -3x^2 + 2x - 1$

For Problems 13–24, use the approach that you think is the most appropriate to graph each quadratic function.

13. $f(x) = -x^2 + 3$
14. $f(x) = (x + 1)^2 + 1$

15. $f(x) = x^2 + x - 1$
16. $f(x) = -x^2 + 3x - 4$

17. $f(x) = -2x^2 + 4x + 1$
18. $f(x) = 4x^2 - 8x + 5$

19. $f(x) = -\left(x + \dfrac{5}{2}\right)^2 + \dfrac{3}{2}$

20. $f(x) = x^2 - 4x$
21. $f(x) = x^2 + 2x$

22. $f(x) = -(x - 3)^2 + 2$

23. $f(x) = (x + 2)^2 - 4$
24. $f(x) = 2x^2 + 2$

For Problems 25–40, find the x intercepts and the vertex of each parabola.

25. $f(x) = x^2 - 8x + 15$
26. $f(x) = x^2 - 16x + 63$

27. $f(x) = 2x^2 - 28x + 96$

28. $f(x) = 3x^2 - 60x + 297$

29. $f(x) = -x^2 + 10x - 24$

30. $f(x) = -2x^2 + 36x - 160$

31. $f(x) = 3x^2 - 2$
32. $f(x) = 5x^2 - 10x$

33. $f(x) = 3x^2 + 9x$ **34.** $f(x) = 6x^2 - 4$

35. $f(x) = x^2 - 14x + 44$

36. $f(x) = x^2 - 18x + 68$

37. $f(x) = -x^2 + 9x - 21$

38. $f(x) = 2x^2 + 3x + 3$

39. $f(x) = -4x^2 + 4x + 4$

40. $f(x) = -2x^2 + 3x + 7$

For Problems 41–52, solve each problem.

41. Suppose that the equation $p(x) = -2x^2 + 280x - 1000$, where x represents the number of items sold, describes the profit function for a certain business. How many items should be sold to maximize the profit?

42. Suppose that the cost function for the production of a particular item is given by the equation $C(x) = 2x^2 - 320x + 12,920$, where x represents the number of items. How many items should be produced to minimize the cost?

43. The height of a projectile fired vertically into the air (neglecting air resistance) at an initial velocity of 96 feet per second is a function of time, x, and is given by the equation $f(x) = 96x - 16x^2$. Find the highest point reached by the projectile.

44. Find two numbers whose sum is 30 and whose product is a maximum.

45. Find two numbers whose sum is 20 such that the sum of their squares is a minimum.

46. Find two numbers whose sum is 30, such that the sum of the square of one number plus ten times the other number is a minimum.

47. Find two numbers whose sum is 50 and whose product is a maximum.

48. Find two numbers whose difference is 40 and whose product is a minimum.

49. Two hundred and forty meters of fencing is available to enclose a rectangular playground. What should be the dimensions of the playground to maximize the area?

50. An outdoor adventure company advertises that they will provide a mountain bike and a picnic lunch for $50 per person. They must have a guarantee of 30 people to run the ride. Furthermore, they agree that for each person in excess of 30, they will reduce the price per person for all riders by $0.50. How many people will it take to maximize the company's revenue?

51. A video rental service has 1000 subscribers, each of whom pays $15 per month. On the basis of a survey, they believe that for each decrease of $0.25 in the monthly rate, they could obtain 20 additional subscribers. At what rate will maximum revenue be obtained and how many subscribers will there be at that rate?

52. A manufacturer finds that for the first 500 units of its product that are produced and sold, the profit is $50 per unit. The profit on each of the units beyond 500 is decreased by $0.10 times the number of additional units sold. What level of output will maximize profit?

■ ■ ■ THOUGHTS INTO WORDS

53. Suppose your friend was absent the day this section was discussed. How would you explain to her the ideas pertaining to x intercepts of the graph of a function, zeros of the function, and solutions of the equation $f(x) = 0$?

54. Give a step-by-step explanation of how to find the x intercepts of the graph of the function $f(x) = 2x^2 + 7x - 4$.

55. Give a step-by-step explanation of how to find the vertex of the parabola determined by the equation $f(x) = -x^2 - 6x - 5$.

■ ■ ■ **FURTHER INVESTIGATIONS**

56. Suppose that an arch is shaped like a parabola. It is 20 feet wide at the base and 100 feet high. How wide is the arch 50 feet above the ground? (See Figure 3.46.)

57. A parabolic arch 27 feet high spans a parkway. The center section of the parkway is 50 feet wide. How wide is the arch if it has a minimum clearance of 15 feet above the center section?

58. A parabolic arch spans a stream 200 feet wide. How high must the arch be above the stream to give a minimum clearance of 40 feet over a 120-foot-wide channel in the center?

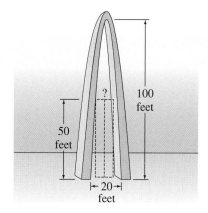

Figure 3.46

GRAPHING CALCULATOR ACTIVITIES

59. Suppose that the viewing window on your graphing calculator is set so that $-15 \le x \le 15$ and $-10 \le y \le 10$. Now try to graph the function $f(x) = x^2 - 8x + 28$. Nothing appears on the screen, so the parabola must be outside the viewing window. We could arbitrarily expand the window until the parabola appeared. However, let's be a little more systematic and use $\left(-\dfrac{b}{2a}, f\left(-\dfrac{b}{2a}\right)\right)$ to find the vertex. We find the vertex is at $(4, 12)$, so let's change the y values of the window so that $0 \le y \le 25$. Now we get a good picture of the parabola.

Graph each of the following parabolas, and keep in mind that you may need to change the dimensions of the viewing window to obtain a good picture.

a. $f(x) = x^2 - 2x + 12$
b. $f(x) = -x^2 - 4x - 16$
c. $f(x) = x^2 + 12x + 44$
d. $f(x) = x^2 - 30x + 229$
e. $f(x) = -2x^2 + 8x - 19$

60. Use a graphing calculator to graph each of the following parabolas, and then use the trace function to help estimate the x intercepts and the vertex. Finally, use the approach of Example 3 to find the x intercepts and the vertex.

a. $f(x) = x^2 - 6x + 3$
b. $f(x) = x^2 - 18x + 66$
c. $f(x) = -x^2 + 8x - 3$
d. $f(x) = -x^2 + 24x - 129$
e. $f(x) = 14x^2 - 7x + 1$
f. $f(x) = -\dfrac{1}{2}x^2 + 5x - \dfrac{17}{2}$

61. In Problems 25–40, you were asked to find the x intercepts and the vertex of some parabolas. Now use a graphing calculator to graph each parabola and visually justify your answers.

3.5   Transformations of Some Basic Curves

Objectives

■   Know the basic shape for the graphs of $f(x) = x^3$, $f(x) = x^4$, $f(x) = \sqrt{x}$, and $f(x) = |x|$.

■   Graph by applying horizontal and vertical translations.

■   Graph $y = -f(x)$ by reflecting the graph of $y = f(x)$ through the x axis.

■   Graph $y = f(-x)$ by reflecting the graph of $y = f(x)$ through the y axis.

■   Understand and apply the concepts of the vertical stretching or shrinking of a graph.

From our work in Section 3.3, we know that the graph of $f(x) = (x - 5)^2$ is the basic parabola $f(x) = x^2$ translated five units to the right. Likewise, we know that the graph of $f(x) = -x^2 - 2$ is the basic parabola reflected across the x axis and translated downward two units. Translations and reflections apply not only to parabolas but to curves in general. Therefore, if we know the shapes of a few basic curves, then it is easy to sketch numerous variations of these curves by using the concepts of translation and reflection.

     Let's begin this section by establishing the graphs of four basic curves and then apply some transformations to these curves. First let's restate, in function vocabulary, the graphing suggestions offered in Chapter 2. Pay special attention to suggestions 2 and 3, where we restate the concepts of intercepts and symmetry using function notation.

1.   Determine the domain of the function.

2.   Find the y intercept [we are labeling the y axis with $f(x)$] by evaluating $f(0)$. Find the x intercept by finding the value(s) of x such that $f(x) = 0$.

3.   Determine any types of symmetry that the equation possesses. If $f(-x) = f(x)$, then the function exhibits y axis symmetry. If $f(-x) = -f(x)$, then the function exhibits origin symmetry. (Note that the definition of a function rules out the possibility that the graph of a function has x axis symmetry.)

4.   Set up a table of ordered pairs that satisfy the equation. The type of symmetry and the domain will affect your choice of values of x in the table.

5. Plot the points associated with the ordered pairs and connect them with a smooth curve. Then, if appropriate, reflect this part of the curve according to any symmetries possessed by the graph.

E X A M P L E   1

Graph $f(x) = x^3$.

Solution

The domain is the set of real numbers. Because $f(0) = 0$, the origin is on the graph. Because $f(-x) = (-x)^3 = -x^3 = -f(x)$, the graph is symmetric with respect to the origin. Therefore we can concentrate our table on the positive values of x. By connecting the points associated with the ordered pairs from the table with a smooth curve and then reflecting it through the origin, we get the graph in Figure 3.47.

x	$f(x) = x^3$
0	0
1	1
2	8
$\frac{1}{2}$	$\frac{1}{8}$

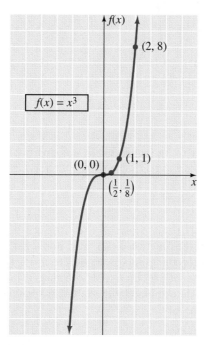

Figure 3.47   ■

E X A M P L E   2

Graph $f(x) = x^4$.

Solution

The domain is the set of real numbers. Because $f(0) = 0$, the origin is on the graph. Because $f(-x) = (-x)^4 = x^4 = f(x)$, the graph has y axis symmetry, and we can concentrate our table of values on the positive values of x. If we connect the points associated with the ordered pairs from the table with a smooth curve and then reflect across the vertical axis, we get the graph in Figure 3.48.

x	$f(x) = x^4$
0	0
1	1
2	16
$\frac{1}{2}$	$\frac{1}{16}$

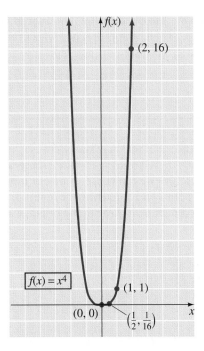

Figure 3.48 ∎

Remark: The curve in Figure 3.48 is not a parabola, even though it resembles one; this curve is flatter at the bottom and steeper.

EXAMPLE 3

Graph $f(x) = \sqrt{x}$.

Solution

The domain of the function is the set of nonnegative real numbers. Because $f(0) = 0$, the origin is on the graph. Because $f(-x) \neq f(x)$ and $f(-x) \neq -f(x)$, there is no symmetry, so let's set up a table of values using nonnegative values for x. Plotting the points determined by the table and connecting them with a smooth curve produces Figure 3.49.

x	$f(x) = \sqrt{x}$
0	0
1	1
4	2
9	3

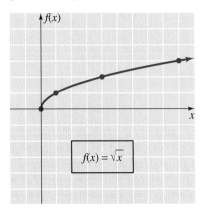

Figure 3.49 ∎

Sometimes a new function is defined in terms of old functions. In such cases the definition plays an important role in the study of the new function. Consider the following example.

E X A M P L E   4

Graph $f(x) = |x|$.

Solution

The concept of absolute value is defined for all real numbers by

$$|x| = x \quad \text{if } x \geq 0$$
$$|x| = -x \quad \text{if } x < 0$$

Therefore the absolute value function can be expressed as

$$f(x) = |x| = \begin{cases} x & \text{if } x \geq 0 \\ -x & \text{if } x < 0 \end{cases}$$

The graph of $f(x) = x$ for $x \geq 0$ is the ray in the first quadrant, and the graph of $f(x) = -x$ for $x < 0$ is the half-line (not including the origin) in the second quadrant, as indicated in Figure 3.50. Note that the graph has y axis symmetry.

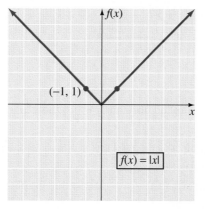

Figure 3.50   ■

■ Translations of the Basic Curves

From our work in Section 3.3, we know that

1. The graph of $f(x) = x^2 + 3$ is the graph of $f(x) = x^2$ moved up three units, and

2. The graph of $f(x) = x^2 - 2$ is the graph of $f(x) = x^2$ moved down two units.

Now we can describe in general the concept of a vertical translation.

Vertical Translation

The graph of $y = f(x) + k$ is the graph of $y = f(x)$ shifted k units upward if $k > 0$ or shifted $|k|$ units downward if $k < 0$.

In Figure 3.51, the graph of $f(x) = |x| + 2$ is obtained by shifting the graph of $f(x) = |x|$ upward two units, and the graph of $f(x) = |x| - 3$ is obtained by shifting the graph of $f(x) = |x|$ downward three units. [Remember that $f(x) = |x| - 3$ can be written as $f(x) = |x| + (-3)$.]

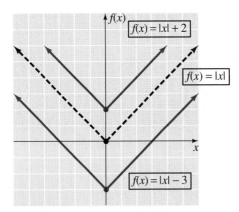

Figure 3.51

We also graphed horizontal translations of the basic parabola in Section 3.3. For example,

1. The graph of $f(x) = (x - 4)^2$ is the graph of $f(x) = x^2$ shifted four units to the right, and

2. The graph of $f(x) = (x + 5)^2$ is the graph of $f(x) = x^2$ shifted five units to the left.

The general concept of a horizontal translation can be described as follows.

Horizontal Translation

The graph of $y = f(x - h)$ is the graph of $y = f(x)$ shifted h units to the right if $h > 0$ or shifted $|h|$ units to the left if $h < 0$.

In Figure 3.52, the graph of $f(x) = (x - 3)^3$ is obtained by shifting the graph of $f(x) = x^3$ three units to the right. Likewise, the graph of $f(x) = (x + 2)^3$ is obtained by shifting the graph of $f(x) = x^3$ two units to the left.

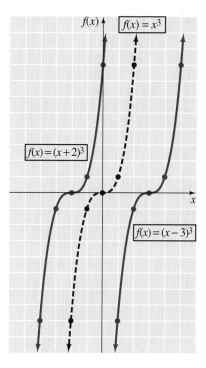

Figure 3.52

■ Reflections of the Basic Curves

From our work in Section 3.3, we know that the graph of $f(x) = -x^2$ is the graph of $f(x) = x^2$ reflected through the x axis. The general concept of an x axis reflection can be described as follows.

x Axis Reflection

The graph of $y = -f(x)$ is the graph of $y = f(x)$ reflected through the x axis.

In Figure 3.53, the graph of $f(x) = -\sqrt{x}$ is obtained by reflecting the graph of $f(x) = \sqrt{x}$ through the x axis. Reflections are sometimes referred to as **mirror images.** Thus if we think of the x axis in Figure 3.53 as a mirror, then the graphs of $f(x) = \sqrt{x}$ and $f(x) = -\sqrt{x}$ are mirror images of each other.

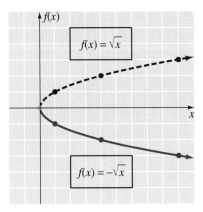

Figure 3.53

In Section 3.3, we did not consider a y axis reflection of the basic parabola $f(x) = x^2$ because it is symmetric with respect to the y axis. In other words, a y axis reflection of $f(x) = x^2$ produces the same figure. However, at this time let's describe the general concept of a y axis reflection.

y Axis Reflection

The graph of $y = f(-x)$ is the graph of $y = f(x)$ reflected through the y axis.

Now suppose that we want to do a y axis reflection of $f(x) = \sqrt{x}$. The domain for the function $f(x) = \sqrt{x}$ is restricted to values of x such that $x \geq 0$. The function for the y axis reflection is $f(x) = \sqrt{-x}$. To find the domain for the reflection, we know the function will be defined if the expression under the radical is greater than or equal to 0. Therefore, the domain for the reflection is $-x \geq 0$. Simplifying $-x \geq 0$ by multiplying both sides by -1 gives $x \leq 0$. Figure 3.54 shows the y axis reflection of $f(x) = \sqrt{x}$.

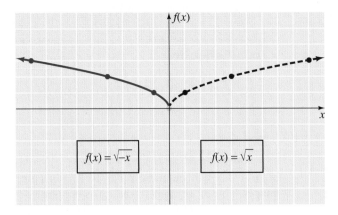

Figure 3.54

■ Vertical Stretching and Shrinking

Translations and reflections are called **rigid transformations** because the basic shape of the curve being transformed is not changed. In other words, only the positions of the graphs change. Now we want to consider some transformations that distort the shape of the original figure somewhat.

In Section 3.3, we graphed the equation $y = 2x^2$ by doubling the y coordinates of the ordered pairs that satisfy the equation $y = x^2$. We obtained a parabola with its vertex at the origin, symmetric to the y axis, but *narrower* than the basic parabola. Likewise, we graphed the equation $y = \frac{1}{2}x^2$ by halving the y coordinates of the ordered pairs that satisfy $y = x^2$. We obtained a parabola with its vertex at the origin, symmetric to the y axis, but *wider* than the basic parabola.

The concepts of *narrower* and *wider* can be used to describe parabolas, but they cannot be used to describe some other curves accurately. Instead, we use the more general concepts of vertical stretching and shrinking.

Vertical Stretching and Shrinking

> The graph of $y = cf(x)$, where $c > 0$, is the graph of $y = f(x)$ either stretched or shrunk by a factor of c. If $c > 1$, the graph is said to be stretched by a factor of c, and if $0 < c < 1$, the graph is said to be shrunk by a factor of c.

In Figure 3.55 the graph of $f(x) = 2\sqrt{x}$ is obtained by doubling the y coordinates of points on the graph of $f(x) = \sqrt{x}$. Likewise, the graph of $f(x) = \frac{1}{2}\sqrt{x}$ is obtained by halving the y coordinates of points on the graph of $f(x) = \sqrt{x}$.

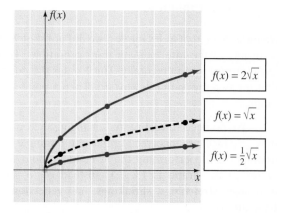

Figure 3.55

■ Successive Transformations

Some curves are the result of performing more than one transformation on a basic curve. Let's consider the graph of a function that involves a stretching, a reflection, a horizontal translation, and a vertical translation of the basic absolute value function.

E X A M P L E   5

Graph $f(x) = -2|x - 3| + 1$.

Solution

This is the basic absolute value curve stretched by a factor of 2, reflected through the x axis, shifted three units to the right, and shifted one unit upward. To sketch the graph, we locate the point (3, 1) and then determine a point on each of the rays. The graph is shown in Figure 3.56.

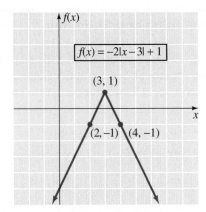

Figure 3.56

▶ **Now work Problem 9.**   ■

Remark: Note that in Example 5 we did not sketch the original basic curve $f(x) = |x|$ or any of the intermediate transformations. However, it is helpful to picture each transformation mentally. This locates the point (3, 1) and establishes the fact that the two rays point downward. Then a point on each ray determines the final graph.

　　We do need to realize that changing the order of doing the transformations may produce an incorrect graph. In Example 5 performing the translations first, and then performing the stretching and x axis reflection would locate the vertex of the graph at (3, −1) instead of (3, 1). **Unless parentheses indicate otherwise, stretchings, shrinkings, and reflections should be performed before translations.**

E X A M P L E   6   Graph $f(x) = \sqrt{-3 - x}$.

Solution

It appears that this function is a y axis reflection and a horizontal translation of the basic function $f(x) = \sqrt{x}$. First let's rewrite the expression under the radical.

$$f(x) = \sqrt{-3 - x} = \sqrt{-(3 + x)} = \sqrt{-(x + 3)}$$

Now to graph $f(x) = \sqrt{-(x + 3)}$, we would first reflect the graph of $f(x) = \sqrt{x}$ across the y axis and then shift the graph 3 units to the left. The graph is shown in Figure 3.57. Because it is always a good idea to check your graph by plotting a few points, we have added the points $(-7, 2)$ and $(-4, 1)$ to the graph.

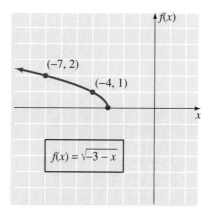

Figure 3.57

▶ **Now work Problem 23.**   ■

Now suppose that we want to graph a function such as

$$f(x) = \frac{2x^2}{x^2 + 4}$$

Because this is neither a basic function that we recognize nor a transformation of a basic function, we must revert to our previous graphing experiences. In other words, we need to find the domain, find the intercepts, check for symmetry, check for any restrictions, set up a table of values, plot the points, and sketch the curve. (If you want to do this now, you can check your result on page 385.) If the new function is defined in terms of an old function, we may be able to apply the definition of the old function and thereby simplify the new function for graphing purposes. For example, Problem 13 in Problem Set 3.5 asks you to graph the function $f(x) = |x| + x$. This function can be simplified by applying the definition of absolute value. We will leave that for you to do later.

Finally, let's use a graphing utility to give another illustration of the concept of stretching and shrinking a curve.

EXAMPLE 7   If $f(x) = \sqrt{25 - x^2}$, sketch a graph of $y = 2(f(x))$ and $y = \dfrac{1}{2}(f(x))$.

Solution

If $y = f(x) = \sqrt{25 - x^2}$, then

$$y = 2(f(x)) = 2\sqrt{25 - x^2} \qquad \text{and} \qquad y = \frac{1}{2}(f(x)) = \frac{1}{2}\sqrt{25 - x^2}$$

Graphing all three of these functions on the same set of axes produces Figure 3.58.

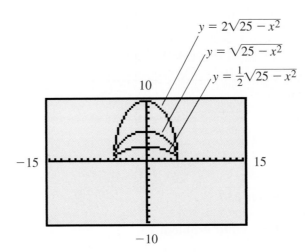

Figure 3.58   ■

CONCEPT QUIZ   For Problems 1–5, select the correct answer.

1. What is the domain of $f(x) = \sqrt{x}$?
   A. $x > 0$ B. $x \geq 0$ C. All real numbers

2. What is the domain of $f(x) = |x|$?
   A. $x > 0$ B. $x \geq 0$ C. All real numbers

3. If a graph is symmetric to the y axis, then which of the following is equal to $f(2)$?
   A. $-f(2)$ B. $f(-2)$ C. $-f(-2)$

4. Which of the following describes the graph of $f(x) = x^4 + 3$?
   A. The graph of $f(x) = x^4$ shifted up 3 units
   B. The graph of $f(x) = x^4$ shifted to the left 3 units
   C. The graph of $f(x) = x^4$ shifted to the right 3 units

5. For the graph of the function $f(x) = -2|x + 1| - 3$, what are the coordinates of the vertex?
   A. $(-1, 3)$ B. $(1, -3)$ C. $(-1, -3)$ D. $(0, 5)$

For Problems 6–8, answer true or false.

6. When the graph of a parabola is stretched, it is said to be narrower than the basic parabola.

7. A horizontal translation is a rigid transformation, and the shape of the graph is not changed.

8. When applying successive transformations to a graph, unless parentheses indicate otherwise, stretchings, shrinkings, or reflections should be performed before translations.

Problem Set 3.5

For Problems 1–30, graph each function.

1. $f(x) = x^4 + 2$

2. $f(x) = -x^4 - 1$

3. $f(x) = (x - 2)^4$

4. $f(x) = (x + 3)^4 + 1$

5. $f(x) = -x^3$

6. $f(x) = x^3 - 2$

7. $f(x) = (x + 2)^3$

8. $f(x) = (x - 3)^3 - 1$

9. $f(x) = |x - 1| + 2$

10. $f(x) = -|x + 2|$

11. $f(x) = |x + 1| - 3$

12. $f(x) = 2|x|$

13. $f(x) = x + |x|$

14. $f(x) = \dfrac{|x|}{x}$

15. $f(x) = -|x - 2| - 1$

16. $f(x) = 2|x + 1| - 4$

17. $f(x) = x - |x|$

18. $f(x) = |x| - x$

19. $f(x) = -2\sqrt{x}$

20. $f(x) = 2\sqrt{x - 1}$

21. $f(x) = \sqrt{x + 2} - 3$

22. $f(x) = -\sqrt{x + 2} + 2$

23. $f(x) = \sqrt{2 - x}$

24. $f(x) = \sqrt{-1 - x}$

25. $f(x) = -2x^4 + 1$

26. $f(x) = 2(x - 2)^4 - 4$

27. $f(x) = -2x^3$

28. $f(x) = 2x^3 + 3$

29. $f(x) = 3(x - 2)^3 - 1$

30. $f(x) = -2(x + 1)^3 + 2$

31. Suppose that the graph of $y = f(x)$ with a domain of $-2 \le x \le 2$ is shown in Figure 3.59.

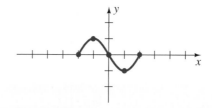

Figure 3.59

Sketch the graph of each of the following transformations of $y = f(x)$.

   a. $y = f(x) + 3$

   b. $y = f(x - 2)$

   c. $y = -f(x)$

   d. $y = f(x + 3) - 4$

■ ■ ■ THOUGHTS INTO WORDS

32. Are the graphs of the two functions $f(x) = \sqrt{x-2}$ and $g(x) = \sqrt{2-x}$ y axis reflections of each other? Defend your answer.

33. Are the graphs of $f(x) = 2\sqrt{x}$ and $g(x) = \sqrt{2x}$ identical? Defend your answer.

34. Are the graphs of $f(x) = \sqrt{x+4}$ and $g(x) = \sqrt{-x+4}$ y axis reflections of each other? Defend your answer.

▦ GRAPHING CALCULATOR ACTIVITIES

35. Use your graphing calculator to check your graphs for Problems 13–30.

36. Graph $f(x) = \sqrt{x^2 + 8}$, $f(x) = \sqrt{x^2 + 4}$, and $f(x) = \sqrt{x^2 + 1}$ on the same set of axes. Look at these graphs and predict the graph of $f(x) = \sqrt{x^2 - 4}$. Now graph it with the calculator to test your prediction.

37. For each of the following, predict the general shape and location of the graph, and then use your calculator to graph the function to check your prediction.

   a. $f(x) = \sqrt{x^2}$ **b.** $f(x) = \sqrt{x^3}$

   c. $f(x) = |x^2|$ **d.** $f(x) = |x^3|$

38. Graph $f(x) = x^4 + x^3$. Now predict the graph for each of the following and check each prediction with your graphing calculator.

   a. $f(x) = x^4 + x^3 - 4$

   b. $f(x) = (x-3)^4 + (x-3)^3$

   c. $f(x) = -x^4 - x^3$ **d.** $f(x) = x^4 - x^3$

39. Graph $f(x) = \sqrt[3]{x}$. Now predict the graph for each of the following, and check each prediction with your graphing calculator.

   a. $f(x) = 5 + \sqrt[3]{x}$ **b.** $f(x) = \sqrt[3]{x+4}$

   c. $f(x) = -\sqrt[3]{x}$ **d.** $f(x) = \sqrt[3]{x-3} - 5$

   e. $f(x) = \sqrt[3]{-x}$

Answers to the Concept Quiz

1. B   **2.** C   **3.** B   **4.** A   **5.** C   **6.** True   **7.** True   **8.** True

3.6   Combining Functions

Objectives

■ Combine functions by finding the sum, difference, product, or quotient.

■ Find the composition of two functions.

In subsequent mathematics courses, it is common to encounter functions that are defined in terms of sums, differences, products, and quotients of simpler functions. For example, if $h(x) = x^2 + \sqrt{x-1}$, then we may consider the function h as the sum of f and g, where $f(x) = x^2$ and $g(x) = \sqrt{x - 1}$. In general, *if f and g are functions, and D is the intersection of their domains,* then the following definitions can be stated:

Sum	$(f + g)(x) = f(x) + g(x)$
Difference	$(f - g)(x) = f(x) - g(x)$
Product	$(f \cdot g)(x) = f(x) \cdot g(x)$
Quotient	$\left(\dfrac{f}{g}\right)(x) = \dfrac{f(x)}{g(x)}, \qquad g(x) \neq 0$

E X A M P L E   1   If $f(x) = 3x - 1$ and $g(x) = x^2 - x - 2$, find (a) $(f + g)(x)$, (b) $(f - g)(x)$, (c) $(f \cdot g)(x)$, and (d) $(f/g)(x)$. Determine the domain of each.

Solutions

a. $(f + g)(x) = f(x) + g(x) = (3x - 1) + (x^2 - x - 2) = x^2 + 2x - 3$

b. $(f - g)(x) = f(x) - g(x)$

$$= (3x - 1) - (x^2 - x - 2)$$

$$= 3x - 1 - x^2 + x + 2$$

$$= -x^2 + 4x + 1$$

c. $(f \cdot g)(x) = f(x) \cdot g(x)$

$$= (3x - 1)(x^2 - x - 2)$$

$$= 3x^3 - 3x^2 - 6x - x^2 + x + 2$$

$$= 3x^3 - 4x^2 - 5x + 2$$

d. $\left(\dfrac{f}{g}\right)(x) = \dfrac{f(x)}{g(x)} = \dfrac{3x - 1}{x^2 - x - 2}$

The domain of both f and g is the set of all real numbers. Therefore the domain of $f + g$, $f - g$, and $f \cdot g$ is the set of all real numbers. For f/g, the denominator $x^2 - x - 2$ cannot equal zero. Solving $x^2 - x - 2 = 0$ produces

$$(x - 2)(x + 1) = 0$$

$$x - 2 = 0 \quad \text{or} \quad x + 1 = 0$$

$$x = 2 \quad \text{or} \quad x = -1$$

Therefore the domain for f/g is the set of all real numbers except 2 and -1.

▶ **Now work Problem 3.**   ■

■ Composition of Functions

Besides adding, subtracting, multiplying, and dividing functions, there is another important operation called *composition*. The composition of two functions can be defined as follows.

Definition 3.2

The **composition** of functions f and g is defined by

$$(f \circ g)(x) = f(g(x))$$

for all x in the domain of g such that $g(x)$ is in the domain of f.

The left side, $(f \circ g)(x)$, of the equation in Definition 3.2 is read "the composition of f and g," and the right side is read "f of g of x." It may also be helpful for you to have a mental picture of Definition 3.2 as two function machines hooked together to produce another function (called the **composite function**), as illustrated in Figure 3.60. Note that what comes out of the g function is substituted into the f function. Thus composition is sometimes called the **substitution of functions.**

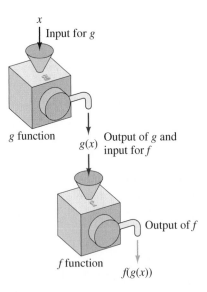

Figure 3.60

Figure 3.60 also illustrates the fact that $f \circ g$ is defined *for all x in the domain of g such that g(x) is in the domain of f.* In other words, what comes out of g must be capable of being fed into f. Let's consider some examples.

E X A M P L E   2

If $f(x) = x^2$ and $g(x) = 3x - 4$, find $(f \circ g)(x)$ and determine its domain.

Solution

We apply Definition 3.2 to obtain

$$(f \circ g)(x) = f(g(x))$$
$$= f(3x - 4)$$
$$= (3x - 4)^2$$
$$= 9x^2 - 24x + 16$$

Because g and f are both defined for all real numbers, so is $f \circ g$. ∎

Definition 3.2, with f and g interchanged, defines the composition of g and f as $(g \circ f)(x) = g(f(x))$.

E X A M P L E   3

If $f(x) = x^2$ and $g(x) = 3x - 4$, find $(g \circ f)(x)$ and determine its domain.

Solution

$$(g \circ f)(x) = g(f(x))$$
$$= g(x^2)$$
$$= 3x^2 - 4$$

Because f and g are defined for all real numbers, so is $g \circ f$.

▶ **Now work Problem 13.** ∎

The results of Examples 2 and 3 demonstrate an important idea: **The composition of functions is not a commutative operation.** In other words, $f \circ g \neq g \circ f$ for all functions f and g. However, as we will see in the next section, there is a special class of functions for which $f \circ g = g \circ f$.

E X A M P L E   4

If $f(x) = \sqrt{x}$ and $g(x) = 2x - 1$, find $(f \circ g)(x)$ and $(g \circ f)(x)$. Also determine the domain of each composite function.

Solution

$$(f \circ g)(x) = f(g(x))$$
$$= f(2x - 1)$$
$$= \sqrt{2x - 1}$$

The domain and range of g are the set of all real numbers, but the domain of f is all *nonnegative* real numbers. Therefore $g(x)$, which is $2x - 1$, must be nonnegative:

$$2x - 1 \geq 0$$
$$2x \geq 1$$
$$x \geq \frac{1}{2}$$

Thus the domain of $f \circ g$ is $D = \left\{ x \middle| x \geq \dfrac{1}{2} \right\}$.

$$(g \circ f)(x) = g(f(x))$$
$$= g(\sqrt{x})$$
$$= 2\sqrt{x} - 1$$

The domain and range of f are the set of nonnegative real numbers. The domain of g is the set of all real numbers. Therefore the domain of $g \circ f$ is $D = \{x | x \geq 0\}$.

▶ **Now work Problem 19.**   ■

EXAMPLE 5

If $f(x) = 2/(x - 1)$ and $g(x) = 1/x$, find $(f \circ g)(x)$ and $(g \circ f)(x)$. Determine the domain for each composite function.

Solution

$$(f \circ g)(x) = f(g(x))$$
$$= f\left(\frac{1}{x} \right)$$
$$= \frac{2}{\dfrac{1}{x} - 1} = \frac{2}{\dfrac{1 - x}{x}}$$
$$= \frac{2x}{1 - x}$$

The domain of g is all real numbers except zero, and the domain of f is all real numbers except 1. Because $g(x)$, which is $1/x$, cannot equal 1, we have

$$\frac{1}{x} \neq 1$$
$$x \neq 1$$

Therefore the domain of $f \circ g$ is $D = \{x | x \neq 0 \text{ and } x \neq 1\}$.

$$(g \circ f)(x) = g(f(x))$$
$$= g\left(\frac{2}{x - 1} \right)$$
$$= \frac{1}{\dfrac{2}{x - 1}}$$
$$= \frac{x - 1}{2}$$

The domain of f is all real numbers except 1, and the domain of g is all real numbers except zero. Because $f(x)$, which is $2/(x-1)$, will never equal zero, the domain of $g \circ f$ is $D = \{x | x \neq 1\}$.

▶ **Now work Problem 21.** ■

A graphing utility can be used to find the graph of a composite function without actually forming the function algebraically. Let's see how this works.

E X A M P L E 6

If $f(x) = x^3$ and $g(x) = x - 4$, use a graphing utility to obtain the graphs of $y = (f \circ g)(x)$ and $y = (g \circ f)(x)$.

Solution

To find the graph of $y = (f \circ g)(x)$, we can make the following assignments:

$$Y_1 = x - 4$$
$$Y_2 = (Y_1)^3$$

(Note that we have substituted Y_1 for x in $f(x)$ and assigned this expression to Y_2, much the same way as we would do it algebraically.) The graph of $y = (f \circ g)(x)$ is shown in Figure 3.61.

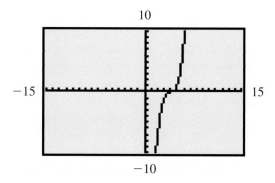

Figure 3.61

To find the graph of $y = (g \circ f)(x)$, we can make the following assignments:

$$Y_1 = x^3$$
$$Y_2 = Y_1 - 4$$

The graph of $y = (g \circ f)(x)$ is shown in Figure 3.62.

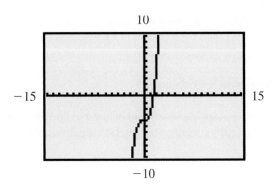

Figure 3.62   ■

Take another look at Figures 3.61 and 3.62. Note that in Figure 3.61 the graph of $y = (f \circ g)(x)$ is the basic cubic curve $f(x) = x^3$ shifted four units to the right. Likewise, in Figure 3.62 the graph of $y = (g \circ f)(x)$ is the basic cubic curve shifted four units downward. These are examples of a more general concept of using composite functions to represent various geometric transformations.

CONCEPT QUIZ   For Problems 1–8, answer true or false.

1. If $f(x) = \sqrt{x}$ and $g(x) = x^2$, then the domain of $(f \circ g)(x)$ is all real numbers.

2. If $f(x) = 2x + 6$ and $g(x) = x - 7$, then the domain of $\dfrac{f}{g}$ is all real numbers.

3. The composition of functions is a commutative operation.

4. The sum of two functions is a commutative operation.

5. The composition of two functions $(f \circ g)(x)$ means that we multiply the functions.

6. When forming the composition of two functions $(f \circ g)(x)$, the range elements of g are members of the domain of f.

7. If $f(x) = \dfrac{1}{x - 6}$ and $g(x) = 2x$, then the domain of $(f \circ g)(x)$ is all real numbers except 3 and 6.

8. If the domain of f is $x > 0$ and the domain of g is $x < 0$, then the sum $(f + g)(x)$ is not defined.

Problem Set 3.6

For Problems 1–8, find $f + g, f - g, f \cdot g$, and $\dfrac{f}{g}$.

1. $f(x) = 3x - 4, \quad g(x) = 5x + 2$

2. $f(x) = -6x - 1, \quad g(x) = -8x + 7$

3. $f(x) = x^2 - 6x + 4, \quad g(x) = -x - 1$

4. $f(x) = 2x^2 - 3x + 5, \quad g(x) = x^2 - 4$

5. $f(x) = x^2 - x - 1, \quad g(x) = x^2 + 4x - 5$

6. $f(x) = x^2 - 2x - 24$, $g(x) = x^2 - x - 30$

7. $f(x) = \sqrt{x - 1}$, $g(x) = \sqrt{x}$

8. $f(x) = \sqrt{x - 2}$, $g(x) = \sqrt{3x - 1}$

For Problems 9–28, find $(f \circ g)(x)$ and $(g \circ f)(x)$. Also specify the domain for each.

9. $f(x) = 2x$, $g(x) = 3x - 1$

10. $f(x) = 4x + 1$, $g(x) = 3x$

11. $f(x) = 5x - 3$, $g(x) = 2x + 1$

12. $f(x) = 3 - 2x$, $g(x) = -4x$

13. $f(x) = 3x + 4$, $g(x) = x^2 + 1$

14. $f(x) = 3$, $g(x) = -3x^2 - 1$

15. $f(x) = 3x - 4$, $g(x) = x^2 + 3x - 4$

16. $f(x) = 2x^2 - x - 1$, $g(x) = x + 4$

17. $f(x) = \dfrac{1}{x}$, $g(x) = 2x + 7$

18. $f(x) = \dfrac{1}{x^2}$, $g(x) = x$

19. $f(x) = \sqrt{x - 2}$, $g(x) = 3x - 1$

20. $f(x) = \dfrac{1}{x}$, $g(x) = \dfrac{1}{x^2}$

21. $f(x) = \dfrac{1}{x - 1}$, $g(x) = \dfrac{2}{x}$

22. $f(x) = \dfrac{4}{x + 2}$, $g(x) = \dfrac{3}{2x}$

23. $f(x) = 2x + 1$, $g(x) = \sqrt{x - 1}$

24. $f(x) = \sqrt{x + 1}$, $g(x) = 5x - 2$

25. $f(x) = \dfrac{1}{x - 1}$, $g(x) = \dfrac{x + 1}{x}$

26. $f(x) = \dfrac{x - 1}{x + 2}$, $g(x) = \dfrac{1}{x}$

27. $f(x) = \sqrt{x - 9}$, $g(x) = x^2$

28. $f(x) = 4x^2 - 9$, $g(x) = \sqrt{x}$

For Problems 29–34, solve each problem.

29. If $f(x) = 3x - 2$ and $g(x) = x^2 + 1$, find $(f \circ g)(-1)$ and $(g \circ f)(3)$.

30. If $f(x) = x^2 - 2$ and $g(x) = x + 4$, find $(f \circ g)(2)$ and $(g \circ f)(-4)$.

31. If $f(x) = 2x - 3$ and $g(x) = x^2 - 3x - 4$, find $(f \circ g)(-2)$ and $(g \circ f)(1)$.

32. If $f(x) = 1/x$ and $g(x) = 2x + 1$, find $(f \circ g)(1)$ and $(g \circ f)(2)$.

33. If $f(x) = \sqrt{x}$ and $g(x) = 3x - 1$, find $(f \circ g)(4)$ and $(g \circ f)(4)$.

34. If $f(x) = x + 5$ and $g(x) = |x|$, find $(f \circ g)(-4)$ and $(g \circ f)(-4)$.

For Problems 35–40, show that $(f \circ g)(x) = x$ and that $(g \circ f)(x) = x$.

35. $f(x) = 2x$, $g(x) = \dfrac{1}{2}x$

36. $f(x) = \dfrac{3}{4}x$, $g(x) = \dfrac{4}{3}x$

37. $f(x) = x - 2$, $g(x) = x + 2$

38. $f(x) = 2x + 1$, $g(x) = \dfrac{x - 1}{2}$

39. $f(x) = 3x + 4$, $g(x) = \dfrac{x - 4}{3}$

40. $f(x) = 4x - 3$, $g(x) = \dfrac{x + 3}{4}$

■■■ THOUGHTS INTO WORDS

41. Discuss whether addition, subtraction, multiplication, and division of functions are commutative operations.

42. Explain why the composition of two functions is not a commutative operation.

43. Explain how to find the domain of

$$\left(\frac{f}{g}\right)(x) \text{ if } f(x) = \frac{x-1}{x+2} \text{ and } g(x) = \frac{x+3}{x-5}.$$

■■■ FURTHER INVESTIGATIONS

44. If $f(x) = 3x - 4$ and $g(x) = ax + b$, find conditions on a and b that will guarantee that $f \circ g = g \circ f$.

45. If $f(x) = x^2$ and $g(x) = \sqrt{x}$, with both having a domain of the set of nonnegative real numbers, then show that $(f \circ g)(x) = x$ and $(g \circ f)(x) = x$.

46. If $f(x) = 3x^2 - 2x - 1$ and $g(x) = x$, find $f \circ g$ and $g \circ f$. (Recall that we have previously named $g(x) = x$ the *identity function*.)

47. In Section 3.1 we defined an *even function* to be a function such that $f(-x) = f(x)$ and an *odd function* to be one such that $f(-x) = -f(x)$. Verify that (a) the sum of two even functions is an even function, and (b) the sum of two odd functions is an odd function.

 ### GRAPHING CALCULATOR ACTIVITIES

48. For each of the following, predict the general shape and location of the graph, and then use your calculator to graph the function to check your prediction. (Your knowledge of the graphs of the basic functions that are being added or subtracted should be helpful when you are making your predictions.)

a. $f(x) = x^4 + x^2$ **b.** $f(x) = x^3 + x^2$

c. $f(x) = x^4 - x^2$ **d.** $f(x) = x^2 - x^4$

e. $f(x) = x^2 - x^3$ **f.** $f(x) = x^3 - x^2$

g. $f(x) = |x| + \sqrt{x}$ **h.** $f(x) = |x| - \sqrt{x}$

49. For each of the following, find the graphs of $y = (f \circ g)(x)$ and $y = (g \circ f)(x)$.

a. $f(x) = x^2$ and $g(x) = x + 5$

b. $f(x) = x^3$ and $g(x) = x + 3$

c. $f(x) = x - 6$ and $g(x) = -x^3$

d. $f(x) = x^2 - 4$ and $g(x) = \sqrt{x}$

e. $f(x) = \sqrt{x}$ and $g(x) = x^2 + 4$

f. $f(x) = \sqrt[3]{x}$ and $g(x) = x^3 - 5$

Answers to the Concept Quiz

1. True   **2.** False   **3.** False   **4.** True   **5.** False   **6.** True   **7.** False   **8.** True

Inverse Functions

Objectives

- Determine if a function is one-to-one.

- Verify that two functions are inverses of each other.

- Find the inverse of a function.

- Find the intervals where a function is increasing (or decreasing).

Recall the *vertical line test*: If each vertical line intersects a graph in no more than one point, then the graph represents a function. There is also a useful distinction between two basic types of functions. Consider the graphs of the two functions in Figure 3.63: (a) $f(x) = 2x - 1$ and (b) $g(x) = x^2$. In part (a), any *horizontal line* will intersect the graph in no more than one point. Therefore every value of $f(x)$ has only one value of x associated with it. Any function that has this property of having exactly one value of x associated with each value of $f(x)$ is called a **one-to-one function**. The function $g(x) = x^2$ is not a one-to-one function because the horizontal line in Figure 3.63(b) intersects the parabola in two points.

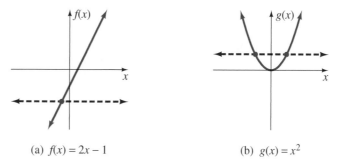

(a) $f(x) = 2x - 1$ (b) $g(x) = x^2$

Figure 3.63

Stated another way, a function f is said to be one-to-one if $x_1 \neq x_2$ implies that $f(x_1) \neq f(x_2)$. In other words, different values for x always result in different values for $f(x)$. Thus without a graph, we can show that $f(x) = 2x - 1$ is a one-to-one function as follows: If $x_1 \neq x_2$, then $2x_1 \neq 2x_2$, and therefore $2x_1 - 1 \neq 2x_2 - 1$. Furthermore, we can show that $f(x) = x^2$ is not a one-to-one function because $f(2) = 4$ and $f(-2) = 4$; that is, different values for x produce the same value for $f(x)$.

Now let's consider a one-to-one function f that assigns the value $f(x)$ in its range R to each x in its domain D (see Figure 3.64(a)). We can define a new function g that goes from R to D; it assigns $f(x)$ in R back to x in D, as indicated in

Figure 3.64(b). The functions *f* and *g* are called *inverse functions* of each other. The following definition states this concept precisely.

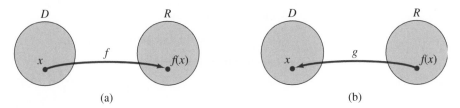

(a)   (b)

Figure 3.64

Definition 3.3

> Let *f* be a one-to-one function with a domain of X and a range of Y. A function *g* with a domain of Y and a range of X is called the **inverse function** of *f* if
>
> $$(f \circ g)(x) = x \qquad \text{for every } x \text{ in } Y$$
>
> and
>
> $$(g \circ f)(x) = x \qquad \text{for every } x \text{ in } X$$

In Definition 3.3, note that for *f* and *g* to be inverses of each other, the domain of *f* must equal the range of *g*, and the range of *f* must equal the domain of *g*. Furthermore, *g* must reverse the correspondences given by *f*, and *f* must reverse the correspondences given by *g*. In other words, inverse functions *undo* each other. Let's use Definition 3.3 to verify that two specific functions are inverses of each other.

E X A M P L E   1   Verify that $f(x) = 4x - 5$ and $g(x) = \dfrac{x + 5}{4}$ are inverse functions.

Solution

Because the set of real numbers is the domain and range of both functions, we know that the domain of *f* equals the range of *g* and that the range of *f* equals the domain of *g*. Furthermore,

$$(f \circ g)(x) = f(g(x))$$

$$= f\left(\frac{x + 5}{4}\right)$$

$$= 4\left(\frac{x + 5}{4}\right) - 5 = x$$

and

$$(g \circ f)(x) = g(f(x))$$

$$= g(4x - 5)$$

$$= \frac{4x - 5 + 5}{4} = x$$

Therefore f and g are inverses of each other.

▶   **Now work Problem 19.**   ∎

EXAMPLE   2

Verify that $f(x) = x^2 + 1$ for $x \geq 0$ and $g(x) = \sqrt{x - 1}$ for $x \geq 1$ are inverse functions.

Solution

First, note that the domain of f equals the range of g—namely, the set of nonnegative real numbers. Also, the range of f equals the domain of g—namely, the set of real numbers greater than or equal to 1. Furthermore,

$$(f \circ g)(x) = f(g(x))$$

$$= f(\sqrt{x - 1})$$

$$= (\sqrt{x - 1})^2 + 1$$

$$= x - 1 + 1 = x$$

and

$$(g \circ f)(x) = g(f(x))$$

$$= g(x^2 + 1)$$

$$= \sqrt{x^2 + 1 - 1}$$

$$= \sqrt{x^2} = x \qquad \sqrt{x^2} = x \text{ because } x \geq 1$$

Therefore f and g are inverses of each other.

▶   **Now work Problem 25.**   ∎

The inverse of a function f is commonly denoted by f^{-1} (read "f inverse" or "the inverse of f"). Do not confuse the -1 in f^{-1} with a negative exponent. The symbol f^{-1} *does not* mean $1/f^1$ but refers to the inverse function of function f.

Remember that a function can also be thought of as a set of ordered pairs no two of which have the same first component. Along those lines, a one-to-one function further requires that no two of the ordered pairs have the same second component. Then, if the components of each ordered pair of a given one-to-one function are interchanged, the resulting function and the given function are inverses of each other. Thus if

$$f = \{(1, 4), (2, 7), (5, 9)\}$$

then

$$f^{-1} = \{(4, 1), (7, 2), (9, 5)\}$$

Graphically, two functions that are inverses of each other are **mirror images with reference to the line $y = x$.** This is because ordered pairs (a, b) and (b, a) are reflections of each other with respect to the line $y = x$, as illustrated in Figure 3.65. (You will verify this in the next set of exercises.) Therefore, if the graph of a function f is known, as in Figure 3.66(a), then the graph of f^{-1} can be determined by reflecting f across the line $y = x$, as in Figure 3.66(b).

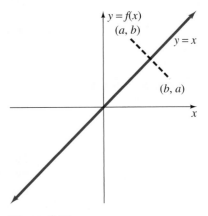

Figure 3.65

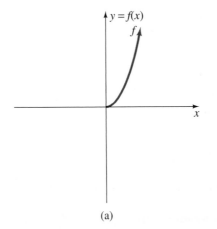

(a)

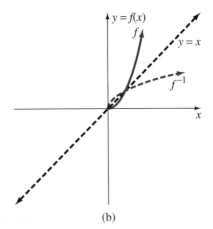

(b)

Figure 3.66

■ Finding Inverse Functions

The idea of inverse functions *undoing each other* provides the basis for an informal approach to finding the inverse of a function. Consider the function

$$f(x) = 2x + 1$$

To each x this function assigns twice x plus 1. To undo this function, we can subtract 1 and divide by 2. Hence the inverse is

$$f^{-1}(x) = \frac{x - 1}{2}$$

Now let's verify that f and f^{-1} are indeed inverses of each other:

$$(f \circ f^{-1})(x) = f(f^{-1}(x)) \qquad\qquad (f^{-1} \circ f)(x) = f^{-1}(f(x))$$

$$= f\left(\frac{x - 1}{2}\right) \qquad\qquad\qquad\qquad = f^{-1}(2x + 1)$$

$$= 2\left(\frac{x - 1}{2}\right) + 1 \qquad\qquad\qquad = \frac{2x + 1 - 1}{2}$$

$$= x - 1 + 1 = x \qquad\qquad\qquad\quad = \frac{2x}{2} = x$$

Thus the inverse of $f(x) = 2x + 1$ is $f^{-1}(x) = \dfrac{x - 1}{2}$.

This informal approach may not work very well with more complex functions, but it does emphasize how inverse functions are related to each other. A more formal and systematic technique for finding the inverse of a function can be described as follows.

1. Replace the symbol $f(x)$ with y.

2. Interchange x and y.

3. Solve the equation for y in terms of x.

4. Replace y with the symbol $f^{-1}(x)$.

The following examples illustrate this technique.

E X A M P L E   3   Find the inverse of $f(x) = \dfrac{2}{3}x + \dfrac{3}{5}$.

Solution

When we replace $f(x)$ with y, the equation becomes $y = \dfrac{2}{3}x + \dfrac{3}{5}$. Interchanging x and y produces $x = \dfrac{2}{3}y + \dfrac{3}{5}$. Now, solving for y, we obtain

$$x = \frac{2}{3}y + \frac{3}{5}$$

$$15(x) = 15\left(\frac{2}{3}y + \frac{3}{5}\right)$$

$$15x = 10y + 9$$
$$15x - 9 = 10y$$
$$\frac{15x - 9}{10} = y$$

Finally, by replacing y with $f^{-1}(x)$, we can express the inverse function as

$$f^{-1}(x) = \frac{15x - 9}{10}$$

The domain of f is equal to the range of f^{-1} (both are the set of real numbers), and the range of f equals the domain of f^{-1} (both are the set of real numbers). Furthermore, we could show that $(f \circ f^{-1})(x) = x$ and $(f^{-1} \circ f)(x) = x$. We leave this for you to complete.

▶ **Now work Problem 41.** ∎

Does $f(x) = x^2 - 2$ have an inverse function? Sometimes a graph of the function helps to answer such a question. In Figure 3.67(a), it should be evident that f is not a one-to-one function and therefore cannot have an inverse. However, it should also be apparent from the graph that if we restrict the domain of f to be the nonnegative real numbers, then it is a one-to-one function and should have an inverse [see Figure 3.67(b)]. The next example illustrates how to find the inverse function.

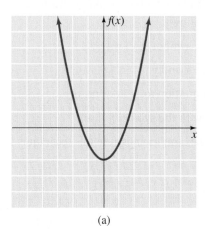
(a)

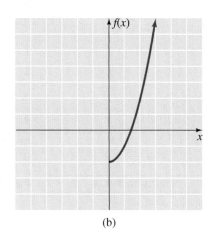
(b)

Figure 3.67

E X A M P L E   4

Find the inverse of $f(x) = x^2 - 2$, where $x \geq 0$.

Solution

When we replace $f(x)$ with y, the equation becomes

$$y = x^2 - 2, \qquad x \geq 0$$

Interchanging x and y produces

$$x = y^2 - 2, \qquad y \geq 0$$

Now let's solve for y; keep in mind that y is to be nonnegative.

$$x = y^2 - 2$$
$$x + 2 = y^2$$
$$\sqrt{x + 2} = y, \qquad x \geq -2$$

Finally, by replacing y with $f^{-1}(x)$, we can express the inverse function as

$$f^{-1}(x) = \sqrt{x + 2}, \qquad x \geq -2$$

The domain of f equals the range of f^{-1} (both are the nonnegative real numbers), and the range of f equals the domain of f^{-1} (both are the real numbers greater than or equal to -2). It can also be shown that $(f \circ f^{-1})(x) = x$ and $(f^{-1} \circ f)(x) = x$. Again, we leave this for you to complete.

▶ **Now work Problem 47.** ■

■ Increasing and Decreasing Functions

Now we can formulate some general ideas that were specifically illustrated in Example 4. In Figure 3.68 the function f is said to be *increasing* on the intervals $(-\infty, x_1]$ and $[x_2, \infty)$, and f is said to be *decreasing* on the interval $[x_1, x_2]$.

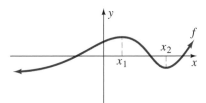

Figure 3.68

More specifically, increasing and decreasing functions are defined as follows:

Definition 3.4

Let f be a function, with the interval I a subset of the domain of f. Let x_1 and x_2 be in I. Then

   1. f is *increasing on I* if $f(x_1) < f(x_2)$ whenever $x_1 < x_2$.

   2. f is *decreasing on I* if $f(x_1) > f(x_2)$ whenever $x_1 < x_2$.

   3. f is *constant on I* if $f(x_1) = f(x_2)$ for every x_1 and x_2.

Apply Definition 3.4 and you will see that the quadratic function $f(x) = x^2$ shown in Figure 3.69 is decreasing on $(-\infty, 0]$ and increasing on $[0, \infty)$. Likewise, the linear function $f(x) = 2x$ in Figure 3.70 is increasing throughout its domain of real numbers, so we say it is increasing on $(-\infty, \infty)$. The function $f(x) = -2x$ in Figure 3.71 is decreasing on $(-\infty, \infty)$. For our purposes in this text, we will rely on our knowledge of the graphs of the functions to determine where functions are increasing and decreasing. More formal techniques for determining where functions increase and decrease will be developed in calculus.

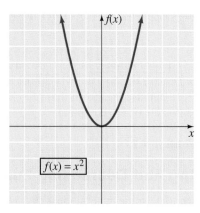

Figure 3.69

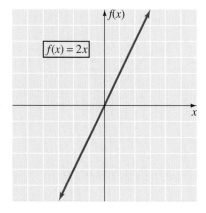

Figure 3.70

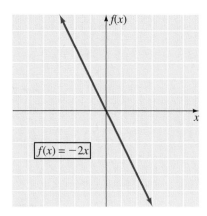

Figure 3.71

A function that is always increasing (or always decreasing) over its entire domain is one-to-one and so has an inverse. Furthermore, as illustrated by Example 4, even if a function is not one-to-one over its entire domain, it may be so over some subset of the domain. It then has an inverse over this restricted domain.

As functions become more complex, a graphing utility can be used to help with the problems we have discussed in this section. For example, suppose that we want to know whether the function $f(x) = \dfrac{3x + 1}{x - 4}$ is a one-to-one function and therefore has an inverse. Using a graphing utility, we can quickly get a sketch of the graph (see Figure 3.72). Then, by applying the horizontal line test to the graph, we can be fairly certain that the function is one-to-one. (Later we will develop some concepts that will allow us to be absolutely certain of this conclusion.)

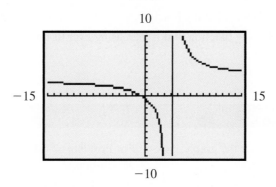

Figure 3.72

A graphing utility can also be used to help determine intervals on which a function is increasing or decreasing. For example, to determine such intervals for the function $f(x) = \sqrt{x^2 + 4}$, let's use a graphing utility to get a sketch of the curve (see Figure 3.73). From this graph we see that the function is decreasing on the interval $(-\infty, 0]$ and increasing on the interval $[0, \infty)$.

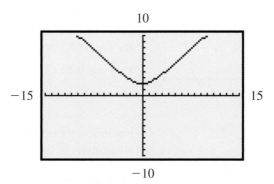

Figure 3.73

CONCEPT QUIZ

For Problems 1–8, answer true or false.

1. Over the domain of all real numbers, a quadratic function is never a one-to-one function.

2. If a horizontal line intersects the graph of a function at more than one point, then the function is one-to-one.

3. If $f(1) = 4$ and $f(-3) = 4$, then f is not a one-to-one function.

4. The only condition for f and g to be inverse functions is that g must reverse the correspondence given by f.

5. Given that f has an inverse function and $f(2) = -8$, then $f^{-1}(-8) = 2$.

6. The graphs of two functions that are inverses of each other are mirror images with reference to the line $y = -x$.

7. The quadratic function $f(x) = (x - 3)^2$ is decreasing on the interval $(-\infty, 0]$.

8. Any function that is increasing over its entire domain is a one-to-one function.

Problem Set 3.7

For Problems 1–6, determine whether the graph represents a one-to-one function.

1.

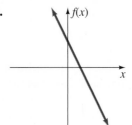

Figure 3.74

2.

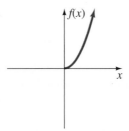

Figure 3.75

3.

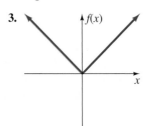

Figure 3.76

4.

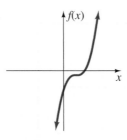

Figure 3.77

5.

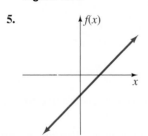

Figure 3.78

6.

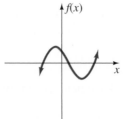

Figure 3.79

For Problems 7–14, determine whether the function f is one-to-one.

7. $f(x) = 5x + 4$

8. $f(x) = -3x + 4$

9. $f(x) = x^3$

10. $f(x) = x^5 + 1$

11. $f(x) = |x| + 1$

12. $f(x) = -|x| - 2$

13. $f(x) = -x^4$

14. $f(x) = x^4 + 1$

For Problems 15–18, list the domain and range of the function, form the inverse function f^{-1}, and list the domain and range of f^{-1}.

15. $f = \{(1, 5), (2, 9), (5, 21)\}$

16. $f = \{(1, 1), (4, 2), (9, 3), (16, 4)\}$

17. $f = \{(0, 0), (2, 8), (-1, -1), (-2, -8)\}$

18. $f = \{(-1, 1), (-2, 4), (-3, 9), (-4, 16)\}$

For Problems 19–26, verify that the two given functions are inverses of each other.

19. $f(x) = 5x - 9$ and $g(x) = \dfrac{x + 9}{5}$

20. $f(x) = -3x + 4$ and $g(x) = \dfrac{4 - x}{3}$

21. $f(x) = -\dfrac{1}{2}x + \dfrac{5}{6}$ and $g(x) = -2x + \dfrac{5}{3}$

22. $f(x) = x^3 + 1$ and $g(x) = \sqrt[3]{x - 1}$

23. $f(x) = \dfrac{1}{x-1}$ for $x > 1$

and $g(x) = \dfrac{x+1}{x}$ for $x > 0$

24. $f(x) = x^2 + 2$ for $x \geq 0$ and $g(x) = \sqrt{x-2}$
for $x \geq 2$

25. $f(x) = \sqrt{2x-4}$ for $x \geq 2$ and $g(x) = \dfrac{x^2+4}{2}$
for $x \geq 0$

26. $f(x) = x^2 - 4$ for $x \geq 0$ and $g(x) = \sqrt{x+4}$
for $x \geq -4$

For Problems 27–36, determine whether f and g are inverse functions.

27. $f(x) = 3x$ and $g(x) = -\dfrac{1}{3}x$

28. $f(x) = \dfrac{3}{4}x - 2$ and $g(x) = \dfrac{4}{3}x + \dfrac{8}{3}$

29. $f(x) = x^3$ and $g(x) = \sqrt[3]{x}$

30. $f(x) = \dfrac{1}{x+1}$ and $g(x) = \dfrac{1-x}{x}$

31. $f(x) = x$ and $g(x) = \dfrac{1}{x}$

32. $f(x) = \dfrac{3}{5}x + \dfrac{1}{3}$ and $g(x) = \dfrac{5}{3}x - 3$

33. $f(x) = x^2 - 3$ for $x \geq 0$ and $g(x) = \sqrt{x+3}$
for $x \geq -3$

34. $f(x) = |x-1|$ for $x \geq 1$ and $g(x) = |x+1|$
for $x \geq 0$

35. $f(x) = \sqrt{x+1}$ and $g(x) = x^2 - 1$
for $x \geq 0$

36. $f(x) = \sqrt{2x-2}$ and $g(x) = \dfrac{1}{2}x^2 + 1$

For Problems 37–50, find f^{-1} and also verify that
$(f \circ f^{-1})(x) = x$ and $(f^{-1} \circ f)(x) = x$.

37. $f(x) = x - 4$ **38.** $f(x) = 2x - 1$

39. $f(x) = -3x - 4$ **40.** $f(x) = -5x + 6$

41. $f(x) = \dfrac{3}{4}x - \dfrac{5}{6}$

42. $f(x) = \dfrac{2}{3}x - \dfrac{1}{4}$

43. $f(x) = -\dfrac{2}{3}x$ **44.** $f(x) = \dfrac{4}{3}x$

45. $f(x) = \sqrt{x}$ for $x \geq 0$

46. $f(x) = \dfrac{1}{x}$ for $x \neq 0$

47. $f(x) = x^2 + 4$ for $x \geq 0$

48. $f(x) = x^2 + 1$ for $x \leq 0$

49. $f(x) = 1 + \dfrac{1}{x}$ for $x > 0$

50. $f(x) = \dfrac{x}{x+1}$ for $x > -1$

For Problems 51–58, find f^{-1} and graph f and f^{-1} on the same set of axes.

51. $f(x) = 3x$ **52.** $f(x) = -x$

53. $f(x) = 2x + 1$ **54.** $f(x) = -3x - 3$

55. $f(x) = \dfrac{2}{x-1}$ for $x > 1$

56. $f(x) = \dfrac{-1}{x-2}$ for $x > 2$

57. $f(x) = x^2 - 4$ for $x \geq 0$

58. $f(x) = \sqrt{x-3}$ for $x \geq 3$

For Problems 59–66, find the intervals on which the given function is increasing and the intervals on which it is decreasing.

59. $f(x) = x^2 + 1$

60. $f(x) = x^3$

61. $f(x) = -3x + 1$

62. $f(x) = (x-3)^2 + 1$

63. $f(x) = -(x + 2)^2 - 1$

64. $f(x) = x^2 - 2x + 6$

65. $f(x) = -2x^2 - 16x - 35$

66. $f(x) = x^2 + 3x - 1$

■ ■ ■ THOUGHTS INTO WORDS

67. Does the function $f(x) = 4$ have an inverse? Explain your answer.

68. Explain why every nonconstant linear function has an inverse.

69. Are the functions $f(x) = x^4$ and $g(x) = \sqrt[4]{x}$ inverses of each other? Explain your answer.

70. What does it mean to say that 2 and -2 are additive inverses of each other? What does it mean to say that 2 and $\dfrac{1}{2}$ are multiplicative inverses of each other? What does it mean to say that the functions $f(x) = x - 2$ and $f(x) = x + 2$ are inverses of each other? Do you think that the concept of "inverse" is being used in a consistent manner? Explain your answer.

■ ■ ■ FURTHER INVESTIGATIONS

71. The function notation and the operation of composition can be used to find inverses as follows: To find the inverse of $f(x) = 5x + 3$, we know that $f(f^{-1}(x))$ must produce x. Therefore

$$f[f^{-1}(x)] = 5[f^{-1}(x)] + 3 = x$$

$$5[f^{-1}(x)] = x - 3$$

$$f^{-1}(x) = \frac{x - 3}{5}$$

Use this approach to find the inverse of each of the following functions.

a. $f(x) = 3x - 9$ **b.** $f(x) = -2x + 6$

c. $f(x) = -x + 1$ **d.** $f(x) = 2x$

e. $f(x) = -5x$ **f.** $f(x) = x^2 + 6$ for $x \geq 0$

72. If $f(x) = 2x + 3$ and $g(x) = 3x - 5$, find

a. $(f \circ g)^{-1}(x)$

b. $(f^{-1} \circ g^{-1})(x)$

c. $(g^{-1} \circ f^{-1})(x)$

GRAPHING CALCULATOR ACTIVITIES

73. For Problems 37–44, graph the given function, the inverse function that you found, and $f(x) = x$ on the same set of axes. In each case the given function and its inverse should produce graphs that are reflections of each other through the line $f(x) = x$.

74. There is another way we can use the graphing calculator to help show that two functions are inverses of each other. Suppose we want to show that $f(x) = x^2 - 2$ for $x \geq 0$ and $g(x) = \sqrt{x + 2}$ for

$x \geq -2$ are inverses of each other. Let's make the following assignments for our graphing calculator:

$$f: \quad Y_1 = x^2 - 2$$

$$g: \quad Y_2 = \sqrt{x + 2}$$

$$f \circ g: \quad Y_3 = (Y_2)^2 - 2$$

$$g \circ f: \quad Y_4 = \sqrt{Y_1 + 2}$$

Now we can proceed as follows:

1. Graph $Y_1 = x^2 - 2$ and note that for $x \geq 0$, the range is greater than or equal to -2.

2. Graph $Y_2 = \sqrt{x + 2}$ and note that for $x \geq -2$, the range is greater than or equal to zero.

> Thus the domain of f equals the range of g, and the range of f equals the domain of g.

3. Graph $Y_3 = (Y_2)^2 - 2$ for $x \geq -2$ and observe the line $y = x$ for $x \geq -2$.

4. Graph $Y_4 = \sqrt{Y_1 + 2}$ for $x \geq 0$ and observe the line $y = x$ for $x \geq 0$.

> Thus $(f \circ g)(x) = x$ and $(g \circ f)(x) = x$, and the two functions are inverses of each other.

Use this approach to check your answers for Problems 45–50.

75. Use the technique demonstrated in Problem 74 to show that

$$f(x) = \frac{x}{\sqrt{x^2 + 1}}$$

and

$$g(x) = \frac{x}{\sqrt{1 - x^2}} \quad \text{for } -1 < x < 1$$

are inverses of each other.

Answers to the Concept Quiz

1. True   **2.** False   **3.** True   **4.** False   **5.** True   **6.** False   **7.** False   **8.** True

Chapter 3 Summary

The function serves as a thread to tie the ideas in this chapter together.

■ Function Concept

Definition 3.1

> A function f is a correspondence between two sets X and Y that assigns to each element x of set X one and only one element y of set Y. The element y being assigned is called the **image** of x. The set X is called the **domain** of the function, and the set of all images is called the **range** of the function.

A function can also be thought of as a set of ordered pairs no two of which have the same first component. If each vertical line intersects a graph in no more than one point, then the graph represents a function.

■ Graphing Functions

Any function that can be written in the form

$$f(x) = ax + b$$

where a and b are real numbers, is a **linear function.** The graph of a linear function is a straight line.

Any function that can be written in the form

$$f(x) = ax^2 + bx + c$$

where a, b, and c are real numbers and $a \neq 0$, is a **quadratic function.** The graph of any quadratic function is a **parabola,** which can be drawn using either one of the following methods:

1. Express the function in the form $f(x) = a(x - h)^2 + k$, and use the values of a, h, and k to determine the parabola.
2. Express the function in the form $f(x) = ax^2 + bx + c$ and use the fact that the vertex is at

$$\left(-\frac{b}{2a}, f\left(-\frac{b}{2a}\right)\right)$$

and the axis of symmetry is

$$x = -\frac{b}{2a}$$

Another important skill in graphing is to be able to recognize equations of the transformations of basic curves. We worked with the following transformations in this chapter.

Vertical Translation   The graph of $y = f(x) + k$ is the graph of $y = f(x)$ shifted k units upward if $k > 0$ or shifted $|k|$ units downward if $k < 0$.

Horizontal Translation   The graph of $y = f(x - h)$ is the graph of $y = f(x)$ shifted h units to the right if $h > 0$ or shifted $|h|$ units to the left if $h < 0$.

x Axis Reflection   The graph of $y = -f(x)$ is the graph of $y = f(x)$ reflected through the x axis.

y Axis Reflection   The graph of $y = f(-x)$ is the graph of $y = f(x)$ reflected through the y axis.

Vertical Stretching and Shrinking   The graph of $y = cf(x)$, where $c > 0$, is the graph of $y = f(x)$ either stretched or shrunk by a factor of c. If $c > 1$, the graph is said to be **stretched** by a factor of c, and if $0 < c < 1$, the graph is said to be **shrunk** by a factor of c.

The following suggestions are helpful for graphing functions that are unfamiliar:

1. Determine the domain of the function.
2. Find the intercepts.
3. Determine what type of symmetry the equation exhibits.
4. Set up a table of values that satisfy the equation. The type of symmetry and the domain will affect your choice of values for x in the table.
5. Plot the points associated with the ordered pairs and connect them with a smooth curve. Then, if appropriate, reflect this part of the curve according to the symmetry possessed by the graph.

■ Operations on Functions

Sum of two functions
$$(f + g)(x) = f(x) + g(x)$$

Difference of two functions
$$(f - g)(x) = f(x) - g(x)$$

Product of two functions
$$(f \cdot g)(x) = f(x) \cdot g(x)$$

Quotient of two functions
$$\left(\frac{f}{g}\right)(x) = \frac{f(x)}{g(x)}, \qquad g(x) \neq 0$$

Definition 3.2

The **composition** of functions f and g is defined by
$$(f \circ g)(x) = f(g(x))$$
for all x in the domain of g such that $g(x)$ is in the domain of f.

Remember that the composition of functions is *not a commutative operation*.

■ Inverse Functions

Definition 3.3

Let f be a one-to-one function with a domain of X and a range of Y. A function g with a domain of Y and a range of X is called the **inverse function** of f if
$$(f \circ g)(x) = x \qquad \text{for every } x \text{ in } Y$$
and
$$(g \circ f)(x) = x \qquad \text{for every } x \text{ in } X$$

The inverse of a function f is denoted by f^{-1}. Graphically, two functions that are inverses of each other are mirror images with reference to the line $y = x$.

A systematic technique for finding the inverse of a function can be described as follows.

1. Let $y = f(x)$.
2. Interchange x and y.
3. Solve the equation for y in terms of x.
4. The inverse function $f^{-1}(x)$ is determined by the equation in step 3.

Don't forget that the domain of f must equal the range of f^{-1}, and the domain of f^{-1} must equal the range of f.

Increasing and decreasing functions are defined as follows:

Definition 3.4

Let f be a function, with the interval I a subset of the domain of f. Let x_1 and x_2 be in I. Then

1. f is *increasing on I* if $f(x_1) < f(x_2)$ whenever $x_1 < x_2$.
2. f is *decreasing on I* if $f(x_1) > f(x_2)$ whenever $x_1 < x_2$.
3. f is *constant on I* if $f(x_1) = f(x_2)$ for every x_1 and x_2.

A function that is always increasing or always decreasing over its entire domain is a one-to-one function and therefore has an inverse.

Chapter 3   Review Problem Set

1. If $f(x)$ is $3x^2 - 2x - 1$, find $f(2), f(-1)$, and $f(-3)$.

2. For each of the following functions, find
$$\frac{f(a + h) - f(a)}{h}.$$

   a. $f(x) = -5x + 4$ **b.** $f(x) = 2x^2 - x + 4$

   c. $f(x) = -3x^2 + 2x - 5$

3. Determine the domain and range of the function $f(x) = x^2 + 5$.

4. Determine the domain of the function $f(x) = \dfrac{2}{2x^2 + 7x - 4}$.

5. Express the domain of $f(x) = \sqrt{x^2 - 7x + 10}$ using interval notation.

For Problems 6–15, graph each function.

6. $f(x) = -2x + 2$ **7.** $f(x) = 2x^2 - 1$

8. $f(x) = -\sqrt{x - 2} + 1$ **9.** $f(x) = x^2 - 8x + 17$

10. $f(x) = -x^3 + 2$ **11.** $f(x) = 2|x - 1| + 3$

12. $f(x) = -2x^2 - 12x - 19$

13. $f(x) = -\dfrac{1}{3}x + 1$ **14.** $f(x) = -\dfrac{2}{x^2}$

15. $f(x) = 2|x| - x$

16. If $f(x) = 2x + 3$ and $g(x) = x^2 - 4x - 3$, find $f + g$, $f - g, f \cdot g$, and f/g.

For Problems 17–20, find $(f \circ g)(x)$ and $(g \circ f)(x)$. Also specify the domain for each.

17. $f(x) = 3x - 9$ and $g(x) = -2x + 7$

18. $f(x) = x^2 - 5$ and $g(x) = 5x - 4$

19. $f(x) = \sqrt{x - 5}$ and $g(x) = x + 2$

20. $f(x) = \dfrac{1}{x - 3}$ and $g(x) = \dfrac{1}{x + 2}$

21. For each of the following, classify the function as even, odd, or neither even nor odd.

   a. $f(x) = 3x^2 - 4x + 6$

   b. $f(x) = -x^3$

   c. $f(x) = -4x^2 + 6$

   d. $f(x) = 2x^3 + x - 2$

22. If $f(x) = \begin{cases} x^2 - 2 & \text{for } x \geq 0 \\ -3x + 4 & \text{for } x < 0 \end{cases}$, find $f(5), f(0)$, and $f(-3)$.

23. If $f(x) = -x^2 - x + 4$ and $g(x) = \sqrt{x - 2}$, find $f(g(6))$ and $g(f(-2))$.

24. If $f(x) = |x|$ and $g(x) = x^2 - x - 1$, find $(f \circ g)(1)$ and $(g \circ f)(-3)$.

25. Determine the linear function whose graph is a line that is parallel to the line determined by
$g(x) = \dfrac{2}{3}x + 4$ and contains the point $(5, -2)$.

26. Determine the linear function whose graph is a line that is perpendicular to the line determined by
$g(x) = -\dfrac{1}{2}x - 6$ and contains the point $(-6, 3)$.

For Problems 27–30, solve each problem.

27. The cost for burning a 100-watt light bulb is given by the function $c(h) = 0.006h$, where h represents the number of hours that the bulb burns. How much, to the nearest cent, does it cost to burn a 100-watt bulb for 4 hours per night for a 30-day month?

28. "All Items 30% Off Marked Price" is a sign in a local department store. Form a function and then use it to determine how much one has to pay for each of the following marked items: a $65 pair of shoes, a $48 pair of slacks, a $15.50 belt.

29. Find two numbers whose sum is 10, such that the sum of the square of one number plus four times the other number is a minimum.

30. A group of students is arranging a chartered flight to Europe. The charge per person is $496 if 100 students go on the flight. If more than 100 students go, the charge per student is reduced by an amount equal to $4 times the number of students above 100. How many students should the airline try to get in order to maximize its revenue?

For Problems 31–34, determine whether f and g are inverse functions.

31. $f(x) = 7x - 1$ and $g(x) = \dfrac{x + 1}{7}$

32. $f(x) = -\dfrac{2}{3}x$ and $g(x) = \dfrac{3}{2}x$

33. $f(x) = x^2 - 6$ for $x \geq 0$ and $g(x) = \sqrt{x + 6}$
for $x \geq -6$

34. $f(x) = 2 - x^2$ for $x \geq 0$ and $g(x) = \sqrt{2 - x}$
for $x \leq 2$

For Problems 35–38, find f^{-1} and verify that $(f \circ f^{-1})(x) = x$ and $(f^{-1} \circ f)(x) = x$.

35. $f(x) = 4x + 5$

36. $f(x) = -3x - 7$

37. $f(x) = \dfrac{5}{6}x - \dfrac{1}{3}$

38. $f(x) = -2 - x^2$ for $x \geq 0$

For Problems 39 and 40, find the intervals on which the function is increasing and the intervals on which it is decreasing.

39. $f(x) = -2x^2 + 16x - 35$

40. $f(x) = 2\sqrt{x - 3}$

1. If $f(x) = -\frac{1}{2}x + \frac{1}{3}$, find $f(-3)$.

2. If $f(x) = -x^2 - 6x + 3$, find $f(-2)$.

3. If $f(x) = 3x^2 + 2x - 5$, find $\dfrac{f(a + h) - f(a)}{h}$.

4. Determine the domain of the function
$$f(x) = \frac{-3}{2x^2 + 7x - 4}.$$

5. Determine the domain of the function $f(x) = \sqrt{5 - 3x}$.

6. If $f(x) = 3x - 1$ and $g(x) = 2x^2 - x - 5$, find $f + g$, $f - g$, and $f \cdot g$.

7. If $f(x) = -3x + 4$ and $g(x) = 7x + 2$, find $(f \circ g)(x)$.

8. If $f(x) = 2x + 5$ and $g(x) = 2x^2 - x + 3$, find $(g \circ f)(x)$.

9. If $f(x) = \dfrac{3}{x - 2}$ and $g(x) = \dfrac{2}{x}$, find $(f \circ g)(x)$.

10. If $f(x) = x^2 - 2x - 3$ and $g(x) = |x - 3|$, find $f(g(-2))$ and $g(f(1))$.

11. Classify each of the following functions as even, odd, or neither even nor odd.

    a. $f(x) = 3x^2 - 10$

    b. $f(x) = -x^5 + x^3$

    c. $f(x) = -x^2 + 6x - 4$

    d. $f(x) = 2x^4 + x^2$

12. If $f(x) = \dfrac{3}{x}$ and $g(x) = \dfrac{2}{x - 1}$, determine the domain of $\left(\dfrac{f}{g}\right)(x)$.

13. If $f(x) = 2x^2 - x + 1$ and $g(x) = x^2 + 3$, find $(f + g)(-2)$, $(f - g)(4)$, and $(g - f)(-1)$.

14. If $f(x) = x^2 + 5x - 6$ and $g(x) = x - 1$, find $(f \cdot g)(x)$ and $\left(\dfrac{f}{g}\right)(x)$.

15. Determine the linear function whose graph is a line that has a slope of $-\dfrac{5}{6}$ and contains the point $(4, -8)$.

16. Find the inverse of the function $f(x) = -3x - 6$.

17. Find the inverse of the function $f(x) = \dfrac{2}{3}x - \dfrac{3}{5}$.

18. A retailer has a number of items that he wants to sell at a profit of 35% of the cost. What linear function can be used to determine the selling prices of the items? What price should he charge for a tie that cost him $13?

19. Find two numbers whose sum is 60, such that the sum of the square of one number plus 12 times the other number is a minimum.

For Problems 20 and 21, use the concepts of translation, reflection, or both to describe how the second curve can be obtained from the first curve.

20. $f(x) = x^3$, $g(x) = (x - 6)^3 - 4$

21. $f(x) = \sqrt{x}$, $g(x) = -\sqrt{x + 5} + 7$

For Problems 22–25, graph each function.

22. $f(x) = -2x^2 - 12x - 14$

23. $f(x) = 3|x - 2| - 1$

24. $f(x) = \sqrt{-x + 2}$

25. $f(x) = -x - 1$

For Problems 1–10, evaluate each expression.

1. $(3^{-2})^{-1}$

2. $\left(\dfrac{7}{9}\right)^{-1}$

3. $\dfrac{1}{\left(\dfrac{1}{2}\right)^{-3}}$

4. $8^{-1} + 2^{-3}$

5. $(3^{-2} + 2^{-3})^{-1}$

6. $-\sqrt{0.16}$

7. $\sqrt[3]{3\dfrac{3}{8}}$

8. $9^{3/2}$

9. $8^{2/3}$

10. $(-27)^{4/3}$

For Problems 11–15, evaluate each algebraic expression for the given values of the variables.

11. $-3(x - 1) + 4(2x + 3) - (3x + 5)$
   for $x = -9$

12. $\dfrac{3}{n} - \dfrac{5}{n} + \dfrac{9}{n}$ for $n = -7$

13. $\dfrac{4}{x - 2} + \dfrac{7}{x + 1}$ for $x = 6$

14. $(2x + 5y)(2x - 5y)$ for $x = 5$ and $y = -1$

15. $\dfrac{\dfrac{2}{x} - \dfrac{3}{y}}{\dfrac{1}{x} + \dfrac{4}{y}}$ for $x = -3$ and $y = 11$

For Problems 16–19, simplify each rational expression.

16. $\dfrac{12x^3y^2}{27xy}$

17. $\dfrac{6x^2 + 11x - 7}{8x^2 - 22x + 9}$

18. $\dfrac{8x^3 + 64}{4x^2 - 16}$

19. $\dfrac{xy + 4y - 2x - 8}{x^2 + 4x}$

For Problems 20–24, perform the indicated operations involving rational expressions. Express final answers in simplest form.

20. $\dfrac{3a^2b}{4a^3b^2} \div \dfrac{6a}{27b}$

21. $\dfrac{x^2 - x}{x + 5} \cdot \dfrac{x^2 + 5x + 4}{x^4 - x^2}$

22. $\dfrac{x + 3}{10} + \dfrac{2x + 1}{15} - \dfrac{x - 2}{18}$

23. $\dfrac{7}{12ab} - \dfrac{11}{15a^2}$

24. $\dfrac{8}{x^2 - 4x} + \dfrac{2}{x}$

For Problems 25–27, simplify each complex fraction.

25. $\dfrac{\dfrac{2}{x} - 3}{\dfrac{3}{y} + 4}$

26. $\dfrac{\dfrac{5}{x^2} - \dfrac{3}{x}}{\dfrac{1}{y} + \dfrac{2}{y^2}}$

27. $\dfrac{\dfrac{3a}{} - 1}{2 - \dfrac{1}{a}}$

For Problems 28–30, perform the indicated operations and simplify. Express final answers using positive exponents only.

28. $(-3x^{-1}y^2)(4x^{-2}y^{-3})$

29. $\dfrac{48x^{-4}y^2}{6xy}$

30. $\left(\dfrac{27a^{-4}b^{-3}}{-3a^{-1}b^{-4}}\right)^{-1}$

For Problems 31–36, express each in simplest radical form. All variables represent positive real numbers.

31. $\sqrt{\dfrac{8}{25}}$

32. $\dfrac{4\sqrt{3}}{7\sqrt{6}}$

33. $\sqrt{48x^3y^7}$

34. $\dfrac{4}{\sqrt{5} - \sqrt{3}}$

35. $\sqrt[3]{48x^4y^5}$

36. $\dfrac{\sqrt[3]{4}}{\sqrt[3]{2}}$

For Problems 37–40, find each of the indicated products or quotients. Express answers in the standard form of a complex number.

37. $(5 - 2i)(6 + 5i)$

38. $(-3 - i)(-2 - 4i)$

39. $\dfrac{5}{4i}$

40. $\dfrac{6 + 2i}{3 - 4i}$

For Problems 41–58, solve each equation.

41. $3(2x - 1) - 2(5x + 1) = 4(3x + 4)$

42. $n + \dfrac{3n - 1}{9} - 4 = \dfrac{3n + 1}{3}$

43. $0.92 + 0.9(x - 0.3) = 2x - 5.95$

44. $|4x - 1| = 11$ **45.** $|2x - 1| = |-x + 4|$

46. $x^3 = 36x$ **47.** $(3x - 1)^2 = 45$

48. $(2x + 5)^2 = -32$ **49.** $2x^2 - 3x + 4 = 0$

50. $(n + 4)(n - 6) = 11$ **51.** $(2n - 1)(n + 6) = 0$

52. $(x + 5)(3x - 1) = (x + 5)(2x + 7)$

53. $(x - 4)(2x + 9) = (2x - 1)(x + 2)$

54. $(3x - 1)(x + 1) = (2x + 1)(x - 3)$

55. $\sqrt{3x} - x = -6$

56. $\sqrt{x + 19} - \sqrt{x + 28} = -1$

57. $12x^4 - 19x^2 + 5 = 0$

58. $x^3 - 4x^2 - 3x + 12 = 0$

For Problems 59–68, solve each inequality and express the solution set using interval notation.

59. $|5x - 2| > 13$ **60.** $(x - 2)(x + 4) \le 0$

61. $|6x + 2| \le 8$ **62.** $x(x + 5) < 24$

63. $-5(y - 1) + 3 > 3y - 4 - 4y$

64. $\dfrac{x - 2}{5} - \dfrac{3x - 1}{4} \le \dfrac{3}{10}$

65. $(2x + 1)(x - 2)(x + 5) > 0$

66. $\dfrac{x - 3}{x - 7} \ge 0$ **67.** $\dfrac{2x}{x + 3} > 4$

68. $2x^3 + 5x^2 - 3x < 0$

69. Find the center and the length of a radius of the circle $x^2 + 4x + y^2 - 8y + 11 = 0$.

70. On a number line, find the coordinate of the point that is three-fourths of the distance from -6 to 10.

71. In a Cartesian plane, find the coordinates of a point that is two-thirds of the distance from $(-1, 2)$ to $(8, 11)$.

72. Find the slope of the line determined by the equation
$-2x + 5y = 7$.

73. Write the equation of the line that contains the two points $(3, -4)$ and $(-2, -1)$.

74. Write the equation of the circle that has a diameter with endpoints at $(-2, 4)$ and $(6, 10)$.

75. If $f(x) = 3x - 2$ and $g(x) = x^2 + 2x$, find $f(g(3))$ and $g(f(2))$.

76. If $f(x) = 2x - 1$ and $g(x) = \sqrt{x + 2}$, find $f(g(x))$ and $g(f(x))$.

77. Express the domain of the function $f(x) = \sqrt{x^2 + 7x - 30}$.

78. If $f(x) = -x^2 + 6x - 1$, find $\dfrac{f(a + h) - f(a)}{h}$.

For Problems 79–83, graph each equation.

79. $2x^2 + y^2 = 8$ **80.** $2x^2 - y^2 = 8$

81. $x^2 + 6x + y^2 - 2y + 6 = 0$

82. $-2x + y = 6$ **83.** $xy^2 = 4$

84. Graph the inequality $-3x + y \ge -6$.

For Problems 85–90, graph each function.

85. $f(x) = -|x - 2| + 4$

86. $f(x) = -x^2 - 6x - 10$

87. $f(x) = x - 2$

88. $f(x) = (x - 2)^2$

89. $f(x) = (x - 2)^3$ **90.** $f(x) = \sqrt{x - 2}$

For Problems 91–105, use an equation or an inequality to help solve each problem.

91. Find three consecutive odd integers whose sum is 57.

92. Eric has a collection of 63 coins consisting of nickels, dimes, and quarters. The number of dimes is 6 more than the number of nickels, and the number of quarters is 1 more than twice the number of nickels. How many coins of each kind are in the collection?

93. One of two supplementary angles is 4° larger than one-third of the other angle. Find the measure of each of the angles.

94. If a ring costs a jeweler $300, at what price should it be sold to make a profit of 50% on the selling price?

95. Beth invested a certain amount of money at 8% interest and $300 more than that amount at 9%. Her total yearly interest was $316. How much did she invest at each rate?

96. Two trains leave the same depot at the same time, one traveling east and the other west. At the end of $4\frac{1}{2}$ hours, the trains are 639 miles apart. If the rate of the train traveling east is 10 miles per hour faster than the rate of the other train, find their rates.

97. A 10-quart radiator contains a 50% solution of antifreeze. How much needs to be drained out and replaced with pure antifreeze to obtain a 70% antifreeze solution?

98. Sam shot rounds of 70, 73, and 76 on the first three days of a golf tournament. What must he shoot on the fourth day of the tournament to average 72 or lower for the four days?

99. The cube of a number equals nine times the same number. Find the number.

100. A strip of uniform width is to be cut off both sides and both ends of a sheet of paper that is 8 inches by 14 inches to reduce the size of the paper to an area of 72 square inches. Find the width of the strip.

101. A sum of $2450 is to be divided between two people in the ratio of 3 to 4. How much does each person receive?

102. Working together, Sue and Dean can complete a task in $1\frac{1}{5}$ hours. Dean can do the task by himself in 2 hours. How long would it take Sue to complete the task by herself?

103. Dudley bought a number of shares of stock for $300. A month later he sold all but 10 shares at a profit of $5 per share and regained his original investment of $300. How many shares did he originally buy and at what price per share?

104. The units digit of a two-digit number is 1 more than twice the tens digit. The sum of the digits is 10. Find the number.

105. The sum of the two smallest angles of a triangle is 40° less than the other angle. The sum of the smallest and largest angles is twice the other angle. Find the measures of the three angles of the triangle.

4

Polynomial and Rational Functions

Many problems that involve maximum and minimum values can be solved by using polynomial and rational functions.

© Terry Oakley/Alamy

Using Descartes' rule of signs, which is introduced in this chapter, we can determine that the equation $3x^4 + 2x^2 + 5 = 0$ contains four nonreal complex solutions. We also know that these solutions will appear in conjugate pairs.

Using the rational root theorem, which is introduced in this chapter, we can determine that the only possible rational solutions for the equation $x^4 - 6x^3 + 22x^2 - 30x + 13 = 0$ are ± 1 and ± 13.

Suppose we have a rectangular piece of cardboard that measures 20 inches by 14 inches. From each corner a square piece is cut out, and then the flaps are turned up to form an open box. The equation $V = x(20 - 2x)(14 - 2x)$, where x is the length of a side of the square pieces to be cut out, represents the volume of the box. Using a graphical approach, we can determine that $x \approx 2.7$; that is, the length of a side of the square piece cut from each corner is approximately 2.7 inches. This will determine a box with a maximum volume.

In earlier chapters we solved linear and quadratic equations and graphed linear and quadratic functions. In this chapter we will extend our equation-solving processes and graphing techniques to include more general polynomial equations and functions. Our knowledge of polynomial functions will then allow us to work with rational functions; the function concept will unify the chapter. To facilitate our study in this chapter, we will first review the concept of dividing polynomials; then we will introduce a special division technique called synthetic division.

4.1   Dividing Polynomials

Objectives

■   Divide a polynomial by a binomial divisor.

■   Use synthetic division to determine the quotient and remainder for division problems.

In Chapter 0 we used the properties

$$\frac{a + b}{c} = \frac{a}{c} + \frac{b}{c} \quad \text{and} \quad \frac{a - b}{c} = \frac{a}{c} - \frac{b}{c}$$

as a basis for dividing a polynomial by a monomial. For example,

$$\frac{18x^3 + 24x^2}{6x} = \frac{18x^3}{6x} + \frac{24x^2}{6x} = 3x^2 + 4x$$

and

$$\frac{35x^2y^3 - 55x^3y^4}{5xy^2} = \frac{35x^2y^3}{5xy^2} - \frac{55x^3y^4}{5xy^2} = 7xy - 11x^2y^2$$

You may recall from a previous algebra course that the format used to divide a polynomial by a binomial resembles the long-division format in arithmetic. Let's work through an example step by step.

Step 1   Use the conventional long-division format and arrange both the dividend and the divisor in descending powers of the variable.

$$3x + 1 \overline{)3x^3 - 5x^2 + 10x + 1}$$

Step 2   Find the first term of the quotient by dividing the first term of the dividend by the first term of the divisor.

$$\begin{array}{r} x^2 \\ 3x + 1 \overline{)3x^3 - 5x^2 + 10x + 1} \end{array}$$

Step 3   Multiply the entire divisor by the quotient term in step 2 and place this product in position to be subtracted from the dividend.

$$\begin{array}{r} x^2 \\ 3x + 1 \overline{)3x^3 - 5x^2 + 10x + 1} \\ \underline{3x^3 + x^2} \end{array}$$

Step 4 Subtract.

$$
\begin{array}{r}
x^2 \qquad\qquad\qquad \\
3x + 1\overline{)3x^3 - 5x^2 + 10x + 1} \\
\underline{3x^3 +\ \ x^2 \qquad\qquad} \\
-6x^2 + 10x + 1
\end{array}
$$

Step 5 Repeat steps 2, 3, and 4, and use $-6x^2 + 10x + 1$ as a new dividend.

$$
\begin{array}{r}
x^2 - 2x \qquad\qquad \\
3x + 1\overline{)3x^3 - 5x^2 + 10x + 1} \\
\underline{3x^3 +\ \ x^2 \qquad\qquad} \\
-6x^2 + 10x + 1 \\
\underline{-6x^2 -\ \ 2x \quad} \\
12x + 1
\end{array}
$$

Step 6 Repeat steps 2, 3, and 4 and use $12x + 1$ as a new dividend.

$$
\begin{array}{r}
x^2 - 2x\ \ + 4 \\
3x + 1\overline{)3x^3 - 5x^2 + 10x + 1} \\
\underline{3x^3 +\ \ x^2 \qquad\qquad} \\
-6x^2 + 10x + 1 \\
\underline{-6x^2 -\ \ 2x \quad} \\
12x + 1 \\
\underline{12x + 4} \\
-3
\end{array}
$$

Therefore $3x^3 - 5x^2 + 10x + 1 = (3x + 1)(x^2 - 2x + 4) + (-3)$, which has the familiar form

$$\text{Dividend} = (\text{Divisor})(\text{Quotient}) + \text{Remainder}$$

This result is commonly called the **division algorithm for polynomials,** which can be stated in general terms as follows.

Division Algorithm for Polynomials

If $f(x)$ and $g(x)$ are polynomials and $g(x) \neq 0$, then unique polynomials $q(x)$ and $r(x)$ exist such that

$$f(x) = g(x)q(x) + r(x)$$

Dividend   Divisor   Quotient   Remainder

where $r(x) = 0$ or the degree of $r(x)$ is less than the degree of $g(x)$.

Let's consider one more example to illustrate this division process further.

E X A M P L E   1

Divide $t^2 - 3t + 2t^4 - 1$ by $t^2 + 4t$.

Solution

Don't forget to arrange both the dividend and the divisor in descending powers of the variable.

$$
\begin{array}{r}
2t^2 - 8t\ + 33 \\
t^2 + 4t \overline{)2t^4 + 0t^3 +\ \ t^2 -\ \ 3t - 1} \\
\underline{2t^4 + 8t^3} \\
-8t^3 +\ \ t^2 -\ \ 3t - 1 \\
\underline{-8t^3 - 32t^2} \\
33t^2 -\ \ 3t - 1 \\
\underline{33t^2 + 132t} \\
-135t - 1
\end{array}
$$

Notice the insertion of a t^3 term with a zero coefficient

The division process is completed when the degree of the remainder is less than the degree of the divisor

▶ **Now work Problem 7.** ■

■ Synthetic Division

If the divisor is of the form $x - c$, where c is a constant, then the typical long-division algorithm can be simplified to a process called **synthetic division.** First, let's consider another division problem and use the regular-division algorithm. Then, in a step-by-step fashion, we will demonstrate some shortcuts that will lead us into the synthetic-division procedure. Consider the division problem $(2x^4 + x^3 - 17x^2 + 13x + 2) \div (x - 2)$.

$$
\begin{array}{r}
2x^3 + 5x^2 -\ 7x\ -\ 1 \\
x - 2 \overline{)2x^4 +\ \ x^3 - 17x^2 + 13x + 2} \\
\underline{2x^4 - 4x^3} \\
5x^3 - 17x^2 \\
\underline{5x^3 - 10x^2} \\
-7x^2 + 13x \\
\underline{-7x^2 + 14x} \\
-x + 2 \\
\underline{-x + 2}
\end{array}
$$

Because the dividend is written in descending powers of x, the quotient is produced in descending powers of x. In other words, the numerical coefficients are the *key issues,* so let's rewrite the problem in terms of its coefficients:

$$
\begin{array}{r}
2\quad\ 5\ \ -7\ -1 \\
1 - 2 \overline{)2\quad\ 1\ -17\ \ 13\ \ 2} \\
②\ -4 \\
5\ \ ⊝17 \\
⑤\ -10 \\
-7\ \ ⑬ \\
⊝7\ \ 14 \\
-1\ \ ② \\
⊝1\ \ 2
\end{array}
$$

Now observe that the circled numbers are simply repetitions of the numbers directly above them in the format. Thus the circled numbers can be omitted, and the format will be as follows (disregard the arrows for the moment):

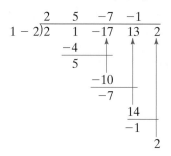

Next, by moving some numbers up (indicated by the arrows) and by not writing the 1 that is the coefficient of x in the divisor, we obtain the following more compact form:

$$
\begin{array}{r}
 \quad 2 \quad\ \ 5 \quad -7 \quad -1 \\
-2)\overline{2 \quad\ \ 1 \quad -17 \quad\ 13 \quad\ 2} \\
\underline{-4 \quad -10 \quad\ 14 \quad\ 2} \\
5 \quad -7 \quad -1
\end{array}
$$

$$\text{(1)}$$
$$\text{(2)}$$
$$\text{(3)}$$
$$\text{(4)}$$

Note that line (4) reveals all of the coefficients of the quotient (line (1)) except for the first coefficient, 2. Thus we can omit line (1), begin line (4) with the first coefficient, and then use the following form:

$$
\begin{array}{r}
-2)\overline{2 \quad\ \ 1 \quad -17 \quad 13 \quad\ 2} \\
\underline{-4 \quad -10 \quad 14 \quad\ 2} \\
2 \quad\ 5 \quad -7 \quad -1 \quad 0
\end{array}
$$

$$\text{(5)}$$
$$\text{(6)}$$
$$\text{(7)}$$

Line (7) contains the coefficients of the quotient, where the zero indicates the remainder. Finally, by changing the constant in the divisor to 2 (instead of -2), which changes the signs of the numbers in line (6), we can *add* the corresponding entries in lines (5) and (6) rather than subtract. Thus the final synthetic division form for this problem is

$$
\begin{array}{r}
2)\overline{2 \quad\ 1 \quad -17 \quad\ \ 13 \quad\ \ 2} \\
\underline{4 \quad\ \ 10 \quad -14 \quad -2} \\
2 \quad 5 \quad -7 \quad -1 \quad\ \ 0
\end{array}
$$

Now we will consider another problem and indicate a step-by-step procedure for setting up and carrying out the synthetic-division process. Suppose that we want to do the following division problem:

$$x + 4)\overline{2x^3 + 5x^2 - 13x - 2}$$

Step 1 Write the coefficients of the dividend as follows:

$$)\overline{2 \quad 5 \quad -13 \quad -2}$$

Step 2 In the divisor use -4 instead of 4 so that later we can add rather than subtract.

$$-4\overline{)2\quad 5\quad -13\quad -2}$$

Step 3 Bring down the first coefficient of the dividend.

$$-4\overline{)2\quad 5\quad -13\quad -2}$$
$$\overline{2}$$

Step 4 Multiply that first coefficient by the divisor, which yields $2(-4) = -8$; add this result to the second coefficient of the dividend.

$$-4\overline{)2\quad\quad 5\quad -13\quad -2}$$
$$\underline{-8}$$
$$2\quad -3$$

Step 5 Multiply $(-3)(-4)$, which yields 12; add this result to the third coefficient of the dividend.

$$-4\overline{)2\quad\quad 5\quad -13\quad -2}$$
$$\underline{-8\quad\ \ 12}$$
$$2\quad -3\quad -1$$

Step 6 Multiply $(-1)(-4)$, which yields 4; add this result to the last term of the dividend.

$$-4\overline{)2\quad\quad 5\quad -13\quad -2}$$
$$\underline{-8\quad\ \ 12\quad\ \ 4}$$
$$2\quad -3\quad -1\quad\ \ 2$$

The last row indicates a quotient of $2x^2 - 3x - 1$ and a remainder of 2.

Now let's consider some examples in which we show only the final compact form of synthetic division.

EXAMPLE 2   Find the quotient and remainder for $(x^3 + 8x^2 + 13x - 6) \div (x + 3)$.

Solution

$$-3\overline{)1\quad\quad 8\quad\quad 13\quad -6}$$
$$\underline{-3\quad -15\quad\ \ 6}$$
$$1\quad\quad 5\quad\ -2\quad\ \ 0$$

Thus the quotient is $x^2 + 5x - 2$, and the remainder is zero.

▶ **Now work Problem 29.**   ∎

EXAMPLE 3

Find the quotient and the remainder for $(3x^4 + 5x^3 - 29x^2 - 45x + 14) \div (x - 3)$.

Solution

$$
\begin{array}{r|rrrrr}
3) & 3 & 5 & -29 & -45 & 14 \\
  &   & 9 & 42 & 39 & -18 \\
\hline
  & 3 & 14 & 13 & -6 & -4 \\
\end{array}
$$

Thus the quotient is $3x^3 + 14x^2 + 13x - 6$, and the remainder is -4.

▶ **Now work Problem 23.**   ■

EXAMPLE 4

Find the quotient and the remainder for $(4x^4 - 2x^3 + 6x - 1) \div (x - 1)$.

Solution

$$
\begin{array}{r|rrrrr}
1) & 4 & -2 & 0 & 6 & -1 \\
  &   & 4 & 2 & 2 & 8 \\
\hline
  & 4 & 2 & 2 & 8 & 7 \\
\end{array}
$$

Note that a zero has been inserted as the coefficient of the missing x^2 term

Thus the quotient is $4x^3 + 2x^2 + 2x + 8$, and the remainder is 7.

▶ **Now work Problem 27.**   ■

EXAMPLE 5

Find the quotient and the remainder for $(x^4 + 16) \div (x + 2)$.

Solution

$$
\begin{array}{r|rrrrr}
-2) & 1 & 0 & 0 & 0 & 16 \\
   &   & -2 & 4 & -8 & 16 \\
\hline
   & 1 & -2 & 4 & -8 & 32 \\
\end{array}
$$

Note that zeros have been inserted as coefficients of the missing terms in the dividend

Thus the quotient is $x^3 - 2x^2 + 4x - 8$ and the remainder is 32.

▶ **Now work Problem 33.**   ■

CONCEPT QUIZ

For Problems 1–3, given $(x^3 + 6x^2 - 5x - 2) \div (x - 1) = x^2 + 7x + 2$, match the mathematical expression with the correct term.

1. $x^2 + 7x + 2$ **2.** $x^3 + 6x^2 - 5x - 1$ **3.** $x - 1$

A. Dividend   B. Quotient   C. Divisor

For Problems 4–8, answer true or false.

4. For long division of polynomials, the degree of the remainder is always less than the degree of the divisor.

5. The polynomial divisor of $(x + 3)$ would become a divisor of 3 for synthetic division.

6. If a synthetic division problem gave a quotient line of 3 −1 4 7 0, we would know that the remainder is zero.

7. If a synthetic division problem gave a quotient of 2 3 −5 6, we would know that the quotient is $2x^2 + 3x - 5$ with a remainder of 6.

8. If a synthetic division problem gave a quotient line of 4 0 −3 7, we would know that the quotient is $4x - 3$ with a remainder of 7.

Problem Set 4.1

For Problems 1–14, find the quotient and remainder for each division problem.

1. $(12x^2 + 7x - 10) \div (3x - 2)$

2. $(20x^2 - 39x + 18) \div (5x - 6)$

3. $(3t^3 + 7t^2 - 10t - 4) \div (3t + 1)$

4. $(4t^3 - 17t^2 + 7t + 10) \div (4t - 5)$

5. $(6x^2 + 19x + 11) \div (3x + 2)$

6. $(20x^2 + 3x - 1) \div (5x + 2)$

7. $(3x^3 + 2x^2 - 5x - 1) \div (x^2 + 2x)$

8. $(4x^3 - 5x^2 + 2x - 6) \div (x^2 - 3x)$

9. $(5y^3 - 6y^2 - 7y - 2) \div (y^2 - y)$

10. $(8y^3 - y^2 - y + 5) \div (y^2 + y)$

11. $(4a^3 - 2a^2 + 7a - 1) \div (a^2 - 2a + 3)$

12. $(5a^3 + 7a^2 - 2a - 9) \div (a^2 + 3a - 4)$

13. $(3x^2 - 2xy - 8y^2) \div (x - 2y)$

14. $(4a^2 - 8ab + 4b^2) \div (a - b)$

For Problems 15–38, use *synthetic division* to determine the quotient and remainder for each division problem.

15. $(3x^2 + x - 4) \div (x - 1)$

16. $(2x^2 - 5x - 3) \div (x - 3)$

17. $(x^2 + 2x - 10) \div (x - 4)$

18. $(x^2 - 10x + 15) \div (x - 8)$

19. $(4x^2 + 5x - 4) \div (x + 2)$

20. $(5x^2 + 18x - 8) \div (x + 4)$

21. $(x^3 - 2x^2 - x + 2) \div (x - 2)$

22. $(x^3 - 5x^2 + 2x + 8) \div (x + 1)$

23. $(3x^4 - x^3 + 2x^2 - 7x - 1) \div (x + 1)$

24. $(2x^3 - 5x^2 - 4x + 6) \div (x - 2)$

25. $(x^3 - 7x - 6) \div (x + 2)$

26. $(x^3 + 6x^2 - 5x - 1) \div (x - 1)$

27. $(x^4 + 4x^3 - 7x - 1) \div (x - 3)$

28. $(2x^4 + 3x^2 + 3) \div (x + 2)$

29. $(x^3 + 6x^2 + 11x + 6) \div (x + 3)$

30. $(x^3 - 4x^2 - 11x + 30) \div (x - 5)$

31. $(x^5 - 1) \div (x - 1)$

32. $(x^5 - 1) \div (x + 1)$

33. $(x^5 + 1) \div (x - 1)$

34. $(x^5 + 1) \div (x + 1)$

35. $(2x^3 + 3x^2 - 2x + 3) \div \left(x + \dfrac{1}{2} \right)$

36. $(9x^3 - 6x^2 + 3x - 4) \div \left(x - \dfrac{1}{3} \right)$

37. $(4x^4 - 5x^2 + 1) \div \left(x - \dfrac{1}{2} \right)$

38. $(3x^4 - 2x^3 + 5x^2 - x - 1) \div \left(x + \dfrac{1}{3} \right)$

39. How would you describe synthetic division to someone who had just completed an elementary algebra course?

40. Why is synthetic division restricted to situations where the divisor is of the form $x - c$?

Answers to the Concept Quiz

1. B   **2.** A   **3.** C   **4.** True   **5.** False   **6.** True   **7.** True   **8.** False

4.2   Remainder and Factor Theorems

Objectives

■ Use the remainder theorem to evaluate a function for a given value.

■ Determine if an expression is a factor of a given polynomial.

■ Find linear factors of a polynomial.

■ Remainder Theorem

Let's consider the division algorithm (stated in the previous section) when the dividend, $f(x)$, is divided by a *linear polynomial* of the form $x - c$. Then the division algorithm

$$f(x) = g(x)q(x) + r(x)$$

Dividend   Divisor   Quotient   Remainder

becomes

$$f(x) = (x - c)q(x) + r(x)$$

Because the degree of the remainder, $r(x)$, must be less than the degree of the divisor, $x - c$, the remainder is a constant. Therefore, letting R represent the remainder, we have

$$f(x) = (x - c)q(x) + R$$

If we evaluate f at c, we obtain

$$\begin{aligned} f(c) &= (c - c)q(c) + R \\ &= 0 \cdot q(c) + R \\ &= R \end{aligned}$$

In other words, if a polynomial is divided by a linear polynomial of the form $x - c$, then the remainder is the value of the polynomial at c. Let's state this more formally as the **remainder theorem.**

Property 4.1 Remainder theorem

If a polynomial $f(x)$ is divided by $x - c$, then the remainder is equal to $f(c)$.

EXAMPLE 1

If $f(x) = x^3 + 2x^2 - 5x - 1$, find $f(2)$ first (a) by using synthetic division and the remainder theorem and then (b) by evaluating $f(2)$ directly.

Solutions

a.
$$
\begin{array}{r|rrrr}
2) & 1 & 2 & -5 & -1 \\
   &   & 2 & 8 & 6 \\
\hline
   & 1 & 4 & 3 & ⑤ \leftarrow R = f(2)
\end{array}
$$

b. $f(2) = 2^3 + 2(2)^2 - 5(2) - 1 = 8 + 8 - 10 - 1 = 5$

⊙ **Now work Problem 1.** ∎

EXAMPLE 2

If $f(x) = x^4 + 7x^3 + 8x^2 + 11x + 5$, find $f(-6)$ first (a) by using synthetic division and the remainder theorem, and then (b) by evaluating $f(-6)$ directly.

Solutions

a.
$$
\begin{array}{r|rrrrr}
-6) & 1 & 7 & 8 & 11 & 5 \\
    &   & -6 & -6 & -12 & 6 \\
\hline
    & 1 & 1 & 2 & -1 & ⑪ \leftarrow R = f(-6)
\end{array}
$$

b. $f(-6) = (-6)^4 + 7(-6)^3 + 8(-6)^2 + 11(-6) + 5$
$$= 1296 - 1512 + 288 - 66 + 5 = 11$$

⊙ **Now work Problem 5.** ∎

In Example 2 note that finding $f(-6)$ by synthetic division and the remainder theorem involves easier computation than evaluating $f(-6)$ directly. This is often the case.

EXAMPLE 3

Find the remainder when $x^3 + 3x^2 - 13x - 15$ is divided by $x + 1$.

Solution

Let $f(x) = x^3 + 3x^2 - 13x - 15$ and write $x + 1$ as $x - (-1)$ so that we can apply the remainder theorem.

$$f(-1) = (-1)^3 + 3(-1)^2 - 13(-1) - 15 = 0$$

Thus the remainder is zero. ∎

Example 3 illustrates an important special case of the remainder theorem in which the remainder is *zero*. In this case, we say that $x + 1$ is a **factor** of $x^3 + 3x^2 - 13x - 15$.

■ Factor Theorem

A general *factor theorem* can be formulated by considering the equation

$$f(x) = (x - c)q(x) + R$$

If $x - c$ is a factor of $f(x)$, then the remainder R, which is also $f(c)$, must be zero. Conversely, if $R = f(c) = 0$, then $f(x) = (x - c)q(x)$; in other words, $x - c$ is a factor of $f(x)$. The **factor theorem** can be stated as follows.

Property 4.2 Factor Theorem

A polynomial $f(x)$ has a factor $x - c$ if and only if $f(c) = 0$.

E X A M P L E 4

Is $x - 1$ a factor of $x^3 + 5x^2 + 2x - 8$?

Solution

Let $f(x) = x^3 + 5x^2 + 2x - 8$ and compute $f(1)$ to obtain

$$f(1) = 1^3 + 5(1)^2 + 2(1) - 8 = 0$$

Therefore, by the factor theorem, $x - 1$ is a factor of $f(x)$.

▶ **Now work Problem 19.** ■

E X A M P L E 5

Is $x + 3$ a factor of $2x^3 + 5x^2 - 6x - 7$?

Solution

Using synthetic division, we obtain

$$
\begin{array}{r|rrrr}
-3 & 2 & 5 & -6 & -7 \\
   &   & -6 & 3 & 9 \\
\hline
   & 2 & -1 & -3 & ②
\end{array}
\quad \longleftarrow R = f(-3)
$$

Because $f(-3) \neq 0$, we know that $x + 3$ is not a factor of the given polynomial.

▶ **Now work Problem 23.** ■

In Examples 4 and 5 we were concerned only with determining whether a linear polynomial of the form $x - c$ was a factor of another polynomial. For such problems, it is reasonable to compute $f(c)$ either directly or by synthetic division, whichever way seems easier. However, if more information is required, such as complete factorization of the given polynomial, then using synthetic division becomes appropriate, as in the next two examples.

E X A M P L E   6

Show that $x - 1$ is a factor of $x^3 - 2x^2 - 11x + 12$ and find the other linear factors of the polynomial.

Solution

Let's use synthetic division to divide $x^3 - 2x^2 - 11x + 12$ by $x - 1$:

$$
\begin{array}{r|rrrr}
1) & 1 & -2 & -11 & 12 \\
  &   & 1 & -1 & -12 \\
\hline
  & 1 & -1 & -12 & 0
\end{array}
$$

The last line indicates a quotient of $x^2 - x - 12$ and a remainder of zero. The zero remainder means that $x - 1$ is a factor. Furthermore, we can write

$$x^3 - 2x^2 - 11x + 12 = (x - 1)(x^2 - x - 12)$$

We can factor the quadratic polynomial $x^2 - x - 12$ as $(x - 4)(x + 3)$ by using our conventional factoring techniques. Thus we obtain

$$x^3 - 2x^2 - 11x + 12 = (x - 1)(x - 4)(x + 3)$$

▶ **Now work Problem 29.**   ■

E X A M P L E   7

Show that $x + 4$ is a factor of $f(x) = x^3 - 5x^2 - 22x + 56$ and complete the factorization of $f(x)$.

Solution

We use synthetic division to divide $x^3 - 5x^2 - 22x + 56$ by $x + 4$:

$$
\begin{array}{r|rrrr}
-4) & 1 & -5 & -22 & 56 \\
  &   & -4 & 36 & -56 \\
\hline
  & 1 & -9 & 14 & 0
\end{array}
$$

The last line indicates a quotient of $x^2 - 9x + 14$ and a remainder of zero. The zero remainder means that $x + 4$ is a factor. Furthermore, we can write

$$x^3 - 5x^2 - 22x + 56 = (x + 4)(x^2 - 9x + 14)$$

and then complete the factoring to obtain

$$f(x) = x^3 - 5x^2 - 22x + 56 = (x + 4)(x - 7)(x - 2)$$

▶ **Now work Problem 31.**   ■

The factor theorem also plays a significant role in determining some general factorization ideas, as the last example of this section illustrates.

EXAMPLE 8

Verify that $x + 1$ is a factor of $x^n + 1$ whenever n is an odd positive integer.

Solution

Let $f(x) = x^n + 1$ and compute $f(-1)$ to obtain

$$f(-1) = (-1)^n + 1$$
$$= -1 + 1 \qquad \text{Any odd power of } -1 \text{ is } -1$$
$$= 0$$

Because $f(-1) = 0$, we know that $x + 1$ is a factor of $f(x)$.

▶ **Now work Problem 41.** ■

CONCEPT QUIZ

For Problems 1–6, answer true or false.

1. When a polynomial is divided by a divisor that is a linear factor, the remainder is a constant term.
2. If $f(3) = -12$, then the remainder, when $f(x)$ is divided by $x - 3$, is -12.
3. If $f(-5) = 0$, then $(x - 5)$ is a factor of $f(x)$.
4. If $(x + 3)$ is a factor of $f(x)$, then the division of $f(x)$ by $(x + 3)$ has a remainder of 0.
5. For any polynomial of the form $x^n + 1$ where n is an odd positive integer, $(x + 1)$ is a factor.
6. If $f(2) = 8$, then $(x - 2)$ is a factor of $f(x)$.

Problem Set 4.2

For Problems 1–10, find $f(c)$ (a) by using synthetic division and the remainder theorem and (b) by evaluating $f(c)$ directly.

1. $f(x) = x^2 + x - 8$ and $c = 2$

2. $f(x) = x^3 + x^2 - 2x - 4$ and $c = -1$

3. $f(x) = 3x^3 + 4x^2 - 5x + 3$ and $c = -4$

4. $f(x) = 2x^4 + x^2 + 6$ and $c = 1$

5. $f(x) = x^4 - 2x^3 - 3x^2 + 5x - 1$ and $c = -2$

6. $f(x) = 2x^4 + x^3 - 4x^2 - x + 1$ and $c = 2$

7. $f(t) = 6t^3 - 35t^2 + 8t - 10$ and $c = 6$

8. $f(t) = 2t^5 - 1$ and $c = -2$

9. $f(n) = 3n^4 - 2n^3 + 4n - 1$ and $c = 3$

10. $f(n) = -2n^4 + 4n - 5$ and $c = -3$

For Problems 11–18, find $f(c)$ *either* by using synthetic division and the remainder theorem *or* by evaluating $f(c)$ directly.

11. $f(x) = 5x^6 - x^3 - 1$ and $c = -1$

12. $f(x) = 2x^3 - 3x^2 - 5x + 4$ and $c = 4$

13. $f(x) = x^4 - 8x^3 + 9x^2 - 15x + 2$ and $c = 7$

14. $f(t) = 5t^3 - 8t^2 + 9t - 4$ and $c = -5$

15. $f(n) = -2n^4 + 2n^2 - n - 5$ and $c = -2$

16. $f(x) = 4x^7 + 3$ and $c = 3$

17. $f(x) = 2x^3 - 5x^2 + 4x - 3$ and $c = \dfrac{1}{2}$

18. $f(x) = 3x^3 + 4x^2 - 5x - 7$ and $c = -\dfrac{1}{3}$

For Problems 19–28, use the factor theorem to help answer each question about factors.

19. Is $x - 2$ a factor of $3x^2 - 4x - 4$?

20. Is $x + 3$ a factor of $6x^2 + 13x - 15$?

21. Is $x + 2$ a factor of $x^3 + x^2 - 7x - 10$?

22. Is $x - 3$ a factor of $2x^3 - 3x^2 - 10x + 3$?

23. Is $x - 1$ a factor of $3x^3 + 5x^2 - x - 2$?

24. Is $x + 4$ a factor of $x^3 - 4x^2 + 2x - 8$?

25. Is $x - 2$ a factor of $x^3 - 8$?

26. Is $x + 2$ a factor of $x^3 + 8$?

27. Is $x - 3$ a factor of $x^4 - 81$?

28. Is $x + 3$ a factor of $x^4 - 81$?

For Problems 29–34, use synthetic division to show that $g(x)$ is a factor of $f(x)$, and complete the factorization of $f(x)$.

29. $g(x) = x + 2$; $f(x) = x^3 + 7x^2 + 4x - 12$

30. $g(x) = x - 1$; $f(x) = 3x^3 + 19x^2 - 38x + 16$

31. $g(x) = x - 3$; $f(x) = 6x^3 - 17x^2 - 5x + 6$

32. $g(x) = x + 2$; $f(x) = 12x^3 + 29x^2 + 8x - 4$

33. $g(x) = x + 1$; $f(x) = x^3 - 2x^2 - 7x - 4$

34. $g(x) = x - 5$; $f(x) = 2x^3 + x^2 - 61x + 30$

For Problems 35–38, find the value(s) of k that make(s) the second polynomial a factor of the first.

35. $x^3 - kx^2 + 5x + k$; $x - 2$

36. $k^2x^4 + 3kx^2 - 4$; $x - 1$

37. $x^3 + 4x^2 - 11x + k$; $x + 2$

38. $kx^3 + 19x^2 + x - 6$; $x + 3$

39. Show that $x + 2$ is a factor of $x^{12} - 4096$.

40. Argue that $f(x) = 2x^4 + x^2 + 3$ has no factor of the form $x - c$, where c is a real number.

41. Verify that $x - 1$ is a factor of $x^n - 1$ for all positive integral values of n.

42. Verify that $x + 1$ is a factor of $x^n - 1$ for all even positive integral values of n.

43. a. Verify that $x - y$ is a factor of $x^n - y^n$ whenever n is a positive integer.

   b. Verify that $x + y$ is a factor of $x^n - y^n$ whenever n is an even positive integer.

   c. Verify that $x + y$ is a factor of $x^n + y^n$ whenever n is an odd positive integer.

■ ■ ■ **THOUGHTS INTO WORDS**

44. In your own words, explain how the remainder theorem is used to prove the factor theorem.

45. It is sometimes said that the factor theorem is a special case of the remainder theorem. What does this statement mean?

■ ■ ■ **FURTHER INVESTIGATIONS**

The remainder and factor theorems are true for any complex value of c. Therefore, for Problems 46–48, find $f(c)$ (a) by using synthetic division and the remainder theorem and (b) by evaluating $f(c)$ directly.

46. $f(x) = x^3 - 5x^2 + 2x + 1$ and $c = i$

47. $f(x) = x^2 + 4x - 2$ and $c = 1 + i$

48. $f(x) = x^3 + 2x^2 + x - 2$ and $c = 2 - 3i$

For Problems 49 and 50, solve each problem.

49. Show that $x - 2i$ is a factor of $f(x) = x^4 + 6x^2 + 8$.

50. Show that $x + 3i$ is a factor of $f(x) = x^4 + 14x^2 + 45$.

51. Consider changing the form of the polynomial $f(x) = x^3 + 4x^2 - 3x + 2$ as follows:

$$f(x) = x^3 + 4x^2 - 3x + 2$$
$$= (x^2 + 4x - 3)x + 2$$
$$= [x(x + 4) - 3]x + 2$$

The final form, $f(x) = [x(x + 4) - 3]x + 2$, is called the **nested form** of the polynomial. It is particularly well suited to evaluating functional values of f either by hand or with a calculator.

For each of the following, find the indicated functional values, using the nested form of the given polynomial.

a. $f(4), f(-5),$ and $f(7)$ for $f(x) = x^3 + 5x^2 - 2x + 1$

b. $f(3), f(6),$ and $f(-7)$ for $f(x) = 2x^3 - 4x^2 - 3x + 2$

c. $f(4), f(5),$ and $f(-3)$ for $f(x) = -2x^3 + 5x^2 - 6x - 7$

d. $f(5), f(6),$ and $f(-3)$ for $f(x) = x^4 + 3x^3 - 2x^2 + 5x - 1$

4.3   Polynomial Equations

Objectives

■ Know the concept of multiplicity of roots.

■ Solve polynomial equations using the rational roots theorem and the factor theorem.

■ Use Descartes' rule of signs to determine the possibilities for the nature of the solutions of a polynomial equation.

■ Given the solutions and an indicated degree, write a polynomial equation using integral coefficients with those solutions.

In Chapter 1 we solved a large variety of *linear equations* of the form $ax + b = 0$ and *quadratic equations* of the form $ax^2 + bx + c = 0$. Linear and quadratic equations are special cases of a general class of equations we refer to as **polynomial equations.** The equation

$$a_n x^n + a_{n-1} x^{n-1} + \cdots + a_1 x + a_0 = 0$$

where the coefficients $a_0, a_1, \ldots, a_n$ are real numbers and n is a positive integer, is called a **polynomial equation of degree n.** The following are examples of polynomial equations:

$\sqrt{2}x - 6 = 0$	Degree 1
$\dfrac{3}{4}x^2 - \dfrac{2}{3}x + 5 = 0$	Degree 2
$4x^3 - 3x^2 - 7x - 9 = 0$	Degree 3
$5x^4 - x + 6 = 0$	Degree 4

Remark: The most general polynomial equation allows complex numbers as coefficients. However, for our purposes in this text, we will restrict the coefficients to real numbers. We refer to such equations as **polynomial equations over the reals.**

In general, solving polynomial equations of degree greater than 2 can be very difficult and often requires mathematics beyond the scope of this text. However, there are some general methods for solving polynomial equations that you should know because certain types of polynomial equations *can* be solved with the techniques available to us at this time.

Let's begin by listing some previously encountered polynomial equations and their solution sets.

Equation	Solution Set
$3x + 4 = 7$	$\{1\}$
$x^2 + x - 6 = 0$	$\{-3, 2\}$
$2x^3 - 3x^2 - 2x + 3 = 0$	$\left\{-1, 1, \dfrac{3}{2}\right\}$
$x^4 - 16 = 0$	$\{-2, 2, -2i, 2i\}$

Note that in each of these examples, the number of solutions corresponds to the degree of the equation. The first-degree equation has one solution, the second-degree equation has two solutions, the third-degree equation has three solutions, and the fourth-degree equation has four solutions. Now consider the equation

$$(x - 4)^2 (x + 5)^3 = 0$$

It can be written

$$(x - 4)(x - 4)(x + 5)(x + 5)(x + 5) = 0$$

which implies that

$$x - 4 = 0 \quad \text{or} \quad x - 4 = 0 \quad \text{or} \quad x + 5 = 0 \quad \text{or}$$

$$x + 5 = 0 \quad \text{or} \quad x + 5 = 0$$

Therefore

$$x = 4 \quad \text{or} \quad x = 4 \quad \text{or} \quad x = -5 \quad \text{or}$$

$$x = -5 \quad \text{or} \quad x = -5$$

We say that the solution set of the original equation is $\{-5, 4\}$, but we also say that the equation has a solution of 4 with a **multiplicity of two** and a solution of -5 with a **multiplicity of three.** Furthermore, note that the sum of the multiplicities is 5, which agrees with the degree of the equation.

We can state the following general property.

Property 4.3

> A polynomial equation of degree n has n solutions, where any solution of multiplicity p is counted p times.

■ Finding Rational Solutions

As we stated earlier, solving polynomial equations of degree greater than 2 can be very difficult. However, *rational solutions* of polynomial equations with integral coefficients can be found by using techniques from this chapter. The following property restricts the possible rational solutions of such an equation.

Property 4.4   Rational Root Theorem

> Consider the polynomial equation
>
> $$a_n x^n + a_{n-1} x^{n-1} + \cdots + a_1 x + a_0 = 0$$
>
> where the coefficients $a_0, a_1, \ldots, a_n$ are integers. If the rational number c/d, reduced to lowest terms, is a solution of the equation, then c is a factor of the constant term a_0, and d is a factor of the leading coefficient a_n.

The *why* behind the rational root theorem is based on some simple factoring ideas, as indicated by the following outline of a proof for the theorem.

Outline of Proof   If c/d is to be a solution, then

$$a_n \left(\frac{c}{d} \right)^n + a_{n-1} \left(\frac{c}{d} \right)^{n-1} + \cdots + a_1 \left(\frac{c}{d} \right) + a_0 = 0$$

We multiply both sides of this equation by d^n and then add $-a_0 d^n$ to both sides:

$$a_n c^n + a_{n-1} c^{n-1} d + \cdots + a_1 c d^{n-1} = -a_0 d^n$$

Because c is a factor of the left side of this equation, c must also be a factor of $-a_0 d^n$. Furthermore, because c/d is in reduced form, c and d have no common factors other than -1 or 1. Thus c must be a factor of a_0. In the same way, from the equation

$$a_{n-1} c^{n-1} d + \cdots + a_1 c d^{n-1} + a_0 d^n = -a_n c^n$$

we can conclude that d is a factor of the left side; therefore d is also a factor of a_n. ∎

The rational root theorem, synthetic division, the factor theorem, and some previous knowledge about solving linear and quadratic equations all merge to form a basis for finding rational solutions. Let's consider some examples.

E X A M P L E   1   Find all rational solutions of $3x^3 + 8x^2 - 15x + 4 = 0$.

Solution

If c/d is a rational solution, then c must be a factor of 4 and d must be a factor of 3. Therefore, the possible values for c and d are the following:

For c	$\pm 1, \pm 2, \pm 4$
For d	$\pm 1, \pm 3$

Thus the possible values for c/d are

$$\pm 1, \pm \frac{1}{3}, \pm 2, \pm \frac{2}{3}, \pm 4, \pm \frac{4}{3}$$

By using synthetic division, we can test $x - 1$:

$$
\begin{array}{r|rrrr}
1) & 3 & 8 & -15 & 4 \\
   &   & 3 & 11 & -4 \\
\hline
   & 3 & 11 & -4 & 0 \\
\end{array}
$$

This shows that $x - 1$ is a factor of the given polynomial; therefore 1 is a rational solution of the equation. Furthermore, the synthetic-division result also indicates how to factor the given polynomial.

$$3x^3 + 8x^2 - 15x + 4 = 0$$
$$(x - 1)(3x^2 + 11x - 4) = 0$$

The quadratic factor can be further factored by using techniques we are familiar with.

$$(x - 1)(3x^2 + 11x - 4) = 0$$
$$(x - 1)(3x - 1)(x + 4) = 0$$
$$x - 1 = 0 \quad\text{or}\quad 3x - 1 = 0 \quad\text{or}\quad x + 4 = 0$$
$$x = 1 \quad\text{or}\quad x = \frac{1}{3} \quad\text{or}\quad x = -4$$

Thus the entire solution set consists of rational numbers and can be listed as $\left\{ -4, \dfrac{1}{3}, 1 \right\}$.

▶ **Now work Problem 1.**   ∎

In Example 1 we were fortunate that the first time we used synthetic division, we got a rational solution. But this often does not happen, and then we need to conduct a little organized search, as the next example illustrates.

EXAMPLE 2

Find all rational solutions of $3x^3 + 7x^2 - 22x - 8 = 0$.

Solution

If c/d is a rational solution, then c must be a factor of -8, and d must be a factor of 3. Therefore the possible values for c and d are the following:

For c $\pm 1, \pm 2, \pm 4, \pm 8$
For d $\pm 1, \pm 3$

Thus the possible values for c/d are

$$\pm 1, \pm\frac{1}{3}, \pm 2, \pm\frac{2}{3}, \pm 4, \pm\frac{4}{3}, \pm 8, \pm\frac{8}{3}$$

Let's begin our search for rational solutions by trying the integers first.

$$\begin{array}{r|rrrr} 1) & 3 & 7 & -22 & -8 \\ & & 3 & 10 & -12 \\ \hline & 3 & 10 & -12 & \fbox{-20} \end{array}$$
⟵ This indicates that $x - 1$ is not a factor and thus 1 is not a solution

$$\begin{array}{r|rrrr} -1) & 3 & 7 & -22 & -8 \\ & & -3 & -4 & 26 \\ \hline & 3 & 4 & -26 & \fbox{18} \end{array}$$
⟵ This indicates that -1 is not a solution

$$\begin{array}{r|rrrr} 2) & 3 & 7 & -22 & -8 \\ & & 6 & 26 & 8 \\ \hline & 3 & 13 & 4 & 0 \end{array}$$
⟵ This indicates that 2 is a solution

Now we know that $x - 2$ is a factor, and we can proceed as follows:

$$3x^3 + 7x^2 - 22x - 8 = 0$$
$$(x - 2)(3x^2 + 13x + 4) = 0$$
$$(x - 2)(3x + 1)(x + 4) = 0$$
$$x - 2 = 0 \quad \text{or} \quad 3x + 1 = 0 \quad \text{or} \quad x + 4 = 0$$
$$x = 2 \quad \text{or} \quad 3x = -1 \quad \text{or} \quad x = -4$$
$$x = 2 \quad \text{or} \quad x = -\frac{1}{3} \quad \text{or} \quad x = -4$$

The solution set is $\left\{-4, -\frac{1}{3}, 2\right\}$.

▶ **Now work Problem 3.**   ■

In Examples 1 and 2 we were solving third-degree equations. Therefore, once we found one linear factor by synthetic division, we were able to factor the remaining quadratic factor in the usual way. However, if the given equation is of degree 4 or more, then we may need to find more than one linear factor by synthetic division, as in the next example.

EXAMPLE 3

Solve $x^4 - 6x^3 + 22x^2 - 30x + 13 = 0$.

Solution

The possible values for c/d are ± 1 and ± 13. By synthetic division we test 1.

$$
\begin{array}{r|rrrrr}
1) & 1 & -6 & 22 & -30 & 13 \\
  &   & 1 & -5 & 17 & -13 \\
\hline
  & 1 & -5 & 17 & -13 & 0
\end{array}
$$

This indicates that $x - 1$ is a factor of the given polynomial. The bottom line of the synthetic division indicates that the given polynomial can now be factored as follows:

$$x^4 - 6x^3 + 22x^2 - 30x + 13 = 0$$
$$(x - 1)(x^3 - 5x^2 + 17x - 13) = 0$$

Therefore,

$$x - 1 = 0 \quad \text{or} \quad x^3 - 5x^2 + 17x - 13 = 0$$

Now we can use the same approach to look for rational solutions of $x^3 - 5x^2 + 17x - 13 = 0$. The possible values for c/d are, again, ± 1 and ± 13. By synthetic division we test 1 again.

$$
\begin{array}{r|rrrr}
1) & 1 & -5 & 17 & -13 \\
  &   & 1 & -4 & 13 \\
\hline
  & 1 & -4 & 13 & 0
\end{array}
$$

This indicates that $x - 1$ is also a factor of $x^3 - 5x^2 + 17x - 13$, and the other factor is $x^2 - 4x + 13$. Now we can solve the original equation.

$$x^4 - 6x^3 + 22x^2 - 30x + 13 = 0$$
$$(x - 1)(x^3 - 5x^2 + 17x - 13) = 0$$
$$(x - 1)(x - 1)(x^2 - 4x + 13) = 0$$
$$x - 1 = 0 \quad \text{or} \quad x - 1 = 0 \quad \text{or} \quad x^2 - 4x + 13 = 0$$
$$x = 1 \quad \text{or} \quad x = 1 \quad \text{or} \quad x^2 - 4x + 13 = 0$$

We use the quadratic formula on $x^2 - 4x + 13 = 0$ to produce

$$x = \frac{4 \pm \sqrt{16 - 52}}{2} = \frac{4 \pm \sqrt{-36}}{2} = \frac{4 \pm 6i}{2} = 2 \pm 3i$$

Thus the original equation has a rational solution of 1 with a multiplicity of two and two complex solutions, $2 + 3i$ and $2 - 3i$. We list the solution set as $\{1, 2 \pm 3i\}$.

▶ **Now work Problem 13.**

■

Example 3 illustrates two general properties. First, note that the coefficient of x^4 is 1, which forces the possible rational solutions to be integers. In general, **the possible rational solutions of $x^n + a_{n-1}x^{n-1} + \cdots + a_1x + a_0 = 0$ are the integral factors of a_0.** Second, note that the complex solutions of Example 3 are conjugates of each other. The following general property can be stated.

Property 4.5

> If a polynomial equation with real coefficients has any nonreal complex solutions, they must occur in conjugate pairs.

Remark: The justification for Property 4.5 is based on some properties of conjugates that were presented in Problem 79 of Problem Set 0.8. We will not show the details of such a proof at this time.

Each of Properties 4.3, 4.4, and 4.5 yields some information about the solutions of a polynomial equation. Before we state one more property that will give us some additional information, we need to illustrate two ideas.

In a polynomial that is arranged in descending powers of x, if two successive terms differ in sign, we say that there is a **variation in sign.** Terms with zero coefficients are disregarded when counting sign variations. For example, the polynomial

$$+3x^3 - 2x^2 + 4x + 7$$

has *two* sign variations, whereas the polynomial

$$+x^5 - 4x^3 + x - 5$$

has *three* variations.

Another idea that we need to understand is that the solutions of

$$a_n(-x)^n + a_{n-1}(-x)^{n-1} + \cdots + a_1(-x) + a_0 = 0$$

are the opposites of the solutions of

$$a_nx^n + a_{n-1}x^{n-1} + \cdots + a_1x + a_0 = 0$$

In other words, if a new equation is formed by replacing x with $-x$ in a given equation, then the solutions of the new equation are the opposites of the solutions of the original equation. For example, the solution set of $x^2 + 7x + 12 = 0$ is $\{-4, -3\}$; the solution set of $(-x)^2 + 7(-x) + 12 = 0$, which simplifies to $x^2 - 7x + 12 = 0$, is $\{3, 4\}$.

Now we can state a property that can help us to determine the nature of the solutions of a polynomial equation without actually solving the equation.

Property 4.6 Descartes' Rule of Signs

Let $a_n x^n + a_{n-1}x^{n-1} + \cdots + a_1 x + a_0 = 0$ be a polynomial equation with real coefficients.

1. The number of **positive real solutions** of the given equation is either equal to the number of variations in sign of the polynomial or less than that number of variations by a positive even integer.

2. The number of **negative real solutions** of the given equation is either equal to the number of variations in sign of the polynomial $a_n(-x)^n + a_{n-1}(-x)^{n-1} + \cdots + a_1(-x) + a_0$ or less than that number of variations by a positive even integer.

Property 4.6, along with Properties 4.3 and 4.5, allows us to acquire some information about the solutions of a polynomial equation without actually solving the equation. Let's consider some equations and indicate how much we know about their solutions without solving them.

$$x^3 + 3x^2 + 5x + 4 = 0$$

1. No variations of sign in $x^3 + 3x^2 + 5x + 4$ means that there are *no positive solutions*.

2. Replace x with $-x$ in the given polynomial to produce $(-x)^3 + 3(-x)^2 + 5(-x) + 4$, which simplifies to $-x^3 + 3x^2 - 5x + 4$. This polynomial contains three variations of sign; thus there are *three or one negative solution(s)*.

Conclusion   The given equation has either three negative real solutions or one negative real solution and two nonreal complex solutions.

$$2x^4 + 3x^2 - x - 1 = 0$$

1. There is one variation of sign in the given polynomial; thus the equation has *one positive solution*.

2. Replace x with $-x$ to produce $2(-x)^4 + 3(-x)^2 - (-x) - 1$, which simplifies to $2x^4 + 3x^2 + x - 1$ and contains one variation of sign. Thus the given equation has *one negative solution*.

Conclusion   The given equation has one positive, one negative, and two nonreal complex solutions.

$$3x^4 + 2x^2 + 5 = 0$$

1. No variations of sign in the given polynomial means that there are *no positive solutions*.

2. Replace x with $-x$ to produce $3(-x)^4 + 2(-x)^2 + 5$, which simplifies to $3x^4 + 2x^2 + 5$ and still contains no variations of sign. Thus there are *no negative solutions*.

Conclusion   The given equation contains four nonreal complex solutions. We also know that these solutions will appear in conjugate pairs.

$$2x^5 - 4x^3 + 2x - 5 = 0$$

1. Three variations of sign in the given polynomial imply that *the number of positive solutions is three or one*.

2. Replace x with $-x$ to produce $2(-x)^5 - 4(-x)^3 + 2(-x) - 5 = -2x^5 + 4x^3 - 2x - 5$, which contains two variations of sign. Thus *the number of negative solutions is two or zero*.

Conclusion   The given equation has

three positive and two negative solutions, or

three positive and two nonreal complex solutions, or

one positive, two negative, and two nonreal complex solutions, or

one positive and four nonreal complex solutions

It should be evident from the previous discussions that sometimes we can truly pinpoint the nature of the solutions of a polynomial equation. However, for some equations (such as the last example), if we use the properties discussed in this section, the best that we can do is to restrict the nature of the solutions to a few possibilities.

Finally, we need to realize that some of the properties presented in these last two sections help us to determine polynomial equations with specified roots. Let's consider some examples.

E X A M P L E   4   Find a polynomial equation with integral coefficients that has the given numbers as solutions and the indicated degree.

a. $1, \dfrac{1}{2}, -2$; degree 3   **b.** 2 of multiplicity four; degree 4

c. $1 + i, -3i$; degree 4

Solutions

a. If $1, \frac{1}{2}$, and -2 are solutions, then $(x - 1)$, $\left(x - \frac{1}{2}\right)$, and $(x + 2)$ are factors of the polynomial. Thus we can form the following third-degree polynomial equation:

$$(x - 1)\left(x - \frac{1}{2}\right)(x + 2) = 0$$

$$(x - 1)(2x - 1)(x + 2) = 0$$

$$2x^3 + x^2 - 5x + 2 = 0$$

b. If 2 is to be a solution with multiplicity four, then the equation $(x - 2)^4 = 0$ can be formed. Using the binomial expansion pattern, we can express the equation as follows:

$$(x - 2)^4 = 0$$

$$x^4 - 8x^3 + 24x^2 - 32x + 16 = 0$$

c. By Property 4.5, if $1 + i$ is a solution, then so is $1 - i$. Likewise, because $-3i$ is a solution, so is $3i$. Therefore we can form the following equation:

$$[x - (1 + i)][x - (1 - i)](x + 3i)(x - 3i) = 0$$

$$[(x - 1) - i][(x - 1) + i](x^2 + 9) = 0$$

$$[(x - 1)^2 - i^2](x^2 + 9) = 0$$

$$(x^2 - 2x + 1 + 1)(x^2 + 9) = 0$$

$$x^4 - 2x^3 + 11x^2 - 18x + 18 = 0$$

▶ **Now work Problem 41.**   ■

A graphing utility can be very helpful when solving polynomial equations, especially if they are of degree greater than 2. Even the search for possible rational solutions can be simplified by looking at a graph. To find the rational solutions of $3x^3 + 8x^2 - 15x + 4 = 0$ (Example 1), we could begin by graphing the equation $y = 3x^3 + 8x^2 - 15x + 4$. This graph is shown in Figure 4.1. From the graph it looks as if 1 and -4 are two of the x intercepts and therefore solutions of the original equation. Let's check them in the equation.

$$3(1)^3 + 8(1)^2 - 15(1) + 4 = 3 + 8 - 15 + 4 = 0$$

$$3(-4)^3 + 8(-4)^2 - 15(-4) + 4 = -192 + 128 + 60 + 4 = 0$$

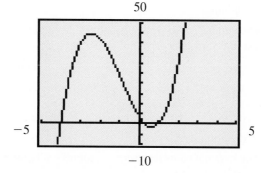

Figure 4.1

Thus $x - 1$ and $x + 4$ are factors of $3x^3 + 8x^2 - 15x + 4$ and the remaining factor could be found by division. We could then determine the solution set as we did in Example 1.

Let's consider an example where we use a graphing utility to approximate the real number solutions of a polynomial equation.

E X A M P L E   5

Find the real number solutions of the equation $x^4 - 2x^3 - 5 = 0$.

Solution

Let's use a graphing utility to get a sketch of the graph of $y = x^4 - 2x^3 - 5$, as in Figure 4.2. From this graph we see that one x intercept is between -1 and -2, and another is between 2 and 3. We can use the zoom-in and trace features to approximate these values at -1.2 and 2.4, to the nearest tenth. Thus the real solutions for the equation $x^4 - 2x^3 - 5 = 0$ are approximately -1.2 and 2.4. (The other two solutions must be conjugate complex numbers.)

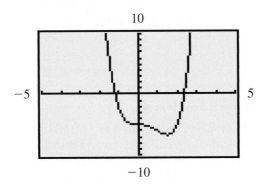

Figure 4.2   ■

C O N C E P T   Q U I Z

For Problems 1–8, answer true or false.

1. For a polynomial equation, the number of solutions is equal to the degree of the polynomial.

2. The equation $(x - 7)^3 (x + 1)^2 = 0$ has a solution of 7 with a multiplicity of 3 and a solution of -1 with a multiplicity of 2.

3. Given $2x^4 - 3x^2 + 8 = 0$, the only possible rational solutions are $\pm\frac{1}{8}, \pm\frac{1}{2}$, and $\pm\frac{1}{4}$.

4. According to the rational roots theorem, $\frac{1}{3}$ is a possible rational solution of the equation $3x^3 - 4x^2 + x + 5 = 0$.

5. The rational roots theorem can identify all possible real number roots of a polynomial equation.

6. If $-3 + 5i$ is a solution to a polynomial equation, then $-3 - 5i$ is also a solution of the equation.

7. The equation $2x^3 + 4x^2 + x + 6 = 0$ has no positive solutions.

8. The equation $x^4 + x^2 + 4 = 0$ has no real number solutions.

Problem Set 4.3

For Problems 1–20, use the rational root theorem and the factor theorem to help solve each equation. Be sure that the number of solutions for each equation agrees with Property 4.3; take into account the multiplicity of solutions.

1. $x^3 + x^2 - 4x - 4 = 0$

2. $x^3 - 2x^2 - 11x + 12 = 0$

3. $6x^3 + x^2 - 10x + 3 = 0$

4. $8x^3 - 2x^2 - 41x - 10 = 0$

5. $3x^3 + 13x^2 - 52x + 28 = 0$

6. $15x^3 + 14x^2 - 3x - 2 = 0$

7. $x^3 - 2x^2 - 7x - 4 = 0$

8. $x^3 - x^2 - 8x + 12 = 0$

9. $x^4 - 4x^3 - 7x^2 + 34x - 24 = 0$

10. $x^4 + 4x^3 - x^2 - 16x - 12 = 0$

11. $x^3 - 10x - 12 = 0$

12. $x^3 - 4x^2 + 8 = 0$

13. $x^4 - 3x^3 - 2x^2 + 6x + 4 = 0$

14. $2x^4 + 3x^3 - 11x^2 - 9x + 15 = 0$

15. $6x^4 - 13x^3 - 19x^2 + 12x = 0$

16. $x^3 - x^2 + x - 1 = 0$

17. $x^4 - 3x^3 + 2x^2 + 2x - 4 = 0$

18. $x^4 + x^3 - 3x^2 - 17x - 30 = 0$

19. $2x^5 - 5x^4 + x^3 + x^2 - x + 6 = 0$

20. $4x^4 + 12x^3 + x^2 - 12x + 4 = 0$

For Problems 21–26, verify that each equation has no rational solutions.

21. $x^4 - x^3 - 8x^2 - 3x + 1 = 0$

22. $x^4 + 3x - 2 = 0$

23. $2x^4 - 3x^3 + 6x^2 - 24x + 5 = 0$

24. $3x^4 - 4x^3 - 10x^2 + 3x - 4 = 0$

25. $x^5 - 2x^4 + 3x^3 + 4x^2 + 7x - 1 = 0$

26. $x^5 + 2x^4 - 2x^3 + 5x^2 - 2x - 3 = 0$

27. The rational root theorem pertains to polynomial equations with integral coefficients. However, if the co-efficients are nonintegral rational numbers, we can first apply the *multiplication property of equality* to produce an equivalent equation with integral coefficients. Use this method to solve each of the following equations.

a. $\dfrac{1}{10}x^3 + \dfrac{1}{2}x^2 + \dfrac{1}{5}x - \dfrac{4}{5} = 0$

b. $\dfrac{1}{10}x^3 + \dfrac{1}{5}x^2 - \dfrac{1}{2}x - \dfrac{3}{5} = 0$

c. $x^3 + \dfrac{9}{2}x^2 - x - 12 = 0$

d. $x^3 - \dfrac{5}{6}x^2 - \dfrac{22}{3}x + \dfrac{5}{2} = 0$

For Problems 28–37, use Descartes' rule of signs (Property 4.6) to determine the possibilities for the nature of the solutions for each of the equations. *Do not solve the equations.*

28. $6x^2 + 7x - 20 = 0$

29. $8x^2 - 14x + 3 = 0$

30. $2x^3 + x - 3 = 0$

31. $4x^3 + 3x + 7 = 0$

32. $3x^3 - 2x^2 + 6x + 5 = 0$

33. $4x^3 + 5x^2 - 6x - 2 = 0$

34. $x^5 - 3x^4 + 5x^3 - x^2 + 2x - 1 = 0$

35. $2x^5 + 3x^3 - x + 1 = 0$

36. $x^5 + 32 = 0$

37. $2x^6 + 3x^4 - 2x^2 - 1 = 0$

For Problems 38–47, find a polynomial equation with integral coefficients that has the given numbers as solutions and the indicated degree.

38. $2, 4, -3$; degree 3

39. $1, -1, 2, -4$; degree 4

40. $-2, \dfrac{1}{2}, \dfrac{2}{3}$; degree 3

41. $3, -\dfrac{2}{3}, \dfrac{3}{4}$; degree 3

42. 1 of multiplicity 5;   degree 5

43. -3 of multiplicity 4;   degree 4

44. $3, 2 + 3i$;   degree 3

45. $-2, 1 - 4i$;   degree 3

46. $1 - i, 2i$;   degree 4

47. $-2 + 3i, -i$;   degree 4

■■■ THOUGHTS INTO WORDS

48. Explain the concept of *multiplicity of roots* of an equation.

49. How would you defend the statement that the equation $2x^4 + 3x^3 + x^2 + 5 = 0$ has no positive solutions?

Does it have any negative solutions? Defend your answer.

50. How do we know by inspection that the equation $2x^4 + 3x^2 + 6 = 0$ has no real number solutions?

■■■ FURTHER INVESTIGATIONS

51. Use the rational root theorem to argue that $\sqrt{2}$ is not a rational number. [*Hint:* The solutions of $x^2 - 2 = 0$ are $\pm\sqrt{2}$.]

52. Use the rational root theorem to argue that $\sqrt{12}$ is not a rational number.

53. Defend the following statement: *Every polynomial equation of odd degree with real coefficients has at least one real number solution.*

54. The following synthetic division shows that 2 is a solution of $x^4 + x^3 + x^2 - 9x - 10 = 0$:

```
2)1   1   1   -9   -10
        2   6   14   10
   1   3   7    5    0   ⟵
```

Note that the new quotient row (indicated by the arrow) consists entirely of nonnegative numbers. This indicates that searching for solutions greater than 2 would be a waste of time because larger divisors would continue to increase each of the numbers (except the 1 on the far left) in the new quotient row. (Try 3 as a divisor!) Thus we say that 2 is an **upper bound** for the real number solutions of the given equation.

Now consider the following synthetic division, which shows that -1 is also a solution of $x^4 + x^3 + x^2 - 9x - 10 = 0$:

```
-1)1   1   1   -9   -10
        -1   0   -1   10
    1   0   1   -10    0   ⟵
```

The new quotient row (indicated by the arrow) shows that there is no need to look for solutions less than -1 because any divisor less than -1 would increase the size (in absolute value) of each number in the new quotient row (except the 1 on the far left). (Try -2 as a divisor!) Thus we say that -1 is a **lower bound** for the real number solutions of the given equation.

The following general property can be stated: If $a_n x^n + a_{n-1} x^{n-1} + \cdots + a_1 x + a_0 = 0$ is a polynomial equation with real coefficients, where $a_n > 0$, and if the polynomial is divided synthetically by $x - c$, then:

1. If $c > 0$ and all numbers in the new quotient row of the synthetic division are nonnegative, then c is an upper bound for the real number solutions of the given equation.

2. If $c < 0$ and the numbers in the new quotient row alternate in sign (with 0 considered either

positive or negative, as needed), then c is a lower bound for the real number solutions of the given equation.

Find the smallest positive integer and the largest negative integer that are upper and lower bounds, respectively, for the real number solutions of each of the following equations. Keep in mind that the integers that serve as bounds do not necessarily have to be solutions of the equation.

a. $x^3 - 3x^2 + 25x - 75 = 0$

b. $x^3 + x^2 - 4x - 4 = 0$

c. $x^4 + 4x^3 - 7x^2 - 22x + 24 = 0$

d. $3x^3 + 7x^2 - 22x - 8 = 0$

e. $x^4 - 2x^3 - 9x^2 + 2x + 8 = 0$

GRAPHING CALCULATOR ACTIVITIES

55. Suppose that we want to solve the equation $x^3 + 2x^2 - 14x - 40 = 0$. Let's graph the function $f(x) = x^3 + 2x^2 - 14x - 40$. The graph has only one x intercept, so the equation must have one real number solution and two nonreal complex solutions. The graph also indicates that the real solution is approximately 4. We can determine that 4 is a solution, and then we can proceed to solve the equation using the ideas of this section.

Solve each of the following equations, using a graphing calculator whenever it seems to be helpful. Express all irrational solutions in lowest radical form.

a. $x^3 + 2x^2 - 14x - 40 = 0$

b. $x^3 + x^2 - 7x + 65 = 0$

c. $x^4 - 6x^3 - 6x^2 + 32x + 24 = 0$

d. $x^4 + 3x^3 - 39x^2 + 11x + 24 = 0$

e. $x^3 - 14x^2 + 26x - 24 = 0$

f. $x^4 + 2x^3 - 3x^2 - 4x + 4 = 0$

56. Use a graphing calculator to help determine the nature of the solutions for each of the following equations. You may also need to use the property stated in Problem 54.

a. $2x^3 - 3x^2 - 3x + 2 = 0$

b. $3x^3 + 7x^2 + 8x + 2 = 0$

c. $2x^4 + 3x^2 + 1 = 0$

d. $4x^5 - 8x^4 - 5x^3 + 10x^2 + x - 2 = 0$

e. $x^4 - x^3 + 2x^2 - x - 1 = 0$

f. $x^5 - x^4 + x^3 - x^2 + x - 3 = 0$

g. $x^4 - 14x^3 + 23x^2 + 14x - 24 = 0$

h. $x^3 + 13x^2 - 28x + 30 = 0$

57. Find approximations, to the nearest hundredth, of the real number solutions of each of the following equations.

a. $x^2 - 4x + 1 = 0$

b. $3x^3 - 2x^2 + 12x - 8 = 0$

c. $x^4 - 8x^3 + 14x^2 - 8x + 13 = 0$

d. $x^4 + 6x^3 - 10x^2 - 22x + 161 = 0$

e. $7x^5 - 5x^4 + 35x^3 - 25x^2 + 28x - 20 = 0$

Answers to the Concept Quiz

1. True   **2.** True   **3.** False   **4.** True   **5.** False   **6.** True   **7.** True   **8.** True

4.4 Graphing Polynomial Functions

Objectives

■ Know the patterns for the graphs of $f(x) = ax^n$.

■ Graph polynomial functions.

Just as we have a vocabulary to deal with linear, quadratic, and polynomial equations, we also have terms that classify functions. In Chapter 3 we defined a **linear function** by means of the equation

$$f(x) = ax + b$$

and a **quadratic function** by means of the equation

$$f(x) = ax^2 + bx + c$$

Both of these are special cases of a general class of functions called **polynomial functions.** Any function of the form

$$f(x) = a_n x^n + a_{n-1} x^{n-1} + \cdots + a_1 x + a_0$$

is called a **polynomial function of degree n,** where a_n is a nonzero real number; $a_{n-1}, \ldots, a_1, a_0$ are real numbers; and n is a nonnegative integer. The following are examples of polynomial functions:

$$f(x) = 5x^3 - 2x^2 + x - 4 \qquad \text{Degree 3}$$
$$f(x) = -2x^4 - 5x^3 + 3x^2 + 4x - 1 \qquad \text{Degree 4}$$
$$f(x) = 3x^5 + 2x^2 - 3 \qquad \text{Degree 5}$$

Remark: Our previous work with polynomial equations is sometimes presented as *finding zeros of polynomial functions.* The *solutions,* or *roots,* of a polynomial equation are also called the **zeros** of the polynomial function. For example, -2 and 2 are solutions of $x^2 - 4 = 0$ and they are zeros of $f(x) = x^2 - 4$. That is, $f(-2) = 0$ and $f(2) = 0$.

For a complete discussion of graphing polynomial functions, we would need some tools from calculus. However, the graphing techniques that we have discussed so far enable us to graph certain kinds of polynomial functions. For example, polynomial functions of the form

$$f(x) = ax^n$$

are quite easy to graph. We know from our previous work that if $n = 1$, then functions such as $f(x) = 2x$, $f(x) = -3x$, and $f(x) = \dfrac{1}{2}x$ are lines through the origin that have slopes of 2, -3, and $\dfrac{1}{2}$, respectively.

Furthermore, if $n = 2$, then we know that the graphs of functions of the form $f(x) = ax^2$ are parabolas that are symmetric with respect to the y axis and have vertices at the origin.

We have also previously graphed the special case of $f(x) = ax^n$, where $a = 1$ and $n = 3$; namely, the function $f(x) = x^3$. This graph is shown in Figure 4.3.

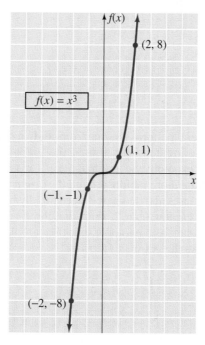

Figure 4.3

From our work with transformations of graphs in Section 3.5, we know that the graphs of functions of the form $f(x) = ax^3$, where $a > 1$, are vertical stretchings of $f(x) = x^3$ and can be easily determined by plotting a few points. Likewise, if $0 < a < 1$, the graph of $f(x) = ax^3$ is a shrinking of $f(x) = x^3$. Furthermore, we know that $f(x) = -x^3$ is an x axis (and also a y axis) reflection of $f(x) = x^3$. Figure 4.4 shows graphs of $f(x) = \dfrac{1}{2}x^3$ and $f(x) = -x^3$.

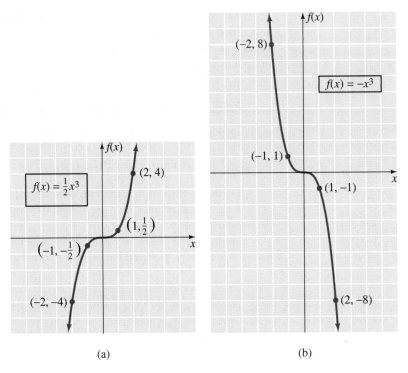

(a)   (b)

Figure 4.4

Two general patterns emerge from studying functions of the form $f(x) = x^n$. If n is odd and greater than 3, then the graph of $f(x) = x^n$ closely resembles Figure 4.3. For example, the graph of $f(x) = x^5$ is shown in Figure 4.5. Note that it *flattens out* a little more rapidly around the origin than the graph of $f(x) = x^3$ does and that it increases and decreases more rapidly because of the larger exponent. If n is even and greater than 2, then the graphs of $f(x) = x^n$ are not parabolas; they resemble the basic parabola, but they are flatter at the bottom and steeper. Figure 4.6 shows the graph of $f(x) = x^4$.

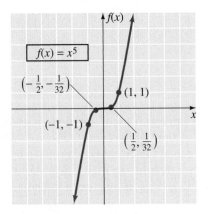

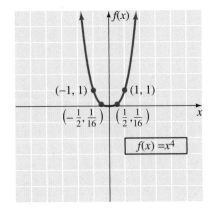

Figure 4.5   **Figure 4.6**

Graphs of functions of the form $f(x) = ax^n$, where n is an integer greater than 2 and $a \neq 1$, are variations of those shown in Figures 4.3 and 4.6. If n is odd, the curve is symmetric about the origin; if n is even, the graph is symmetric about the y axis.

Transformations of these basic curves are easy to sketch. For example, in Figure 4.7 we translated the graph of $f(x) = x^3$ upward two units to produce the graph of $f(x) = x^3 + 2$. In Figure 4.8 we obtained the graph of $f(x) = (x - 1)^5$ by translating the graph of $f(x) = x^5$ one unit to the right. In Figure 4.9 we sketched the graph of $f(x) = -x^4$ as the x axis reflection of $f(x) = x^4$.

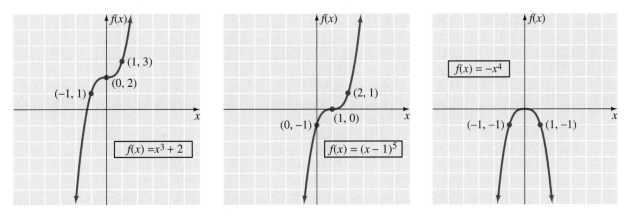

Figure 4.7   **Figure 4.8**   **Figure 4.9**

■ Graphing Polynomial Functions in Factored Form

As we mentioned earlier, a complete discussion of graphing polynomials of degree greater than 2 requires some tools from calculus. In fact, as the degree increases, the graphs often become more complicated. We do know that polynomial functions produce smooth continuous curves with a number of turning points, as illustrated in Figures 4.10 and 4.11. Figure 4.10 shows some graphs of polynomial functions of odd degree. As suggested by the graphs, every polynomial function of odd degree has at least *one real zero*—that is, at least one real number c such that $f(c) = 0$. Geometrically, the zeros of the function are the x intercepts of the graph. Figure 4.11 shows some graphs of polynomial functions of even degree.

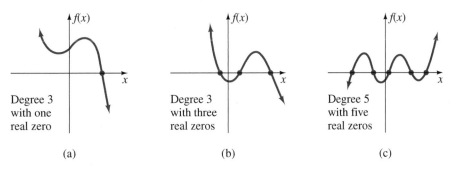

Degree 3 with one real zero (a)   Degree 3 with three real zeros (b)   Degree 5 with five real zeros (c)

Figure 4.10

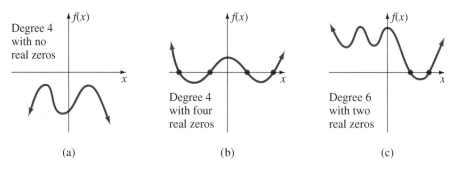

Figure 4.11

As indicated by the graphs in Figures 4.10 and 4.11, polynomial functions usually have **turning points** where the function changes either from increasing to decreasing or from decreasing to increasing. In calculus we are able to verify that *a polynomial function of degree n has at most n − 1 turning points*. Now let's illustrate how this information, along with some other techniques, can be used to graph polynomial functions that are expressed in factored form.

E X A M P L E   1

Graph $f(x) = (x + 2)(x - 1)(x - 3)$.

Solution

First, let's find the x intercepts (zeros of the function) by setting each factor equal to zero and solving for x:

$$x + 2 = 0 \qquad \text{or} \qquad x - 1 = 0 \qquad \text{or} \qquad x - 3 = 0$$
$$x = -2 \qquad \text{or} \qquad x = 1 \qquad \text{or} \qquad x = 3$$

| $x < -2$ | $-2 < x < 1$ | $1 < x < 3$ | $x > 3$ |

$$-2 \qquad\qquad 1 \qquad\qquad 3$$

Figure 4.12

Thus the points $(-2, 0)$, $(1, 0)$, and $(3, 0)$ are on the graph. Second, the points associated with the x intercepts divide the x axis into four intervals, as we see in Figure 4.12. In each of these intervals, $f(x)$ is either always positive or always negative. That is, the graph is either completely above or completely below the x axis. The sign can be determined by selecting a *test value* for x in each of the intervals. Any additional points that are easily obtained improve the accuracy of the graph. The following table summarizes these results.

Interval	Test Value	Sign of $f(x)$	Location of Graph
$x < -2$	$f(-3) = -24$	Negative	Below x axis
$-2 < x < 1$	$f(0) = 6$	Positive	Above x axis
$1 < x < 3$	$f(2) = -4$	Negative	Below x axis
$x > 3$	$f(4) = 18$	Positive	Above x axis

Additional values: $f(-1) = 8$

Making use of the x intercepts and the information in the table, we sketched the graph in Figure 4.13. [The points $(-3, -24)$ and $(4, 18)$ are not shown, but we used them to indicate a rapid decrease and increase of the curve in those regions.]

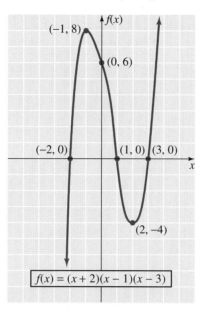

Figure 4.13

▶ **Now work Problem 11.**   ∎

Remark: In Figure 4.13 we indicated turning points of the graph at $(2, -4)$ and $(-1, 8)$. Keep in mind that these are only approximations; again, the tools of calculus are needed to find the exact turning points.

EXAMPLE 2

Graph $f(x) = -x^4 + 3x^3 - 2x^2$.

Solution

The polynomial can be factored as follows:

$$\begin{aligned} f(x) &= -x^4 + 3x^3 - 2x^2 \\ &= -x^2(x^2 - 3x + 2) \\ &= -x^2(x - 1)(x - 2) \end{aligned}$$

Now we can find the x intercepts.

$$\begin{array}{ccccc} -x^2 = 0 & \text{or} & x - 1 = 0 & \text{or} & x - 2 = 0 \\ x = 0 & \text{or} & x = 1 & \text{or} & x = 2 \end{array}$$

$x < 0$	$0 < x < 1$	$1 < x < 2$	$x > 2$
	0	1	2

Figure 4.14

Thus the points $(0, 0)$, $(1, 0)$, and $(2, 0)$ are on the graph and divide the x axis into four intervals (see Figure 4.14). The following table determines some points and summarizes the sign behavior of $f(x)$.

Interval	Test Value	Sign of $f(x)$	Location of Graph
$x < 0$	$f(-1) = -6$	Negative	Below x axis
$0 < x < 1$	$f\left(\dfrac{1}{2}\right) = -\dfrac{3}{16}$	Negative	Below x axis
$1 < x < 2$	$f\left(\dfrac{3}{2}\right) = \dfrac{9}{16}$	Positive	Above x axis
$x > 2$	$f(3) = -18$	Negative	Below x axis

We can use the table and the x intercepts to graph Figure 4.15.

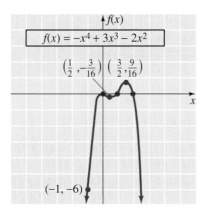

Figure 4.15

▶ **Now work Problem 25.** ∎

E X A M P L E   3

Graph $f(x) = x^3 + 3x^2 - 4$.

Solution

By using the rational root theorem, synthetic division, and the factor theorem, we can factor the given polynomial as follows:

$$f(x) = x^3 + 3x^2 - 4$$
$$= (x - 1)(x^2 + 4x + 4)$$
$$= (x - 1)(x + 2)^2$$

Now we can find the x intercepts.

$$x - 1 = 0 \quad \text{or} \quad (x + 2)^2 = 0$$
$$x = 1 \quad \text{or} \quad x = -2$$

Thus the points $(-2, 0)$ and $(1, 0)$ are on the graph and divide the x axis into three intervals (see Figure 4.16). The following table determines some points and summarizes the sign behavior of $f(x)$.

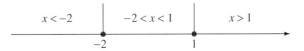

Figure 4.16

Interval	Test Value	Sign of f(x)	Location of Graph
$x < -2$	$f(-3) = -4$	Negative	Below x axis
$-2 < x < 1$	$f(0) = -4$	Negative	Below x axis
$x > 1$	$f(2) = 16$	Positive	Above x axis

Additional values: $f(-1) = -2; f(-4) = -20$

With the results of the table and the x intercepts, we sketched the graph in Figure 4.17.

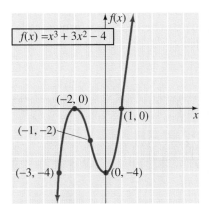

Figure 4.17

▶ **Now work Problem 31.**   ∎

Finally, let's use a graphical approach to solve a problem involving a polynomial function.

PROBLEM 1

Suppose that we have a rectangular piece of cardboard that measures 20 inches by 14 inches. From each corner a square piece is cut out, and then the flaps are turned up to form an open box (see Figure 4.18). Determine the length of a side of the square pieces to be cut out so that the volume of the box is as large as possible.

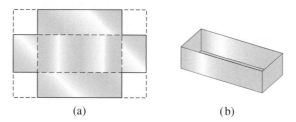

(a)   (b)

Figure 4.18

Solution

Let x represent the length of a side of the square to be cut from each corner. Then $20 - 2x$ represents the length of the open box and $14 - 2x$ represents the width. The volume of a rectangular box is given by the formula $V = lwh$, so the volume of this box can be represented by $V = x(20 - 2x)(14 - 2x)$. Now we let $y = V$ and graph the function $y = x(20 - 2x)(14 - 2x)$ as shown in Figure 4.19. For this problem we are interested only in the part of the graph between $x = 0$ and $x = 7$ because the length of a side of the square has to be less than 7 inches for a box to be formed. Figure 4.20 gives us a view of that part of the graph. Now we can use the zoom-in and trace features to determine that as x equals approximately 2.7, the value of y is a maximum of approximately 339.0. Thus square pieces of length approximately 2.7 inches on a side should be cut from each corner of the rectangular piece of cardboard. The open box formed will have a volume of approximately 339.0 cubic inches.

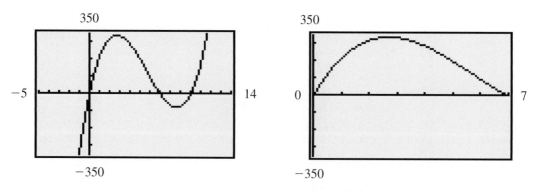

Figure 4.19   **Figure 4.20**

▶ **Now work Problem 55 using a graphing utility.**   ■

CONCEPT QUIZ   For Problems 1–4, match the function with its graph (Figures 4.21–4.24).

1. $f(x) = -x^4$ **3.** $f(x) = (x + 2)^3$

2. $f(x) = 2x^3$ **4.** $f(x) = x^3 + 2$

A.

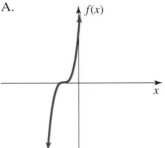

Figure 4.21

B.

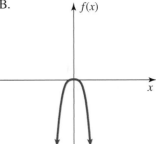

Figure 4.22

C.

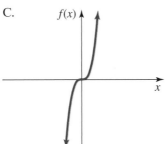

Figure 4.23

D.

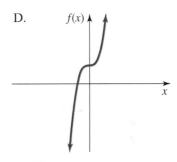

Figure 4.24

For Problems 5–8, answer true or false.

5. The solutions of a polynomial equation are called the zeros of the polynomial function.

6. The graphs of $f(x) = x^4$ and $f(x) = x^6$ are parabolas.

7. Every polynomial function of odd degree has at least one real number zero.

8. Turning points of polynomial functions are where the function crosses the x axis.

Problem Set 4.4

For Problems 1–22, graph each polynomial function.

1. $f(x) = x^3 - 3$

2. $f(x) = (x + 1)^3$

3. $f(x) = (x - 2)^3 + 1$

4. $f(x) = -(x - 3)^3$

5. $f(x) = x^4 - 2$

6. $f(x) = (x + 3)^4$

7. $f(x) = (x + 1)^4 + 3$ **8.** $f(x) = -x^5$

9. $f(x) = (x - 1)^5 + 2$ **10.** $f(x) = -(x - 2)^4$

11. $f(x) = (x - 1)(x + 1)(x - 3)$

12. $f(x) = (x - 2)(x + 1)(x + 3)$

13. $f(x) = (x + 4)(x + 1)(1 - x)$

14. $f(x) = x(x + 2)(2 - x)$

15. $f(x) = -x(x + 3)(x - 2)$

16. $f(x) = -x^2(x - 1)(x + 1)$

17. $f(x) = (x + 3)(x + 1)(x - 1)(x - 2)$

18. $f(x) = (2x - 1)(x - 2)(x - 3)$

19. $f(x) = (x - 1)^2(x + 2)$

20. $f(x) = (x + 2)^3(x - 4)$

21. $f(x) = (x + 1)^2(x - 1)^2$

22. $f(x) = x(x - 2)^2(x + 1)$

For Problems 23–34, graph each polynomial function by first factoring the given polynomial. You may need to use some factoring techniques from Chapter 0, as well as the rational root theorem and the factor theorem.

23. $f(x) = x^3 + x^2 - 2x$

24. $f(x) = -x^3 - x^2 + 6x$

25. $f(x) = -x^4 - 3x^3 - 2x^2$

26. $f(x) = x^4 - 6x^3 + 8x^2$

27. $f(x) = x^3 - x^2 - 4x + 4$

28. $f(x) = x^3 + 2x^2 - x - 2$

29. $f(x) = x^3 - 13x + 12$

30. $f(x) = x^3 - x^2 - 9x + 9$

31. $f(x) = x^3 - 4x^2 - 3x + 18$

32. $f(x) = 2x^3 - 3x^2 - 3x + 2$

33. $f(x) = -x^3 + 6x^2 - 11x + 6$

34. $f(x) = x^4 - 5x^2 + 4$

For Problems 35–41, find (a) the y intercepts, (b) the x intercepts, and (c) the intervals of x where $f(x) > 0$ and where $f(x) < 0$. *Do not* sketch the graph.

35. $f(x) = (x - 5)(x + 4)(x - 3)$

36. $f(x) = (x + 3)(x - 6)(8 - x)$

37. $f(x) = (x - 4)^2(x + 3)^3$

38. $f(x) = (x + 3)^4(x - 1)^3$

39. $f(x) = (x + 2)^2(x - 1)^3(x - 2)$

40. $f(x) = x(x - 6)^2(x + 4)$

41. $f(x) = (x + 2)^5(x - 4)^2$

■ ■ ■ THOUGHTS INTO WORDS

42. The graph of $f(x) = x^3 - 3$ is the graph of $f(x) = x^3$ translated three units downward. Describe each of the following as transformations of the basic cubic function $f(x) = x^3$.

    a. $f(x) = (x + 4)^3$ **b.** $f(x) = -3x^3$

    c. $f(x) = (x - 2)^3 + 6$ **d.** $f(x) = 2(x + 1)^3 - 4$

43. How would you go about graphing $f(x) = -(x - 1)(x + 2)^3$?

44. Give a general description of how to graph polynomial functions that are in factored form.

■ ■ ■ FURTHER INVESTIGATIONS

45. A polynomial function with real coefficients is **continuous everywhere;** that is, its graph has no holes or breaks. This is the basis for the following property: **If $f(x)$ is a polynomial with real coefficients, and if $f(a)$ and $f(b)$ are of opposite sign, then there is at least one real zero between a and b.** This property, along with what we already know about polynomial functions, provides the basis for locating and approximating irrational solutions of a polynomial equation.

Consider the equation $x^3 + 2x - 4 = 0$. Apply Descartes' rule of signs to determine that this equation has one positive real solution and two nonreal complex solutions. (You may want to confirm this!) The rational root theorem indicates that the only possible positive *rational* solutions are 1, 2, and 4. Use a little more compact format for synthetic division to obtain the following results when testing for 1 and 2 as possible solutions:

	1	0	2	-4
1	1	1	3	-1
2	1	2	6	8

Because $f(1) = -1$ (negative) and $f(2) = 8$ (positive), there must be an *irrational* solution between 1 and 2. Furthermore, because -1 is closer to 0 than to 8, our guess is that the solution is closer to 1 than to 2. Let's start looking at 1.0, 1.1, 1.2, and so on, until we can clamp the solution between two numbers.

	1	0	2	-4
1.0	1	1	3	-1
1.1	1	1.1	3.21	-0.469
1.2	1	1.2	3.44	0.128

A calculator is very helpful at this time

Because $f(1.1) = -0.469$ and $f(1.2) = 0.128$, the irrational solution must be between 1.1 and 1.2. Furthermore, because 0.128 is closer to 0 than to -0.469, our guess is that the solution is closer to 1.2 than to 1.1. Let's start looking at 1.15, 1.16, and so on.

	1	0	2	-4
1.15	1	1.15	3.3225	-0.179
1.16	1	1.16	3.3456	-0.119
1.17	1	1.17	3.3689	-0.058
1.18	1	1.18	3.3924	0.003

Because $f(1.17) = -0.058$ and $f(1.18) = 0.003$, the irrational solution must be between 1.17 and 1.18. Therefore we can use 1.2 as a rational approximation to the nearest tenth.

For each of the following equations, find an approximation, to the nearest tenth, of each irrational solution.

a. $x^3 + x - 6 = 0$
b. $x^3 - 6x - 4 = 0$
c. $x^3 - 27x + 18 = 0$
d. $x^3 - x^2 - x - 1 = 0$
e. $x^3 - 24x - 32 = 0$
f. $x^3 - 5x^2 + 3 = 0$

GRAPHING CALCULATOR ACTIVITIES

46. Graph $f(x) = x^3$. Now predict the graphs for $f(x) = x^3 + 2$, $f(x) = -x^3 + 2$, and $f(x) = -x^3 - 2$. Graph these three functions on the same set of axes with the graph of $f(x) = x^3$.

47. Draw a rough sketch of the graphs of the functions $f(x) = x^3 - x^2$, $f(x) = -x^3 + x^2$, and $f(x) = -x^3 - x^2$. Now graph these three functions to check your sketches.

48. Graph $f(x) = x^4 + x^3 + x^2$. What should the graphs of $f(x) = x^4 - x^3 + x^2$ and $f(x) = -x^4 - x^3 - x^2$ look like? Graph them to see whether you were right.

49. How should the graphs of $f(x) = x^3$, $f(x) = x^5$, and $f(x) = x^7$ compare? Graph these three functions on the same set of axes.

50. How should the graphs of $f(x) = x^2$, $f(x) = x^4$, and $f(x) = x^6$ compare? Graph these three functions on the same set of axes.

51. For each of the following functions, find the x intercepts and find the intervals of x where $f(x) > 0$ and those where $f(x) < 0$.

a. $f(x) = x^3 - 3x^2 - 6x + 8$

b. $f(x) = x^3 - 8x^2 - x + 8$

c. $f(x) = x^3 - 7x^2 + 16x - 12$

d. $f(x) = x^3 - 19x^2 + 90x - 72$

e. $f(x) = x^4 + 3x^3 - 3x^2 - 11x - 6$

f. $f(x) = x^4 + 12x^2 - 64$

52. Find the coordinates of the turning points of each of the following graphs. Express x and y values to the nearest integer.

a. $f(x) = 2x^3 - 3x^2 - 12x + 40$

b. $f(x) = 2x^3 - 33x^2 + 60x + 1050$

c. $f(x) = -2x^3 - 9x^2 + 24x + 100$

d. $f(x) = x^4 - 4x^3 - 2x^2 + 12x + 3$

e. $f(x) = x^3 - 30x^2 + 288x - 900$

f. $f(x) = x^5 - 2x^4 - 3x^3 - 2x^2 + x - 1$

53. For each of the following functions, find the x intercepts and find the turning points. Express your answers to the nearest tenth.

a. $f(x) = x^3 + 2x^2 - 3x + 4$

b. $f(x) = 42x^3 - x^2 - 246x - 35$

c. $f(x) = x^4 - 4x^2 - 4$

54. A rectangular piece of cardboard is 13 inches long and 9 inches wide. From each corner a square piece is cut out, and then the flaps are turned up to form an open box. Determine the length of a side of the square pieces so that the volume of the box is as large as possible.

55. A company determines that its weekly profit from manufacturing and selling x units of a certain item is given by $P(x) = -x^3 + 3x^2 + 2880x - 500$. What weekly production rate will maximize the profit?

Answers to the Concept Quiz

1. B   **2.** C   **3.** A   **4.** D   **5.** True   **6.** False   **7.** True   **8.** False

---

4.5   Graphing Rational Functions

Objectives

■ Determine the domain for a rational function.

■ Find the vertical asymptote(s) for a rational function.

■ Find the horizontal asymptote for a rational function.

■ Graph rational functions.

A function of the form

$$f(x) = \frac{p(x)}{q(x)}, \qquad q(x) \neq 0$$

where $p(x)$ and $q(x)$ are polynomial functions with no common factors other than -1 and 1, is called a **rational function.** The following are examples of rational functions:

$$f(x) = \frac{2}{x - 1} \qquad f(x) = \frac{x}{x - 2}$$

$$f(x) = \frac{x^2}{x^2 - x - 6} \qquad f(x) = \frac{x^3 - 8}{x + 4}$$

In each example the domain of the rational function is the set of all real numbers except those that make the denominator zero. For example, the domain of $f(x) = \frac{2}{x - 1}$ is the set of all real numbers except 1. As you will see, these exclusions from the domain are important numbers from a graphing standpoint. They represent breaks in an otherwise continuous curve.

Let's set the stage for graphing rational functions by considering in detail the function $f(x) = 1/x$. First, note that at $x = 0$, the function is undefined. Second, let's consider an extensive table of values to show some number trends and to build a basis for defining the concept of an *asymptote*.

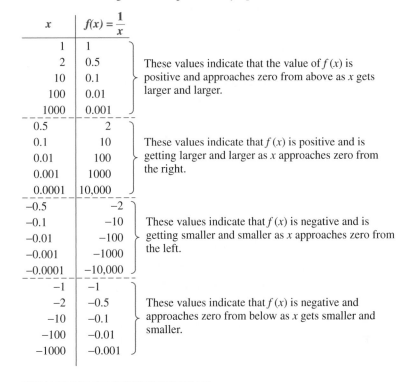

x	$f(x) = \dfrac{1}{x}$	
1	1	
2	0.5	These values indicate that the value of $f(x)$ is positive and approaches zero from above as x gets larger and larger.
10	0.1	
100	0.01	
1000	0.001	
0.5	2	
0.1	10	These values indicate that $f(x)$ is positive and is getting larger and larger as x approaches zero from the right.
0.01	100	
0.001	1000	
0.0001	10,000	
−0.5	−2	
−0.1	−10	These values indicate that $f(x)$ is negative and is getting smaller and smaller as x approaches zero from the left.
−0.01	−100	
−0.001	−1000	
−0.0001	−10,000	
−1	−1	
−2	−0.5	These values indicate that $f(x)$ is negative and approaches zero from below as x gets smaller and smaller.
−10	−0.1	
−100	−0.01	
−1000	−0.001	

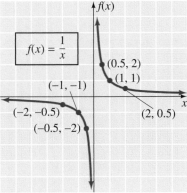

$$f(x) = \frac{1}{x}$$

(0.5, 2)
(1, 1)
(−1, −1)
(−2, −0.5)
(2, 0.5)
(−0.5, −2)

Figure 4.25

Using some points from the table and the patterns discussed, we can sketch the graph of $f(x) = \dfrac{1}{x}$ as shown in Figure 4.25. [Because the equation $f(x) = \dfrac{1}{x}$ exhibits origin symmetry, we could concentrate our point plotting in the first quadrant and then reflect that portion of the curve through the origin.] Note that the graph approaches both axes but does not touch either axis. We say that the y axis [or the $f(x)$ axis] is a *vertical asymptote,* and the x axis is a *horizontal asymptote.* In general, the following definitions can be given:

Vertical Asymptote   A line $x = a$ is a **vertical asymptote** for the graph of a function f if it satisfies either of the following two properties.

1. $f(x)$ either increases or decreases without bound as x approaches the number a from the right, as in Figure 4.26, *or*

2. $f(x)$ either increases or decreases without bound as x approaches the number a from the left, as in Figure 4.27,

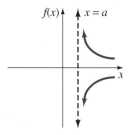

Figure 4.26

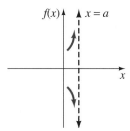

Figure 4.27

Horizontal Asymptote   A line $y = b$ [or $f(x) = b$] is a **horizontal asymptote** for the graph of a function f if it satisfies either of the following two properties.

1. $f(x)$ approaches the number b from above or below as x gets infinitely small, as in Figure 4.28, *or*

2. $f(x)$ approaches the number b from above or below as x gets infinitely large, as in Figure 4.29,

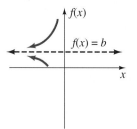

Figure 4.28

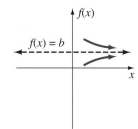

Figure 4.29

The following suggestions will help you graph rational functions of the type we are considering in this section.

1. Check for y axis symmetry and origin symmetry.

2. Find any vertical asymptote(s) by setting the denominator equal to zero and solving for x.

3. Find any horizontal asymptote(s) by studying the behavior of $f(x)$ as x gets infinitely large or as x gets infinitely small. (This suggestion will take on more meaning as you study the next four examples.)

4. Study the behavior of the graph when it is close to the asymptotes.

5. Plot as many points as necessary to determine the shape of the graph. The number may be affected by whether the graph has any symmetry.

Keep these suggestions in mind as you study the following examples.

EXAMPLE 1

Graph $f(x) = \dfrac{-2}{x - 1}$.

Solution

Because $x = 1$ makes the denominator zero, the line $x = 1$ is a vertical asymptote; we have indicated this with a dashed line in Figure 4.30. Now let's look for a horizontal asymptote by checking some large and small values of x in the following table:

x	$f(x)$	
10	$-\dfrac{2}{9}$	
100	$-\dfrac{2}{99}$	This portion of the table shows that as x gets very large, the value of $f(x)$ approaches 0 from below.
1000	$-\dfrac{2}{999}$	
-10	$\dfrac{2}{11}$	
-100	$\dfrac{2}{101}$	This portion shows that as x gets very small, the value of $f(x)$ approaches 0 from above.
-1000	$\dfrac{2}{1001}$	

Therefore the x axis is a horizontal asymptote. Finally, let's check the behavior of the graph near the vertical asymptote.

x	$f(x)$	
2	-2	
1.5	-4	
1.1	-20	As x approaches 1 from the right side, the value of $f(x)$ gets smaller and smaller.
1.01	-200	
1.001	-2000	
0	2	
0.5	4	
0.9	20	As x approaches 1 from the left side, the value of $f(x)$ gets larger and larger.
0.99	200	
0.999	2000	

The graph of $f(x) = \dfrac{-2}{x-1}$ is shown in Figure 4.30.

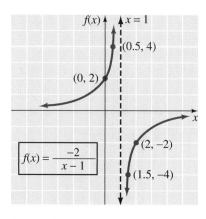

▶ **Now work Problem 3.**

Figure 4.30   ■

EXAMPLE 2

Graph $f(x) = \dfrac{x}{x+2}$.

Solution

Because $x = -2$ makes the denominator zero, the line $x = -2$ is a vertical asymptote. To study the behavior of $f(x)$ as x gets very large or very small, let's change the form of the rational expression by dividing both the numerator and the denominator by x:

$$f(x) = \frac{x}{x+2} = \frac{\dfrac{x}{x}}{\dfrac{x+2}{x}} = \frac{1}{\dfrac{x}{x}+\dfrac{2}{x}} = \frac{1}{1+\dfrac{2}{x}}$$

Now we can see that (1) as x gets larger and larger, the value of $f(x)$ approaches 1 from below, and (2) as x gets smaller and smaller, the value of $f(x)$ approaches 1 from above. (Perhaps you should check these claims by plugging in some values for x.) Thus the line $f(x) = 1$ is a horizontal asymptote. Drawing the asymptotes (dashed lines), and plotting a few points, enables us to complete the graph shown in Figure 4.31.

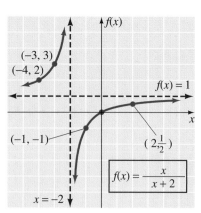

▶ **Now work Problem 7.**

Figure 4.31   ■

In the next two examples pay special attention to the role of symmetry. It will allow us to focus on the portion of a curve in quadrants I and IV and then to reflect that portion of the curve across the vertical axis to complete the graph.

E X A M P L E   3

Graph $f(x) = \dfrac{2x^2}{x^2 + 4}$.

Solution

First, note that $f(-x) = f(x)$; therefore this graph is symmetric with respect to the y axis. Second, the denominator $x^2 + 4$ cannot equal zero for any real number x. Thus there is no vertical asymptote. Third, dividing both the numerator and the denominator of the rational expression by x^2 produces

$$f(x) = \frac{2x^2}{x^2 + 4} = \frac{\dfrac{2x^2}{x^2}}{\dfrac{x^2 + 4}{x^2}} = \frac{2}{\dfrac{x^2}{x^2} + \dfrac{4}{x^2}} = \frac{2}{1 + \dfrac{4}{x^2}}$$

Now we can see that as x gets larger and larger, the value of $f(x)$ approaches 2 from below. Therefore the line $f(x) = 2$ is a horizontal asymptote. We can plot a few points using positive values for x, sketch this part of the curve, and then reflect across the $f(x)$ axis to obtain the complete graph shown in Figure 4.32.

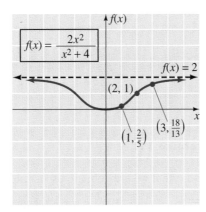

Figure 4.32

▶ **Now work Problem 21.**   ■

E X A M P L E   4

Graph $f(x) = \dfrac{3}{x^2 - 4}$.

Solution

First, note that $f(-x) = f(x)$; therefore this graph is symmetric about the $f(x)$ axis. Second, by setting the denominator equal to zero and solving for x, we obtain

$$x^2 - 4 = 0$$
$$x^2 = 4$$
$$x = \pm 2$$

The lines $x = 2$ and $x = -2$ are vertical asymptotes. Next, we can see that $\dfrac{3}{x^2 - 4}$ approaches zero from above as x gets larger and larger. Finally, we can plot a few points using positive values for x (not 2), sketch this part of the curve, and then reflect it across the $f(x)$ axis to obtain the complete graph shown in Figure 4.33.

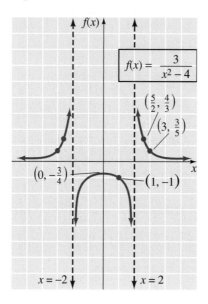

Figure 4.33

▶ **Now work Problem 11.** ■

Now suppose that we are going to use a graphing utility to obtain a graph of the function $f(x) = \dfrac{4x^2}{x^4 - 16}$. Before we enter this function into a graphing utility, let's analyze what we know about the graph.

1. Because $f(0) = 0$, the origin is a point on the graph.
2. Because $f(-x) = f(x)$, the graph is symmetric with respect to the y axis.
3. By setting the denominator equal to zero and solving for x, we can determine the vertical asymptotes:

$$x^4 - 16 = 0$$
$$(x^2 + 4)(x^2 - 4) = 0$$
$$x^2 + 4 = 0 \quad \text{or} \quad x^2 - 4 = 0$$
$$x^2 = -4 \quad \text{or} \quad x^2 = 4$$
$$x = \pm 2i \quad \text{or} \quad x = \pm 2$$

Remember that we are working with ordered pairs of real numbers. Thus the lines $x = -2$ and $x = 2$ are vertical asymptotes.

4. Dividing both the numerator and the denominator of the rational expression by x^4 produces

$$\frac{4x^2}{x^4 - 16} = \frac{\dfrac{4x^2}{x^4}}{\dfrac{x^4 - 16}{x^4}} = \frac{\dfrac{4}{x^2}}{1 - \dfrac{16}{x^4}}$$

From the last expression, we see that as $|x|$ gets larger and larger, the value of $f(x)$ approaches zero from above. Therefore the x axis is a horizontal asymptote.

We can enter the function in a graphing utility and obtain the graph shown in Figure 4.34. Note that the graph is consistent with all of the information we determined before we used the graphing utility. In other words, our knowledge of graphing techniques enhances our use of a graphing utility.

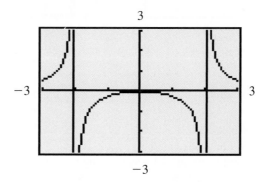

Figure 4.34

Remark: In Figure 4.34 the origin is a point of the graph that is on the horizontal asymptote. More will be said about such situations in the next section.

Back in Problem Set 1.4 you were asked to solve the following problem: How much pure alcohol should be added to 6 liters of a 40% alcohol solution to raise it to a 60% alcohol solution? The answer of 3 liters can be found by solving the following equation, where x represents the amount of pure alcohol to be added:

Pure alcohol   +   Pure alcohol   =   Pure alcohol in
to start with   added   final solution

$$0.40(6) \quad + \quad x \quad = 0.60(6 + x)$$

Now let's consider this problem in a more general setting. Again, x represents the amount of pure alcohol to be added, and the rational expression $\dfrac{2.4 + x}{6 + x}$ represents the concentration of pure alcohol in the final solution. Let's graph the rational function $y = \dfrac{2.4 + x}{6 + x}$ as shown in Figure 4.35. For this particular problem, x is non-negative, so we are interested only in the part of the graph that is in the first quadrant. We can change the boundaries of the viewing rectangle so that $0 \le x \le 15$ and $0 \le y \le 2$ to obtain Figure 4.36. Now we are ready to answer questions about this situation.

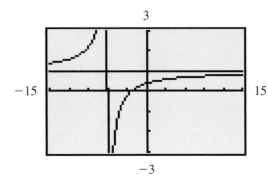

Figure 4.35   **Figure 4.36**

1. How much pure alcohol needs to be added to raise the 40% solution to a 60% alcohol solution? (*Answer:* Using the trace feature of the graphing utility, we find that $y = 0.6$ when $x = 3$. Therefore 3 liters of pure alcohol must be added.)

2. How much pure alcohol needs to be added to raise the 40% solution to a 70% alcohol solution? [*Answer:* Using the trace feature, we find that $y = 0.7$ when $x = 6$. Therefore 6 liters of pure alcohol must be added.]

3. What percent of alcohol do we have if we add 9 liters of pure alcohol to the 6 liters of a 40% solution? [*Answer:* Using the trace feature, we find that $y = 0.76$ when $x = 9$. Therefore adding 9 liters of pure alcohol will give a 76% alcohol solution.]

CONCEPT QUIZ   For Problems 1–6, answer true or false.

1. The domain of $f(x) = \dfrac{x - 4}{2x + 1}$ is all real numbers expect 4 and $-\dfrac{1}{2}$.

2. If the graph of a rational function has an asymptote at $x = -3$, then -3 is not in the domain of the function.

3. Vertical asymptotes can be found by setting the denominator of the function equal to 0.

4. Horizontal asymptotes can be found by setting the numerator of the function equal to 0.

5. If the numerator of a rational function is a constant, then the horizontal asymptote of the graph of the function is always the line $f(x) = 0$.

6. The graph of a rational function always has a vertical asymptote.

Problem Set 4.5

For Problems 1–22, identify the vertical asymptotes, if any, and graph each rational function.

1. $f(x) = \dfrac{-1}{x}$

2. $f(x) = \dfrac{1}{x^2}$

3. $f(x) = \dfrac{3}{x + 1}$

4. $f(x) = \dfrac{-1}{x - 3}$

5. $f(x) = \dfrac{2}{(x - 1)^2}$

6. $f(x) = \dfrac{-3}{(x + 2)^2}$

7. $f(x) = \dfrac{x}{x - 3}$

8. $f(x) = \dfrac{2x}{x - 1}$

9. $f(x) = \dfrac{-3x}{x + 2}$

10. $f(x) = \dfrac{-x}{x + 1}$

11. $f(x) = \dfrac{1}{x^2 - 1}$

12. $f(x) = \dfrac{-2}{x^2 - 4}$

13. $f(x) = \dfrac{-2}{(x + 1)(x - 2)}$

14. $f(x) = \dfrac{3}{(x + 2)(x - 4)}$

15. $f(x) = \dfrac{2}{x^2 + x - 2}$

16. $f(x) = \dfrac{-1}{x^2 + x - 6}$

17. $f(x) = \dfrac{x + 2}{x}$

18. $f(x) = \dfrac{2x - 1}{x}$

19. $f(x) = \dfrac{4}{x^2 + 2}$

20. $f(x) = \dfrac{4x^2}{x^2 + 1}$

21. $f(x) = \dfrac{2x^4}{x^4 + 1}$

22. $f(x) = \dfrac{x^2 - 4}{x^2}$

■ ■ ■ THOUGHTS INTO WORDS

23. How would you explain the concept of an asymptote to an elementary algebra student?

24. Give a step-by-step description of how you would go about graphing $f(x) = \dfrac{-2}{x^2 - 9}$.

■ ■ ■ FURTHER INVESTIGATIONS

25. The function $f(x) = \dfrac{(x - 2)(x + 3)}{x - 2}$ has a domain of all the real numbers except 2 and can be simplified to $f(x) = x + 3$. Thus its graph is a straight line with a hole at (2, 5). Graph each of the following functions.

a. $f(x) = \dfrac{(x + 4)(x - 1)}{x + 4}$

b. $f(x) = \dfrac{x^2 - 5x + 6}{x - 2}$

c. $f(x) = \dfrac{x - 1}{x^2 - 1}$

d. $f(x) = \dfrac{x + 2}{x^2 + 6x + 8}$

GRAPHING CALCULATOR ACTIVITIES

26. Use a graphing calculator to check your graphs for Problem 25. What feature of the graph does not show up on the calculator?

27. Each of the following graphs is a transformation of $f(x) = 1/x$. First predict the general shape and location of the graph, and then check your prediction with a graphing calculator.

a. $f(x) = \dfrac{1}{x} - 2$

b. $f(x) = \dfrac{1}{x + 3}$

c. $f(x) = -\dfrac{1}{x}$

d. $f(x) = \dfrac{1}{x - 2} + 3$

e. $f(x) = \dfrac{2x + 1}{x}$

28. Graph $f(x) = \dfrac{1}{x^2}$. How should the graphs of $f(x) =$

$\dfrac{1}{(x-4)^2}$, $f(x) = \dfrac{1 + 3x^2}{x^2}$, and $f(x) = -\dfrac{1}{x^2}$ com-

pare to the graph of $f(x) = \dfrac{1}{x^2}$? Graph the three

functions on the same set of axes with the graph of

$f(x) = \dfrac{1}{x^2}$.

29. Graph $f(x) = \dfrac{1}{x^3}$. How should the graphs of $f(x) =$

$\dfrac{2x^3 + 1}{x^3}$, $f(x) = \dfrac{1}{(x+2)^3}$, and $f(x) = -\dfrac{1}{x^3}$ com-

pare to the graph of $f(x) = \dfrac{1}{x^3}$? Graph the three

functions on the same set of axes with the graph of

$f(x) = \dfrac{1}{x^3}$.

30. Use a graphing calculator to check your graphs for Problems 19–22.

31. Graph each of the following functions. Be sure that you get a complete graph for each one. Sketch each graph on a sheet of paper and keep them all handy as you study the next section.

a. $f(x) = \dfrac{x^2}{x^2 - x - 2}$ **b.** $f(x) = \dfrac{x}{x^2 - 4}$

c. $f(x) = \dfrac{3x}{x^2 + 1}$ **d.** $f(x) = \dfrac{x^2 - 1}{x - 2}$

32. Suppose that x ounces of pure acid has been added to 14 ounces of a 15% acid solution.

a. Set up the rational expression that represents the concentration of pure acid in the final solution.
b. Graph the rational function that displays the level of concentration.
c. How many ounces of pure acid must be added to the 14 ounces of a 15% solution to raise it to a 40.5% solution? Check your answer.
d. How many ounces of pure acid must be added to the 14 ounces of a 15% solution to raise it to a 50% solution? Check your answer.
e. What percent of acid do we obtain if we add 12 ounces of pure acid to the 14 ounces of a 15% solution? Check your answer.

33. Solve the following problem both algebraically and graphically: One solution contains 50% alcohol, and another solution contains 80% alcohol. How many liters of each solution should be mixed to produce 10.5 liters of a 70% alcohol solution? Check your answer.

Answers to the Concept Quiz

1. False   **2.** True   **3.** True   **4.** False   **5.** True   **6.** False

---

4.6   More on Graphing Rational Functions

Objectives

■ Determine if any points of the graph of a rational function are on the horizontal asymptote.

■ Find the equation for an oblique asymptote of a rational function.

■ Graph rational functions with vertical, horizontal, or oblique asymptotes.

The rational functions that we studied in the previous section "behaved" rather well. In fact, once we established the vertical and horizontal asymptotes, a little bit of point plotting usually determined the graph rather easily. Such is not always the case with rational functions. In this section we will investigate some rational functions that behave a little differently.

Vertical asymptotes occur at values of x where the denominator is zero, so there can be no points of a graph on a vertical asymptote. However, recall that horizontal asymptotes are created by the behavior of $f(x)$ as x gets infinitely large or infinitely small. This does not restrict the possibility that for some values of x, there will be points of the graph on the horizontal asymptote. Let's consider some examples.

E X A M P L E   1   Graph $f(x) = \dfrac{x^2}{x^2 - x - 2}$.

Solution

First, let's identify the vertical asymptotes by setting the denominator equal to zero and solving for x.

$$x^2 - x - 2 = 0$$
$$(x - 2)(x + 1) = 0$$
$$x - 2 = 0 \quad \text{or} \quad x + 1 = 0$$
$$x = 2 \quad \text{or} \quad x = -1$$

Thus the lines $x = 2$ and $x = -1$ are vertical asymptotes. Next, we can divide both the numerator and the denominator of the rational expression by x^2.

$$f(x) = \frac{x^2}{x^2 - x - 2} = \frac{\dfrac{x^2}{x^2}}{\dfrac{x^2 - x - 2}{x^2}} = \frac{1}{1 - \dfrac{1}{x} - \dfrac{2}{x^2}}$$

Now we can see that as x gets larger and larger, the value of $f(x)$ approaches 1 from above. Thus the line $f(x) = 1$ is a horizontal asymptote. To determine whether any points of the graph are *on* the horizontal asymptote, we can see whether the equation

$$\frac{x^2}{x^2 - x - 2} = 1$$

has any solutions:

$$\frac{x^2}{x^2 - x - 2} = 1$$
$$x^2 = x^2 - x - 2$$
$$0 = -x - 2$$
$$x = -2$$

Therefore the point $(-2, 1)$ is on the graph. Now, by drawing the asymptotes, plotting a few points [including $(-2, 1)$], and studying the behavior of the function close to the asymptotes, we can sketch the curve shown in Figure 4.37.

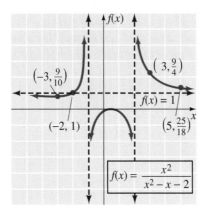

Figure 4.37

▶ **Now work Problem 1.** ■

E X A M P L E   2

Graph $f(x) = \dfrac{x}{x^2 - 4}$.

Solution

First, note that $f(-x) = -f(x)$; therefore this graph has origin symmetry. Second, let's identify the vertical asymptotes:

$$x^2 - 4 = 0$$
$$x^2 = 4$$
$$x = \pm 2$$

Thus the lines $x = -2$ and $x = 2$ are vertical asymptotes. Next, by dividing the numerator and the denominator of the rational expression by x^2, we obtain

$$f(x) = \frac{x}{x^2 - 4} = \frac{\dfrac{x}{x^2}}{\dfrac{x^2 - 4}{x^2}} = \frac{\dfrac{1}{x}}{1 - \dfrac{4}{x^2}}$$

From this form, we can see that as x gets larger and larger, the value of $f(x)$ approaches zero from above. Therefore the x axis is a horizontal asymptote. Because $f(0) = 0$, we know that the origin is a point of the graph. Finally, by concentrating our point plotting on positive values of x, we can sketch the portion of the curve to the right of the vertical axis and then use the fact that the graph is symmetric with respect to the origin to complete the graph. Figure 4.38 shows the completed graph.

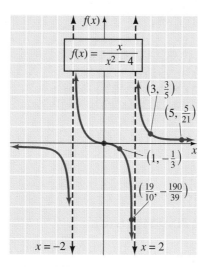

$$f(x) = \frac{x}{x^2 - 4}$$

$\left(3, \frac{3}{5}\right)$

$\left(5, \frac{5}{21}\right)$

$\left(1, -\frac{1}{3}\right)$

$\left(\frac{19}{10}, -\frac{190}{39}\right)$

$x = -2$ $x = 2$

Figure 4.38

▶ **Now work Problem 5.** ■

EXAMPLE 3

Graph $f(x) = \dfrac{3x}{x^2 + 1}$.

Solution

First, observe that $f(-x) = -f(x)$; therefore this graph is symmetric with respect to the origin. Second, because $x^2 + 1$ is a positive number for all real number values of x, there are no vertical asymptotes for this graph. Next, by dividing the numerator and denominator of the rational expression by x^2, we obtain

$$f(x) = \frac{3x}{x^2 + 1} = \frac{\dfrac{3x}{x^2}}{\dfrac{x^2 + 1}{x^2}} = \frac{\dfrac{3}{x}}{1 + \dfrac{1}{x^2}}$$

From this form, we see that as x gets larger and larger, the value of $f(x)$ approaches zero from above. Thus the x axis is a horizontal asymptote. Because $f(0) = 0$, the origin is a point of the graph. Finally, by concentrating our point plotting on positive values of x, we can sketch the portion of the curve to the right of the vertical axis and then use origin symmetry to complete the graph, as shown in Figure 4.39.

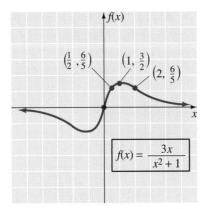

Figure 4.39

▶ **Now work Problem 11.** ∎

■ Oblique Asymptotes

Thus far we have restricted our study of rational functions to those where the degree of the numerator is less than or equal to the degree of the denominator. As our final examples of graphing rational functions, we will consider functions where the degree of the numerator is one greater than the degree of the denominator.

EXAMPLE 4   Graph $f(x) = \dfrac{x^2 - 1}{x - 2}$.

Solution

First, let's observe that $x = 2$ is a vertical asymptote. Second, because the degree of the numerator is greater than the degree of the denominator, we can change the form of the rational expression by division. We use synthetic division:

$$
\begin{array}{r}
2)\overline{\,1 \quad 0 \quad -1\,} \\
\underline{\quad 2 \quad 4} \\
1 \quad 2 \quad 3
\end{array}
$$

Therefore the original function can be rewritten

$$f(x) = \frac{x^2 - 1}{x - 2} = x + 2 + \frac{3}{x - 2}$$

Now, for very large values of $|x|$, the fraction $\dfrac{3}{x - 2}$ is close to zero. Therefore, as $|x|$ gets larger and larger, the graph of $f(x) = x + 2 + \dfrac{3}{x - 2}$ gets closer and closer to the line $f(x) = x + 2$. We call this line an **oblique asymptote** and indicate it with a dashed line in Figure 4.36. Finally, because this is a new situation, it may be necessary to plot a large number of points on both sides of the vertical asymptote, so let's make an extensive table of values. The graph of the function is shown in Figure 4.40.

x	$f(x) = \dfrac{x^2 - 1}{x - 2}$	
2.1	34.1	
2.5	10.5	
3	8	
4	7.5	These values indicate the behavior of $f(x)$ to the right of the vertical asymptote $x = 2$.
5	8	
6	8.75	
10	12.375	
1.9	−26.1	
1.5	−2.5	
1	0	
0	0.5	These values indicate the behavior of $f(x)$ to the left of the vertical asymptote $x = 2$.
−1	0	
−3	−1.6	
−5	−3.4	
−10	−8.25	

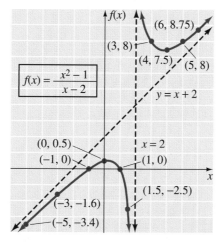

Figure 4.40

▶ **Now work Problem 15.**

If the degree of the numerator of a rational function is *exactly one more* than the degree of its denominator, then the graph of the function has an oblique asymptote. [If the graph is a line, as is the case with $f(x) = \dfrac{(x - 2)(x + 1)}{x - 2}$, then we consider it to be its own asymptote.] As in Example 4, we find the equation of the oblique asymptote by changing the form of the function using long division. Let's consider another example:

Graph $f(x) = \dfrac{x^2 - x - 2}{x - 1}$.

Solution

From the given form of the function, we see that $x = 1$ is a vertical asymptote. Then, by factoring the numerator, we can change the form to

$$f(x) = \frac{(x - 2)(x + 1)}{(x - 1)}$$

which indicates x intercepts of 2 and -1. By long division, we can change the original form of the function to

$$f(x) = x - \frac{2}{x - 1}$$

which indicates an oblique asymptote $f(x) = x$. Finally, by plotting a few additional points, we can determine the graph as shown in Figure 4.41.

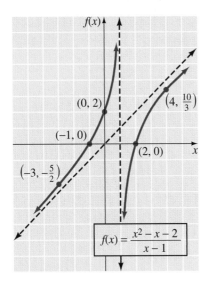

Figure 4.41

▶ **Now work Problem 17.**   ■

Finally, let's combine our knowledge of rational functions with the use of a graphing utility to obtain the graph of a fairly complex rational function.

Graph the rational function $f(x) = \dfrac{x^3 - 2x^2 - x - 1}{x^2 - 36}$.

Solution

Before entering this function into a graphing utility, let's analyze what we know about the graph.

1. Because $f(0) = \dfrac{1}{36}$, the point $\left(0, \dfrac{1}{36}\right)$ is on the graph.

2. Because $f(-x) \neq f(x)$ and $f(-x) \neq -f(x)$, there is no symmetry with respect to the origin or the y axis.

3. The denominator is zero at $x = \pm 6$. Thus the lines $x = 6$ and $x = -6$ are vertical asymptotes.

4. Let's change the form of the rational expression by division.

$$
\begin{array}{r}
x - 2 \\
x^2 - 36{\overline{\smash{\big)}\,x^3 - 2x^2 - x - 1}} \\
\underline{x^3 - 36x} \\
-2x^2 + 35x - 1 \\
\underline{-2x^2 + 72} \\
35x - 73
\end{array}
$$

Thus the original function can be rewritten as

$$f(x) = x - 2 + \frac{35x - 73}{x^2 - 36}$$

Therefore the line $y = x - 2$ is an oblique asymptote. Now we let $Y_1 = x - 2$ and $Y_2 = \dfrac{x^3 - 2x^2 - x - 1}{x^2 - 36}$ and use a viewing rectangle where $-15 \leq x \leq 15$ and $-30 \leq y \leq 30$ (see Figure 4.42).

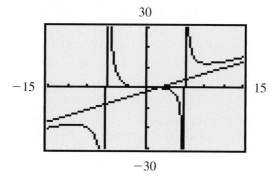

Figure 4.42

▶ **Now work Problem 26h using a graphing utility.** ■

Note that the graph in Figure 4.42 is consistent with the information we had before we used the graphing utility. (The graph may appear to have origin symmetry, but remember that the point $\left(0, \dfrac{1}{36}\right)$ is on the graph, whereas the point $\left(0, -\dfrac{1}{36}\right)$ is not.) Also note that the curve does intersect the oblique asymptote. We can use the zoom-in and trace features of the graphing utility to find that point of intersection, or we can do it algebraically as follows: Because $y = \dfrac{x^3 - 2x^2 - x - 1}{x^2 - 36}$ and $y = x - 2$, we can equate the two expressions for y and solve the resulting equation for x:

$$\frac{x^3 - 2x^2 - x - 1}{x^2 - 36} = x - 2$$

$$x^3 - 2x^2 - x - 1 = (x - 2)(x^2 - 36)$$

$$x^3 - 2x^2 - x - 1 = x^3 - 2x^2 - 36x + 72$$

$$35x = 73$$

$$x = \frac{73}{35}$$

If $x = \dfrac{73}{35}$, then $y = x - 2 = \dfrac{73}{35} - 2 = \dfrac{3}{35}$. The point of intersection of the curve and the oblique asymptote is $\left(\dfrac{73}{35}, \dfrac{3}{35}\right)$.

CONCEPT QUIZ

For Problems 1–6, answer true or false.

1. Every graph of a rational function has horizontal asymptotes.

2. For rational functions where the degree of the numerator is one more than the degree of the denominator, the graph of the function will have an oblique asymptote.

3. The graph of a rational function can cross a horizontal asymptote.

4. The graph of a rational function can cross a vertical asymptote.

5. If the graph of a rational function is symmetric with respect to the origin, then the line $f(x) = 0$ is a vertical asymptote.

6. For rational functions where the degree of the numerator is less than or equal to the degree of the denominator, the graph of the function will have a horizontal asymptote.

Problem Set 4.6

For Problems 1–20, graph each rational function. Check first for symmetry and identify the asymptotes.

1. $f(x) = \dfrac{x^2}{x^2 + x - 2}$

2. $f(x) = \dfrac{x^2}{x^2 + 2x - 3}$

3. $f(x) = \dfrac{2x^2}{x^2 - 2x - 8}$

4. $f(x) = \dfrac{-x^2}{x^2 + 3x - 4}$

5. $f(x) = \dfrac{-x}{x^2 - 1}$

6. $f(x) = \dfrac{2x}{x^2 - 9}$

7. $f(x) = \dfrac{x}{x^2 + x - 6}$

8. $f(x) = \dfrac{-x}{x^2 - 2x - 8}$

9. $f(x) = \dfrac{x^2}{x^2 - 4x + 3}$

10. $f(x) = \dfrac{1}{x^3 + x^2 - 6x}$

11. $f(x) = \dfrac{x}{x^2 + 2}$

12. $f(x) = \dfrac{6x}{x^2 + 1}$

13. $f(x) = \dfrac{-4x}{x^2 + 1}$

14. $f(x) = \dfrac{-5x}{x^2 + 2}$

15. $f(x) = \dfrac{x^2 + 2}{x - 1}$

16. $f(x) = \dfrac{x^2 - 3}{x + 1}$

17. $f(x) = \dfrac{x^2 - x - 6}{x + 1}$

18. $f(x) = \dfrac{x^2 + 4}{x + 2}$

19. $f(x) = \dfrac{x^2 + 1}{1 - x}$

20. $f(x) = \dfrac{x^3 + 8}{x^2}$

■ ■ ■ THOUGHTS INTO WORDS

21. Explain the concept of an oblique asymptote.

22. Explain why it is possible for curves to intersect horizontal and oblique asymptotes but not to intersect vertical asymptotes.

23. Give a step-by-step description of how you would go about graphing $f(x) = \dfrac{x^2 - x - 12}{x - 2}$.

24. Your friend is having difficulty finding the point of intersection of a curve and the oblique asymptote. How would you help?

GRAPHING CALCULATOR ACTIVITIES

25. First check for symmetry and identify the asymptotes for the graphs of the following rational functions. Then use your graphing utility to graph each function.

a. $f(x) = \dfrac{4x^2}{x^2 + x - 2}$

b. $f(x) = \dfrac{-2x}{x^2 - 5x - 6}$

c. $f(x) = \dfrac{x^2}{x^2 - 9}$

d. $f(x) = \dfrac{x^2 - 4}{x^2 - 9}$

e. $f(x) = \dfrac{x^2 - 9}{x^2 - 4}$

f. $f(x) = \dfrac{x^2 + 2x + 1}{x^2 - 5x + 6}$

26. For each of the following rational functions, first determine and graph any oblique asymptotes. Then, on the same set of axes, graph the function.

a. $f(x) = \dfrac{x^2 - 1}{x - 2}$

b. $f(x) = \dfrac{x^2 + 1}{x + 2}$

c. $f(x) = \dfrac{2x^2 + x + 1}{x + 1}$

d. $f(x) = \dfrac{x^2 + 4}{x - 3}$

e. $f(x) = \dfrac{3x^2 - x - 2}{x - 2}$

f. $f(x) = \dfrac{4x^2 + x + 1}{x + 1}$

g. $f(x) = \dfrac{x^3 + x^2 - x - 1}{x^2 + 2x + 3}$

h. $f(x) = \dfrac{x^3 + 2x^2 + x - 3}{x^2 - 4}$

Answers to the Concept Quiz

1. False   **2.** True   **3.** True   **4.** False   **5.** False   **6.** True

4.7 Direct and Inverse Variation

Objectives

- Translate statements of variation into equations.
- Find the constant of variation for stated conditions.
- Solve application problems for direct, inverse, or joint variation.

The amount of simple interest earned by a fixed amount of money invested at a certain rate *varies directly* as the time.

At a constant temperature, the volume of an enclosed gas *varies inversely* as the pressure.

These statements illustrate two basic types of functional relationships, **direct variation** and **inverse variation,** that are widely used, especially in the physical sciences. These relationships can be expressed by equations that determine functions. The purpose of this section is to investigate these special functions.

■ Direct Variation

The statement *y varies directly as x* means

$$y = kx$$

where k is a nonzero constant, called the **constant of variation.** The phrase *y is directly proportional to x* is also used to indicate direct variation; k is then referred to as the **constant of proportionality.**

Remark: Note that the equation $y = kx$ defines a function and can be written $f(x) = kx$. However, in this section it is more convenient not to use function notation but instead to use variables that are meaningful in terms of the physical entities involved in the particular problem.

Statements that indicate direct variation may also involve powers of a variable. For example, *y varies directly as the square of x* can be written $y = kx^2$. In general, *y varies directly as the nth power of x* $(n > 0)$ means

$$y = kx^n$$

There are basically three types of problems in which we deal with direct variation: (1) translating an English statement into an equation expressing the direct variation, (2) finding the constant of variation from given values of the variables, and (3) finding additional values of the variables once the constant of variation has been determined. Let's consider an example of each of these types of problems.

EXAMPLE 1

Translate the statement *the tension on a spring varies directly as the distance it is stretched* into an equation, using k as the constant of variation.

Solution

Let t represent the tension and d the distance; the equation is

$$t = kd$$

⊙ **Now work Problem 1.** ∎

EXAMPLE 2

If A varies directly as the square of e and if $A = 96$ when $e = 4$, find the constant of variation.

Solution

Because A varies directly as the square of e, we have

$$A = ke^2$$

We substitute 96 for A and 4 for e to obtain

$$96 = k(4)^2$$
$$96 = 16k$$
$$6 = k$$

The constant of variation is 6.

⊙ **Now work Problem 11.** ∎

EXAMPLE 3

If y is directly proportional to x and if $y = 6$ when $x = 8$, find the value of y when $x = 24$.

Solution

The statement "y is directly proportional to x" translates into

$$y = kx$$

Let $y = 6$ and $x = 8$; the constant of variation becomes

$$6 = k(8)$$
$$\frac{6}{8} = k$$
$$\frac{3}{4} = k$$

Thus the specific equation is

$$y = \frac{3}{4}x$$

Now we let $x = 24$ to obtain

$$y = \frac{3}{4}(24) = 18$$

▶ **Now work Problem 19.** ■

■ Inverse Variation

The second basic type of variation, *inverse variation*, is defined as follows. The statement "*y* varies inversely as *x*" means

$$y = \frac{k}{x}$$

where k is a nonzero constant, which is again referred to as the constant of variation. The phrase "*y* is inversely proportional to *x*" is also used to express inverse variation. As with direct variation, statements indicating variation may involve powers of x. For example, "*y* varies inversely as the square of *x*" can be written $y = k/x^2$. In general, "*y* varies inversely as the nth power of $x(n > 0)$" means

$$y = \frac{k}{x^n}$$

The following examples illustrate the three basic kinds of problems that involve inverse variation.

EXAMPLE 4

Translate the statement "the length of a rectangle of fixed area varies inversely as the width" into an equation, using k as the constant of variation.

Solution

Let l represent the length and w the width; the equation is

$$l = \frac{k}{w}$$

▶ **Now work Problem 5.** ■

EXAMPLE 5

If y is inversely proportional to x, and if $y = 14$ when $x = 4$, find the constant of variation.

Solution

Because y is inversely proportional to x, we have

$$y = \frac{k}{x}$$

We substitute 4 for x and 14 for y to obtain

$$14 = \frac{k}{4}$$

Solving this equation yields

$$k = 56$$

The constant of variation is 56.

▶ **Now work Problem 10.** ■

E X A M P L E 6

The time required for a car to travel a certain distance varies inversely as the rate at which it travels. If it takes 4 hours at 50 miles per hour to travel the distance, how long will it take at 40 miles per hour?

Solution

Let t represent time and r rate. The phrase "time required . . . varies inversely as the rate" translates into

$$t = \frac{k}{r}$$

We substitute 4 for t and 50 for r to find the constant of variation:

$$4 = \frac{k}{50}$$

$$k = 200$$

Thus the specific equation is

$$t = \frac{200}{r}$$

Now we substitute 40 for r to get

$$t = \frac{200}{40}$$

$$= 5$$

It will take 5 hours at 40 miles per hour.

▶ **Now work Problem 23.** ■

The terms "direct" and "inverse," as applied to variation, refer to the relative behavior of the variables involved in the equation. That is, in *direct variation* ($y = kx$), an assignment of **increasing absolute values for x** produces **increasing absolute values for y.** However, in *inverse variation* ($y = k/x$), an assignment of **increasing absolute values for x** produces **decreasing absolute values for y.**

■ Joint Variation

Variation may involve more than two variables. The following table illustrates some different types of variation statements and their equivalent algebraic equations that use k as the constant of variation. Statements 1, 2, and 3 illustrate the concept of *joint variation*. Statements 4 and 5 show that both direct and inverse variation may occur in the same problem. Statement 6 combines joint variation with inverse variation.

Variation Statement	Algebraic Equation
1. y varies jointly as x and z	$y = kxz$
2. y varies jointly as x, z, and w	$y = kxzw$
3. V varies jointly as h and the square of r	$V = khr^2$
4. h varies directly as V and inversely as w	$h = \dfrac{kV}{w}$
5. y is directly proportional to x and inversely proportional to the square of z	$y = \dfrac{kx}{z^2}$
6. y varies jointly as w and z and inversely as x	$y = \dfrac{kwz}{x}$

The final two examples of this section illustrate different kinds of problems involving some of these variation situations.

E X A M P L E   7

The volume of a pyramid varies jointly as its altitude and the area of its base. If a pyramid with an altitude of 9 feet and a base with an area of 17 square feet has a volume of 51 cubic feet, find the volume of a pyramid with an altitude of 14 feet and a base with an area of 45 square feet.

Solution

Let's use the following variables:

V = volume h = altitude

B = area of base k = constant of variation

The fact that the volume varies jointly as the altitude and the area of the base can be represented by the equation

$V = kBh$

We substitute 51 for V, 17 for B, and 9 for h to obtain

$51 = k(17)(9)$

$51 = 153k$

$$\frac{51}{153} = k$$

$$\frac{1}{3} = k$$

Therefore the specific equation is $V = \frac{1}{3}Bh$. Now we can substitute 45 for B and 14 for h to get

$$V = \frac{1}{3}(45)(14) = (15)(14) = 210$$

The volume is 210 cubic feet.

▶ **Now work Problem 31.**   ■

EXAMPLE 8

Suppose that y varies jointly as x and z and inversely as w. If $y = 154$ when $x = 6$, $z = 11$, and $w = 3$, find y when $x = 8$, $z = 9$, and $w = 6$.

Solution

The statement "y varies jointly as x and z and inversely as w" translates into the equation

$$y = \frac{kxz}{w}$$

We substitute 154 for y, 6 for x, 11 for z, and 3 for w to produce

$$154 = \frac{(k)(6)(11)}{3}$$

$$154 = 22k$$

$$7 = k$$

Thus the specific equation is

$$y = \frac{7xz}{w}$$

Now we can substitute 8 for x, 9 for z, and 6 for w to get

$$y = \frac{7(8)(9)}{6} = 84$$

▶ **Now work Problem 29.**   ■

CONCEPT QUIZ For Problems 1–4, match the statement of variation with its equation.

1. y varies inversely as the cube of x.

   A. $y = \dfrac{kw^2}{x^3}$

2. y varies directly as the cube of x.

   B. $y = \dfrac{k}{x^3}$

3. y varies directly as the square of w and inversely as the cube of x.

   C. $y = kw^2x^3$

4. y varies jointly as the square of w and the cube of x.

   D. $y = kx^3$

For Problems 5–8, answer true or false.

5. The statement y varies jointly as x and w means that y varies directly as x and inversely as w.

6. The constant of variation is always a positive number.

7. If a worker gets paid \$8.50 for each hour worked, we would say that his pay varies directly with the number of hours worked.

8. If a fast food restaurant loses \$0.25 for each special burger sold, we would say that the amount of money lost varies inversely as the number of special burgers sold.

Problem Set 4.7

For Problems 1–8, translate each statement of variation into an equation; use k as the constant of variation.

1. y varies directly as the cube of x.

2. a varies inversely as the square of b.

3. A varies jointly as l and w.

4. s varies jointly as g and the square of t.

5. At a constant temperature, the volume, V, of a gas varies inversely as the pressure, P.

6. y varies directly as the square of x and inversely as the cube of w.

7. The volume, V, of a cone varies jointly as its height, h, and the square of a radius, r.

8. I is directly proportional to r and t.

For Problems 9–18, find the constant of variation for each stated condition.

9. y varies directly as x, and $y = 72$ when $x = 3$.

10. y varies inversely as the square of x, and $y = 4$ when $x = 2$.

11. A varies directly as the square of r, and $A = 154$ when $r = 7$.

12. V varies jointly as B and h, and $V = 104$ when $B = 24$ and $h = 13$.

13. A varies jointly as b and h, and $A = 81$ when $b = 9$ and $h = 18$.

14. s varies jointly as g and the square of t, and $s = -108$ when $g = 24$ and $t = 3$.

15. y varies jointly as x and z and inversely as w, and $y = 154$ when $x = 6$, $z = 11$, and $w = 3$.

16. V varies jointly as h and the square of r, and $V = 1100$ when $h = 14$ and $r = 5$.

17. y is directly proportional to the square of x and inversely proportional to the cube of w, and $y = 18$ when $x = 9$ and $w = 3$.

18. y is directly proportional to x and inversely proportional to the square root of w, and $y = \dfrac{1}{5}$ when $x = 9$ and $w = 10$.

For Problems 19–32, solve each problem.

19. If y is directly proportional to x and $y = 5$ when $x = -15$, find the value of y when $x = -24$.

20. If y is inversely proportional to the square of x and $y = \dfrac{1}{8}$ when $x = 4$, find y when $x = 8$.

21. If V varies jointly as B and h, and $V = 96$ when $B = 36$ and $h = 8$, find V when $B = 48$ and $h = 6$.

22. If A varies directly as the square of e and $A = 150$ when $e = 5$, find A when $e = 10$.

23. The time required for a car to travel a certain distance varies inversely as the rate at which it travels. If it takes 3 hours to travel the distance at 50 miles per hour, how long will it take at 30 miles per hour?

24. The distance that a freely falling body falls varies directly as the square of the time it falls. If a body falls 144 feet in 3 seconds, how far will it fall in 5 seconds?

25. The period (the time required for one complete oscillation) of a simple pendulum varies directly as the square root of its length. If a pendulum 12 feet long has a period of 4 seconds, find the period of a pendulum of length 3 feet.

26. Suppose the number of days it takes to complete a construction job varies inversely as the number of people assigned to the job. If it takes 7 people 8 days to do the job, how long will it take 10 people to complete the job?

27. The number of days needed to assemble some machines varies directly as the number of machines and inversely as the number of people working. If it takes 4 people 32 days to assemble 16 machines, how many days will it take 8 people to assemble 24 machines?

28. The volume of a gas at a constant temperature varies inversely as the pressure. What is the volume of a gas under a pressure of 25 pounds if the gas occupies 15 cubic centimeters under a pressure of 20 pounds?

29. The volume, V, of a gas varies directly as the temperature, T, and inversely as the pressure, P. If $V = 48$ when $T = 320$ and $P = 20$, find V when $T = 280$ and $P = 30$.

30. The volume of a cylinder varies jointly as its altitude and the square of the radius of its base. If the volume of a cylinder is 1386 cubic centimeters when the radius of the base is 7 centimeters and its altitude is 9 centimeters, find the volume of a cylinder that has a base of radius 14 centimeters if the altitude of the cylinder is 5 centimeters.

31. The cost of labor varies jointly as the number of workers and the number of days that they work. If it costs $900 to have 15 people work for 5 days, how much will it cost to have 20 people work for 10 days?

32. The cost of publishing pamphlets varies directly as the number of pamphlets produced. If it costs $96 to publish 600 pamphlets, how much does it cost to publish 800 pamphlets?

■■■ **THOUGHTS INTO WORDS**

33. How would you explain the difference between direct variation and inverse variation?

34. Suppose that y varies directly as the square of x. Does doubling the value of x also double the value of y? Explain your answer.

35. Suppose that y varies inversely as x. Does doubling the value of x also double the value of y? Explain your answer.

■■■ **FURTHER INVESTIGATIONS**

C In the previous problems, we chose numbers to make computations reasonable without the use of a calculator. However, variation-type problems often involve messy computations and the calculator becomes a very useful tool. Use your calculator to help solve the following problems.

The symbol, **C**, signals a problem that requires a calculator.

36. The simple interest earned by a certain amount of money varies jointly as the rate of interest and the time (in years) that the money is invested.

    a. If some money invested at 11% for 2 years earns $385, how much would the same amount earn at 12% for 1 year?

    b. If some money invested at 12% for 3 years earns $819, how much would the same amount earn at 14% for 2 years?

    c. If some money invested at 14% for 4 years earns $1960, how much would the same amount earn at 15% for 2 years?

37. The period (the time required for one complete oscillation) of a simple pendulum varies directly as the square root of its length. If a pendulum 9 inches long has a period of 2.4 seconds, find the period of a pendulum of length 12 inches. Express the answer to the nearest tenth of a second.

38. The volume of a cylinder varies jointly as its altitude and the square of the radius of its base. If the volume of a cylinder is 549.5 cubic meters when the radius of the base is 5 meters and its altitude is 7 meters, find the volume of a cylinder that has a base of radius 9 meters and an altitude of 14 meters.

39. If y is directly proportional to x and inversely proportional to the square of z, and if $y = 0.336$ when $x = 6$ and $z = 5$, find the constant of variation.

40. If y is inversely proportional to the square root of x and $y = 0.08$ when $x = 225$, find y when $x = 625$.

Answers to the Concept Quiz

1. B   **2.** D   **3.** A   **4.** C   **5.** False   **6.** False   **7.** True   **8.** False

Two themes unify this chapter: (1) solving polynomial equations and (2) graphing polynomial and rational functions.

■ Solving Polynomial Equations

The following concepts and properties provide the basis for solving polynomial equations.

1. Synthetic division.

2. The factor theorem: A polynomial $f(x)$ has a factor $x - c$ if and only if $f(c) = 0$.

3. Property 4.3: A polynomial equation of degree n has n solutions, where any solution of multiplicity p is counted p times.

4. The rational root theorem: Consider the polynomial equation

$$a_n x^n + a_{n-1} x^{n-1} + \cdots + a_1 x + a_0 = 0$$

where *the coefficients are integers*. If the rational number c/d, reduced to lowest terms, is a solution of the equation, then c is a factor of the constant term, a_0, and d is a factor of the leading coefficient, a_n.

5. Property 4.5: If a polynomial equation with real coefficients has any nonreal complex solutions, they must occur in conjugate pairs.

6. Property 4.6, Descartes' rule of signs: Let $a_n x^n + a_{n-1} x^{n-1} + \cdots + a_1 x + a_0 = 0$ be a polynomial equation with real coefficients.

   a. The number of *positive real solutions* either is equal to the number of sign variations in the given polynomial or is less than the number of sign variations by a positive even integer.

   b. The number of *negative real solutions* either is equal to the number of sign variations in

   $$a_n(-x)^n + a_{n-1}(-x)^{n-1} + \cdots + a_1(-x) + a_0$$

   or is less than that number of sign variations by a positive even integer.

■ Graphing Polynomial and Rational Functions

Graphs of polynomial functions of the form $f(x) = ax^n$, where n is an integer greater than 2 and $a \neq 1$, are variations of the graphs shown in Figures 4.3 and 4.6. If n is odd, the curve is symmetric about the origin, and if n is even, the graph is symmetric about the vertical axis.

Graphs of polynomial functions of the form $f(x) = ax^n$ can be translated horizontally and vertically and reflected across the x axis. For example:

1. The graph of $f(x) = 2(x - 4)^3$ is the graph of $f(x) = 2x^3$ moved four units to the right.

2. The graph of $f(x) = 3x^4 + 4$ is the graph of $f(x) = 3x^4$ moved up four units.

3. The graph of $f(x) = -x^5$ is the graph of $f(x) = x^5$ reflected across the x axis.

To graph a polynomial function that is expressed in factored form, the following steps are helpful.

1. Find the x intercepts, which are also called the *zeros* of the polynomial.

2. Use a test value in each of the intervals determined by the x intercepts to find out whether the function is positive or negative over that interval.

3. Plot any additional points that are needed to determine the graph.

To graph a rational function, the following steps are useful.

1. Check for vertical axis and origin symmetry.

2. Find any vertical asymptotes by setting the denominator equal to zero and solving it for x.

3. Find any horizontal asymptotes by studying the behavior of $f(x)$ as x gets very large or very small. This may require changing the form of the original rational expression.

4. If the degree of the numerator is one larger than the degree of the denominator, determine the equation of the oblique asymptote.

5. Study the behavior of the graph when it is close to the asymptotic lines.

6. Plot as many points as necessary to determine the graph. This may be affected by whether the graph has any symmetries.

■ Applications of Functions

Relationships that involve **direct** and **inverse variation** can be expressed by equations that determine functions. The statement *y varies directly as x* means

$$y = kx$$

where k is the **constant of variation.** The statement *y varies directly as the nth power of x* $(n > 0)$ means

$$y = kx^n$$

The statement *y varies inversely as x* means $y = \dfrac{k}{x}$.

The statement *y varies inversely as the nth power of x* $(n > 0)$ means $y = \dfrac{k}{x^n}$.

The statement *y varies jointly as x and w* means $y = kxw$.

Chapter 4 Review Problem Set

For Problems 1 and 2, find the quotient and remainder of each division problem.

1. $(6x^3 + 11x^2 - 27x + 32) \div (2x + 7)$

2. $(2a^3 - 3a^2 + 13a - 1) \div (a^2 - a + 6)$

For Problems 3–6, use synthetic division to determine the quotient and remainder.

3. $(3x^3 - 4x^2 + 6x - 2) \div (x - 1)$

4. $(5x^3 + 7x^2 - 9x + 10) \div (x + 2)$

5. $(-2x^4 + x^3 - 2x^2 - x - 1) \div (x + 4)$

6. $(-3x^4 - 5x^2 + 9) \div (x + 3)$

For Problems 7–10, find $f(c)$ either by using synthetic division and the remainder theorem or by evaluating $f(c)$ directly.

7. $f(x) = 4x^5 - 3x^3 + x^2 - 1$ and $c = 1$

8. $f(x) = 4x^3 - 7x^2 + 6x - 8$ and $c = -3$

9. $f(x) = -x^4 + 9x^2 - x - 2$ and $c = -2$

10. $f(x) = x^4 - 9x^3 + 9x^2 - 10x + 16$ and $c = 8$

For Problems 11–14, use the factor theorem to help answer some questions about factors.

11. Is $x + 2$ a factor of $2x^3 + x^2 - 7x - 2$?

12. Is $x - 3$ a factor of $x^4 + 5x^3 - 7x^2 - x + 3$?

13. Is $x - 4$ a factor of $x^5 - 1024$?

14. Is $x + 1$ a factor of $x^5 + 1$?

For Problems 15–18, use the rational root theorem and the factor theorem to help solve each equation.

15. $x^3 - 3x^2 - 13x + 15 = 0$

16. $8x^3 + 26x^2 - 17x - 35 = 0$

17. $x^4 - 5x^3 + 34x^2 - 82x + 52 = 0$

18. $x^3 - 4x^2 - 10x + 4 = 0$

For Problems 19 and 20, use Descartes' rule of signs (Property 4.6) to list the possibilities for the nature of the solutions. *Do not* solve the equations.

19. $4x^4 - 3x^3 + 2x^2 + x + 4 = 0$

20. $x^5 + 3x^3 + x + 7 = 0$

For Problems 21–24, graph each polynomial function.

21. $f(x) = -(x - 2)^3 + 3$

22. $f(x) = (x + 3)(x - 1)(3 - x)$

23. $f(x) = x^4 - 4x^2$

24. $f(x) = x^3 - 4x^2 + x + 6$

For Problems 25–28, graph each rational function. Be sure to identify the asymptotes.

25. $f(x) = \dfrac{2x}{x - 3}$

26. $f(x) = \dfrac{-3}{x^2 + 1}$

27. $f(x) = \dfrac{-x^2}{x^2 - x - 6}$

28. $f(x) = \dfrac{x^2 + 3}{x + 1}$

29. If y varies directly as x and inversely as w, and if $y = 27$ when $x = 18$ and $w = 6$, find the constant of variation.

30. If y varies jointly as x and the square root of w, and if $y = 140$ when $x = 5$ and $w = 16$, find y when $x = 9$ and $w = 49$.

31. The weight of a body above the surface of the earth varies inversely as the square of its distance from the center of the earth. Assuming the radius of the earth to be 4000 miles, determine how much a man would weigh 1000 miles above the earth's surface if he weighs 200 pounds on the surface.

32. The number of hours needed to assemble some furniture varies directly as the number of pieces of furniture and inversely as the number of people working. If it takes three people 10 hours to assemble 20 pieces of furniture, how many hours will it take four people to assemble 40 pieces of furniture?

1. Find the quotient and remainder for the division problem $(6x^3 - 19x^2 + 3x + 20) \div (3x - 5)$.

2. Find the quotient and remainder for the division problem $(3x^4 + 8x^3 - 5x^2 - 12x - 15) \div (x + 3)$.

3. Find the quotient and remainder for the division problem $(4x^4 - 7x^2 + 4) \div (x - 2)$.

4. If $f(x) = x^5 - 8x^4 + 9x^3 - 13x^2 - 9x - 10$, find $f(7)$.

5. If $f(x) = 3x^4 + 20x^3 - 6x^2 + 9x + 19$, find $f(-7)$.

6. If $f(x) = x^5 - 35x^3 - 32x + 15$, find $f(6)$.

7. Is $x - 5$ a factor of $3x^3 - 11x^2 - 22x - 20$?

8. Is $x + 2$ a factor of $5x^3 + 9x^2 - 9x - 17$?

9. Is $x + 3$ a factor of $x^4 - 16x^2 - 17x + 12$?

10. Is $x - 6$ a factor of $x^4 - 2x^2 + 3x - 12$?

11. Use Descartes' rule of signs to determine the nature of the roots of $5x^4 + 3x^3 - x^2 - 9 = 0$.

12. Find the x intercepts of the graph of the function $f(x) = 3x^3 + 19x^2 - 14x$.

13. Find the equation of the vertical asymptote for the graph of the function $f(x) = \dfrac{5x}{x + 3}$.

14. Find the equation of the horizontal asymptote for the graph of the function $f(x) = \dfrac{5x^2}{x^2 - 4}$.

15. What type of symmetry does the equation $f(x) = \dfrac{x^2}{x^2 + 2}$ exhibit?

16. What type of symmetry does the equation $f(x) = \dfrac{-3x}{x^2 + 1}$ exhibit?

17. Find the equation of the oblique asymptote for the graph of the function $f(x) = \dfrac{4x^2 + x + 1}{x + 1}$.

18. If y varies inversely as x and $y = \dfrac{1}{2}$ when $x = -8$, find the constant of variation.

19. If V varies jointly as h and the square of r, and $V = 396$ when $h = 14$ and $r = 3$, find V when $h = 11$ and $r = 7$.

20. The simple interest earned by a certain amount of money varies jointly as the rate of interest and the time (in years) that the money is invested. If $140 is earned for the money invested at 7% for 5 years, how much is earned if the same amount is invested at 8% for 3 years?

For Problems 21–25, graph each of the functions. Be sure to identify the asymptotes for the rational functions.

21. $f(x) = (2 - x)(x - 1)(x + 1)$

22. $f(x) = \dfrac{-x}{x - 3}$

23. $f(x) = (x + 2)^2(x - 1)$

24. $f(x) = \dfrac{-2}{x^2 - 4}$

25. $f(x) = \dfrac{4x^2 + x + 1}{x + 1}$

For Problems 1–10, evaluate each numerical expression.

1. $\left(\dfrac{3}{4}\right)^{-3}$

2. $\sqrt[3]{-\dfrac{8}{27}}$

3. -5^{-2}

4. $8^{4/3}$

5. $9^{-(3/2)}$

6. $\dfrac{1}{\left(\dfrac{2}{3}\right)^{-3}}$

7. $\left(\dfrac{8}{27}\right)^{-2/3}$

8. -6^{-2}

9. $(-64)^{2/3}$

10. $\sqrt[3]{-\dfrac{64}{27}}$

For Problems 11–33, solve each problem.

11. Express the domain of the function $f(x) = \sqrt{2x^2 + 11x - 6}$ using interval notation.

12. If $f(x) = 3x - 1$ and $g(x) = x^2 - x + 3$, find $(f \circ g)(-2)$ and $(g \circ f)(3)$.

13. If $f(x) = -\dfrac{2}{x}$ and $g(x) = \dfrac{1}{x - 4}$, find $(f \circ g)(x)$ and $(g \circ f)(x)$. Also indicate the domain of each composite function.

14. If $f(x) = -2x + 7$, find the inverse of f.

15. If $f(x) = x^2 + 7x - 2$, find $\dfrac{f(a + h) - f(a)}{h}$.

16. If $f(x) = 2x^4 - 17x^3 - 10x^2 + 11x + 15$, find $f(9)$.

17. Find the quotient for $(3x^5 - 25x^3 - 7x^2 + x + 6) \div (x - 3)$.

18. Is $x + 2$ a factor of $2x^4 + 3x^3 + x^2 + 2x - 16$?

19. Find the remainder when $x^4 + 2x^3 - x^2 + 3x - 4$ is divided by $x - 2$.

20. Find the center and the length of a radius of the circle $x^2 + y^2 + 6x - 4y + 4 = 0$.

21. Write the equation of the line that contains the points $(-4, 2)$ and $(5, -1)$.

22. Write the equation of the perpendicular bisector of the line segment determined by $(-2, -4)$ and $(6, 2)$.

23. Find the length of the major axis of the ellipse $16x^2 + y^2 = 64$.

24. Find the equations of the asymptotes of the hyperbola $x^2 - 9y^2 = 18$.

25. If y varies directly as x and if $y = 3$ when $x = 4$, find y when $x = 16$.

26. If y varies inversely as the square root of x and if $y = \dfrac{2}{5}$ when $x = 25$, find y when $x = 49$.

27. Suppose the number of days it takes to complete a construction job varies inversely as the number of people assigned to the job. If it takes 8 people 10 days to do the job, how long will it take 12 people to complete the job?

28. Find the equation of the oblique asymptote for the graph of the function $f(x) = \dfrac{x^2 + 2x - 1}{x - 1}$.

29. Sandy has a collection of 57 coins worth $10. They consist of nickels, dimes, and quarters, and the number of quarters is 2 more than three times the number of nickels. How many coins of each kind does she have?

30. A retailer bought a dress for $75 and wants to sell it at a profit of 40% of the selling price. What price should she ask for the dress?

31. A container has 8 quarts of a 30% alcohol solution. How much pure alcohol should be added to raise it to a 40% solution?

32. Claire rode her bicycle out into the country at a speed of 15 miles per hour and returned along the same route at 10 miles per hour. If the entire trip took $7\dfrac{1}{2}$ hours, how far out did she ride?

33. Adam can do a job in 2 hours less time than it takes Carl to do the same job. Working together, they can do the job in 2 hours and 24 minutes. How long would it take Adam to do the job by himself?

For Problems 34–45, solve each equation.

34. $(2x - 5)(6x + 1) = (3x + 2)(4x - 7)$

35. $(2x + 1)(x - 2) = (3x - 2)(x + 4)$

36. $4x^3 + 20x^2 - 56x = 0$

37. $6x^3 + 17x^2 + x - 10 = 0$

38. $|4x - 3| = 7$

39. $\dfrac{2x - 1}{3} - \dfrac{3x + 2}{4} = -\dfrac{5}{6}$

40. $(3x - 2)(3x + 4) = 0$

41. $\sqrt{3x + 1} + 2 = 4$

42. $\sqrt{n - 2} - 6 = -3$

43. $x^4 + 3x^2 - 54 = 0$

44. $(2x - 1)(x + 3) = 49$

45. $x^4 - 2x^3 + 2x^2 - 7x + 6 = 0$

For Problems 46–53, solve each inequality and express the solution set using interval notation.

46. $3(x - 1) - 5(x + 2) > 3(x + 4)$

47. $\dfrac{x - 1}{2} + \dfrac{2x + 1}{5} \geq \dfrac{x - 2}{3}$

48. $x^2 - 3x < 18$

49. $(x - 1)(x + 3)(2 - x) \leq 0$

50. $|2x - 1| > 6$

51. $|3x + 2| \leq 8$

52. $\dfrac{4x - 3}{x - 2} \geq 0$

53. $\dfrac{x + 3}{x - 4} < 3$

For Problems 54–64, graph each function.

54. $f(x) = -2x + 4$

55. $f(x) = 2x^2 - 3$

56. $f(x) = 2x^2 + 4x$

57. $f(x) = 2\sqrt{x + 2} - 1$

58. $f(x) = \dfrac{2x}{x + 1}$

59. $f(x) = -|x - 2| + 1$

60. $f(x) = 2\sqrt{x} + 1$

61. $f(x) = 3x^2 + 12x + 9$

62. $f(x) = -(x - 3)^3 + 1$

63. $f(x) = (x + 1)(x - 2)(x - 4)$

64. $f(x) = x^4 - x^2$

5

Exponential and Logarithmic Functions

© Dynamic Graphics / Jupiter Images

Compound interest is a good illustration of exponential growth.

Is it better to invest money at 6% interest compounded quarterly or at 5.75% compounded continuously? Questions of this type can be answered using the concept of *effective annual rate of interest*, sometimes called *effective yield*. We will present this concept in this chapter.

The formula $A = Pe^{rt}$ yields the accumulated value, A, of a sum of money, P, that has been invested for t years at a rate of interest of r% compounded continuously. Using this formula and logarithms, we can determine that it will take approximately 13.7 years for a sum of money to triple in value if it is invested at 8% interest compounded continuously.

Richter numbers are commonly used to report the magnitude of earthquakes. We will define a Richter number and use it in some problem-solving situations in this chapter.

Generally speaking, in this chapter we will continue our study of exponents in several ways: We will (1) extend the meaning of an exponent, (2) work with some exponential functions, (3) introduce the concept of a logarithm, (4) work with some logarithmic functions, and (5) use the concepts of exponent and logarithm to develop more problem-solving skills. Your calculator will be a valuable tool throughout this chapter.

415

5.1 Exponents and Exponential Functions

Objectives

- Review the laws of exponents.
- Solve exponential equations.
- Graph exponential functions.

■ Exponents

In Chapter 0 we defined the expression b^n to mean n factors of b, where n is any positive integer, and b is any real number. For example,

$$2^3 = 2 \cdot 2 \cdot 2 = 8 \qquad\qquad \left(\frac{1}{3}\right)^4 = \left(\frac{1}{3}\right)\left(\frac{1}{3}\right)\left(\frac{1}{3}\right)\left(\frac{1}{3}\right) = \frac{1}{81}$$

$$(-4)^2 = (-4)(-4) = 16 \qquad -(0.5)^3 = -[(0.5)(0.5)(0.5)] = -0.125$$

Also in Chapter 0, by defining $b^0 = 1$ and $b^{-n} = 1/b^n$, where n is any positive integer, and b is any nonzero real number, we extended the concept of an exponent to include all integers. For example,

$$(0.76)^0 = 1 \qquad\qquad 2^{-3} = \frac{1}{2^3} = \frac{1}{8}$$

$$\left(\frac{2}{3}\right)^{-2} = \frac{1}{\left(\frac{2}{3}\right)^2} = \frac{1}{\frac{4}{9}} = \frac{9}{4} \qquad\qquad (0.4)^{-1} = \frac{1}{(0.4)^1} = \frac{1}{0.4} = 2.5$$

Finally, in Chapter 0 we provided for the use of any rational number as an exponent by defining

$$b^{m/n} = \sqrt[n]{b^m} = (\sqrt[n]{b})^m$$

where n is a positive integer greater than 1, and b is a real number such that $\sqrt[n]{b}$ exists. For example,

$$27^{2/3} = (\sqrt[3]{27})^2 = 9 \qquad 16^{1/4} = \sqrt[4]{16^1} = 2$$

$$\left(\frac{1}{9}\right)^{1/2} = \sqrt{\frac{1}{9}} = \frac{1}{3} \qquad 32^{-1/5} = \frac{1}{32^{1/5}} = \frac{1}{\sqrt[5]{32}} = \frac{1}{2}$$

If we were to make a formal extension of the concept of an exponent to include the use of irrational numbers, we would require some ideas from calculus, which is beyond the scope of this text. However, we can give you a brief glimpse at the general idea involved. Consider the number $2^{\sqrt{3}}$. By using the nonterminating and nonrepeating decimal representation $1.73205\ldots$ for $\sqrt{3}$, we can form the sequence of numbers 2^1, $2^{1.7}$, $2^{1.73}$, $2^{1.732}$, $2^{1.7320}$, $2^{1.73205}$, $\ldots$. It is a reasonable idea that each successive power gets closer to $2^{\sqrt{3}}$. This is precisely what happens if n is irrational and b^n is properly defined by using the concept of a *limit*. Furthermore, this process ensures that an expression such as 2^x will yield exactly one value for each value of x.

From now on, we can use any real number as an exponent, and the basic properties stated in Chapter 0 can be extended to include all real numbers as exponents. Let's restate those properties with the restriction that the bases a and b are to be positive numbers to avoid expressions such as $(-4)^{1/2}$, which do not represent real numbers.

Property 5.1

If a and b are positive real numbers and m and n are any real numbers, then the following properties hold:

1. $b^n \cdot b^m = b^{n+m}$ Product of two powers

2. $(b^n)^m = b^{mn}$ Power of a power

3. $(ab)^n = a^n b^n$ Power of a product

4. $\left(\dfrac{a}{b}\right)^n = \dfrac{a^n}{b^n}$ Power of a quotient

5. $\dfrac{b^n}{b^m} = b^{n-m}$ Quotient of two powers

Another property that can be used to solve certain types of equations involving exponents can be stated as follows.

Property 5.2

If $b > 0$ but $b \neq 1$, and if m and n are real numbers, then

$$b^n = b^m \quad \text{if and only if } n = m$$

The following examples illustrate the use of Property 5.2.

EXAMPLE 1

Solve $2^x = 32$.

Solution

$$2^x = 32$$
$$2^x = 2^5 \qquad 32 = 2^5$$
$$x = 5 \qquad \text{Apply Property 5.2}$$

The solution set is $\{5\}$.

▶ **Now work Problem 1.** ∎

EXAMPLE 2

Solve $2^{3x} = \dfrac{1}{64}$.

Solution

$$2^{3x} = \frac{1}{64}$$

$$2^{3x} = \frac{1}{2^6}$$

$$2^{3x} = 2^{-6}$$

$$3x = -6 \qquad \text{Apply Property 5.2}$$

$$x = -2$$

The solution set is $\{-2\}$.

▶ **Now work Problem 5.** ■

EXAMPLE 3

Solve $\left(\dfrac{1}{5}\right)^{x-4} = \dfrac{1}{125}$.

Solution

$$\left(\frac{1}{5}\right)^{x-4} = \frac{1}{125}$$

$$\left(\frac{1}{5}\right)^{x-4} = \left(\frac{1}{5}\right)^3$$

$$x - 4 = 3 \qquad \text{Apply Property 5.2}$$

$$x = 7$$

The solution set is $\{7\}$.

▶ **Now work Problem 13.** ■

EXAMPLE 4

Solve $9^x = 243$.

Solution

$$9^x = 243$$

$$(3^2)^x = 3^5$$

$$3^{2x} = 3^5$$

$$2x = 5 \qquad \text{Apply Property 5.2}$$

$$x = \frac{5}{2}$$

The solution set is $\left\{\dfrac{5}{2}\right\}$.

▶ **Now work Problem 19.** ■

E X A M P L E   5   Solve $(8^{2x})(4^{2x-1}) = 16$.

Solution

$$(8^{2x})(4^{2x-1}) = 16$$
$$(2^3)^{2x}(2^2)^{2x-1} = 2^4$$
$$(2^{6x})(2^{4x-2}) = 2^4$$
$$2^{6x+4x-2} = 2^4$$
$$2^{10x-2} = 2^4$$
$$10x - 2 = 4 \qquad \text{Apply Property 5.2}$$
$$10x = 6$$
$$x = \frac{6}{10}$$
$$x = \frac{3}{5}$$

The solution set is $\left\{\dfrac{3}{5}\right\}$.

▶ **Now work Problem 23.** ■

■ Exponential Functions

If b is any positive number, then the expression b^x designates exactly one real number for every real value of x. Therefore the equation $f(x) = b^x$ defines a function whose domain is the set of real numbers. Furthermore, if we add the restriction $b \neq 1$, then any equation of the form $f(x) = b^x$ describes what we will call later a one-to-one function and is called an **exponential function.** This leads to the following definition:

Definition 5.1

If $b > 0$ and $b \neq 1$, then the function f defined by
$$f(x) = b^x$$
where x is any real number, is called the **exponential function with base b.**

Remark: The function $f(x) = 1^x$ is a constant function (its graph is a horizontal line), and therefore it is not an exponential function.

Now let's graph some exponential functions.

EXAMPLE 6   Graph the function $f(x) = 2^x$.

Solution

Let's set up a table of values. Keep in mind that the domain is the set of real numbers and the equation $f(x) = 2^x$ exhibits no symmetry. We can plot the points and connect them with a smooth curve to produce Figure 5.1.

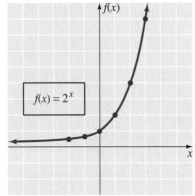

x	2^x
-2	$\frac{1}{4}$
-1	$\frac{1}{2}$
0	1
1	2
2	4
3	8

$f(x) = 2^x$

Figure 5.1

▶ **Now work Problem 27.**   ■

In the table for Example 6 we chose integral values for x to keep the computation simple. However, with a calculator, we could easily acquire functional values by using nonintegral exponents. Consider the following additional values for $f(x) = 2^x$:

$$f(0.5) \approx 1.41 \qquad f(1.7) \approx 3.25 \qquad \approx \text{ means "is approximately equal to"}$$

$$f(-0.5) \approx 0.71 \qquad f(-2.6) \approx 0.16$$

Use your calculator to check these results. Also note that the points generated by these values fit the graph in Figure 5.1.

EXAMPLE 7   Graph $f(x) = \left(\dfrac{1}{2}\right)^x$.

Solution

Again, let's set up a table of values, plot the points, and connect them with a smooth curve. The graph is shown in Figure 5.2.

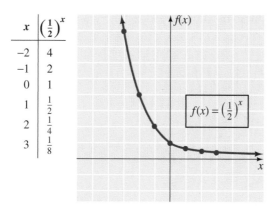

x	$\left(\frac{1}{2}\right)^x$
-2	4
-1	2
0	1
1	$\frac{1}{2}$
2	$\frac{1}{4}$
3	$\frac{1}{8}$

$$f(x) = \left(\tfrac{1}{2}\right)^x$$

Figure 5.2

▶ **Now work Problem 31.** ∎

Remark: Because $\left(\dfrac{1}{2}\right)^x = 1/2^x = 2^{-x}$, the graphs of $f(x) = 2^x$ and $f(x) = \left(\dfrac{1}{2}\right)^x$ are reflections of each other across the y axis. Therefore Figure 5.2 could have been drawn by reflecting Figure 5.1 across the y axis.

The graphs in Figures 5.1 and 5.2 illustrate a general behavior pattern of exponential functions. That is, if $b > 1$, then the graph of $f(x) = b^x$ goes up to the right, and the function is called an **increasing function.** If $0 < b < 1$, then the graph of $f(x) = b^x$ goes down to the right, and the function is called a **decreasing function.** These facts are illustrated in Figure 5.3. Notice that $b^0 = 1$ for any $b > 0$; thus, **all graphs of $f(x) = b^x$ contain the point (0, 1).** Note that the x axis is a horizontal asymptote of the graphs of $f(x) = b^x$.

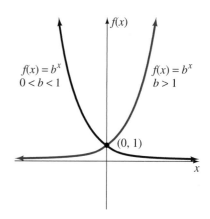

Figure 5.3

As you graph exponential functions, don't forget to use your previous graphing experience. For example, consider the following functions.

1. The graph of $f(x) = 2^x + 3$ is the graph of $f(x) = 2^x$ *moved up three units.* The line $f(x) = 3$ is the horizontal asymptote.
2. The graph of $f(x) = 2^{x-4}$ is the graph of $f(x) = 2^x$ *moved to the right four units.*
3. The graph of $f(x) = -2^x$ is the graph of $f(x) = 2^x$ *reflected across the x axis.*
4. The graph of $f(x) = 2^x + 2^{-x}$ is symmetric with respect to the y axis because $f(-x) = 2^{-x} + 2^x = f(x)$.

The graphs of these functions are shown in Figure 5.4.

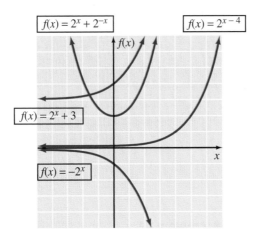

Figure 5.4

If you are faced with an exponential function that is not of the form $f(x) = b^x$ or a variation of it, don't forget the graphing suggestions offered in Chapter 2. Let's consider one such example.

EXAMPLE 8

Graph $f(x) = 2^{-x^2}$.

Solution

Because $f(-x) = 2^{-(-x)^2} = 2^{-x^2} = f(x)$, we know that this curve is symmetric with respect to the y axis. Therefore let's set up a table of values using nonnegative values for x. We can plot these points, connect them with a smooth curve, and reflect this portion of the curve across the y axis to produce the graph in Figure 5.5.

x	2^{-x^2}
0	1
$\frac{1}{2}$	0.84
1	0.5
$\frac{3}{2}$	0.21
2	0.06

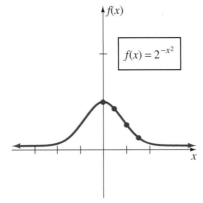

$f(x) = 2^{-x^2}$

Figure 5.5

▶ **Now work Problem 35.**

EXAMPLE   9

Use a graphing utility to obtain a graph of $f(x) = 50(2^x)$, and find an approximate value for x when $f(x) = 15,000$.

Solution

First, we must find an appropriate viewing rectangle. Because $50(2^{10}) = 51,200$, let's set the boundaries so that $0 \le x \le 10$ and $0 \le y \le 50,000$ with a scale of 10,000 on the y axis. (Certainly other boundaries could be used, but these will give us a graph that we can work with for this problem.) The graph of $f(x) = 50(2^x)$ is shown in Figure 5.6. Now we can use the trace and zoom-in features of the graphing utility to find that $x \approx 8.2$ at $y = 15,000$.

50,000

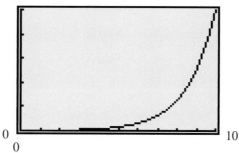

0   10
0

Figure 5.6   ■

Remark: In Example 9 we used a graphical approach to solve the equation $50(2^x) = 15,000$. In Section 5.4 we will use an algebraic approach to solve that same kind of equation.

CONCEPT QUIZ   For Problems 1–8, answer true or false.

1. $2^2 \cdot 2^3 = 4^5$.

2. $2^2 \cdot 2^3 = 2^6$.

3. $2^{-3} = -8$.

4. $-2^{-3} = 8$.

5. $5^{\frac{3}{4}} = \sqrt[4]{5^3}$.

6. The function $f(x) = 1^x$ is an exponential function.

7. The exponential function $f(x) = b^x$ where $b > 1$ is an increasing function.

8. All graphs of $f(x) = b^x$ where $b > 0$ and $b \ne 1$ contain the point $(0, 1)$.

Problem Set 5.1

For Problems 1–26, solve each equation.

1. $3^x = 27$

2. $2^x = 64$

3. $\left(\dfrac{1}{2}\right)^x = \dfrac{1}{8}$

4. $\left(\dfrac{1}{2}\right)^n = 4$

5. $3^{-x} = \dfrac{1}{81}$

6. $3^{x+1} = 9$

7. $5^{2n-1} = 125$

8. $2^{3-n} = 8$

9. $9^{2n+1} = 27^n$

10. $4^n = 32^{n-1}$

11. $\left(\dfrac{1}{2}\right)^x = 8$

12. $\left(\dfrac{2}{5}\right)^x = \dfrac{25}{4}$

13. $\left(\dfrac{2}{3}\right)^t = \dfrac{9}{4}$

14. $\left(\dfrac{3}{4}\right)^n = \dfrac{64}{27}$

15. $4^{3x-1} = 256$

16. $16^x = 64$

17. $4^n = 8$

18. $27^{4x} = 9^{x+1}$

19. $4^{2x+3} = 8^{x-1}$

20. $3^{n+1} = \left(\dfrac{1}{9}\right)^{2n-3}$

21. $32^x = 16^{1-x}$

22. $\left(\dfrac{1}{8}\right)^{-2t} = 2^{t+3}$

23. $(2^{2x-1})(2^{x+2}) = 32$

24. $(27)(3^x) = 9^x$

25. $(3^x)(3^{5x}) = 81$

26. $(4^x)(16^{3x-1}) = 8$

For Problems 27–52, graph each function.

27. $f(x) = 3^x$

28. $f(x) = \left(\dfrac{1}{3}\right)^x$

29. $f(x) = 4^x$

30. $f(x) = \left(\dfrac{1}{4}\right)^x$

31. $f(x) = \left(\dfrac{2}{3}\right)^x$

32. $f(x) = \left(\dfrac{3}{2}\right)^x$

33. $f(x) = 2^x + 1$

34. $f(x) = 2^x - 3$

35. $f(x) = 2^{x-1}$

36. $f(x) = 2^{x+2}$

37. $f(x) = 3^x - 2$

38. $f(x) = 3^x + 1$

39. $f(x) = 3^{x+1}$

40. $f(x) = 3^{x-3}$

41. $f(x) = -3^x$

42. $f(x) = -2^x$

43. $f(x) = 2^{-x+1}$

44. $f(x) = 2^{-x-2}$

45. $f(x) = 3 \cdot 2^x$

46. $f(x) = \dfrac{1}{3} \cdot 2^x$

47. $f(x) = 2^x + 2^{-x}$

48. $f(x) = 2^{x^2}$

49. $f(x) = 3^{1-x^2}$

50. $f(x) = 2^{|x|}$

51. $f(x) = 2^{-|x|}$

52. $f(x) = 2^x - 2^{-x}$

53. Graph $f(x) = -\left(\dfrac{1}{2}\right)^x$. Then, on the same set of axes,

graph $f(x) = -\left(\dfrac{1}{2}\right)^x + 2$, $f(x) = -\left(\dfrac{1}{2}\right)^{x+3}$,

and $f(x) = -\left(\dfrac{1}{2}\right)^{-x}$.

■ ■ ■ THOUGHTS INTO WORDS

54. Why is the base of an exponential function restricted to positive numbers not including 1?

55. How would you go about graphing the function

$f(x) = -\left(\dfrac{1}{3}\right)^x$?

56. Explain how you would solve the equation

$$(4^{x-1})(8^{2x+3}) = 128$$

GRAPHING CALCULATOR ACTIVITIES

57. Use your graphing calculator to check your graphs for Problems 36–40.

58. Graph $f(x) = 4^x$. Where should the graphs of $f(x) = 4^{x-2}$, $f(x) = 4^{x-4}$, and $f(x) = 4^{x+3}$ be located? Graph all three functions on the same set of axes with $f(x) = 4^x$.

59. Graph $f(x) = \left(\dfrac{1}{4}\right)^x$. Where should the graphs of $f(x) = \left(\dfrac{1}{4}\right)^x - 2$, $f(x) = \left(\dfrac{1}{4}\right)^x + 3$, and $f(x) = \left(\dfrac{1}{4}\right)^x - 4$ be located? Graph all three functions on the same set of axes with $f(x) = \left(\dfrac{1}{4}\right)^x$.

60. Graph $f(x) = \left(\dfrac{3}{4}\right)^x$. Now predict the graphs for $f(x) = -\left(\dfrac{3}{4}\right)^x$, $f(x) = \left(\dfrac{3}{4}\right)^{-x}$, and $f(x) = -\left(\dfrac{3}{4}\right)^{-x}$.

Graph all three functions on the same set of axes with $f(x) = \left(\dfrac{3}{4}\right)^x$.

61. Graph $f(x) = (-2)^x$. Explain your result.

62. What is the solution for $3^x = 5$? Do you agree that it is between 1 and 2 because $3^1 = 3$ and $3^2 = 9$? Now graph $f(x) = 3^x - 5$ and use the zoom-in and trace features of your graphing calculator to find an approximation, to the nearest hundredth, for the x intercept. You should get the answer 1.46. This should be an approximation for the solution for $3^x = 5$. Try it; raise 3 to the 1.46 power.

Find an approximate solution, to the nearest hundredth, for each of the following equations by graphing the appropriate function and finding the x intercept.

a. $2^x = 19$ **b.** $3^x = 50$
c. $4^x = 47$ **d.** $5^x = 120$
e. $2^x = 1500$ **f.** $3^{x-1} = 34$

Answers to the Concept Quiz

1. False   **2.** False   **3.** False   **4.** False   **5.** True   **6.** False   **7.** True   **8.** True

---

5.2   Applications of Exponential Functions

Objectives

■ Solve compound interest problems.

■ Solve exponential growth and decay problems.

Many real-world situations that exhibit growth or decay can be represented by equations that describe exponential functions. For example, suppose an economist predicts an annual inflation rate of 5% per year for the next 10 years. This means that an item that presently costs $8 will cost $8(105%) = $8(1.05) = $8.40 a year from now. The same item will cost [$8(105%)] × (105%) = $8(1.05)² = $8.82 in 2 years. In general, the equation

$$P = P_0(1.05)^t$$

yields the predicted price P of an item in t years if the present cost is P_0 and the annual inflation rate is 5%. By using this equation, we can look at some future prices based on the prediction of a 5% inflation rate.

A $1.29 jar of mustard will cost 1.29(1.05)^3 = 1.49 in 3 years.

A $3.29 bag of potato chips will cost 3.29(1.05)^5 = 4.20 in 5 years.

A $8.69 can of coffee will cost 8.69(1.05)^7 = 12.23 in 7 years.

■ Compound Interest

Compound interest provides another illustration of exponential growth. Suppose that $500 (called the **principal**) is invested at an interest rate of 8% **compounded annually.** The interest earned the first year is $500(0.08) = $40, and this amount is added to the original $500 to form a new principal of $540 for the second year. The interest earned during the second year is $540(0.08) = $43.20, and this amount is added to $540 to form a new principal of $583.20 for the third year. Each year a new principal is formed by reinvesting the interest earned during that year.

In general, suppose that a sum of money P (called the principal) is invested at an interest rate of r% compounded annually. The interest earned the first year is Pr, and the new principal for the second year is $P + Pr$ or $P(1 + r)$. Note that the new principal for the second year can be found by multiplying the original principal P by $(1 + r)$. In like fashion, we can find the new principal for the third year by multiplying the previous principal, $P(1 + r)$, by $1 + r$, thus obtaining $P(1 + r)^2$. If this process is continued, then after t years the total amount of money accumulated, A, is given by

$$A = P(1 + r)^t$$

Consider the following examples of investments made at a certain rate of interest compounded annually.

1. $750 invested for 5 years at 4% compounded annually produces

$$A = \$750(1.04)^5 = \$912.49$$

2. $1000 invested for 10 years at 6% compounded annually produces

$$A = \$1000(1.06)^{10} = \$1790.85$$

3. $5000 invested for 20 years at 9% compounded annually produces

$$A = \$5000(1.09)^{20} = \$28,022.05$$

The compound interest formula can be used to determine what rate of interest is needed to accumulate a certain amount of money based on a given initial investment. The next problem illustrates this idea.

PROBLEM 1

What rate of interest is needed for an investment of $1000 to yield $3000 in 10 years if the interest is compounded annually?

Solution

Let's substitute $1000 for P, $3000 for A, and 10 years for t in the compound interest formula and solve for r:

$$A = P(1 + r)^t$$
$$3000 = 1000(1 + r)^{10}$$
$$3 = (1 + r)^{10}$$
$$3^{0.1} = [(1 + r)^{10}]^{0.1} \qquad \text{Raise both sides to the 0.1 power}$$
$$1.116123174 \approx 1 + r$$
$$0.116123174 \approx r$$
$$r = 11.6\% \quad \text{to the nearest tenth of a percent}$$

Therefore a rate of interest of approximately 11.6% is needed. (Perhaps you should check this answer.)

▶ **Now work Problem 28.**   ■

If money invested at a certain rate of interest is to be compounded more than once a year, then the basic formula $A = P(1 + r)^t$ can be adjusted according to the number of compounding periods in a year. For example, for **compounding semiannually,** the formula becomes

$$A = P\left(1 + \frac{r}{2}\right)^{2t}$$

and for **compounding quarterly,** the formula becomes

$$A = P\left(1 + \frac{r}{4}\right)^{4t}$$

In general, if n represents the number of **compounding periods** in a year, then the formula becomes

$$A = P\left(1 + \frac{r}{n}\right)^{nt}$$

The following examples illustrate the use of the formula.

1. $750 invested for 5 years at 4% compounded semiannually produces

$$A = \$750\left(1 + \frac{0.04}{2}\right)^{2(5)} = \$750(1.02)^{10} = \$914.25$$

2. $1000 invested for 10 years at 6% compounded quarterly produces

$$A = \$1000\left(1 + \frac{0.06}{4}\right)^{4(10)} = \$1000(1.015)^{40} = \$1814.02$$

3. $5000 invested for 20 years at 9% compounded monthly produces

$$A = \$5000\left(1 + \frac{0.09}{12}\right)^{12(20)} = \$5000(1.0075)^{240} = \$30,045.76$$

You may find it interesting to compare these results with those obtained earlier for compounding annually.

■ Exponential Decay

Suppose it is estimated that the value of a car depreciates 15% per year for the first 5 years. Therefore a car that costs $19,000 will be worth $19,000 (100% – 15%) = $19,000 (85%) = $19,000 (0.85) = $16,150 in 1 year. In 2 years the value of the car will have depreciated to $19,000(0.85)^2 = \$13,727.50$. The equation

$$V = V_0(0.85)^t$$

yields the value V of a car in t years if the initial cost is V_0 and the car's value depreciates 15% per year. Therefore we can estimate some car values to the nearest dollar.

A \$17,000 car will be worth $\$17,000(0.85)^3 = \$10,440$ in 3 years.
A \$25,000 car will be worth $\$25,000(0.85)^5 = \$11,093$ in 5 years.
A \$40,000 car will be worth $\$40,000(0.85)^4 = \$20,880$ in 4 years.

Another example of exponential decay involves radioactive substances. The rate of decay can be described exponentially and is based on the half-life of a substance. The **half-life** of a radioactive substance is the amount of time that it takes for one-half of an initial amount of the substance to disappear as the result of decay. For example, suppose we have 200 grams of a certain substance that has a half-life of 5 days. After 5 days, $200\left(\frac{1}{2}\right) = 100$ grams remain. After 10 days, $200\left(\frac{1}{2}\right)^2 = 50$ grams remain. After 15 days, $200\left(\frac{1}{2}\right)^3 = 25$ grams remain. In general, after t days, $200\left(\frac{1}{2}\right)^{t/5}$ grams remain.

The previous discussion leads into the following half-life formula. Suppose there is an initial amount, Q_0, of a radioactive substance with a half-life of h. The amount of substance remaining, Q, after a time period of t is given by the formula

$$Q = Q_0\left(\frac{1}{2}\right)^{t/h}$$

The units of measure for t and h must be the same.

PROBLEM 2

Barium-140 has a half-life of 13 days. If there are 500 milligrams of barium initially, how many milligrams remain after 26 days? After 100 days?

Solution

With $Q_0 = 500$ and $h = 13$, the half-life formula becomes

$$Q = 500\left(\frac{1}{2}\right)^{t/13}$$

If $t = 26$, then

$$Q = 500\left(\frac{1}{2}\right)^{26/13}$$

$$= 500\left(\frac{1}{2}\right)^{2}$$

$$= 500\left(\frac{1}{4}\right)$$

$$= 125$$

Thus 125 milligrams remain after 26 days. If $t = 100$, then

$$Q = 500\left(\frac{1}{2}\right)^{100/13}$$

$$= 500(0.5)^{100/13}$$

$$= 2.4 \quad \text{to the nearest tenth of a milligram}$$

Approximately 2.4 milligrams remain after 100 days.

▶ **Now work Problem 35.**   ■

Remark: The solution to Problem 2 clearly demonstrates the useful role of the calculator in the application of mathematics. We solved the first part of the problem easily without the calculator, but the calculator certainly was helpful for the second part of the problem.

■ Number *e*

An interesting situation occurs if we consider the compound interest formula for $P = \$1$, $r = 100\%$, and $t = 1$ year. The formula becomes $A = 1\left(1 + \dfrac{1}{n}\right)^{n}$. The following table shows some values, rounded to eight decimal places, of $\left(1 + \dfrac{1}{n}\right)^{n}$ for different values of n.

n	$\left(1 + \dfrac{1}{n}\right)^n$
1	2.00000000
10	2.59374246
100	2.70481383
1000	2.71692393
10,000	2.71814593
100,000	2.71826824
1,000,000	2.71828047
10,000,000	2.71828169
100,000,000	2.71828181
1,000,000,000	2.71828183

The table suggests that as n increases, the value of $\left(1 + \dfrac{1}{n}\right)^n$ gets closer and closer to some fixed number. This does happen, and the fixed number is called e. To five decimal places, $e = 2.71828$.

The function defined by the equation $f(x) = e^x$ is the **natural exponential function.** It has a great many real-world applications, some of which we will look at in a moment. First, however, let's get a picture of the natural exponential function. Because $2 < e < 3$, the graph of $f(x) = e^x$ must fall between the graphs of $f(x) = 2^x$ and $f(x) = 3^x$. To be more specific, let's use our calculator to determine a table of values. We use the $\boxed{e^x}$ key and round the results to the nearest tenth to obtain the table. Plotting the points determined by this table and connecting them with a smooth curve produces Figure 5.7. Again, note that the x axis is a horizontal asymptote.

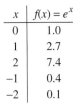

x	$f(x) = e^x$
0	1.0
1	2.7
2	7.4
−1	0.4
−2	0.1

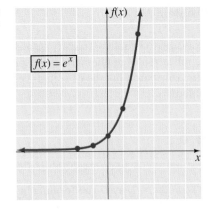

Figure 5.7

■ Back to Compound Interest

Let's return to the concept of compound interest. If the number of compounding periods in a year is increased indefinitely, we arrive at the concept of **compounding continuously.** Mathematically, we can do this by applying the limit concept to the expression

$$P\left(1 + \frac{r}{n}\right)^{nt}$$

We will not show the details here, but the following result is obtained. The formula

$$A = Pe^{rt}$$

yields the accumulated value, A, of a sum of money, P, that has been invested for t years at a rate of $r\%$ compounded continuously. The following examples illustrate the use of this formula.

1. $750 invested for 5 years at 4% compounded continuously produces

$$A = 750e^{(0.04)(5)} = 750e^{0.2} = \$916.05$$

2. $1000 invested for 10 years at 6% compounded continuously produces

$$A = 1000e^{(0.06)(10)} = 1000e^{0.6} = \$1822.12$$

3. $5000 invested for 20 years at 9% compounded continuously produces

$$A = 5000e^{(0.09)(20)} = 5000e^{1.8} = \$30{,}248.24$$

Again, you may find it interesting to compare these results with those we obtained earlier using a different number of compounding periods.

Is it better to invest at 6% interest compounded quarterly or at 5.75% compounded continuously? To answer such a question, we can use the concept of **effective yield** (sometimes called *effective annual rate of interest*). The effective yield of an investment is the simple interest rate that would yield the same amount in 1 year. Thus, for the *6% compounded quarterly* investment, we can calculate the effective yield as follows:

$$P(1 + r) = P\left(1 + \frac{0.06}{4}\right)^4$$

$$1 + r = \left(1 + \frac{0.06}{4}\right)^4 \qquad \text{Multiply both sides by } \frac{1}{P}$$

$$1 + r = (1.015)^4$$

$$r = (1.015)^4 - 1$$

$$r \approx 0.0613635506$$

$$r = 6.14\% \quad \text{to the nearest hundredth of a percent}$$

Likewise, for the 5.75% *compounded continuously* investment we can calculate the effective yield as follows:

$$P(1 + r) = Pe^{0.0575}$$
$$1 + r = e^{0.0575}$$
$$r = e^{0.0575} - 1$$
$$r \approx 0.0591852707$$
$$r = 5.92\% \qquad \text{to the nearest hundredth of a percent}$$

Therefore, comparing the two effective yields, we see that it is better to invest at 6% compounded quarterly than to invest at 5.75% compounded continuously.

■ Law of Exponential Growth

The ideas behind compounded continuously carry over to other growth situations. The law of exponential growth,

$$Q(t) = Q_0 e^{kt}$$

is used as a mathematical model for numerous growth-and-decay applications. In this equation, $Q(t)$ represents the quantity of a given substance at any time t; Q_0 is the initial amount of the substance (when $t = 0$), and k is a constant that depends on the particular application. If $k < 0$, then $Q(t)$ decreases as t increases, and we refer to the model as the **law of decay**.

Let's consider some growth-and-decay applications.

PROBLEM 3

Suppose that in a certain culture, the equation $Q(t) = 15{,}000e^{0.3t}$ expresses the number of bacteria present as a function of the time t, where t is expressed in hours. Find (a) the initial number of bacteria, and (b) the number of bacteria after 3 hours.

Solutions

a. The initial number of bacteria is produced when $t = 0$.

$$Q(0) = 15{,}000e^{0.3(0)}$$
$$= 15{,}000e^{0}$$
$$= 15{,}000 \qquad e^0 = 1$$

b. $Q(3) = 15{,}000e^{0.3(3)}$
$$= 15{,}000e^{0.9}$$
$$= 36{,}894 \qquad \text{to the nearest whole number}$$

There should be approximately 36,894 bacteria present after 3 hours.

▶ **Now work Problem 37.**   ■

PROBLEM 4

Suppose the number of bacteria present in a certain culture after t minutes is given by the equation $Q(t) = Q_0 e^{0.05t}$, where Q_0 represents the initial number of bacteria. If 5000 bacteria are present after 20 minutes, how many bacteria were present initially?

Solution

If 5000 bacteria are present after 20 minutes, then $Q(20) = 5000$.

$$5000 = Q_0 e^{0.05(20)}$$
$$5000 = Q_0 e^1$$
$$\frac{5000}{e} = Q_0$$
$$1839 = Q_0 \quad \text{to the nearest whole number}$$

Thus approximately 1839 bacteria were present initially.

▶ **Now work Problem 39.** ∎

PROBLEM 5

The number of grams of a certain radioactive substance present after t seconds is given by the equation $Q(t) = 200e^{-0.3t}$. How many grams remain after 7 seconds?

Solution

Use $Q(t) = 200e^{-0.3t}$ to obtain

$$Q(7) = 200e^{(-0.3)(7)}$$
$$= 200e^{-2.1} = 24.5 \quad \text{to the nearest tenth}$$

Thus approximately 24.5 grams remain after 7 seconds.

▶ **Now work Problem 40.** ∎

Finally, let's use the graphical approach to solve two problems.

PROBLEM 6

Suppose that $1000 is invested at 6% interest compounded quarterly. How long will it take for the money to double?

Solution

Substitute $1000 for P, 0.06 for r, and 4 for n in the formula $A = P\left(1 + \dfrac{r}{n}\right)^{nt}$ to produce $A = 1000\left(1 + \dfrac{0.06}{4}\right)^{4t}$. If we let $y = A$ and $x = t$, we can graph the equation $y = 1000(1.015)^{4x}$. By letting $x = 20$, we obtain $y = 1000(1.015)^{4(20)} = 1000(1.015)^{80} \approx 3291$. Therefore let's set the boundaries of the viewing rectangle so that $0 \le x \le 20$ and $0 \le y \le 3300$ with a y scale of 1000. Then we obtain Figure 5.8. Now we want to find the value of x so that $y = 2000$. (The money is to double.)

Using the zoom-in and trace features of the graphing utility, we can determine that an x value of approximately 11.6 will produce a y value of 2000. Thus it will take approximately 11.6 years for the $1000 investment to double.

3300

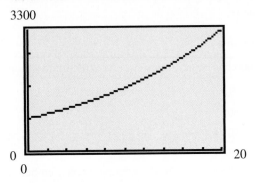

0

0   20

Figure 5.8

▷ **Now work Problem 69.** ■

E X A M P L E   1

Graph the function $y = \dfrac{1}{\sqrt{2\pi}} e^{-x^2/2}$ and find its maximum value.

Solution

If $x = 0$, then

$$y = \frac{1}{\sqrt{2\pi}} e^0 = \frac{1}{\sqrt{2\pi}} \approx 0.4$$

Let's set the boundaries of the viewing rectangle so that $-5 \le x \le 5$ and $0 \le y \le 1$ with a y scale of 0.1; the graph of the function is shown in Figure 5.9. From the graph, we see that the maximum value of the function occurs at $x = 0$, which we have already determined to be approximately 0.4.

1

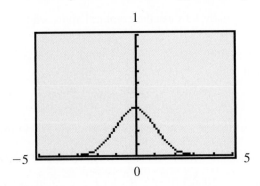

−5   5

0

Figure 5.9

▷ **Now work Problem 58 using a graphing utility.** ■

Remark: The curve in Figure 5.9 is called the **normal distribution curve.** You may want to ask your instructor to explain what it means to assign grades on the basis of the normal distribution curve.

For Problems 1–6, answer true or false.

1. In the formula for compound interest $A = P(1 + r)^t$, the P represents the amount of money invested and is called the principal.

2. In the formula for compound interest $A = P\left(1 + \dfrac{r}{n}\right)^{nt}$, the n represents the number of compounding periods in year.

3. In an exponential problem using the formula $Q = Q_0 b^x$ the problem will be about decay or depreciation if $0 < b < 1$.

4. The natural exponential function $f(x) = e^x$ is a decreasing function.

5. If 400 grams of a radioactive substance has a half-life of 12 days, then 100 grams remain after 18 days.

6. If an amount P is invested at 5% compounded quarterly, then the effective yield will be greater than 5%.

Problem Set 5.2

1. Assuming that the rate of inflation is 4% per year, the equation $P = P_0(1.04)^t$ yields the predicted price P of an item in t years if it presently costs P_0. Find the predicted price of each of the following items for the indicated years ahead.

    a. $1.29 pack of chewing gum in 3 years

    b. $4.29 chicken sandwich meal in 5 years

    c. $2.29 gallon of gasoline in 4 years

    d. $1.59 soft drink in 10 years

    e. $22,000 car in 5 years (nearest dollar)

    f. $210,000 house in 8 years

    g. $59.00 pair of jeans in 7 years

2. Suppose it is estimated that the value of a new car declines, or the car depreciates, 20% per year for the first 5 years. The equation $A = P_0(0.8)^t$ yields the value A of a new car after t years if the original price is P_0. Find the value (to the nearest dollar) of each of the following new cars after the indicated time.

    a. $35,000 car after 4 years

    b. $24,000 car after 2 years

    c. $18,000 car after 5 years

    d. $25,000 car after 3 years

For Problems 3–14, use the formula

$$A = P\left(1 + \frac{r}{n}\right)^{nt}$$

to find the total amount of money accumulated at the end of the indicated time period for each of the following investments. Estimate to the nearest cent.

3. $250 for 5 years at 6% compounded annually

4. $350 for 7 years at 5% compounded annually

5. $300 for 6 years at 8% compounded semiannually

6. $450 for 10 years at 4% compounded semiannually

7. $600 for 12 years at 7% compounded quarterly

8. $750 for 15 years at 9% compounded quarterly

9. $1000 for 5 years at 4% compounded monthly

10. $1250 for 8 years at 7% compounded monthly

11. $600 for 10 years at $8\frac{1}{2}$% compounded annually

12. $1500 for 15 years at $6\frac{1}{4}$% compounded semiannually

13. $8000 for 10 years at 4.5% compounded quarterly

14. $10,000 for 25 years at 5.25% compounded monthly

For Problems 15–22, use the formula $A = Pe^{rt}$ to find the total amount of money accumulated at the end of the indicated time period by continuous compounding.

15. $400 for 5 years at 7%

16. $500 for 7 years at 6%

17. $750 for 8 years at 8%

18. $1000 for 10 years at 9%

19. $2000 for 15 years at 4%

20. $5000 for 20 years at 7%

21. $7500 for 10 years at 5.5%

22. $10,000 for 25 years at 9.25%

For Problems 23–43, solve the problem.

23. Rueben has a finance plan with the furniture store where the $4830 he spent accrues finance charges at an annual interest rate of 10.9% compounded monthly for 3 years before he starts to make payments. What will be the balance on the account at the end of those three years?

24. In a certain balloon mortgage loan, the borrower pays the lender all of the principal and interest for the loan at the end of the five years. What will be the payoff amount for a loan of $185,000 at 6% annual interest rate where the interest is compounded monthly?

25. Jody took out a $3200 student loan her freshman year of college. The loan was at a 2.5% annual interest rate and accrued interest quarterly. Jody is obligated to begin repaying the loan back in five years. At that time what will be the amount she needs to repay?

26. To pay the tution for medical school Melissa borrowed $8400 in student loans her first year. The loan is for seven years at an annual interest rate of 3.4%, and interest is compounded semiannually. What will be the amount of principal and interest in seven years?

27. Mark became overextended in his gambling debt and could not pay $500 he owed. The loan person said he could have three weeks to pay off the $500 at 10% interest per week compounded continuously. How much will Mark have to pay at the end of the three weeks?

28. What rate of interest, to the nearest tenth of a percent, compounded annually is needed for an investment of $200 to grow to $350 in 5 years?

29. What rate of interest, to the nearest tenth of a percent, compounded quarterly is needed for an investment of $1500 to grow to $2700 in 10 years?

30. Find the effective yield, to the nearest tenth of a percent, of an investment at 7.5% compounded monthly.

31. Find the effective yield, to the nearest hundredth of a percent, of an investment at 7.75% compounded continuously.

32. Which investment yields the greater return: 7% compounded monthly or 6.85% compounded continuously?

33. Which investment yields the greater return: 8.25% compounded quarterly or 8.3% compounded semiannually?

34. Suppose that a certain radioactive substance has a half-life of 20 years. If there are presently 2500 milligrams of the substance, how much, to the nearest milligram, will remain after 40 years? After 50 years?

35. Strontium-90 has a half-life of 29 years. If there are 400 grams of strontium initially, how much, to the nearest gram, will remain after 87 years? After 100 years?

36. The half-life of radium is approximately 1600 years. If the present amount of radium in a certain location is 500 grams, how much will remain after 800 years? Express your answer to the nearest gram.

37. Suppose that in a certain culture, the equation $Q(t) = 1000e^{0.4t}$ expresses the number of bacteria present as a function of the time t, where t is expressed in hours. How many bacteria are present at the end of 2 hours? 3 hours? 5 hours?

38. The number of bacteria present at a given time under certain conditions is given by the equation $Q = 5000e^{0.05t}$, where t is expressed in minutes. How many bacteria are present at the end of 10 minutes? 30 minutes? 1 hour?

39. The number of bacteria present in a certain culture after t hours is given by the equation $Q = Q_0e^{0.3t}$, where

Q_0 represents the initial number of bacteria. If 6640 bacteria are present after 4 hours, how many bacteria were present initially?

40. The number of grams Q of a certain radioactive substance present after t seconds is given by the equation $Q = 1500e^{-0.4t}$. How many grams remain after 5 seconds? 10 seconds? 20 seconds?

41. The atmospheric pressure, measured in pounds per square inch, is a function of the altitude above sea level. The equation $P(a) = 14.7e^{-0.21a}$, where a is the altitude measured in miles, can be used to approximate atmospheric pressure. Find the atmospheric pressure at each of the following locations:

 a. Mount McKinley in Alaska — altitude of 3.85 miles
 b. Denver, Colorado — the mile-high city
 c. Asheville, North Carolina — altitude of 1985 feet
 d. Phoenix, Arizona — altitude of 1090 feet

42. Suppose that the present population of a city is 75,000. Using the equation $P(t) = 75,000e^{0.01t}$ to estimate future growth, estimate the population 10 years from now, 15 years from now, and 25 years from now.

43. The brightness of a star viewed from Earth is measured in magnitudes. A star of any given magnitude is 2.512 times as bright as a star of the next higher magnitude. Therefore, to determine how many times brighter one star is than another, we can use the exponential function $f(x) = 2.512^x$, where x is the higher magnitude minus the lower magnitude.

 a. How many times brighter is a star of magnitude 1 than a star of magnitude 6?
 b. The star Altair has a magnitude of 0.9, and the Kapteyn's star has a magnitude of 8.8. How many times brighter than Kapteyn's star is Altair?
 c. The sun has a magnitude of -26.7, and Sirius has a magnitude of -1.6. How many times brighter is the sun than Sirius?

For Problems 44–49, graph each exponential function.

44. $f(x) = e^x + 1$ **45.** $f(x) = e^x - 2$

46. $f(x) = 2e^x$ **47.** $f(x) = -e^x$

48. $f(x) = e^{2x}$ **49.** $f(x) = e^{-x}$

■■■ THOUGHTS INTO WORDS

50. Explain the difference between simple interest and compound interest.

51. How would you explain the concept of effective yield to someone who missed class the day it was discussed?

52. How would you explain the half-life formula to someone who missed class the day it was discussed?

■■■ FURTHER INVESTIGATIONS

53. Complete the following chart that illustrates what happens to $1000 invested at various rates of interest for different lengths of time but always compounded continuously. Round your answers to the nearest dollar.

$1000 compounded continuously

	4%	6%	8%	10%
5 years				
10 years				
15 years				
20 years				
25 years				

54. Complete the following chart, which illustrates what happens to $1000 invested at 6% for different lengths of time and different numbers of compounding periods. Round all of your answers to the nearest dollar.

$1000 at 6%

	1 year	5 years	10 years	20 years
Compounded annually				
Compounded semiannually				
Compounded quarterly				
Compounded monthly				
Compounded continuously				

55. Complete the following chart that illustrates what happens to $1000 in 10 years based on different rates of interest and different numbers of compounding periods. Round your answers to the nearest dollar.

$1000 for 10 years

	4%	6%	8%	10%
Compounded annually				
Compounded semiannually				
Compounded quarterly				
Compounded monthly				
Compounded continuously				

For Problems 56–60, graph each function.

56. $f(x) = x(2^x)$

57. $f(x) = \dfrac{e^x + e^{-x}}{2}$

58. $f(x) = \dfrac{2}{e^x + e^{-x}}$

59. $f(x) = \dfrac{e^x - e^{-x}}{2}$

60. $f(x) = \dfrac{2}{e^x - e^{-x}}$

GRAPHING CALCULATOR ACTIVITIES

61. Use your graphing calculator to check your graphs for Problems 44–49 and 56–60.

62. How should the graphs of $f(x) = 2^x$, $f(x) = e^x$, and $f(x) = 3^x$ compare? Graph them on the same set of axes.

63. Graph $f(x) = e^x$. Where should the graphs of $f(x) = e^{x-2}$, $f(x) = e^{x+4}$, and $f(x) = e^{x-6}$ be located? Graph all three functions on the same set of axes.

64. Graph $f(x) = e^x$ again. Now predict the graphs for $f(x) = -e^x$, $f(x) = e^{-x}$, and $f(x) = -e^{-x}$. Graph these three functions on the same set of axes.

65. How do you think the graphs of $f(x) = e^x$, $f(x) = e^{2x}$, and $f(x) = 2e^x$ will compare? Graph them on the same set of axes to see whether you were right.

66. Find an approximate solution, to the nearest hundredth, for each of the following equations by graphing the appropriate function and finding the x intercept.

a. $e^x = 7$
b. $e^x = 21$
c. $e^x = 53$
d. $2e^x = 60$
e. $e^{x+1} = 150$
f. $e^{x-2} = 300$

67. Use a graphing approach to argue that it is better to invest money at 6% compounded quarterly than at 5.75% compounded continuously.

68. How long will it take $500 to be worth $1500 if it is invested at 7.5% interest compounded semiannually?

69. How long will it take $5000 to triple if it is invested at 6.75% interest compounded quarterly?

5.3 Logarithms

Objectives

- Switch between exponential and logarithmic form of equations.
- Evaluate logarithmic expressions.
- Solve logarithmic equations.
- Apply the properties of logarithms to simplify expressions.

In Sections 5.1 and 5.2 we gave meaning to exponential expressions of the form b^n, where b is any positive real number and n is any real number; we next used exponential expressions of the form b^n to define exponential functions; and then we used exponential functions to help solve problems. In the next three sections we will follow the same basic pattern with respect to a new concept, a *logarithm*. Let's begin with the following definition.

Definition 5.2

If r is any positive real number, then the unique exponent t such that $b^t = r$ is called the **logarithm of r with base b** and is denoted by $\log_b r$.

According to Definition 5.2, the logarithm of 16 base 2 is the exponent t such that $2^t = 16$; thus we can write $\log_2 16 = 4$. Likewise, we can write $\log_{10} 1000 = 3$ because $10^3 = 1000$. In general, Definition 5.2 can be remembered in terms of the statement

$$\log_b r = t \quad \text{is equivalent to} \quad b^t = r$$

Therefore we can easily switch back and forth between exponential and logarithmic forms of equations, as the next examples illustrate:

$$\log_2 8 = 3 \qquad \text{is equivalent to } 2^3 = 8$$

$$\log_{10} 100 = 2 \qquad \text{is equivalent to } 10^2 = 100$$

$$\log_3 81 = 4 \qquad \text{is equivalent to } 3^4 = 81$$

$$\log_{10} 0.001 = -3 \qquad \text{is equivalent to } 10^{-3} = 0.001$$

$$2^7 = 128 \qquad \text{is equivalent to } \log_2 128 = 7$$

$$5^3 = 125 \qquad \text{is equivalent to } \log_5 125 = 3$$

$$\left(\frac{1}{2}\right)^4 = \frac{1}{16} \qquad \text{is equivalent to } \log_{1/2} \frac{1}{16} = 4$$

$$10^{-2} = 0.01 \qquad \text{is equivalent to } \log_{10} 0.01 = -2$$

Some logarithms can be determined by changing to exponential form and using the properties of exponents, as in the next two examples.

EXAMPLE 1   Evaluate $\log_{10} 0.0001$.

Solution

Let $\log_{10} 0.0001 = x$. Changing to exponential form yields $10^x = 0.0001$, which can be solved as follows:

$$10^x = 0.0001$$
$$10^x = 10^{-4} \qquad 0.0001 = \frac{1}{10{,}000} = \frac{1}{10^4} = 10^{-4}$$
$$x = -4$$

Thus we have $\log_{10} 0.0001 = -4$.

▶ **Now work Problem 15.**   ■

EXAMPLE 2   Evaluate $\log_9(\sqrt[5]{27}/3)$.

Solution

Let $\log_9(\sqrt[5]{27}/3) = x$. Changing to exponential form yields $9^x = \sqrt[5]{27}/3$, which can be solved as follows:

$$9^x = \frac{(27)^{1/5}}{3}$$

$$(3^2)^x = \frac{(3^3)^{1/5}}{3}$$

$$3^{2x} = \frac{3^{3/5}}{3}$$

$$3^{2x} = 3^{-2/5}$$

$$2x = -\frac{2}{5}$$

$$x = -\frac{1}{5}$$

Therefore we have $\log_9 \dfrac{\sqrt[5]{27}}{3} = -\dfrac{1}{5}$.

▶ **Now work Problem 23.**   ■

Some equations that involve logarithms can also be solved by changing them to exponential form and using our knowledge of exponents.

E X A M P L E   3

Solve $\log_8 x = \dfrac{2}{3}$.

Solution

We can change $\log_8 x = \dfrac{2}{3}$ to exponential form to obtain

$$8^{2/3} = x$$

Therefore

$$x = (\sqrt[3]{8})^2 = 2^2 = 4$$

The solution set is $\{4\}$.

▶ **Now work Problem 37.**   ■

E X A M P L E   4

Solve $\log_b \dfrac{27}{64} = 3$.

Solution

We can change $\log_b \dfrac{27}{64} = 3$ to exponential form to obtain

$$b^3 = \frac{27}{64}$$

Therefore

$$b = \sqrt[3]{\frac{27}{64}} = \frac{3}{4}$$

The solution set is $\left\{\dfrac{3}{4}\right\}$.

▶ **Now work Problem 39.**   ■

■ Properties of Logarithms

Some properties of logarithms are a direct consequence of Definition 5.4 and the properties of exponents. For example, by writing the exponential equations $b^1 = b$ and $b^0 = 1$ in logarithmic form, we obtain the following property.

Property 5.3

For $b > 0$ and $b \neq 1$,

$$\log_b b = 1 \quad \text{and} \quad \log_b 1 = 0$$

Therefore, according to Property 5.3, we can write

$$\log_{10} 10 = 1 \qquad \log_4 4 = 1$$
$$\log_{10} 1 = 0 \qquad \log_5 1 = 0$$

Also, from Definition 5.2 we know that $\log_b r$ is the exponent t such that $b^t = r$. Therefore raising b to the $\log_b r$ power must produce r. This fact is stated in Property 5.4.

Property 5.4

For $b > 0$, $b \neq 1$, and $r > 0$,

$$b^{\log_b r} = r$$

Therefore, according to Property 5.4, we can write

$$10^{\log_{10} 72} = 72 \qquad 3^{\log_3 85} = 85 \qquad e^{\log_e 7} = 7$$

Because a logarithm is by definition an exponent, it is reasonable to predict that logarithms will have some properties that correspond to the basic exponential properties. This is an accurate prediction; these properties provide a basis for computational work with logarithms. Let's state the first of these properties and show how we can verify it by using our knowledge of exponents.

Property 5.5

For positive numbers b, r, and s, where $b \neq 1$,

$$\log_b rs = \log_b r + \log_b s$$

To verify Property 5.5, we can proceed as follows. Let $m = \log_b r$ and $n = \log_b s$. Change each of these equations to exponential form.

$$m = \log_b r \qquad \text{becomes } r = b^m$$
$$n = \log_b s \qquad \text{becomes } s = b^n$$

Thus the product rs becomes

$$rs = b^m \cdot b^n = b^{m+n}$$

Now, by changing $rs = b^{m+n}$ back to logarithmic form, we obtain

$$\log_b rs = m + n$$

Replacing m with $\log_b r$ and n with $\log_b s$ yields

$$\log_b rs = \log_b r + \log_b s$$

The following two examples demonstrate a use of Property 5.5.

E X A M P L E   5   If $\log_2 5 = 2.3219$ and $\log_2 3 = 1.5850$, evaluate $\log_2 15$.

Solution

Because $15 = 5 \cdot 3$, we can apply Property 5.5 as follows:

$$\log_2 15 = \log_2(5 \cdot 3)$$
$$= \log_2 5 + \log_2 3$$
$$= 2.3219 + 1.5850 = 3.9069$$

▶ **Now work Problem 47.**   ∎

E X A M P L E   6   If $\log_{10} 11 = 1.0414$ and $\log_{10} 17 = 1.2304$, evaluate $\log_{10} 187$.

Solution

Because $187 = 11 \cdot 17$, we can use Property 5.5 as follows:

$$\log_{10} 187 = \log_{10}(11 \cdot 17)$$
$$= \log_{10} 11 + \log_{10} 17$$
$$= 1.0414 + 1.2304 = 2.2718$$

▶ **Now work Problem 53.**   ∎

Because $b^m/b^n = b^{m-n}$, we would expect a corresponding property that pertains to logarithms. Property 5.6 is that property. It can be verified by using an approach similar to the one we used for Property 5.5. This verification is left for you to do as an exercise in the next problem set.

Property 5.6

For positive numbers b, r, and s, where $b \neq 1$,

$$\log_b\left(\frac{r}{s}\right) = \log_b r - \log_b s$$

Property 5.6 can be used to change a division problem into an equivalent subtraction problem, as the next two examples illustrate.

EXAMPLE 7

If $\log_5 36 = 2.2266$ and $\log_5 4 = 0.8614$, evaluate $\log_5 9$.

Solution

Because $9 = \dfrac{36}{4}$, we can use Property 5.6 as follows:

$$\log_5 9 = \log_5\left(\frac{36}{4}\right)$$
$$= \log_5 36 - \log_5 4$$
$$= 2.2266 - 0.8614 = 1.3652 \qquad \blacksquare$$

EXAMPLE 8

Evaluate $\log_{10}\left(\dfrac{379}{86}\right)$, given that $\log_{10} 379 = 2.5786$ and $\log_{10} 86 = 1.9345$.

Solution

$$\log_{10}\left(\frac{379}{86}\right) = \log_{10} 379 - \log_{10} 86$$
$$= 2.5786 - 1.9345 = 0.6441$$

▶ **Now work Problem 57.** ∎

Another property of exponents states that $(b^n)^m = b^{mn}$. The corresponding property of logarithms is stated in Property 5.7. Again, we leave the verification of this property as an exercise for you to do in the next set of problems.

Property 5.7

If r is a positive real number, b is a positive real number other than 1, and p is any real number, then

$$\log_b r^p = p(\log_b r)$$

The next two examples demonstrate a use of Property 5.7.

EXAMPLE 9

Evaluate $\log_2 22^{1/3}$, given that $\log_2 22 = 4.4594$.

Solution

$$\log_2 22^{1/3} = \frac{1}{3}\log_2 22 \qquad\qquad \text{Property 5.7}$$

$$= \frac{1}{3}(4.4594) = 1.4865$$

▶ **Now work Problem 51.**   ■

EXAMPLE 10

Evaluate $\log_{10}(8540)^{3/5}$, given that $\log_{10} 8540 = 3.9315$.

Solution

$$\log_{10}(8540)^{3/5} = \frac{3}{5}\log_{10} 8540$$

$$= \frac{3}{5}(3.9315) = 2.3589$$ ■

The properties of logarithms can be used to change the forms of various logarithmic expressions, as we will see in the next two examples.

EXAMPLE 11

Express $\log_b \sqrt{\dfrac{x}{yz}}$ in terms of the logarithms of x, y, and z.

Solution

$$\log_b \sqrt{\frac{x}{yz}} = \log_b\left(\frac{x}{yz}\right)^{1/2}$$

$$= \frac{1}{2}\log_b\left(\frac{x}{yz}\right) \qquad\qquad \text{Property 5.7}$$

$$= \frac{1}{2}(\log_b x - \log_b yz) \qquad\qquad \text{Property 5.6}$$

$$= \frac{1}{2}(\log_b x - (\log_b y + \log_b z)) \qquad\quad \text{Property 5.5}$$

$$= \frac{1}{2}(\log_b x - \log_b y - \log_b z) \qquad\quad \text{Distributive property}$$

▶ **Now work Problem 71.**   ■

Remark: The final expression, $\frac{1}{2}(\log_b x - \log_b y - \log_b z)$, in Example 11 could also be written as

$$\frac{1}{2}\log_b x - \frac{1}{2}\log_b y - \frac{1}{2}\log_b z$$

EXAMPLE 12

Express $2 \log_b x + 3 \log_b y - 4 \log_b z$ as one logarithm.

Solution

$$2 \log_b x + 3 \log_b y - 4 \log_b z = \log_b x^2 + \log_b y^3 - \log_b z^4 \qquad \text{Property 5.7}$$

$$= \log_b x^2 y^3 - \log_b z^4 \qquad \text{Property 5.5}$$

$$= \log_b \left(\frac{x^2 y^3}{z^4} \right) \qquad \text{Property 5.6}$$

▶ **Now work Problem 81.** ■

Sometimes we need to change from an indicated sum or difference of logarithmic quantities to an indicated product or quotient. This is especially helpful when we are solving certain kinds of equations that involve logarithms. Note in these next two examples how we can use the properties, along with the process of changing from logarithmic form to exponential form, to solve some equations.

EXAMPLE 13

Solve $\log_{10} x + \log_{10}(x + 9) = 1$.

Solution

$$\log_{10} x + \log_{10}(x + 9) = 1$$

$$\log_{10}[x(x + 9)] = 1 \qquad \text{Property 5.5}$$

$$x(x + 9) = 10^1 \qquad \text{Change to exponential form}$$

$$x^2 + 9x = 10$$

$$x^2 + 9x - 10 = 0$$

$$(x + 10)(x - 1) = 0$$

$$x + 10 = 0 \qquad \text{or} \qquad x - 1 = 0$$

$$x = -10 \qquad \text{or} \qquad x = 1$$

Because the left-hand side of the original equation is meaningful only if $x > 0$ and $x + 9 > 0$, the solution -10 must be discarded. Thus the solution set is $\{1\}$.

▶ **Now work Problem 87.** ■

EXAMPLE 14

Solve $\log_5(x + 4) - \log_5 x = 2$.

Solution

$$\log_5(x + 4) - \log_5 x = 2$$

$$\log_5 \left(\frac{x + 4}{x} \right) = 2 \qquad \text{Property 5.6}$$

$$5^2 = \frac{x+4}{x} \qquad \text{Change to exponential form}$$

$$25 = \frac{x+4}{x}$$

$$25x = x + 4$$

$$24x = 4$$

$$x = \frac{4}{24} = \frac{1}{6}$$

The solution set is $\left\{\dfrac{1}{6}\right\}$.

▶ **Now work Problem 97.** ∎

E X A M P L E   1 5

Solve $\log_2 3 + \log_2(x+4) = 3$.

Solution

$$\log_2 3 + \log_2(x+4) = 3$$

$$\log_2 3(x+4) = 3 \qquad \text{Property 5.5}$$

$$3(x+4) = 2^3 \qquad \text{Change to exponential form}$$

$$3x + 12 = 8$$

$$3x = -4$$

$$x = -\frac{4}{3}$$

The only restriction is that $x + 4 > 0$ or $x > -4$. Therefore, the solution set is $\left\{-\dfrac{4}{3}\right\}$. Perhaps you should check this answer.

▶ **Now work Problem 91.** ∎

C O N C E P T   Q U I Z

For Problems 1–5, answer true or false.

1. For $a > 0$, $\log_b a = c$ is equivalent to $b^c = a$.

2. For $b > 0$, $\log_b (\log_b b) = 0$.

3. The expression $\log_{\frac{1}{3}} 12$ is undefined because the base of the logarithmic expression must be greater than 1.

4. The logarithm of -8 base 2 denoted $\log_2(-8)$ is undefined.

5. For the equation $\log_2(x+3) + \log_2(x+5) = 1$, the solutions are restricted to values of x that are greater than -3.

For Problems 6–8, match the expression with its equivalent form.

6. $\log_3 2x$

7. $\log_3 \dfrac{1}{2} x$

8. $\log_3 \sqrt{x}$

A. $\log_3 \dfrac{1}{2} + \log_3 x$

B. $\dfrac{1}{2} \log_3 x$

C. $\log_3 2 + \log_3 x$

Problem Set 5.3

For Problems 1–8, write each equation in logarithmic form. For example, $2^4 = 16$ becomes $\log_2 16 = 4$.

1. $3^2 = 9$

2. $2^5 = 32$

3. $5^3 = 125$

4. $10^1 = 10$

5. $2^{-4} = \dfrac{1}{16}$

6. $\left(\dfrac{2}{3}\right)^{-3} = \dfrac{27}{8}$

7. $10^{-2} = 0.01$

8. $10^5 = 100{,}000$

For Problems 9–14, write each equation in exponential form. For example, $\log_2 8 = 3$ becomes $2^3 = 8$.

9. $\log_2 64 = 6$

10. $\log_3 27 = 3$

11. $\log_{10} 0.1 = -1$

12. $\log_5 \left(\dfrac{1}{25}\right) = -2$

13. $\log_2 \left(\dfrac{1}{16}\right) = -4$

14. $\log_{10} 0.00001 = -5$

For Problems 15–34, evaluate each expression.

15. $\log_6 36$

16. $\log_3 243$

17. $\log_5 \left(\dfrac{1}{5}\right)$

18. $\log_4 \left(\dfrac{1}{64}\right)$

19. $\log_{10} 10$

20. $\log_{10} 1$

21. $\log_3 \sqrt{3}$

22. $\log_5 \sqrt[3]{25}$

23. $\log_3 \left(\dfrac{\sqrt{27}}{3}\right)$

24. $\log_{1/2} \left(\dfrac{\sqrt[4]{8}}{2}\right)$

25. $\log_{1/4} \left(\dfrac{\sqrt[4]{32}}{2}\right)$

26. $\log_2 \left(\dfrac{\sqrt[3]{16}}{4}\right)$

27. $\log_9 27$

28. $\log_8 32$

29. $\log_4 \left(\dfrac{1}{8}\right)$

30. $\log_4 32$

31. $10^{\log_{10} 7}$

32. $5^{\log_5 13}$

33. $\log_2 (\log_5 5)$

34. $\log_6 (\log_2 64)$

For Problems 35–46, solve each equation.

35. $\log_5 x = 2$

36. $\log_{10} x = 3$

37. $\log_8 t = \dfrac{5}{3}$

38. $\log_4 m = \dfrac{3}{2}$

39. $\log_b 3 = \dfrac{1}{2}$

40. $\log_b 2 = \dfrac{1}{2}$

41. $\log_{10} x = 0$

42. $\log_{10} x = 1$

43. $\log_{1/2} x = 3$

44. $\log_{2/3} x = 2$

45. $\log_3 x = -2$

46. $\log_2 x = -5$

For Problems 47–55, given that $\log_2 5 = 2.3219$ and $\log_2 7 = 2.8074$, evaluate each expression by using Properties 5.5–5.7.

47. $\log_2 35$

48. $\log_2 \left(\dfrac{7}{5}\right)$

49. $\log_2 125$

50. $\log_2 49$

51. $\log_2 \sqrt{7}$

52. $\log_2 \sqrt[3]{5}$

53. $\log_2 175$

54. $\log_2 56$

55. $\log_2 80$

For Problems 56–64, given that $\log_8 5 = 0.7740$ and $\log_8 11 = 1.1531$, evaluate each expression using Properties 5.5–5.7.

56. $\log_8 55$

57. $\log_8\left(\dfrac{5}{11}\right)$

58. $\log_8 25$

59. $\log_8 \sqrt{11}$

60. $\log_8(5)^{2/3}$

61. $\log_8 88$

62. $\log_8 320$

63. $\log_8\left(\dfrac{25}{11}\right)$

64. $\log_8\left(\dfrac{121}{25}\right)$

For Problems 65–76, express each as the sum or difference of simpler logarithmic quantities. (Assume that all variables represent positive real numbers.) For example,

$$\log_b\left(\frac{x^3}{y^2}\right) = \log_b x^3 - \log_b y^2$$

$$= 3 \log_b x - 2 \log_b y$$

65. $\log_b xyz$

66. $\log_b\left(\dfrac{x^2}{y}\right)$

67. $\log_b x^2 y^3$

68. $\log_b x^{2/3} y^{3/4}$

69. $\log_b \sqrt{xy}$

70. $\log_b \sqrt[3]{x^2 z}$

71. $\log_b \sqrt{\dfrac{x}{y}}$

72. $\log_b\left[x\left(\sqrt{\dfrac{x}{y}}\right)\right]$

73. $\log_b\left(\dfrac{x^2 y}{z}\right)$

74. $\log_b\left(\dfrac{x}{y^2 z}\right)$

75. $\log_b \sqrt[3]{\dfrac{x^2}{yz}}$

76. $\log_b \sqrt[4]{\dfrac{xy}{z}}$

For Problems 77–84, express each as a single logarithm. (Assume that all variables represent positive numbers.) For example,

$$3 \log_b x + 5 \log_b y = \log_b x^3 y^5$$

77. $\log_b x + \log_b y - \log_b z$

78. $2 \log_b x - 4 \log_b y$

79. $(\log_b x - \log_b y) - \log_b z$

80. $\log_b x - (\log_b y - \log_b z)$

81. $\log_b x + \dfrac{1}{2} \log_b y$

82. $2 \log_b x + 4 \log_b y - 3 \log_b z$

83. $2 \log_b x + \dfrac{1}{2} \log_b(x - 1) - 4 \log_b(2x + 5)$

84. $\dfrac{1}{2} \log_b x - 3 \log_b x + 4 \log_b y$

For Problems 85–102, solve each equation.

85. $\log_3 x + \log_3 4 = 2$

86. $\log_7 5 + \log_7 x = 1$

87. $\log_{10} x + \log_{10}(x - 21) = 2$

88. $\log_{10} x + \log_{10}(x - 3) = 1$

89. $\log_2 x + \log_2(x - 3) = 2$

90. $\log_3 x + \log_3(x - 2) = 1$

91. $\log_2 5 + \log_2(x + 6) = 3$

92. $\log_2(x - 1) - \log_2(x + 3) = 2$

93. $\log_5 x = \log_5(x + 2) + 1$

94. $\log_3(x + 3) + \log_3(x + 5) = 1$

95. $\log_2(x + 2) = 1 - \log_2(x + 3)$

96. $\log_4 7 + \log_4(x + 3) = 2$

97. $\log_{10}(2x - 1) - \log_{10}(x - 2) = 1$

98. $\log_{10}(9x - 2) = 1 + \log_{10}(x - 4)$

99. $\log_5(3x - 2) = 1 + \log_5(x - 4)$

100. $\log_6 x + \log_6(x + 5) = 2$

101. $\log_8(x + 7) + \log_8 x = 1$

102. $\log_6(x + 1) + \log_6(x - 4) = 2$

103. Verify Property 5.6.

104. Verify Property 5.7.

■ ■ ■ **THOUGHTS INTO WORDS**

105. How would you explain the concept of a logarithm to someone who has never studied algebra?

106. Explain, without using Property 5.4, why $4^{\log_4 9}$ equals 9.

107. In the next section we are going to show that the logarithmic function $f(x) = \log_2 x$ is the inverse of the exponential function $f(x) = 2^x$. From that information, how could you sketch a graph of $f(x) = \log_2 x$?

Answers to the Concept Quiz

1. True **2.** True **3.** False **4.** True **5.** True **6.** C **7.** A **8.** B

5.4 Logarithmic Functions

Objectives

■ Graph logarithmic functions.

■ Evaluate common logarithms.

■ Evaluate natural logarithms.

The concept of a logarithm can now be used to define a logarithmic function.

Definition 5.3

If $b > 0$ and $b \neq 1$, then the function defined by

$$f(x) = \log_b x$$

where x is any positive real number, is called the **logarithmic function with base b.**

We can obtain the graph of a specific logarithmic function in various ways. For example, we can change the equation $y = \log_2 x$ to the exponential equation $2^y = x$, from which we can determine a table of values. We will instruct you to use this approach to graph some logarithmic functions in the next set of exercises.

We can also obtain the graph of a logarithmic function by setting up a table of values directly from the logarithmic equation. We will demonstrate this approach.

EXAMPLE 1

Graph $f(x) = \log_2 x$.

Solution

Let's choose some values for x where the corresponding values for $\log_2 x$ are easily determined. (Remember that logarithms are defined only for the positive real numbers.) We plot the points determined by the table and connect them with a smooth curve to produce Figure 5.10.

x	$f(x)$	
$\frac{1}{8}$	-3	$\log_2 \frac{1}{8} = -3$ because $2^{-3} = \frac{1}{2^3} = \frac{1}{8}$
$\frac{1}{4}$	-2	
$\frac{1}{2}$	-1	
1	0	$\log_2 1 = 0$ because $2^0 = 1$
2	1	
4	2	
8	3	

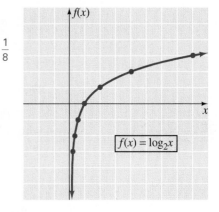

Note that the $f(x)$ axis is a vertical asymptote.

▶ **Now work Problem 41.**

Figure 5.10 ∎

Now suppose that we consider two functions f and g as follows:

$f(x) = b^x$ Domain: all real numbers
       Range: positive real numbers

$g(x) = \log_b x$ Domain: positive real numbers
         Range: all real numbers

Furthermore, suppose that we consider the composition of f and g and the composition of g and f:

$$(f \circ g)(x) = f(g(x)) = f(\log_b x) = b^{\log_b x} = x$$
$$(g \circ f)(x) = g(f(x)) = g(b^x) = \log_b b^x = x \log_b b = x(1) = x$$

Therefore, because the domain of f is the range of g, and the range of f is the domain of g, and because $f(g(x)) = x$ and $g(f(x)) = x$, the two functions f and g are *inverses of each other*.

Remember that the graphs of a function and its inverse are reflections of each other through the line $y = x$. Thus the graph of a logarithmic function can be determined by reflecting the graph of its inverse exponential function through the line $y = x$. This idea is illustrated in Figure 5.11, where the graph of $y = 2^x$ has been reflected across the line $y = x$ to produce the graph of $y = \log_2 x$.

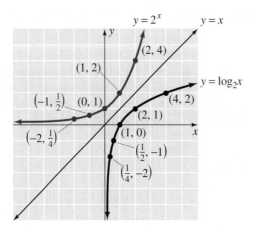

Note that the x axis is an asymptote for the graph of $y = 2^x$ and the y axis is an asymptote for the graph of $y = \log_2 x$.

Figure 5.11

Figure 5.3 illustrated the general behavior patterns of exponential functions with two graphs. We can now reflect each of these graphs through the line $y = x$ and observe the general behavior patterns of logarithmic functions, as shown in Figure 5.12.

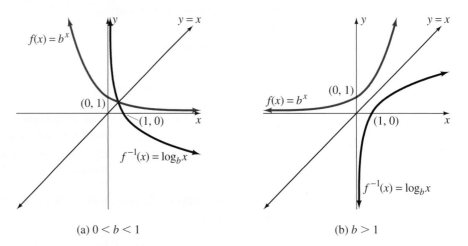

(a) $0 < b < 1$ (b) $b > 1$

Figure 5.12

Finally, when graphing logarithmic functions, don't forget about transformations of the basic curves.

1. The graph of $f(x) = 3 + \log_2 x$ is the graph of $f(x) = \log_2 x$ *moved up three units.* (Because $\log_2 x + 3$ is easily confused with $\log_2(x + 3)$, we commonly write $3 + \log_2 x$.)

2. The graph of $f(x) = \log_2(x - 4)$ is the graph of $f(x) = \log_2 x$ *moved four units to the right.*

3. The graph of $f(x) = -\log_2 x$ is the graph of $f(x) = \log_2 x$ *reflected across the x axis.*

■ Common Logarithms: Base 10

The properties of logarithms that we discussed in Section 5.3 are true for any valid base. However, because the Hindu–Arabic numeration system that we use is a base-10 system, logarithms to base 10 have historically been used for computational purposes. Base-10 logarithms are called **common logarithms.**

Originally, common logarithms were developed to aid in complicated numerical calculations that involve products, quotients, and powers of real numbers. Today they are seldom used for that purpose because the calculator and computer can much more effectively handle the messy computational problems. However, common logarithms do still occur in applications, so they deserve our attention.

As we know from earlier work, the definition of a logarithm allows us to evaluate $\log_{10} x$ for values of x that are integral powers of 10. Consider the following examples:

$$\log_{10} 1000 = 3 \qquad \text{because } 10^3 = 1000$$
$$\log_{10} 100 = 2 \qquad \text{because } 10^2 = 100$$
$$\log_{10} 10 = 1 \qquad \text{because } 10^1 = 10$$
$$\log_{10} 1 = 0 \qquad \text{because } 10^0 = 1$$
$$\log_{10} 0.1 = -1 \qquad \text{because } 10^{-1} = \frac{1}{10} = 0.1$$
$$\log_{10} 0.01 = -2 \qquad \text{because } 10^{-2} = \frac{1}{10^2} = 0.01$$
$$\log_{10} 0.001 = -3 \qquad \text{because } 10^{-3} = \frac{1}{10^3} = 0.001$$

When we work exclusively with base-10 logarithms, it is customary to omit writing the numeral 10 to designate the base. Thus the expression $\log_{10} x$ is written as $\log x$, and a statement such as $\log_{10} 1000 = 3$ becomes $\log 1000 = 3$. We will follow this practice from now on in this chapter, but don't forget that the base is understood to be 10:

$$\log_{10} x = \log x$$

To find the common logarithm of a positive number that is not an integral power of 10, we can use an appropriately equipped calculator. We used a calculator equipped with a common logarithmic function (ordinarily a key labeled $\boxed{\log}$ is used) to obtain the following results rounded to four decimal places:

$$\log 1.75 = 0.2430$$
$$\log 23.8 = 1.3766$$
$$\log 134 = 2.1271$$
$$\log 0.192 = -0.7167$$
$$\log 0.0246 = -1.6091$$

Be sure you can use a calculator and obtain these results

In order to use logarithms to solve problems, we sometimes need to be able to determine a number when the logarithm of the number is known. That is, we may need to determine x when $\log x$ is known. Let's consider an example.

E X A M P L E   2   Find x if $\log x = 0.2430$.

Solution

If $\log x = 0.2430$, then changing to exponential form yields $10^{0.2430} = x$; use the $\boxed{10^x}$ key to find x:

$$x = 10^{0.2430} \approx 1.749846689$$

Therefore $x = 1.7498$ rounded to five significant digits.

▶ **Now work Problem 11.**   ∎

Be sure that you can use your calculator and obtain the following results. We have rounded the values for x to five significant digits.

If $\log x = 0.7629$, then $x = 10^{0.7629} = 5.7930$.
If $\log x = 1.4825$, then $x = 10^{1.4825} = 30.374$.
If $\log x = 4.0214$, then $x = 10^{4.0214} = 10{,}505$.
If $\log x = -1.5162$, then $x = 10^{-1.5162} = 0.030465$.
If $\log x = -3.8921$, then $x = 10^{-3.8921} = 0.00012820$.

The **common logarithmic function** is defined by the equation $f(x) = \log x$. It should now be a simple matter to set up a table of values and sketch the function. You will do this in the next set of exercises. Remember that $g(x) = 10^x$ and $f(x) = \log x$ are inverses of each other. Therefore we could also get the graph of $f(x) = \log x$ by reflecting the exponential curve $g(x) = 10^x$ across the line $y = x$, as shown in Figure 5.13.

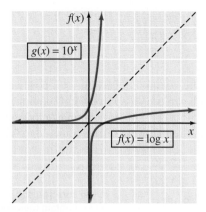

Figure 5.13

■ Natural Logarithms — Base *e*

In many practical applications of logarithms, the number *e* (remember $e \approx 2.71828$) is used as a base. Logarithms with a base of *e* are called **natural logarithms,** and the symbol ln *x* is commonly used instead of $\log_e x$:

$$\log_e x = \ln x$$

Natural logarithms can also be found with an appropriately equipped calculator. Use a calculator with a natural logarithm function (ordinarily a key labeled $\boxed{\ln x}$) to obtain the following results rounded to four decimal places:

$\ln 3.21 = 1.1663$

$\ln 47.28 = 3.8561$

$\ln 842 = 6.7358$

$\ln 0.21 = -1.5606$

$\ln 0.0046 = -5.3817$

$\ln 10 = 2.3026$

Be sure that you can use your calculator to obtain these results. Keep in mind the significance of a statement such as $\ln 3.21 = 1.1663$. By changing to exponential form, we are claiming that *e* raised to the 1.1663 power is approximately 3.21. Using a calculator, we obtain $e^{1.1663} = 3.210093293$.

Two special cases come from comparing exponential form to logarithmic form. Because $e^1 = e$, we know that

$$\log_e e = \ln e = 1$$

and because $e^0 = 1$, we have

$$\log_e 1 = \ln 1 = 0$$

Let's do a few more problems finding *x* when given ln *x*. Be sure that you agree with these results.

If $\ln x = 2.4156$, then $x = e^{2.4156} = 11.196$.

If $\ln x = 0.9847$, then $x = e^{0.9847} = 2.6770$.

If $\ln x = 4.1482$, then $x = e^{4.1482} = 63.320$.

If $\ln x = -1.7654$, then $x = e^{-1.7654} = 0.17112$.

The **natural logarithmic function** is defined by the equation $f(x) = \ln x$. It is the inverse of the natural exponential function $g(x) = e^x$. Therefore one way to graph $f(x) = \ln x$ is to reflect the graph of $g(x) = e^x$ across the line $y = x$, as shown in Figure 5.14.

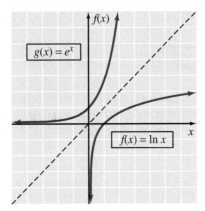

Figure 5.14

Now based on our previous work with transformations, we should be able to make the following statements:

1. The graph of $f(x) = e^{-x}$ is the graph of $f(x) = e^x$ reflected through the y axis.

2. The graph of $f(x) = e^x + 4$ is the graph of $f(x) = e^x$ shifted upward four units.

3. The graph of $f(x) = \ln(x + 2)$ is the graph of $f(x) = \ln x$ shifted two units to the left.

4. The graph of $f(x) = -\ln x$ is the graph of $f(x) = \ln x$ reflected through the x axis.

The graphs of these four functions are shown in Figures 5.15 and 5.16.

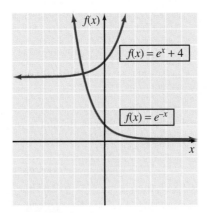

Figure 5.15

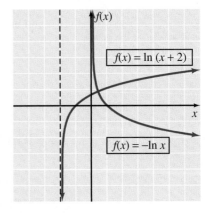

Figure 5.16

CONCEPT QUIZ For Problems 1–8, answer true or false.

1. For $f(x) = \log_b x$, the domain is $[0, \infty)$.

2. The x axis is an asymptote for the graph of $f(x) = \log_b x$.

3. For all $b > 0$ and $b \neq 1$, the graph of $f(x) = \log_b x$ passes through the point $(1, 0)$.

4. The functions $f(x) = \log_2 x$ and $f(x) = \log_{\frac{1}{2}} x^{-1}$ are inverse functions.

5. The base for $\log x$ is 10.

6. The base for $\ln x$ is 2.

7. The inverse of the function $f(x) = \ln x$ is the function $f(x) = e^x$.

8. The graph of $f(x) = \ln(x + 2)$ is the graph of $f(x) = \ln x$ shifted up two units.

Problem Set 5.4

For Problems 1–10, use a calculator to find each **common logarithm.** Express answers to four decimal places.

1. log 7.24

2. log 2.05

3. log 52.23

4. log 825.8

5. log 3214.1

6. log 14,189

7. log 0.729

8. log 0.04376

9. log 0.00034

10. log 0.000069

For Problems 11–20, use your calculator to find x when given log x. Express answers to five significant digits.

11. $\log x = 2.6143$

12. $\log x = 1.5263$

13. $\log x = 4.9547$

14. $\log x = 3.9335$

15. $\log x = 1.9006$

16. $\log x = 0.5517$

17. $\log x = -1.3148$

18. $\log x = -0.1452$

19. $\log x = -2.1928$

20. $\log x = -2.6542$

For Problems 21–30, use your calculator to find each **natural logarithm.** Express answers to four decimal places.

21. ln 5

22. ln 18

23. ln 32.6

24. ln 79.5

25. ln 430

26. ln 371.8

27. ln 0.46

28. ln 0.524

29. ln 0.0314

30. ln 0.008142

For Problems 31–40, use your calculator to find x when given ln x. Express answers to five significant digits.

31. $\ln x = 0.4721$

32. $\ln x = 0.9413$

33. $\ln x = 1.1425$

34. $\ln x = 2.7619$

35. $\ln x = 4.6873$

36. $\ln x = 3.0259$

37. $\ln x = -0.7284$

38. $\ln x = -1.6246$

39. $\ln x = -3.3244$

40. $\ln x = -2.3745$

41. Complete the following table and then graph $f(x) = \log x$. (Express the values for log x to the nearest tenth.)

x	0.1	0.5	1	2	4	8	10
$\log x$							

42. Complete the following table and then graph $f(x) = \ln x$. (Express the values for ln x to the nearest tenth.)

x	0.1	0.5	1	2	4	8	10
$\ln x$							

43. Graph $y = \log_{1/2} x$ by graphing $\left(\dfrac{1}{2}\right)^y = x$.

44. Graph $y = \log_2 x$ by graphing $2^y = x$.

45. Graph $f(x) = \log_3 x$ by reflecting the graph of $g(x) = 3^x$ across the line $y = x$.

46. Graph $f(x) = \log_4 x$ by reflecting the graph of $g(x) = 4^x$ across the line $y = x$.

For Problems 47–53, graph each function. Remember that the graph of $f(x) = \log_2 x$ is given in Figure 5.10.

47. $f(x) = 3 + \log_2 x$ **48.** $f(x) = -2 + \log_2 x$

49. $f(x) = \log_2(x + 3)$ **50.** $f(x) = \log_2(x - 2)$

51. $f(x) = \log_2(2x)$ **52.** $f(x) = -\log_2 x$

53. $f(x) = 2 \log_2 x$

For Problems 54–61, use your knowledge of transformations to help sketch a graph of each of the functions.

54. $f(x) = 10^x - 1$ **55.** $f(x) = 10^{x-1}$

56. $f(x) = -\log(x - 2)$ **57.** $f(x) = -\log(-x)$

58. $f(x) = e^{x+3}$ **59.** $f(x) = -e^x$

60. $f(x) = \ln(-x)$ **61.** $f(x) = \ln(x + 4)$

62. In chemistry the term pH, meaning "hydrogen power," is defined as the negative base-10 logarithm of the concentration, in moles per liter, of H^+ ions. In other words, pH is a function of the number of H^+ ions and can be expressed as

$$f(x) = -\log x$$

where x is the number of H^+ ions in moles per liter of the solution. A solution with a pH below 7 is called an *acid solution*, and a solution with a pH above 7 is called a *basic solution*.

Find, to the nearest tenth, the pH of each of the following solutions with the given H^+ concentrations, and identify each as an acid or basic solution.

a. $2(10)^{-9}$ **b.** $7.1(10)^{-4}$
c. $8(10)^{-2}$ **d.** $6(10)^{-7}$
e. $1.8(10)^{-11}$

For Problems 63–70, perform the following calculations and express answers to the nearest hundredth. (These calculations are in preparation for our work in the next section.)

63. $\dfrac{\log 7}{\log 3}$ **64.** $\dfrac{\ln 2}{\ln 7}$

65. $\dfrac{2 \ln 3}{\ln 8}$ **66.** $\dfrac{\ln 5}{2 \ln 3}$

67. $\dfrac{\ln 3}{0.04}$ **68.** $\dfrac{\ln 2}{0.03}$

69. $\dfrac{\log 2}{5 \log 1.02}$ **70.** $\dfrac{\log 5}{3 \log 1.07}$

■ ■ ■ **THOUGHTS INTO WORDS**

71. Describe three ways in which the graph of $f(x) = \log_3 x$ can be obtained.

72. How do we know that $\log_2 6$ is between 2 and 3?

■ ■ ■ **FURTHER INVESTIGATIONS**

73. Graph the function $f(x) = \log_2 x^2$.

74. Graph the function $f(x) = 2 \log_2 x$.

75. According to Property 5.7, $\log_2 x^2 = 2 \log_2 x$. Why are the graphs for Problems 73 and 74 different?

GRAPHING CALCULATOR ACTIVITIES

76. Graph $f(x) = x$, $f(x) = e^x$, and $f(x) = \ln x$ on the same set of axes.

77. Graph $f(x) = x$, $f(x) = 10^x$, and $f(x) = \log x$ on the same set of axes.

78. Graph $f(x) = \ln x$. How should the graphs of $f(x) = 2 \ln x$, $f(x) = 4 \ln x$, and $f(x) = 6 \ln x$ compare to this basic curve? Graph the three functions on the same set of axes with the graph of $f(x) = \ln x$.

79. Graph $f(x) = \log x$. Now predict the graphs for $f(x) = 3 + \log x$, $f(x) = -2 + \log x$, and $f(x) = -4 + \log x$. Graph them on the same set of axes with the graph of $f(x) = \log x$.

80. Graph $\ln x$. Now predict the graphs for $f(x) = \ln (x - 2)$, $f(x) = \ln (x - 6)$, and $f(x) = \ln (x + 4)$. Graph the three functions on the same set of axes with $f(x) = \ln x$.

81. For each of the following, predict the general shape and location of the graph, and use your graphing calculator to graph the function to check your prediction.

a. $f(x) = \log x + \ln x$ **b.** $f(x) = \log x - \ln x$
c. $f(x) = \ln x - \log x$ **d.** $f(x) = \ln x^2$

5.5 Exponential and Logarithmic Equations: Problem Solving

Objectives

- Solve logarithmic equations.
- Use logarithms to solve application problems involving exponential growth and decay.
- Solve application problems involving Richter numbers.
- Use a change-of-base formula to evaluate a logarithm.

In Section 5.1 we solved exponential equations such as $3^x = 81$ by expressing both sides of the equation as a power of 3 and then applying the property *If $b^n = b^m$, then $n = m$*. However, if we try this same approach with an equation such as $3^x = 5$, we face the difficulty of expressing 5 as a power of 3. We can solve this type of problem by using the properties of logarithms and the following property of equality.

Property 5.8

If $x > 0$, $y > 0$, $b > 0$, and $b \neq 1$, then

$$x = y \quad \text{if and only if } \log_b x = \log_b y$$

Property 5.8 is stated in terms of any valid base b; however, for most applications we use either common logarithms or natural logarithms. Let's consider some examples.

EXAMPLE 1   Solve $3^x = 5$ to the nearest hundredth.

Solution

By using common logarithms, we can proceed as follows:

$$3^x = 5$$

$$\log 3^x = \log 5 \qquad\qquad\qquad \text{Property 5.8}$$

$$x \log 3 = \log 5 \qquad\qquad\qquad \log r^p = p \log r$$

$$x = \frac{\log 5}{\log 3}$$

$$x = 1.46 \quad \text{nearest hundredth}$$

✔   **Check**   Because $3^{1.46} \approx 4.972754647$, we say that, to the nearest hundredth, the solution set for $3^x = 5$ is $\{1.46\}$.

▶   **Now work Problem 1.**   ■

A WORD OF CAUTION!   The expression $\dfrac{\log 5}{\log 3}$ means that we must *divide*, not subtract, the logarithms. That is, $\dfrac{\log 5}{\log 3}$ *does not* mean $\log\left(\dfrac{5}{3}\right)$. Remember that $\log\left(\dfrac{5}{3}\right) = \log 5 - \log 3$.

EXAMPLE 2   Solve $e^{x+1} = 5$ to the nearest hundredth.

Solution

Because base e is used in the exponential expression, let's use natural logarithms to help solve this equation.

$$e^{x+1} = 5$$

$$\ln e^{x+1} = \ln 5 \qquad\qquad\qquad \text{Property 5.8}$$

$$(x + 1) \ln e = \ln 5 \qquad\qquad\qquad \ln r^p = p \ln r$$

$$(x + 1)(1) = \ln 5 \qquad\qquad\qquad \ln e = 1$$

$$x = \ln 5 - 1$$

$$x = 0.61 \quad \text{nearest hundredth}$$

The solution set is $\{0.61\}$. Check it!

▶   **Now work Problem 11.**   ■

EXAMPLE 3   Solve $2^{3x-2} = 3^{2x+1}$ to the nearest hundredth.

Solution

$$2^{3x-2} = 3^{2x+1}$$

$$\log 2^{3x-2} = \log 3^{2x+1}$$

$$(3x - 2)\log 2 = (2x + 1)\log 3$$

$$3x \log 2 - 2 \log 2 = 2x \log 3 + \log 3$$

$$3x \log 2 - 2x \log 3 = \log 3 + 2 \log 2$$

$$x(3 \log 2 - 2 \log 3) = \log 3 + 2 \log 2$$

$$x = \frac{\log 3 + 2 \log 2}{3 \log 2 - 2 \log 3}$$

$$x = -21.10 \quad \text{nearest hundredth}$$

The solution set is $\{-21.10\}$. Check it!

▶ **Now work Problem 15.**   ■

Remark: In Example 3 the expression $\dfrac{\log 3 + 2 \log 2}{3 \log 2 - 2 \log 3}$ can be simplified to $-\dfrac{\log 12}{\log 1.125}$ using the properties of logarithms. Perhaps it would be worth your time to do that simplification and then arrive at the same value for x.

■ Logarithmic Equations

In Example 13 of Section 5.3 we solved the logarithmic equation

$$\log_{10} x + \log_{10}(x + 9) = 1$$

by simplifying the left side of the equation to $\log_{10}[x(x + 9)]$ and then changing the equation to exponential form to complete the solution. At this time we can use Property 5.8 to solve this type of logarithmic equation another way, and we can also expand our equation-solving capabilities. Let's consider some examples.

EXAMPLE 4   Solve $\log x + \log(x - 15) = 2$.

Solution

Because $\log 100 = 2$, the given equation becomes

$$\log x + \log(x - 15) = \log 100$$

Now we can simplify the left side, apply Property 5.8, and proceed as follows:

$$\log[(x)(x - 15)] = \log 100$$

$$x(x - 15) = 100$$

$$x^2 - 15x - 100 = 0$$
$$(x - 20)(x + 5) = 0$$
$$x - 20 = 0 \quad \text{or} \quad x + 5 = 0$$
$$x = 20 \quad \text{or} \quad x = -5$$

The domain of a logarithmic function must contain only positive numbers, so x and $x - 15$ must be positive in this problem. Therefore we discard the solution of -5, and the solution set is $\{20\}$.

▶ **Now work Problem 19.**   ■

E X A M P L E   5

Solve $\ln(x + 2) = \ln(x + 10) - \ln 3$.

Solution

$$\ln(x + 2) = \ln(x + 10) - \ln 3$$
$$\ln(x + 2) = \ln\frac{(x + 10)}{3}$$
$$(x + 2) = \frac{(x + 10)}{3}$$
$$3(x + 2) = x + 10$$
$$3x + 6 = x + 10$$
$$2x = 4$$
$$x = 2$$

The solution set is $\{2\}$.

▶ **Now work Problem 31.**   ■

E X A M P L E   6

Solve $\log_b(x + 2) + \log_b(2x - 1) = \log_b x$.

Solution

$$\log_b(x + 2) + \log_b(2x - 1) = \log_b x$$
$$\log_b[(x + 2)(2x - 1)] = \log_b x$$
$$(x + 2)(2x - 1) = x$$
$$2x^2 + 3x - 2 = x$$
$$2x^2 + 2x - 2 = 0$$
$$x^2 + x - 1 = 0$$

We can use the quadratic formula to obtain

$$x = \frac{-1 \pm \sqrt{1 + 4}}{2}$$
$$= \frac{-1 \pm \sqrt{5}}{2}$$

Because $x + 2$, $2x - 1$, and x all have to be positive, the solution of $(-1 - \sqrt{5})/2$ has to be discarded, and the solution set is

$$\left\{ \frac{-1 + \sqrt{5}}{2} \right\}$$

▶ **Now work Problem 25.** ■

■ Problem Solving

In Section 5.2 we used the compound interest formula

$$A = P\left(1 + \frac{r}{n} \right)^{nt}$$

to determine the amount of money A accumulated at the end of t years if P dollars is invested at rate r of interest compounded n times per year. Now let's use this formula to solve other types of problems that deal with compound interest.

PROBLEM 1

How long will it take $500 to double if it is invested at 8% compounded quarterly?

Solution

To *double* $500 means that the $500 will grow to $1000. We want to find out how long it will take; that is, what is t? Thus

$$1000 = 500\left(1 + \frac{0.08}{4} \right)^{4t}$$
$$= 500(1 + 0.02)^{4t}$$
$$= 500(1.02)^{4t}$$

We multiply both sides of $1000 = 500(1.02)^{4t}$ by $\dfrac{1}{500}$ to get

$$2 = (1.02)^{4t}$$

Therefore

$$\log 2 = \log(1.02)^{4t} \qquad \text{Property 5.8}$$
$$= 4t \log 1.02 \qquad \log r^p = p \log r$$

Now let's solve for t:

$$4t \log 1.02 = \log 2$$
$$t = \frac{\log 2}{4 \log 1.02}$$
$$t = 8.8 \quad \text{nearest tenth}$$

Therefore we are claiming that $500 invested at 8% interest compounded quarterly will double in approximately 8.8 years.

✔ **Check** $500 invested at 8% compounded quarterly for 8.8 years will produce

$$A = \$500\left(1 + \frac{0.08}{4}\right)^{4(8.8)}$$

$$= \$500(1.02)^{35.2}$$

$$= \$1003.91$$

▶ **Now work Problem 43.** ∎

PROBLEM 2

Suppose that the number of bacteria present in a certain culture after t minutes is given by the equation $Q(t) = Q_0 e^{0.04t}$, where Q_0 represents the initial number of bacteria. How long would it take for the bacteria count to grow from 500 to 2000?

Solution

Substituting into $Q(t) = Q_0 e^{0.04t}$ and solving for t, we obtain

$$2000 = 500 e^{0.04t}$$

$$4 = e^{0.04t}$$

$$\ln 4 = \ln e^{0.04t}$$

$$\ln 4 = 0.04t \ln e$$

$$\ln 4 = 0.04t \qquad\qquad \ln e = 1$$

$$\frac{\ln 4}{0.04} = t$$

$$34.7 = t \quad \text{nearest tenth}$$

It should take approximately 34.7 minutes.

▶ **Now work Problem 53.** ∎

∎ Richter Numbers

Seismologists use the Richter scale to measure and report the magnitude of earthquakes. The equation

$$R = \log \frac{I}{I_0} \qquad R \text{ is called a Richter number}$$

compares the intensity I of an earthquake to a minimal or reference intensity I_0. The reference intensity is the smallest earth movement that can be recorded on a seismograph. Suppose that the intensity of an earthquake was determined to be 50,000 times the reference intensity. In this case $I = 50,000 I_0$ and the Richter number would be calculated as follows:

$$R = \log \frac{50{,}000\, I_0}{I_0}$$

$$= \log 50{,}000$$

$$\approx 4.698970004$$

Thus a Richter number of 4.7 would be reported. Let's consider two more examples that involve Richter numbers.

PROBLEM   3

An earthquake that occurred in San Francisco in 1989 was reported to have a Richter number of 6.9. How did its intensity compare to the reference intensity?

Solution

$$6.9 = \log \frac{I}{I_0}$$

$$10^{6.9} = \frac{I}{I_0}$$

$$I = (10^{6.9})(I_0)$$

$$I \approx 7{,}943{,}282\, I_0$$

Thus its intensity was a little less than 8 million times the reference intensity.

▶ **Now work Problem 55.**   ■

PROBLEM   4

An earthquake in Iran in 1990 had a Richter number of 7.7. Compare the intensity of this earthquake to that of the one in San Francisco (see Problem 3).

Solution

From Problem 3 we have $I = (10^{6.9})(I_0)$ for the earthquake in San Francisco. Then, using a Richter number of 7.7, we obtain $I = (10^{7.7})(I_0)$ for the earthquake in Iran. Therefore, by comparison,

$$\frac{(10^{7.7})(I_0)}{(10^{6.9})(I_0)} = 10^{7.7-6.9} = 10^{0.8} \approx 6.3$$

The earthquake in Iran was about 6 times as intense as the one in San Francisco.

▶ **Now work Problem 57.**   ■

■ Logarithms with Base Other Than 10 or *e*

The basic approach of applying Property 5.8 and using either common or natural logarithms can also be used to evaluate a logarithm to some base other than 10 or *e*. The next example illustrates this idea.

EXAMPLE 7

Evaluate $\log_3 41$.

Solution

Let $x = \log_3 41$. Changing to exponential form, we obtain

$$3^x = 41$$

Now we can apply Property 5.8.

$$\log 3^x = \log 41$$

$$x \log 3 = \log 41$$

$$x = \frac{\log 41}{\log 3}$$

$$x = 3.3802 \quad \text{rounded to four decimal places}$$

Therefore we are claiming that 3 raised to the 3.3802 power will produce approximately 41. Check it!

▶ **Now work Problem 35.**   ■

Using the method of Example 7 to evaluate $\log_a r$ produces the following formula, which is often referred to as the **change-of-base formula** for logarithms.

Property 5.9

If a, b, and r are positive numbers, with $a \neq 1$ and $b \neq 1$, then

$$\log_a r = \frac{\log_b r}{\log_b a}$$

Property 5.9 provides us with another way to express logarithms with bases other than 10 or e in terms of common or natural logarithms. For example, $\log_4 78$ is of the form $\log_a r$ with $r = 78$ and $a = 4$. Therefore, in terms of common logarithms (base 10), we have the following:

$$\log_4 78 = \frac{\log_{10} 78}{\log_{10} 4}$$

$$= \frac{\log 78}{\log 4}$$

$$= 3.1427 \quad \text{rounded to four decimal places}$$

In a similar fashion, we can use natural logarithms to evaluate expressions such as $\log_4 78$.

$$\log_4 78 = \frac{\ln 78}{\ln 4}$$

$$= 3.1427 \quad \text{rounded to four decimal places}$$

Property 5.9 also provides us with another way of solving equations such as $3^x = 5$.

$$3^x = 5$$

$$x = \log_3 5 \qquad \text{Changed to logarithmic form}$$

$$x = \frac{\log 5}{\log 3} \qquad \text{Applied Property 5.9}$$

$$x = 1.46 \qquad \text{to the nearest hundredth}$$

Finally, by using Property 5.9, we can obtain a relationship between common and natural logarithms by letting $a = 10$ and $b = e$. Then

$$\log_a r = \frac{\log_b r}{\log_b a}$$

becomes

$$\log_{10} r = \frac{\log_e r}{\log_e 10}$$

$$\log_e r = (\log_e 10)(\log_{10} r)$$

$$\log_e r = (2.3026)(\log_{10} r)$$

Thus the natural logarithm of any positive number is approximately equal to 2.3026 times the common logarithm of the number.

Now we can use a graphing utility to graph logarithmic functions such as $f(x) = \log_2 x$. Using the change-of-base formula, we can express this function as $f(x) = \frac{\log x}{\log 2}$ or as $f(x) = \frac{\ln x}{\ln 2}$. The graph of $f(x) = \log_2 x$ is shown in Figure 5.17.

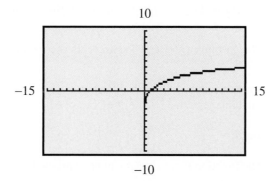

Figure 5.17

Finally, let's use a graphical approach to solve an equation that is cumbersome to solve with an algebraic approach.

E X A M P L E   8

Solve the equation $(5^x - 5^{-x})/2 = 3$.

Solution

First, we need to recognize that the solutions for the equation $(5^x - 5^{-x})/2 = 3$ are the x intercepts of the graph of the equation $y = (5^x - 5^{-x})/2 - 3$. Thus let's use a graphing utility to obtain the graph of this equation as shown in Figure 5.18. We use the zoom-in and trace features to determine that the graph crosses the x axis at approximately 1.13. Thus the solution set of the original equation is {1.13}.

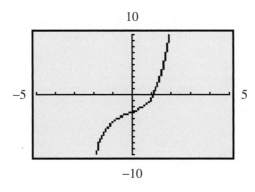

Figure 5.18

▶ **Now work Problem 75.**   ■

C O N C E P T   Q U I Z

For Problems 1–8, answer true or false.

1. For $x > 0$ and $y > 0$, if $\log_2 x = \log_2 y$, then $x = y$.

2. For $x > 0$ and $y > 0$, if $x = y$, then $\ln x = \log y$.

3. The formula $Q = Q_0 e^{at}$ represents the exponential decay of a substance if $0 < a < 1$.

4. The formula $Q = Q_0 e^{at}$ represents the exponential growth of a substance if $a > 0$.

5. An earthquake with a Richter number of 7.0 is 10 times more intense than an earthquake with a Richter number of 6.0.

6. $\log_3 7 = \dfrac{\ln 7}{\ln 3}$.

7. The function $f(x) = \log_4 x$ can be expressed as $f(x) = \dfrac{\log x}{\log 4}$.

8. The solution to a logarithmic equation cannot be an irrational number.

Problem Set 5.5

For Problems 1–18, solve each exponential equation. Express approximate solutions to the nearest hundredth.

1. $2^x = 9$ **2.** $3^x = 20$

3. $5^t = 123$ **4.** $4^t = 12$

5. $2^{x+1} = 7$ **6.** $3^{x-2} = 11$

7. $7^{2t-1} = 35$ **8.** $5^{3t+1} = 9$

9. $e^x = 4.1$ **10.** $e^x = 30$

11. $e^{x-1} = 8.2$ **12.** $e^{x-2} = 13.1$

13. $2e^x = 12.4$ **14.** $3e^x - 1 = 17$

15. $3^{x-1} = 2^{x+3}$ **16.** $5^{2x+1} = 7^{x+3}$

17. $5^{x-1} = 2^{2x+1}$ **18.** $3^{2x+1} = 2^{3x+2}$

For Problems 19–34, solve each logarithmic equation. Express irrational solutions in simplest radical form.

19. $\log x + \log(x + 3) = 1$

20. $\log x + \log(x + 21) = 2$

21. $\log(2x - 1) - \log(x - 3) = 1$

22. $\log(3x - 1) = 1 + \log(5x - 2)$

23. $\log(x - 2) = 1 - \log(x + 3)$

24. $\log(x + 1) = \log 3 - \log(2x - 1)$

25. $\log_b (x + 4) + \log_b (x - 2) = \log_b 2x$

26. $\log_b 3x + \log_b (x - 1) = \log_b (x + 4)$

27. $\ln (x + 4) - \ln (x + 3) = \ln x$

28. $\log_8 3x = \log_8 (x + 5) - \log_8 (x + 4)$

29. $\log(x + 1) - \log(x + 2) = \log \dfrac{1}{x}$

30. $\log(x + 2) - \log(2x + 1) = \log x$

31. $\ln(3t - 4) - \ln(t + 1) = \ln 2$

32. $\ln(2t + 5) = \ln 3 + \ln(t - 1)$

33. $\log(x^2) = (\log x)^2$

34. $\log \sqrt{x} = \sqrt{\log x}$

For Problems 35–42, evaluate each logarithm to three decimal places.

35. $\log_3 14$ **36.** $\log_4 94$

37. $\log_5 2.1$ **38.** $\log_6 0.345$

39. $\log_7 176$ **40.** $\log_8 296$

41. $\log_9 14.32$ **42.** $\log_7 0.024$

For Problems 43–61, solve each problem.

43. How long will it take $1000 to double if it is invested at 6% interest compounded semiannually?

44. How long will it take $750 to be worth $1000 if it is invested at 5% interest compounded quarterly?

45. How long will it take $500 to triple if it is invested at 4% interest compounded continuously?

46. How long will it take $2000 to double if it is invested at 8% interest compounded continuously?

47. At what rate of interest (to the nearest tenth of a percent) compounded annually will an investment of $200 grow to $350 in 5 years?

48. At what rate of interest (to the nearest tenth of a percent) compounded continuously will an investment of $500 grow to $900 in 10 years?

49. A piece of machinery valued at $30,000 depreciates at a rate of 10% yearly. How long will it take until the machinery has a value of $15,000?

50. For a certain strain of bacteria, the number present after t hours is given by the equation $Q = Q_0 e^{0.34t}$, where Q_0 represents the initial number of bacteria. How long will it take 400 bacteria to increase to 4000 bacteria?

51. The number of grams of a certain radioactive substance present after t hours is given by the equation $Q = Q_0 e^{-0.45t}$, where Q_0 represents the initial number of grams. How long will it take 2500 grams to be reduced to 1250 grams?

52. The atmospheric pressure in pounds per square inch is expressed by the equation $P(a) = 14.7e^{-0.21a}$, where a is the altitude above sea level measured in miles. If the atmospheric pressure in Cheyenne, Wyoming, is approximately 11.53 pounds per square inch, find its altitude above sea level. Express your answer to the nearest hundred feet.

53. Suppose you are given the equation $P(t) = P_0 e^{0.02t}$ to predict population growth, where P_0 represents an initial population and t is the time in years. How long does this equation predict it will take a city of 50,000 to have a population of 75,000?

54. In a certain bacterial culture, the equation $Q(t) = Q_0 e^{0.4t}$ yields the number of bacteria as a function of the time, where Q_0 is an initial number of bacteria and t is time measured in hours. How long will it take 500 bacteria to increase to 2000?

55. An earthquake in San Francisco in 1906 was reported to have a Richter number of 8.3. How did its intensity compare to the reference intensity?

56. An earthquake in Los Angeles in 1971 had an intensity of approximately 5 million times the reference intensity. What Richter number was associated with that earthquake?

57. Calculate how many times more intense an earthquake with a Richter number of 7.3 is than an earthquake with a Richter number of 6.4.

58. Calculate how many times more intense an earthquake with a Richter number of 8.9 is than an earthquake with a Richter number of 6.2.

59. In Problem 43 of Problem Set 5.2 we used the function $f(x) = 2.512^x$, where x is the higher magnitude minus the lower magnitude, to compare the relative brightness of stars. Suppose star A is 212 times brighter than star B. Find the difference, to the nearest tenth, of their magnitudes.

60. See Problem 59. If star C has a magnitude of 7 and is 100 times brighter than star D, find the magnitude of star D. Express your answer to the nearest whole number.

61. See Problem 59. If star E is 10,000 times brighter than star F, and if star F has a magnitude of 20, find the magnitude of star E. Express your answer to the nearest whole number.

■ ■ ■ THOUGHTS INTO WORDS

62. Explain the concept of a Richter number.

63. Explain how you would solve the equation $7^x = 134$.

64. Explain how you would evaluate $\log_4 79$.

65. How do logarithms with a base of 9 compare to logarithms with a base of 3?

■ ■ ■ FURTHER INVESTIGATIONS

66. Use the approach of Example 7 to develop Property 5.9.

67. Let $r = b$ in Property 5.9 and verify that
$$\log_a b = \frac{1}{\log_b a}.$$

68. To solve the equation $(5^x - 5^{-x})/2 = 3$, let's begin as follows:
$$\frac{5^x - 5^{-x}}{2} = 3$$
$$5^x - 5^{-x} = 6$$

$$5^x(5^x - 5^{-x}) = 6(5^x) \qquad \text{Multiply both sides by } 5^x.$$
$$5^{2x} - 1 = 6(5^x)$$

$$5^{2x} - 6(5^x) - 1 = 0$$

This final equation is of quadratic form. Finish the solution and check your answer against the answer in Example 8.

69. Solve the equation $y = (10^x + 10^{-x})/2$ for x in terms of y.

70. Solve the equation $y = (e^x - e^{-x})/2$ for x in terms of y.

GRAPHING CALCULATOR ACTIVITIES

71. Check your answers for Problems 15–18 by graphing the appropriate function and finding the x intercept.

72. Graph $f(x) = \log_2 x$. Then predict the graphs for $f(x) = \log_3 x$, $f(x) = \log_4 x$, and $f(x) = \log_8 x$. Now graph these three functions on the same set of axes with the graph of $f(x) = \log_2 x$.

73. Graph $f(x) = x$, $f(x) = 2^x$, and $f(x) = \log_2 x$ on the same set of axes.

74. Graph $f(x) = x$, $f(x) = \left(\dfrac{1}{2}\right)^x$, and $f(x) = \log_{1/2} x$ on the same set of axes.

75. Use both a graphical and an algebraic approach to solve the equation $(2^x - 2^{-x})/3 = 4$.

This chapter can be summarized in terms of three main topics: (1) exponents and exponential functions, (2) logarithms and logarithmic functions, and (3) applications of exponential and logarithmic functions.

■ Exponents and Exponential Functions

If a and b are positive numbers, and m and n are real numbers, then the following properties hold:

1. $b^n \cdot b^m = b^{n+m}$ Product of two powers

2. $(b^n)^m = b^{mn}$ Power of a power

3. $(ab)^n = a^n b^n$ Power of a product

4. $\left(\dfrac{a}{b}\right)^n = \dfrac{a^n}{b^n}$ Power of a quotient

5. $\dfrac{b^n}{b^m} = b^{n-m}$ Quotient of two powers

A function defined by an equation of the form

$$f(x) = b^x \qquad b > 0 \text{ and } b \neq 1$$

is called an **exponential function.** Figure 5.19 illustrates the general behavior of the graph of an exponential function of the form $f(x) = b^x$.

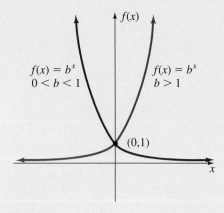

$f(x) = b^x$
$0 < b < 1$

$f(x) = b^x$
$b > 1$

$(0,1)$

Figure 5.19

■ Logarithms and Logarithmic Functions

If r is any positive real number, then the unique exponent t such that $b^t = r$ is called the **logarithm of r with base b;** it is denoted by $\log_b r$.

The following properties of logarithms are used frequently:

1. $\log_b b = 1$

2. $\log_b 1 = 0$

3. $b^{\log_b r} = r$

4. $\log_b rs = \log_b r + \log_b s$

5. $\log_b\left(\dfrac{r}{s}\right) = \log_b r - \log_b s$

6. $\log_b(r^p) = p \log_b r$

Logarithms with a base of 10 are called **common logarithms.** The expression $\log_{10} x$ is usually written $\log x$.

Many calculators are equipped with a common logarithmic function. Often a key labeled $\boxed{\log}$ is used to find common logarithms.

Natural logarithms are logarithms that have a base of e, where e is an irrational number whose decimal approximation to eight digits is 2.7182818. Natural logarithms are denoted by $\log_e x$ or $\ln x$.

Many calculators are also equipped with a natural logarithmic function. Often a key labeled $\boxed{\ln x}$ is used for this purpose.

A function defined by an equation of the form

$$f(x) = \log_b x \qquad b > 0 \text{ and } b \neq 1$$

is called a **logarithmic function.**

The graph of a logarithmic function (such as $y = \log_2 x$) can be determined by either changing the equation to exponential form ($2^y = x$) and plotting points, or by reflecting the graph of the inverse function ($y = 2^x$) across the line $y = x$. This last approach is based on the fact that exponential and logarithmic functions are inverses of each other.

Figure 5.20 illustrates the general behavior of the graph of a logarithmic function of the form $f(x) = \log_b x$.

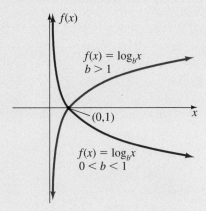

Figure 5.20

■ Applications

We use the following properties of equality frequently when solving exponential and logarithmic equations:

1. If $b > 0$, $b \neq 1$, and m and n are real numbers, then

$$b^n = b^m \quad \text{if and only if } n = m$$

2. If $x > 0, y > 0, b > 0$, and $b \neq 1$, then

$$x = y \quad \text{if and only if } \log_b x = \log_b y$$

A general formula for any principal, P, that is compounded n times per year for any number, t, of years at a given rate, r, is

$$A = P\left(1 + \frac{r}{n}\right)^{nt}$$

where A represents the total amount of money accumulated at the end of the t years. As n gets infinitely large, the value of $[1 + (1/n)]^n$ approaches the number e, where e equals 2.71828 to five decimal places.

The formula

$$A = Pe^{rt}$$

yields the accumulated value, A, of a sum of money, P, that has been invested for t years at a rate of $r\%$ **compounded continuously.**

The formula

$$Q = Q_0\left(\frac{1}{2}\right)^{t/h}$$

is referred to as the **half-life** formula.

The equation

$$Q(t) = Q_0 e^{kt}$$

is used as a mathematical model for exponential growth and decay problems.

The formula

$$R = \log \frac{I}{I_0}$$

yields the Richter number associated with the magnitude of an earthquake.

The formula

$$\log_a r = \frac{\log_b r}{\log_b a}$$

is often called the **change-of-base formula.**

Chapter 5   Review Problem Set

For Problems 1–10, evaluate each expression.

1. $8^{5/3}$

2. $-25^{3/2}$

3. $(-27)^{4/3}$

4. $\log_6 216$

5. $\log_7\left(\dfrac{1}{49}\right)$

6. $\log_2 \sqrt[3]{2}$

7. $\log_2\left(\dfrac{\sqrt[4]{32}}{2}\right)$

8. $\log_{10} 0.00001$

9. $\ln e$

10. $7^{\log_7 12}$

For Problems 11–24, solve each equation. Express approximate solutions to the nearest hundredth.

11. $\log_{10} 2 + \log_{10} x = 1$

12. $\log_3 x = -2$

13. $4^x = 128$

14. $3^t = 42$

15. $\log_2 x = 3$

16. $\left(\dfrac{1}{27}\right)^{3x} = 3^{2x-1}$

17. $2e^x = 14$

18. $2^{2x+1} = 3^{x+1}$

19. $\ln(x + 4) - \ln(x + 2) = \ln x$

20. $\log x + \log(x - 15) = 2$

21. $\log(\log x) = 2$

22. $\log(7x - 4) - \log(x - 1) = 1$

23. $\ln(2t - 1) = \ln 4 + \ln(t - 3)$

24. $64^{2t+1} = 8^{-t+2}$

For Problems 25–28, if $\log 3 = 0.4771$ and $\log 7 = 0.8451$, evaluate each expression.

25. $\log\left(\dfrac{7}{3}\right)$

26. $\log 21$

27. $\log 27$

28. $\log 7^{2/3}$

29. Express each of the following as the sum or difference of simpler logarithmic quantities. Assume that all variables represent positive real numbers.

a. $\log_b\left(\dfrac{x}{y^2}\right)$ **b.** $\log_b \sqrt[4]{xy^2}$ **c.** $\log_b\left(\dfrac{\sqrt{x}}{y^3}\right)$

30. Express each of the following as a single logarithm. Assume that all variables represent positive real numbers.

a. $3 \log_b x + 2 \log_b y$ **b.** $\dfrac{1}{2}\log_b y - 4 \log_b x$

c. $\dfrac{1}{2}(\log_b x + \log_b y) - 2 \log_b z$

For Problems 31–34, approximate each of the logarithms to two decimal places.

31. $\log_2 3$

32. $\log_3 2$

33. $\log_4 191$

34. $\log_2 0.23$

For Problems 35–42, graph each function.

35. $f(x) = \left(\dfrac{3}{4}\right)^x$

36. $f(x) = 2^{x+2}$

37. $f(x) = e^{x-1}$

38. $f(x) = -1 + \log x$

39. $f(x) = 3^x - 3^{-x}$

40. $f(x) = e^{-x^2/2}$

41. $f(x) = \log_2(x - 3)$

42. $f(x) = 3 \log_3 x$

For Problems 43–48, find the total amount of money accumulated at the end of the indicated time period for each of the investments.

43. $750 for 10 years at 8% compounded quarterly

44. $1250 for 15 years at 7% compounded monthly

45. $2500 for 20 years at 6.5% compounded semi-annually

46. $2500 for 10 years at 8% compounded continuously

47. $5000 for 15 years at 6% compounded continuously

48. $7500 for 20 years at 5% compounded continuously

For Problems 49–55, solve each problem.

49. How long will it take $100 to double if it is invested at 10% interest compounded annually?

50. How long will it take $1000 to be worth $3500 if it is invested at 7.5% interest compounded quarterly?

51. At what rate of interest (to the nearest tenth of a percent) compounded continuously will an investment of $500 grow to $1000 in 8 years?

52. Suppose that the present population of a city is 50,000 and suppose that the equation $P(t) = P_0 e^{0.02t}$, where P_0 represents an initial population, can be used to estimate future populations. Estimate the population of that city in 10 years, 15 years, and 20 years.

53. The number of bacteria present in a certain culture after t hours is given by the equation $Q = Q_0 e^{0.29t}$, where Q_0 represents the initial number of bacteria. How long will it take 500 bacteria to increase to 2000 bacteria?

54. Suppose that a certain radioactive substance has a half-life of 40 days. If there are presently 750 grams of the substance, how much, to the nearest gram, will remain after 100 days?

55. An earthquake occurred in Mexico City in 1985 that had an intensity level about 125,000,000 times the reference intensity. Find the Richter number for that earthquake.

For Problems 1–4, evaluate each expression.

1. $\log_3 \sqrt{3}$ **2.** $\log_2(\log_2 4)$

3. $-2 + \ln e^3$ **4.** $\log_2(0.5)$

For Problems 5–10, solve each equation.

5. $4^x = \dfrac{1}{64}$ **6.** $9^x = \dfrac{1}{27}$

7. $2^{3x-1} = 128$ **8.** $\log_9 x = \dfrac{5}{2}$

9. $\log x + \log(x + 48) = 2$

10. $\ln x = \ln 2 + \ln(3x - 1)$

For Problems 11–13, given that $\log_3 4 = 1.2619$ and $\log_3 5 = 1.4650$, evaluate each expression.

11. $\log_3 100$

12. $\log_3 1.25$

13. $\log_3 \sqrt{5}$

14. Solve $e^x = 176$ to the nearest hundredth.

15. Solve $2^{x-2} = 314$ to the nearest hundredth.

16. Determine $\log_5 632$ to four decimal places.

For Problems 17–23, solve each problem.

17. If \$4500 is invested at 6% interest compounded continuously, how much money has accumulated at the end of 10 years?

18. How many times more intense is an earthquake with a Richter number of 8.5 than an earthquake with a Richter number of 6.8? Express your answer to the nearest whole number.

19. A piece of farm machinery valued at \$45,000 depreciates at a rate of 15% per year. How long (to the nearest tenth of a year) will it take until the machinery has a value of \$10,000?

20. If \$3500 is invested at 7.5% interest compounded quarterly, how much money has accumulated at the end of 8 years?

21. How long will it take \$5000 to be worth \$12,500 if it is invested at 7% compounded annually? Express your answer to the nearest tenth of a year.

22. The number of bacteria present in a certain culture after t hours is given by $Q(t) = Q_0 e^{0.23t}$, where Q_0 represents the initial number of bacteria. How long will it take 400 bacteria to increase to 2400 bacteria? Express your answer to the nearest tenth of an hour.

23. Suppose that a certain radioactive substance has a half-life of 50 years. If there are presently 7500 grams of the substance, how much will remain after 32 years? Express your answer to the nearest gram.

For Problems 24 and 25, graph each of the functions.

24. $f(x) = e^x - 2$

25. $f(x) = \log_2(x - 2)$

For Problems 1–8, evaluate each expression.

1. $\sqrt[3]{0.008}$

2. $\log_2 64$

3. $\ln e^4$

4. $(4^{-1} + 3^{-2})^{-1}$

5. $\left(\dfrac{2^{-1}}{3^{-3}}\right)^{-2}$

6. $\log_3\left(\dfrac{1}{9}\right)$

7. $\log_2 \sqrt[5]{2}$

8. $\sqrt{1\dfrac{7}{9}}$

For Problems 9–12, perform the indicated operations involving rational expressions. Express final answers in simplest form.

9. $\dfrac{x}{x^2 - 9} \div \dfrac{x^2 + x}{x^2 + 4x + 3}$

10. $\dfrac{4}{x^2 + 3x} + \dfrac{7}{x}$

11. $\dfrac{x^3 - 8}{x^3 + 2x^2 + 4x} \cdot \dfrac{x^2 + 4}{x^2 + 5x - 14}$

12. $\dfrac{3n + 2}{6} - \dfrac{n + 1}{14}$

13. Simplify the radical expression
$3\sqrt{8} - \sqrt{18} + 2\sqrt{32}$.

14. Simplify the complex fraction $\dfrac{\dfrac{3}{xy} - \dfrac{2}{x^2}}{\dfrac{4}{y} + \dfrac{7}{x}}$.

15. Write the equation of the line that has a slope of $-\dfrac{5}{6}$ and contains the point $(4, -3)$.

16. Find the slope of the line $-2x + 9y = 14$.

17. Determine the linear function whose graph is a line that contains the points $(3, 5)$ and $(-2, -4)$.

18. Determine the equation of the line that contains the point $(2, 8)$ and is parallel to the line $3x + 5y = 10$.

19. Determine the linear function whose graph is a line perpendicular to the line $g(x) = \dfrac{1}{2}x - 4$ and contains the point $(-3, 7)$.

20. If $f(x) = 4x - 1$ and $g(x) = x^2 - x + 7$, find $f(g(-1))$ and $g(f(2))$.

21. If $f(x) = |x - 2|$ and $g(x) = \sqrt{x + 3}$, find $f(g(-3))$ and $g(f(8))$.

22. If $f(x) = 7x - 3$ and $g(x) = 2x^2 - 3x - 4$, find $f(g(x))$.

23. If $f(x) = x^2 - 1$ and $g(x) = 3x^2 + 4x - 2$, find $g(f(x))$.

For Problems 24–28, express each in simplest radical form. All variables represent positive real numbers.

24. $\dfrac{5\sqrt{2}}{7\sqrt{6}}$

25. $\sqrt{50x^3y^2}$

26. $\sqrt[3]{56x^5y^4}$

27. $\dfrac{\sqrt[3]{3}}{\sqrt[3]{4}}$

28. $\dfrac{\sqrt{2}}{2\sqrt{3} - \sqrt{6}}$

For Problems 29–32, find each of the indicated products or quotients. Express answers in the standard form of a complex number.

29. $(-2 - 3i)(5 + 6i)$

30. $\dfrac{4 - 2i}{1 + 3i}$

31. $(6 + 9i)(6 - 9i)$

32. $-\dfrac{3}{2i}$

33. Find the value of $\log_3 94$ to the nearest hundredth.

34. Find the inverse of the function $f(x) = \dfrac{1}{2}x - 6$.

35. Find the quotient when $3x^3 + x^2 - 14x - 8$ is divided by $x + 2$.

36. Find the quotient when $2x^3 + 13x^2 - 11x + 2$ is divided by $2x - 1$.

For Problems 37–61, solve each equation.

37. $(x - 2)(3x + 7) = 0$ **38.** $(x + 4)(x - 1) = 66$

39. $\frac{2}{3}(x - 1) - \frac{3}{4}(2x + 3) = -\frac{5}{6}(3x + 4)$

40. $\frac{4x - 1}{7} - \frac{5x - 2}{4} = -2$

41. $(2x - 1)(3x + 6) = (5x - 2)(2x - 1)$

42. $(x - 3)(x + 5) = (2x - 3)(x + 6)$

43. $\frac{-4}{7x - 2} = \frac{3}{6x + 1}$

44. $0.08(x + 200) = 0.07x + 20$

45. $0.05(x - 2.5) = 15.4$ **46.** $\frac{x}{x - 2} + \frac{2}{3} = \frac{2}{x - 2}$

47. $6x^2 + x - 15 = 0$ **48.** $x^2 + 4x + 5 = 0$

49. $x^2 - 6x - 216 = 0$ **50.** $|3x - 7| = 14$

51. $|2x - 1| = |4x + 5|$ **52.** $2x^3 + 15x^2 - 27x = 0$

53. $\frac{3x}{x^2 + x - 6} + \frac{2}{x^2 + 4x + 3} = \frac{x}{x^2 - x - 2}$

54. $8^{x-1} = 4^{3x+2}$

55. $\log_3(x + 2) + \log_3(x + 1) = 2$

56. $\sqrt{x + 6} = x$

57. $x^{2/3} + x^{1/3} - 20 = 0$

58. $2x^4 + 11x^2 - 63 = 0$

59. $x^3 - 3x^2 - 18x + 40 = 0$

60. $x^4 - 2x^3 - 23x^2 - 12x + 36 = 0$

61. $2x^3 - 19x^2 + 49x - 20 = 0$

For Problems 62–73, solve each inequality and express the solution set using interval notation.

62. $(2x + 3)(x - 4) > 0$ **63.** $x(x - 5)(x + 7) \leq 0$

64. $\frac{x + 6}{x - 4} \geq 0$ **65.** $\frac{x - 1}{x + 3} + 1 < 0$

66. $|8x - 3| \geq 6$ **67.** $|10x + 1| < 4$

68. $-3(2x - 1) - (x + 4) \leq 2$

69. $2(x + 3) > 2(x - 4)$

70. $\frac{x + 1}{3} - \frac{2x + 3}{5} \geq 1$ **71.** $-4 \leq \frac{4x - 3}{2} \leq 4$

72. $2x^2 + x - 21 > 0$ **73.** $\left|\frac{x - 1}{x + 2}\right| > 3$

74. Find the length of the major axis of the ellipse $2x^2 + y^2 = 16$.

75. Find the equations of the asymptotes of the hyperbola $4x^2 - y^2 = 16$.

76. Find the length of a line segment whose endpoints are at $(-1, -3)$ and $(4, 2)$. Express the answer in simplest radical form.

77. In the xy plane, find the coordinates of the point that is three-fourths of the distance from $(2, -6)$ to $(-2, 4)$.

78. What kind of symmetry does the graph of $y^2 = 4x$ possess?

79. What kind of symmetry does the graph of $xy = -10$ possess?

80. Find the center of the circle $x^2 + 16x + y^2 - 14y + 109 = 0$.

81. If V varies jointly as B and h, and $V = 90$ when $B = 27$ and $h = 10$, find V when $B = 18$ and $h = 5$.

82. y is directly proportional to x and inversely proportional to w. If $y = \frac{6}{5}$ when $x = 2$ and $w = 5$, find y when $x = 7$ and it $w = 4$.

83. Find the x intercepts of the graph of the function $f(x) = x^3 - 8x^2 + x + 42$.

84. Find the x intercepts of the graph of the function $f(x) = x^4 - 2x^3 - 8x^2$.

85. Find the equations of the asymptotes for the graph of the function $f(x) = \dfrac{2x}{x-3}$.

86. Find the equations of the asymptotes for the graph of the function $f(x) = \dfrac{x^2 + 2x - 4}{x - 3}$.

For Problems 87–94, graph each function.

87. $f(x) = x^2 - 2x + 4$ **88.** $f(x) = -3x^2 + 6x - 4$

89. $f(x) = -2x - 1$ **90.** $f(x) = 2\sqrt{x - 1} + 3$

91. $f(x) = -|x + 2| - 1$ **92.** $f(x) = -(x - 2)^3 - 2$

93. $f(x) = \log_3 x$ **94.** $f(x) = 5^{x-2}$

For Problems 95–104, use an equation or an inequality to help solve each problem.

95. Find four consecutive whole numbers such that the product of the smallest and largest is 2 less than the product of the other two numbers.

96. Find four consecutive whole numbers such that the product of the two largest is 34 more than the product of the two smallest numbers.

97. In a class of 200 students, the number of females is 20 more than twice the number of males. How many males and how many females are in the class?

98. Ramon is taking a college algebra course in which four tests determine his final grade. He has scores of 88, 91, and 92 on the first three tests. To get an A for the course, he must average at least 90 on the four tests. What must he score on the last test to get an A?

99. Lian is paid double-time for each hour worked over 40 hours in a week. Last week she worked 45 hours and earned $425. What is her normal hourly rate?

100. If a ring costs a jeweler $780, at what price should it be sold to make a profit of 40% of the selling price?

101. In Problem 100, at what price should the ring be sold if the jeweler is satisfied to make a profit of 40% of the cost of the ring?

102. Jose bought a pair of slacks at a 30% discount sale for $55.30. What was the original price of the slacks?

103. A container has 4 liters of a 30% alcohol solution in it. How much pure alcohol should be added to raise it to a 50% solution?

104. Justin contracted to paint a small one-story house for $560. It took him 10 hours longer than he had anticipated, so he earned $2.80 per hour less than he originally calculated. How long had he anticipated it would take him to paint the house?

6

Systems of Equations

*A system of two linear
equations in two variables can
be used to approximate the
effect of the jet stream on
airline schedules.*

© Imagestopshop/Alamy

A 10% salt solution is to be mixed with a 20% salt solution to produce 20 gallons of a 17.5% salt solution. How many gallons of the 10% solution and how many gallons of the 20% solution should be mixed? The two equations $x + y = 20$ and $0.10x + 0.20y = 0.175(20)$ algebraically represent the conditions of the problem, where x represents the number of gallons of the 10% solution, and y represents the number of gallons of the 20% solution. The two equations considered together form a system of linear equations, and the problem can be solved by solving the system of equations.

In this chapter we will begin by reviewing some techniques for solving systems of linear equations that involve two or three variables. Then, because many applications of mathematics require the use of large numbers of variables and equations, we will introduce some additional techniques for solving such extensive systems. These techniques also form a basis for solving systems by using a computer.

6.1   Systems of Two Linear Equations in Two Variables

Objectives

■ Solve systems of two linear equations using either the substitution method or the elimination-by-addition method.

■ Solve application problems using a system of equations.

In Chapter 2 we stated that any equation of the form $Ax + By = C$, where A, B, and C are real numbers (A and B not both zero), is a **linear equation** in the two variables x and y, and its graph is a straight line. Two linear equations in two variables considered together form a **system of two linear equations in two variables,** as illustrated by the following examples:

$$\begin{pmatrix} x + y = 6 \\ x - y = 2 \end{pmatrix} \qquad \begin{pmatrix} 3x + 2y = 1 \\ 5x - 2y = 23 \end{pmatrix} \qquad \begin{pmatrix} 4x - 5y = 21 \\ -3x + y = -7 \end{pmatrix}$$

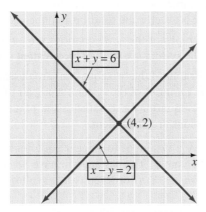

Figure 6.1

To solve a system (such as any of these three examples) means to find all of the ordered pairs that simultaneously satisfy both equations in the system. For example, if we graph the two equations $x + y = 6$ and $x - y = 2$ on the same set of axes, as in Figure 6.1, then the ordered pair associated with the point of intersection of the two lines is the **solution of the system.** Thus we say that $\{(4, 2)\}$ is the solution set of the system

$$\begin{pmatrix} x + y = 6 \\ x - y = 2 \end{pmatrix}$$

To check the solution, we substitute 4 for x and 2 for y in the two equations:

$x + y = 6$ becomes $4 + 2 = 6$, a true statement

$x - y = 2$ becomes $4 - 2 = 2$, a true statement

Because the graph of a linear equation in two variables is a straight line, there are three possible situations that can occur when we are solving a system of two linear equations in two variables. Each situation is shown in Figure 6.2.

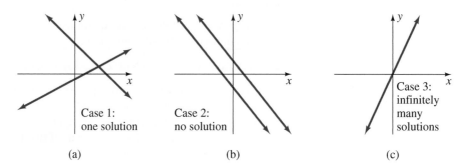

Case 1:
one solution

Case 2:
no solution

Case 3:
infinitely
many
solutions

(a)   (b)   (c)

Figure 6.2

Case 1 The graphs of the two equations are two lines that intersect in *one* point. There is exactly one solution, and the system is called a **consistent system.**

Case 2 The graphs of the two equations are parallel lines. There is *no solution*, and the system is called an **inconsistent system.**

Case 3 The graphs of the two equations are the same line, and there are *infinitely many solutions* of the system. Any pair of real numbers that satisfies one of the equations also satisfies the other equation, and we say that the equations are **dependent.**

Thus as we solve a system of two linear equations in two variables, we can expect one of three outcomes: The system will have *no* solutions, *one* ordered pair as a solution, or *infinitely many* ordered pairs as solutions.

■ The Substitution Method

Solving specific systems of equations by graphing requires accurate graphs. However, unless the solutions are integers, it is difficult to obtain exact solutions from a graph. Therefore we will consider some other techniques for solving systems of equations.

The **substitution method,** which works especially well with systems of two equations in two unknowns, can be described as follows.

Step 1 Solve one of the equations for one variable in terms of the other. (If possible, make a choice that will avoid fractions.)

Step 2 Substitute the expression obtained in step 1 into the other equation, producing an equation in one variable.

Step 3 Solve the equation obtained in step 2.

Step 4 Use the solution obtained in step 3, along with the expression obtained in step 1, to determine the solution of the system.

E X A M P L E   1

Solve the system $\begin{pmatrix} x - 3y = -25 \\ 4x + 5y = 19 \end{pmatrix}$.

Solution

Let's solve the first equation for x in terms of y to produce

$$x = 3y - 25$$

Now we substitute $3y - 25$ for x in the second equation and solve for y.

$$4x + 5y = 19$$
$$4(3y - 25) + 5y = 19$$
$$12y - 100 + 5y = 19$$
$$17y = 119$$
$$y = 7$$

Next we substitute 7 for y in the equation $x = 3y - 25$ to obtain

$$x = 3(7) - 25 = -4$$

The solution set of the given system is $\{(-4, 7)\}$. (You should check this solution in both of the original equations.)

▶ **Now work Problem 3.**   ∎

E X A M P L E   2

Solve the system $\begin{pmatrix} 5x + 9y = -2 \\ 2x + 4y = -1 \end{pmatrix}$.

Solution

A glance at the system should tell us that solving either equation for either variable will produce a fractional form, so let's just use the first equation and solve for x in terms of y.

$$5x + 9y = -2$$
$$5x = -9y - 2$$
$$x = \frac{-9y - 2}{5}$$

Now we can substitute this value for x into the second equation and solve for y.

$$2x + 4y = -1$$
$$2\left(\frac{-9y - 2}{5}\right) + 4y = -1$$
$$2(-9y - 2) + 20y = -5 \qquad \text{Multiplied both sides by 5}$$
$$-18y - 4 + 20y = -5$$
$$2y - 4 = -5$$
$$2y = -1$$
$$y = -\frac{1}{2}$$

Now we can substitute $-\dfrac{1}{2}$ for y in $x = \dfrac{-9y - 2}{5}$:

$$x = \frac{-9\left(-\dfrac{1}{2}\right) - 2}{5} = \frac{\dfrac{9}{2} - 2}{5} = \frac{1}{2}$$

The solution set is $\left\{\left(\dfrac{1}{2}, -\dfrac{1}{2}\right)\right\}$.

▶ **Now work Problem 13.** ∎

EXAMPLE 3

Solve the system $\begin{pmatrix} 6x - 4y = 18 \\ y = \dfrac{3}{2}x - \dfrac{9}{2} \end{pmatrix}$.

Solution

The second equation is given in appropriate form for us to begin the substitution process. We substitute $\dfrac{3}{2}x - \dfrac{9}{2}$ for y into the first equation to yield

$$6x - 4y = 18$$
$$6x - 4\left(\frac{3}{2}x - \frac{9}{2}\right) = 18$$
$$6x - 6x + 18 = 18$$
$$18 = 18$$

Our obtaining a true numerical statement ($18 = 18$) indicates that the system has infinitely many solutions. Any ordered pair that satisfies one of the equations will also satisfy the other equation. Thus in the second equation of the original system, if we let $x = k$, then $y = \dfrac{3}{2}k - \dfrac{9}{2}$. Therefore the solution set can be expressed $\left\{\left(k, \dfrac{3}{2}k - \dfrac{9}{2}\right) \,\middle|\, k \text{ is a real number}\right\}$. If some specific solutions are needed, they can be generated by the ordered pair $\left(k, \dfrac{3}{2}k - \dfrac{9}{2}\right)$. For example, if we let $k = 1$, then we get $\dfrac{3}{2}(1) - \dfrac{9}{2} = -\dfrac{6}{2} = -3$. Thus the ordered pair $(1, -3)$ is a member of the solution set of the given system.

▶ **Now work Problem 9.** ∎

■ The Elimination-by-Addition Method

Now let's consider the **elimination-by-addition method** for solving a system of equations. This is a very important method because it is the basis for developing other techniques for solving systems that contain many equations and variables. The method involves replacing systems of equations with *simpler equivalent systems* until we obtain a system where the solutions are obvious. **Equivalent systems of equations are systems that have exactly the same solution set.** The following operations or transformations can be applied to a system of equations to produce an equivalent system:

1. Any two equations of the system can be interchanged.
2. Both sides of any equation of the system can be multiplied by any nonzero real number.
3. Any equation of the system can be replaced by the sum of that equation and a nonzero multiple of another equation.

EXAMPLE 4

Solve the system $\begin{pmatrix} 3x + 5y = -9 \\ 2x - 3y = 13 \end{pmatrix}$. $\qquad$ (1) (2)

Solution

We can replace the given system with an equivalent system by multiplying equation (2) by -3:

$$\begin{pmatrix} 3x + 5y = -9 \\ -6x + 9y = -39 \end{pmatrix} \qquad \begin{matrix} (3) \\ (4) \end{matrix}$$

Now let's replace equation (4) with an equation formed by multiplying equation (3) by 2 and adding this result to equation (4):

$$\begin{pmatrix} 3x + 5y = -9 \\ 19y = -57 \end{pmatrix} \qquad \begin{matrix} (5) \\ (6) \end{matrix}$$

From equation (6) we can easily determine that $y = -3$. Then, substituting -3 for y in equation (5) produces

$$3x + 5(-3) = -9$$
$$3x - 15 = -9$$
$$3x = 6$$
$$x = 2$$

The solution set for the given system is $\{(2, -3)\}$.

▶ **Now work Problem 23.** ∎

Remark: We are using a format for the elimination-by-addition method that highlights the use of equivalent systems. In Section 6.3 this format will lead naturally to an approach using matrices. Thus it is beneficial to stress the use of equivalent systems at this time.

EXAMPLE 5

Solve the system

$$\begin{pmatrix} \dfrac{1}{2}x + \dfrac{2}{3}y = -4 \\ \dfrac{1}{4}x - \dfrac{3}{2}y = 20 \end{pmatrix} \qquad \begin{matrix} (7) \\ \\ (8) \end{matrix}$$

Solution

The given system can be replaced with an equivalent system by multiplying equation (7) by 6 and equation (8) by 4:

$$\left(\begin{matrix} 3x + 4y = -24 \\ x - 6y = 80 \end{matrix} \right)$$

$$(9)$$
$$(10)$$

Now let's exchange equations (9) and (10):

$$\left(\begin{matrix} x - 6y = 80 \\ 3x + 4y = -24 \end{matrix} \right)$$

$$(11)$$
$$(12)$$

We can replace equation (12) with an equation formed by multiplying equation (11) by -3 and adding this result to equation (12):

$$\left(\begin{matrix} x - 6y = 80 \\ 22y = -264 \end{matrix} \right)$$

$$(13)$$
$$(14)$$

From equation (14) we can determine that $y = -12$. Then, substituting -12 for y in equation (13) produces

$$x - 6(-12) = 80$$
$$x + 72 = 80$$
$$x = 8$$

The solution set of the given system is $\{(8, -12)\}$. (Check this!)

▶ **Now work Problem 29.** ■

EXAMPLE 6

Solve the system $\left(\begin{matrix} x - 4y = 9 \\ 2x - 8y = 7 \end{matrix} \right)$.

$$(15)$$
$$(16)$$

Solution

We can replace equation (16) with an equation formed by multiplying equation (15) by -2 and adding this result to equation (16):

$$\left(\begin{matrix} x - 4y = 9 \\ 0 - 0 = -11 \end{matrix} \right)$$

$$(17)$$
$$(18)$$

The statement $0 = -11$ is a contradiction, and therefore the original system is *inconsistent;* it has no solution. The solution set is $\varnothing$. ■

Both the elimination-by-addition and substitution methods can be used to obtain exact solutions for any system of two linear equations in two unknowns. Sometimes the issue is deciding which method to use on a particular system. Some systems lend themselves to one or the other methods by virtue of the original format of the equations. We will illustrate this idea when we solve some word problems.

■ Using Systems to Solve Problems

Many word problems that we solved earlier in this text with one variable and one equation can also be solved by using a system of two linear equations in two variables. In fact, in many of these problems you may find it more natural to use two variables and two equations.

PROBLEM 1

BJ always runs his pontoon boat at full throttle, which results in the boat traveling at a constant speed. Going up the river against the current, the boat traveled 72 miles in 4.5 hours. The return trip down the river with the current only took three hours. Find the speed of the boat and the speed of the current.

Solution

Let x represent the speed of the boat and let y represent the speed of the current. Going up the river, the rate of the boat against the current is $\dfrac{72 \text{ miles}}{4.5 \text{ hours}} = 16$ miles per hour. Going down the river, the rate of the boat going with the current is $\dfrac{72 \text{ miles}}{3 \text{ hours}} = 24$ miles per hour. The problem translates into the following system of equations.

$$\begin{pmatrix} x + y = 24 \\ x - y = 16 \end{pmatrix} \begin{matrix} \longleftarrow \text{ Rate of boat going with the current} \\ \longleftarrow \text{ Rate of boat going against the current} \end{matrix}$$

Let's use the elimination-by-addition method to solve the system. The second equation can be replaced by an equation formed by adding the two equations.

$$\begin{pmatrix} x + y = 24 \\ 2x = 40 \end{pmatrix}$$

Solving the second equation we can determine that $x = 20$. Now substituting 20 for x in the first equation produces

$$20 + y = 24$$
$$y = 4$$

Therefore, the speed of the boat is 20 miles per hour, and the speed of the current is 4 miles per hour.

⊙ **Now work Problem 65.**   ■

PROBLEM 2

Lucinda invested $950, part of it at 6% interest and the remainder at 8%. Her total yearly income from the two investments was $71.00. How much did she invest at each rate?

Solution

Let x represent the amount invested at 6% and y the amount invested at 8%. The problem translates into the following system:

$$\begin{pmatrix} x + y = 950 \\ 0.06x + 0.08y = 71.00 \end{pmatrix}$$ ⟵ The two investments total $950
⟵ The yearly interest from the two investments totals $71.00

We can multiply the second equation by 100 to produce an equivalent system:

$$\begin{pmatrix} x + y = 950 \\ 6x + 8y = 7100 \end{pmatrix}$$

Because neither equation is solved for one variable in terms of the other, let's use the elimination-by-addition method to solve the system. The second equation can be replaced by an equation formed by multiplying the first equation by -6 and adding this result to the second equation:

$$\begin{pmatrix} x + y = 950 \\ 2y = 1400 \end{pmatrix}$$

Solving the second equation we can determine that $y = 700$. Now we substitute 700 for y in the equation $x + y = 950$:

$$x + 700 = 950$$
$$x = 250$$

Therefore Lucinda must have invested $250 at 6% and $700 at 8%.

Now work Problem 63. ■

The two-variable expression $10t + u$ can be used to represent any two-digit whole number. The t represents the tens digit, and the u represents the units digit. For example, if $t = 4$ and $u = 8$, then $10t + u$ becomes $10(4) + 8 = 48$. Now let's use this general representation for a two-digit number to help solve a problem.

PROBLEM 3

The units digit of a two-digit number is one more than twice the tens digit. The number with the digits reversed is 45 larger than the original number. Find the original number.

Solution

Let u represent the units digit of the original number, and let t represent the tens digit. Then $10t + u$ represents the original number, and $10u + t$ represents the new number with the digits reversed. The problem translates into the following system:

$$\left(\begin{array}{l} u = 2t + 1 \\ 10u + t = 10t + u + 45 \end{array} \right)$$

The units digit is 1 more than twice the tens digit

The number with the digits reversed is 45 larger than the original number

When we simplify the second equation, the system becomes

$$\left(\begin{array}{l} u = 2t + 1 \\ u - t = 5 \end{array} \right)$$

Because of the form of the first equation, this system lends itself to solution by the substitution method. We substitute $2t + 1$ for u in the second equation to produce

$$(2t + 1) - t = 5$$
$$t + 1 = 5$$
$$t = 4$$

Now we substitute 4 for t in the equation $u = 2t + 1$ to get

$$u = 2(4) + 1 = 9$$

The tens digit is 4 and the units digit is 9, so the number is 49.

▶ **Now work Problem 57.**   ∎

In our final example of this section we will use a graphing utility to help solve a system of equations.

EXAMPLE 7

Solve the system $\left(\begin{array}{l} 1.14x + 2.35y = -7.12 \\ 3.26x - 5.05y = 26.72 \end{array} \right)$.

Solution

We began this section by graphing the equations in a system to find the solution. For this problem let's use a graphing utility to help find the solution of the system of equations. First, we need to solve each equation for y in terms of x. Thus the system becomes

$$\left(\begin{array}{l} y = \dfrac{-7.12 - 1.14x}{2.35} \\[3mm] y = \dfrac{3.26x - 26.72}{5.05} \end{array} \right)$$

Now we can enter both of these equations into a graphing utility and obtain Figure 6.3. It appears that the point of intersection is at approximately $x = 2$ and $y = -4$. By direct substitution into the given equations, we can verify that the point of intersection is exactly $(2, -4)$.

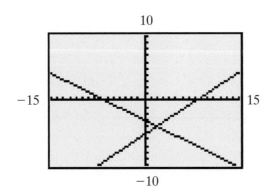

Figure 6.3

⊳ **Now work Problem 84f.**   ■

For Problems 1–8, answer true or false.

1. To *solve a system of equations* means to find all the ordered pairs that satisfy every equation in the system.

2. A consistent system of two linear equations will have more than one solution.

3. If the graph of a system of two distinct linear equations results in two distinct parallel lines, then the system has no solution.

4. If the graphs of the two equations in a system are the same line, then the equations in the system are dependent.

5. Every system of equations has a solution.

6. For the system $\left(\begin{array}{c} 2x + y = 4 \\ x + 5y = 10 \end{array} \right)$, the ordered pair $(1, 2)$ is a solution.

7. Graphing a system of equations is the most accurate method for finding the solution of the system.

8. The only possibilities for the solution set of a system of two linear equations are no solutions, one solution, or two solutions.

Problem Set 6.1

For Problems 1–18, solve each system by using the substitution method.

1. $\left(\begin{array}{c} x + y = 16 \\ y = x + 2 \end{array} \right)$

2. $\left(\begin{array}{c} 2x + 3y = -5 \\ y = 2x + 9 \end{array} \right)$

3. $\left(\begin{array}{c} x = 3y - 25 \\ 4x + 5y = 19 \end{array} \right)$

4. $\left(\begin{array}{c} 3x - 5y = 25 \\ x = y + 7 \end{array} \right)$

5. $\left(\begin{array}{c} y = \dfrac{2}{3}x - 1 \\ 5x - 7y = 9 \end{array} \right)$

6. $\left(\begin{array}{c} y = \dfrac{3}{4}x + 5 \\ 4x - 3y = -1 \end{array} \right)$

7. $\left(\begin{array}{c} a = 4b + 13 \\ 3a + 6b = -33 \end{array} \right)$

8. $\left(\begin{array}{c} 9a - 2b = 28 \\ b = -3a + 1 \end{array} \right)$

9. $\begin{pmatrix} 2x - 3y = 4 \\ y = \dfrac{2}{3}x - \dfrac{4}{3} \end{pmatrix}$ **10.** $\begin{pmatrix} t + u = 11 \\ t = u + 7 \end{pmatrix}$

11. $\begin{pmatrix} u = t - 2 \\ t + u = 12 \end{pmatrix}$ **12.** $\begin{pmatrix} y = 5x - 9 \\ 5x - y = 9 \end{pmatrix}$

13. $\begin{pmatrix} 4x + 3y = -7 \\ 3x - 2y = 16 \end{pmatrix}$

14. $\begin{pmatrix} 5x - 3y = -34 \\ 2x + 7y = -30 \end{pmatrix}$ **15.** $\begin{pmatrix} 5x - y = 4 \\ y = 5x + 9 \end{pmatrix}$

16. $\begin{pmatrix} 2x + 3y = 3 \\ 4x - 9y = -4 \end{pmatrix}$ **17.** $\begin{pmatrix} 4x - 5y = 3 \\ 8x + 15y = -24 \end{pmatrix}$

18. $\begin{pmatrix} 4x + y = 9 \\ y = 15 - 4x \end{pmatrix}$

For Problems 19–34, solve each system by using the elimination-by-addition method.

19. $\begin{pmatrix} 3x + 2y = 1 \\ 5x - 2y = 23 \end{pmatrix}$ **20.** $\begin{pmatrix} 4x + 3y = -22 \\ 4x - 5y = 26 \end{pmatrix}$

21. $\begin{pmatrix} x - 3y = -22 \\ 2x + 7y = 60 \end{pmatrix}$ **22.** $\begin{pmatrix} 6x - y = 3 \\ 5x + 3y = -9 \end{pmatrix}$

23. $\begin{pmatrix} 4x - 5y = 21 \\ 3x + 7y = -38 \end{pmatrix}$ **24.** $\begin{pmatrix} 5x - 3y = -34 \\ 2x + 7y = -30 \end{pmatrix}$

25. $\begin{pmatrix} 5x - 2y = 19 \\ 5x - 2y = 7 \end{pmatrix}$ **26.** $\begin{pmatrix} 3a - 2b = 5 \\ 2a + 7b = 9 \end{pmatrix}$

27. $\begin{pmatrix} 6a - 3b = 4 \\ 5a + 2b = -1 \end{pmatrix}$ **28.** $\begin{pmatrix} 7x + 2y = 11 \\ 7x + 2y = -4 \end{pmatrix}$

29. $\begin{pmatrix} \dfrac{2}{3}s + \dfrac{1}{4}t = -1 \\ \dfrac{1}{2}s - \dfrac{1}{3}t = -7 \end{pmatrix}$ **30.** $\begin{pmatrix} \dfrac{1}{4}s - \dfrac{2}{3}t = -3 \\ \dfrac{1}{3}s + \dfrac{1}{3}t = 7 \end{pmatrix}$

31. $\begin{pmatrix} \dfrac{x}{2} - \dfrac{2y}{5} = \dfrac{-23}{60} \\ \dfrac{2x}{3} + \dfrac{y}{4} = \dfrac{-1}{4} \end{pmatrix}$ **32.** $\begin{pmatrix} \dfrac{2x}{3} - \dfrac{y}{2} = \dfrac{3}{5} \\ \dfrac{x}{4} + \dfrac{y}{2} = \dfrac{7}{80} \end{pmatrix}$

33. $\begin{pmatrix} \dfrac{4x}{5} - \dfrac{3y}{2} = \dfrac{1}{5} \\ -2x + y = -1 \end{pmatrix}$ **34.** $\begin{pmatrix} \dfrac{3x}{2} - \dfrac{2y}{7} = -1 \\ 4x + y = 2 \end{pmatrix}$

For Problems 35–50, solve each system by either the substitution method or the elimination-by-addition method, whichever seems more appropriate.

35. $\begin{pmatrix} 5x - y = -22 \\ 2x + 3y = -2 \end{pmatrix}$ **36.** $\begin{pmatrix} 4x + 5y = -41 \\ 3x - 2y = 21 \end{pmatrix}$

37. $\begin{pmatrix} x = 3y - 10 \\ x = -2y + 15 \end{pmatrix}$ **38.** $\begin{pmatrix} y = 4x - 24 \\ 7x + y = 42 \end{pmatrix}$

39. $\begin{pmatrix} 3x - 5y = 9 \\ 6x - 10y = -1 \end{pmatrix}$ **40.** $\begin{pmatrix} y = \dfrac{2}{5}x - 3 \\ 4x - 7y = 33 \end{pmatrix}$

41. $\begin{pmatrix} \dfrac{1}{2}x - \dfrac{2}{3}y = 22 \\ \dfrac{1}{2}x + \dfrac{1}{4}y = 0 \end{pmatrix}$ **42.** $\begin{pmatrix} \dfrac{x}{2} + \dfrac{y}{3} = 4 \\ 3x + 2y = 24 \end{pmatrix}$

43. $\begin{pmatrix} t = 2u + 2 \\ 9u - 9t = -45 \end{pmatrix}$ **44.** $\begin{pmatrix} 9u - 9t = 36 \\ u = 2t + 1 \end{pmatrix}$

45. $\begin{pmatrix} x + y = 1000 \\ 0.12x + 0.14y = 136 \end{pmatrix}$

46. $\begin{pmatrix} x + y = 10 \\ 0.3x + 0.7y = 4 \end{pmatrix}$

47. $\begin{pmatrix} y = 2x \\ 0.09x + 0.12y = 132 \end{pmatrix}$

48. $\begin{pmatrix} y = 3x \\ 0.1x + 0.11y = 64.5 \end{pmatrix}$

49. $\begin{pmatrix} x + y = 10.5 \\ 0.5x + 0.8y = 7.35 \end{pmatrix}$ **50.** $\begin{pmatrix} 2x + y = 7.75 \\ 3x + 2y = 12.5 \end{pmatrix}$

For Problems 51–70, solve each problem by using a system of equations.

51. The sum of two numbers is 53, and their difference is 19. Find the numbers.

52. The sum of two numbers is −3 and their difference is 25. Find the numbers.

53. The measure of the larger of two complementary angles is 15° more than four times the measure of the smaller angle. Find the measures of both angles.

54. Assume that a plane is flying at a constant speed under unvarying wind conditions. Traveling against a head wind, the plane takes 4 hours to travel 1540 miles. Traveling with a tail wind, the plane flies 1365 miles in 3 hours. Find the speed of the plane and the speed of the wind.

55. The tens digit of a two-digit number is 1 more than three times the units digit. If the sum of the digits is 9, find the number.

56. The units digit of a two-digit number is 1 less than twice the tens digit. The sum of the digits is 8. Find the number.

57. The sum of the digits of a two-digit number is 7. If the digits are reversed, the newly formed number is 9 larger than the original number. Find the original number.

58. The units digit of a two-digit number is 1 less than twice the tens digit. If the digits are reversed, the newly formed number is 27 larger than the original number. Find the original number.

59. A car rental agency rents sedans at $45 a day and convertibles at $65 a day. If, on one day, 32 cars were rented for a total of $1680, how many convertibles were rented?

60. A video store rents new release movies for $5 and old favorites for $2.75. One day the number of new release movies rented was twice the number of old favorites. If the total income from those rentals was $956.25, how many movies of each type were rented?

61. The income from a student concert was $32,500. The price of a student ticket was $10, and nonstudent tickets were sold at $15 each. Three thousand tickets were sold. How many tickets of each kind were sold?

62. Michelle can enter a small business as a full partner and receive a salary of $4,000 a year and 15% of the year's profit, or she can be a sales manager for a salary of $6,000 plus 5% of the year's profit. What must the year's profit be for her total earnings to be the same whether she is a full partner or a sales manager?

63. Sam invested $1950, part of it at 6% and the rest at 8% yearly interest. The yearly income on the 8% investment was $6 more than twice the income from the 6% investment. How much did he invest at each rate?

64. Melinda invested three times as much money at 9% yearly interest as she did at 7%. Her total yearly interest from the two investments was $170. How much did she invest at each rate?

65. One day last summer Jim went kayaking on the Little Susitna River in Alaska. Paddling upstream against the current, he traveled 20 miles in 4 hours. Then he turned around and paddled twice as fast downstream and, with the help of the current, traveled 19 miles in 1 hour. Find the rate of the current.

66. One solution contains 30% alcohol and a second solution contains 70% alcohol. How many liters of each solution should be mixed to make 10 liters containing 40% alcohol?

67. Santo bought 4 gallons of green latex paint and 2 gallons of primer for a total of $116. Not having enough paint to finish the project, Santo returned to the same store and bought 3 gallons of green latex paint and 1 gallon of primer for a total of $80. What is the price of a gallon of green latex paint?

68. Four bottles of Fiji water and 2 bagels cost $10.54. At the same prices, 3 bottles of water and 5 bagels cost $11.02. Find the price per bottle of water and the price per bagel.

69. A cash drawer contains only five- and ten-dollar bills. There are 12 more five-dollar bills than ten-dollar bills. If the drawer contains $330, find the number of each kind of bill.

70. Brad has a collection of dimes and quarters totaling $47.50. The number of quarters is 10 more than twice the number of dimes. How many coins of each kind does he have?

■ ■ ■ THOUGHTS INTO WORDS

71. Give a general description of how to use the substitution method to solve a system of two linear equations in two variables.

72. Give a general description of how to use the elimination-by-addition method to solve a system of two linear equations in two variables.

73. Which method would you use to solve the system $\begin{pmatrix} 9x + 4y = 7 \\ 3x + 2y = 6 \end{pmatrix}$? Why?

74. Which method would you use to solve the system $\begin{pmatrix} 5x + 3y = 12 \\ 3x - y = 10 \end{pmatrix}$? Why?

■ ■ ■ FURTHER INVESTIGATIONS

A system such as

$$\begin{pmatrix} \dfrac{2}{x} + \dfrac{3}{y} = \dfrac{19}{15} \\ -\dfrac{2}{x} + \dfrac{1}{y} = -\dfrac{7}{15} \end{pmatrix}$$

is not a linear system, but it can be solved using the elimination-by-addition method as follows. Add the first equation to the second to produce the equivalent system

$$\begin{pmatrix} \dfrac{2}{x} + \dfrac{3}{y} = \dfrac{19}{15} \\ \dfrac{4}{y} = \dfrac{12}{15} \end{pmatrix}$$

Now solve $\dfrac{4}{y} = \dfrac{12}{15}$ to produce $y = 5$. Substitute 5 for y in the first equation, and solve for x to produce

$$\dfrac{2}{x} + \dfrac{3}{5} = \dfrac{19}{15}$$
$$\dfrac{2}{x} = \dfrac{10}{15}$$
$$10x = 30$$
$$x = 3$$

The solution set of the original system is $\{(3, 5)\}$.

For Problems 75–80, solve each system.

75. $\begin{pmatrix} \dfrac{1}{x} + \dfrac{2}{y} = \dfrac{7}{12} \\ \dfrac{3}{x} - \dfrac{2}{y} = \dfrac{5}{12} \end{pmatrix}$

76. $\begin{pmatrix} \dfrac{3}{x} + \dfrac{2}{y} = 2 \\ \dfrac{2}{x} - \dfrac{3}{y} = \dfrac{1}{4} \end{pmatrix}$

77. $\begin{pmatrix} \dfrac{3}{x} - \dfrac{2}{y} = \dfrac{13}{6} \\ \dfrac{2}{x} + \dfrac{3}{y} = 0 \end{pmatrix}$

78. $\begin{pmatrix} \dfrac{4}{x} + \dfrac{1}{y} = 11 \\ \dfrac{3}{x} - \dfrac{5}{y} = -9 \end{pmatrix}$

79. $\begin{pmatrix} \dfrac{5}{x} - \dfrac{2}{y} = 23 \\ \dfrac{4}{x} + \dfrac{3}{y} = \dfrac{23}{2} \end{pmatrix}$

80. $\begin{pmatrix} \dfrac{2}{x} - \dfrac{7}{y} = \dfrac{9}{10} \\ \dfrac{5}{x} + \dfrac{4}{y} = -\dfrac{41}{20} \end{pmatrix}$

81. Consider the linear system $\begin{pmatrix} a_1x + b_1y = c_1 \\ a_2x + b_2y = c_2 \end{pmatrix}$.

a. Prove that this system has exactly one solution if and only if $\dfrac{a_1}{a_2} \neq \dfrac{b_1}{b_2}$.

b. Prove that this system has no solutions if and only if $\dfrac{a_1}{a_2} = \dfrac{b_1}{b_2} \neq \dfrac{c_1}{c_2}$.

c. Prove that this system has infinitely many solutions if and only if $\dfrac{a_1}{a_2} = \dfrac{b_1}{b_2} = \dfrac{c_1}{c_2}$.

82. For each of the following systems, use the results from Problem 81 to determine whether the system is consistent or inconsistent or the equations are dependent.

a. $\begin{pmatrix} 5x + y = 9 \\ x - 5y = 4 \end{pmatrix}$

b. $\begin{pmatrix} 3x - 2y = 14 \\ 2x + 3y = 9 \end{pmatrix}$

c. $\begin{pmatrix} x - 7y = 4 \\ x - 7y = 9 \end{pmatrix}$

d. $\begin{pmatrix} 3x - 5y = 10 \\ 6x - 10y = 1 \end{pmatrix}$

e. $\begin{pmatrix} 3x + 6y = 2 \\ \dfrac{3}{5}x + \dfrac{6}{5}y = \dfrac{2}{5} \end{pmatrix}$ **f.** $\begin{pmatrix} \dfrac{2}{3}x - \dfrac{3}{4}y = 2 \\ \dfrac{1}{2}x + \dfrac{2}{5}y = 9 \end{pmatrix}$ **g.** $\begin{pmatrix} 7x + 9y = 14 \\ 8x - 3y = 12 \end{pmatrix}$ **h.** $\begin{pmatrix} 4x - 5y = 3 \\ 12x - 15y = 9 \end{pmatrix}$

GRAPHING CALCULATOR ACTIVITIES

83. For each of the systems of equations in Problem 82, use your graphing calculator to help determine whether the system is consistent or inconsistent or the equations are dependent.

84. Use your graphing calculator to help determine the solution set for each of the following systems. Be sure to check your answers.

a. $\begin{pmatrix} y = 3x - 1 \\ y = 9 - 2x \end{pmatrix}$ **b.** $\begin{pmatrix} 5x + y = -9 \\ 3x - 2y = 5 \end{pmatrix}$

c. $\begin{pmatrix} 4x - 3y = 18 \\ 5x + 6y = 3 \end{pmatrix}$ **d.** $\begin{pmatrix} 2x - y = 20 \\ 7x + y = 79 \end{pmatrix}$

e. $\begin{pmatrix} 13x - 12y = 37 \\ 15x + 13y = -11 \end{pmatrix}$

f. $\begin{pmatrix} 1.98x + 2.49y = 13.92 \\ 1.19x + 3.45y = 16.18 \end{pmatrix}$

Answers to the Concept Quiz

1. True   **2.** False   **3.** True   **4.** True   **5.** False   **6.** False   **7.** False   **8.** False

6.2   Systems of Three Linear Equations in Three Variables

Objectives

■ Solve systems of three linear equations.

■ Solve application problems using a system of three linear equations.

Consider a linear equation in three variables x, y, and z, such as $3x - 2y + z = 7$. Any **ordered triple** (x, y, z) that makes the equation a true numerical statement is said to be a **solution** of the equation. For example, the ordered triple $(2, 1, 3)$ is a solution because $3(2) - 2(1) + 3 = 7$. However, the ordered triple $(5, 2, 4)$ is not a solution because $3(5) - 2(2) + 4 \neq 7$. There are infinitely many solutions in the solution set.

Remark: The idea of a linear equation is generalized to include equations of more than two variables. Thus an equation such as $5x - 2y + 9z = 8$ is called a *linear equation in three variables;* the equation $5x - 7y + 2z - 11w = 1$ is called a *linear equation in four variables,* and so on.

To *solve* a system of three linear equations in three variables, such as

$$\begin{pmatrix} 3x - y + 2z = 13 \\ 4x + 2y + 5z = 30 \\ 5x - 3y - z = 3 \end{pmatrix}$$

means to find all of the ordered triples that satisfy all three equations. In other words, the solution set of the system is the intersection of the solution sets of all three equations in the system.

The graph of a linear equation in three variables is a *plane*, not a line. In fact, graphing equations in three variables requires the use of a three-dimensional coordinate system. Thus using a graphing approach to solve systems of three linear equations in three variables is not at all practical. However, a simple graphical analysis does provide us with some direction as to what we can expect as we begin solving such systems.

In general, because each linear equation in three variables produces a plane, a system of three such equations produces three planes. There are various ways in which three planes can be related. For example, they may be mutually parallel, or two of the planes may be parallel with the third intersecting the other two. (You may want to analyze all of the other possibilities for the three planes!) However, for our purposes at this time, we need to realize that from a solution set viewpoint, a system of three linear equations in three variables produces one of the following possibilities.

1. There is *one ordered triple* that satisfies all three equations. The three planes have a common point of intersection, as indicated in Figure 6.4.

Figure 6.4

2. There are *infinitely many ordered triples* in the solution set, all of which are coordinates of *points on a line* common to the three planes. This can happen if the three planes have a common line of intersection, as in Figure 6.5(a), or if two of the planes coincide, and the third plane intersects them, as in Figure 6.5(b).

3. There are *infinitely many ordered triples* in the solution set, all of which are coordinates of *points on a plane*. This can happen if the three planes coincide, as illustrated in Figure 6.6.

Figure 6.6

4. The solution set is *empty*; thus we write $\varnothing$. This can happen in various ways, as illustrated in Figure 6.7. Notice that in each situation there are no points common to all three planes.

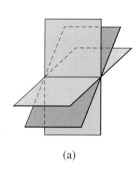

(a)

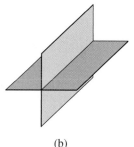

(b)

Figure 6.5

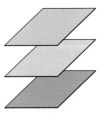

(a) Three planes are parallel.

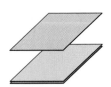

(b) Two planes coincide and the third one is parallel to the coinciding planes.

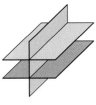

(c) Two planes are parallel and the third intersects them in parallel lines.

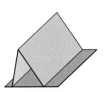

(d) No two planes are parallel, but two of them intersect in a line that is parallel to the third plane.

Figure 6.7

Now that we know what possibilities exist, let's consider finding the solution sets for some systems. Our approach will be the elimination-by-addition method, in which systems are replaced with equivalent systems until a system is obtained where we can easily determine the solution set. The details of this approach will become apparent as we work a few examples.

EXAMPLE 1

Solve the system

$$\left(\begin{array}{ll} 4x - 3y - 2z = 5 & \quad(1) \\ 5y + z = -11 & \quad(2) \\ 3z = 12 & \quad(3) \end{array} \right)$$

Solution

The form of this system makes it easy to solve. From equation (3) we obtain $z = 4$. Then, substituting 4 for z in equation (2), we get

$$5y + 4 = -11$$
$$5y = -15$$
$$y = -3$$

Finally, substituting 4 for z and -3 for y in equation (1) yields

$$4x - 3(-3) - 2(4) = 5$$
$$4x + 1 = 5$$
$$4x = 4$$
$$x = 1$$

Thus the solution set of the given system is $\{(1, -3, 4)\}$.

▶   **Now work Problem 1.**   ■

E X A M P L E   2   Solve the system

$$\begin{pmatrix} x - 2y + 3z = 22 \\ 2x - 3y - z = 5 \\ 3x + y - 5z = -32 \end{pmatrix} \qquad \begin{matrix} (4) \\ (5) \\ (6) \end{matrix}$$

Solution

Equation (5) can be replaced with the equation formed by multiplying equation (4) by -2 and adding this result to equation (5). Equation (6) can be replaced with the equation formed by multiplying equation (4) by -3 and adding this result to equation (6). The following equivalent system is produced, in which equations (8) and (9) contain only the two variables y and z:

$$\begin{pmatrix} x - 2y + 3z = 22 \\ y - 7z = -39 \\ 7y - 14z = -98 \end{pmatrix} \qquad \begin{matrix} (7) \\ (8) \\ (9) \end{matrix}$$

Equation (9) can be replaced with the equation formed by multiplying equation (8) by -7 and adding this result to equation (9). This produces the following equivalent system:

$$\begin{pmatrix} x - 2y + 3z = 22 \\ y - 7z = -39 \\ 35z = 175 \end{pmatrix} \qquad \begin{matrix} (10) \\ (11) \\ (12) \end{matrix}$$

From equation (12) we obtain $z = 5$. Then, substituting 5 for z in equation (11), we obtain

$$y - 7(5) = -39$$
$$y - 35 = -39$$
$$y = -4$$

Finally, substituting -4 for y and 5 for z in equation (10) produces

$$x - 2(-4) + 3(5) = 22$$
$$x + 8 + 15 = 22$$

$$x + 23 = 22$$
$$x = -1$$

The solution set of the original system is $\{(-1, -4, 5)\}$. (You should check this ordered triple in all three of the original equations.)

▶ **Now work Problem 7.** ∎

EXAMPLE 3

Solve the system

$$\begin{pmatrix} 3x - y + 2z = 13 \\ 5x - 3y - z = 3 \\ 4x + 2y + 5z = 30 \end{pmatrix}$$

  (13)
(14)
(15)

Solution

Equation (14) can be replaced with the equation formed by multiplying equation (13) by -3 and adding this result to equation (14). Equation (15) can be replaced with the equation formed by multiplying equation (13) by 2 and adding this result to equation (15). Thus we produce the following equivalent system, in which equations (17) and (18) contain only the two variables x and z:

$$\begin{pmatrix} 3x - y + 2z = 13 \\ -4x - 7z = -36 \\ 10x + 9z = 56 \end{pmatrix}$$

(16)
(17)
(18)

Now if we multiply equation (17) by 5 and equation (18) by 2, we get the following equivalent system:

$$\begin{pmatrix} 3x - y + 2z = 13 \\ -20x - 35z = -180 \\ 20x + 18z = 112 \end{pmatrix}$$

(19)
(20)
(21)

Equation (21) can be replaced with the equation formed by adding equation (20) to equation (21).

$$\begin{pmatrix} 3x - y + 2z = 13 \\ -20x - 35z = -180 \\ -17z = -68 \end{pmatrix}$$

(22)
(23)
(24)

From equation (24), we obtain $z = 4$. Then we can substitute 4 for z in equation (23):

$$-20x - 35(4) = -180$$
$$-20x - 140 = -180$$
$$-20x = -40$$
$$x = 2$$

Now we can substitute 2 for x and 4 for z in equation (22):

$$3(2) - y + 2(4) = 13$$
$$6 - y + 8 = 13$$
$$-y + 14 = 13$$
$$-y = -1$$
$$y = 1$$

The solution set of the original system is $\{(2, 1, 4)\}$.

▶ **Now work Problem 9.**   ■

E X A M P L E   4   Solve the system

$$\left(\begin{array}{l} 2x + 3y + z = 14 \\ 3x - 4y - 2z = -30 \\ 5x + 7y + 3z = 32 \end{array}\right)$$

(25)
(26)
(27)

Solution

Equation (26) can be replaced with the equation formed by multiplying equation (25) by 2 and adding this result to equation (26). Equation (27) can be replaced with the equation formed by multiplying equation (25) by -3 and adding this result to equation (27). The following equivalent system is produced, in which equations (29) and (30) contain only the two variables x and y:

$$\left(\begin{array}{l} 2x + 3y + z = 14 \\ 7x + 2y = -2 \\ -x - 2y = -10 \end{array}\right)$$

(28)
(29)
(30)

Now equation (30) can be replaced with the equation formed by adding equation (29) to equation (30).

$$\left(\begin{array}{l} 2x + 3y + z = 14 \\ 7x + 2y = -2 \\ 6x = -12 \end{array}\right)$$

(31)
(32)
(33)

From equation (33) we obtain $x = -2$. Then, substituting -2 for x in equation (32), we obtain

$$7(-2) + 2y = -2$$
$$2y = 12$$
$$y = 6$$

Finally, substituting 6 for y and -2 for x in equation (31) yields

$$2(-2) + 3(6) + z = 14$$
$$14 + z = 14$$
$$z = 0$$

The solution set of the original system is $\{(-2, 6, 0)\}$.

▶ **Now work Problem 11.**   ■

The ability to solve systems of three linear equations in three unknowns enhances our problem-solving capabilities. Let's conclude this section with a problem that we can solve using such a system.

PROBLEM 1

A small company that manufactures sporting equipment produces three different styles of golf shirts. Each style of shirt requires the services of three departments, as indicated by the following table.

	Style A	Style B	Style C
Cutting department	0.1 hour	0.1 hour	0.3 hour
Sewing department	0.3 hour	0.2 hour	0.4 hour
Packaging department	0.1 hour	0.2 hour	0.1 hour

The cutting, sewing, and packaging departments have available a maximum of 340, 580, and 255 work-hours per week, respectively. How many of each style of golf shirt should be produced each week so that the company is operating at full capacity?

Solution

Let a represent the number of shirts of style A produced per week, b the number of style B per week, and c the number of style C per week. Then the problem translates into the following system of equations:

$$\begin{pmatrix} 0.1a + 0.1b + 0.3c = 340 \\ 0.3a + 0.2b + 0.4c = 580 \\ 0.1a + 0.2b + 0.1c = 255 \end{pmatrix} \begin{matrix} \longleftarrow \text{Cutting department} \\ \longleftarrow \text{Sewing department} \\ \longleftarrow \text{Packaging department} \end{matrix}$$

Solving this system (we will leave the details for you to carry out) produces $a = 500$, $b = 650$, and $c = 750$. Thus the company should produce 500 golf shirts of style A, 650 of style B, and 750 of style C per week.

▶ **Now work Problem 29.** ∎

CONCEPT QUIZ

For Problems 1–8, answer true or false.

1. For a system of three linear equations, any ordered triple that satisfies one of the equations is a solution of the system.

2. The solution set of a system of equations is the intersection of the solution sets of all the equations in the system.

3. The graph of a linear equation in three variables is a plane.

4. For a system of three linear equations, the only way for the solution set to be the empty set is if the equations represents three planes that are parallel.

5. The ordered triple $(0, 0, 0)$ could not be a solution for a system of three linear equations.

6. The solution set for a system of three linear equations could be two ordered triples.

7. It is not possible for the solution set of a system of three linear equations to have an infinite number of solutions.

8. Graphing is a practical way to solve a system of three linear equations.

Problem Set 6.2

For Problems 1–20, solve each system.

1. $\begin{pmatrix} 2x - 3y + 4z = 10 \\ 5y - 2z = -16 \\ 3z = 9 \end{pmatrix}$

2. $\begin{pmatrix} -3x + 2y + z = -9 \\ 4x - 3z = 18 \\ 4z = -8 \end{pmatrix}$

3. $\begin{pmatrix} x + 2y - 3z = 2 \\ 3y - z = 13 \\ 3y + 5z = 25 \end{pmatrix}$

4. $\begin{pmatrix} 2x + 3y - 4z = -10 \\ 2y + 3z = 16 \\ 2y - 5z = -16 \end{pmatrix}$

5. $\begin{pmatrix} 3x + 2y - 2z = 14 \\ x - 6z = 16 \\ 2x + 5z = -2 \end{pmatrix}$

6. $\begin{pmatrix} 3x + 2y - z = -11 \\ 2x - 3y = -1 \\ 4x + 5y = -13 \end{pmatrix}$

7. $\begin{pmatrix} x - 2y + 3z = 7 \\ 2x + y + 5z = 17 \\ 3x - 4y - 2z = 1 \end{pmatrix}$

8. $\begin{pmatrix} x - 2y + z = -4 \\ 2x + 4y - 3z = -1 \\ -3x - 6y + 7z = 4 \end{pmatrix}$

9. $\begin{pmatrix} 2x - y + z = 0 \\ 3x - 2y + 4z = 11 \\ 5x + y - 6z = -32 \end{pmatrix}$

10. $\begin{pmatrix} 2x - y + 3z = -14 \\ 4x + 2y - z = 12 \\ 6x - 3y + 4z = -22 \end{pmatrix}$

11. $\begin{pmatrix} 3x + 2y - z = -11 \\ 2x - 3y + 4z = 11 \\ 5x + y - 2z = -17 \end{pmatrix}$

12. $\begin{pmatrix} 9x + 4y - z = 0 \\ 3x - 2y + 4z = 6 \\ 6x - 8y - 3z = 3 \end{pmatrix}$

13. $\begin{pmatrix} 2x + 3y - 4z = -10 \\ 4x - 5y + 3z = 2 \\ 2y + z = 8 \end{pmatrix}$

14. $\begin{pmatrix} x + 2y - 3z = 2 \\ 3x - z = -8 \\ 2x - 3y + 5z = -9 \end{pmatrix}$

15. $\begin{pmatrix} 3x + 2y - 2z = 14 \\ 2x - 5y + 3z = 7 \\ 4x - 3y + 7z = 5 \end{pmatrix}$

16. $\begin{pmatrix} 4x + 3y - 2z = -11 \\ 3x - 7y + 3z = 10 \\ 9x - 8y + 5z = 9 \end{pmatrix}$

17. $\begin{pmatrix} 2x - 3y + 4z = -12 \\ 4x + 2y - 3z = -13 \\ 6x - 5y + 7z = -31 \end{pmatrix}$

18. $\begin{pmatrix} 3x + 5y - 2z = -27 \\ 5x - 2y + 4z = 27 \\ 7x + 3y - 6z = -55 \end{pmatrix}$

19. $\begin{pmatrix} 5x - 3y - 6z = 22 \\ x - y + z = -3 \\ -3x + 7y - 5z = 23 \end{pmatrix}$

20. $\begin{pmatrix} 4x + 3y - 5z = -29 \\ 3x - 7y - z = -19 \\ 2x + 5y + 2z = -10 \end{pmatrix}$

For Problems 21–31, solve each problem by setting up and solving a system of three linear equations in three variables.

21. A gift store is making a mixture of almonds, pecans, and peanuts, which cost $3.50 per pound, $4 per pound, and $2 per pound, respectively. The store-keeper wants to make 20 pounds of the mix to sell at $2.70 per pound. The number of pounds of peanuts is to be three times the number of pounds of pecans. Find the number of pounds of each to be used in the mixture.

22. The organizer for a church picnic ordered coleslaw, potato salad, and beans amounting to 50 pounds. There was to be three times as much potato salad as coleslaw. The number of pounds of beans was to be 6 less than the number of pounds of potato salad. Find the number of pounds of each.

23. A box contains $7.15 in nickels, dimes, and quarters. There are 42 coins in all, and the sum of the numbers of nickels and dimes is 2 less than the number of quarters. How many coins of each kind are there?

24. A handful of 65 coins consists of pennies, nickels, and dimes. The number of nickels is 4 less than twice the number of pennies, and there are 13 more dimes than nickels. How many coins of each kind are there?

25. The measure of the largest angle of a triangle is twice the smallest angle. The sum of the smallest angle and the largest angle is twice the other angle. Find the measure of each angle.

26. The perimeter of a triangle is 45 centimeters. The longest side is 4 centimeters less than twice the shortest side. The sum of the lengths of the shortest and longest sides is 7 centimeters less than three times the length of the remaining side. Find the lengths of all three sides of the triangle.

27. Part of $30,000 is invested at 6%, another part at 7%, and the remainder at 8% yearly interest. The total yearly income from the three investments is $2200.

The sum of the amounts invested at 6% and 7% equals the amount invested at 8%. How much is invested at each rate?

28. Different amounts are invested at 10%, 11%, and 12% yearly interest. The amount invested at 11% is $300 more than what is invested at 10%, and the total yearly income from all three investments is $324. A total of $2900 is invested. Find the amount invested at each rate.

29. A small company makes three different types of bird houses. Each type requires the services of three different departments, as indicated in the following table.

	Style A	Style B	Style C
Cutting department	0.1 hour	0.2 hour	0.1 hour
Finishing department	0.4 hour	0.4 hour	0.3 hour
Assembly department	0.2 hour	0.1 hour	0.3 hour

The cutting, finishing, and assembly departments have available a maximum of 35, 95, and 62.5 work-hours per week, respectively. How many bird houses of each type should be made per week so that the company is operating at full capacity?

30. A certain diet consists of dishes A, B, and C. Each serving of A has 1 gram of fat, 2 grams of carbohydrate, and 4 grams of protein. Each serving of B has 2 grams of fat, 1 gram of carbohydrate, and 3 grams of protein. Each serving of C has 2 grams of fat, 4 grams of carbohydrate, and 3 grams of protein. The diet allows 15 grams of fat, 24 grams of carbohydrate, and 30 grams of protein. How many servings of each dish can be eaten?

31. Recall that one form of the equation of a circle is $x^2 + y^2 + Dx + Ey + F = 0$. Find the equation of the circle that passes through the points $(-3, 1)$, $(7, 1)$, and $(-7, 5)$.

32. Give a general description of how to solve a system of three linear equations in three variables.

33. Give a step-by-step description of how to solve the system

$$\begin{pmatrix} x - 2y + 3z = -23 \\ 5y - 2z = 32 \\ 4z = -24 \end{pmatrix}$$

34. Give a step-by-step description of how to solve the system

$$\begin{pmatrix} 3x - 2y + 7z = 9 \\ x - 3z = 4 \\ 2x + z = 9 \end{pmatrix}$$

Answers to the Concept Quiz

1. False   **2.** True   **3.** True   **4.** False   **5.** False   **6.** False   **7.** False   **8.** False

6.3   Matrix Approach to Solving Systems

Objectives

■ Represent a system of equations with an augmented matrix.

■ Perform elementary row operations to change an augmented matrix to reduced echelon form.

■ Use a matrix approach to solve systems of equations.

In the first two sections of this chapter we found that the techniques of substitution and elimination-by-addition worked effectively with two equations and two unknowns, but they started to get a bit cumbersome with three equations and three unknowns. Therefore we shall now begin to analyze some techniques that lend themselves to use with larger systems of equations. Furthermore, some of these techniques form the basis for using a computer to solve systems. Even though these techniques are designed primarily for large systems of equations, we shall study them in the context of small systems so that we won't get bogged down with the computational aspects of the techniques.

■ Matrices

A **matrix** is an array of numbers arranged in horizontal rows and vertical columns and enclosed in brackets. For example, the matrix

$$2 \text{ rows} \longrightarrow \begin{bmatrix} 2 & 3 & -1 \\ -4 & 7 & 12 \end{bmatrix}$$

$$3 \text{ columns}$$

has 2 rows and 3 columns and is called a 2×3 (read "two by three") matrix. Each number in a matrix is called an **element** of the matrix. Some additional examples of matrices ("matrices" is the plural of matrix) follow:

$$
\begin{array}{cccc}
3 \times 2 & 2 \times 2 & 1 \times 2 & 4 \times 1 \\
\begin{bmatrix} 2 & 1 \\ 1 & -4 \\ \dfrac{1}{2} & \dfrac{2}{3} \end{bmatrix} &
\begin{bmatrix} 17 & 18 \\ -14 & 16 \end{bmatrix} &
\begin{bmatrix} 7 & 14 \end{bmatrix} &
\begin{bmatrix} 3 \\ -2 \\ 1 \\ 19 \end{bmatrix}
\end{array}
$$

In general, a matrix of m rows and n columns is called a matrix of **dimension $m \times n$** or **order $m \times n$.**

With every system of linear equations we can associate a matrix that consists of the coefficients and constant terms. For example, with the system

$$
\begin{pmatrix}
a_1 x + b_1 y + c_1 z = d_1 \\
a_2 x + b_2 y + c_2 z = d_2 \\
a_3 x + b_3 y + c_3 z = d_3
\end{pmatrix}
$$

we can associate the matrix

$$
\left[\begin{array}{ccc:c}
a_1 & b_1 & c_1 & d_1 \\
a_2 & b_2 & c_2 & d_2 \\
a_3 & b_3 & c_3 & d_3
\end{array}\right]
$$

which is commonly called the **augmented matrix** of the system of equations. The dashed line simply separates the coefficients from the constant terms and reminds us that we are working with an augmented matrix.

On page 485 we listed the operations or transformations that can be applied to a system of equations to produce an equivalent system. Because augmented matrices are essentially abbreviated forms of systems of linear equations, there are analogous transformations that can be applied to augmented matrices. These transformations are usually referred to as **elementary row operations** and can be stated as follows:

For any augmented matrix of a system of linear equations, the following elementary row operations will produce a matrix of an equivalent system.

1. Any two rows of the matrix can be interchanged.

2. Any row of the matrix can be multiplied by a nonzero real number.

3. Any row of the matrix can be replaced by the sum of a nonzero multiple of another row plus that row.

Let's illustrate the use of augmented matrices and elementary row operations to solve a system of two linear equations in two variables.

Solve the system $\begin{pmatrix} x - 3y = -17 \\ 2x + 7y = 31 \end{pmatrix}$.

Solution

The augmented matrix of the system is

$$\begin{bmatrix} 1 & -3 & \vdots & -17 \\ 2 & 7 & \vdots & 31 \end{bmatrix}$$

We would like to change this matrix to one of the form

$$\begin{bmatrix} 1 & 0 & \vdots & a \\ 0 & 1 & \vdots & b \end{bmatrix}$$

where we can easily determine that the solution is $x = a$ and $y = b$. Let's begin by adding -2 times row 1 to row 2 to produce a new row 2.

$$\begin{bmatrix} 1 & -3 & \vdots & -17 \\ 0 & 13 & \vdots & 65 \end{bmatrix}$$

Now we can multiply row 2 by $\dfrac{1}{13}$.

$$\begin{bmatrix} 1 & -3 & \vdots & -17 \\ 0 & 1 & \vdots & 5 \end{bmatrix}$$

Finally, we can add 3 times row 2 to row 1 to produce a new row 1.

$$\begin{bmatrix} 1 & 0 & \vdots & -2 \\ 0 & 1 & \vdots & 5 \end{bmatrix}$$

From this last matrix we see that $x = -2$ and $y = 5$. In other words, the solution set of the original system is $\{(-2, 5)\}$.

▶   **Now work Problem 11.**   ■

It may seem that the matrix approach does not provide us with much extra power for solving systems of two linear equations in two unknowns. However, as the systems get larger, the compactness of the matrix approach becomes more convenient. Let's consider a system of three equations in three variables.

Solve the system

$$\begin{pmatrix} x + 2y - 3z = 15 \\ -2x - 3y + z = -15 \\ 4x + 9y - 4z = 49 \end{pmatrix}$$

Solution

The augmented matrix of this system is

$$\begin{bmatrix} 1 & 2 & -3 & \vdots & 15 \\ -2 & -3 & 1 & \vdots & -15 \\ 4 & 9 & -4 & \vdots & 49 \end{bmatrix}$$

If the system has a unique solution, then we will be able to change the augmented matrix to the form

$$\begin{bmatrix} 1 & 0 & 0 & \vdots & a \\ 0 & 1 & 0 & \vdots & b \\ 0 & 0 & 1 & \vdots & c \end{bmatrix}$$

where we will be able to read the solution $x = a$, $y = b$, and $z = c$. We add 2 times row 1 to row 2 to produce a new row 2. Likewise, we add -4 times row 1 to row 3 to produce a new row 3.

$$\begin{bmatrix} 1 & 2 & -3 & \vdots & 15 \\ 0 & 1 & -5 & \vdots & 15 \\ 0 & 1 & 8 & \vdots & -11 \end{bmatrix}$$

Now we add -2 times row 2 to row 1 to produce a new row 1. Also, we add -1 times row 2 to row 3 to produce a new row 3.

$$\begin{bmatrix} 1 & 0 & 7 & \vdots & -15 \\ 0 & 1 & -5 & \vdots & 15 \\ 0 & 0 & 13 & \vdots & -26 \end{bmatrix}$$

Now let's multiply row 3 by $\dfrac{1}{13}$.

$$\begin{bmatrix} 1 & 0 & 7 & \vdots & -15 \\ 0 & 1 & -5 & \vdots & 15 \\ 0 & 0 & 1 & \vdots & -2 \end{bmatrix}$$

Finally, we can add -7 times row 3 to row 1 to produce a new row 1, and we can add 5 times row 3 to row 2 for a new row 2.

$$\begin{bmatrix} 1 & 0 & 0 & \vdots & -1 \\ 0 & 1 & 0 & \vdots & 5 \\ 0 & 0 & 1 & \vdots & -2 \end{bmatrix}$$

From this last matrix, we can see that the solution set of the original system is $\{(-1, 5, -2)\}$.

▶ **Now work Problem 19.**   ■

The final matrices of Examples 1 and 2,

$$\begin{bmatrix} 1 & 0 & \vdots & -2 \\ 0 & 1 & \vdots & 5 \end{bmatrix} \quad \text{and} \quad \begin{bmatrix} 1 & 0 & 0 & \vdots & -1 \\ 0 & 1 & 0 & \vdots & 5 \\ 0 & 0 & 1 & \vdots & -2 \end{bmatrix}$$

are said to be in **reduced echelon form.** In general, a matrix is in reduced echelon form if the following conditions are satisfied:

1. As we read from left to right, the first nonzero entry of each row is 1.

2. In the *column* containing the leftmost 1 of a row, all the remaining entries are zeros.

3. The leftmost 1 of any row is to the right of the leftmost 1 of the preceding row.

4. Rows containing only zeros are below all the rows containing nonzero entries.

Like the final matrices of Examples 1 and 2, the following are in reduced echelon form:

$$\begin{bmatrix} 1 & 2 & \vdots & -3 \\ 0 & 0 & \vdots & 0 \end{bmatrix} \quad \begin{bmatrix} 1 & 0 & -2 & \vdots & 5 \\ 0 & 1 & 4 & \vdots & 7 \\ 0 & 0 & 0 & \vdots & 0 \end{bmatrix} \quad \begin{bmatrix} 1 & 0 & 0 & 0 & \vdots & 8 \\ 0 & 1 & 0 & 0 & \vdots & -9 \\ 0 & 0 & 1 & 0 & \vdots & -2 \\ 0 & 0 & 0 & 1 & \vdots & 12 \end{bmatrix}$$

In contrast, the following matrices are *not* in reduced echelon form for the reason indicated below each matrix:

$$\begin{bmatrix} 1 & 0 & 0 & \vdots & 11 \\ 0 & 3 & 0 & \vdots & -1 \\ 0 & 0 & 1 & \vdots & -2 \end{bmatrix} \quad \begin{bmatrix} 1 & 2 & -3 & \vdots & 5 \\ 0 & 1 & 7 & \vdots & 9 \\ 0 & 0 & 1 & \vdots & -6 \end{bmatrix}$$

Violates condition 1   Violates condition 2

$$\begin{bmatrix} 1 & 0 & 0 & \vdots & 7 \\ 0 & 0 & 1 & \vdots & -8 \\ 0 & 1 & 0 & \vdots & 14 \end{bmatrix} \quad \begin{bmatrix} 1 & 0 & 0 & 0 & \vdots & -1 \\ 0 & 0 & 0 & 0 & \vdots & 0 \\ 0 & 0 & 1 & 0 & \vdots & 7 \\ 0 & 0 & 0 & 0 & \vdots & 0 \end{bmatrix}$$

Violates condition 3   Violates condition 4

Once we have an augmented matrix in reduced echelon form, it is easy to determine the solution set of the system. Furthermore, the procedure for changing a given augmented matrix to reduced echelon form can be described in a very systematic way. For example, if an augmented matrix of a system of three linear equations in three unknowns has a unique solution, then it can be changed to reduced echelon form as follows:

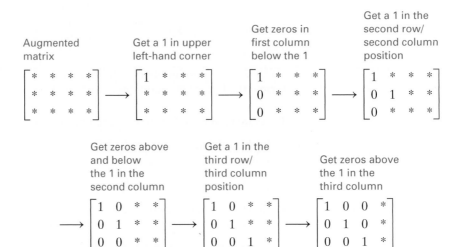

We can identify inconsistent and dependent systems while we are changing a matrix to reduced echelon form. We will show some examples of such cases in a moment, but first let's consider another example of a system of three linear equations in three unknowns where there is a unique solution.

EXAMPLE 3

Solve the system

$$\begin{pmatrix} 2x + 4y - 5z = 37 \\ x + 3y - 4z = 29 \\ 5x - y + 3z = -20 \end{pmatrix}$$

Solution

The augmented matrix

$$\begin{bmatrix} 2 & 4 & -5 & \vdots & 37 \\ 1 & 3 & -4 & \vdots & 29 \\ 5 & -1 & 3 & \vdots & -20 \end{bmatrix}$$

does not have a 1 in the upper left-hand corner, but this can be remedied by exchanging rows 1 and 2.

$$\begin{bmatrix} 1 & 3 & -4 & \vdots & 29 \\ 2 & 4 & -5 & \vdots & 37 \\ 5 & -1 & 3 & \vdots & -20 \end{bmatrix}$$

Now we can get zeros in the first column below the 1 by adding -2 times row 1 to row 2 and by adding -5 times row 1 to row 3.

$$\begin{bmatrix} 1 & 3 & -4 & \vdots & 29 \\ 0 & -2 & 3 & \vdots & -21 \\ 0 & -16 & 23 & \vdots & -165 \end{bmatrix}$$

Next, we can get a 1 for the first nonzero entry of the second row by multiplying the second row by $-\dfrac{1}{2}$.

$$\begin{bmatrix} 1 & 3 & -4 & \vdots & 29 \\ 0 & 1 & -\dfrac{3}{2} & \vdots & \dfrac{21}{2} \\ 0 & -16 & 23 & \vdots & -165 \end{bmatrix}$$

Now we can get zeros above and below the 1 in the second column by adding -3 times row 2 to row 1 and by adding 16 times row 2 to row 3.

$$\begin{bmatrix} 1 & 0 & \dfrac{1}{2} & \vdots & -\dfrac{5}{2} \\ 0 & 1 & -\dfrac{3}{2} & \vdots & \dfrac{21}{2} \\ 0 & 0 & -1 & \vdots & 3 \end{bmatrix}$$

Next, we can get a 1 in the first nonzero entry of the third row by multiplying the third row by -1.

$$\begin{bmatrix} 1 & 0 & \dfrac{1}{2} & \vdots & -\dfrac{5}{2} \\ 0 & 1 & -\dfrac{3}{2} & \vdots & \dfrac{21}{2} \\ 0 & 0 & 1 & \vdots & -3 \end{bmatrix}$$

Finally, we can get zeros above the 1 in the third column by adding $-\dfrac{1}{2}$ times row 3 to row 1 and by adding $\dfrac{3}{2}$ times row 3 to row 2.

$$\begin{bmatrix} 1 & 0 & 0 & \vdots & -1 \\ 0 & 1 & 0 & \vdots & 6 \\ 0 & 0 & 1 & \vdots & -3 \end{bmatrix}$$

From this last matrix, we see that the solution set of the original system is $\{(-1, 6, -3)\}$.

Now work Problem 21.

Example 3 illustrates that even though the process of changing to reduced echelon form can be systematically described, it can involve some rather messy calculations. However, with the aid of a computer, such calculations are not troublesome. For our purposes in this text, the examples and problems involve systems that minimize messy calculations. This will allow us to concentrate on the procedures.

We want to call your attention to another issue in the solution of Example 3. Consider the matrix

$$\begin{bmatrix} 1 & 3 & -4 & \vdots & 29 \\ 0 & 1 & -\dfrac{3}{2} & \vdots & \dfrac{21}{2} \\ 0 & -16 & 23 & \vdots & -165 \end{bmatrix}$$

which was obtained about halfway through the solution. At this step it seems evident that the calculations are getting a little messy. Therefore, instead of continuing toward the reduced echelon form, let's add 16 times row 2 to row 3 to produce a new row 3.

$$\begin{bmatrix} 1 & 3 & -4 & \vdots & 29 \\ 0 & 1 & -\dfrac{3}{2} & \vdots & \dfrac{21}{2} \\ 0 & 0 & -1 & \vdots & 3 \end{bmatrix}$$

The system represented by this matrix is

$$\left(\begin{array}{l} x + 3y - 4z = 29 \\ y - \dfrac{3}{2}z = \dfrac{21}{2} \\ \phantom{y - \dfrac{3}{2}}-z = 3 \end{array} \right)$$

and it is said to be in **triangular form.** The last equation determines the value for z; then we can use the process of back-substitution to determine the values for y and x.

Finally, let's consider two examples to illustrate what happens when we use the matrix approach on inconsistent and dependent systems.

EXAMPLE 4   Solve the system

$$\left(\begin{array}{l} x - 2y + 3z = 3 \\ 5x - 9y + 4z = 2 \\ 2x - 4y + 6z = -1 \end{array} \right)$$

Solution

The augmented matrix of the system is

$$\left[\begin{array}{ccc:c} 1 & -2 & 3 & 3 \\ 5 & -9 & 4 & 2 \\ 2 & -4 & 6 & -1 \end{array}\right]$$

We can get zeros below the 1 in the first column by adding -5 times row 1 to row 2 and by adding -2 times row 1 to row 3.

$$\left[\begin{array}{ccc:c} 1 & -2 & 3 & 3 \\ 0 & 1 & -11 & -13 \\ 0 & 0 & 0 & -7 \end{array}\right]$$

At this step we can stop because the bottom row of the matrix represents the statement $0(x) + 0(y) + 0(z) = -7$, which is obviously false for all values of x, y, and z. Thus the original system is inconsistent; its solution set is $\varnothing$.

⊳   **Now work Problem 26.**   ∎

EXAMPLE 5

Solve the system

$$\left(\begin{array}{l} x + 2y + 2z = 9 \\ x + 3y - 4z = 5 \\ 2x + 5y - 2z = 14 \end{array}\right)$$

Solution

The augmented matrix of the system is

$$\left[\begin{array}{ccc:c} 1 & 2 & 2 & 9 \\ 1 & 3 & -4 & 5 \\ 2 & 5 & -2 & 14 \end{array}\right]$$

We can get zeros in the first column below the 1 in the upper left-hand corner by adding -1 times row 1 to row 2 and adding -2 times row 1 to row 3.

$$\left[\begin{array}{ccc:c} 1 & 2 & 2 & 9 \\ 0 & 1 & -6 & -4 \\ 0 & 1 & -6 & -4 \end{array}\right]$$

Now we can get zeros in the second column above and below the 1 in the second row by adding -2 times row 2 to row 1 and adding -1 times row 2 to row 3.

$$\left[\begin{array}{ccc:c} 1 & 0 & 14 & 17 \\ 0 & 1 & -6 & -4 \\ 0 & 0 & 0 & 0 \end{array}\right]$$

The bottom row of zeros represents the statement $0(x) + 0(y) + 0(z) = 0$, which is true for all values of x, y, and z. The second row represents the statement $y - 6z = -4$, which can be rewritten $y = 6z - 4$. The top row represents the statement $x + 14z = 17$, which can be rewritten $x = -14z + 17$. Therefore, if we let $z = k$, where k is any real number, the solution set of infinitely many ordered triples can be represented by $\{(-14k + 17, 6k - 4, k)|k \text{ is a real number}\}$. Specific solutions can be generated by letting k take on a value. For example, if $k = 2$, then $6k - 4$ becomes $6(2) - 4 = 8$ and $-14k + 17$ becomes $-14(2) + 17 = -11$. Thus the ordered triple $(-11, 8, 2)$ is a member of the solution set.

▶ **Now work Problem 25.** ■

CONCEPT QUIZ   For Problems 1–8, answer true or false.

1. A matrix with dimension $m \times n$ has m rows and n columns.

2. The augmented matrix of a system of equations consists of just the coefficients of the variables.

3. If two columns of a matrix are interchanged, the result is an equivalent matrix.

4. The matrix $\begin{bmatrix} a_1 & b_1 & c_1 \\ a_2 & b_2 & c_2 \\ a_3 & b_3 & c_3 \end{bmatrix}$ is equivalent to the matrix $\begin{bmatrix} a_1 & b_1 & c_1 \\ da_2 & db_2 & dc_2 \\ a_3 & b_3 & c_3 \end{bmatrix}$.

5. The matrix $\left[\begin{array}{ccc:c} 1 & 0 & 0 & -4 \\ 0 & 1 & 0 & 8 \\ 1 & 0 & 0 & 2 \end{array}\right]$ is in reduced echelon form.

6. If the augmented matrix for a system of equations is $\left[\begin{array}{ccc:c} 1 & 0 & 0 & -1 \\ 0 & 1 & 0 & 5 \\ 0 & 0 & 1 & 2 \end{array}\right]$, then the solution of the system of equations is $(-1, 5, 2)$.

7. The system of equations $\begin{pmatrix} 2x + y - z = 25 \\ 3y - 4z = 10 \\ z = 3 \end{pmatrix}$ is said to be in triangular form.

8. If the solution set of a system of equations is represented by $\{(2k + 3, k) \mid k$ is a real number$\}$, then $(11, 4)$ is a specific solution of the system of equations.

Problem Set 6.3

For Problems 1–10, indicate whether each matrix is in reduced echelon form.

1. $\begin{bmatrix} 1 & 0 & \vdots & -4 \\ 0 & 1 & \vdots & 14 \end{bmatrix}$

2. $\begin{bmatrix} 1 & 2 & \vdots & 8 \\ 0 & 0 & \vdots & 0 \end{bmatrix}$

3. $\begin{bmatrix} 1 & 0 & 2 & \vdots & 5 \\ 0 & 1 & 3 & \vdots & 7 \\ 0 & 0 & 0 & \vdots & 0 \end{bmatrix}$

4. $\begin{bmatrix} 1 & 0 & 0 & \vdots & 5 \\ 0 & 3 & 0 & \vdots & 8 \\ 0 & 0 & 1 & \vdots & -11 \end{bmatrix}$

5. $\begin{bmatrix} 1 & 0 & 0 & \vdots & 17 \\ 0 & 0 & 0 & \vdots & 0 \\ 0 & 1 & 0 & \vdots & -14 \end{bmatrix}$

6. $\begin{bmatrix} 1 & 0 & 0 & \vdots & -7 \\ 0 & 1 & 0 & \vdots & 0 \\ 0 & 0 & 1 & \vdots & 9 \end{bmatrix}$

7. $\begin{bmatrix} 1 & 1 & 0 & \vdots & -3 \\ 0 & 1 & 2 & \vdots & 5 \\ 0 & 0 & 1 & \vdots & 7 \end{bmatrix}$

8. $\begin{bmatrix} 1 & 0 & 3 & \vdots & 8 \\ 0 & 1 & 2 & \vdots & -6 \\ 0 & 0 & 0 & \vdots & 0 \end{bmatrix}$

9. $\begin{bmatrix} 1 & 0 & 0 & 3 & \vdots & 4 \\ 0 & 1 & 0 & 5 & \vdots & -3 \\ 0 & 0 & 1 & -1 & \vdots & 7 \\ 0 & 0 & 0 & 0 & \vdots & 0 \end{bmatrix}$

10. $\begin{bmatrix} 1 & 0 & 0 & 0 & \vdots & 2 \\ 0 & 0 & 1 & 0 & \vdots & 4 \\ 0 & 1 & 0 & 0 & \vdots & -3 \\ 0 & 0 & 0 & 1 & \vdots & 9 \end{bmatrix}$

For Problems 11–30, use a matrix approach to solve each system.

11. $\left(\begin{array}{l} x - 3y = 14 \\ 3x + 2y = -13 \end{array} \right)$

12. $\left(\begin{array}{l} x + 5y = -18 \\ -2x + 3y = -16 \end{array} \right)$

13. $\left(\begin{array}{l} 3x - 4y = 33 \\ x + 7y = -39 \end{array} \right)$

14. $\left(\begin{array}{l} 2x + 7y = -55 \\ x - 4y = 25 \end{array} \right)$

15. $\left(\begin{array}{l} x - 6y = -2 \\ 2x - 12y = 5 \end{array} \right)$

16. $\left(\begin{array}{l} 2x - 3y = -12 \\ 3x + 2y = 8 \end{array} \right)$

17. $\left(\begin{array}{l} 3x - 5y = 39 \\ 2x + 7y = -67 \end{array} \right)$

18. $\left(\begin{array}{l} 3x + 9y = -1 \\ x + 3y = 10 \end{array} \right)$

19. $\left(\begin{array}{l} x - 2y - 3z = -6 \\ 3x - 5y - z = 4 \\ 2x + y + 2z = 2 \end{array} \right)$

20. $\left(\begin{array}{l} x + 3y - 4z = 13 \\ 2x + 7y - 3z = 11 \\ -2x - y + 2z = -8 \end{array} \right)$

21. $\left(\begin{array}{l} -2x - 5y + 3z = 11 \\ x + 3y - 3z = -12 \\ 3x - 2y + 5z = 31 \end{array} \right)$

22. $\left(\begin{array}{l} -3x + 2y + z = 17 \\ x - y + 5z = -2 \\ 4x - 5y - 3z = -36 \end{array} \right)$

23. $\left(\begin{array}{l} x - 3y - z = 2 \\ 3x + y - 4z = -18 \\ -2x + 5y + 3z = 2 \end{array} \right)$

24. $\left(\begin{array}{l} x - 4y + 3z = 16 \\ 2x + 3y - 4z = -22 \\ -3x + 11y - z = -36 \end{array} \right)$

25. $\left(\begin{array}{l} x - y + 2z = 1 \\ -3x + 4y - z = 4 \\ -x + 2y + 3z = 6 \end{array} \right)$

26. $\left(\begin{array}{l} x + 2y - 5z = -1 \\ 2x + 3y - 2z = 2 \\ 3x + 5y - 7z = 4 \end{array} \right)$

27. $\left(\begin{array}{l} -2x + y + 5z = -5 \\ 3x + 8y - z = -34 \\ x + 2y + z = -12 \end{array} \right)$

28. $\left(\begin{array}{l} 4x - 10y + 3z = -19 \\ 2x + 5y - z = -7 \\ x - 3y - 2z = -2 \end{array} \right)$

29. $\left(\begin{array}{l} 2x + 3y - z = 7 \\ 3x + 4y + 5z = -2 \\ 5x + y + 3z = 13 \end{array} \right)$

30. $\left(\begin{array}{l} 4x + 3y - z = 0 \\ 3x + 2y + 5z = 6 \\ 5x - y - 3z = 3 \end{array} \right)$

Subscript notation is frequently used for working with larger systems of equations. For Problems 31–34, use a matrix approach to solve each system. Express the solutions as 4-tuples of the form (x_1, x_2, x_3, x_4).

31. $\begin{pmatrix} x_1 - 3x_2 - 2x_3 + x_4 = -3 \\ -2x_1 + 7x_2 + x_3 - 2x_4 = -1 \\ 3x_1 - 7x_2 - 3x_3 + 3x_4 = -5 \\ 5x_1 + x_2 + 4x_3 - 2x_4 = 18 \end{pmatrix}$

32. $\begin{pmatrix} x_1 - 2x_2 + 2x_3 - x_4 = -2 \\ -3x_1 + 5x_2 - x_3 - 3x_4 = 2 \\ 2x_1 + 3x_2 + 3x_3 + 5x_4 = -9 \\ 4x_1 - x_2 - x_3 - 2x_4 = 8 \end{pmatrix}$

33. $\begin{pmatrix} x_1 + 3x_2 - x_3 + 2x_4 = -2 \\ 2x_1 + 7x_2 + 2x_3 - x_4 = 19 \\ -3x_1 - 8x_2 + 3x_3 + x_4 = -7 \\ 4x_1 + 11x_2 - 2x_3 - 3x_4 = 19 \end{pmatrix}$

34. $\begin{pmatrix} x_1 + 2x_2 - 3x_3 + x_4 = -2 \\ -2x_1 - 3x_2 + x_3 - x_4 = 5 \\ 4x_1 + 9x_2 - 2x_3 - 2x_4 = -28 \\ -5x_1 - 9x_2 + 2x_3 - 3x_4 = 14 \end{pmatrix}$

In Problems 35–42, each matrix is the reduced echelon matrix for a system with variables x_1, x_2, x_3, and x_4. Find the solution set of each system.

35. $\left[\begin{array}{cccc|c} 1 & 0 & 0 & 0 & -2 \\ 0 & 1 & 0 & 0 & 4 \\ 0 & 0 & 1 & 0 & -3 \\ 0 & 0 & 0 & 1 & 0 \end{array}\right]$

36. $\left[\begin{array}{cccc|c} 1 & 0 & 0 & 0 & 0 \\ 0 & 1 & 0 & 0 & -5 \\ 0 & 0 & 1 & 0 & 0 \\ 0 & 0 & 0 & 1 & 4 \end{array}\right]$

37. $\left[\begin{array}{cccc|c} 1 & 0 & 0 & 0 & -8 \\ 0 & 1 & 0 & 0 & 5 \\ 0 & 0 & 1 & 0 & -2 \\ 0 & 0 & 0 & 0 & 1 \end{array}\right]$

38. $\left[\begin{array}{cccc|c} 1 & 0 & 0 & 0 & 2 \\ 0 & 1 & 0 & 2 & -3 \\ 0 & 0 & 1 & 3 & 4 \\ 0 & 0 & 0 & 0 & 0 \end{array}\right]$

39. $\left[\begin{array}{cccc|c} 1 & 0 & 0 & 3 & 5 \\ 0 & 1 & 0 & 0 & -1 \\ 0 & 0 & 1 & 4 & 2 \\ 0 & 0 & 0 & 0 & 0 \end{array}\right]$

40. $\left[\begin{array}{ccc|c} 1 & 3 & 0 & 0 \\ 0 & 0 & 1 & 0 \\ 0 & 0 & 0 & 1 \\ 0 & 0 & 0 & 0 \end{array}\right]$

41. $\left[\begin{array}{cccc|c} 1 & 3 & 0 & 0 & 9 \\ 0 & 0 & 1 & 0 & 2 \\ 0 & 0 & 0 & 1 & -3 \\ 0 & 0 & 0 & 0 & 0 \end{array}\right]$

42. $\left[\begin{array}{cccc|c} 1 & 0 & 0 & 0 & 7 \\ 0 & 1 & 0 & 0 & -3 \\ 0 & 0 & 1 & -2 & 5 \\ 0 & 0 & 0 & 0 & 0 \end{array}\right]$

■ ■ ■ THOUGHTS INTO WORDS

43. What is a matrix? What is an augmented matrix of a system of linear equations?

44. Describe how to use matrices to solve the system $\begin{pmatrix} x - 2y = 5 \\ 2x + 7y = 9 \end{pmatrix}$.

■ ■ ■ **FURTHER INVESTIGATIONS**

For Problems 45–50, change each augmented matrix of the system to reduced echelon form and then indicate the solutions of the system.

45. $\begin{pmatrix} x - 2y + 3z = 4 \\ 3x - 5y - z = 7 \end{pmatrix}$ **46.** $\begin{pmatrix} x + 3y - 2z = -1 \\ 2x - 5y + 7z = 4 \end{pmatrix}$

47. $\begin{pmatrix} 2x - 4y + 3z = 8 \\ 3x + 5y - z = 7 \end{pmatrix}$ **48.** $\begin{pmatrix} 3x + 6y - z = 9 \\ 2x - 3y + 4z = 1 \end{pmatrix}$

49. $\begin{pmatrix} x - 2y + 4z = 9 \\ 2x - 4y + 8z = 3 \end{pmatrix}$

50. $\begin{pmatrix} x + y - 2z = -1 \\ 3x + 3y - 6z = -3 \end{pmatrix}$

 GRAPHING CALCULATOR ACTIVITIES

51. If your graphing calculator has the capability of manipulating matrices, this is a good time to become familiar with those operations. Also if your calculator has a Catalog, look for a rref key under the Catalog. The rref stands for "reduced-row-echelon form." The rref key transforms an augmented matrix into reduced echelon form. Refer to your user's manual for the key-punching instructions. To begin the familiarization process, load your calculator with the three augmented matrices in Examples 1, 2, and 3. Then, for each one, carry out the row operations as described in the text.

Answers to the Concept Quiz

1. True   **2.** False   **3.** False   **4.** True   **5.** False   **6.** True   **7.** True   **8.** True

6.4 Determinants

Objectives

■ Know the notation used to identify elements of a matrix.

■ Evaluate the determinant of a matrix.

■ Determine the minor and cofactor of an element in a matrix.

■ Apply properties of determinants to help find the determinant.

Before we introduce the concept of a determinant, let's agree on some convenient new notation. A **general $m \times n$ (m-by-n) matrix** can be represented by

$$A = \begin{bmatrix} a_{11} & a_{12} & a_{13} & \cdots & a_{1n} \\ a_{21} & a_{22} & a_{23} & \cdots & a_{2n} \\ \vdots & \vdots & \vdots & & \vdots \\ a_{m1} & a_{m2} & a_{m3} & \cdots & a_{mn} \end{bmatrix}$$

where the double subscripts are used to identify the number of the row and the number of the column, in that order. For example, a_{23} is the entry at the intersection of the second row and the third column. In general, the entry at the intersection of row i and column j is denoted by a_{ij}.

A **square matrix** is one that has the same number of rows as columns. Each square matrix A with real number entries can be associated with a real number called the **determinant** of the matrix, denoted by $|A|$. We will first define $|A|$ for a 2×2 matrix.

Definition 6.1

If $A = \begin{bmatrix} a_{11} & a_{12} \\ a_{21} & a_{22} \end{bmatrix}$, then

$$|A| = \begin{vmatrix} a_{11} & a_{12} \\ a_{21} & a_{22} \end{vmatrix} = a_{11}a_{22} - a_{12}a_{21}$$

EXAMPLE 1

If $A = \begin{bmatrix} 3 & -2 \\ 5 & 8 \end{bmatrix}$, find $|A|$.

Solution

We use Definition 6.1 to obtain

$$|A| = \begin{vmatrix} 3 & -2 \\ 5 & 8 \end{vmatrix} = 3(8) - (-2)(5)$$

$$= 24 + 10$$

$$= 34$$

▶ **Now work Problem 1.** ∎

Finding the determinant of a square matrix is commonly called **evaluating the determinant,** and the matrix notation is often omitted.

EXAMPLE 2

Evaluate $\begin{vmatrix} -3 & 6 \\ 2 & 8 \end{vmatrix}$.

Solution

$$\begin{vmatrix} -3 & 6 \\ 2 & 8 \end{vmatrix} = (-3)(8) - (6)(2)$$

$$= -24 - 12$$

$$= -36$$

▶ **Now work Problem 3.** ∎

To find the determinants of 3×3 and larger square matrices, it is convenient to introduce some additional terminology.

Definition 6.2

If A is a 3×3 matrix, then the **minor** (denoted by M_{ij}) of the a_{ij} element is the determinant of the 2×2 matrix obtained by deleting row i and column j of A.

E X A M P L E   3

If $A = \begin{bmatrix} 2 & 1 & 4 \\ -6 & 3 & -2 \\ 4 & 2 & 5 \end{bmatrix}$, find M_{11} and M_{23}.

Solution

To find M_{11} we first delete row 1 and column 1 of matrix A.

$$\begin{bmatrix} 2 & 1 & 4 \\ -6 & 3 & -2 \\ 4 & 2 & 5 \end{bmatrix}$$

Thus

$$M_{11} = \begin{vmatrix} 3 & -2 \\ 2 & 5 \end{vmatrix} = 3(5) - (-2)(2) = 19$$

To find M_{23} we first delete row 2 and column 3 of matrix A.

$$\begin{bmatrix} 2 & 1 & 4 \\ 6 & 3 & -2 \\ 4 & 2 & 5 \end{bmatrix}$$

Thus

$$M_{23} = \begin{vmatrix} 2 & 1 \\ 4 & 2 \end{vmatrix} = 2(2) - (1)(4) = 0 \qquad \blacksquare$$

The following definition will also be used.

Definition 6.3

If A is a 3×3 matrix, then the **cofactor** (denoted by C_{ij}) of the element a_{ij} is defined by

$$C_{ij} = (-1)^{i+j} M_{ij}$$

According to Definition 6.3, to find the cofactor of any element a_{ij} of a square matrix A, we find the minor of a_{ij} and multiply it by 1 if $i + j$ is even, or multiply it by -1 if $i + j$ is odd.

EXAMPLE 4   If $A = \begin{bmatrix} 3 & 2 & -4 \\ 1 & 5 & 4 \\ 2 & 3 & 1 \end{bmatrix}$, find C_{32}.

Solution

First, let's find M_{32} by deleting row 3 and column 2 of matrix A.

$$\begin{bmatrix} 3 & 2 & -4 \\ 1 & 5 & 4 \\ 2 & 3 & 1 \end{bmatrix}$$

Thus

$$M_{32} = \begin{vmatrix} 3 & -4 \\ 1 & 4 \end{vmatrix} = 3(4) - (-4)(1) = 16$$

Therefore

$$C_{32} = (-1)^{3+2}M_{32} = (-1)^5(16) = -16 \qquad \blacksquare$$

The concept of a cofactor can be used to define the determinant of a 3×3 matrix as follows.

Definition 6.4

If $A = \begin{bmatrix} a_{11} & a_{12} & a_{13} \\ a_{21} & a_{22} & a_{23} \\ a_{31} & a_{32} & a_{33} \end{bmatrix}$, then

$$|A| = a_{11}C_{11} + a_{21}C_{21} + a_{31}C_{31}$$

Definition 6.4 simply states that the determinant of a 3×3 matrix can be found by multiplying each element of the first column by its corresponding cofactor and then adding the three results. Let's illustrate this procedure.

EXAMPLE 5   Find $|A|$ if $A = \begin{bmatrix} -2 & 1 & 4 \\ 3 & 0 & 5 \\ 1 & -4 & -6 \end{bmatrix}$.

Solution

$$|A| = a_{11}C_{11} + a_{21}C_{21} + a_{31}C_{31}$$

$$= (-2)(-1)^{1+1}\begin{vmatrix} 0 & 5 \\ -4 & -6 \end{vmatrix} + (3)(-1)^{2+1}\begin{vmatrix} 1 & 4 \\ -4 & -6 \end{vmatrix} + (1)(-1)^{3+1}\begin{vmatrix} 1 & 4 \\ 0 & 5 \end{vmatrix}$$

$$= (-2)(1)(20) + (3)(-1)(10) + (1)(1)(5)$$

$$= -40 - 30 + 5$$

$$= -65$$

Now work Problem 13. ∎

When we use Definition 6.4, we often say that *the determinant is being expanded about the first column*. It can also be shown that **any row or column can be used to expand a determinant.** For example, for matrix A in Example 5, the expansion of the determinant about the *second row* is as follows:

$$\begin{vmatrix} -2 & 1 & 4 \\ 3 & 0 & 5 \\ 1 & -4 & -6 \end{vmatrix} = (3)(-1)^{2+1}\begin{vmatrix} 1 & 4 \\ -4 & -6 \end{vmatrix} + (0)(-1)^{2+2}\begin{vmatrix} -2 & 4 \\ 1 & -6 \end{vmatrix} + (5)(-1)^{2+3}\begin{vmatrix} -2 & 1 \\ 1 & -4 \end{vmatrix}$$

$$= (3)(-1)(10) + (0)(1)(8) + (5)(-1)(7)$$

$$= -30 + 0 - 35$$

$$= -65$$

Note that when we expanded about the second row, the computation was simplified by the presence of a zero. In general, it is helpful to expand about the row or column that contains the most zeros.

The concepts of minor and cofactor have been defined in terms of 3×3 matrices. Analogous definitions can be given for any square matrix (that is, any $n \times n$ matrix with $n \geq 2$), and the determinant can then be expanded about any row or column. Certainly as the matrices become larger than 3×3, the computations get more tedious. We will concentrate most of our efforts in this text on 2×2 and 3×3 matrices.

■ Properties of Determinants

Determinants have several interesting properties, some of which are important primarily from a theoretical standpoint. But some of the properties are also very useful when evaluating determinants. We will state these properties for square matrices in general, but we will use 2×2 or 3×3 matrices as examples. We can illustrate some of the proofs of these properties by evaluating the determinants involved, and some of the proofs for 3×3 matrices will be left for you to verify in the next problem set.

Property 6.1

If any row (or column) of a square matrix A contains only zeros, then $|A| = 0$.

If every element of a row (or column) of a square matrix A is zero, then it should be evident that expanding the determinant about that row (or column) of zeros will produce zero.

Property 6.2

> If square matrix B is obtained from square matrix A by interchanging two rows (or two columns), then $|B| = -|A|$.

Property 6.2 states that **interchanging two rows (or columns) changes the sign of the determinant.** As an example of this property, suppose that

$$A = \begin{bmatrix} 2 & 5 \\ -1 & 6 \end{bmatrix}$$

and that rows 1 and 2 are interchanged to form

$$B = \begin{bmatrix} -1 & 6 \\ 2 & 5 \end{bmatrix}$$

Calculating $|A|$ and $|B|$ yields

$$|A| = \begin{vmatrix} 2 & 5 \\ -1 & 6 \end{vmatrix} = 2(6) - (5)(-1) = 17$$

and

$$|B| = \begin{vmatrix} -1 & 6 \\ 2 & 5 \end{vmatrix} = (-1)(5) - (6)(2) = -17$$

Property 6.3

> If square matrix B is obtained from square matrix A by multiplying each element of any row (or column) of A by some real number k, then $|B| = k|A|$.

Property 6.3 states that **multiplying any row (or column) by a factor of k affects the value of the determinant by a factor of k.** As an example of this property, suppose that

$$A = \begin{bmatrix} 1 & -2 & 8 \\ 2 & 1 & 12 \\ 3 & 2 & -16 \end{bmatrix}$$

and that B is formed by multiplying each element of the third column by $\dfrac{1}{4}$:

$$B = \begin{bmatrix} 1 & -2 & 2 \\ 2 & 1 & 3 \\ 3 & 2 & -4 \end{bmatrix}$$

Now let's calculate $|A|$ and $|B|$ by expanding about the third column in each case.

$$|A| = \begin{vmatrix} 1 & -2 & 8 \\ 2 & 1 & 12 \\ 3 & 2 & -16 \end{vmatrix} = (8)(-1)^{1+3}\begin{vmatrix} 2 & 1 \\ 3 & 2 \end{vmatrix} + (12)(-1)^{2+3}\begin{vmatrix} 1 & -2 \\ 3 & 2 \end{vmatrix} + (-16)(-1)^{3+3}\begin{vmatrix} 1 & -2 \\ 2 & 1 \end{vmatrix}$$

$$= (8)(1)(1) + (12)(-1)(8) + (-16)(1)(5)$$

$$= -168$$

$$|B| = \begin{vmatrix} 1 & -2 & 2 \\ 2 & 1 & 3 \\ 3 & 2 & -4 \end{vmatrix} = (2)(-1)^{1+3}\begin{vmatrix} 2 & 1 \\ 3 & 2 \end{vmatrix} + (3)(-1)^{2+3}\begin{vmatrix} 1 & -2 \\ 3 & 2 \end{vmatrix} + (-4)(-1)^{3+3}\begin{vmatrix} 1 & -2 \\ 2 & 1 \end{vmatrix}$$

$$= (2)(1)(1) + (3)(-1)(8) + (-4)(1)(5)$$

$$= -42$$

We see that $|B| = \dfrac{1}{4}|A|$. This example also illustrates the usual computational use of Property 6.3: We can factor out a common factor from a row or column and then adjust the value of the determinant by that factor. For example,

$$\begin{vmatrix} 2 & 6 & 8 \\ -1 & 2 & 7 \\ 5 & 2 & 1 \end{vmatrix} = 2\begin{vmatrix} 1 & 3 & 4 \\ -1 & 2 & 7 \\ 5 & 2 & 1 \end{vmatrix}$$

Factor a 2 from the top row

Property 6.4

If square matrix B is obtained from square matrix A by adding k times a row (or column) of A to another row (or column) of A, then $|B| = |A|$.

Property 6.4 states that **adding the product of k times a row (or column) to another row (or column) does not affect the value of the determinant.** As an example of this property, suppose that

$$A = \begin{bmatrix} 1 & 2 & 4 \\ 2 & 4 & 7 \\ -1 & 3 & 5 \end{bmatrix}$$

Now let's form B by replacing row 2 with the result of adding -2 times row 1 to row 2.

$$B = \begin{bmatrix} 1 & 2 & 4 \\ 0 & 0 & -1 \\ -1 & 3 & 5 \end{bmatrix}$$

Next, let's evaluate $|A|$ and $|B|$ by expanding about the second row in each case.

$$|A| = \begin{vmatrix} 1 & 2 & 4 \\ 2 & 4 & 7 \\ -1 & 3 & 5 \end{vmatrix} = (2)(-1)^{2+1}\begin{vmatrix} 2 & 4 \\ 3 & 5 \end{vmatrix} + (4)(-1)^{2+2}\begin{vmatrix} 1 & 4 \\ -1 & 5 \end{vmatrix} + (7)(-1)^{2+3}\begin{vmatrix} 1 & 2 \\ -1 & 3 \end{vmatrix}$$

$$= 2(-1)(-2) + (4)(1)(9) + (7)(-1)(5)$$

$$= 5$$

$$|B| = \begin{vmatrix} 1 & 2 & 4 \\ 0 & 0 & -1 \\ -1 & 3 & 5 \end{vmatrix} = (0)(-1)^{2+1}\begin{vmatrix} 2 & 4 \\ 3 & 5 \end{vmatrix} + (0)(-1)^{2+2}\begin{vmatrix} 1 & 4 \\ -1 & 5 \end{vmatrix} + (-1)(-1)^{2+3}\begin{vmatrix} 1 & 2 \\ -1 & 3 \end{vmatrix}$$

$$= 0 + 0 + (-1)(-1)(5)$$

$$= 5$$

Note that $|B| = |A|$. Furthermore, note that because of the zeros in the second row, evaluating $|B|$ is much easier than evaluating $|A|$. Property 6.4 can often be used to obtain some zeros before evaluating a determinant.

A word of caution is in order at this time. Be careful not to confuse Properties 6.2, 6.3, and 6.4 with the three elementary row transformations of augmented matrices that were used in Section 6.3. The statements of the two sets of properties do resemble each other, but the properties pertain to *two different concepts,* so be sure you understand the distinction between them.

One final property of determinants should be mentioned.

Property 6.5

> If two rows (or columns) of a square matrix A are identical, then $|A| = 0$.

Property 6.5 can be verified by using Property 6.1. Consider a square matrix A that has two identical rows (or columns). Multiply one of the two identical rows (or columns) by -1 and add that result to the other row (or column). This will produce a row (or column) of zeros, and therefore $|A| = 0$.

Let's conclude this section by evaluating a 4×4 determinant, using Properties 6.3 and 6.4 to facilitate the computation.

EXAMPLE 6
Evaluate $\begin{vmatrix} 6 & 2 & 1 & -2 \\ 9 & -1 & 4 & 1 \\ 12 & -2 & 3 & -1 \\ 0 & 0 & 9 & 3 \end{vmatrix}$.

Solution

First, let's add -3 times the fourth column to the third column.

$$\begin{vmatrix} 6 & 2 & 7 & -2 \\ 9 & -1 & 1 & 1 \\ 12 & -2 & 6 & -1 \\ 0 & 0 & 0 & 3 \end{vmatrix}$$

Now if we expand about the fourth row, we get only one nonzero product.

$$(3)(-1)^{4+4}\begin{vmatrix} 6 & 2 & 7 \\ 9 & -1 & 1 \\ 12 & -2 & 6 \end{vmatrix}$$

Factoring a 3 out of the first column of the 3×3 determinant, we obtain

$$(3)(-1)^8(3)\begin{vmatrix} 2 & 2 & 7 \\ 3 & -1 & 1 \\ 4 & -2 & 6 \end{vmatrix}$$

Now, working with the 3×3 determinant, we can first add column 3 to column 2 and then add -3 times column 3 to column 1.

$$(3)(-1)^8(3)\begin{vmatrix} -19 & 9 & 7 \\ 0 & 0 & 1 \\ -14 & 4 & 6 \end{vmatrix}$$

Finally, by expanding this 3×3 determinant about the second row, we obtain

$$(3)(-1)^8(3)(1)(-1)^{2+3}\begin{vmatrix} -19 & 9 \\ -14 & 4 \end{vmatrix}$$

Our final result is

$$(3)(-1)^8(3)(1)(-1)^5(50) = -450$$

▶ **Now work Problem 29.**   ∎

CONCEPT QUIZ   For Problems 1–8, answer true or false.

1. The element of a matrix a_{15} is located at the intersection of the first row and 5th column.

2. If $A = \begin{vmatrix} a_{11} & a_{12} \\ a_{21} & a_{22} \end{vmatrix}$, then $|A| = a_{11}a_{12} - a_{21}a_{22}$.

3. The determinant of a matrix is never a negative value.

4. If $A = \begin{vmatrix} 2 & -1 & 4 \\ 7 & 0 & 0 \\ 8 & -3 & 9 \end{vmatrix}$, the easiest calculation to find the determinant would be

to expand about the second row.

5. The only way for a determinant to be equal to zero is if a row (or column) contains all zeros.

6. If $\begin{vmatrix} 3 & -4 \\ 2 & 6 \end{vmatrix}$ and $\begin{vmatrix} 2 & 6 \\ 3 & -4 \end{vmatrix}$, then $-|A| = |B|$.

7. Multiplying a row of a matrix by 3 affects the determinant by a factor of $\dfrac{1}{3}$.

8. Interchanging two columns of a matrix has no effect on the determinant.

Problem Set 6.4

For Problems 1–12, evaluate each 2×2 determinant by using Definition 6.1.

1. $\begin{vmatrix} 4 & 3 \\ 2 & 7 \end{vmatrix}$

2. $\begin{vmatrix} 3 & 5 \\ 6 & 4 \end{vmatrix}$

3. $\begin{vmatrix} -3 & 2 \\ 7 & 5 \end{vmatrix}$

4. $\begin{vmatrix} 5 & 3 \\ 6 & -1 \end{vmatrix}$

5. $\begin{vmatrix} 2 & -3 \\ 8 & -2 \end{vmatrix}$

6. $\begin{vmatrix} -5 & 5 \\ -6 & 2 \end{vmatrix}$

7. $\begin{vmatrix} -2 & -3 \\ -1 & -4 \end{vmatrix}$

8. $\begin{vmatrix} -4 & -3 \\ -5 & -7 \end{vmatrix}$

9. $\begin{vmatrix} \dfrac{1}{2} & \dfrac{1}{3} \\ -3 & -6 \end{vmatrix}$

10. $\begin{vmatrix} \dfrac{2}{3} & \dfrac{3}{4} \\ 8 & 6 \end{vmatrix}$

11. $\begin{vmatrix} \dfrac{1}{2} & \dfrac{2}{3} \\ \dfrac{3}{4} & -\dfrac{1}{3} \end{vmatrix}$

12. $\begin{vmatrix} \dfrac{2}{3} & \dfrac{1}{5} \\ -\dfrac{1}{4} & \dfrac{3}{2} \end{vmatrix}$

For Problems 13–28, evaluate each 3×3 determinant. Use the properties of determinants to your advantage.

13. $\begin{vmatrix} 1 & 2 & -1 \\ 3 & 1 & 2 \\ 2 & 4 & 3 \end{vmatrix}$

14. $\begin{vmatrix} 1 & -2 & 1 \\ 2 & 1 & -1 \\ 3 & 2 & 4 \end{vmatrix}$

15. $\begin{vmatrix} 1 & -4 & 1 \\ 2 & 5 & -1 \\ 3 & 3 & 4 \end{vmatrix}$

16. $\begin{vmatrix} 3 & -2 & 1 \\ 2 & 1 & 4 \\ -1 & 3 & 5 \end{vmatrix}$

17. $\begin{vmatrix} 6 & 12 & 3 \\ -1 & 5 & 1 \\ -3 & 6 & 2 \end{vmatrix}$

18. $\begin{vmatrix} 2 & 35 & 5 \\ 1 & -5 & 1 \\ -4 & 15 & 2 \end{vmatrix}$

19. $\begin{vmatrix} 2 & -1 & 3 \\ 0 & 3 & 1 \\ 1 & -2 & -1 \end{vmatrix}$

20. $\begin{vmatrix} 2 & -17 & 3 \\ 0 & 5 & 1 \\ 1 & -3 & -1 \end{vmatrix}$

21. $\begin{vmatrix} -3 & -2 & 1 \\ 5 & 0 & 6 \\ 2 & 1 & -4 \end{vmatrix}$

22. $\begin{vmatrix} -5 & 1 & -1 \\ 3 & 4 & 2 \\ 0 & 2 & -3 \end{vmatrix}$

23. $\begin{vmatrix} 3 & -4 & -2 \\ 5 & -2 & 1 \\ 1 & 0 & 0 \end{vmatrix}$

24. $\begin{vmatrix} -6 & 5 & 3 \\ 2 & 0 & -1 \\ 4 & 0 & 7 \end{vmatrix}$

25. $\begin{vmatrix} 24 & -1 & 4 \\ 40 & 2 & 0 \\ -16 & 6 & 0 \end{vmatrix}$

26. $\begin{vmatrix} 2 & -1 & 3 \\ 0 & 3 & 1 \\ 4 & -8 & -4 \end{vmatrix}$

27. $\begin{vmatrix} 2 & 3 & -4 \\ 4 & 6 & -1 \\ -6 & 1 & -2 \end{vmatrix}$

28. $\begin{vmatrix} 1 & 2 & -3 \\ -3 & -1 & 1 \\ 4 & 5 & 4 \end{vmatrix}$

For Problems 29–32, evaluate each 4×4 determinant. Use the properties of determinants to your advantage.

29. $\begin{vmatrix} 1 & -2 & 3 & 2 \\ 2 & -1 & 0 & 4 \\ -3 & 4 & 0 & -2 \\ -1 & 1 & 1 & 5 \end{vmatrix}$

30. $\begin{vmatrix} 1 & 2 & 5 & 7 \\ -6 & 3 & 0 & 9 \\ -3 & 5 & 2 & 7 \\ 2 & 1 & 4 & 3 \end{vmatrix}$

31. $\begin{vmatrix} 3 & -1 & 2 & 3 \\ 1 & 0 & 2 & 1 \\ 2 & 3 & 0 & 1 \\ 5 & 2 & 4 & -5 \end{vmatrix}$

32. $\begin{vmatrix} 1 & 2 & 0 & 0 \\ 3 & -1 & 4 & 5 \\ -2 & 4 & 1 & 6 \\ 2 & -1 & -2 & -3 \end{vmatrix}$

For Problems 33–42, use the appropriate property of determinants from this section to justify each true statement. *Do not* evaluate the determinants.

33. $(-4) \begin{vmatrix} 2 & 1 & -1 \\ 3 & 2 & 1 \\ 2 & 1 & 3 \end{vmatrix} = \begin{vmatrix} 2 & -4 & -1 \\ 3 & -8 & 1 \\ 2 & -4 & 3 \end{vmatrix}$

34. $\begin{vmatrix} 1 & -2 & 3 \\ 4 & -6 & -8 \\ 0 & 2 & 7 \end{vmatrix} = (-2) \begin{vmatrix} 1 & -2 & 3 \\ -2 & 3 & 4 \\ 0 & 2 & 7 \end{vmatrix}$

35. $\begin{vmatrix} 4 & 7 & 9 \\ 6 & -8 & 2 \\ 4 & 3 & -1 \end{vmatrix} = -\begin{vmatrix} 4 & 9 & 7 \\ 6 & 2 & -8 \\ 4 & -1 & 3 \end{vmatrix}$

36. $\begin{vmatrix} 3 & -1 & 4 \\ 5 & 2 & 7 \\ 3 & -1 & 4 \end{vmatrix} = 0$

37. $\begin{vmatrix} 1 & 3 & 4 \\ -2 & 5 & 7 \\ -3 & -1 & 2 \end{vmatrix} = \begin{vmatrix} 1 & 3 & 4 \\ -2 & 5 & 7 \\ 0 & 8 & 14 \end{vmatrix}$

38. $\begin{vmatrix} 3 & 2 & 0 \\ 1 & 4 & 1 \\ -4 & 9 & 2 \end{vmatrix} = \begin{vmatrix} 3 & 2 & -3 \\ 1 & 4 & 0 \\ -4 & 9 & 6 \end{vmatrix}$

39. $\begin{vmatrix} 6 & 2 & 2 \\ 3 & -1 & 4 \\ 9 & -3 & 6 \end{vmatrix} = 6 \begin{vmatrix} 2 & 2 & 1 \\ 1 & -1 & 2 \\ 3 & -3 & 3 \end{vmatrix} = 18 \begin{vmatrix} 2 & 2 & 1 \\ 1 & -1 & 2 \\ 1 & -1 & 1 \end{vmatrix}$

40. $\begin{vmatrix} 2 & 1 & -3 \\ 0 & 2 & -4 \\ -5 & 1 & 3 \end{vmatrix} = -\begin{vmatrix} 2 & 1 & -3 \\ -5 & 1 & 3 \\ 0 & 2 & -4 \end{vmatrix}$

41. $\begin{vmatrix} 2 & -3 & 2 \\ 1 & -4 & 1 \\ 7 & 8 & 7 \end{vmatrix} = 0$

42. $\begin{vmatrix} 3 & 1 & 2 \\ -4 & 5 & -1 \\ 2 & -2 & -4 \end{vmatrix} = \begin{vmatrix} 3 & 1 & 0 \\ -4 & 5 & -11 \\ 2 & -2 & 0 \end{vmatrix}$

■■■ THOUGHTS INTO WORDS

43. Explain the difference between a matrix and a determinant.

44. Explain the concept of a cofactor and how it is used to help expand a determinant.

45. What does it mean to say that any row or column can be used to expand a determinant?

46. Give a step-by-step explanation of how to evaluate the determinant

$$\begin{vmatrix} 3 & 0 & 2 \\ 1 & -2 & 5 \\ 6 & 0 & 9 \end{vmatrix}$$

■■■ FURTHER INVESTIGATIONS

For Problems 47–50, use

$$A = \begin{bmatrix} a_{11} & a_{12} & a_{13} \\ a_{21} & a_{22} & a_{23} \\ a_{31} & a_{32} & a_{33} \end{bmatrix}$$

as a general representation for any 3×3 matrix.

47. Verify Property 6.2 for 3×3 matrices.

48. Verify Property 6.3 for 3×3 matrices.

49. Verify Property 6.4 for 3×3 matrices.

50. Show that $|A| = a_{11}a_{22}a_{33}a_{44}$ if

$$A = \begin{bmatrix} a_{11} & a_{12} & a_{13} & a_{14} \\ 0 & a_{22} & a_{23} & a_{24} \\ 0 & 0 & a_{33} & a_{34} \\ 0 & 0 & 0 & a_{44} \end{bmatrix}$$

GRAPHING CALCULATOR ACTIVITIES

51. Use a calculator to check your answers for Problems 29–32.

52. Consider the following matrix:

$$A = \begin{bmatrix} 2 & 5 & 7 & 9 \\ -4 & 6 & 2 & 4 \\ 6 & 9 & 12 & 3 \\ 5 & 4 & -2 & 8 \end{bmatrix}$$

Form matrix B by interchanging rows 1 and 3 of matrix A. Now use your calculator to show that $|B| = -|A|$.

53. Consider the following matrix:

$$A = \begin{bmatrix} 2 & 1 & 7 & 6 & 8 \\ 3 & -2 & 4 & 5 & -1 \\ 6 & 7 & 9 & 12 & 13 \\ -4 & -7 & 6 & 2 & 1 \\ 9 & 8 & 12 & 14 & 17 \end{bmatrix}$$

Form matrix B by multiplying each element of the second row of matrix A by 3. Now use your calculator to show that $|B| = 3|A|$.

54. Consider the following matrix:

$$A = \begin{bmatrix} 4 & 3 & 2 & 1 & 5 & -3 \\ 5 & 2 & 7 & 8 & 6 & 3 \\ 0 & 9 & 1 & 4 & 7 & 2 \\ 4 & 3 & 2 & 1 & 5 & -3 \\ -4 & -6 & 7 & 12 & 11 & 9 \\ 5 & 8 & 6 & -3 & 2 & -1 \end{bmatrix}$$

Use your calculator to show that $|A| = 0$.

6.5   Cramer's Rule

Objectives

■ Use Cramer's Rule to solve 2×2 and 3×3 systems of equations.

Determinants provide the basis for another method of solving linear systems. Consider the following linear system of two equations and two unknowns:

$$\begin{pmatrix} a_1x + b_1y = c_1 \\ a_2x + b_2y = c_2 \end{pmatrix}$$

The augmented matrix of this system is

$$\begin{bmatrix} a_1 & b_1 & \vdots & c_1 \\ a_2 & b_2 & \vdots & c_2 \end{bmatrix}$$

Using the elementary row transformations of augmented matrices, we can change this matrix to the following reduced echelon form. (The details of this are left for you to do as an exercise.)

$$\begin{bmatrix} 1 & 0 & \vdots & \dfrac{c_1 b_2 - c_2 b_1}{a_1 b_2 - a_2 b_1} \\ 0 & 1 & \vdots & \dfrac{a_1 c_2 - a_2 c_1}{a_1 b_2 - a_2 b_1} \end{bmatrix}, \qquad a_1 b_2 - a_2 b_1 \neq 0$$

The solution for x and y can be expressed in determinant form as follows:

$$x = \frac{c_1 b_2 - c_2 b_1}{a_1 b_2 - a_2 b_1} = \frac{\begin{vmatrix} c_1 & b_1 \\ c_2 & b_2 \end{vmatrix}}{\begin{vmatrix} a_1 & b_1 \\ a_2 & b_2 \end{vmatrix}}$$

$$y = \frac{a_1 c_2 - a_2 c_1}{a_1 b_2 - a_2 b_1} = \frac{\begin{vmatrix} a_1 & c_1 \\ a_2 & c_2 \end{vmatrix}}{\begin{vmatrix} a_1 & b_1 \\ a_2 & b_2 \end{vmatrix}}$$

This method of using determinants to solve a system of two linear equations in two variables is called **Cramer's rule** and can be stated as follows.

Cramer's Rule (2 × 2 case)

Given the system

$$\begin{pmatrix} a_1 x + b_1 y = c_1 \\ a_2 x + b_2 y = c_2 \end{pmatrix}$$

with

$$D = \begin{vmatrix} a_1 & b_1 \\ a_2 & b_2 \end{vmatrix} \neq 0$$

$$D_x = \begin{vmatrix} c_1 & b_1 \\ c_2 & b_2 \end{vmatrix} \qquad \text{and} \qquad D_y = \begin{vmatrix} a_1 & c_1 \\ a_2 & c_2 \end{vmatrix}$$

the solution for this system is given by

$$x = \frac{D_x}{D} \qquad \text{and} \qquad y = \frac{D_y}{D}$$

Note that the elements of D are the coefficients of the variables in the given system. In D_x the coefficients of x are replaced by the corresponding constants, and in D_y the coefficients of y are replaced by the corresponding constants. Let's illustrate the use of Cramer's rule to solve some systems.

E X A M P L E   1

Solve the system $\begin{pmatrix} 6x + 3y = 2 \\ 3x + 2y = -4 \end{pmatrix}$.

Solution

The system is in the proper form for us to apply Cramer's rule, so let's determine D, D_x, and D_y.

$$D = \begin{vmatrix} 6 & 3 \\ 3 & 2 \end{vmatrix} = 12 - 9 = 3$$

$$D_x = \begin{vmatrix} 2 & 3 \\ -4 & 2 \end{vmatrix} = 4 + 12 = 16$$

$$D_y = \begin{vmatrix} 6 & 2 \\ 3 & -4 \end{vmatrix} = -24 - 6 = -30$$

Therefore

$$x = \frac{D_x}{D} = \frac{16}{3} \quad \text{and} \quad y = \frac{D_y}{D} = \frac{-30}{3} = -10$$

The solution set is $\left\{ \left(\dfrac{16}{3}, -10 \right) \right\}$.

▶ **Now work Problem 1.**   ■

E X A M P L E   2

Solve the system $\begin{pmatrix} y = -2x - 2 \\ 4x - 5y = 17 \end{pmatrix}$.

Solution

To begin, we must change the form of the first equation so that the system fits the form given in Cramer's rule. The equation $y = -2x - 2$ can be rewritten $2x + y = -2$. The system now becomes

$$\begin{pmatrix} 2x + y = -2 \\ 4x - 5y = 17 \end{pmatrix}$$

and we can proceed to determine D, D_x, and D_y.

$$D = \begin{vmatrix} 2 & 1 \\ 4 & -5 \end{vmatrix} = -10 - 4 = -14$$

$$D_x = \begin{vmatrix} -2 & 1 \\ 17 & -5 \end{vmatrix} = 10 - 17 = -7$$

$$D_y = \begin{vmatrix} 2 & -2 \\ 4 & 17 \end{vmatrix} = 34 - (-8) = 42$$

Thus

$$x = \frac{D_x}{D} = \frac{-7}{-14} = \frac{1}{2} \quad \text{and} \quad y = \frac{D_y}{D} = \frac{42}{-14} = -3$$

The solution set is $\left\{\left(\frac{1}{2}, -3\right)\right\}$, which can be verified, as always, by substituting back into the original equations.

▶ **Now work Problem 11.**   ■

E X A M P L E 3   Solve the system

$$\left(\begin{array}{l} \frac{1}{2}x + \frac{2}{3}y = -4 \\ \frac{1}{4}x - \frac{3}{2}y = 20 \end{array}\right)$$

Solution

With such a system, either we can first produce an equivalent system with integral coefficients and then apply Cramer's rule, or we can apply the rule immediately. Let's avoid some work with fractions by multiplying the first equation by 6 and the second equation by 4 to produce the following equivalent system:

$$\left(\begin{array}{l} 3x + 4y = -24 \\ x - 6y = 80 \end{array}\right)$$

Now we can proceed as before.

$$D = \begin{vmatrix} 3 & 4 \\ 1 & -6 \end{vmatrix} = -18 - 4 = -22$$

$$D_x = \begin{vmatrix} -24 & 4 \\ 80 & -6 \end{vmatrix} = 144 - 320 = -176$$

$$D_y = \begin{vmatrix} 3 & -24 \\ 1 & 80 \end{vmatrix} = 240 - (-24) = 264$$

Therefore

$$x = \frac{D_x}{D} = \frac{-176}{-22} = 8$$

and

$$y = \frac{D_y}{D} = \frac{264}{-22} = -12$$

The solution set is $\{(8, -12)\}$.

▶ **Now work Problem 13.**   ■

In the statement of Cramer's rule, the condition that $D \neq 0$ was imposed. If $D = 0$ and either D_x or D_y (or both) is nonzero, then the system is inconsistent and has no solution. If $D = 0$, $D_x = 0$, and $D_y = 0$, then the equations are dependent and there are infinitely many solutions.

■ Cramer's Rule Extended

Without showing the details, we will simply state that Cramer's rule also applies to solving systems of three linear equations in three variables. It can be stated as follows.

Cramer's Rule (3 × 3 case)

Given the system

$$\begin{pmatrix} a_1x + b_1y + c_1z = d_1 \\ a_2x + b_2y + c_2z = d_2 \\ a_3x + b_3y + c_3z = d_3 \end{pmatrix}$$

with

$$D = \begin{vmatrix} a_1 & b_1 & c_1 \\ a_2 & b_2 & c_2 \\ a_3 & b_3 & c_3 \end{vmatrix} \neq 0 \qquad D_x = \begin{vmatrix} d_1 & b_1 & c_1 \\ d_2 & b_2 & c_2 \\ d_3 & b_3 & c_3 \end{vmatrix}$$

$$D_y = \begin{vmatrix} a_1 & d_1 & c_1 \\ a_2 & d_2 & c_2 \\ a_3 & d_3 & c_3 \end{vmatrix} \qquad D_z = \begin{vmatrix} a_1 & b_1 & d_1 \\ a_2 & b_2 & d_2 \\ a_3 & b_3 & d_3 \end{vmatrix}$$

we have

$$x = \frac{D_x}{D} \qquad y = \frac{D_y}{D} \qquad \text{and} \qquad z = \frac{D_z}{D}$$

Again, note the restriction that $D \neq 0$. If $D = 0$ and at least one of D_x, D_y, and D_z is not zero, then the system is inconsistent. If D, D_x, D_y, and D_z are all zero, then the equations are dependent, and there are infinitely many solutions.

E X A M P L E   4   Solve the system

$$\begin{pmatrix} x - 2y + z = -4 \\ 2x + y - z = 5 \\ 3x + 2y + 4z = 3 \end{pmatrix}$$

Solution

We will simply indicate the values of D, D_x, D_y, and D_z and leave the computations for you to check.

$$D = \begin{vmatrix} 1 & -2 & 1 \\ 2 & 1 & -1 \\ 3 & 2 & 4 \end{vmatrix} = 29 \qquad D_x = \begin{vmatrix} -4 & -2 & 1 \\ 5 & 1 & -1 \\ 3 & 2 & 4 \end{vmatrix} = 29$$

$$D_y = \begin{vmatrix} 1 & -4 & 1 \\ 2 & 5 & -1 \\ 3 & 3 & 4 \end{vmatrix} = 58 \qquad D_z = \begin{vmatrix} 1 & -2 & -4 \\ 2 & 1 & 5 \\ 3 & 2 & 3 \end{vmatrix} = -29$$

Therefore

$$x = \frac{D_x}{D} = \frac{29}{29} = 1$$

$$y = \frac{D_y}{D} = \frac{58}{29} = 2$$

$$z = \frac{D_z}{D} = \frac{-29}{29} = -1$$

The solution set is $\{(1, 2, -1)\}$. (Be sure to check it!)

▶ **Now work Problem 17.** ∎

E X A M P L E 5 Solve the system

$$\begin{pmatrix} x + 3y - z = 4 \\ 3x - 2y + z = 7 \\ 2x + 6y - 2z = 1 \end{pmatrix}$$

Solution

$$D = \begin{vmatrix} 1 & 3 & -1 \\ 3 & -2 & 1 \\ 2 & 6 & -2 \end{vmatrix} = 2 \begin{vmatrix} 1 & 3 & -1 \\ 3 & -2 & 1 \\ 1 & 3 & -1 \end{vmatrix} = 2(0) = 0$$

$$D_x = \begin{vmatrix} 4 & 3 & -1 \\ 7 & -2 & 1 \\ 1 & 6 & -2 \end{vmatrix} = -7$$

Therefore, because $D = 0$ and at least one of D_x, D_y, and D_z is not zero, the system is inconsistent. The solution set is $\varnothing$.

▶ **Now work Problem 26.** ∎

EXAMPLE 6

Solve the system

$$\begin{pmatrix} x - y + 2z = 1 \\ -3x + 4y - z = 4 \\ -x + 2y + 3z = 6 \end{pmatrix}$$

Solution

$$D = \begin{vmatrix} 1 & -1 & 2 \\ -3 & 4 & -1 \\ -1 & 2 & 3 \end{vmatrix} = 0 \qquad D_x = \begin{vmatrix} 1 & -1 & 2 \\ 4 & 4 & -1 \\ 6 & 2 & 3 \end{vmatrix} = 0$$

$$D_y = \begin{vmatrix} 1 & 1 & 2 \\ -3 & 4 & -1 \\ -1 & 6 & 3 \end{vmatrix} = 0 \qquad D_z = \begin{vmatrix} 1 & -1 & 1 \\ -3 & 4 & 4 \\ -1 & 2 & 6 \end{vmatrix} = 0$$

Because $D = D_x = D_y = D_z = 0$, the solution set has infinitely many solutions. By adding the first and third equations, we obtain $y + 5z = 7$ or $y = 7 - 5z$. Substituting $7 - 5z$ for y in the second equation and solving for x produces $x = 8 - 7z$. Therefore if we let z be any real number k, then the ordered triple $(8 - 7k, 7 - 5k, k)$ will generate infinitely many solutions of the given system of equations. For example, let $k = 2$; then the ordered triple $(8 - 7k, 7 - 5k, k)$ becomes $(-6, -3, 2)$, which is a solution of the system. (You may want to try some additional values for k.)

▶ **Now work Problem 27.** ∎

Example 5 illustrates why D should be determined first. Once we found that $D = 0$ and $D_x \neq 0$, we knew that the system was inconsistent and there was no need to find D_y and D_z. However, in Example 6 we had to show that D, D_x, D_y, and D_z were all equal to zero.

Finally, it should be noted that Cramer's rule can be extended to systems of n linear equations in n variables; however, that method is not considered to be a very efficient way of solving a large system of linear equations.

CONCEPT QUIZ

For Problems 1–8, answer true or false.

1. When using Cramer's rule, the elements of D are the coefficients of the variables in the given system.

2. The system $\begin{pmatrix} 2x - y + 3z = 9 \\ -x + z - 4y = 2 \\ 5x + 2y - z = 8 \end{pmatrix}$ is in the proper form to apply Cramer's rule.

3. When using Cramer's rule for D_y, the coefficients of y are replaced by the corresponding constants from the system of equations.

4. Applying Cramer's rule to the system $\begin{pmatrix} \dfrac{1}{4}x + \dfrac{5}{12}y = -1 \\ \dfrac{1}{3}x - \dfrac{2}{3}y = 7 \end{pmatrix}$ produces the same

   solution set as applying Cramer's rule to the system $\begin{pmatrix} 3x + 5y = -12 \\ x - 2y = 21 \end{pmatrix}$.

5. When using Cramer's rule, if $D = 0$, then the system either has no solution or an infinite number of solutions.

6. When using Cramer's rule, if $D = 0$ and $D_y = 4$, then the system is inconsistent.

7. When using Cramer's Rule, if $D_x \neq 0$ then the system of equations always has a solution.

8. Cramer's rule can be extended to solve systems of n linear equations in n variables.

Problem Set 6.5

For Problems 1–32, use Cramer's rule to find the solution set for each system. If the equations are dependent, simply indicate that there are infinitely many solutions.

15. $\begin{pmatrix} 2x + 7y = -1 \\ x = 2 \end{pmatrix}$

16. $\begin{pmatrix} 5x - 3y = 2 \\ y = 4 \end{pmatrix}$

1. $\begin{pmatrix} 2x - y = -2 \\ 3x + 2y = 11 \end{pmatrix}$

2. $\begin{pmatrix} 3x + y = -9 \\ 4x - 3y = 1 \end{pmatrix}$

17. $\begin{pmatrix} x - y + 2z = -8 \\ 2x + 3y - 4z = 18 \\ -x + 2y - z = 7 \end{pmatrix}$

3. $\begin{pmatrix} 5x + 2y = 5 \\ 3x - 4y = 29 \end{pmatrix}$

4. $\begin{pmatrix} 4x - 7y = -23 \\ 2x + 5y = -3 \end{pmatrix}$

18. $\begin{pmatrix} x - 2y + z = 3 \\ 3x + 2y + z = -3 \\ 2x - 3y - 3z = -5 \end{pmatrix}$

5. $\begin{pmatrix} 5x - 4y = 14 \\ -x + 2y = -4 \end{pmatrix}$

6. $\begin{pmatrix} -x + 2y = 10 \\ 3x - y = -10 \end{pmatrix}$

19. $\begin{pmatrix} 2x - 3y + z = -7 \\ -3x + y - z = -7 \\ x - 2y - 5z = -45 \end{pmatrix}$

7. $\begin{pmatrix} y = 2x - 4 \\ 6x - 3y = 1 \end{pmatrix}$

8. $\begin{pmatrix} -3x - 4y = 14 \\ -2x + 3y = -19 \end{pmatrix}$

9. $\begin{pmatrix} -4x + 3y = 3 \\ 4x - 6y = -5 \end{pmatrix}$

10. $\begin{pmatrix} x = 4y - 1 \\ 2x - 8y = -2 \end{pmatrix}$

20. $\begin{pmatrix} 3x - y - z = 18 \\ 4x + 3y - 2z = 10 \\ -5x - 2y + 3z = -22 \end{pmatrix}$

11. $\begin{pmatrix} 9x - y = -2 \\ 8x + y = 4 \end{pmatrix}$

12. $\begin{pmatrix} 6x - 5y = 1 \\ 4x - 7y = 2 \end{pmatrix}$

21. $\begin{pmatrix} 4x + 5y - 2z = -14 \\ 7x - y + 2z = 42 \\ 3x + y + 4z = 28 \end{pmatrix}$

13. $\begin{pmatrix} -\dfrac{2}{3}x + \dfrac{1}{2}y = -7 \\ \dfrac{1}{3}x - \dfrac{3}{2}y = 6 \end{pmatrix}$

14. $\begin{pmatrix} \dfrac{1}{2}x + \dfrac{2}{3}y = -6 \\ \dfrac{1}{4}x - \dfrac{1}{3}y = -1 \end{pmatrix}$

22. $\begin{pmatrix} -5x + 6y + 4z = -4 \\ -7x - 8y + 2z = -2 \\ 2x + 9y - z = 1 \end{pmatrix}$

23. $\begin{pmatrix} 2x - y + 3z = -17 \\ 3y + z = 5 \\ x - 2y - z = -3 \end{pmatrix}$

28. $\begin{pmatrix} 3x - 2y + z = 11 \\ 5x + 3y = 17 \\ x + y - 2z = 6 \end{pmatrix}$

24. $\begin{pmatrix} 2x - y + 3z = -5 \\ 3x + 4y - 2z = -25 \\ -x + z = 6 \end{pmatrix}$

29. $\begin{pmatrix} x - 2y + 3z = 1 \\ -2x + 4y - 3z = -3 \\ 5x - 6y + 6z = 10 \end{pmatrix}$

25. $\begin{pmatrix} x + 3y - 4z = -1 \\ 2x - y + z = 2 \\ 4x + 5y - 7z = 0 \end{pmatrix}$

30. $\begin{pmatrix} 2x - y + 2z = -1 \\ 4x + 3y - 4z = 2 \\ x + 5y - z = 9 \end{pmatrix}$

26. $\begin{pmatrix} x - 2y + z = 1 \\ 3x + y - z = 2 \\ 2x - 4y + 2z = -1 \end{pmatrix}$

31. $\begin{pmatrix} -x - y + 3z = -2 \\ -2x + y + 7z = 14 \\ 3x + 4y - 5z = 12 \end{pmatrix}$

27. $\begin{pmatrix} 3x - 2y - 3z = -5 \\ x + 2y + 3z = -3 \\ -x + 4y - 6z = 8 \end{pmatrix}$

32. $\begin{pmatrix} -2x + y - 3z = -4 \\ x + 5y - 4z = 13 \\ 7x - 2y - z = 37 \end{pmatrix}$

■ ■ ■ **THOUGHTS INTO WORDS**

33. Give a step-by-step description of how you would solve the system

$$\begin{pmatrix} 2x - y + 3z = 31 \\ x - 2y - z = 8 \\ 3x + 5y + 8z = 35 \end{pmatrix}$$

34. Give a step-by-step description of how you would find the value of x in the solution for the system

$$\begin{pmatrix} x + 5y - z = -9 \\ 2x - y + z = 11 \\ -3x - 2y + 4z = 20 \end{pmatrix}$$

■ ■ ■ **FURTHER INVESTIGATIONS**

35. A linear system in which the constant terms are all zero is called a **homogeneous system.**

   a. Verify that for a 3×3 homogeneous system, if $D \neq 0$, then $(0, 0, 0)$ is the only solution for the system.

   b. Verify that for a 3×3 homogeneous system, if $D = 0$, then the equations are dependent.

For Problems 36–39, solve each of the homogeneous systems (see Problem 35). If the equations are dependent, indicate that the system has infinitely many solutions.

36. $\begin{pmatrix} x - 2y + 5z = 0 \\ 3x + y - 2z = 0 \\ 4x - y + 3z = 0 \end{pmatrix}$

37. $\begin{pmatrix} 2x - y + z = 0 \\ 3x + 2y + 5z = 0 \\ 4x - 7y + z = 0 \end{pmatrix}$

38. $\begin{pmatrix} 3x + y - z = 0 \\ x - y + 2z = 0 \\ 4x - 5y - 2z = 0 \end{pmatrix}$

39. $\begin{pmatrix} 2x - y + 2z = 0 \\ x + 2y + z = 0 \\ x - 3y + z = 0 \end{pmatrix}$

GRAPHING CALCULATOR ACTIVITIES

40. Use determinants and your calculator to solve each of the following systems.

a.
$$\begin{pmatrix} 4x - 3y + z = 10 \\ 8x + 5y - 2z = -6 \\ -12x - 2y + 3z = -2 \end{pmatrix}$$

b.
$$\begin{pmatrix} 2x + y - z + w = -4 \\ x + 2y + 2z - 3w = 6 \\ 3x - y - z + 2w = 0 \\ 2x + 3y + z + 4w = -5 \end{pmatrix}$$

c.
$$\begin{pmatrix} x - 2y + z - 3w = 4 \\ 2x + 3y - z - 2w = -4 \\ 3x - 4y + 2z - 4w = 12 \\ 2x - y - 3z + 2w = -2 \end{pmatrix}$$

d.
$$\begin{pmatrix} 1.98x + 2.49y + 3.45z = 80.10 \\ 2.15x + 3.20y + 4.19z = 97.16 \\ 1.49x + 4.49y + 2.79z = 83.92 \end{pmatrix}$$

Answers to the Concept Quiz

1. True   **2.** False   **3.** True   **4.** True   **5.** True   **6.** True   **7.** False   **8.** True

6.6   Partial Fractions

Objectives

■ Find partial fraction decompositions for rational expressions.

In Chapter 0 we reviewed the process of adding rational expressions. For example,

$$\frac{3}{x - 2} + \frac{2}{x + 3} = \frac{3(x + 3) + 2(x - 2)}{(x - 2)(x + 3)} = \frac{3x + 9 + 2x - 4}{(x - 2)(x + 3)} = \frac{5x + 5}{(x - 2)(x + 3)}$$

Now suppose that we want to reverse the process. That is, suppose we are given the rational expression

$$\frac{5x + 5}{(x - 2)(x + 3)}$$

and we want to express it as the sum of two simpler rational expressions called **partial fractions.** This process, called **partial fraction decomposition,** has several applications in calculus and differential equations. The following property provides the basis for partial fraction decomposition:

Property 6.6

Let $f(x)$ and $g(x)$ be polynomials with real coefficients, such that the degree of $f(x)$ is less than the degree of $g(x)$. The indicated quotient $f(x)/g(x)$ can be decomposed into partial fractions as follows.

1. If $g(x)$ has a linear factor of the form $ax + b$, then the partial fraction decomposition will contain a term of the form

$$\frac{A}{ax + b}, \qquad \text{where } A \text{ is a constant}$$

2. If $g(x)$ has a linear factor of the form $ax + b$ raised to the kth power, then the partial fraction decomposition will contain terms of the form

$$\frac{A_1}{ax + b} + \frac{A_2}{(ax + b)^2} + \cdots + \frac{A_k}{(ax + b)^k}$$

where $A_1, A_2, \ldots, A_k$ are constants.

3. If $g(x)$ has a quadratic factor of the form $ax^2 + bx + c$, where $b^2 - 4ac < 0$, then the partial fraction decomposition will contain a term of the form

$$\frac{Ax + B}{ax^2 + bx + c}, \qquad \text{where } A \text{ and } B \text{ are constants}$$

4. If $g(x)$ has a quadratic factor of the form $ax^2 + bx + c$ raised to the kth power, where $b^2 - 4ac < 0$, then the partial fraction decomposition will contain terms of the form

$$\frac{A_1 x + B_1}{ax^2 + bx + c} + \frac{A_2 x + B_2}{(ax^2 + bx + c)^2} + \cdots + \frac{A_k x + B_k x}{(ax^2 + bx + c)^k}$$

where $A_1, A_2, \ldots, A_k$ and $B_1, B_2, \ldots, B_k$ are constants.

In parts 3 and 4 of Property 6.6, the condition that $b^2 - 4ac < 0$ means that the quadratic factor cannot be expressed as a product of linear factors with real coefficients.

Note that Property 6.6 applies only to **proper fractions**—that is, fractions where the degree of the numerator is less than the degree of the denominator. If the numerator is not of lower degree, we can divide and then apply Property 6.6 to the remainder, which will be a proper fraction. For example,

$$\frac{x^3 - 3x^2 - 3x - 5}{x^2 - 4} = x - 3 + \frac{x - 17}{x^2 - 4}$$

and the proper fraction $\dfrac{x - 17}{x^2 - 4}$ can be decomposed into partial fractions by apply-

ing Property 6.6. Now let's consider some examples to illustrate the four cases in Property 6.6.

E X A M P L E   1   Find the partial fraction decomposition of $\dfrac{11x + 2}{2x^2 + x - 1}$.

Solution

The denominator can be expressed as $(x + 1)(2x - 1)$. Therefore, according to part 1 of Property 6.6, each of the linear factors produces a partial fraction of the form *constant over linear factor*. In other words, we can write

$$\frac{11x + 2}{(x + 1)(2x - 1)} = \frac{A}{x + 1} + \frac{B}{2x - 1} \tag{1}$$

for some constants A and B. To find A and B, we multiply both sides of equation (1) by the least common denominator $(x + 1)(2x - 1)$:

$$11x + 2 = A(2x - 1) + B(x + 1) \tag{2}$$

Equation (2) is an **identity:** *It is true for all values of x.* Therefore let's choose some convenient values for x that will determine the values for A and B. If we let $x = -1$, then equation (2) becomes an equation only in A.

$$11(-1) + 2 = A[2(-1) - 1] + B(-1 + 1)$$
$$-9 = -3A$$
$$3 = A$$

If we let $x = \dfrac{1}{2}$, then equation (2) becomes an equation only in B.

$$11\left(\frac{1}{2}\right) + 2 = A\left[2\left(\frac{1}{2}\right) - 1\right] + B\left(\frac{1}{2} + 1\right)$$
$$\frac{15}{2} = \frac{3}{2}B$$
$$5 = B$$

Therefore the given rational expression can now be written

$$\frac{11x + 2}{2x^2 + x - 1} = \frac{3}{x + 1} + \frac{5}{2x - 1}$$

▶ **Now work Problem 5.**   ■

The key idea in Example 1 is the statement that equation (2) is true for all values of x. If we had chosen *any* two values for x, we still would have been able to determine the values for A and B. For example, letting $x = 1$ and then $x = 2$ produces the equations $13 = A + 2B$ and $24 = 3A + 3B$. Solving this system of two equations in two unknowns produces $A = 3$ and $B = 5$. In Example 1 our choices of letting $x = -1$ and then $x = \dfrac{1}{2}$ simply eliminated the need for solving a system of equations to find A and B.

<hr>

E X A M P L E   2   Find the partial fraction decomposition of $\dfrac{-2x^2 + 7x + 2}{x(x-1)^2}$.

Solution

We can apply part 1 of Property 6.6 to determine that there is a partial fraction of the form A/x corresponding to the factor of x. Next, applying part 2 of Property 6.6, we see that the squared factor $(x-1)^2$ gives rise to a sum of partial fractions of the form

$$\frac{B}{x-1} + \frac{C}{(x-1)^2}$$

Therefore the complete partial fraction decomposition is of the form

$$\frac{-2x^2 + 7x + 2}{x(x-1)^2} = \frac{A}{x} + \frac{B}{x-1} + \frac{C}{(x-1)^2} \tag{1}$$

We can multiply both sides of equation (1) by $x(x-1)^2$ to produce

$$-2x^2 + 7x + 2 = A(x-1)^2 + Bx(x-1) + Cx \tag{2}$$

which is true for all values of x. If we let $x = 1$, then equation (2) becomes an equation only in C.

$$-2(1)^2 + 7(1) + 2 = A(1-1)^2 + B(1)(1-1) + C(1)$$
$$7 = C$$

If we let $x = 0$, then equation (2) becomes an equation just in A.

$$-2(0)^2 + 7(0) + 2 = A(0-1)^2 + B(0)(0-1) + C(0)$$
$$2 = A$$

If we let $x = 2$, then equation (2) becomes an equation in A, B, and C.

$$-2(2)^2 + 7(2) + 2 = A(2-1)^2 + B(2)(2-1) + C(2)$$
$$8 = A + 2B + 2C$$

But we already know that $A = 2$ and $C = 7$, so we can easily determine B.

$$8 = 2 + 2B + 14$$
$$-8 = 2B$$
$$-4 = B$$

Therefore the original rational expression can be written

$$\frac{-2x^2 + 7x + 2}{x(x-1)^2} = \frac{2}{x} - \frac{4}{x-1} + \frac{7}{(x-1)^2}$$

▶ **Now work Problem 13.**   ■

EXAMPLE 3   Find the partial fraction decomposition of $\dfrac{4x^2 + 6x - 10}{(x + 3)(x^2 + x + 2)}$.

Solution

We can apply part 1 of Property 6.6 to determine that there is a partial fraction of the form $A/(x + 3)$ that corresponds to the factor $x + 3$. Furthermore, because $b^2 - 4ac < 0$ for the quadratic factor $x^2 + x + 2$, there is a partial fraction of the form $(Bx + C)/(x^2 + x + 2)$. Therefore we have the following decomposition form:

$$\frac{4x^2 + 6x - 10}{(x + 3)(x^2 + x + 2)} = \frac{A}{x + 3} + \frac{Bx + C}{x^2 + x + 2} \tag{1}$$

We multiply both sides of equation (1) by $(x + 3)(x^2 + x + 2)$ to produce

$$4x^2 + 6x - 10 = A(x^2 + x + 2) + (Bx + C)(x + 3) \tag{2}$$

which is true for all values of x. If we let $x = -3$, then equation (2) becomes an equation in A alone.

$$4(-3)^2 + 6(-3) - 10 = A[(-3)^2 + (-3) + 2] + [B(-3) + C][(-3) + 3]$$
$$8 = 8A$$
$$1 = A$$

If we let $x = 0$, then equation (2) becomes an equation in A and C.

$$4(0)^2 + 6(0) - 10 = A(0^2 + 0 + 2) + [B(0) + C](0 + 3)$$
$$-10 = 2A + 3C$$

Because $A = 1$, we obtain the value of C.

$$-10 = 2 + 3C$$
$$-12 = 3C$$
$$-4 = C$$

If we let $x = 1$, then equation (2) becomes an equation in A, B, and C.

$$4(1)^2 + 6(1) - 10 = A(1^2 + 1 + 2) + [B(1) + C](1 + 3)$$
$$0 = 4A + 4B + 4C$$
$$0 = A + B + C$$

But because $A = 1$ and $C = -4$, we obtain the value of B.

$$0 = A + B + C$$
$$0 = 1 + B + (-4)$$
$$3 = B$$

Therefore the original rational expression can now be written

$$\frac{4x^2 + 6x - 10}{(x + 3)(x^2 + x + 2)} = \frac{1}{x + 3} + \frac{3x - 4}{x^2 + x + 2}$$

▶ **Now work Problem 15.**   ■

E X A M P L E   4   Find the partial fraction decomposition of $\dfrac{x^3 + x^2 + x + 3}{(x^2 + 1)^2}$.

Solution

Because $b^2 - 4ac < 0$ for the quadratic factor $x^2 + 1$, we can use the following decomposition form:

$$\frac{x^3 + x^2 + x + 3}{(x^2 + 1)^2} = \frac{Ax + B}{x^2 + 1} + \frac{Cx + D}{(x^2 + 1)^2} \tag{1}$$

We multiply both sides of equation (1) by $(x^2 + 1)^2$ to produce

$$x^3 + x^2 + x + 3 = (Ax + B)(x^2 + 1) + Cx + D \tag{2}$$

which is true for all values of x. Equation (2) is an identity, so we know that the coefficients of similar terms on both sides of the equation must be equal. Therefore let's collect similar terms on the right side of equation (2).

$$x^3 + x^2 + x + 3 = Ax^3 + Ax + Bx^2 + B + Cx + D$$
$$= Ax^3 + Bx^2 + (A + C)x + B + D$$

Now we can equate coefficients from both sides:

$$1 = A \qquad 1 = B \qquad 1 = A + C \qquad \text{and} \qquad 3 = B + D$$

From these equations we can determine that $A = 1$, $B = 1$, $C = 0$, and $D = 2$. Therefore the original rational expression can be written

$$\frac{x^3 + x^2 + x + 3}{(x^2 + 1)^2} = \frac{x + 1}{x^2 + 1} + \frac{2}{(x^2 + 1)^2}$$

▶   **Now work Problem 21.** ■

C O N C E P T   Q U I Z   For Problems 1–8, answer true or false.

1. The process of partial fraction decomposition expresses a rational expression as the sum of two or more simpler rational expressions.

2. A rational expression is considered a proper fraction if the degree of the numerator is equal to or less than the degree of the denominator.

3. The process of partial fraction decomposition applies only to proper fractions.

4. To apply partial fraction decomposition to a rational expression that is not a proper fraction, long division is used to obtain a remainder that is a proper fraction.

5. If an equation is an identity, any value of x substituted into the equation produces an equivalent equation.

6. A quadratic expression such as $ax^2 + bx + c$ is not factorable over the real numbers if $b^2 - 4ac < 0$.

7. Given that $5x + 3 = A(x - 4) + B(x + 7)$ is an identity, the value of A can be determined by substituting any value of x into the identity.

8. Given that $3x - 1 = A(x - 2) + B(x + 5)$ is an identity, the value of B can be determined by substituting -5 for x into the identity.

Problem Set 6.6

For Problems 1–22, find the partial fraction decomposition for each rational expression.

1. $\dfrac{11x - 10}{(x - 2)(x + 1)}$

2. $\dfrac{11x - 2}{(x + 3)(x - 4)}$

3. $\dfrac{-2x - 8}{x^2 - 1}$

4. $\dfrac{-2x + 32}{x^2 - 4}$

5. $\dfrac{20x - 3}{6x^2 + 7x - 3}$

6. $\dfrac{-2x - 8}{10x^2 - x - 2}$

7. $\dfrac{x^2 - 18x + 5}{(x - 1)(x + 2)(x - 3)}$

8. $\dfrac{-9x^2 + 7x - 4}{x^3 - 3x^2 - 4x}$

9. $\dfrac{-6x^2 + 7x + 1}{x(2x - 1)(4x + 1)}$

10. $\dfrac{15x^2 + 20x + 30}{(x + 3)(3x + 2)(2x + 3)}$

11. $\dfrac{2x + 1}{(x - 2)^2}$

12. $\dfrac{-3x + 1}{(x + 1)^2}$

13. $\dfrac{-6x^2 + 19x + 21}{x^2(x + 3)}$

14. $\dfrac{10x^2 - 73x + 144}{x(x - 4)^2}$

15. $\dfrac{-2x^2 - 3x + 10}{(x^2 + 1)(x - 4)}$

16. $\dfrac{8x^2 + 15x + 12}{(x^2 + 4)(3x - 4)}$

17. $\dfrac{3x^2 + 10x + 9}{(x + 2)^3}$

18. $\dfrac{2x^3 + 8x^2 + 2x + 4}{(x + 1)^2(x^2 + 3)}$

19. $\dfrac{5x^2 + 3x + 6}{x(x^2 - x + 3)}$

20. $\dfrac{x^3 + x^2 + 2}{(x^2 + 2)^2}$

21. $\dfrac{2x^3 + x + 3}{(x^2 + 1)^2}$

22. $\dfrac{4x^2 + 3x + 14}{x^3 - 8}$

■ ■ ■ THOUGHTS INTO WORDS

23. Give a general description of partial fraction decomposition for someone who missed class the day it was discussed.

24. Give a step-by-step explanation of how to find the partial fraction decomposition of $\dfrac{11x + 5}{2x^2 + 5x - 3}$.

Answers to the Concept Quiz

1. True **2.** False **3.** True **4.** True **5.** True **6.** True **7.** False **8.** False

The primary focus of this entire chapter is the development of different techniques for solving systems of linear equations.

■ Substitution Method

With the aid of an example, we can describe the substitution method as follows. Suppose we want to solve the system

$$\begin{pmatrix} x - 2y = 22 \\ 3x + 4y = -24 \end{pmatrix}$$

Step 1 Solve the first equation for x in terms of y.

$$x - 2y = 22$$
$$x = 2y + 22$$

Step 2 Substitute $2y + 22$ for x in the second equation.

$$3(2y + 22) + 4y = -24$$

Step 3 Solve the equation obtained in step 2.

$$6y + 66 + 4y = -24$$
$$10y + 66 = -24$$
$$10y = -90$$
$$y = -9$$

Step 4 Substitute -9 for y in the equation of step 1.

$$x = 2(-9) + 22 = 4$$

The solution set is $\{(4, -9)\}$.

■ Elimination-by-Addition Method

This method allows us to replace systems of equations with *simpler equivalent systems* until we obtain a system where we can easily determine the solution. The following operations produce equivalent systems.

1. Any two equations of a system can be interchanged.

2. Both sides of any equation of the system can be multiplied by any nonzero real number.

3. Any equation of the system can be replaced by the sum of a nonzero multiple of another equation plus that equation.

For example, through a sequence of operations, we can transform the system

$$\begin{pmatrix} 5x + 3y = -28 \\ \dfrac{1}{2}x - y = -8 \end{pmatrix}$$

to the equivalent system

$$\begin{pmatrix} x - 2y = -16 \\ 13y = 52 \end{pmatrix}$$

where we can easily determine the solution set $\{(-8, 4)\}$.

■ Matrix Approach

We can change the augmented matrix of a system to reduced echelon form by applying the following elementary row operations.

1. Any two rows of the matrix can be interchanged.

2. Any row of the matrix can be multiplied by a nonzero real number.

3. Any row of the matrix can be replaced by the sum of a nonzero multiple of another row plus that row.

For example, the augmented matrix of the system

$$\begin{pmatrix} x - 2y + 3z = 4 \\ 2x + y - 4z = 3 \\ -3x + 4y - z = -2 \end{pmatrix}$$

is

$$\begin{bmatrix} 1 & -2 & 3 & \vdots & 4 \\ 2 & 1 & -4 & \vdots & 3 \\ -3 & 4 & -1 & \vdots & -2 \end{bmatrix}$$

We can change this matrix to the reduced echelon form

$$\begin{bmatrix} 1 & 0 & 0 & \vdots & 4 \\ 0 & 1 & 0 & \vdots & 3 \\ 0 & 0 & 1 & \vdots & 2 \end{bmatrix}$$

where the solution set $\{(4, 3, 2)\}$ is obvious.

■ Cramer's Rule

Cramer's rule for solving systems of linear equations involves the use of determinants. It is stated for the 2×2 case on page 527 and for the 3×3 case on page 530. For example the solution set of the system

$$\left(\begin{array}{l} 3x - y - z = 2 \\ 2x + y + 3z = 9 \\ -x + 5y - 6z = -29 \end{array} \right)$$

is determined by

$$x = \frac{\begin{vmatrix} 2 & -1 & -1 \\ 9 & 1 & 3 \\ -29 & 5 & -6 \end{vmatrix}}{\begin{vmatrix} 3 & -1 & -1 \\ 2 & 1 & 3 \\ -1 & 5 & -6 \end{vmatrix}} = \frac{-83}{-83} = 1$$

$$y = \frac{\begin{vmatrix} 3 & 2 & -1 \\ 2 & 9 & 3 \\ -1 & -29 & -6 \end{vmatrix}}{\begin{vmatrix} 3 & -1 & -1 \\ 2 & 1 & 3 \\ -1 & 5 & -6 \end{vmatrix}} = \frac{166}{-83} = -2$$

$$z = \frac{\begin{vmatrix} 3 & -1 & 2 \\ 2 & 1 & 9 \\ -1 & 5 & -29 \end{vmatrix}}{\begin{vmatrix} 3 & -1 & -1 \\ 2 & 1 & 3 \\ -1 & 5 & -6 \end{vmatrix}} = \frac{-249}{-83} = 3$$

Be sure that you understand the process of partial fraction decomposition outlined in Property 6.6.

Chapter 6   Review Problem Set

For Problems 1–4, solve each system by using the *substitution* method.

1. $\left(\begin{array}{l} 3x - y = 16 \\ 5x + 7y = -34 \end{array} \right)$ **2.** $\left(\begin{array}{l} 6x + 5y = -21 \\ x - 4y = 11 \end{array} \right)$

3. $\left(\begin{array}{l} 2x - 3y = 12 \\ 3x + 5y = -20 \end{array} \right)$ **4.** $\left(\begin{array}{l} 5x + 8y = 1 \\ 4x + 7y = -2 \end{array} \right)$

For Problems 5–8, solve each system by using the *elimination-by-addition* method.

5. $\left(\begin{array}{l} 4x - 3y = 34 \\ 3x + 2y = 0 \end{array} \right)$ **6.** $\left(\begin{array}{l} \dfrac{1}{2}x - \dfrac{2}{3}y = 1 \\ \dfrac{3}{4}x + \dfrac{1}{6}y = -1 \end{array} \right)$

7. $\left(\begin{array}{l} 2x - y + 3z = -19 \\ 3x + 2y - 4z = 21 \\ 5x - 4y - z = -8 \end{array} \right)$

8. $\left(\begin{array}{l} 3x + 2y - 4z = 4 \\ 5x + 3y - z = 2 \\ 4x - 2y + 3z = 11 \end{array} \right)$

For Problems 9–12, solve each system by *changing the augmented matrix to reduced echelon form*.

9. $\left(\begin{array}{l} x - 3y = 17 \\ -3x + 2y = -23 \end{array} \right)$

10. $\left(\begin{array}{l} 2x + 3y = 25 \\ 3x - 5y = -29 \end{array} \right)$

11. $\left(\begin{array}{l} x - 2y + z = -7 \\ 2x - 3y + 4z = -14 \\ -3x + y - 2z = 10 \end{array} \right)$

12. $\left(\begin{array}{l} -2x - 7y + z = 9 \\ x + 3y - 4z = -11 \\ 4x + 5y - 3z = -11 \end{array} \right)$

For Problems 13–16, solve each system using *Cramer's rule.*

13. $\begin{pmatrix} 5x + 3y = -18 \\ 4x - 9y = -3 \end{pmatrix}$ **14.** $\begin{pmatrix} 0.2x + 0.3y = 2.6 \\ 0.5x - 0.1y = 1.4 \end{pmatrix}$

15. $\begin{pmatrix} 2x - 3y - 3z = 25 \\ 3x + y + 2z = -5 \\ 5x - 2y - 4z = 32 \end{pmatrix}$

16. $\begin{pmatrix} 3x - y + z = -10 \\ 6x - 2y + 5z = -35 \\ 7x + 3y - 4z = 19 \end{pmatrix}$

For Problems 17–24, solve each system by using the method you think is most appropriate.

17. $\begin{pmatrix} 4x + 7y = -15 \\ 3x - 2y = 25 \end{pmatrix}$ **18.** $\begin{pmatrix} \dfrac{3}{4}x - \dfrac{1}{2}y = -15 \\ \dfrac{2}{3}x + \dfrac{1}{4}y = -5 \end{pmatrix}$

19. $\begin{pmatrix} x + 4y = 3 \\ 3x - 2y = 1 \end{pmatrix}$ **20.** $\begin{pmatrix} 7x - 3y = -49 \\ y = \dfrac{3}{5}x - 1 \end{pmatrix}$

21. $\begin{pmatrix} x - y - z = 4 \\ -3x + 2y + 5z = -21 \\ 5x - 3y - 7z = 30 \end{pmatrix}$

22. $\begin{pmatrix} 2x - y + z = -7 \\ -5x + 2y - 3z = 17 \\ 3x + y + 7z = -5 \end{pmatrix}$

23. $\begin{pmatrix} 3x - 2y - 5z = 2 \\ -4x + 3y + 11z = 3 \\ 2x - y + z = -1 \end{pmatrix}$

24. $\begin{pmatrix} 7x - y + z = -4 \\ -2x + 9y - 3z = -50 \\ x - 5y + 4z = 42 \end{pmatrix}$

For Problems 25–30, evaluate each determinant.

25. $\begin{vmatrix} -2 & 6 \\ 3 & 8 \end{vmatrix}$ **26.** $\begin{vmatrix} 5 & -4 \\ 7 & -3 \end{vmatrix}$

27. $\begin{vmatrix} 2 & 3 & -1 \\ 3 & 4 & -5 \\ 6 & 4 & 2 \end{vmatrix}$ **28.** $\begin{vmatrix} 3 & -2 & 4 \\ 1 & 0 & 6 \\ 3 & -3 & 5 \end{vmatrix}$

29. $\begin{vmatrix} 5 & 4 & 3 \\ 2 & -7 & 0 \\ 3 & -2 & 0 \end{vmatrix}$ **30.** $\begin{vmatrix} 5 & -4 & 2 & 1 \\ 3 & 7 & 6 & -2 \\ 2 & 1 & -5 & 0 \\ 3 & -2 & 4 & 0 \end{vmatrix}$

For Problems 31 and 32, find the partial fraction decomposition.

31. $\dfrac{5x^2 - 4}{x^2(x + 2)}$ **32.** $\dfrac{x^2 - x - 21}{(x^2 + 4)(2x - 1)}$

For Problems 33–36, solve each problem by setting up and solving a system of linear equations.

33. The sum of the digits of a two-digit number is 9. If the digits are reversed, the newly formed number is 45 less than the original number. Find the original number.

34. Sara invested $2500, part of it at 10% and the rest at 12% yearly interest. The yearly income on the 12% investment was $102 more than the income on the 10% investment. How much money did she invest at each rate?

35. A box contains $17.70 in nickels, dimes, and quarters. The number of dimes is 8 less than twice the number of nickels. The number of quarters is 2 more than the sum of the numbers of nickels and dimes. How many coins of each kind are there in the box?

36. The measure of the largest angle of a triangle is 10° more than four times the smallest angle. The sum of the smallest and largest angles is three times the measure of the other angle. Find the measure of each angle of the triangle.

For Problems 1–4, refer to the following systems of equations.

$$\text{I.} \begin{pmatrix} 3x - 2y = 4 \\ 9x - 6y = 12 \end{pmatrix} \quad \text{II.} \begin{pmatrix} 5x - y = 4 \\ 3x + 7y = 9 \end{pmatrix}$$

$$\text{III.} \begin{pmatrix} 2x - y = 4 \\ 2x - y = -6 \end{pmatrix}$$

1. For which system are the graphs parallel lines?

2. For which system are the equations dependent?

3. For which system is the solution set $\varnothing$?

4. Which system is consistent?

For Problems 5–8, evaluate each determinant.

5. $\begin{vmatrix} -2 & 4 \\ -5 & 6 \end{vmatrix}$

6. $\begin{vmatrix} \dfrac{1}{2} & \dfrac{1}{3} \\ \dfrac{3}{4} & -\dfrac{2}{3} \end{vmatrix}$

7. $\begin{vmatrix} -1 & 2 & 1 \\ 3 & 1 & -2 \\ 2 & -1 & 1 \end{vmatrix}$

8. $\begin{vmatrix} 2 & 4 & -5 \\ -4 & 3 & 0 \\ -2 & 6 & 1 \end{vmatrix}$

9. How many ordered pairs of real numbers are in the solution set for the system
$$\begin{pmatrix} y = 3x - 4 \\ 9x - 3y = 12 \end{pmatrix}?$$

10. Solve the system $\begin{pmatrix} 3x - 2y = -14 \\ 7x + 2y = -6 \end{pmatrix}$.

11. Solve the system $\begin{pmatrix} 4x - 5y = 17 \\ y = -3x + 8 \end{pmatrix}$.

12. Find the value of x in the solution for the system
$$\begin{pmatrix} \dfrac{3}{4}x - \dfrac{1}{2}y = -21 \\ \dfrac{2}{3}x + \dfrac{1}{6}y = -4 \end{pmatrix}$$

13. Find the value of y in the solution for the system
$$\begin{pmatrix} 4x - y = 7 \\ 3x + 2y = 2 \end{pmatrix}.$$

14. Is $(1, -1, 4)$ a solution of the following system?
$$\begin{pmatrix} 2x - y + z = 7 \\ 3x - 2y + 2z = 13 \\ x - 4y + 5z = 17 \end{pmatrix}$$

15. Suppose that the augmented matrix of a system of three linear equations in the three variables x, y, and z can be changed to the matrix
$$\begin{bmatrix} 1 & 1 & -4 & \vdots & 3 \\ 0 & 1 & 4 & \vdots & 5 \\ 0 & 0 & 3 & \vdots & 6 \end{bmatrix}$$
Find the value of x in the solution for the system.

16. Suppose that the augmented matrix of a system of three linear equations in the three variables x, y, and z can be changed to the matrix
$$\begin{bmatrix} 1 & 2 & -3 & \vdots & 4 \\ 0 & 1 & 2 & \vdots & 5 \\ 0 & 0 & 2 & \vdots & -8 \end{bmatrix}$$
Find the value of y in the solution for the system.

17. How many ordered triples are there in the solution set for the following system?
$$\begin{pmatrix} x + 3y - z = 5 \\ 2x - y - z = 7 \\ 5x + 8y - 4z = 22 \end{pmatrix}$$

18. How many ordered triples are there in the solution set for the following system?
$$\begin{pmatrix} 3x - y - 2z = 1 \\ 4x + 2y + z = 5 \\ 6x - 2y - 4z = 9 \end{pmatrix}$$

19. Solve the following system:
$$\begin{pmatrix} 5x - 3y - 2z = -1 \\ 4y + 7z = 3 \\ 4z = -12 \end{pmatrix}$$

20. Solve the following system:

$$\begin{pmatrix} x - 2y + z = 0 \\ y - 3z = -1 \\ 2y + 5z = -2 \end{pmatrix}$$

21. Find the value of x in the solution for the system

$$\begin{pmatrix} x - 4y + z = 12 \\ -2x + 3y - z = -11 \\ 5x - 3y + 2z = 17 \end{pmatrix}$$

22. Find the partial fraction decomposition of

$$\frac{11x - 22}{(2x - 1)(x - 6)}$$

23. A car wash charges $5.00 for an express wash and $15.00 for a full wash. On a recent day there were 75 car washes of these two types, which brought in $825.00. Find the number of express washes.

24. One solution is 30% alcohol and another solution is 70% alcohol. Some of each of the two solutions is mixed to produce 8 liters of a 40% solution. How many liters of the 70% solution should be used?

25. A box contains $7.25 in nickels, dimes, and quarters. There are 43 coins, and the number of quarters is 1 more than three times the number of nickels. Find the number of quarters in the box.

For Problems 1–10, evaluate each expression.

1. $(3^{-2} + 2^{-1})^{-2}$

2. $\left(\dfrac{4^{-2}}{2^{-3}}\right)^{-1}$

3. $\ln e^5$

4. $\log 0.001$

5. $\sqrt{2\dfrac{1}{4}}$

6. $(-27)^{4/3}$

7. $\log_2 64$

8. $\log_4 64$

9. $\left(\dfrac{3}{2}\right)^{-3}$

10. $\sqrt[3]{-0.008}$

For Problems 11–14, evaluate each algebraic expression for the given values of the variables.

11. $5(2x - 1) - 3(x + 4) - 2(3x + 11)$ for $x = -13$

12. $\dfrac{14x^3y^2}{28x^2y^3}$ for $x = 7$ and $y = -6$

13. $\dfrac{7}{n} - \dfrac{9}{2n} + \dfrac{5}{3n}$ for $n = 15$

14. $\dfrac{6x^2 + 7x - 20}{3x^2 - 25x + 28}$ for $x = 8$

For Problems 15–18, perform the indicated operations involving rational expressions and express final answers in simplest form.

15. $\dfrac{3a^2b}{7a} \div \dfrac{9ab^3}{14b^2}$

16. $\dfrac{3}{x^2 - 4} - \dfrac{2}{x^2 + 2x}$

17. $\dfrac{x^3 + 1}{x^2 + 5x + 6} \cdot \dfrac{x^2 - 9}{x^2 - 2x - 3}$

18. $\dfrac{2x - 1}{4} - \dfrac{x + 2}{6} + \dfrac{3x - 1}{8}$

For Problems 19–21, perform the indicated operations and simplify. Express final answers using positive exponents only.

19. $\dfrac{36x^{-2}y^{-4}}{24xy^{-2}}$

20. $\left(\dfrac{28a^2b^{-4}}{7a^{-1}b^3}\right)^{-1}$

21. $(-5x^{-1}y^2)(6x^{-2}y^{-5})$

For Problems 22–24, express each in simplest radical form. All variables represent positive real numbers.

22. $8\sqrt{72}$

23. $\sqrt[3]{54x^3y^5}$

24. $\dfrac{3\sqrt{6}}{4\sqrt{27}}$

25. Write the equation of the line that is perpendicular to the line $2x - y = 4$ and contains the point $(6, 2)$.

26. Determine the linear function whose graph is a line that contains the points $(-2, -3)$ and $(4, 8)$.

27. If $f(x) = -x^2 - 2x - 4$ and $g(x) = -4x + 5$, find $f(g(-2))$ and $g(f(3))$.

28. If $f(x) = x + 7$ and $g(x) = 2x^2 + x - 3$, find $f(g(x))$ and $g(f(x))$.

29. Find the indicated product $(4 - 3i)(-2 + 6i)$ and express the answer in the standard form of a complex number.

30. If $f(x) = -3x + 10$, find the inverse of f.

31. Find the length of the minor axis of the ellipse $3x^2 + y^2 = 27$.

32. Find the remainder when $3x^3 - 2x^2 + x - 4$ is divided by $x - 2$.

33. Find the remainder when $2x^3 + 11x^2 - 2x - 21$ is divided by $2x + 3$.

34. Find the quotient and the remainder when $3x^3 + 22x^2 + 4x - 1$ is divided by $3x + 1$.

35. Find the quotient and the remainder when $2x^4 - 3x^3 + 2x^2 - x + 6$ is divided by $x + 1$.

36. Is $x + 3$ a factor of $x^3 + 9x^2 + 23x + 15$?

37. Is $x - 1$ a factor of $x^6 + 1$?

38. Is $x + 1$ a factor of $x^7 + 1$?

39. Find the equations of the asymptotes of the hyperbola $y^2 - 9x^2 = 16$.

40. If y varies directly as x, and if $y = -2$ when $x = 3$, find y when $x = 9$.

41. If y is inversely proportional to x, and $y = 2$ when $x = -2$, find y when $x = 4$.

42. Evaluate $\log_2 101$ to the nearest hundredth.

43. Find the equations of the vertical and horizontal asymptotes for the graph of the function
$$f(x) = \frac{2x}{x - 1}.$$

44. What type of symmetry does the graph of
$$f(x) = \frac{x^2}{x^2 - 3} \text{ possess?}$$

45. Find the equation of the oblique asymptote for the graph of the function $f(x) = \dfrac{2x^2 + 3}{x + 1}$.

For Problems 46–58, solve each equation or system of equations.

46. $\dfrac{3}{2}(x - 1) - \dfrac{2}{3}(x + 2) = \dfrac{3}{4}(2x - 5)$

47. $\dfrac{2x - 1}{6} + \dfrac{x + 3}{8} = -2$

48. $|2x + 5| = 6$

49. $\begin{pmatrix} 3x - 2y = -11 \\ 4x + 5y = 16 \end{pmatrix}$

50. $14x^3 + 45x^2 - 14x = 0$

51. $2x^3 + x^2 - 3x + 6 = 0$

52. $12x^2 - 14x - 10 = 0$

53. $(x + 1)(3x - 2) = (4x + 5)(x + 1)$

54. $\begin{pmatrix} x - y + z = -4 \\ 2x + y + 3z = -5 \\ -3x - 2y - z = -4 \end{pmatrix}$

55. $9^{x-2} = 27^{x+3}$

56. $4^{x+1} = 16^{2x-1}$

57. $2^{x+3} = 7^{x-1}$ (to the nearest hundredth)

58. $\log(x + 3) + \log(x - 1) = 1$

For Problems 59–66, solve each inequality and express the solution set using interval notation.

59. $\dfrac{x - 1}{4} - \dfrac{x + 2}{12} \geq 1$

60. $|2x - 1| < 4$ **61.** $|3x + 2| > 6$

62. $-3 \leq 2(x - 1) - (x + 4)$

63. $(x - 2)(x + 4)(x - 7) \geq 0$

64. $(x - 1)(x + 5) < (x - 1)(2x + 3)$

65. $\dfrac{x + 1}{x - 4} - 2 > 0$ **66.** $x^2 + 5x \leq 24$

For Problems 67–74, graph each function.

67. $f(x) = 2|x + 3| + 1$ **68.** $f(x) = \log_3(x - 1)$

69. $f(x) = -x^2 + 4x - 5$

70. $f(x) = 2\sqrt{x + 1} + 2$

71. $f(x) = 2^{x-2}$

72. $f(x) = (x - 1)(x + 1)(x - 4)$

73. $f(x) = \dfrac{-1}{x^2 - 9}$ **74.** $f(x) = \dfrac{x^2 + x + 1}{x - 1}$

For Problems 75–83, use an equation, a system of equations, or an inequality to help solve each problem.

75. The ratio of female students to male students at a certain university is 5 to 6. If there are 13,200 students at the university, find the number of female students and the number of male students.

76. A retailer of sporting goods bought a driver for $170. He wants to sell the driver to make a 15% profit based on the selling price. What price should he mark on the driver?

77. In Problem 76, suppose the retailer wanted a profit of 15% based on the cost of the driver. What price should he mark on the driver?

78. The perimeter of a rectangle is 38 centimeters, and the area is 84 square centimeters. Find the length and width of the rectangle.

79. Maria can do a certain task in 30 minutes. Dana can accomplish that same task in 20 minutes. One day

Maria started the task, then after 10 minutes she was joined by Dana and they finished the task together. How long did it take them to finish the task once they started working together?

80. How many liters of a 50% acid solution must be added to 5 liters of a 10% acid solution to produce a 30% acid solution?

81. A car can be rented from agency A at $30 per day plus $0.15 a mile or from agency B at $20 per day plus $0.20 a mile. If the car is driven m miles in one day, for what values of m does it cost less to rent from agency A?

82. The cost of labor varies jointly as the number of workers and the number of days they work. If it costs $12,500 to have 25 people work for 5 days, how much will it cost to have 30 people work for 6 days?

83. How long will it take for $500 to grow to $1000 if it is invested at 6% interest compounded semiannually?

Algebra of Matrices

© John Sann /Getty Images /Taxi

Cryptography, considered both a branch of mathematics and of computer science, is the practice and study of hiding information. Matrices are often used to encrypt data for coded messages, computer passwords, and secure financial transactions.

Matrices can be used to code and decode messages. By setting up a one-to-one correspondence between the letters of the alphabet and the first 26 counting numbers, we can convert a message into a numerical matrix. Then by multiplying this matrix by some 2 × 2 matrix, we can give the message a numerical code of numbers. The receiver of the message can decode it by multiplying by the multiplicative inverse of the previously agreed upon 2 × 2 matrix. The details of this process are described in Problem 51 of Problem Set 7.3.

In Section 6.3 we used matrices strictly as a device to help solve systems of linear equations. Our primary objective was to develop techniques for solving systems of equations, not the study of matrices. However, matrices can be studied from an algebraic viewpoint, much as we study the set of real numbers. That is, we can define certain operations on matrices and verify properties of those operations. This algebraic approach to matrices is the focal point of this chapter. In order to get a simplified view of the algebra of matrices, we will begin by studying 2 × 2 matrices, and then later we will enlarge our discussion to include *m* × *n* matrices. As a bonus, another technique for solving systems of linear equations will emerge from our study.

7.1 Algebra of 2 × 2 Matrices

Objectives

- Add and subtract matrices.

- Multiply a matrix by a scalar.

- Multiply 2 × 2 matrices.

Throughout these next two sections, we will be working primarily with 2 × 2 matrices; therefore any reference to matrices means 2 × 2 matrices unless stated otherwise. The following 2 × 2 matrix notation will be used frequently:

$$A = \begin{bmatrix} a_{11} & a_{12} \\ a_{21} & a_{22} \end{bmatrix} \qquad B = \begin{bmatrix} b_{11} & b_{12} \\ b_{21} & b_{22} \end{bmatrix} \qquad C = \begin{bmatrix} c_{11} & c_{12} \\ c_{21} & c_{22} \end{bmatrix}$$

Two matrices are equal if and only if all elements in corresponding positions are equal. Thus $A = B$ if and only if $a_{11} = b_{11}$, $a_{12} = b_{12}$, $a_{21} = b_{21}$, and $a_{22} = b_{22}$.

■ Addition of Matrices

To add two matrices, we add the elements that appear in corresponding positions. Therefore the sum of matrix A and matrix B is defined as follows.

Definition 7.1

$$A + B = \begin{bmatrix} a_{11} & a_{12} \\ a_{21} & a_{22} \end{bmatrix} + \begin{bmatrix} b_{11} & b_{12} \\ b_{21} & b_{22} \end{bmatrix}$$

$$= \begin{bmatrix} a_{11} + b_{11} & a_{12} + b_{12} \\ a_{21} + b_{21} & a_{22} + b_{22} \end{bmatrix}$$

For example,

$$\begin{bmatrix} 2 & -1 \\ -3 & 4 \end{bmatrix} + \begin{bmatrix} -5 & 4 \\ -1 & 7 \end{bmatrix} = \begin{bmatrix} -3 & 3 \\ -4 & 11 \end{bmatrix}$$

It is not difficult to show that **the commutative and associative properties are valid for the addition of matrices.** Thus we can state that

$$A + B = B + A \qquad \text{and} \qquad (A + B) + C = A + (B + C)$$

Because

$$\begin{bmatrix} a_{11} & a_{12} \\ a_{21} & a_{22} \end{bmatrix} + \begin{bmatrix} 0 & 0 \\ 0 & 0 \end{bmatrix} = \begin{bmatrix} a_{11} & a_{12} \\ a_{21} & a_{22} \end{bmatrix}$$

we see that $\begin{bmatrix} 0 & 0 \\ 0 & 0 \end{bmatrix}$, which is called the **zero matrix** and represented by O, is the **additive identity element.** Thus we can state that

$$A + O = O + A = A$$

Because every real number has an additive inverse, it follows that any matrix A has an **additive inverse,** $-A$, that is formed by taking the additive inverse of each element of A. For example, if

$$A = \begin{bmatrix} 4 & -2 \\ -1 & 0 \end{bmatrix} \qquad \text{then} \qquad -A = \begin{bmatrix} -4 & 2 \\ 1 & 0 \end{bmatrix}$$

and

$$A + (-A) = \begin{bmatrix} 4 & -2 \\ -1 & 0 \end{bmatrix} + \begin{bmatrix} -4 & 2 \\ 1 & 0 \end{bmatrix} = \begin{bmatrix} 0 & 0 \\ 0 & 0 \end{bmatrix}$$

In general, we can state that **every matrix A has an additive inverse $-A$** such that

$$A + (-A) = (-A) + A = O$$

■ Subtraction of Matrices

Again, paralleling the algebra of real numbers, *subtraction* of matrices can be defined in terms of *adding the additive inverse*. Therefore we can define subtraction as follows.

Definition 7.2

$$A - B = A + (-B)$$

For example,

$$\begin{bmatrix} 2 & -7 \\ -6 & 5 \end{bmatrix} - \begin{bmatrix} 3 & 4 \\ -2 & -1 \end{bmatrix} = \begin{bmatrix} 2 & -7 \\ -6 & 5 \end{bmatrix} + \begin{bmatrix} -3 & -4 \\ 2 & 1 \end{bmatrix}$$

$$= \begin{bmatrix} -1 & -11 \\ -4 & 6 \end{bmatrix}$$

■ Scalar Multiplication

When we work with matrices, we commonly refer to a single real number as a **scalar** to distinguish it from a matrix. Then taking the **product** of a scalar and a matrix (often referred to as a **scalar multiplication**) can be done by multiplying each element of the matrix by the scalar. For example,

$$3\begin{bmatrix} -4 & -6 \\ 1 & -2 \end{bmatrix} = \begin{bmatrix} 3(-4) & 3(-6) \\ 3(1) & 3(-2) \end{bmatrix} = \begin{bmatrix} -12 & -18 \\ 3 & -6 \end{bmatrix}$$

In general, scalar multiplication can be defined as follows.

Definition 7.3

$$kA = k\begin{bmatrix} a_{11} & a_{12} \\ a_{21} & a_{22} \end{bmatrix} = \begin{bmatrix} ka_{11} & ka_{12} \\ ka_{21} & ka_{22} \end{bmatrix}$$

where k is any real number.

EXAMPLE 1

If $A = \begin{bmatrix} -4 & 3 \\ 2 & -5 \end{bmatrix}$ and $B = \begin{bmatrix} 2 & -3 \\ 7 & -6 \end{bmatrix}$, find $-2A$, $3A + 2B$, and $A - 4B$.

Solutions

$$-2A = -2\begin{bmatrix} -4 & 3 \\ 2 & -5 \end{bmatrix} = \begin{bmatrix} 8 & -6 \\ -4 & 10 \end{bmatrix}$$

$$3A + 2B = 3\begin{bmatrix} -4 & 3 \\ 2 & -5 \end{bmatrix} + 2\begin{bmatrix} 2 & -3 \\ 7 & -6 \end{bmatrix}$$

$$= \begin{bmatrix} -12 & 9 \\ 6 & -15 \end{bmatrix} + \begin{bmatrix} 4 & -6 \\ 14 & -12 \end{bmatrix}$$

$$= \begin{bmatrix} -8 & 3 \\ 20 & -27 \end{bmatrix}$$

$$A - 4B = \begin{bmatrix} -4 & 3 \\ 2 & -5 \end{bmatrix} - 4\begin{bmatrix} 2 & -3 \\ 7 & -6 \end{bmatrix}$$

$$= \begin{bmatrix} -4 & 3 \\ 2 & -5 \end{bmatrix} - \begin{bmatrix} 8 & -12 \\ 28 & -24 \end{bmatrix}$$

$$= \begin{bmatrix} -4 & 3 \\ 2 & -5 \end{bmatrix} + \begin{bmatrix} -8 & 12 \\ -28 & 24 \end{bmatrix}$$

$$= \begin{bmatrix} -12 & 15 \\ -26 & 19 \end{bmatrix}$$

▶ **Now work Problem 3.**

The following properties, which are easy to check, pertain to scalar multiplication and matrix addition (where k and l represent any real numbers):

$$k(A + B) = kA + kB$$
$$(k + l)A = kA + lA$$
$$(kl)A = k(lA)$$

■ Multiplication of Matrices

At this time, it probably would seem quite natural to define matrix multiplication by multiplying the corresponding elements of two matrices. However, it turns out that such a definition does not have many worthwhile applications. Therefore we use a special type of **matrix multiplication,** sometimes referred to as a *row-by-column* multiplication. We will state the definition, paraphrase what it says, and then give some examples.

Definition 7.4

$$AB = \begin{bmatrix} a_{11} & a_{12} \\ a_{21} & a_{22} \end{bmatrix}\begin{bmatrix} b_{11} & b_{12} \\ b_{21} & b_{22} \end{bmatrix}$$

$$= \begin{bmatrix} a_{11}b_{11} + a_{12}b_{21} & a_{11}b_{12} + a_{12}b_{22} \\ a_{21}b_{11} + a_{22}b_{21} & a_{21}b_{12} + a_{22}b_{22} \end{bmatrix}$$

Notice the row-by-column pattern of Definition 7.4. We multiply the rows of A by the columns of B in a pairwise entry fashion, adding the results. For example, the element in the first row and second column of the product is obtained by multiplying the elements of the first row of A by the elements of the second column of B and adding the results.

$$\begin{bmatrix} \boxed{a_{11} \quad a_{12}} \\ a_{21} \quad a_{22} \end{bmatrix}\begin{bmatrix} b_{11} & \boxed{b_{12}} \\ b_{21} & \boxed{b_{22}} \end{bmatrix} = \begin{bmatrix} & a_{11}b_{12} + a_{12}b_{22} \\ & \end{bmatrix}$$

Now let's look at some specific examples.

EXAMPLE 2

If $A = \begin{bmatrix} -2 & 1 \\ 4 & 5 \end{bmatrix}$ and $B = \begin{bmatrix} 3 & -2 \\ -1 & 7 \end{bmatrix}$, find AB and BA.

Solutions

$$AB = \begin{bmatrix} -2 & 1 \\ 4 & 5 \end{bmatrix}\begin{bmatrix} 3 & -2 \\ -1 & 7 \end{bmatrix}$$

$$= \begin{bmatrix} (-2)(3) + (1)(-1) & (-2)(-2) + (1)(7) \\ (4)(3) + (5)(-1) & (4)(-2) + (5)(7) \end{bmatrix}$$

$$= \begin{bmatrix} -7 & 11 \\ 7 & 27 \end{bmatrix}$$

$$BA = \begin{bmatrix} 3 & -2 \\ -1 & 7 \end{bmatrix}\begin{bmatrix} -2 & 1 \\ 4 & 5 \end{bmatrix}$$

$$= \begin{bmatrix} (3)(-2) + (-2)(4) & (3)(1) + (-2)(5) \\ (-1)(-2) + (7)(4) & (-1)(1) + (7)(5) \end{bmatrix} = \begin{bmatrix} -14 & -7 \\ 30 & 34 \end{bmatrix}$$

▶ **Now work Problem 13.**   ■

Example 2 makes it immediately apparent that **matrix multiplication is not a commutative operation.**

EXAMPLE 3

If $A = \begin{bmatrix} 2 & -6 \\ -3 & 9 \end{bmatrix}$ and $B = \begin{bmatrix} -3 & 6 \\ -1 & 2 \end{bmatrix}$, find AB.

Solution

Once you feel comfortable with Definition 7.4, you can do the addition mentally.

$$AB = \begin{bmatrix} 2 & -6 \\ -3 & 9 \end{bmatrix}\begin{bmatrix} -3 & 6 \\ -1 & 2 \end{bmatrix} = \begin{bmatrix} 0 & 0 \\ 0 & 0 \end{bmatrix}$$ ■

Example 3 illustrates that the product of two matrices can be the zero matrix even though neither of the two matrices is the zero matrix. This is different from the property of real numbers, which states $ab = 0$ *if and only if a* $= 0$ *or b* $= 0$.

As we illustrated and stated earlier, matrix multiplication is *not* a commutative operation. However, it is an **associative operation**, and it does abide by two **distributive properties.** These properties can be stated as follows:

$$(AB)C = A(BC)$$
$$A(B + C) = AB + AC$$
$$(B+C)A = BA+CA$$

We will ask you to verify these properties in the next set of problems.

CONCEPT QUIZ

For the following problems, given that A, B, and C are 2 × 2 matrices and k and l are real numbers, answer true or false.

1. The matrix $\begin{bmatrix} 0 & 0 \\ 0 & 0 \end{bmatrix}$ is the additive identity element.

2. If $A + B = \begin{bmatrix} 0 & 0 \\ 0 & 0 \end{bmatrix}$, then A and B are additive inverses.

3. $A + B = B + A$.

4. $AB = BA$.

5. If $AB = \begin{bmatrix} 0 & 0 \\ 0 & 0 \end{bmatrix}$, then either $A = \begin{bmatrix} 0 & 0 \\ 0 & 0 \end{bmatrix}$ or $B = \begin{bmatrix} 0 & 0 \\ 0 & 0 \end{bmatrix}$.

6. The product of A times B can never equal $\begin{bmatrix} 0 & 0 \\ 0 & 0 \end{bmatrix}$.

7. To perform the scalar multiplication kA, only the elements in the first row of A are multiplied by k.

8. If $A = \begin{bmatrix} 1 & 2 \\ 3 & 4 \end{bmatrix}$, then $A^2 = AA = \begin{bmatrix} 1 & 4 \\ 9 & 16 \end{bmatrix}$.

Problem Set 7.1

For Problems 1–12, compute the indicated matrix by using the following matrices:

$$A = \begin{bmatrix} 1 & -2 \\ 3 & 4 \end{bmatrix} \qquad B = \begin{bmatrix} 2 & -3 \\ 5 & -1 \end{bmatrix}$$

$$C = \begin{bmatrix} 0 & 6 \\ -4 & 2 \end{bmatrix} \qquad D = \begin{bmatrix} -2 & 3 \\ 5 & -4 \end{bmatrix}$$

$$E = \begin{bmatrix} 2 & 5 \\ 7 & 3 \end{bmatrix}$$

1. $A + B$

2. $B - C$

3. $3C + D$

4. $2D - E$

5. $4A - 3B$

6. $2B + 3D$

7. $(A - B) - C$

8. $B - (D - E)$

9. $2D - 4E$

10. $3A - 4E$

11. $B - (D + E)$

12. $A - (B + C)$

For Problems 13–26, compute AB and BA.

13. $A = \begin{bmatrix} 1 & -1 \\ 2 & -2 \end{bmatrix}$, $B = \begin{bmatrix} 3 & -4 \\ -1 & 2 \end{bmatrix}$

14. $A = \begin{bmatrix} -3 & 4 \\ 2 & 1 \end{bmatrix}$, $B = \begin{bmatrix} -2 & 5 \\ 6 & -1 \end{bmatrix}$

15. $A = \begin{bmatrix} 1 & -3 \\ -4 & 6 \end{bmatrix}$, $B = \begin{bmatrix} 7 & -3 \\ 4 & 5 \end{bmatrix}$

16. $A = \begin{bmatrix} 5 & 0 \\ -2 & 3 \end{bmatrix}$, $B = \begin{bmatrix} -3 & 6 \\ 4 & 1 \end{bmatrix}$

17. $A = \begin{bmatrix} 2 & -4 \\ 1 & -2 \end{bmatrix}$, $B = \begin{bmatrix} 1 & -2 \\ -3 & 6 \end{bmatrix}$

18. $A = \begin{bmatrix} 1 & 2 \\ 1 & 2 \end{bmatrix}$, $B = \begin{bmatrix} 2 & 2 \\ -1 & -1 \end{bmatrix}$

19. $A = \begin{bmatrix} -3 & -2 \\ -4 & -1 \end{bmatrix}$, $B = \begin{bmatrix} 2 & -1 \\ 4 & 5 \end{bmatrix}$

20. $A = \begin{bmatrix} -2 & 3 \\ -1 & 7 \end{bmatrix}$, $B = \begin{bmatrix} -1 & -3 \\ -5 & -7 \end{bmatrix}$

21. $A = \begin{bmatrix} 2 & -1 \\ -5 & 3 \end{bmatrix}$, $B = \begin{bmatrix} 3 & 1 \\ 5 & 2 \end{bmatrix}$

22. $A = \begin{bmatrix} -8 & -5 \\ 3 & 2 \end{bmatrix}$, $B = \begin{bmatrix} -2 & -5 \\ 3 & 8 \end{bmatrix}$

23. $A = \begin{bmatrix} \frac{1}{2} & -\frac{1}{3} \\ \frac{1}{3} & \frac{1}{4} \end{bmatrix}$, $B = \begin{bmatrix} 4 & -6 \\ 6 & -4 \end{bmatrix}$

24. $A = \begin{bmatrix} \frac{1}{3} & \frac{1}{2} \\ \frac{3}{2} & -\frac{2}{3} \end{bmatrix}$, $B = \begin{bmatrix} -6 & -18 \\ 12 & -12 \end{bmatrix}$

25. $A = \begin{bmatrix} 5 & 6 \\ 2 & 3 \end{bmatrix}$, $B = \begin{bmatrix} 1 & -2 \\ -\frac{2}{3} & \frac{5}{3} \end{bmatrix}$

26. $A = \begin{bmatrix} -3 & -5 \\ 2 & 4 \end{bmatrix}$, $B = \begin{bmatrix} -2 & -\frac{5}{2} \\ 1 & \frac{3}{2} \end{bmatrix}$

For Problems 27–30, use the following matrices:

$$A = \begin{bmatrix} -2 & 3 \\ 5 & 4 \end{bmatrix} \qquad B = \begin{bmatrix} 0 & 1 \\ 1 & 0 \end{bmatrix}$$

$$C = \begin{bmatrix} 1 & 0 \\ 1 & 0 \end{bmatrix} \qquad D = \begin{bmatrix} 1 & 1 \\ 1 & 1 \end{bmatrix}$$

$$I = \begin{bmatrix} 1 & 0 \\ 0 & 1 \end{bmatrix}$$

27. Compute AB and BA.

28. Compute AC and CA.

29. Compute AD and DA.

30. Compute AI and IA.

For Problems 31–34, use the following matrices:

$$A = \begin{bmatrix} 2 & 4 \\ 5 & -3 \end{bmatrix} \qquad B = \begin{bmatrix} -2 & 3 \\ -1 & 2 \end{bmatrix}$$

$$C = \begin{bmatrix} 2 & 1 \\ 3 & 7 \end{bmatrix}$$

31. Show that $(AB)C = A(BC)$.

32. Show that $A(B + C) = AB + AC$.

33. Show that $(A + B)C = AC + BC$.

34. Show that $(3 + 2)A = 3A + 2A$.

For Problems 35–43, use the following matrices:

$$A = \begin{bmatrix} a_{11} & a_{12} \\ a_{21} & a_{22} \end{bmatrix} \qquad B = \begin{bmatrix} b_{11} & b_{12} \\ b_{21} & b_{22} \end{bmatrix}$$

$$C = \begin{bmatrix} c_{11} & c_{12} \\ c_{21} & c_{22} \end{bmatrix} \qquad O = \begin{bmatrix} 0 & 0 \\ 0 & 0 \end{bmatrix}$$

35. Show that $A + B = B + A$.

36. Show that $(A + B) + C = A + (B + C)$.

37. Show that $A + (-A) = O$.

38. Show that $k(A + B) = kA + kB$ for any real number k.

39. Show that $(k + l)A = kA + lA$ for any real numbers k and l.

40. Show that $(kl)A = k(lA)$ for any real numbers k and l.

41. Show that $(AB)C = A(BC)$.

42. Show that $A(B + C) = AB + AC$.

43. Show that $(A + B)C = AC + BC$.

■■■ **THOUGHTS INTO WORDS**

44. How would you show that addition of 2 × 2 matrices is a commutative operation?

45. How would you show that subtraction of 2 × 2 matrices is not a commutative operation?

46. How would you explain matrix multiplication to someone who missed class the day it was discussed?

47. Your friend says that because multiplication of real numbers is a commutative operation, it seems reasonable that multiplication of matrices should also be a commutative operation. How would you react to that statement?

■■■ **FURTHER INVESTIGATIONS**

48. If $A = \begin{bmatrix} 2 & 0 \\ 0 & 3 \end{bmatrix}$, calculate A^2 and A^3, where A^2 means AA and A^3 means AAA.

49. If $A = \begin{bmatrix} 1 & -1 \\ 2 & 3 \end{bmatrix}$, calculate A^2 and A^3.

50. Does $(A + B)(A - B) = A^2 - B^2$ for all 2 × 2 matrices? Defend your answer.

GRAPHING CALCULATOR ACTIVITIES

51. Use your calculator to check the answers to all three parts of Example 1.

52. Use a calculator to check your answers for Problems 21–26.

53. Use the following matrices:

$$A = \begin{bmatrix} 7 & -4 \\ 6 & 9 \end{bmatrix} \qquad B = \begin{bmatrix} -3 & 8 \\ -5 & 7 \end{bmatrix} \qquad C = \begin{bmatrix} 8 & -2 \\ 4 & -7 \end{bmatrix}$$

a. Show that $(AB)C = A(BC)$.

b. Show that $A(B + C) = AB + AC$.

c. Show that $(B + C)A = BA + CA$.

Answers to the Concept Quiz

1. True **2.** True **3.** True **4.** False **5.** False **6.** False **7.** False **8.** False

7.2 Multiplicative Inverses

Objectives

- Find the multiplicative inverse of a matrix.
- Solve a 2×2 system of equations by using matrices.

We know that 1 is a multiplicative identity element for the set of real numbers. That is, $a(1) = 1(a) = a$ for any real number a. Is there a multiplicative identity element for 2×2 matrices? Yes. The matrix

$$I = \begin{bmatrix} 1 & 0 \\ 0 & 1 \end{bmatrix}$$

is the **multiplicative identity element** because

$$\begin{bmatrix} 1 & 0 \\ 0 & 1 \end{bmatrix}\begin{bmatrix} a_{11} & a_{12} \\ a_{21} & a_{22} \end{bmatrix} = \begin{bmatrix} a_{11} & a_{12} \\ a_{21} & a_{22} \end{bmatrix}$$

and

$$\begin{bmatrix} a_{11} & a_{12} \\ a_{21} & a_{22} \end{bmatrix}\begin{bmatrix} 1 & 0 \\ 0 & 1 \end{bmatrix} = \begin{bmatrix} a_{11} & a_{12} \\ a_{21} & a_{22} \end{bmatrix}$$

Therefore we can state that

$$AI = IA = A$$

for all 2×2 matrices.

Again, refer to the real numbers, where every nonzero real number a has a multiplicative inverse $1/a$ such that $a(1/a) = (1/a)a = 1$. Does every 2×2 matrix have a multiplicative inverse? To help answer this question, let's think about finding the multiplicative inverse (if one exists) for a specific matrix. This should give us some clues about a general approach.

E X A M P L E   1

Find the multiplicative inverse of $A = \begin{bmatrix} 3 & 5 \\ 2 & 4 \end{bmatrix}$.

Solution

We are looking for a matrix A^{-1} such that $AA^{-1} = A^{-1}A = I$. In other words, we want to solve the following matrix equation

$$\begin{bmatrix} 3 & 5 \\ 2 & 4 \end{bmatrix}\begin{bmatrix} x & y \\ z & w \end{bmatrix} = \begin{bmatrix} 1 & 0 \\ 0 & 1 \end{bmatrix}$$

We need to multiply the two matrices on the left side of this equation and then set the elements of the product matrix equal to the corresponding elements of the identity matrix. We obtain the following system of equations:

$$\left(\begin{array}{l} 3x + 5z = 1 \\ 3y + 5w = 0 \\ 2x + 4z = 0 \\ 2y + 4w = 1 \end{array} \right)$$

                (1)
                (2)
                (3)
                (4)

Solving equations (1) and (3) simultaneously produces values for x and z as follows:

$$x = \frac{\begin{vmatrix} 1 & 5 \\ 0 & 4 \end{vmatrix}}{\begin{vmatrix} 3 & 5 \\ 2 & 4 \end{vmatrix}} = \frac{1(4) - 5(0)}{3(4) - 5(2)} = \frac{4}{2} = 2$$

$$z = \frac{\begin{vmatrix} 3 & 1 \\ 2 & 0 \end{vmatrix}}{\begin{vmatrix} 3 & 5 \\ 2 & 4 \end{vmatrix}} = \frac{3(0) - 1(2)}{3(4) - 5(2)} = \frac{-2}{2} = -1$$

Likewise, solving equations (2) and (4) simultaneously produces values for y and w:

$$y = \frac{\begin{vmatrix} 0 & 5 \\ 1 & 4 \end{vmatrix}}{\begin{vmatrix} 3 & 5 \\ 2 & 4 \end{vmatrix}} = \frac{0(4) - 5(1)}{3(4) - 5(2)} = \frac{-5}{2} = -\frac{5}{2}$$

$$w = \frac{\begin{vmatrix} 3 & 0 \\ 2 & 1 \end{vmatrix}}{\begin{vmatrix} 3 & 5 \\ 2 & 4 \end{vmatrix}} = \frac{3(1) - 0(2)}{3(4) - 5(2)} = \frac{3}{2}$$

Therefore

$$A^{-1} = \begin{bmatrix} x & y \\ z & w \end{bmatrix} = \begin{bmatrix} 2 & -\dfrac{5}{2} \\ -1 & \dfrac{3}{2} \end{bmatrix}$$

To check this, we perform the following multiplication:

$$\begin{bmatrix} 3 & 5 \\ 2 & 4 \end{bmatrix} \begin{bmatrix} 2 & -\dfrac{5}{2} \\ -1 & \dfrac{3}{2} \end{bmatrix} = \begin{bmatrix} 2 & -\dfrac{5}{2} \\ -1 & \dfrac{3}{2} \end{bmatrix} \begin{bmatrix} 3 & 5 \\ 2 & 4 \end{bmatrix} = \begin{bmatrix} 1 & 0 \\ 0 & 1 \end{bmatrix} \quad \blacksquare$$

Now let's use the approach in Example 1 on the general matrix

$$A = \begin{bmatrix} a_{11} & a_{12} \\ a_{21} & a_{22} \end{bmatrix}$$

We want to find

$$A^{-1} = \begin{bmatrix} x & y \\ z & w \end{bmatrix}$$

such that $AA^{-1} = I$. Therefore we need to solve the matrix equation

$$\begin{bmatrix} a_{11} & a_{12} \\ a_{21} & a_{22} \end{bmatrix} \begin{bmatrix} x & y \\ z & w \end{bmatrix} = \begin{bmatrix} 1 & 0 \\ 0 & 1 \end{bmatrix}$$

for x, y, z, and w. Once again, we multiply the two matrices on the left side of the equation and set the elements of this product matrix equal to the corresponding elements of the identity matrix. We then obtain the following system of equations:

$$\left(\begin{aligned} a_{11}x + a_{12}z &= 1 \\ a_{11}y + a_{12}w &= 0 \\ a_{21}x + a_{22}z &= 0 \\ a_{21}y + a_{22}w &= 1 \end{aligned} \right)$$

Solving this system produces

$$x = \frac{a_{22}}{a_{11}a_{22} - a_{12}a_{21}} \qquad y = \frac{-a_{12}}{a_{11}a_{22} - a_{12}a_{21}}$$

$$z = \frac{-a_{21}}{a_{11}a_{22} - a_{12}a_{21}} \qquad w = \frac{a_{11}}{a_{11}a_{22} - a_{12}a_{21}}$$

Note that the number in each denominator, $a_{11}a_{22} - a_{12}a_{21}$, is the determinant of the matrix A. Thus, if $|A| \neq 0$, then

$$A^{-1} = \frac{1}{|A|} \begin{bmatrix} a_{22} & -a_{12} \\ -a_{21} & a_{11} \end{bmatrix}$$

Matrix multiplication will show that $AA^{-1} = A^{-1}A = I$. **If $|A| = 0$, then the matrix A has no multiplicative inverse.**

EXAMPLE 2   Find A^{-1} (if it exists) if $A = \begin{bmatrix} 3 & 5 \\ -2 & -4 \end{bmatrix}$.

Solution

First, let's find $|A|$:

$$|A| = (3)(-4) - (5)(-2) = -2$$

Therefore

$$A^{-1} = \frac{1}{-2} \begin{bmatrix} -4 & -5 \\ 2 & 3 \end{bmatrix} = -\frac{1}{2} \begin{bmatrix} -4 & -5 \\ 2 & 3 \end{bmatrix} = \begin{bmatrix} 2 & \frac{5}{2} \\ -1 & -\frac{3}{2} \end{bmatrix}$$

It is easy to check that $AA^{-1} = A^{-1}A = I$.

▶ **Now work Problem 5.**   ■

EXAMPLE 3   Find A^{-1} (if it exists) if $A = \begin{bmatrix} 8 & -2 \\ -12 & 3 \end{bmatrix}$.

Solution

$$|A| = (8)(3) - (-2)(-12) = 0$$

Therefore A has no multiplicative inverse.

▶ **Now work Problem 7.**   ■

■ More About Multiplication of Matrices

Thus far we have found the products of only 2×2 matrices. The *row-by-column* multiplication pattern can be applied to many different kinds of matrices, which we shall see in the next section. For now let's find the product of a 2×2 matrix and a 2×1 matrix, with the 2×2 matrix on the left, as follows:

$$\begin{bmatrix} a_{11} & a_{12} \\ a_{21} & a_{22} \end{bmatrix} \begin{bmatrix} b_{11} \\ b_{21} \end{bmatrix} = \begin{bmatrix} a_{11}b_{11} + a_{12}b_{21} \\ a_{21}b_{11} + a_{22}b_{21} \end{bmatrix}$$

Note that the product matrix is a 2×1 matrix. The following example illustrates this pattern:

$$\begin{bmatrix} -2 & 3 \\ 1 & -4 \end{bmatrix} \begin{bmatrix} 5 \\ 7 \end{bmatrix} = \begin{bmatrix} (-2)(5) + (3)(7) \\ (1)(5) + (-4)(7) \end{bmatrix} = \begin{bmatrix} 11 \\ -23 \end{bmatrix}$$

■ Back to Solving Systems of Equations

The linear system of equations

$$\begin{pmatrix} a_{11}x + a_{12}y = d_1 \\ a_{21}x + a_{22}y = d_2 \end{pmatrix}$$

can be represented by the matrix equation

$$\begin{bmatrix} a_{11} & a_{12} \\ a_{21} & a_{22} \end{bmatrix} \begin{bmatrix} x \\ y \end{bmatrix} = \begin{bmatrix} d_1 \\ d_2 \end{bmatrix}$$

If we let

$$A = \begin{bmatrix} a_{11} & a_{12} \\ a_{21} & a_{22} \end{bmatrix} \qquad X = \begin{bmatrix} x \\ y \end{bmatrix} \qquad \text{and} \qquad B = \begin{bmatrix} d_1 \\ d_2 \end{bmatrix}$$

then the previous matrix equation can be written $AX = B$.

If A^{-1} exists, then we can multiply both sides of $AX = B$ by A^{-1} (on the left) and simplify as follows:

$$AX = B$$
$$A^{-1}(AX) = A^{-1}(B)$$
$$(A^{-1}A)X = A^{-1}B$$
$$IX = A^{-1}B$$
$$X = A^{-1}B$$

Therefore the product $A^{-1}B$ is the solution of the system.

E X A M P L E   4

Solve the system $\begin{pmatrix} 5x + 4y = 10 \\ 6x + 5y = 13 \end{pmatrix}$.

Solution

If we let

$$A = \begin{bmatrix} 5 & 4 \\ 6 & 5 \end{bmatrix} \qquad X = \begin{bmatrix} x \\ y \end{bmatrix} \qquad \text{and} \qquad B = \begin{bmatrix} 10 \\ 13 \end{bmatrix}$$

then the given system can be represented by the matrix equation $AX = B$. From our previous discussion we know that the solution of this equation is $X = A^{-1}B$, so we need to find A^{-1} and the product $A^{-1}B$.

$$A^{-1} = \frac{1}{|A|}\begin{bmatrix} 5 & -4 \\ -6 & 5 \end{bmatrix} = \frac{1}{1}\begin{bmatrix} 5 & -4 \\ -6 & 5 \end{bmatrix} = \begin{bmatrix} 5 & -4 \\ -6 & 5 \end{bmatrix}$$

Therefore

$$A^{-1}B = \begin{bmatrix} 5 & -4 \\ -6 & 5 \end{bmatrix}\begin{bmatrix} 10 \\ 13 \end{bmatrix} = \begin{bmatrix} -2 \\ 5 \end{bmatrix}$$

The solution set of the given system is $\{(-2, 5)\}$.

▶ **Now work Problem 27.**   ■

EXAMPLE 5   Solve the system $\begin{pmatrix} 3x - 2y = 9 \\ 4x + 7y = -17 \end{pmatrix}$.

Solution

If we let

$$A = \begin{bmatrix} 3 & -2 \\ 4 & 7 \end{bmatrix} \qquad X = \begin{bmatrix} x \\ y \end{bmatrix} \qquad \text{and} \qquad B = \begin{bmatrix} 9 \\ -17 \end{bmatrix}$$

then the system is represented by $AX = B$, where $X = A^{-1}B$ and

$$A^{-1} = \frac{1}{|A|}\begin{bmatrix} 7 & 2 \\ -4 & 3 \end{bmatrix} = \frac{1}{29}\begin{bmatrix} 7 & 2 \\ -4 & 3 \end{bmatrix} = \begin{bmatrix} \frac{7}{29} & \frac{2}{29} \\ -\frac{4}{29} & \frac{3}{29} \end{bmatrix}$$

Therefore

$$A^{-1}B = \begin{bmatrix} \frac{7}{29} & \frac{2}{29} \\ -\frac{4}{29} & \frac{3}{29} \end{bmatrix}\begin{bmatrix} 9 \\ -17 \end{bmatrix} = \begin{bmatrix} 1 \\ -3 \end{bmatrix}$$

The solution set of the given system is $\{(1, -3)\}$.

▶ **Now work Problem 31.**   ■

This technique of using matrix inverses to solve systems of linear equations is especially useful when there are many systems to be solved that have the same coefficients but different constant terms.

CONCEPT QUIZ For the following problems, answer true or false.

1. Every 2×2 matrix has a multiplicative inverse.

2. If $A = \begin{bmatrix} 4 & 7 \\ 2 & 5 \end{bmatrix}$, then $A^{-1} = \begin{bmatrix} \dfrac{1}{4} & \dfrac{1}{7} \\ \dfrac{1}{2} & \dfrac{1}{5} \end{bmatrix}$.

3. If $|A| = 0$, then A does not have a multiplicative inverse.

4. If $|A| = 1$, then A does have a multiplicative inverse.

5. The multiplicative identity element for 2×2 matrices is $\begin{bmatrix} 1 & 1 \\ 1 & 1 \end{bmatrix}$.

6. If A has an inverse A^{-1}, then $AA^{-1} = \begin{bmatrix} 0 & 1 \\ 1 & 0 \end{bmatrix}$.

7. The commutative property holds for the multiplication of a matrix and its inverse.

8. The commutative property holds for the multiplication of a matrix and the multiplicative identity element.

Problem Set 7.2

For Problems 1–18, find the multiplicative inverse (if one exists) of each matrix.

1. $\begin{bmatrix} 5 & 7 \\ 2 & 3 \end{bmatrix}$

2. $\begin{bmatrix} 3 & 4 \\ 2 & 3 \end{bmatrix}$

3. $\begin{bmatrix} 3 & 8 \\ 2 & 5 \end{bmatrix}$

4. $\begin{bmatrix} 2 & 9 \\ 3 & 13 \end{bmatrix}$

5. $\begin{bmatrix} -1 & 2 \\ 3 & 4 \end{bmatrix}$

6. $\begin{bmatrix} 1 & -2 \\ 4 & -3 \end{bmatrix}$

7. $\begin{bmatrix} -2 & -3 \\ 4 & 6 \end{bmatrix}$

8. $\begin{bmatrix} 5 & -1 \\ 3 & 4 \end{bmatrix}$

9. $\begin{bmatrix} -3 & 2 \\ -4 & 5 \end{bmatrix}$

10. $\begin{bmatrix} 3 & -4 \\ 6 & -8 \end{bmatrix}$

11. $\begin{bmatrix} 0 & 1 \\ 5 & 3 \end{bmatrix}$

12. $\begin{bmatrix} -2 & 0 \\ -3 & 5 \end{bmatrix}$

13. $\begin{bmatrix} -2 & -3 \\ -1 & -4 \end{bmatrix}$

14. $\begin{bmatrix} -2 & -5 \\ -3 & -6 \end{bmatrix}$

15. $\begin{bmatrix} -2 & 5 \\ -3 & 6 \end{bmatrix}$

16. $\begin{bmatrix} -3 & 4 \\ 1 & -2 \end{bmatrix}$

17. $\begin{bmatrix} 1 & 1 \\ 1 & -1 \end{bmatrix}$

18. $\begin{bmatrix} 1 & -1 \\ 1 & 1 \end{bmatrix}$

For Problems 19–26, compute AB.

19. $A = \begin{bmatrix} 4 & 3 \\ 2 & 5 \end{bmatrix}$, $B = \begin{bmatrix} 3 \\ 6 \end{bmatrix}$

20. $A = \begin{bmatrix} 5 & -2 \\ 3 & 1 \end{bmatrix}$, $B = \begin{bmatrix} 5 \\ 8 \end{bmatrix}$

21. $A = \begin{bmatrix} -3 & -4 \\ 2 & 1 \end{bmatrix}$, $B = \begin{bmatrix} 4 \\ -3 \end{bmatrix}$

22. $A = \begin{bmatrix} 5 & 2 \\ -1 & -3 \end{bmatrix}$, $B = \begin{bmatrix} 3 \\ -5 \end{bmatrix}$

23. $A = \begin{bmatrix} -4 & 2 \\ 7 & -5 \end{bmatrix}$, $B = \begin{bmatrix} -1 \\ -4 \end{bmatrix}$

24. $A = \begin{bmatrix} 0 & -3 \\ 2 & 9 \end{bmatrix}$, $B = \begin{bmatrix} -3 \\ -6 \end{bmatrix}$

25. $A = \begin{bmatrix} -2 & -3 \\ -5 & -6 \end{bmatrix}$, $B = \begin{bmatrix} 5 \\ -2 \end{bmatrix}$

26. $A = \begin{bmatrix} -3 & -5 \\ 4 & -7 \end{bmatrix}$, $B = \begin{bmatrix} -3 \\ -10 \end{bmatrix}$

For Problems 27–40, use the method of matrix inverses to solve each system.

27. $\left(\begin{array}{l} 2x + 3y = 13 \\ x + 2y = 8 \end{array} \right)$

28. $\left(\begin{array}{l} 3x + 2y = 10 \\ 7x + 5y = 23 \end{array} \right)$

29. $\left(\begin{array}{l} 4x - 3y = -23 \\ -3x + 2y = 16 \end{array} \right)$

30. $\left(\begin{array}{l} 6x - y = -14 \\ 3x + 2y = -17 \end{array} \right)$

31. $\left(\begin{array}{l} x - 7y = 7 \\ 6x + 5y = -5 \end{array} \right)$

32. $\left(\begin{array}{l} x + 9y = -5 \\ 4x - 7y = -20 \end{array} \right)$

33. $\left(\begin{array}{l} 3x - 5y = 2 \\ 4x - 3y = -1 \end{array} \right)$

34. $\left(\begin{array}{l} 5x - 2y = 6 \\ 7x - 3y = 8 \end{array} \right)$

35. $\left(\begin{array}{l} y = 19 - 3x \\ 9x - 5y = 1 \end{array} \right)$

36. $\left(\begin{array}{l} 4x + 3y = 31 \\ x = 5y + 2 \end{array} \right)$

37. $\left(\begin{array}{l} 3x + 2y = 0 \\ 30x - 18y = -19 \end{array} \right)$

38. $\left(\begin{array}{l} 12x + 30y = 23 \\ 12x - 24y = -13 \end{array} \right)$

39. $\left(\begin{array}{l} \dfrac{1}{3}x + \dfrac{3}{4}y = 12 \\ \dfrac{2}{3}x + \dfrac{1}{5}y = -2 \end{array} \right)$

40. $\left(\begin{array}{l} \dfrac{3}{2}x + \dfrac{1}{6}y = 11 \\ \dfrac{2}{3}x - \dfrac{1}{4}y = 1 \end{array} \right)$

■ ■ ■ **THOUGHTS INTO WORDS**

41. Describe how to solve the system $\left(\begin{array}{l} x - 2y = -10 \\ 3x + 5y = 14 \end{array} \right)$ using each of the following techniques:

a. Substitution method

b. Elimination-by-addition method

c. Reduced echelon form of the augmented matrix

d. Determinants

e. The method of matrix inverses

GRAPHING CALCULATOR ACTIVITIES

42. Use your calculator to find the multiplicative inverse (if it exists) of each of the following matrices. Be sure to check your answers by showing that $A^{-1}A = I$.

a. $\begin{bmatrix} 7 & 6 \\ 8 & 7 \end{bmatrix}$

b. $\begin{bmatrix} -12 & 5 \\ -19 & 8 \end{bmatrix}$

c. $\begin{bmatrix} -7 & 9 \\ 6 & -8 \end{bmatrix}$

d. $\begin{bmatrix} -6 & -11 \\ -4 & -8 \end{bmatrix}$

e. $\begin{bmatrix} 13 & 12 \\ 4 & 4 \end{bmatrix}$

f. $\begin{bmatrix} 15 & -8 \\ -9 & 5 \end{bmatrix}$

g. $\begin{bmatrix} 9 & 36 \\ 3 & 12 \end{bmatrix}$

h. $\begin{bmatrix} 1.2 & 1.5 \\ 7.6 & 4.5 \end{bmatrix}$

43. Use your calculator to find the multiplicative inverse of $\begin{bmatrix} \dfrac{1}{2} & \dfrac{2}{5} \\ \dfrac{3}{4} & \dfrac{1}{4} \end{bmatrix}$. What difficulty did you encounter?

44. Use your calculator and the method of matrix inverses to solve each of the following systems. Be sure to check your solutions.

a. $\begin{pmatrix} 5x + 7y = 82 \\ 7x + 10y = 116 \end{pmatrix}$

b. $\begin{pmatrix} 9x - 8y = -150 \\ -10x + 9y = 168 \end{pmatrix}$

c. $\begin{pmatrix} 15x - 8y = -15 \\ -9x + 5y = 12 \end{pmatrix}$

d. $\begin{pmatrix} 1.2x + 1.5y = 5.85 \\ 7.6x + 4.5y = 19.55 \end{pmatrix}$

e. $\begin{pmatrix} 12x - 7y = -34.5 \\ 8x + 9y = 79.5 \end{pmatrix}$

f. $\begin{pmatrix} \dfrac{3x}{2} + \dfrac{y}{6} = 11 \\ \dfrac{2x}{3} - \dfrac{y}{4} = 1 \end{pmatrix}$

g. $\begin{pmatrix} 114x + 129y = 2832 \\ 127x + 214y = 4139 \end{pmatrix}$

h. $\begin{pmatrix} \dfrac{x}{2} + \dfrac{2y}{5} = 14 \\ \dfrac{3x}{4} + \dfrac{y}{4} = 14 \end{pmatrix}$

Answers to the Concept Quiz

1. False   **2.** False   **3.** True   **4.** True   **5.** False   **6.** False   **7.** True   **8.** True

7.3   *m* × *n* Matrices

Objectives

■ Add and subtract general $m \times n$ matrices.

■ Determine the additive inverse of an $m \times n$ matrix.

■ Multiply an $m \times n$ matrix by a scalar.

■ Find the inverse of a square matrix.

■ Solve a 3×3 system of equations using matrices.

Now let's see how much of the algebra of 2×2 matrices extends to $m \times n$ matrices — that is, to matrices of any dimension. In Section 6.4 we represented a general $m \times n$ matrix by

$$A = \begin{bmatrix} a_{11} & a_{12} & a_{13} & \cdots & a_{1n} \\ a_{21} & a_{22} & a_{23} & \cdots & a_{2n} \\ \vdots & \vdots & \vdots & & \vdots \\ a_{m1} & a_{m2} & a_{m3} & \cdots & a_{mn} \end{bmatrix}$$

We denote the element at the intersection of row i and column j by a_{ij}. It is also customary to denote a matrix A with the abbreviated notation (a_{ij}).

Addition of matrices can be extended to matrices of any dimensions by the following definition.

Definition 7.5

> Let $A = (a_{ij})$ and $B = (b_{ij})$ be two matrices *of the same dimension*. Then
> $$A + B = (a_{ij}) + (b_{ij}) = (a_{ij} + b_{ij})$$

Definition 7.5 states that to add two matrices, we add the elements that appear in corresponding positions in the matrices. For this to work, the matrices must be of the same dimension. An example of the sum of two 3×2 matrices is

$$\begin{bmatrix} 3 & 2 \\ 4 & -1 \\ -3 & 8 \end{bmatrix} + \begin{bmatrix} -2 & 1 \\ -3 & -7 \\ 5 & 9 \end{bmatrix} = \begin{bmatrix} 1 & 3 \\ 1 & -8 \\ 2 & 17 \end{bmatrix}$$

The **commutative** and **associative properties** hold for any matrices that can be added. The $m \times n$ **zero matrix,** denoted by O**,** is the matrix that contains all zeros. It is the **identity element for addition.** For example,

$$\begin{bmatrix} 2 & 3 & -1 & -5 \\ -7 & 6 & 2 & 8 \end{bmatrix} + \begin{bmatrix} 0 & 0 & 0 & 0 \\ 0 & 0 & 0 & 0 \end{bmatrix} = \begin{bmatrix} 2 & 3 & -1 & -5 \\ -7 & 6 & 2 & 8 \end{bmatrix}$$

Every matrix A has an **additive inverse,** $-A$, that can be found by changing the sign of each element of A. For example, if

$$A = \begin{bmatrix} 2 & -3 & 0 & 4 & -7 \end{bmatrix}$$

then

$$-A = \begin{bmatrix} -2 & 3 & 0 & -4 & 7 \end{bmatrix}$$

Furthermore, $A + (-A) = O$ for all matrices.

The definition we gave earlier for subtraction, $A - B = A + (-B)$, can be extended to any two matrices of the same dimension. For example,

$$\begin{bmatrix} -4 & 3 & -5 \end{bmatrix} - \begin{bmatrix} 7 & -4 & -1 \end{bmatrix} = \begin{bmatrix} -4 & 3 & -5 \end{bmatrix} + \begin{bmatrix} -7 & 4 & 1 \end{bmatrix}$$
$$= \begin{bmatrix} -11 & 7 & -4 \end{bmatrix}$$

The **scalar product** of any real number k and any $m \times n$ matrix $A = (a_{ij})$ is defined by

$$kA = (ka_{ij})$$

In other words, to find kA, we simply multiply each element of A by k. For example,

$$(-4)\begin{bmatrix} 1 & -1 \\ -2 & 3 \\ 4 & 5 \\ 0 & -8 \end{bmatrix} = \begin{bmatrix} -4 & 4 \\ 8 & -12 \\ -16 & -20 \\ 0 & 32 \end{bmatrix}$$

The properties $k(A + B) = kA + kB$, $(k + l)A = kA + lA$, and $(kl)A = k(lA)$ hold for all matrices. The matrices A and B must be of the same dimension to be added.

The *row-by-column* definition for multiplying two matrices can be extended, but we must take care. In order for us to define the product AB of two matrices A and B, **the number of columns of A must equal the number of rows of B.** Therefore suppose $A = (a_{ij})$ is $m \times n$ and $B = (b_{ij})$ is $n \times p$. Then

$$AB = \begin{bmatrix} a_{11} & a_{12} & \cdots & a_{1n} \\ \vdots & \vdots & \cdots & \vdots \\ \boxed{a_{i1}} & a_{i2} & & a_{in} \\ \vdots & \vdots & & \vdots \\ a_{m1} & a_{m2} & \cdots & a_{mn} \end{bmatrix}\begin{bmatrix} b_{11} & \cdots & \boxed{b_{1j}} & \cdots & b_{1p} \\ b_{21} & \cdots & b_{2j} & \cdots & b_{2p} \\ \vdots & & \vdots & & \vdots \\ b_{n1} & \cdots & \boxed{b_{nj}} & \cdots & b_{np} \end{bmatrix} = C$$

The product matrix C is of dimension $m \times p$, and the general element, c_{ij}, is determined as follows:

$$c_{ij} = a_{i1}b_{1j} + a_{i2}b_{2j} + \cdots + a_{in}b_{nj}$$

A specific element of the product matrix, such as c_{23}, is the result of multiplying the elements in row 2 of matrix A by the elements in column 3 of matrix B and adding the results. Therefore

$$c_{23} = a_{21}b_{13} + a_{22}b_{23} + \cdots + a_{2n}b_{n3}$$

The following example illustrates the product of a 2×3 matrix and a 3×2 matrix:

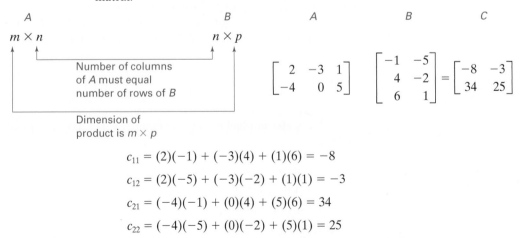

$$c_{11} = (2)(-1) + (-3)(4) + (1)(6) = -8$$

$$c_{12} = (2)(-5) + (-3)(-2) + (1)(1) = -3$$

$$c_{21} = (-4)(-1) + (0)(4) + (5)(6) = 34$$

$$c_{22} = (-4)(-5) + (0)(-2) + (5)(1) = 25$$

Recall that **matrix multiplication is not commutative.** In fact, it may be that AB is defined and BA is not defined. For example, if A is a 2×3 matrix and B is a 3×4 matrix, then the product AB is a 2×4 matrix, but the product BA is not defined because the number of columns of B does not equal the number of rows of A.

The **associative property for multiplication** and the two **distributive properties** hold if the matrices have the proper number of rows and columns for the operations to be defined. In that case, we have $(AB)C = A(BC), A(B + C) = AB + AC$, and $(A + B)C = AC + BC$.

■ Square Matrices

Now let's extend some of the algebra of 2×2 matrices to all square matrices (where the number of rows equals the number of columns). For example, the general **multiplicative identity element** for square matrices contains 1s in the main diagonal from the upper left-hand corner to the lower right-hand corner and zeros elsewhere. Therefore, for 3×3 and 4×4 matrices, the multiplicative identity elements are as follows:

$$I_3 = \begin{bmatrix} 1 & 0 & 0 \\ 0 & 1 & 0 \\ 0 & 0 & 1 \end{bmatrix} \qquad I_4 = \begin{bmatrix} 1 & 0 & 0 & 0 \\ 0 & 1 & 0 & 0 \\ 0 & 0 & 1 & 0 \\ 0 & 0 & 0 & 1 \end{bmatrix}$$

We saw in Section 7.2 that some, but not all, 2×2 matrices have multiplicative inverses. In general, some, but not all, square matrices of a particular dimension have multiplicative inverses. If an $n \times n$ square matrix A does have a multiplicative inverse A^{-1}, then

$$AA^{-1} = A^{-1}A = I_n$$

The technique used in Section 7.2 for finding multiplicative inverses of 2×2 matrices does generalize, but it becomes quite complicated. Therefore we shall now describe another technique that works for all square matrices. Given an $n \times n$ matrix A, we begin by forming the $n \times 2n$ matrix

$$\begin{bmatrix} a_{11} & a_{12} & \cdots & a_{1n} & \vdots & 1 & 0 & 0 & \cdots & 0 \\ a_{21} & a_{22} & \cdots & a_{2n} & \vdots & 0 & 1 & 0 & \cdots & 0 \\ \vdots & \vdots & & \vdots & \vdots & \vdots & \vdots & \vdots & & \vdots \\ a_{n1} & a_{n2} & \cdots & a_{nn} & \vdots & 0 & 0 & 0 & \cdots & 1 \end{bmatrix}$$

where the identity matrix I_n appears to the right of A. Now we apply a succession of elementary row transformations to this double matrix until we obtain a matrix of the form

$$\begin{bmatrix} 1 & 0 & 0 & \cdots & 0 & \vdots & b_{11} & b_{12} & \cdots & b_{1n} \\ 0 & 1 & 0 & \cdots & 0 & \vdots & b_{21} & b_{22} & \cdots & b_{2n} \\ \vdots & \vdots & \vdots & & \vdots & \vdots & \vdots & \vdots & & \vdots \\ 0 & 0 & 0 & \cdots & 1 & \vdots & b_{n1} & b_{n2} & \cdots & b_{nn} \end{bmatrix}$$

The B matrix in this matrix is the desired inverse A^{-1}. If A does not have an inverse, then it is impossible to change the original matrix to this final form.

E X A M P L E   1

Find A^{-1} if $A = \begin{bmatrix} 2 & 4 \\ 3 & 5 \end{bmatrix}$.

Solution

First, we form the matrix

$$\left[\begin{array}{cc:cc} 2 & 4 & 1 & 0 \\ 3 & 5 & 0 & 1 \end{array}\right]$$

Now we multiply row 1 by $\dfrac{1}{2}$.

$$\left[\begin{array}{cc:cc} 1 & 2 & \dfrac{1}{2} & 0 \\ 3 & 5 & 0 & 1 \end{array}\right]$$

Next, we add -3 times row 1 to row 2 to form a new row 2.

$$\left[\begin{array}{cc:cc} 1 & 2 & \dfrac{1}{2} & 0 \\ 0 & -1 & -\dfrac{3}{2} & 1 \end{array}\right]$$

Then we multiply row 2 by -1.

$$\left[\begin{array}{cc:cc} 1 & 2 & \dfrac{1}{2} & 0 \\ 0 & 1 & \dfrac{3}{2} & -1 \end{array}\right]$$

Finally, we add -2 times row 2 to row 1 to form a new row 1.

$$\left[\begin{array}{cc:cc} 1 & 0 & \boxed{\begin{array}{cc} -\dfrac{5}{2} & 2 \\ \dfrac{3}{2} & -1 \end{array}} \end{array}\right]$$

The matrix inside the box is A^{-1}; that is,

$$A^{-1} = \begin{bmatrix} -\dfrac{5}{2} & 2 \\ \dfrac{3}{2} & -1 \end{bmatrix}$$

This can be checked, as always, by showing that $AA^{-1} = A^{-1}A = I_2$.

▶ **Now work Problem 21.**

■

EXAMPLE 2

Find A^{-1} if $A = \begin{bmatrix} 1 & 1 & 2 \\ 2 & 3 & -1 \\ -3 & 1 & -2 \end{bmatrix}$.

Solution

We first form the matrix

$$\begin{bmatrix} 1 & 1 & 2 & \vdots & 1 & 0 & 0 \\ 2 & 3 & -1 & \vdots & 0 & 1 & 0 \\ -3 & 1 & -2 & \vdots & 0 & 0 & 1 \end{bmatrix}$$

We add -2 times row 1 to row 2 and add 3 times row 1 to row 3.

$$\begin{bmatrix} 1 & 1 & 2 & \vdots & 1 & 0 & 0 \\ 0 & 1 & -5 & \vdots & -2 & 1 & 0 \\ 0 & 4 & 4 & \vdots & 3 & 0 & 1 \end{bmatrix}$$

We add -1 times row 2 to row 1 and add -4 times row 2 to row 3.

$$\begin{bmatrix} 1 & 0 & 7 & \vdots & 3 & -1 & 0 \\ 0 & 1 & -5 & \vdots & -2 & 1 & 0 \\ 0 & 0 & 24 & \vdots & 11 & -4 & 1 \end{bmatrix}$$

We multiply row 3 by $\frac{1}{24}$.

$$\begin{bmatrix} 1 & 0 & 7 & \vdots & 3 & -1 & 0 \\ 0 & 1 & -5 & \vdots & -2 & 1 & 0 \\ 0 & 0 & 1 & \vdots & \frac{11}{24} & -\frac{1}{6} & \frac{1}{24} \end{bmatrix}$$

We add -7 times row 3 to row 1 and add 5 times row 3 to row 2.

$$\begin{bmatrix} 1 & 0 & 0 & \vdots & -\frac{5}{24} & \frac{1}{6} & -\frac{7}{24} \\ 0 & 1 & 0 & \vdots & \frac{7}{24} & \frac{1}{6} & \frac{5}{24} \\ 0 & 0 & 1 & \vdots & \frac{11}{24} & -\frac{1}{6} & \frac{1}{24} \end{bmatrix}$$

Therefore

$$A^{-1} = \begin{bmatrix} -\frac{5}{24} & \frac{1}{6} & -\frac{7}{24} \\ \frac{7}{24} & \frac{1}{6} & \frac{5}{24} \\ \frac{11}{24} & -\frac{1}{6} & \frac{1}{24} \end{bmatrix}$$

Be sure to check this!

▶ **Now work Problem 27.**

∎

■ Systems of Equations

In Section 7.2 we used the concept of the multiplicative inverse to solve systems of two linear equations in two variables. This same technique can be applied to general systems of n linear equations in n variables. Let's consider one such example involving three equations in three variables.

EXAMPLE 3

Solve the system

$$\left(\begin{array}{l} x + y + 2z = -8 \\ 2x + 3y - z = 3 \\ -3x + y - 2z = 4 \end{array} \right)$$

Solution

If we let

$$A = \begin{bmatrix} 1 & 1 & 2 \\ 2 & 3 & -1 \\ -3 & 1 & -2 \end{bmatrix} \qquad X = \begin{bmatrix} x \\ y \\ z \end{bmatrix} \qquad \text{and} \qquad B = \begin{bmatrix} -8 \\ 3 \\ 4 \end{bmatrix}$$

then the given system can be represented by the matrix equation $AX = B$. Therefore we know that $X = A^{-1}B$, so we need to find A^{-1} and the product $A^{-1}B$. We found the matrix A^{-1} in Example 2, so let's use that result and find $A^{-1}B$.

$$X = A^{-1}B = \begin{bmatrix} -\dfrac{5}{24} & \dfrac{1}{6} & -\dfrac{7}{24} \\ \dfrac{7}{24} & \dfrac{1}{6} & \dfrac{5}{24} \\ \dfrac{11}{24} & -\dfrac{1}{6} & \dfrac{1}{24} \end{bmatrix} \begin{bmatrix} -8 \\ 3 \\ 4 \end{bmatrix} = \begin{bmatrix} 1 \\ -1 \\ -4 \end{bmatrix}$$

The solution set of the given system is $\{(1, -1, -4)\}$.

▶ **Now work Problem 41.**   ■

CONCEPT QUIZ

For the following problems, answer true or false.

1. If A is a 5×2 matrix, then it has 5 rows and 2 columns of elements.

2. If $B = \begin{bmatrix} 3 & 2 & 5 & 7 \\ 4 & 0 & 8 & -1 \end{bmatrix}$, then B is a 4×2 matrix.

3. Only square matrices have an additive inverse.

4. For matrices that can be added, the commutative property holds.

5. If A is a 3×3 matrix, then $AA^{-1} = \begin{bmatrix} 1 & 1 & 1 \\ 1 & 1 & 1 \\ 1 & 1 & 1 \end{bmatrix}$.

6. Every square matrix has a multiplicative inverse matrix.

7. Given that a_{14} is an element of matrix A, then the element is in the first row and fourth column.

8. If $A = \begin{bmatrix} 2 & -4 & 5 \\ -1 & 0 & 3 \end{bmatrix}$, then $3A = \begin{bmatrix} 6 & -4 & 5 \\ -3 & 0 & 3 \end{bmatrix}$.

Problem Set 7.3

For Problems 1–8, find $A + B$, $A - B$, $2A + 3B$, and $4A - 2B$.

1. $A = \begin{bmatrix} 2 & -1 & 4 \\ -2 & 0 & 5 \end{bmatrix}$, $B = \begin{bmatrix} -1 & 4 & -7 \\ 5 & -6 & 2 \end{bmatrix}$

2. $A = \begin{bmatrix} 3 & -6 \\ 2 & -1 \\ -4 & 5 \end{bmatrix}$, $B = \begin{bmatrix} 1 & 0 \\ 5 & -7 \\ -6 & 9 \end{bmatrix}$

3. $A = \begin{bmatrix} 2 & -1 & 4 & 12 \end{bmatrix}$, $B = \begin{bmatrix} -3 & -6 & 9 & -5 \end{bmatrix}$

4. $A = \begin{bmatrix} 3 \\ -9 \\ 7 \end{bmatrix}$, $B = \begin{bmatrix} -6 \\ 12 \\ 9 \end{bmatrix}$

5. $A = \begin{bmatrix} 3 & -2 & 1 \\ -1 & 4 & -7 \\ 0 & 5 & 9 \end{bmatrix}$, $B = \begin{bmatrix} 5 & -1 & -3 \\ 10 & -2 & 4 \\ 7 & 0 & 12 \end{bmatrix}$

6. $A = \begin{bmatrix} 7 & -4 \\ -5 & 9 \\ -1 & 2 \end{bmatrix}$, $B = \begin{bmatrix} 12 & 3 \\ -2 & -4 \\ -6 & 7 \end{bmatrix}$

7. $A = \begin{bmatrix} -1 & 0 \\ 2 & 3 \\ -5 & -4 \\ -7 & 11 \end{bmatrix}$, $B = \begin{bmatrix} 1 & 2 \\ -3 & 7 \\ 6 & -5 \\ 9 & -2 \end{bmatrix}$

8. $A = \begin{bmatrix} 0 & -1 & -2 \\ 3 & -4 & 6 \\ 5 & 4 & -9 \end{bmatrix}$, $B = \begin{bmatrix} 2 & 1 & -7 \\ -6 & 4 & 5 \\ 3 & -2 & -1 \end{bmatrix}$

For Problems 9–20, find AB and BA, whenever they exist.

9. $A = \begin{bmatrix} 2 & -1 \\ 0 & -4 \\ -5 & 3 \end{bmatrix}$, $B = \begin{bmatrix} 5 & -2 & 6 \\ -1 & 4 & -2 \end{bmatrix}$

10. $A = \begin{bmatrix} -2 & 3 & -1 \\ 7 & -4 & 5 \end{bmatrix}$, $B = \begin{bmatrix} 1 & -1 \\ -2 & 3 \\ -5 & -6 \end{bmatrix}$

11. $A = \begin{bmatrix} 2 & -1 & -3 \\ 0 & -4 & 7 \end{bmatrix}$, $B = \begin{bmatrix} 2 & 1 & -1 & 4 \\ 0 & -2 & 3 & 5 \\ -6 & 4 & -2 & 0 \end{bmatrix}$

12. $A = \begin{bmatrix} 3 & -1 & -4 \\ -5 & 2 & 2 \end{bmatrix}$, $B = \begin{bmatrix} 3 & -2 \\ -4 & -1 \end{bmatrix}$

13. $A = \begin{bmatrix} 1 & -1 & 2 \\ 0 & 1 & -2 \\ 3 & 1 & 4 \end{bmatrix}$, $B = \begin{bmatrix} 2 & 3 & -1 \\ 4 & 0 & 2 \\ -5 & 1 & -1 \end{bmatrix}$

14. $A = \begin{bmatrix} 1 & 0 & 1 \\ 0 & 1 & 1 \\ -1 & 2 & 3 \end{bmatrix}$, $B = \begin{bmatrix} -1 & -1 & 1 \\ 0 & 1 & 0 \\ 2 & -3 & 1 \end{bmatrix}$

15. $A = \begin{bmatrix} 2 & -1 & 3 & 4 \end{bmatrix}$, $B = \begin{bmatrix} -1 \\ -3 \\ 2 \\ -4 \end{bmatrix}$

16. $A = \begin{bmatrix} -2 \\ 3 \\ -5 \end{bmatrix}$, $B = \begin{bmatrix} 3 & -4 & -5 \end{bmatrix}$

17. $A = \begin{bmatrix} 2 \\ -7 \end{bmatrix}, \quad B = \begin{bmatrix} 3 & -2 \\ 1 & 0 \\ -1 & 4 \end{bmatrix}$

18. $A = \begin{bmatrix} 3 & -2 & 2 & -4 \\ 1 & 0 & -1 & 2 \end{bmatrix}, \quad B = \begin{bmatrix} 3 & -2 & 1 \\ -3 & 1 & 4 \\ 5 & 2 & 0 \\ -4 & -1 & -2 \end{bmatrix}$

19. $A = \begin{bmatrix} 3 \\ -4 \\ 2 \end{bmatrix}, \quad B = \begin{bmatrix} 3 & -4 \end{bmatrix}$

20. $A = \begin{bmatrix} 3 & -7 \end{bmatrix}, \quad B = \begin{bmatrix} 8 \\ -9 \end{bmatrix}$

For Problems 21–36, use the technique discussed in this section to find the multiplicative inverse (if it exists) of each matrix.

21. $\begin{bmatrix} 1 & 3 \\ 4 & 2 \end{bmatrix}$

22. $\begin{bmatrix} 1 & 2 \\ 2 & -3 \end{bmatrix}$

23. $\begin{bmatrix} 2 & 1 \\ 7 & 4 \end{bmatrix}$

24. $\begin{bmatrix} 3 & 7 \\ 2 & 5 \end{bmatrix}$

25. $\begin{bmatrix} -2 & 1 \\ 3 & -4 \end{bmatrix}$

26. $\begin{bmatrix} -3 & 1 \\ 3 & -2 \end{bmatrix}$

27. $\begin{bmatrix} 1 & 2 & 3 \\ 1 & 3 & 4 \\ 1 & 4 & 3 \end{bmatrix}$

28. $\begin{bmatrix} 1 & 3 & -2 \\ 1 & 4 & -1 \\ -2 & -7 & 5 \end{bmatrix}$

29. $\begin{bmatrix} 1 & -2 & 1 \\ -2 & 5 & 3 \\ 3 & -5 & 7 \end{bmatrix}$

30. $\begin{bmatrix} 1 & 4 & -2 \\ -3 & -11 & 1 \\ 2 & 7 & 3 \end{bmatrix}$

31. $\begin{bmatrix} 2 & 3 & -4 \\ 3 & -1 & -2 \\ 1 & -4 & 2 \end{bmatrix}$

32. $\begin{bmatrix} -2 & 2 & 3 \\ 1 & -1 & 0 \\ 0 & 1 & 4 \end{bmatrix}$

33. $\begin{bmatrix} 1 & 2 & 3 \\ -3 & -4 & 3 \\ 2 & 4 & -1 \end{bmatrix}$

34. $\begin{bmatrix} 1 & -2 & 3 \\ -1 & 3 & -2 \\ -2 & 6 & 1 \end{bmatrix}$

35. $\begin{bmatrix} 2 & 0 & 0 \\ 0 & 4 & 0 \\ 0 & 0 & 10 \end{bmatrix}$

36. $\begin{bmatrix} 1 & -3 & 5 \\ 0 & 1 & 2 \\ 0 & 0 & 1 \end{bmatrix}$

For Problems 37–46, use the method of matrix inverses to solve each system. The required multiplicative inverses were found in Problems 21–36.

37. $\left(\begin{array}{l} 2x + y = -4 \\ 7x + 4y = -13 \end{array} \right)$

38. $\left(\begin{array}{l} 3x + 7y = -38 \\ 2x + 5y = -27 \end{array} \right)$

39. $\left(\begin{array}{l} -2x + y = 1 \\ 3x - 4y = -14 \end{array} \right)$

40. $\left(\begin{array}{l} -3x + y = -18 \\ 3x - 2y = 15 \end{array} \right)$

41. $\left(\begin{array}{l} x + 2y + 3z = -2 \\ x + 3y + 4z = -3 \\ x + 4y + 3z = -6 \end{array} \right)$

42. $\left(\begin{array}{l} x + 3y - 2z = 5 \\ x + 4y - z = 3 \\ -2x - 7y + 5z = -12 \end{array} \right)$

43. $\left(\begin{array}{l} x - 2y + z = -3 \\ -2x + 5y + 3z = 34 \\ 3x - 5y + 7z = 14 \end{array} \right)$

44. $\left(\begin{array}{l} x + 4y - 2z = 2 \\ -3x - 11y + z = -2 \\ 2x + 7y + 3z = -2 \end{array} \right)$

45. $\left(\begin{array}{l} x + 2y + 3z = 2 \\ -3x - 4y + 3z = 0 \\ 2x + 4y - z = 4 \end{array} \right)$

46. $\left(\begin{array}{l} x - 2y + 3z = -39 \\ -x + 3y - 2z = 40 \\ -2x + 6y + z = 45 \end{array} \right)$

47. We can generate five systems of linear equations from the system

$$\left(\begin{array}{l} x + y + 2z = a \\ 2x + 3y - z = b \\ -3x + y - 2z = c \end{array} \right)$$

by letting a, b, and c assume five different sets of values. Solve the system for each set of values. The inverse of the coefficient matrix of these systems is given in Example 2 of this section.

a. $a = 7, b = 1$, and $c = -1$
b. $a = -7, b = 5$, and $c = 1$
c. $a = -9, b = -8$, and $c = 19$
d. $a = -1, b = -13$, and $c = -17$
e. $a = -2, b = 0$, and $c = -2$

■ ■ ■ **THOUGHTS INTO WORDS**

48. How would you describe row-by-column multiplication of matrices?

49. Give a step-by-step explanation of how to find the multiplicative inverse of the matrix $\begin{bmatrix} 1 & 3 \\ -2 & 4 \end{bmatrix}$ by using the technique of Section 7.3.

50. Explain how to find the multiplicative inverse of the matrix in Problem 49 by using the technique of Section 7.2.

■ ■ ■ **FURTHER INVESTIGATIONS**

51. Matrices can be used to code and decode messages. For example, suppose that we set up a one-to-one correspondence between the letters of the alphabet and the first 26 counting numbers, as follows:

A   B   C   Z
↕   ↕   ↕   ⋯   ↕
1   2   3   26

Now suppose that we want to code the message PLAY IT BY EAR. We can partition the letters of the message into groups of two. Because the last group will contain only one letter, let's arbitrarily stick in a Z to form a group of two. Let's also assign a number to each letter on the basis of the letter/number association we exhibited.

P   L   A   Y   I   T   B   Y   E   A   R   Z
↕   ↕   ↕   ↑   ↑   ↕   ↕   ↕   ↕   ↕   ↑   ↕
16   12   1   25   9   20   2   25   5   1   18   26

Each pair of numbers can be recorded as columns in a 2×6 matrix B.

$$B = \begin{bmatrix} 16 & 1 & 9 & 2 & 5 & 18 \\ 12 & 25 & 20 & 25 & 1 & 26 \end{bmatrix}$$

Now let's choose a 2×2 matrix such that the matrix contains only integers and its inverse also contains only integers. For example, we can use $A = \begin{bmatrix} 3 & 1 \\ 5 & 2 \end{bmatrix}$;

then $A^{-1} = \begin{bmatrix} 2 & -1 \\ -5 & 3 \end{bmatrix}$.

Next, let's find the product AB:

$$AB = \begin{bmatrix} 3 & 1 \\ 5 & 2 \end{bmatrix} \begin{bmatrix} 16 & 1 & 9 & 2 & 5 & 18 \\ 12 & 25 & 20 & 25 & 1 & 26 \end{bmatrix}$$

$$= \begin{bmatrix} 60 & 28 & 47 & 31 & 16 & 80 \\ 104 & 55 & 85 & 60 & 27 & 142 \end{bmatrix}$$

Now we have our **coded message:**

60   104   28   55   47   85   31   60   16   27   80   142

A person decoding the message would put the numbers back into a 2×6 matrix, multiply it on the left by A^{-1}, and convert the numbers back to letters.

Each of the following coded messages was formed by using the matrix $A = \begin{bmatrix} 2 & 3 \\ 1 & 2 \end{bmatrix}$. Decode each of the messages.

a. 53   34   48   25   39   22   35   20   78   47
56   37   83   54

b. 62   40   78   47   64   36   19   11   93   57   93
56   88   57

c. 64   36   58   37   63   36   21   13   75   47   63
36   38   23   118   72

d. 29   15   96   58   60   37   75   47   19   10
37   21   70   42   90   55   98   59   72   45
51   28   86   56

52. Suppose that the ordered pair (x, y) of a rectangular coordinate system is recorded as a 2×1 matrix and then multiplied on the left by the matrix

$\begin{bmatrix} 1 & 0 \\ 0 & -1 \end{bmatrix}$. We would obtain

$$\begin{bmatrix} 1 & 0 \\ 0 & -1 \end{bmatrix} \begin{bmatrix} x \\ y \end{bmatrix} = \begin{bmatrix} x \\ -y \end{bmatrix}$$

The point $(x, -y)$ is an x axis reflection of the point (x, y). Therefore the matrix $\begin{bmatrix} 1 & 0 \\ 0 & -1 \end{bmatrix}$ performs an

x axis reflection. What type of geometric transformation is performed by each of the following matrices?

a. $\begin{bmatrix} -1 & 0 \\ 0 & 1 \end{bmatrix}$

b. $\begin{bmatrix} -1 & 0 \\ 0 & -1 \end{bmatrix}$

c. $\begin{bmatrix} 0 & -1 \\ 1 & 0 \end{bmatrix}$ [*Hint:* Check the slopes of lines through the origin.]

d. $\begin{bmatrix} 0 & 1 \\ -1 & 0 \end{bmatrix}$

GRAPHING CALCULATOR ACTIVITIES

53. Use your calculator to check your answers for Problems 14, 18, 28, 30, 32, 34, 36, 42, 44, 46, and 47.

54. Use your calculator and the method of matrix inverses to solve each of the following systems. Be sure to check your solutions.

a. $\begin{pmatrix} 2x - 3y + 4z = 54 \\ 3x + y - z = 32 \\ 5x - 4y + 3z = 58 \end{pmatrix}$

b. $\begin{pmatrix} 17x + 15y - 19z = 10 \\ 18x - 14y + 16z = 94 \\ 13x + 19y - 14z = -23 \end{pmatrix}$

c. $\begin{pmatrix} 1.98x + 2.49y + 3.15z = 45.72 \\ 2.29x + 1.95y + 2.75z = 42.05 \\ 3.15x + 3.20y + 1.85z = 42 \end{pmatrix}$

d. $\begin{pmatrix} x_1 + 2x_2 - 4x_3 + 7x_4 = -23 \\ 2x_1 - 3x_2 + 5x_3 - x_4 = -22 \\ 5x_1 + 4x_2 - 2x_3 - 8x_4 = 59 \\ 3x_1 - 7x_2 + 8x_3 + 9x_4 = -103 \end{pmatrix}$

e. $\begin{pmatrix} 2x_1 - x_2 + 3x_3 - 4x_4 + 12x_5 = 98 \\ x_1 + 2x_2 - x_3 - 7x_4 + 5x_5 = 41 \\ 3x_1 + 4x_2 - 7x_3 + 6x_4 - 9x_5 = -41 \\ 4x_1 - 3x_2 + x_3 - x_4 + x_5 = 4 \\ 7x_1 + 8x_2 - 4x_3 - 6x_4 - 6x_5 = 12 \end{pmatrix}$

Answers to the Concept Quiz

1. True   **2.** False   **3.** False   **4.** True   **5.** False   **6.** False   **7.** True   **8.** False

---

7.4 Systems Involving Linear Inequalities: Linear Programming

Objectives

■ Solve a system of linear inequalities.

■ Find the minimum and maximum value of linear functions for a specified region.

■ Solve problems using linear programming.

Finding solution sets for **systems of linear inequalities** relies heavily on the graphing approach. (Recall that we discussed graphing of linear inequalities in Section 2.3.) The solution set of the system

$$\begin{pmatrix} x + y > 2 \\ x - y < 2 \end{pmatrix}$$

is the intersection of the solution sets of the individual inequalities. In Figure 7.1(a) we indicate the solution set for $x + y > 2$, and in Figure 7.1(b) we indicate the solution set for $x - y < 2$. The shaded region in Figure 7.1(c) represents the intersection of the two solution sets; therefore it is the graph of the system. Remember that dashed lines are used to indicate that the points on the lines are not included in the solution set. In the following examples we indicate only the final solution set for the system.

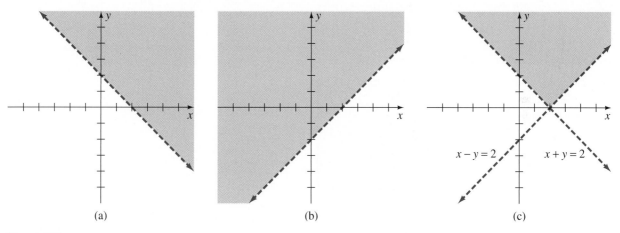

(a)   (b)   (c)

Figure 7.1

E X A M P L E   1

Solve the following system by graphing: $\begin{pmatrix} 2x - y \geq 4 \\ x + 2y < 2 \end{pmatrix}$.

Solution

The graph of $2x - y \geq 4$ consists of all points *on or below* the line $2x - y = 4$. The graph of $x + 2y < 2$ consists of all points *below* the line $x + 2y = 2$. The graph of the system is indicated by the shaded region in Figure 7.2. Note that all points in the shaded region are on or below the line $2x - y = 4$ and below the line $x + 2y = 2$.

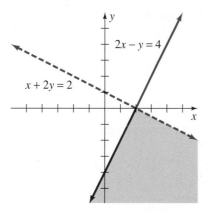

Figure 7.2

▶   **Now work Problem 3.**   ■

E X A M P L E   2

Solve the following system by graphing: $\begin{pmatrix} x \le 2 \\ y \ge -1 \end{pmatrix}$.

Solution

Remember that even though each inequality contains only one variable, we are working in a rectangular coordinate system involving ordered pairs. That is, the system could also be written

$$\begin{pmatrix} x + 0(y) \le 2 \\ 0(x) + y \ge -1 \end{pmatrix}$$

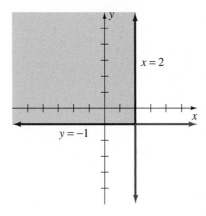

Figure 7.3

The graph of this system is the shaded region in Figure 7.3. Note that all points in the shaded region are *on or to the left of* the line $x = 2$ and *on or above* the line $y = -1$.

⊙ **Now work Problem 19.**   ∎

A system may contain more than two inequalities, as the next example illustrates.

E X A M P L E   3

Solve the following system by graphing:

$$\begin{pmatrix} x \ge 0 \\ y \ge 0 \\ 2x + 3y \le 12 \\ 3x + y \le 6 \end{pmatrix}$$

Solution

The solution set for the system is the intersection of the solution sets of the four inequalities. The shaded region in Figure 7.4 indicates the solution set for the system. Note that all points in the shaded region are *on or to the right of* the y axis, *on or above* the x axis, *on or below* the line $2x + 3y = 12$, and *on or below* the line $3x + y = 6$.

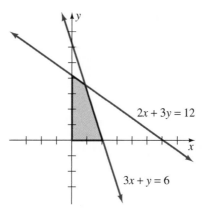

$2x + 3y = 12$

$3x + y = 6$

Figure 7.4

▶ **Now work Problem 21.** ■

■ Linear Programming: Another Look at Problem Solving

Throughout this text problem solving is a unifying theme. Therefore it seems appropriate at this time to give you a brief glimpse of an area of mathematics that was developed in the 1940s specifically as a problem-solving tool. Many applied problems involve the idea of *maximizing* or *minimizing* a certain function that is subject to various constraints; these can be expressed as linear inequalities. **Linear programming** was developed as one method for solving such problems.

Remark: The term "programming" refers to the distribution of limited resources in order to maximize or minimize a certain function, such as cost, profit, distance, and so on. Thus it does not mean the same thing it means in computer programming. The constraints that govern the distribution of resources determine the linear inequalities and equations; thus the term "linear programming" is used.

Before we introduce a linear programming type of problem, we need to extend one mathematical concept a bit. A **linear function in two variables**, x and y, is a function of the form $f(x, y) = ax + by + c$, where a, b, and c are real numbers. In other words, with each ordered pair (x, y) we associate a third number by the rule $ax + by + c$. For example, suppose the function f is described by $f(x, y) = 4x + 3y + 5$. Then $f(2, 1) = 4(2) + 3(1) + 5 = 16$.

First, let's take a look at some mathematical ideas that form the basis for solving a linear programming problem. Consider the shaded region in Figure 7.5 and the following linear functions in two variables:

$$f(x, y) = 4x + 3y + 5$$

$$f(x, y) = 2x + 7y - 1$$

$$f(x, y) = x - 2y$$

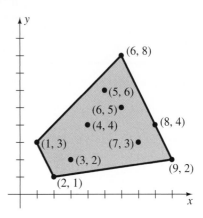

Figure 7.5

Suppose that we need to find the maximum value and the minimum value achieved by each of the functions in the indicated region. The following chart summarizes the values for the ordered pairs indicated in Figure 7.5. Note that for each function, the maximum and minimum values are obtained at vertices of the region.

	Ordered Pairs	Value of $f(x, y) = 4x + 3y + 5$	Value of $f(x, y) = 2x + 7y - 1$	Value of $f(x, y) = x - 2y$
Vertex	(2, 1)	16 (*minimum*)	10 (*minimum*)	0
	(3, 2)	23	19	−1
Vertex	(9, 2)	47	31	5 (*maximum*)
Vertex	(1, 3)	18	22	−5
	(7, 3)	42	34	1
	(4, 4)	33	35	−4
	(8, 4)	49	43	0
	(6, 5)	44	46	−4
	(5, 6)	43	51	−7
Vertex	(6, 8)	53 (*maximum*)	67 (*maximum*)	−10 (*minimum*)

We claim that for linear functions, maximum and minimum functional values are *always* obtained at vertices of the region. To substantiate this, let's consider the family of lines $x - 2y = k$, where k is an arbitrary constant. (We are now working only with the function $f(x, y) = x - 2y$.) In slope–intercept form, $x - 2y = k$ becomes $y = \dfrac{1}{2}x - \dfrac{1}{2}k$, so we have a family of parallel lines each having a slope of $\frac{1}{2}$. In Figure 7.6 we sketched some of these lines so that each line has at least one point in common with the given region. Note that $x - 2y$ reaches a minimum value of -10 at the vertex $(6, 8)$ and a maximum value of 5 at the vertex $(9, 2)$.

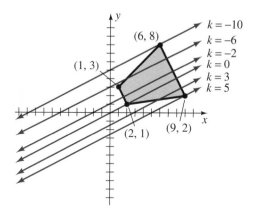

Figure 7.6

In general, suppose that f is a **linear function in two variables**, x and y, and that S is a region of the xy plane. If f **attains a maximum (minimum)** value in S, then that maximum (minimum) value is obtained at a vertex of S.

Remark: A subset of the xy plane is said to be **bounded** if there is a circle that contains all of its points; otherwise the subset is said to be **unbounded.** A bounded set will contain maximum and minimum values for a function, but an unbounded set may not contain such values.

Now we will consider two examples that illustrate a general graphing approach to solving a linear programming problem in two variables. The first example gives us the general makeup of such a problem; the second example illustrates the type of setting from which the function and inequalities evolve.

E X A M P L E   4

Find the maximum value and the minimum value of the function $f(x, y) = 9x + 13y$ in the region determined by the following system of inequalities:

$$\begin{pmatrix} x \geq 0 \\ y \geq 0 \\ 2x + 3y \leq 18 \\ 2x + y \leq 10 \end{pmatrix}$$

Solution

First, let's graph the inequalities to determine the region, as indicated in Figure 7.7. (Such a region is called the **set of feasible solutions,** and the inequalities are referred to as **constraints.**) The point $(3, 4)$ is determined by solving the system

$$\begin{pmatrix} 2x + 3y = 18 \\ 2x + y = 10 \end{pmatrix}$$

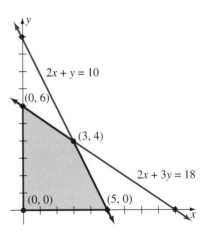

Figure 7.7

Next we can determine the values of the given function at the vertices of the region. (Such a function to be maximized or minimized is called the **objective function.**)

Vertices	Value of $f(x, y) = 9x + 13y$
$(0, 0)$	0 *(minimum)*
$(5, 0)$	45
$(3, 4)$	79 *(maximum)*
$(0, 6)$	78

A minimum value of 0 is obtained at $(0, 0)$, and a maximum value of 79 is obtained at $(3, 4)$.

▶ **Now work Problem 29.**   ■

P R O B L E M   1

A company that manufactures gidgets and gadgets has the following production information available.

1. To produce a gidget requires 3 hours of working time on machine A and 1 hour on machine B.

2. To produce a gadget requires 2 hours on machine A and 1 hour on machine B.

3. Machine A is available for no more than 120 hours per week, and machine B is available for no more than 50 hours per week.

4. Gidgets can be sold at a profit of $3.75 each, and a profit of $3 can be realized on a gadget.

How many gidgets and how many gadgets should the company produce each week to maximize its profit? What would the maximum profit be?

Solution

Let x be the number of gidgets and y be the number of gadgets. Thus the profit function is $P(x, y) = 3.75x + 3y$. The constraints for the problem can be represented by the following inequalities:

$$3x + 2y \leq 120 \quad \text{Machine A is available for no more than 120 hours.}$$
$$x + y \leq 50 \quad \text{Machine B is available for no more than 50 hours.}$$
$$\left. \begin{array}{l} x \geq 0 \\ y \geq 0 \end{array} \right\} \quad \text{The number of gidgets and gadgets must be represented by a nonnegative numer.}$$

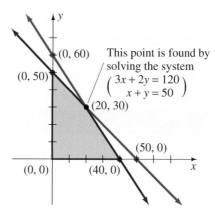

Figure 7.8

When we graph these inequalities, we obtain the set of feasible solutions indicated by the shaded region in Figure 7.8. Next we find the value of the profit function at each of the vertices; this produces the chart that follows.

Vertices	Value of $P(x, y) = 3.75x + 3y$
$(0, 0)$	0
$(40, 0)$	150
$(20, 30)$	165 *(maximum)*
$(0, 50)$	150

Thus a maximum profit of $165 is realized by producing 20 gidgets and 30 gadgets.

▶ **Now work Problem 41.**   ∎

CONCEPT QUIZ

1. Write a system of a two linear inequalities that has the empty set as the solution.

For problems 2–8, answer true or false.

2. The point $(2, -5)$ is a solution of the system of inequalities $\begin{pmatrix} 2x - y > 9 \\ x + 3y < 0 \end{pmatrix}$.

3. The coordinates of every point in the rectangular coordinate plane satisfy the system of inequalities $\begin{pmatrix} 2x - y > 4 \\ 2x - y < 4 \end{pmatrix}$.

4. Given $f(x, y) = 2x - y + 3$, then $f(4, -1) = 12$.

5. A subset of the rectangular coordinate plane is bounded if there can be circle drawn that contains all of its points.

6. The region determined by the system of inequalities $\begin{pmatrix} x \geq 0 \\ y \geq 0 \end{pmatrix}$ is a bounded region.

7. For linear programming, maximum and minimum function values are always obtained at the vertices of a bounded region.

8. For linear programming problems, the region determined by the system of inequalities is called the set of feasible solutions.

Problem Set 7.4

For Problems 1–24, indicate the solution set for each system of inequalities by graphing the system and shading the appropriate region.

1. $\begin{pmatrix} x + y > 3 \\ x - y > 0 \end{pmatrix}$

2. $\begin{pmatrix} x - y < 2 \\ x + y < 1 \end{pmatrix}$

3. $\begin{pmatrix} x - 2y \leq 4 \\ x + 2y > 4 \end{pmatrix}$

4. $\begin{pmatrix} 3x - y > 6 \\ 2x + y \leq 4 \end{pmatrix}$

5. $\begin{pmatrix} 2x + 3y \leq 6 \\ 3x - 2y \leq 6 \end{pmatrix}$

6. $\begin{pmatrix} 4x + 3y \geq 12 \\ 3x - 4y \geq 12 \end{pmatrix}$

7. $\begin{pmatrix} 2x - y \geq 4 \\ x + 3y < 3 \end{pmatrix}$

8. $\begin{pmatrix} 3x - y < 3 \\ x + y \geq 1 \end{pmatrix}$

9. $\begin{pmatrix} x + 2y > -2 \\ x - y < -3 \end{pmatrix}$

10. $\begin{pmatrix} x - 3y < -3 \\ 2x - 3y > -6 \end{pmatrix}$

11. $\begin{pmatrix} y > x - 4 \\ y < x \end{pmatrix}$

12. $\begin{pmatrix} y \leq x + 2 \\ y \geq x \end{pmatrix}$

13. $\begin{pmatrix} x - y > 2 \\ x - y > -1 \end{pmatrix}$

14. $\begin{pmatrix} x + y > 1 \\ x + y > 3 \end{pmatrix}$

15. $\begin{pmatrix} y \geq x \\ x > -1 \end{pmatrix}$

16. $\begin{pmatrix} y < x \\ y \leq 2 \end{pmatrix}$

17. $\begin{pmatrix} y < x \\ y > x + 3 \end{pmatrix}$

18. $\begin{pmatrix} x \leq 3 \\ y \leq -1 \end{pmatrix}$

19. $\begin{pmatrix} y > -2 \\ x > 1 \end{pmatrix}$

20. $\begin{pmatrix} x + 2y > 4 \\ x + 2y < 2 \end{pmatrix}$

21. $\begin{pmatrix} x \geq 0 \\ y \geq 0 \\ x + y \leq 4 \\ 2x + y \leq 6 \end{pmatrix}$

22. $\begin{pmatrix} x \geq 0 \\ y \geq 0 \\ x - y \leq 5 \\ 4x + 7y \leq 28 \end{pmatrix}$

23. $\begin{pmatrix} x \geq 0 \\ y \geq 0 \\ 2x + y \leq 4 \\ 2x - 3y \leq 6 \end{pmatrix}$

24. $\begin{pmatrix} x \geq 0 \\ y \geq 0 \\ 3x + 5y \geq 15 \\ 5x + 3y \geq 15 \end{pmatrix}$

For Problems 25–28 (Figures 7.9–7.12), find the maximum value and the minimum value of the given function in the indicated region.

25. $f(x, y) = 3x + 5y$

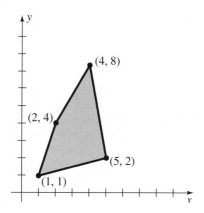

Figure 7.9

26. $f(x, y) = 8x + 3y$

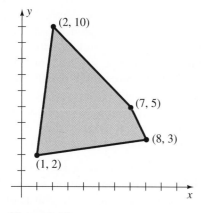

Figure 7.10

27. $f(x, y) = x + 4y$

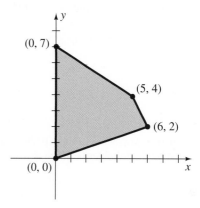

Figure 7.11

28. $f(x, y) = 2.5x + 3.5y$

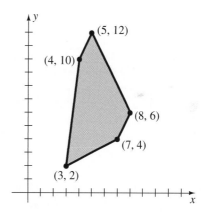

Figure 7.12

29. Maximize the function $f(x, y) = 3x + 7y$ in the region determined by the following constraints:

$$3x + 2y \leq 18$$
$$3x + 4y \geq 12$$
$$x \geq 0$$
$$y \geq 0$$

30. Maximize the function $f(x, y) = 1.5x + 2y$ in the region determined by the following constraints:

$$3x + 2y \leq 36$$
$$3x + 10y \leq 60$$
$$x \geq 0$$
$$y \geq 0$$

31. Maximize the function $f(x, y) = 40x + 55y$ in the region determined by the following constraints:

$$2x + y \leq 10$$
$$x + y \leq 7$$
$$2x + 3y \leq 18$$
$$x \geq 0$$
$$y \geq 0$$

32. Maximize the function $f(x, y) = 0.08x + 0.09y$ in the region determined by the following constraints:

$$x + y \leq 8000$$
$$y \leq \frac{1}{3}x$$
$$y \geq 500$$
$$x \leq 7000$$
$$x \geq 0$$

33. Minimize the function $f(x, y) = 0.2x + 0.5y$ in the region determined by the following constraints:

$$2x + y \geq 12$$
$$2x + 5y \geq 20$$
$$x \geq 0$$
$$y \geq 0$$

34. Minimize the function $f(x, y) = 3x + 7y$ in the region determined by the following constraints:

$$x + y \geq 9$$
$$6x + 11y \geq 84$$
$$x \geq 0$$
$$y \geq 0$$

35. Maximize the function $f(x, y) = 9x + 2y$ in the region determined by the following constraints:

$$5y - 4x \leq 20$$
$$4x + 5y \leq 60$$
$$x \geq 0$$
$$x \leq 10$$
$$y \geq 0$$

36. Maximize the function $f(x, y) = 3x + 4y$ in the region determined by the following constraints:

$$2y - x \leq 6$$
$$x + y \leq 12$$
$$x \geq 2$$
$$x \leq 8$$
$$y \geq 0$$

For Problems 37–42, solve each linear programming problem by using the graphing method illustrated in Problem 1 on page 582.

37. Suppose that an investor wants to invest up to $10,000. She plans to buy one speculative type of stock and one conservative type. The speculative stock is paying a 12% return, and the conservative stock is paying a 9% return. She has decided to invest at least $2000 in the conservative stock and no more than $6000 in the speculative stock. Furthermore, she does not want the speculative investment to exceed the conservative one. How much should she invest at each rate to maximize her return?

38. A manufacturer of golf clubs makes a profit of $50 per set on a model A set and $45 per set on a model B set. Daily production of the model A clubs is between 30 and 50 sets, inclusive, and that of the model B clubs is between 10 and 20 sets, inclusive. The total daily production is not to exceed 50 sets. How many sets of each model should be manufactured per day to maximize the profit?

39. A company makes two types of calculators. Type A sells for $12, and type B sells for $10. It costs the company $9 to produce one type A calculator and $8 to produce one type B calculator. In one month the company is equipped to produce between 200 and 300, inclusive, of the type A calculator and between 100 and 250, inclusive, of the type B calculator, but not more than 500 altogether. How many calculators of each type should be produced per month to maximize the difference between the total selling price and the total cost of production?

40. A manufacturer of small copiers makes a profit of $200 on a deluxe model and $250 on a standard model. The company wants to produce at least 50 deluxe models per week and at least 75 standard models per week. However, the weekly production is not to exceed 150 copiers. How many copiers of each kind should be produced in order to maximize the profit?

41. Products A and B are manufactured by a company according to the following production information.

a. To produce one unit of product A requires 1 hour of working time on machine I, 2 hours on machine II, and 1 hour on machine III.

b. To produce one unit of product B requires 1 hour of working time on machine I, one hour on machine II, and 3 hours on machine III.

c. Machine I is available for no more than 40 hours per week, machine II is available for no more than 40 hours per week, and machine III for no more than 60 hours per week.

d. Product A can be sold at a profit of $2.75 per unit and product B at a profit of $3.50 per unit.

How many units each of product A and product B should be produced per week to maximize profit?

42. Suppose that the company we refer to in Problem 1 also manufactures widgets and wadgets and has the following production information available.

a. To produce a widget requires 4 hours of working time on machine A and 2 hours on machine B.

b. To produce a wadget requires 5 hours of working time on machine A and 5 hours on machine B.

c. Machine A is available for no more than 200 hours per month, and machine B is available for no more than 150 hours per month.

d. Widgets can be sold at a profit of $7 each and wadgets at a profit of $8 each.

How many widgets and how many wadgets should be produced per month in order to maximize profit?

■ ■ ■ THOUGHTS INTO WORDS

43. Describe in your own words the process of solving a system of inequalities.

44. What is linear programming? Write a paragraph or two answering this question in a way that elementary algebra students could understand.

Answers to the Concept Quiz

1. $\begin{pmatrix} x + y > 5 \\ x + y < -2 \end{pmatrix}$ Answers may vary   **2.** False   **3.** False   **4.** True   **5.** True   **6.** False

7. True   **8.** True

Be sure that you understand the following ideas pertaining to the algebra of matrices.

1. Matrices of the same dimension are added by adding the elements in corresponding positions.

2. Matrix addition is a commutative and an associative operation.

3. Matrices of any specific dimension have an additive identity element, which is the matrix of that same dimension containing all zeros.

4. Every matrix A has an additive inverse, $-A$, that can be found by changing the sign of each element of A.

5. Matrices of the same dimension can be subtracted by the definition $A - B = A + (-B)$.

6. The scalar product of a real number k and a matrix A can be found by multiplying each element of A by k.

7. The following properties hold for scalar multiplication and matrix addition:

$$k(A + B) = kA + kB$$
$$(k + l)A = kA + lA$$
$$(kl)A = k(lA)$$

8. If A is an $m \times n$ matrix and B is an $n \times p$ matrix, then the product AB is an $m \times p$ matrix. The general term, c_{ij}, of the product matrix $C = AB$ is determined by the equation

$$c_{ij} = a_{i1}b_{1j} + a_{i2}b_{2j} + \cdots + a_{in}b_{nj}$$

9. Matrix multiplication is not a commutative operation, but it is an associative operation.

10. Matrix multiplication has two distributive properties:

$$A(B + C) = AB + AC \quad \text{and} \quad (A + B)C = AC + BC$$

11. The general multiplicative identity element, I_n, for square $n \times n$ matrices contains only 1s in the main diagonal and zeros elsewhere. For example,

$$I_2 = \begin{bmatrix} 1 & 0 \\ 0 & 1 \end{bmatrix} \quad \text{and} \quad I_3 = \begin{bmatrix} 1 & 0 & 0 \\ 0 & 1 & 0 \\ 0 & 0 & 1 \end{bmatrix}$$

12. If a square matrix A has a multiplicative inverse A^{-1}, then $AA^{-1} = A^{-1}A = I_n$.

13. The multiplicative inverse of the 2×2 matrix

$$A = \begin{bmatrix} a_{11} & a_{12} \\ a_{21} & a_{22} \end{bmatrix}$$

is

$$A^{-1} = \frac{1}{|A|} \begin{bmatrix} a_{22} & -a_{12} \\ -a_{21} & a_{11} \end{bmatrix}$$

for $|A| \neq 0$. If $|A| = 0$, then the matrix A has no inverse.

14. A general technique for finding the inverse of a square matrix, when it exists, is described on page 569.

15. The solution set of a system of n linear equations in n variables can be found by multiplying the inverse of the coefficient matrix by the column matrix consisting of the constant terms. For example, the solution set of the system

$$\begin{pmatrix} 2x + 3y - z = 4 \\ 3x - y + 2z = 5 \\ 5x - 7y - 4z = -1 \end{pmatrix}$$

can be found by the product

$$\begin{bmatrix} 2 & 3 & -1 \\ 3 & -1 & 2 \\ 5 & -7 & -4 \end{bmatrix}^{-1} \begin{bmatrix} 4 \\ 5 \\ -1 \end{bmatrix}$$

The solution set of a system of linear inequalities is the intersection of the solution sets of the individual inequalities. Such solution sets are easily determined by the graphing approach.

Linear programming problems deal with the idea of maximizing or minimizing a certain linear function that is subject to various constraints. The constraints are expressed as linear inequalities. Example 4 and Problem 1 of Section 7.4 are a good summary of the general approach to linear programming problems in this chapter.

Chapter 7   Review Problem Set

For Problems 1–10, compute the indicated matrix, if it exists, using the following matrices:

$$A = \begin{bmatrix} 2 & -4 \\ -3 & 8 \end{bmatrix} \quad B = \begin{bmatrix} 5 & -1 \\ 0 & 2 \end{bmatrix}$$

$$C = \begin{bmatrix} 3 & -1 \\ -2 & 4 \\ 5 & -6 \end{bmatrix} \quad D = \begin{bmatrix} -2 & -1 & 4 \\ 5 & 0 & -3 \end{bmatrix}$$

$$E = \begin{bmatrix} 1 \\ -3 \\ -7 \end{bmatrix} \quad F = \begin{bmatrix} 1 & -2 \\ 4 & -4 \\ 7 & -8 \end{bmatrix}$$

1. $A + B$

2. $B - A$

3. $C - F$

4. $2A + 3B$

5. $3C - 2F$

6. CD

7. DC

8. $DC + AB$

9. DE

10. EF

11. Use A and B from the preceding problems and show that $AB \neq BA$.

12. Use C, D, and F from the preceding problems and show that $D(C + F) = DC + DF$.

13. Use C, D, and F from the preceding problems and show that $(C + F)D = CD + FD$.

For each matrix in Problems 14–23, find the multiplicative inverse, if it exists.

14. $\begin{bmatrix} 9 & 5 \\ 7 & 4 \end{bmatrix}$

15. $\begin{bmatrix} 9 & 4 \\ 7 & 3 \end{bmatrix}$

16. $\begin{bmatrix} -2 & 1 \\ 2 & 3 \end{bmatrix}$

17. $\begin{bmatrix} 4 & -6 \\ 2 & -3 \end{bmatrix}$

18. $\begin{bmatrix} -1 & -3 \\ -4 & -5 \end{bmatrix}$

19. $\begin{bmatrix} 0 & -3 \\ 7 & 6 \end{bmatrix}$

20. $\begin{bmatrix} 1 & -2 & 1 \\ 2 & -5 & 2 \\ -3 & 7 & 5 \end{bmatrix}$

21. $\begin{bmatrix} 1 & 3 & -2 \\ 4 & 13 & -7 \\ 5 & 16 & -8 \end{bmatrix}$

22. $\begin{bmatrix} -2 & 4 & 7 \\ 1 & -3 & 5 \\ 1 & -5 & 22 \end{bmatrix}$

23. $\begin{bmatrix} -1 & 2 & 3 \\ 2 & -5 & -7 \\ -3 & 5 & 11 \end{bmatrix}$

For Problems 24–28, use the multiplicative inverse matrix approach to solve each system. The required inverses were found in Problems 14–23.

24. $\begin{pmatrix} 9x + 5y = 12 \\ 7x + 4y = 10 \end{pmatrix}$

25. $\begin{pmatrix} -2x + y = -9 \\ 2x + 3y = 5 \end{pmatrix}$

26. $\begin{pmatrix} x - 2y + z = 7 \\ 2x - 5y + 2z = 17 \\ -3x + 7y + 5z = -32 \end{pmatrix}$

27. $\begin{pmatrix} x + 3y - 2z = -7 \\ 4x + 13y - 7z = -21 \\ 5x + 16y - 8z = -23 \end{pmatrix}$

28. $\begin{pmatrix} -x + 2y + 3z = 22 \\ 2x - 5y - 7z = -51 \\ -3x + 5y + 11z = 71 \end{pmatrix}$

For Problems 29–32, indicate the solution set for each system of linear inequalities by graphing the system and shading the appropriate region.

29. $\begin{pmatrix} x - y \geq 0 \\ x + y \leq 0 \end{pmatrix}$

30. $\begin{pmatrix} 3x - 2y < 6 \\ 2x - 3y < 6 \end{pmatrix}$

31. $\begin{pmatrix} x - 4y < 4 \\ 2x + y \geq 2 \end{pmatrix}$

32. $\begin{pmatrix} x \geq 0 \\ y \geq 0 \\ x + 2y \leq 4 \\ 2x - y \leq 4 \end{pmatrix}$

33. Maximize the function $f(x, y) = 8x + 5y$ in the region determined by the following constraints:

$$y \leq 4x$$

$$x + y \leq 5$$

$$x \geq 0$$

$$y \geq 0$$

$$x \leq 4$$

34. Maximize the function $f(x, y) = 2x + 7y$ in the region determined by the following constraints:

$$x \geq 0$$
$$y \geq 0$$
$$x + 2y \leq 16$$
$$x + y \leq 9$$
$$3x + 2y \leq 24$$

35. Maximize the function $f(x, y) = 7x + 5y$ in the region determined by the constraints of Problem 34.

36. Maximize the function $f(x, y) = 150x + 200y$ in the region determined by the constraints of Problem 34.

37. A manufacturer of electric ice cream freezers makes a profit of \$4.50 on a one-gallon freezer and a profit of \$5.25 on a two-gallon freezer. The company wants to produce at least 75 one-gallon and at least 100 two-gallon freezers per week. However, the weekly production is not to exceed a total of 250 freezers. How many freezers of each type should be produced per week in order to maximize the profit?

For Problems 1–10, compute the indicated matrix, if it exists, using the following matrices.

$$A = \begin{bmatrix} -1 & 3 \\ 4 & -2 \end{bmatrix} \qquad B = \begin{bmatrix} 3 & -2 \\ 4 & -1 \end{bmatrix}$$

$$C = \begin{bmatrix} -3 \\ 5 \\ -6 \end{bmatrix} \qquad D = \begin{bmatrix} 2 & -1 \\ 3 & -2 \\ 6 & 5 \end{bmatrix}$$

$$E = \begin{bmatrix} 2 & -1 & 4 \\ 5 & 1 & -3 \end{bmatrix} \qquad F = \begin{bmatrix} -1 & 6 \\ 2 & -5 \\ 3 & 4 \end{bmatrix}$$

1. AB

2. BA

3. DE

4. BC

5. EC

6. $2A - B$

7. $3D + 2F$

8. $-3A - 2B$

9. EF

10. $AB - EF$

For Problems 11–16, find the multiplicative inverse, if it exists.

11. $\begin{bmatrix} 3 & -2 \\ 5 & -3 \end{bmatrix}$

12. $\begin{bmatrix} -2 & 5 \\ 3 & -7 \end{bmatrix}$

13. $\begin{bmatrix} 1 & -3 \\ -2 & 8 \end{bmatrix}$

14. $\begin{bmatrix} 3 & 5 \\ 1 & 4 \end{bmatrix}$

15. $\begin{bmatrix} -2 & 2 & 3 \\ 1 & -1 & 0 \\ 0 & 1 & 4 \end{bmatrix}$

16. $\begin{bmatrix} 1 & -2 & 4 \\ 0 & 1 & 3 \\ 0 & 0 & 1 \end{bmatrix}$

For Problems 17–19, use the multiplicative inverse matrix approach to solve each system.

17. $\begin{pmatrix} 3x - 2y = 48 \\ 5x - 3y = 76 \end{pmatrix}$

18. $\begin{pmatrix} x - 3y = 36 \\ -2x + 8y = -100 \end{pmatrix}$

19. $\begin{pmatrix} 3x + 5y = 92 \\ x + 4y = 61 \end{pmatrix}$

20. Solve the system

$$\begin{pmatrix} -x + 3y + z = 1 \\ 2x + 5y = 3 \\ 3x + y - 2z = -2 \end{pmatrix}$$

where the inverse of the coefficient matrix is

$$\begin{bmatrix} -\dfrac{10}{9} & \dfrac{7}{9} & -\dfrac{5}{9} \\ \dfrac{4}{9} & -\dfrac{1}{9} & \dfrac{2}{9} \\ -\dfrac{13}{9} & \dfrac{10}{9} & -\dfrac{11}{9} \end{bmatrix}$$

21. Solve the system

$$\begin{pmatrix} x + y + 2z = 3 \\ 2x + 3y - z = 3 \\ -3x + y - 2z = 3 \end{pmatrix}$$

where the inverse of the coefficient matrix is

$$\begin{bmatrix} -\dfrac{5}{24} & \dfrac{1}{6} & -\dfrac{7}{24} \\ \dfrac{7}{24} & \dfrac{1}{6} & \dfrac{5}{24} \\ \dfrac{11}{24} & -\dfrac{1}{6} & \dfrac{1}{24} \end{bmatrix}$$

For Problems 22–24, indicate the solution set for each system of inequalities by graphing the system and shading the appropriate region.

22. $\begin{pmatrix} 2x - y > 4 \\ x + 3y < 3 \end{pmatrix}$

23. $\begin{pmatrix} 2x - 3y \leq 6 \\ x + 4y > 4 \end{pmatrix}$

24. $\begin{pmatrix} y \leq 2x - 2 \\ y \geq x + 1 \end{pmatrix}$

25. Maximize the function $f(x, y) = 500x + 350y$ in the region determined by the following constraints:

$$3x + 2y \leq 24$$

$$x + 2y \leq 16$$

$$x + y \leq 9$$

$$x \geq 0$$

$$y \geq 0$$

Conic Sections

Parabolic surfaces are used in the construction of satellite dishes.

© Imagebroker/Alamy

Parabolas, circles, ellipses, and hyperbolas can be formed by causing a right circular conical surface and a plane to intersect, as shown in Figure 8.1; these figures are often referred to as **conic sections.** The conic sections are not new to you. You graphed some circles, parabolas, ellipses, and hyperbolas in Chapters 2 and 3. At that time, however, except for the circle, we did not present any formal definitions or standard forms of equations. In Chapter 2 we developed the standard form for the equation of a circle, $(x - h)^2 + (y - k)^2 = r^2$. We used this equation to solve a variety of problems that pertain to circles. It is now time to study the other conic sections in the same manner. We will define each conic section and derive the standard form of an equation. Then we will use the standard forms to study specific conic sections.

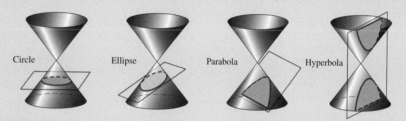

Figure 8.1

The conic sections have a wide variety of practical applications. Some of these will be presented in this chapter, but you may also find it interesting to search the Internet for additional applications.

8.1   Parabolas

Objectives

- Find the vertex, focus, and directrix of a parabola.

- Sketch the graph of a parabola.

- Write the equation of a parabola in standard form.

- Determine the equation of a parabola.

- Solve application problems involving parabolas.

We discussed parabolas as the graphs of quadratic functions in Sections 3.3 and 3.4. All parabolas in those sections had vertical lines as axes of symmetry. Furthermore, we did not state the definition for a parabola at that time. We shall now define a parabola and derive standard forms of equations for those that have either vertical or horizontal axes of symmetry.

Definition 8.1

> A **parabola** is the set of all points in a plane such that the distance of each point from a fixed point F (the **focus**) is equal to its distance from a fixed line d (the **directrix**) in the plane.

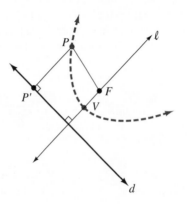

Figure 8.2

Using Definition 8.1, we can sketch a parabola by starting with a fixed line d and a fixed point F, not on d. Then a point P is on the parabola if and only if $PF = PP'$, where $\overline{PP'}$ is perpendicular to the directrix d (see Figure 8.2). The dashed curved line in Figure 8.2 indicates the possible positions of P; it is the parabola. The line l, through F and perpendicular to the directrix d, is called the **axis of symmetry.** The point V, on the axis of symmetry halfway from F to the directrix d, is the **vertex** of the parabola.

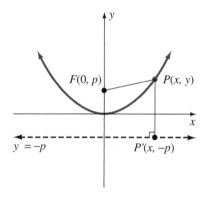

Figure 8.3

We can derive a standard form for the equation of a parabola by super-imposing coordinates on the plane such that the origin is at the vertex of the parabola and the y axis is the axis of symmetry, as in Figure 8.3. If the focus is at $(0, p)$, where $p \neq 0$, then the equation of the directrix is $y = -p$. There-fore, for any point P on the parabola, $PF = PP'$, and using the distance formula yields

$$\sqrt{(x - 0)^2 + (y - p)^2} = \sqrt{(x - x)^2 + (y + p)^2}$$

Squaring both sides and simplifying, we obtain

$$(x - 0)^2 + (y - p)^2 = (x - x)^2 + (y + p)^2$$

$$x^2 + y^2 - 2py + p^2 = y^2 + 2py + p^2$$

$$x^2 = 4py$$

Thus the **standard form for the equation of a parabola,** with its vertex at the origin and the y axis as its axis of symmetry, is

$$x^2 = 4py$$

If $p > 0$, the parabola opens upward; if $p < 0$, the parabola opens downward.

A line segment that contains the focus and has endpoints on the parabola is called a **focal chord**. The specific focal chord that is parallel to the directrix is called the **primary focal chord;** this is line segment $\overline{QP}$ in Figure 8.4. Because $FP = PP' = |2p|$, the entire length of the primary focal chord is $|4p|$ units. You will see how we can use this fact when graphing parabolas.

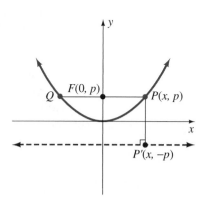

Figure 8.4

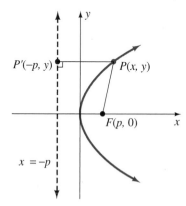

Figure 8.5

In a similar fashion, we can develop the standard form for the equation of a parabola with its vertex at the origin and the x axis as its axis of symmetry. By choosing a focus at $F(p, 0)$ and a directrix with an equation of $x = -p$ (see Figure 8.5) and by applying the definition of a parabola, we obtain the standard form for the equation:

$$y^2 = 4px$$

If $p > 0$, the parabola opens to the right, as in Figure 8.5; if $p < 0$, the parabola opens to the left.

The concept of symmetry can be used to decide which of the two equations, $x^2 = 4py$ or $y^2 = 4px$, is to be used. The graph of $x^2 = 4py$ is symmetric with respect to the y axis, because replacing x with $-x$ does not change the equation. Likewise, the graph of $y^2 = 4px$ is symmetric with respect to the x axis because replacing y with $-y$ leaves the equation unchanged. The following property summarizes our previous discussion.

Property 8.1

The graph of each of the following equations is a parabola that has its vertex at the origin and has the indicated focus, directrix, and symmetry.

1. $x^2 = 4py$ focus $(0, p)$, directrix $y = -p$, y axis symmetry

2. $y^2 = 4px$ focus $(p, 0)$, directrix $x = -p$, x axis symmetry

Now let's illustrate some uses of the equations $x^2 = 4py$ and $y^2 = 4px$.

EXAMPLE 1

Find the focus and directrix of the parabola $x^2 = -8y$, and sketch its graph.

Solution

If we compare $x^2 = -8y$ to the standard form $x^2 = 4py$, we get $4p = -8$. Therefore $p = -2$, and the parabola opens downward. The focus is at $(0, -2)$, and the equation of the directrix is $y = -(-2) = 2$. The primary focal chord is $|4p| = |-8| = 8$ units long. Therefore, the endpoints of the primary focal chord are at $(4, -2)$ and $(-4, -2)$. The graph is sketched in Figure 8.6.

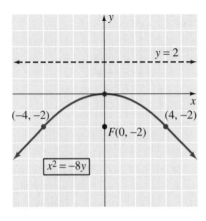

Figure 8.6

▶ **Now work Problem 3.** ∎

EXAMPLE 2

Write the equation of the parabola that is symmetric with respect to the y axis, has its vertex at the origin, and contains the point $P(6, 3)$.

Solution

The standard form of the parabola is $x^2 = 4py$. Because P is on the parabola, the ordered pair $(6, 3)$ must satisfy the equation. Therefore

$$6^2 = 4p(3)$$
$$36 = 12p$$
$$3 = p$$

If $p = 3$, the equation becomes

$$x^2 = 4(3)y$$
$$x^2 = 12y$$

▶ **Now work Problem 33.** ∎

EXAMPLE 3

Find the focus and directrix of the parabola $y^2 = 6x$ and sketch its graph.

Solution

If we compare $y^2 = 6x$ to the standard form $y^2 = 4px$, we see that $4p = 6$ and therefore $p = \dfrac{3}{2}$. Thus the focus is at $\left(\dfrac{3}{2}, 0\right)$, and the equation of the directrix is $x = -\dfrac{3}{2}$. The parabola opens to the right. The primary focal chord is $|4p| = |6| = 6$ units long. Thus the endpoints of the primary focal chord are at $\left(\dfrac{3}{2}, 3\right)$ and $\left(\dfrac{3}{2}, -3\right)$. The graph is sketched in Figure 8.7.

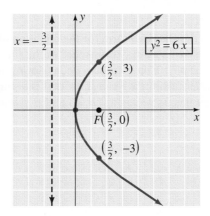

Figure 8.7

▶ **Now work Problem 1.** ∎

■ Other Parabolas

In much the same way, we can develop the standard form for an equation of a parabola that is symmetric with respect to a line parallel to a coordinate axis. In Figure 8.8 we have taken the vertex V at (h, k) and the focus F at $(h, k + p)$; the equation of the directrix is $y = k - p$. By the definition of a parabola, we know that $FP = PP'$. Therefore, applying the distance formula, we obtain

$$\sqrt{(x - h)^2 + (y - (k + p))^2} = \sqrt{(x - x)^2 + [y - (k - p)]^2}$$

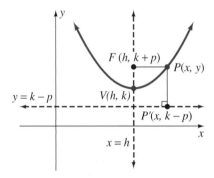

Figure 8.8

We leave it to you to show that this equation simplifies to

$$(x - h)^2 = 4p(y - k)$$

which is called the **standard form for the equation of a parabola** that has its vertex at (h, k) and is symmetric with respect to the line $x = h$. If $p > 0$, the parabola opens upward; if $p < 0$, the parabola opens downward.

In a similar fashion, we can show that the **standard form for the equation of a parabola** that has its vertex at (h, k) and is symmetric with respect to the line $y = k$ is

$$(y - k)^2 = 4p(x - h)$$

If $p > 0$, the parabola opens to the right; if $p < 0$, it opens to the left.

Let's summarize our discussion of parabolas that have lines of symmetry parallel to the x axis or y axis by stating the following property:

Property 8.2

The graph of each of the following equations is a parabola that has its vertex at (h, k) and has the indicated focus, directrix, and symmetry:

1. $(x - h)^2 = 4p(y - k)$ focus $(h, k + p)$, directrix $y = k - p$,
line of symmetry $x = h$

2. $(y - k)^2 = 4p(x - h)$ focus $(h + p, k)$, directrix $x = h - p$,
line of symmetry $y = k$

E X A M P L E   4

Find the vertex, focus, and directrix of the parabola $y^2 + 4y - 4x + 16 = 0$ and sketch its graph.

Solution

We write the given equation as $y^2 + 4y = 4x - 16$, and we can complete the square on the left side by adding 4 to both sides:

$$y^2 + 4y + 4 = 4x - 16 + 4$$
$$(y + 2)^2 = 4x - 12$$
$$(y + 2)^2 = 4(x - 3)$$

Now let's compare this final equation to the form $(y - k)^2 = 4p(x - h)$:

$$(y - (-2))^2 = 4(x - 3)$$

$$k = -2 \qquad 4p = 4 \qquad h = 3$$
$$p = 1$$

The vertex is at $(3, -2)$, and because $p > 0$, the parabola opens to the right, and the focus is at $(4, -2)$. The equation of the directrix is $x = 2$. The primary focal chord is $|4p| = |4| = 4$ units long, and its endpoints are at $(4, 0)$ and $(4, -4)$. The graph is shown in Figure 8.9.

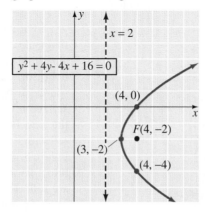

Figure 8.9

▶ **Now work Problem 21.** ∎

Remark: If we were using a graphing calculator to graph the parabola in Example 4, then after the step $(y + 2)^2 = 4x - 12$, we would solve for y to obtain $y = -2 \pm \sqrt{4x - 12}$. Then we could enter the two functions $Y_1 = -2 + \sqrt{4x - 12}$ and $Y_2 = -2 - \sqrt{4x - 12}$ and obtain a figure that closely resembles Figure 8.9. (You are asked to do this in the next Graphing Calculator Activities.) Some computer programs can graph the equation in Example 4 without changing its form.

EXAMPLE 5

Write the equation of the parabola if its focus is at $(-4, 1)$, and the equation of its directrix is $y = 5$.

Solution

Because the directrix is a horizontal line, we know that the equation of the parabola is of the form $(x - h)^2 = 4p(y - k)$. The vertex is halfway between the focus and the directrix, so the vertex is at $(-4, 3)$. This means that $h = -4$ and $k = 3$. The parabola opens downward because the focus is below the directrix, and the distance between the focus and the vertex is 2 units; thus $p = -2$. We substitute -4 for h, 3 for k, and -2 for p in the equation $(x - h)^2 = 4p(y - k)$ to obtain

$$(x - (-4))^2 = 4(-2)(y - 3)$$

which simplifies to

$$(x + 4)^2 = -8(y - 3)$$
$$x^2 + 8x + 16 = -8y + 24$$
$$x^2 + 8x + 8y - 8 = 0$$

▶ **Now work Problem 29.**   ∎

Remark: For a problem such as Example 5, you may find it helpful to put the given information on a set of axes and draw a rough sketch of the parabola to help you with the analysis of the problem.

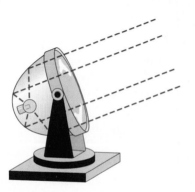

Parabolas possess various properties that make them very useful. For example, if a parabola is rotated about its axis, a parabolic surface is formed. The rays from a source of light placed at the focus of this surface reflect from the surface parallel to the axis. It is for this reason that parabolic reflectors are used on searchlights as in Figure 8.10. Likewise, rays of light coming into a parabolic surface parallel to the axis are reflected through the focus. This property of parabolas is useful in the design of mirrors for telescopes (see Figure 8.11) and in the construction of radar antennas.

Figure 8.10

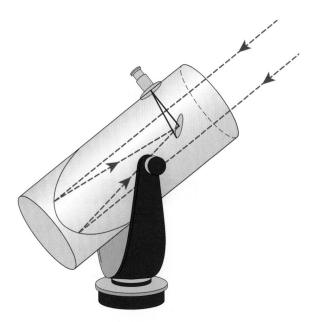

Figure 8.11

Cables hung between structures supporting suspension bridges, such as the Golden Gate Bridge in San Francisco, form parabolas. A bullet fired into the air follows the curvature of a parabola if air resistance and other outside factors are ignored — in other words, if only the force of gravity is considered (see Figure 8.12).

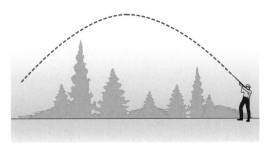

Figure 8.12

CONCEPT QUIZ

For Problems 1–8, answer true or false.

1. For a parabola, the axis of symmetry is parallel to the directrix.

2. For the parabola whose equation is $x^2 = 8y$, the length of the primary focal chord is 8 units.

3. The graph of $y^2 = -4x$ is symmetric with respect to the x axis.

4. For a parabola whose equation is $y^2 = 3x$, the directrix is the point $\left(\frac{4}{3}, 0\right)$.

5. The parabola whose equation is $x^2 = -6y$ opens downward.

6. For the graph of a parabola whose equation is $(x - 4)^2 = 8(y - 1)$, the vertex is located at $(-4, 1)$.

7. The vertex of a parabola always lies on the axis of symmetry.

8. A parabola has an infinite number of focal chords but only one primary focal chord.

Problem Set 8.1

For Problems 1–22, find the vertex, focus, and directrix of the given parabola, and sketch its graph.

1. $y^2 = 8x$

2. $y^2 = -4x$

3. $x^2 = -12y$

4. $x^2 = 8y$

5. $y^2 = -2x$

6. $y^2 = 6x$

7. $x^2 = 6y$

8. $x^2 = -7y$

9. $x^2 - 4y + 8 = 0$

10. $x^2 - 8y - 24 = 0$

11. $x^2 + 8y + 16 = 0$

12. $x^2 + 4y - 4 = 0$

13. $y^2 - 12x + 24 = 0$

14. $y^2 + 8x - 24 = 0$

15. $x^2 - 2x - 4y + 9 = 0$

16. $x^2 + 4x - 8y - 4 = 0$

17. $x^2 + 6x + 8y + 1 = 0$

18. $x^2 - 4x + 4y - 4 = 0$

19. $y^2 - 2y + 12x - 35 = 0$

20. $y^2 + 4y + 8x - 4 = 0$

21. $y^2 + 6y - 4x + 1 = 0$

22. $y^2 - 6y - 12x + 21 = 0$

For Problems 23–42, find an equation of the parabola that satisfies the given conditions.

23. Focus $(0, 3)$, directrix $y = -3$

24. Focus $\left(0, -\dfrac{1}{2}\right)$, directrix $y = \dfrac{1}{2}$

25. Focus $(-1, 0)$, directrix $x = 1$

26. Focus $(5, 0)$, directrix $x = 1$

27. Focus $(0, 1)$, directrix $y = 7$

28. Focus $(0, -2)$, directrix $y = -10$

29. Focus $(3, 4)$, directrix $y = -2$

30. Focus $(-3, -1)$, directrix $y = 7$

31. Focus $(-4, 5)$, directrix $x = 0$

32. Focus $(5, -2)$, directrix $x = -1$

33. Vertex $(0, 0)$, symmetric with respect to the y axis, and contains the point $(-2, -4)$

34. Vertex $(0, 0)$, symmetric with respect to the x axis, and contains the point $(-3, 5)$

35. Vertex $(0, 0)$, focus $\left(\dfrac{5}{2}, 0\right)$

36. Vertex $(0, 0)$, focus $\left(0, -\dfrac{7}{2}\right)$

37. Vertex $(7, 3)$, focus $(7, 5)$, and symmetric with respect to the line $x = 7$

38. Vertex $(-4, -6)$, focus $(-7, -6)$, and symmetric with respect to the line $y = -6$

39. Vertex $(8, -3)$, focus $(11, -3)$, and symmetric with respect to the line $y = -3$

40. Vertex $(-2, 9)$, focus $(-2, 5)$, and symmetric with respect to the line $x = -2$

41. Vertex $(-9, 1)$, symmetric with respect to the line $x = -9$, and contains the point $(-8, 0)$

42. Vertex $(6, -4)$, symmetric with respect to the line $y = -4$, and contains the point $(8, -3)$

For Problems 43–47, solve each problem.

43. One section of a suspension bridge hangs between two towers that are 40 feet above the surface and 300 feet apart, as in Figure 8.13. A cable strung between the tops of the two towers is in the shape of a parabola with its vertex 10 feet above the surface. With axes drawn as indicated in the figure, find the equation of the parabola.

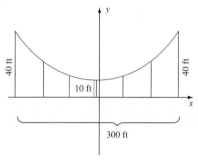

Figure 8.13

44. Suppose that five equally spaced vertical cables are used to support the bridge in Figure 8.13. Find the total length of these supports.

45. Suppose that an arch is shaped like a parabola. It is 20 feet wide at the base and 100 feet high. How wide is the arch 50 feet above the ground?

46. A parabolic arch 27 feet high spans a parkway. How wide is the arch if the center section of the parkway, a section that is 50 feet wide, has a minimum clearance of 15 feet?

47. A parabolic arch spans a stream 200 feet wide. How high must the arch be above the stream to give a minimum clearance of 40 feet over a channel in the center that is 120 feet wide?

■ ■ ■ **THOUGHTS INTO WORDS**

48. Give a step-by-step description of how you would go about graphing the parabola $x^2 - 2x - 4y - 7 = 0$.

49. Suppose that someone graphed the equation $y^2 - 6y - 2x + 11 = 0$ and obtained the graph in Figure 8.14. How do you know by looking at the equation that this graph is incorrect?

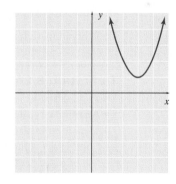

Figure 8.14

GRAPHING CALCULATOR ACTIVITIES

50. The parabola determined by the equation $x^2 + 4x - 8y - 4 = 0$ (Problem 16) is easy to graph using a graphing calculator because it can be expressed as a function of x without much computation. Let's solve the equation for y.

$$8y = x^2 + 4x - 4$$

$$y = \frac{x^2 + 4x - 4}{8}$$

Use your graphing calculator to graph this function.

As noted in the Remark following Example 4, solving the equation $y^2 + 4y - 4x + 16 = 0$ for y produces two functions; namely, $Y_1 = -2 + \sqrt{4x - 12}$ and $Y_2 = -2 - \sqrt{4x - 12}$. Graph these two functions on the same set of axes. Your result should resemble Figure 8.9.

Use your graphing calculator to check your graphs for Problems 1–22.

Answers to the Concept Quiz

1. False   **2.** True   **3.** True   **4.** False   **5.** True   **6.** False   **7.** True   **8.** True

8.2   Ellipses

Objectives

- Find the vertices, endpoints of the minor axis, and the foci of an ellipse.

- Sketch the graph of an ellipse.

- Write the equation of an ellipse in standard form.

- Determine the equation of an ellipse.

- Solve application problems involving ellipses.

Let's begin by defining an ellipse.

Definition 8.2

> An **ellipse** is the set of all points in a plane such that the sum of the distances of each point from two fixed points F and F' (the **foci**) in the plane is constant. (*Foci* is the plural of *focus*.)

Using two thumbtacks, a piece of string, and a pencil, you can easily draw an ellipse by satisfying the conditions of Definition 8.2. First, insert two thumbtacks in a piece of cardboard at points F and F', and fasten the ends of the piece of string to the thumbtacks, as in Figure 8.15. Then loop the string around the point of a pencil and hold the pencil so that the string is taut. Finally, move the pencil around the tacks, always keeping the string taut: You will draw an ellipse. The two points F and F' are the foci referred to in Definition 8.2, and the sum of the distances FP and $F'P$ is constant because it represents the length of the piece of string. With the same piece of string, you can vary the shape of the ellipse by changing the positions of the foci. Moving F and F' farther apart will make the ellipse flatter. Likewise, moving F and F' closer together will cause the ellipse to resemble a circle. In fact, if $F = F'$, you will obtain a circle.

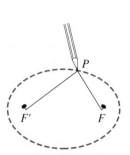

Figure 8.15   **Figure 8.16**

   We can derive a standard form for the equation of an ellipse by superimposing coordinates on the plane such that the foci are on the x axis, equidistant from the origin, as in Figure 8.16. If F has coordinates $(c, 0)$, where $c > 0$, then F' has coordinates $(-c, 0)$, and the distance between F and F' is $2c$ units. We will let $2a$ represent the constant sum of $FP + F'P$. Note that $2a > 2c$ and therefore $a > c$. For any point P on the ellipse,

$$FP + F'P = 2a$$

We can use the distance formula to write this as

$$\sqrt{(x - c)^2 + (y - 0)^2} + \sqrt{(x + c)^2 + (y - 0)^2} = 2a$$

Let's change the form of this equation to

$$\sqrt{(x - c)^2 + y^2} = 2a - \sqrt{(x + c)^2 + y^2}$$

and square both sides:

$$(x - c)^2 + y^2 = 4a^2 - 4a\sqrt{(x + c)^2 + y^2} + (x + c)^2 + y^2$$

This can be simplified to

$$a^2 + cx = a\sqrt{(x + c)^2 + y^2}$$

Again we square both sides to produce

$$a^4 + 2a^2cx + c^2x^2 = a^2[(x + c)^2 + y^2]$$

which can be written in the form

$$x^2(a^2 - c^2) + a^2y^2 = a^2(a^2 - c^2)$$

Dividing both sides by $a^2(a^2 - c^2)$ leads to the form

$$\frac{x^2}{a^2} + \frac{y^2}{a^2 - c^2} = 1$$

Letting $b^2 = a^2 - c^2$, where $b > 0$, produces the equation

$$\frac{x^2}{a^2} + \frac{y^2}{b^2} = 1 \tag{1}$$

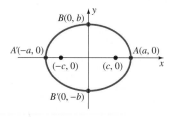

Figure 8.17

Because $c > 0$, $a > c$, and $b^2 = a^2 - c^2$, it follows that $a^2 > b^2$ and hence $a > b$. This equation that we have derived is called the **standard form for the equation of an ellipse** with its foci on the x axis and its center at the origin.

The x intercepts of equation (1) can be found by letting $y = 0$. Doing this produces $x^2/a^2 = 1$, or $x^2 = a^2$; consequently, the x intercepts are a and $-a$. The corresponding points on the graph (see Figure 8.17) are $A(a, 0)$ and $A'(-a, 0)$, and the line segment $\overline{A'A}$, which is of length $2a$, is called the **major axis** of the ellipse. The endpoints of the major axis are also referred to as the **vertices** of the ellipse. Similarly, letting $x = 0$ produces $y^2/b^2 = 1$, or $y^2 = b^2$; consequently, the y intercepts are b and $-b$. The corresponding points on the graph are $B(0, b)$ and $B'(0, -b)$, and the line segment $\overline{BB'}$, which is of length $2b$, is called the **minor axis.** Because $a > b$, **the major axis is always longer than the minor axis.** The point of intersection of the major and minor axes is called the **center** of the ellipse.

Let's summarize this discussion by stating the following property.

Property 8.3

The graph of the equation

$$\frac{x^2}{a^2} + \frac{y^2}{b^2} = 1$$

for $a^2 > b^2$ is an ellipse with the endpoints of its major axis (the vertices) at $(a, 0)$ and $(-a, 0)$ and the endpoints of its minor axis at $(0, b)$ and $(0, -b)$. The foci are at $(c, 0)$ and $(-c, 0)$, where $c^2 = a^2 - b^2$.

Note that replacing y with $-y$, or x with $-x$, or both x and y with $-x$ and $-y$, leaves the equation unchanged. Thus the graph of

$$\frac{x^2}{a^2} + \frac{y^2}{b^2} = 1$$

is symmetric with respect to the x axis, the y axis, and the origin.

E X A M P L E   1

Find the vertices, the endpoints of the minor axis, and the foci of the ellipse $4x^2 + 9y^2 = 36$, and sketch the ellipse.

Solution

The given equation can be changed to standard form by dividing both sides by 36:

$$\frac{4x^2}{36} + \frac{9y^2}{36} = \frac{36}{36}$$

$$\frac{x^2}{9} + \frac{y^2}{4} = 1$$

Therefore $a^2 = 9$ and $b^2 = 4$; hence the vertices are at $(3, 0)$ and $(-3, 0)$, and the endpoints of the minor axis are at $(0, 2)$ and $(0, -2)$. Because $c^2 = a^2 - b^2$, we have

$$c^2 = 9 - 4 = 5$$

Thus the foci are at $(\sqrt{5}, 0)$ and $(\sqrt{-5}, 0)$. The ellipse is sketched in Figure 8.18.

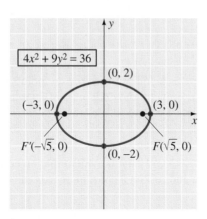

Figure 8.18

▶ **Now work Problem 7.**   ■

Remark: For a problem such as Example 1, it is not necessary to change to standard form to find the values for a and b. After all, $\pm a$ are the x intercepts and $\pm b$ are the y intercepts. These values can be found quite easily from the given form of the equation.

E X A M P L E   2

Find the equation of the ellipse with vertices at $(\pm 6, 0)$ and foci at $(\pm 4, 0)$.

Solution

From the given information, we know that $a = 6$ and $c = 4$. Therefore

$$b^2 = a^2 - c^2 = 36 - 16 = 20$$

We can substitute 36 for a^2 and 20 for b^2 in the standard form to produce

$$\frac{x^2}{36} + \frac{y^2}{20} = 1$$

We multiply both sides by 180 to get

$$5x^2 + 9y^2 = 180$$

▶ **Now work Problem 23.**   ■

■ Ellipses with Foci on the *y* Axis

We can develop a standard form for the equation of an ellipse with foci on the y axis in a similar fashion. The following property summarizes the results of such a development with the foci at $(0, c)$ and $(0, -c)$, where $c > 0$.

Property 8.4

The graph of the equation

$$\frac{x^2}{b^2} + \frac{y^2}{a^2} = 1$$

where $a^2 > b^2$, is an ellipse with the endpoints of its major axis (vertices) at $(0, a)$ and $(0, -a)$ and the endpoints of its minor axis at $(b, 0)$ and $(-b, 0)$. The foci are at $(0, c)$ and $(0, -c)$, where $c^2 = a^2 - b^2$.

From Properties 8.3 and 8.4 it is evident that an equation of an ellipse with its center at the origin and foci on a coordinate axis can be written in the form

$$\frac{x^2}{p} + \frac{y^2}{q} = 1 \qquad \text{or} \qquad qx^2 + py^2 = pq$$

where p and q are positive. If $p > q$, the major axis lies on the x axis, and if $q > p$, the major axis is on the y axis. It is not necessary to memorize these facts because for any specific problem the endpoints of the major and minor axes are determined by the x and y intercepts. However, it is necessary to remember the relationship $c^2 = a^2 - b^2$.

EXAMPLE 3   Find the vertices, the endpoints of the minor axis, and the foci of the ellipse $18x^2 + 4y^2 = 36$, and sketch the ellipse.

Solution

To find the x intercepts, we let $y = 0$ and we obtain

$$18x^2 = 36$$
$$x^2 = 2$$
$$x = \pm\sqrt{2}$$

Similarly, to find the y intercepts, we let $x = 0$ and we obtain

$$4y^2 = 36$$
$$y^2 = 9$$
$$y = \pm 3$$

Because $3 > \sqrt{2}$, we know that $a = 3$ and $b = \sqrt{2}$. Therefore the vertices are at $(0, 3)$ and $(0, -3)$, and the endpoints of the minor axes are at $(\sqrt{2}, 0)$ and $(-\sqrt{2}, 0)$. From the relationship $c^2 = a^2 - b^2$, we get

$$c^2 = 9 - 2 = 7$$

Thus the foci are at $(0, \sqrt{7})$ and $(0, -\sqrt{7})$. The ellipse is sketched in Figure 8.19.

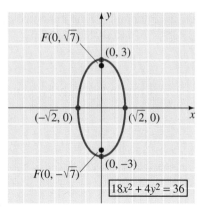

Figure 8.19

⊳   **Now work Problem 9.**   ∎

∎ Other Ellipses

In the same way, we can develop the standard form for an equation of an ellipse that is symmetric with respect to a line parallel to a coordinate axis. We will not show such developments in this text, but Figures 8.20 and 8.21 indicate the basic facts we need in order to develop and use the resulting equations. Note that in each case the center of the ellipse is at a point (h, k). Furthermore, the physical significance of a, b, and c is the same as before. However, these values are used relative to the center (h, k) to find the endpoints of the major and minor axes and to find the foci. Let's see how this works in a specific example.

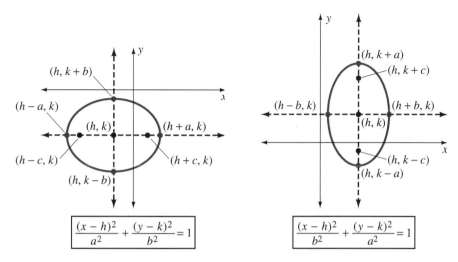

Figure 8.20   **Figure 8.21**

EXAMPLE 4

Find the vertices, the endpoints of the minor axis, and the foci of the ellipse $9x^2 + 54x + 4y^2 - 8y + 49 = 0$, and sketch the ellipse.

Solution

First, we need to change to standard form by completing the square on both x and y.

$$9(x^2 + 6x + \underline{}) + 4(y^2 - 2y + \underline{}) = -49$$

$$9(x^2 + 6x + 9) + 4(y^2 - 2y + 1) = -49 + 81 + 4$$

$$9(x + 3)^2 + 4(y - 1)^2 = 36$$

$$\frac{(x + 3)^2}{4} + \frac{(y - 1)^2}{9} = 1$$

Because $a > b$, this last equation is of the form

$$\frac{(x - h)^2}{b^2} + \frac{(y - k)^2}{a^2} = 1$$

where $h = -3$, $k = 1$, $a = 3$, and $b = 2$. Thus the endpoints of the major axis (vertices) are up three units and down three units from the center, $(-3, 1)$, so they are at $(-3, 4)$ and $(-3, -2)$. Likewise, the endpoints of the minor axis are two units to the right and two units to the left of the center. Thus they are at $(-1, 1)$ and $(-5, 1)$. From the relationship $c^2 = a^2 - b^2$, we get

$$c^2 = 9 - 4 = 5$$

Thus the foci are at $(-3, 1 + \sqrt{5})$ and $(-3, 1 - \sqrt{5})$. The ellipse is sketched in Figure 8.22.

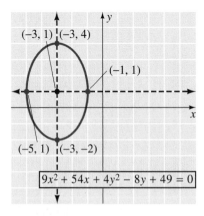

Figure 8.22

▶ **Now work Problem 15.**   ∎

E X A M P L E   5

Write the equation of the ellipse that has vertices at $(-3, -5)$ and $(7, -5)$ and foci at $(-1, -5)$ and $(5, -5)$.

Solution

Because the vertices and foci are on the same horizontal line ($y = -5$), this ellipse has an equation of the form

$$\frac{(x - h)^2}{a^2} + \frac{(y - k)^2}{b^2} = 1$$

The center of the ellipse is at the midpoint of the major axis. Therefore

$$h = \frac{-3 + 7}{2} = 2 \quad \text{and} \quad k = \frac{-5 + (-5)}{2} = -5$$

The distance between the center $(2, -5)$ and a vertex $(7, -5)$ is 5 units; thus $a = 5$. The distance between the center $(2, -5)$ and a focus $(5, -5)$ is 3 units; thus $c = 3$. Using the relationship $c^2 = a^2 - b^2$, we obtain

$$b^2 = a^2 - c^2 = 25 - 9 = 16$$

Now let's substitute 2 for h, -5 for k, 25 for a^2, and 16 for b^2 in the general form, and then we can simplify.

$$\frac{(x - 2)^2}{25} + \frac{(y + 5)^2}{16} = 1$$

$$16(x - 2)^2 + 25(y + 5)^2 = 400$$

$$16(x^2 - 4x + 4) + 25(y^2 + 10y + 25) = 400$$

$$16x^2 - 64x + 64 + 25y^2 + 250y + 625 = 400$$

$$16x^2 - 64x + 25y^2 + 250y + 289 = 0$$

⊳ **Now work Problem 33.**   ∎

Remark: Again, for a problem such as Example 5, it might be helpful to start by recording the given information on a set of axes and drawing a rough sketch of the figure.

Like parabolas, ellipses possess properties that make them very useful. For example, the elliptical surface formed by rotating an ellipse about its major axis has the following property: Light or sound waves emitted at one focus reflect off the surface and converge at the other focus. This is the principle behind "whispering galleries," such as the rotunda of the Capitol Building in Washington, D.C. In such buildings, two people standing at two specific spots that are the foci of the elliptical ceiling can whisper and yet hear each other clearly, even though they may be quite far apart.

Ellipses also play an important role in astronomy. Johannes Kepler (1571–1630) showed that the orbit of a planet is an ellipse with the sun at one focus. For example, the orbit of Earth is elliptical but nearly circular; at the same time, the moon moves about Earth in an elliptical path (see Figure 8.23).

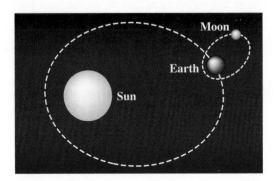

Figure 8.23

The arches for concrete bridges are sometimes elliptical. (One example is shown in Figure 8.25 in the next set of problems.) Also, elliptical gears are used in certain kinds of machinery that require a slow but powerful force at impact, such as a heavy-duty punch (see Figure 8.24).

Figure 8.24

CONCEPT QUIZ

For Problems 1–8, answer true or false.

1. For an ellipse, the endpoints of its major axis are called the vertices.

2. For an ellipse, the major axis is always longer than the minor axis.

3. The foci of an ellipse always lie on the major axis.

4. The point of intersection of the axes of an ellipse is called the vertex.

5. The graph of an ellipse, whose equation is $\dfrac{x^2}{25} + \dfrac{y^2}{9} = 1$, is symmetric to both the x and y axes.

6. For the ellipse whose equation $\dfrac{x^2}{16} + \dfrac{y^2}{25} = 1$, the endpoints of the minor axis are at $(4, 0)$ and $(-4, 0)$.

7. For the ellipse whose equation is $\dfrac{x^2}{16} + \dfrac{y^2}{25} = 1$, the foci are at $(3, 0)$ and $(-3, 0)$.

8. For the ellipse whose equation is $\dfrac{x^2}{49} + \dfrac{y^2}{36} = 1$, the length of the major axis is 7 units.

Problem Set 8.2

For Problems 1–22, find the vertices, the endpoints of the minor axis, and the foci of the given ellipse, and sketch its graph.

1. $\dfrac{x^2}{4} + \dfrac{y^2}{1} = 1$ **2.** $\dfrac{x^2}{16} + \dfrac{y^2}{1} = 1$

3. $\dfrac{x^2}{4} + \dfrac{y^2}{9} = 1$ **4.** $\dfrac{x^2}{4} + \dfrac{y^2}{16} = 1$

5. $9x^2 + 3y^2 = 27$ **6.** $4x^2 + 3y^2 = 36$

7. $2x^2 + 5y^2 = 50$ **8.** $5x^2 + 36y^2 = 180$

9. $12x^2 + y^2 = 36$ **10.** $8x^2 + y^2 = 16$

11. $7x^2 + 11y^2 = 77$ **12.** $4x^2 + y^2 = 12$

13. $4x^2 - 8x + 9y^2 - 36y + 4 = 0$

14. $x^2 + 6x + 9y^2 - 36y + 36 = 0$

15. $4x^2 + 16x + y^2 + 2y + 1 = 0$

16. $9x^2 - 36x + 4y^2 + 16y + 16 = 0$

17. $x^2 - 6x + 4y^2 + 5 = 0$

18. $16x^2 + 9y^2 + 36y - 108 = 0$

19. $9x^2 - 72x + 2y^2 + 4y + 128 = 0$

20. $5x^2 + 10x + 16y^2 + 160y + 325 = 0$

21. $2x^2 + 12x + 11y^2 - 88y + 172 = 0$

22. $9x^2 + 72x + y^2 + 6y + 135 = 0$

For Problems 23–36, find an equation of the ellipse that satisfies the given conditions.

23. Vertices $(\pm 5, 0)$, foci $(\pm 3, 0)$

24. Vertices $(\pm 4, 0)$, foci $(\pm 2, 0)$

25. Vertices $(0, \pm 6)$, foci $(0, \pm 5)$

26. Vertices $(0, \pm 3)$, foci $(0, \pm 2)$

27. Vertices $(\pm 3, 0)$, length of minor axis is 2

28. Vertices $(0, \pm 5)$, length of minor axis is 4

29. Foci $(0, \pm 2)$, length of minor axis is 3

30. Foci $(\pm 1, 0)$, length of minor axis is 2

31. Vertices $(0, \pm 5)$, contains the point $(3, 2)$

32. Vertices $(\pm 6, 0)$, contains the point $(5, 1)$

33. Vertices $(5, 1)$ and $(-3, 1)$, foci $(3, 1)$ and $(-1, 1)$

34. Vertices $(2, 4)$ and $(2, -6)$, foci $(2, 3)$ and $(2, -5)$

35. Center $(0, 1)$ one focus at $(-4, 1)$, length of minor axis is 6

36. Center $(3, 0)$, one focus at $(3, 2)$, length of minor axis is 4

For Problems 37–40, solve each problem.

37. Find an equation of the set of points in a plane such that the sum of the distances between each point of the set and the points $(2, 0)$ and $(-2, 0)$ is 8 units.

38. Find an equation of the set of points in a plane such that the sum of the distances between each point of the set and the points $(0, 3)$ and $(0, -3)$ is 10 units.

39. An arch of the bridge shown in Figure 8.25 (see page 614) is semielliptical and the major axis is horizontal. The arch is 30 feet wide and 10 feet high. Find the height of the arch 10 feet from the center of the base.

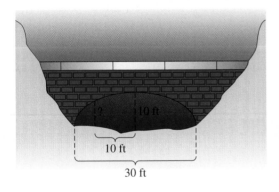

Figure 8.25

40. In Figure 8.25, how much clearance is there 10 feet from the bank?

■ ■ ■ **THOUGHTS INTO WORDS**

41. What type of figure is the graph of the equation $x^2 + 6x + 2y^2 - 20y + 59 = 0$? Explain your answer.

42. Suppose that someone graphed the equation $4x^2 - 16x + 9y^2 + 18y - 11 = 0$ and obtained the graph shown in Figure 8.26. How do you know by looking at the equation that this is an incorrect graph?

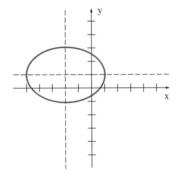

Figure 8.26

GRAPHING CALCULATOR ACTIVITIES

43. Use your graphing calculator to check your graphs for Problems 13–22.

44. Use your graphing calculator to graph each of the following ellipses.

    a. $2x^2 - 40x + y^2 + 2y + 185 = 0$
    b. $x^2 - 4x + 2y^2 - 48y + 272 = 0$
    c. $4x^2 - 8x + y^2 - 4y - 136 = 0$
    d. $x^2 + 6x + 2y^2 + 56y + 301 = 0$

Answers to the Concept Quiz

1. True   **2.** True   **3.** True   **4.** False   **5.** True   **6.** True   **7.** False   **8.** False

8.3 **Hyperbolas**

Objectives

- Find the vertices, the foci, and the equations of the asymptotes for a hyperbola.

- Sketch the graph of a hyperbola.

- Write the equation of a hyperbola in standard form.

- Solve application problems involving hyperbolas.

A hyperbola and an ellipse are similar by definition; however, an ellipse involves the *sum* of distances and a hyperbola involves the *difference* of distances.

Definition 8.3

A **hyperbola** is the set of all points in a plane such that the difference of the distances of each point from two fixed points F and F' (the **foci**) in the plane is a positive constant.

Using Definition 8.3, we can sketch a hyperbola by starting with two fixed points F and F', as shown in Figure 8.27. Then we locate all points P such that $PF' - PF$ is a positive constant. Likewise, as shown in Figure 8.27, all points Q are located such that $QF - QF'$ is the same positive constant. The two dashed curved lines in Figure 8.27 make up the hyperbola. The two curves are sometimes referred to as the *branches* of the hyperbola.

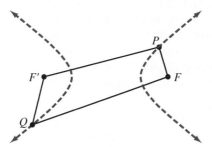

Figure 8.27

To develop a standard form for the equation of a hyperbola, let's superimpose coordinates on the plane such that the foci are located at $F(c, 0)$ and $F'(-c, 0)$, as indicated in Figure 8.28. Using the distance formula and setting $2a$ equal to the difference of the distances from any point P on the hyperbola to the foci, we have the following equation:

$$\left| \sqrt{(x - c)^2 + (y - 0)^2} - \sqrt{(x + c)^2 + (y - 0)^2} \right| = 2a$$

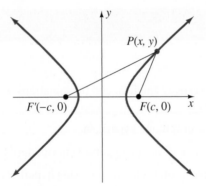

Figure 8.28

(The absolute value sign is used to allow the point P to be on either branch of the hyperbola.) Using the same type of simplification procedure that we used for deriving the standard form for the equation of an ellipse, we find that this equation simplifies to

$$\frac{x^2}{a^2} - \frac{y^2}{c^2 - a^2} = 1$$

Letting $b^2 = c^2 - a^2$, where $b > 0$, we obtain the standard form

$$\frac{x^2}{a^2} - \frac{y^2}{b^2} = 1 \qquad\qquad (1)$$

Equation (1) indicates that this hyperbola is symmetric with respect to both axes and the origin. Furthermore, by letting $y = 0$, we obtain $x^2/a^2 = 1$, or $x^2 = a^2$, so the x intercepts are a and $-a$. The corresponding points $A(a, 0)$ and A' $(-a, 0)$ are the **vertices** of the hyperbola, and the line segment $\overline{AA'}$ is called the **transverse axis;** it is of length $2a$ (see Figure 8.29). The midpoint of the transverse axis is called the **center** of the hyperbola; it is located at the origin. By letting $x = 0$ in equation (1), we obtain $-y^2/b^2 = 1$, or $y^2 = -b^2$. This implies that there are no y intercepts, as indicated in Figure 8.29.

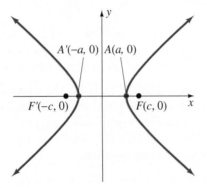

Figure 8.29

The following property summarizes the previous discussion.

Property 8.5

The graph of the equation

$$\frac{x^2}{a^2} - \frac{y^2}{b^2} = 1$$

is a hyperbola with vertices at $(a, 0)$ and $(-a, 0)$. The foci are at $(c, 0)$ and $(-c, 0)$, where $c^2 = a^2 + b^2$.

In conjunction with every hyperbola there are two intersecting lines that pass through the center of the hyperbola. These lines, referred to as *asymptotes,* are very helpful when we are sketching a hyperbola. Their equations are easily determined by using the following type of reasoning. Solving the equation

$$\frac{x^2}{a^2} - \frac{y^2}{b^2} = 1$$

for y produces $y = \pm\frac{b}{a}\sqrt{x^2 - a^2}$. From this form, it is evident that there are no points on the graph for $x^2 - a^2 < 0$; that is, if $-a < x < a$. However, there are points on the graph if $x \geq a$ or $x \leq -a$. If $x \geq a$, then $y = \pm\frac{b}{a}\sqrt{x^2 - a^2}$ can be written

$$y = \pm\frac{b}{a}\sqrt{x^2\left(1 - \frac{a^2}{x^2}\right)}$$

$$= \pm\frac{b}{a}\sqrt{x^2}\sqrt{1 - \frac{a^2}{x^2}}$$

$$= \pm\frac{b}{a}x\sqrt{1 - \frac{a^2}{x^2}}$$

Now suppose that we are going to determine some y values for very large values of x. (Remember that a and b are arbitrary constants; they have specific values for a particular hyperbola.) When x is very large, a^2/x^2 will be close to zero, so the radicand will be close to 1. Therefore the y value will be close to either $(b/a)x$ or $-(b/a)x$. In other words, as x becomes larger and larger, the point $P(x, y)$ gets closer and closer to either the line $y = (b/a)x$ or the line $y = -(b/a)x$. A corresponding situation occurs when $x \leq -a$. The lines with equations

$$y = \pm\frac{b}{a}x$$

are called the **asymptotes** of the hyperbola.

As we mentioned earlier, the asymptotes are very helpful for sketching hyperbolas. An easy way to sketch the asymptotes is first to plot the vertices $A(a, 0)$ and $A'(-a, 0)$ and the points $B(0, b)$ and $B'(0, -b)$, as in Figure 8.30. The line segment $\overline{BB'}$ is of length $2b$ and is called the **conjugate axis** of the hyperbola. The horizontal line segments drawn through B and B', together with the vertical line segments drawn through A and A', form a rectangle. The diagonals of this rectangle have slopes b/a and $-(b/a)$. Therefore, by extending the diagonals, we obtain the asymptotes $y = (b/a)x$ and $y = -(b/a)x$. The two branches of the hyperbola can be sketched by using the asymptotes as guidelines, as shown in Figure 8.30.

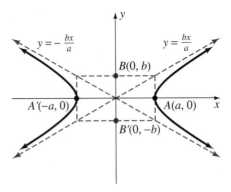

Figure 8.30

EXAMPLE 1

Find the vertices, the foci, and the equations of the asymptotes of the hyperbola $9x^2 - 4y^2 = 36$, and sketch the hyperbola.

Solution

Dividing both sides of the given equation by 36 and simplifying, we change the equation to the standard form:

$$\frac{x^2}{4} - \frac{y^2}{9} = 1$$

where $a^2 = 4$ and $b^2 = 9$. Hence $a = 2$ and $b = 3$. The vertices are $(\pm 2, 0)$ and the endpoints of the conjugate axis are $(0, \pm 3)$; these points determine the rectangle whose diagonals extend to become the asymptotes. Using $a = 2$ and $b = 3$, the equations of the asymptotes are $y = \frac{3}{2}x$ and $y = -\frac{3}{2}x$. Then, using the relationship $c^2 = a^2 + b^2$, we obtain $c^2 = 4 + 9 = 13$. Thus the foci are at $(\sqrt{13}, 0)$ and $(-\sqrt{13}, 0)$. Using the vertices and the asymptotes, we have sketched the hyperbola in Figure 8.31.

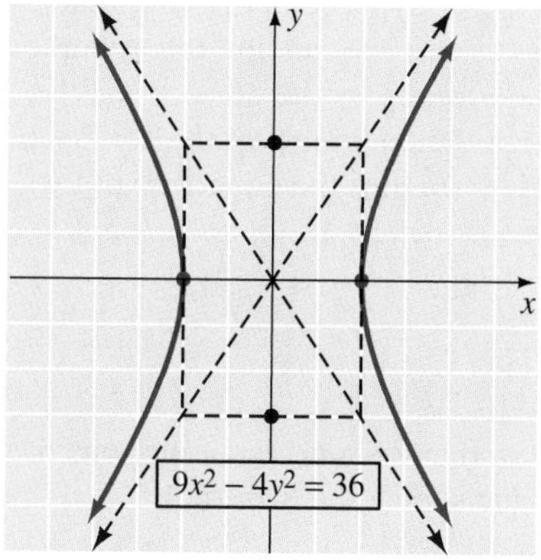

Figure 8.31

⊙ **Now work Problem 7.**   ∎

EXAMPLE 2   Find the equation of the hyperbola with vertices at $(\pm 4, 0)$ and foci at $(\pm 2\sqrt{5}, 0)$.

Solution

From the given information, we know that $a = 4$ and $c = 2\sqrt{5}$. Then using the relationship $b^2 = c^2 - a^2$, we obtain

$$b^2 = (2\sqrt{5})^2 - 4^2 = 20 - 16 = 4$$

Substituting 16 for a^2 and 4 for b^2 in the standard form produces

$$\frac{x^2}{16} - \frac{y^2}{4} = 1$$

Multiplying both sides of this equation by 16 yields

$$x^2 - 4y^2 = 16$$

⊙ **Now work Problem 23.**   ∎

■ **Hyperbolas with Foci on the *y* Axis**

In a similar fashion, we can develop a standard form for the equation of a hyperbola with foci on the *y* axis. The following property summarizes the results of such a development, where the foci are at $(0, c)$ and $(0, -c)$.

Property 8.6

The graph of the equation

$$\frac{y^2}{a^2} - \frac{x^2}{b^2} = 1$$

is a hyperbola with vertices at $(0, a)$ and $(0, -a)$. The foci are at $(0, c)$ and $(0, -c)$, where $c^2 = a^2 + b^2$.

For this type of hyperbola, the endpoints of the conjugate axis are at $(b, 0)$ and $(-b, 0)$. In this case we can find the asymptotes by extending the diagonals of the rectangle determined by the horizontal lines through the vertices and the vertical lines through the endpoints of the conjugate axis. The slopes of these diagonals are a/b and $-a/b$; thus the equations of these asymptotes are

$$y = \frac{a}{b}x \quad \text{and} \quad y = -\frac{a}{b}x$$

EXAMPLE 3

Find the vertices, the foci, and the equations of the asymptotes of the hyperbola $4y^2 - x^2 = 12$, and sketch the hyperbola.

Solution

We can divide both sides of the given equation by 12 to change the equation to the standard form:

$$\frac{y^2}{3} - \frac{x^2}{12} = 1$$

where $a^2 = 3$ and $b^2 = 12$. Hence $a = \sqrt{3}$ and $b = 2\sqrt{3}$. The vertices, $(0, \pm\sqrt{3})$, and the endpoints of the conjugate axis, $(\pm 2\sqrt{3}, 0)$, determine the rectangle whose diagonals extend to become the asymptotes. Using $a = 3$ and $b = 2\sqrt{3}$, we find the equations of the asymptotes are $y = (\sqrt{3}/2\sqrt{3})x = \frac{1}{2}x$ and $y = -\frac{1}{2}x$. Then, using the relationship $c^2 = a^2 + b^2$, we obtain $c^2 = 3 + 12 = 15$. So the foci are at $(0, \sqrt{15})$ and $(0, -\sqrt{15})$. The hyperbola is sketched in Figure 8.32.

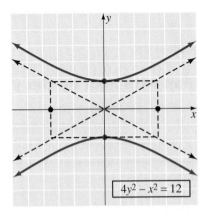

$$4y^2 - x^2 = 12$$

Figure 8.32

▶ **Now work Problem 5.**

■ Other Hyperbolas

In the same way, we can develop the standard form for an equation of a hyperbola that is symmetric with respect to a line parallel to a coordinate axis. We will not show such developments in this text but will simply state and use the results.

$$\frac{(x - h)^2}{a^2} - \frac{(y - k)^2}{b^2} = 1 \qquad \text{a hyperbola with center at } (h, k) \text{ and transverse axis on the horizontal line } y = k$$

$$\frac{(y - k)^2}{a^2} - \frac{(x - h)^2}{b^2} = 1 \qquad \text{a hyperbola with center at } (h, k) \text{ and transverse axis on the vertical line } x = h$$

The relationship $c^2 = a^2 + b^2$ still holds, and the physical significance of a, b, and c remains the same. However, these values are used relative to the center (h, k) to find the endpoints of the transverse and conjugate axes and to find the foci. Furthermore, the slopes of the asymptotes are as before, but these lines now contain the new center, (h, k). Let's see how all of this works in a specific example.

E X A M P L E   4

Find the vertices, the foci, and the equations of the asymptotes of the hyperbola $9x^2 - 36x - 16y^2 + 96y - 252 = 0$, and sketch the hyperbola.

Solution

First, we need to change to a standard form by completing the square on both x and y.

$$9(x^2 - 4x + \underline{\ \ }) - 16(y^2 - 6y + \underline{\ \ }) = 252$$

$$9(x^2 - 4x + 4) - 16(y^2 - 6y + 9) = 252 + 36 - 144$$

$$9(x - 2)^2 - 16(y - 3)^2 = 144$$

$$\frac{(x - 2)^2}{16} - \frac{(y - 3)^2}{9} = 1$$

The center is at $(2, 3)$, and the transverse axis is on the line $y = 3$. Because $a^2 = 16$, we know that $a = 4$. Therefore, the vertices are four units to the right and four units to the left of the center, $(2, 3)$, so they are at $(6, 3)$ and $(-2, 3)$. Likewise, because $b^2 = 9$, or $b = 3$, the endpoints of the conjugate axis are three units up and three units down from the center, so they are at $(2, 6)$ and $(2, 0)$. With $a = 4$ and $b = 3$, the slopes of the asymptotes are $\frac{3}{4}$ and $-\frac{3}{4}$. Then, using the slopes, the center $(2, 3)$, and the point–slope form for writing the equation of a line, we can determine the equations of the asymptotes to be $3x - 4y = -6$ and $3x + 4y = 18$. From the relationship $c^2 = a^2 + b^2$ we obtain $c^2 = 16 + 9 = 25$. Thus the foci are at $(7, 3)$ and $(-3, 3)$. The hyperbola is sketched in Figure 8.33.

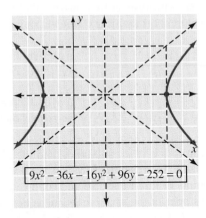

$$9x^2 - 36x - 16y^2 + 96y - 252 = 0$$

Figure 8.33

▶ **Now work Problem 13.** ∎

E X A M P L E   5

Find the equation of the hyperbola with vertices at $(-4, 2)$ and $(-4, -4)$ and with foci at $(-4, 3)$ and $(-4, -5)$.

Solution

Because the vertices and foci are on the same vertical line ($x = -4$), this hyperbola has an equation of the form

$$\frac{(y - k)^2}{a^2} - \frac{(x - h)^2}{b^2} = 1$$

The center of the hyperbola is at the midpoint of the transverse axis. Therefore

$$h = \frac{-4 + (-4)}{2} = -4 \qquad \text{and} \qquad k = \frac{2 + (-4)}{2} = -1$$

The distance between the center, $(-4, -1)$ and a vertex, $(-4, 2)$ is three units, so $a = 3$. The distance between the center, $(-4, -1)$, and a focus, $(-4, 3)$, is four units, so $c = 4$. Then, using the relationship $c^2 = a^2 + b^2$, we obtain

$$b^2 = c^2 - a^2 = 16 - 9 = 7$$

Now we can substitute -4 for h, -1 for k, 9 for a^2, and 7 for b^2 in the general form and simplify.

$$\frac{(y + 1)^2}{9} - \frac{(x + 4)^2}{7} = 1$$

$$7(y + 1)^2 - 9(x + 4)^2 = 63$$

$$7(y^2 + 2y + 1) - 9(x^2 + 8x + 16) = 63$$
$$7y^2 + 14y + 7 - 9x^2 - 72x - 144 = 63$$
$$7y^2 + 14y - 9x^2 - 72x - 200 = 0$$

▶ **Now work Problem 35.** ■

The hyperbola also has numerous applications, including many you may not be aware of. For example, one method of artillery range-finding is based on the concept of a hyperbola. If each of two listening posts, P_1 and P_2 in Figure 8.34 records the time that an artillery blast is heard, then the difference between the times multiplied by the speed of sound gives the difference of the distances of the gun from the two fixed points. Thus the gun is located somewhere on the hyperbola whose foci are the two listening posts. By bringing in a third listening post, P_3, we can form another hyperbola with foci at P_2 and P_3. Then the location of the gun must be at one of the intersections of the two hyperbolas.

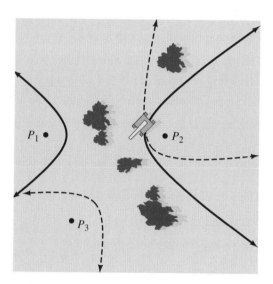

Figure 8.34

This same principle of intersecting hyperbolas is used in a long-range navigation system known as LORAN. Radar stations serve as the foci of the hyperbolas, and of course computers are used for the many calculations that are necessary to fix the location of a plane or ship. At the present time, LORAN is probably used mostly for coastal navigation in connection with small pleasure boats.

Some unique architectural creations have used the concept of a hyperbolic paraboloid, pictured in Figure 8.35. For example, the TWA building at Kennedy Airport is so designed. Some comets, upon entering the sun's gravitational field, follow a hyperbolic path, with the sun as one of the foci (see Figure 8.36).

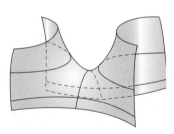

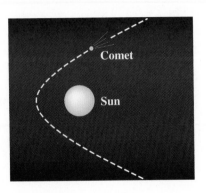

Figure 8.35   **Figure 8.36**

CONCEPT QUIZ   For Problems 1–8, answer true or false.

1. The graph of a hyperbola consists of two curves called the branches of the hyperbola.

2. The asymptotes of a hyperbola are always straight lines.

3. The graph of a hyperbola whose center is at the origin is symmetric with both the x and y axes.

4. The slopes of the asymptotes of a hyperbola are negative reciprocals.

5. A hyperbola has both a transverse axis and a conjugate axis.

6. For a hyperbola's asymptote whose slope is positive, the slope of the asymptote is the ratio of the length of the conjugate axis to the length of the transverse axis.

7. The foci of the hyperbola whose equation is $\dfrac{x^2}{9} - \dfrac{y^2}{16} = 1$ are located at the points $(0, 5)$ and $(0, -5)$.

8. For the hyperbola whose equation is $\dfrac{y^2}{36} - \dfrac{x^2}{49} = 1$, the asymptotes are the lines $y = \pm\dfrac{6}{7}x$.

Problem Set 8.3

For Problems 1–22, find the vertices, the foci, and the equations of the asymptotes, and sketch each hyperbola.

1. $\dfrac{x^2}{9} - \dfrac{y^2}{4} = 1$

2. $\dfrac{x^2}{4} - \dfrac{y^2}{16} = 1$

3. $\dfrac{y^2}{4} - \dfrac{x^2}{9} = 1$

4. $\dfrac{y^2}{16} - \dfrac{x^2}{4} = 1$

5. $9y^2 - 16x^2 = 144$

6. $4y^2 - x^2 = 4$

7. $x^2 - y^2 = 9$

8. $x^2 - y^2 = 1$

9. $5y^2 - x^2 = 25$

10. $y^2 - 2x^2 = 8$

11. $y^2 - 9x^2 = -9$

12. $16y^2 - x^2 = -16$

13. $4x^2 - 24x - 9y^2 - 18y - 9 = 0$

14. $9x^2 + 72x - 4y^2 - 16y + 92 = 0$

15. $y^2 - 4y - 4x^2 - 24x - 36 = 0$

16. $9y^2 + 54y - x^2 + 6x + 63 = 0$

17. $2x^2 - 8x - y^2 + 4 = 0$

18. $x^2 + 6x - 3y^2 = 0$

19. $y^2 + 10y - 9x^2 + 16 = 0$

20. $4y^2 - 16y - x^2 + 12 = 0$

21. $x^2 + 4x - y^2 - 4y - 1 = 0$

22. $y^2 + 8y - x^2 + 2x + 14 = 0$

For Problems 23–38, find an equation of the hyperbola that satisfies the given conditions.

23. Vertices $(\pm 2, 0)$, foci $(\pm 3, 0)$

24. Vertices $(\pm 1, 0)$, foci $(\pm 4, 0)$

25. Vertices $(0, \pm 3)$, foci $(0, \pm 5)$

26. Vertices $(0, \pm 2)$, foci $(0, \pm 6)$

27. Vertices $(\pm 1, 0)$, contains the point $(2, 3)$

28. Vertices $(0, \pm 1)$, contains the point $(-3, 5)$

29. Vertices $(0, \pm\sqrt{3})$, length of conjugate axis is 4

30. Vertices $(\pm\sqrt{5}, 0)$, length of conjugate axis is 6

31. Foci $(\pm\sqrt{23}, 0)$, length of transverse axis is 8

32. Foci $(0, \pm 3\sqrt{2})$, length of conjugate axis is 4

33. Vertices $(6, -3)$ and $(2, -3)$, foci $(7, -3)$ and $(1, -3)$

34. Vertices $(-7, -4)$ and $(-5, -4)$, foci $(-8, -4)$ and $(-4, -4)$

35. Vertices $(-3, 7)$ and $(-3, 3)$, foci $(-3, 9)$ and $(-3, 1)$

36. Vertices $(7, 5)$ and $(7, -1)$, foci $(7, 7)$ and $(7, -3)$

37. Vertices $(0, 0)$ and $(4, 0)$, foci $(5, 0)$ and $(-1, 0)$

38. Vertices $(0, 0)$ and $(0, -6)$, foci $(0, 2)$ and $(0, -8)$

For Problems 39–48, identify the graph of each of the equations as a straight line, a circle, a parabola, an ellipse, or a hyperbola. Do not sketch the graphs.

39. $x^2 - 7x + y^2 + 8y - 2 = 0$

40. $x^2 - 7x - y^2 + 8y - 2 = 0$

41. $5x - 7y = 9$

42. $4x^2 - x + y^2 + 2y - 3 = 0$

43. $10x^2 + y^2 = 8$

44. $-3x - 2y = 9$

45. $5x^2 + 3x - 2y^2 - 3y - 1 = 0$

46. $x^2 + y^2 - 3y - 6 = 0$

47. $x^2 - 3x + y - 4 = 0$

48. $5x + y^2 - 2y - 1 = 0$

■ ■ ■ **THOUGHTS INTO WORDS**

49. What is the difference between the graphs of the equations $x^2 + y^2 = 0$ and $x^2 - y^2 = 0$?

50. What is the difference between the graphs of the equations $4x^2 + 9y^2 = 0$ and $9x^2 + 4y^2 = 0$?

51. A flashlight produces a "cone of light" that can be cut by the plane of a wall to illustrate the conic sections.

Try shining a flashlight against a wall (stand within a couple of feet of the wall) at different angles to produce a circle, an ellipse, a parabola, and one branch of a hyperbola. (You may find it difficult to distinguish between a parabola and a branch of a hyperbola.) Write a paragraph explaining this experiment to someone else.

GRAPHING CALCULATOR ACTIVITIES

52. Use a graphing calculator to check your graphs for Problems 13–22. Be sure to graph the asymptotes for each hyperbola.

53. Use a graphing calculator to check your answers for Problems 39–48.

Answers to the Concept Quiz

1. True   **2.** True   **3.** True   **4.** False   **5.** True   **6.** True   **7.** False   **8.** True

---

8.4   Systems Involving Nonlinear Equations

Objectives

■   Sketch graphs of nonlinear systems of equations.

■   Using a graph, approximate the real number solutions of nonlinear systems of equations.

■   Solve nonlinear systems of equations using substitution.

■   Solve nonlinear systems of equations using elimination.

In Chapters 6 and 7 we used several techniques to solve systems of linear equations. We will use two of those techniques in this section to solve some systems that contain at least one nonlinear equation. Furthermore, we will use our knowledge of graphing lines, circles, parabolas, ellipses, and hyperbolas to get a pictorial view of the systems. That will give us a basis for predicting approximate real number solutions if there are any. We have once again arrived at a topic that vividly illustrates the merging of mathematical ideas. Let's begin by considering a system that contains one linear and one nonlinear equation.

E X A M P L E   1

Solve the system $\begin{pmatrix} x^2 + y^2 = 13 \\ 3x + 2y = 0 \end{pmatrix}$.

Solution

From our previous graphing experiences, we should recognize that $x^2 + y^2 = 13$ is a circle and $3x + 2y = 0$ is a straight line. Thus the system can be pictured as in Figure 8.37. The graph indicates that the solution set of this system should consist of two ordered pairs of real numbers that represent the points of intersection in the second and fourth quadrants.

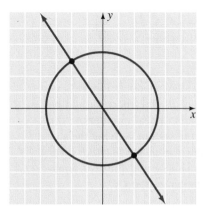

Figure 8.37

Now let's solve the system analytically by using the *substitution method.* We can change the form of $3x + 2y = 0$ to $y = -3x/2$ and then substitute $-3x/2$ for y in the other equation to produce

$$x^2 + \left(-\frac{3x}{2}\right)^2 = 13$$

This equation can now be solved for x.

$$x^2 + \frac{9x^2}{4} = 13$$

$$4x^2 + 9x^2 = 52$$

$$13x^2 = 52$$

$$x^2 = 4$$

$$x = \pm 2$$

We substitute 2 for x and then -2 for x in the second equation of the system to produce two values for y.

$3x + 2y = 0$	$3x + 2y = 0$
$3(2) + 2y = 0$	$3(-2) + 2y = 0$
$2y = -6$	$2y = 6$
$y = -3$	$y = 3$

Therefore the solution set of the system is $\{(2, -3), (-2, 3)\}$.

▶ **Now work Problem 1.**   ■

Remark: Don't forget that, as always, you can check the solutions by substituting them back into the original equations. Graphing the system permits you to approximate any possible real number solutions before solving the system. Then, after solving the system, you can use the graph again to check that the answers are reasonable.

EXAMPLE 2 Solve the system $\left(\begin{array}{l} x^2 + y^2 = 16 \\ y^2 - x^2 = 4 \end{array}\right)$.

Solution

Graphing the system produces Figure 8.38. This figure indicates that there should be four ordered pairs of real numbers in the solution set of the system. Solving the system by using the *elimination method* works nicely. We can simply add the two equations, which eliminates the x's.

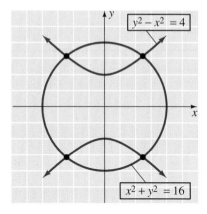

$$x^2 + y^2 = 16$$
$$\underline{-x^2 + y^2 = 4}$$
$$2y^2 = 20$$
$$y^2 = 10$$
$$y = \pm\sqrt{10}$$

Figure 8.38

Substituting $\sqrt{10}$ for y in the first equation yields

$$x^2 + y^2 = 16$$
$$x^2 + (\sqrt{10})^2 = 16$$
$$x^2 + 10 = 16$$
$$x^2 = 6$$
$$x = \pm\sqrt{6}$$

Thus $(\sqrt{6}, \sqrt{10})$ and $(-\sqrt{6}, \sqrt{10})$ are solutions. Substituting $-\sqrt{10}$ for y in the first equation yields

$$x^2 + y^2 = 16$$
$$x^2 + (-\sqrt{10})^2 = 16$$
$$x^2 + 10 = 16$$
$$x^2 = 6$$
$$x = \pm\sqrt{6}$$

Thus $(\sqrt{6}, -\sqrt{10})$ and $(-\sqrt{6}, -\sqrt{10})$ are also solutions. The solution set is $\{(-\sqrt{6}, \sqrt{10}), (-\sqrt{6}, -\sqrt{10}), (\sqrt{6}, \sqrt{10}), (\sqrt{6}, -\sqrt{10})\}$.

▶ **Now work Problem 23.** ■

Sometimes a sketch of the graph of a system may not clearly indicate whether the system contains any real number solutions. The next example illustrates such a situation.

E X A M P L E   3

Solve the system $\left(\begin{array}{l} y = x^2 + 2 \\ 6x - 4y = -5 \end{array} \right)$.

Solution

From previous graphing experiences, we recognize that $y = x^2 + 2$ is the basic parabola shifted upward two units and $6x - 4y = -5$ is a straight line (see Figure 8.39). Because of the close proximity of the curves, it is difficult to tell whether they intersect. In other words, the graph does not definitely indicate any real number solutions for the system.

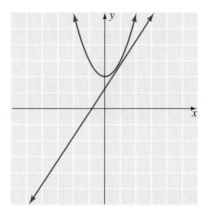

Figure 8.39

Let's solve the system by using the substitution method. We can substitute $x^2 + 2$ for y in the second equation, which produces two values for x.

$$6x - 4(x^2 + 2) = -5$$

$$6x - 4x^2 - 8 = -5$$

$$-4x^2 + 6x - 3 = 0$$

$$4x^2 - 6x + 3 = 0$$

$$x = \frac{6 \pm \sqrt{36 - 48}}{8}$$

$$x = \frac{6 \pm \sqrt{-12}}{8}$$

$$x = \frac{6 \pm 2i\sqrt{3}}{8}$$

$$x = \frac{3 \pm i\sqrt{3}}{4}$$

It is now obvious that the system has no real number solutions. That is, the line and the parabola do not intersect in the real number plane. However, there will be two pairs of complex numbers in the solution set. We can substitute $(3 + i\sqrt{3})/4$ for x in the first equation.

$$y = \left(\frac{3 + i\sqrt{3}}{4}\right)^2 + 2$$

$$y = \frac{6 + 6i\sqrt{3}}{16} + 2$$

$$y = \frac{6 + 6i\sqrt{3} + 32}{16}$$

$$y = \frac{38 + 6i\sqrt{3}}{16} = \frac{19 + 3i\sqrt{3}}{8}$$

Likewise, we can substitute $(3 - i\sqrt{3})/4$ for x in the first equation.

$$y = \left(\frac{3 - i\sqrt{3}}{4}\right)^2 + 2$$

$$y = \frac{6 - 6i\sqrt{3}}{16} + 2$$

$$y = \frac{6 - 6i\sqrt{3} + 32}{16}$$

$$y = \frac{38 - 6i\sqrt{3}}{16} = \frac{19 - 3i\sqrt{3}}{8}$$

The solution set is $\left\{ \left(\dfrac{3 + i\sqrt{3}}{4}, \dfrac{19 + 3i\sqrt{3}}{8}\right), \left(\dfrac{3 - i\sqrt{3}}{4}, \dfrac{19 - 3i\sqrt{3}}{8}\right) \right\}$.

▶ **Now work Problem 11.**   ■

In Example 3 the use of a graphing utility may not, at first, indicate whether the system has any real number solutions. Suppose that we graph the system using a viewing rectangle such that $-15 \leq x \leq 15$ and $-10 \leq y \leq 10$. As shown in the display in Figure 8.40, we cannot tell whether the line and the parabola intersect. However, if we change the viewing rectangle so that $0 \leq x \leq 2$ and $0 \leq y \leq 4$, as shown in Figure 8.41, then it becomes apparent that the two graphs do not intersect.

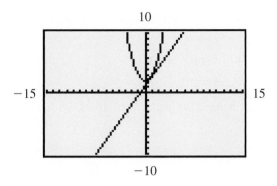

Figure 8.40

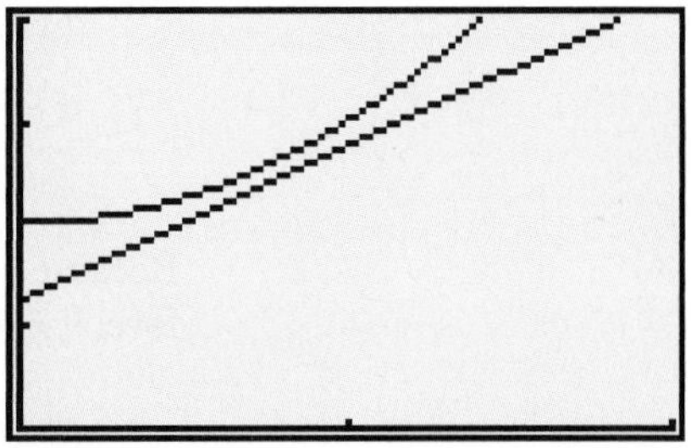

Figure 8.41

E X A M P L E   4

Find the real number solutions for the system $\left(\begin{array}{l} y = \log_2(x - 3) - 2 \\ y = -\log_2 x \end{array} \right)$.

Solution

First, let's use a graphing calculator to obtain a graph of the system as shown in Figure 8.42. The two curves appear to intersect at approximately $x = 4$ and $y = -2$. To solve the system algebraically, we can equate the two expressions for y and solve the resulting equation for x.

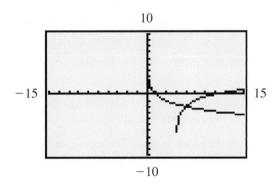

Figure 8.42

$$\log_2(x - 3) - 2 = -\log_2 x$$
$$\log_2 x + \log_2(x - 3) = 2$$
$$\log_2 x(x - 3) = 2$$

At this step we can either change to exponential form or rewrite 2 as $\log_2 4$.

$$\log_2 x(x - 3) = \log_2 4$$
$$x(x - 3) = 4$$

$$x^2 - 3x - 4 = 0$$

$$(x - 4)(x + 1) = 0$$

$$x - 4 = 0 \quad \text{or} \quad x + 1 = 0$$

$$x = 4 \quad \text{or} \quad x = -1$$

Because logarithms are not defined for negative numbers, -1 is discarded. Therefore, if $x = 4$, then

$$y = -\log_2 x$$

becomes

$$y = -\log_2 4$$

$$= -2$$

Therefore the solution set is $\{(4, -2)\}$.

▶ **Now work Problem 31.**   ■

CONCEPT QUIZ   For Problems 1–5, answer true or false.

1. A system of nonlinear equations could have a solution set that consists of one ordered pair of real numbers.

2. A graph can be used to approximate both real number and complex number solutions for nonlinear systems.

3. Every nonlinear system will have at least one real number ordered pair solution.

4. The elimination method for solving nonlinear equations can be used to solve any nonlinear system of equations.

5. The substitution method for solving nonlinear equations can be used to solve any nonlinear system of equations.

6. By visualizing the graphs, determine how many real number solutions there are for the system of equations $\left(\begin{matrix} x^2 + y^2 = 4 \\ x^2 + y^2 = 9 \end{matrix} \right)$?

7. By visualizing the graphs, determine how many real number solutions there are for the system of equations $\left(\begin{matrix} y = x^2 + 3 \\ y = -x^2 + 3 \end{matrix} \right)$?

8. By visualizing the graphs, determine how many real number solutions there are for the system of equations $\left(\begin{matrix} y = 2x \\ x^2 + y^2 = 9 \end{matrix} \right)$?

Problem Set 8.4

For Problems 1–30, (a) graph the system so that approximate real number solutions (if there are any) can be predicted, and (b) solve the system by the substitution or elimination method.

1. $\begin{pmatrix} x^2 + y^2 = 5 \\ x + 2y = 5 \end{pmatrix}$

2. $\begin{pmatrix} x^2 + y^2 = 13 \\ 2x + 3y = 13 \end{pmatrix}$

3. $\begin{pmatrix} x^2 + y^2 = 26 \\ x + y = -4 \end{pmatrix}$

4. $\begin{pmatrix} x^2 + y^2 = 10 \\ x + y = -2 \end{pmatrix}$

5. $\begin{pmatrix} x^2 + y^2 = 2 \\ x - y = 4 \end{pmatrix}$

6. $\begin{pmatrix} x^2 + y^2 = 3 \\ x - y = -5 \end{pmatrix}$

7. $\begin{pmatrix} y = x^2 + 6x + 7 \\ 2x + y = -5 \end{pmatrix}$

8. $\begin{pmatrix} y = x^2 - 4x + 5 \\ y - x = 1 \end{pmatrix}$

9. $\begin{pmatrix} 2x + y = -2 \\ y = x^2 + 4x + 7 \end{pmatrix}$

10. $\begin{pmatrix} 2x + y = 0 \\ y = -x^2 + 2x - 4 \end{pmatrix}$

11. $\begin{pmatrix} y = x^2 - 3 \\ x + y = -4 \end{pmatrix}$

12. $\begin{pmatrix} y = -x^2 + 1 \\ x + y = 2 \end{pmatrix}$

13. $\begin{pmatrix} x^2 + 2y^2 = 9 \\ x - 4y = -9 \end{pmatrix}$

14. $\begin{pmatrix} 2x - y = 7 \\ 3x^2 + y^2 = 21 \end{pmatrix}$

15. $\begin{pmatrix} x + y = -3 \\ x^2 + 2y^2 - 12y - 18 = 0 \end{pmatrix}$

16. $\begin{pmatrix} 4x^2 + 9y^2 = 25 \\ 2x + 3y = 7 \end{pmatrix}$

17. $\begin{pmatrix} x - y = 2 \\ x^2 - y^2 = 16 \end{pmatrix}$

18. $\begin{pmatrix} x^2 - 4y^2 = 16 \\ 2y - x = 2 \end{pmatrix}$

19. $\begin{pmatrix} y = -x^2 + 3 \\ y = x^2 + 1 \end{pmatrix}$

20. $\begin{pmatrix} y = x^2 \\ y = x^2 - 4x + 4 \end{pmatrix}$

21. $\begin{pmatrix} y = x^2 + 2x - 1 \\ y = x^2 + 4x + 5 \end{pmatrix}$

22. $\begin{pmatrix} y = -x^2 + 1 \\ y = x^2 - 2 \end{pmatrix}$

23. $\begin{pmatrix} x^2 - y^2 = 4 \\ x^2 + y^2 = 4 \end{pmatrix}$

24. $\begin{pmatrix} 2x^2 + y^2 = 8 \\ x^2 + y^2 = 4 \end{pmatrix}$

25. $\begin{pmatrix} 8y^2 - 9x^2 = 6 \\ 8x^2 - 3y^2 = 7 \end{pmatrix}$

26. $\begin{pmatrix} 2x^2 + y^2 = 11 \\ x^2 - y^2 = 4 \end{pmatrix}$

27. $\begin{pmatrix} 2x^2 - 3y^2 = -1 \\ 2x^2 + 3y^2 = 5 \end{pmatrix}$

28. $\begin{pmatrix} 4x^2 + 3y^2 = 9 \\ y^2 - 4x^2 = 7 \end{pmatrix}$

29. $\begin{pmatrix} xy = 3 \\ 2x + 2y = 7 \end{pmatrix}$

30. $\begin{pmatrix} x^2 + 4y^2 = 25 \\ xy = 6 \end{pmatrix}$

For Problems 31–36, solve each system for all real number solutions.

31. $\begin{pmatrix} y = \log_3(x - 6) - 3 \\ y = -\log_3 x \end{pmatrix}$

32. $\begin{pmatrix} y = \log_{10}(x - 9) - 1 \\ y = -\log_{10} x \end{pmatrix}$

33. $\begin{pmatrix} y = e^x - 1 \\ y = 2e^{-x} \end{pmatrix}$

34. $\begin{pmatrix} y = 28 - 11e^x \\ y = -e^{2x} \end{pmatrix}$

35. $\begin{pmatrix} y = x^3 \\ y = x^3 + 2x^2 + 5x - 3 \end{pmatrix}$

36. $\begin{pmatrix} y = 3(4^x) - 8 \\ y = 4^{2x} - 2(4^x) - 4 \end{pmatrix}$

■■■ THOUGHTS INTO WORDS

37. What happens if you try to graph the system
$$\begin{pmatrix} 7x^2 + 8y^2 = 36 \\ 11x^2 + 5y^2 = -4 \end{pmatrix}?$$

38. For what value(s) of k will the line $x + y = k$ touch the ellipse $x^2 + 2y^2 = 6$ in one and only one point? Defend your answer.

39. The system

$$\begin{pmatrix} x^2 - 6x + y^2 - 4y + 4 = 0 \\ x^2 - 4x + y^2 + 8y - 5 = 0 \end{pmatrix}$$

represents two circles that intersect in two points. An equivalent system can be formed by replacing the second equation with the result of adding -1 times the

first equation to the second equation. Thus we obtain the system

$$\begin{pmatrix} x^2 - 6x + y^2 - 4y + 4 = 0 \\ 2x + 12y - 9 = 0 \end{pmatrix}$$

Explain why the linear equation in this system is the equation of the common chord of the original two intersecting circles.

GRAPHING CALCULATOR ACTIVITIES

40. Graph the system of equations $\begin{pmatrix} y = x^2 + 2 \\ 6x - 4y = -5 \end{pmatrix}$ and use the trace and zoom-in features of your calculator to show that this system has no real number solutions.

41. Use a graphing calculator to graph the systems in Problems 31–36 and check the reasonableness of your answers to those problems.

For Problems 42–47, use a graphing calculator to approximate, to the nearest tenth, the real number solutions for each system of equations.

42. $\begin{pmatrix} y = e^x + 1 \\ y = x^3 + x^2 - 2x - 1 \end{pmatrix}$

43. $\begin{pmatrix} y = x^3 + 2x^2 - 3x + 2 \\ y = -x^3 - x^2 + 1 \end{pmatrix}$

44. $\begin{pmatrix} y = 2^x + 1 \\ y = 2^{-x} + 2 \end{pmatrix}$

45. $\begin{pmatrix} y = \ln(x - 1) \\ y = x^2 - 16x + 64 \end{pmatrix}$

46. $\begin{pmatrix} x = y^2 - 2y + 3 \\ x^2 + y^2 = 25 \end{pmatrix}$

47. $\begin{pmatrix} y^2 - x^2 = 16 \\ 2y^2 - x^2 = 8 \end{pmatrix}$

Answers to the Concept Quiz

1. True   **2.** False   **3.** False   **4.** False   **5.** True   **6.** No solution   **7.** 1   **8.** 2

The following standard forms for the equations of conic sections were developed in this chapter.

■ Parabolas

$x^2 = 4py$

    focus $(0, p)$, directrix $y = -p$, y axis symmetry

$y^2 = 4px$

    focus $(p, 0)$, directrix $x = -p$, x axis symmetry

$(x - h)^2 = 4p(y - k)$

    focus $(h, k + p)$, directrix $y = k - p$, symmetric with respect to the line $x = h$

$(y - k)^2 = 4p(x - h)$

    focus $(h + p, k)$, directrix $x = h - p$, symmetric with respect to the line $y = k$

■ Ellipses

$\dfrac{x^2}{a^2} + \dfrac{y^2}{b^2} = 1, \qquad a^2 > b^2$

    center $(0, 0)$, vertices $(\pm a, 0)$, endpoints of minor axis $(0, \pm b)$, foci $(\pm c, 0)$, $c^2 = a^2 - b^2$

$\dfrac{x^2}{b^2} + \dfrac{y^2}{a^2} = 1, \qquad a^2 > b^2$

    center $(0, 0)$, vertices $(0, \pm a)$, endpoints of minor axis $(\pm b, 0)$, foci $(0, \pm c)$, $c^2 = a^2 - b^2$

$\dfrac{(x - h)^2}{a^2} + \dfrac{(y - k)^2}{b^2} = 1, \qquad a^2 > b^2$

    center (h, k), vertices $(h \pm a, k)$, endpoints of minor axis $(h, k \pm b)$, foci $(h \pm c, k)$, $c^2 = a^2 - b^2$

$\dfrac{(x - h)^2}{b^2} + \dfrac{(y - k)^2}{a^2} = 1, \qquad a^2 > b^2$

    center (h, k), vertices $(h, k \pm a)$, endpoints of minor axis $(h \pm b, k)$, foci $(h, k \pm c)$, $c^2 = a^2 - b^2$

■ Hyperbolas

$\dfrac{x^2}{a^2} - \dfrac{y^2}{b^2} = 1$

    center $(0, 0)$, vertices $(\pm a, 0)$, endpoints of conjugate axis $(0, \pm b)$, foci $(\pm c, 0)$,

    $c^2 = a^2 + b^2$, asymptotes $y = \pm \dfrac{b}{a} x$

$\dfrac{y^2}{a^2} - \dfrac{x^2}{b^2} = 1$

    center $(0, 0)$, vertices $(0, \pm a)$, endpoints of conjugate axis $(\pm b, 0)$, foci $(0, \pm c)$,

    $c^2 = a^2 + b^2$, asymptotes $y = \pm \dfrac{a}{b} x$

$\dfrac{(x - h)^2}{a^2} - \dfrac{(y - k)^2}{b^2} = 1$

    center (h, k), vertices $(h \pm a, k)$, endpoints of conjugate axis $(h, k \pm b)$, foci $(h \pm c, k)$,

    $c^2 = a^2 + b^2$, asymptotes $y - k = \pm \dfrac{b}{a}(x - h)$

$\dfrac{(y - k)^2}{a^2} - \dfrac{(x - h)^2}{b^2} = 1$

    center (h, k), vertices $(h, k \pm a)$, endpoints of conjugate axis $(h \pm b, k)$, foci $(h, k \pm c)$,

    $c^2 = a^2 + b^2$, asymptotes $y - k = \pm \dfrac{a}{b}(x - h)$

Systems that contain at least one nonlinear equation can often be solved by substitution or by the elimination method. Graphing the system will often provide a basis for predicting approximate real number solutions if there are any.

Chapter 8   Review Problem Set

For Problems 1–12, (a) identify the conic section as a parabola, an ellipse, or a hyperbola, (b) if it is a parabola, find its vertex, focus, and directrix; if it is an ellipse, find its vertices, the endpoints of its minor axis, and its foci; if it is a hyperbola, find its vertices, the endpoints of its conjugate axis, its foci, and its asymptotes; and (c) sketch each of the curves.

1. $x^2 + 2y^2 = 32$ **2.** $y^2 = -12x$

3. $3y^2 - x^2 = 9$ **4.** $2x^2 - 3y^2 = 18$

5. $5x^2 + 2y^2 = 20$ **6.** $x^2 = 2y$

7. $x^2 - 8x - 2y^2 + 4y + 10 = 0$

8. $9x^2 - 54x + 2y^2 + 8y + 71 = 0$

9. $y^2 - 2y + 4x + 9 = 0$

10. $x^2 + 2x + 8y + 25 = 0$

11. $x^2 + 10x + 4y^2 - 16y + 25 = 0$

12. $3y^2 + 12y - 2x^2 - 8x - 8 = 0$

For Problems 13–24, find the equation of the indicated conic section that satisfies the given conditions.

13. Parabola with vertex $(0, 0)$, focus $(-5, 0)$, directrix $x = 5$

14. Ellipse with vertices $(0, \pm 4)$, foci $(0, \pm\sqrt{15})$

15. Hyperbola with vertices $(\pm\sqrt{2}, 0)$, length of conjugate axis 10

16. Ellipse with vertices $(\pm 2, 0)$, contains the point $(1, -2)$

17. Parabola with vertex $(0, 0)$, symmetric with respect to the y axis, contains the point $(2, 6)$

18. Hyperbola with vertices $(0, \pm 1)$, foci $(0, \pm\sqrt{10})$

19. Ellipse with vertices $(6, 1)$ and $(6, 7)$, length of minor axis 2 units

20. Parabola with vertex $(4, -2)$, focus $(6, -2)$

21. Hyperbola with vertices $(-5, -3)$ and $(-5, -5)$, foci $(-5, -2)$ and $(-5, -6)$

22. Parabola with vertex $(-6, -3)$, symmetric with respect to the line $x = -6$, contains the point $(-5, -2)$

23. Ellipse with endpoints of minor axis $(-5, 2)$ and $(-5, -2)$, length of major axis 10 units

24. Hyperbola with vertices $(2, 0)$ and $(6, 0)$, length of conjugate axis 8 units

For Problems 25–30, (a) graph the system and (b) solve the system by using the substitution or elimination method.

25. $\begin{pmatrix} x^2 + y^2 = 17 \\ x - 4y = -17 \end{pmatrix}$ **26.** $\begin{pmatrix} x^2 - y^2 = 8 \\ 3x - y = 8 \end{pmatrix}$

27. $\begin{pmatrix} x - y = 1 \\ y = x^2 + 4x + 1 \end{pmatrix}$

28. $\begin{pmatrix} 4x^2 - y^2 = 16 \\ 9x^2 + 9y^2 = 16 \end{pmatrix}$

29. $\begin{pmatrix} x^2 + 2y^2 = 8 \\ 2x^2 + 3y^2 = 12 \end{pmatrix}$

30. $\begin{pmatrix} y^2 - x^2 = 1 \\ 4x^2 + y^2 = 4 \end{pmatrix}$

1. Find the focus of the parabola $x^2 = -20y$.

2. Find the vertex of the parabola $y^2 - 4y - 8x - 20 = 0$.

3. Find the equation of the directrix for the parabola $2y^2 = 24x$.

4. Find the focus of the parabola $y^2 = 24x$.

5. Find the vertex of the parabola $x^2 + 4x - 12y - 8 = 0$.

6. Find the equation of the directrix for the parabola $x^2 = -16y$.

7. Find the equation of the parabola that has its vertex at the origin, is symmetric with respect to the x axis, and contains the point $(-2, 4)$.

8. Find the equation of the parabola that has its vertex at $(3, 4)$ and its focus at $(3, 1)$.

9. Find the endpoints of the major axis of the ellipse $4x^2 + y^2 = 36$.

10. Find the length of the major axis of the ellipse $x^2 - 4x + 9y^2 - 18y + 4 = 0$.

11. Find the endpoints of the minor axis of the ellipse $9x^2 + 90x + 4y^2 - 8y + 193 = 0$.

12. Find the foci of the ellipse $x^2 + 4y^2 = 16$.

13. Find the center of the ellipse $3x^2 + 30x + y^2 - 16y + 79 = 0$.

14. Find the equation of the ellipse that has the endpoints of its major axis at $(0, \pm 10)$ and its foci at $(0, \pm 8)$.

15. Find the equation of the ellipse that has the endpoints of its major axis at $(2, -2)$ and $(10, -2)$ and the endpoints of its minor axis at $(6, 0)$ and $(6, -4)$.

16. Find the equations of the asymptotes of the hyperbola $4y^2 - 9x^2 = 32$.

17. Find the vertices of the hyperbola $y^2 - 6y - 3x^2 - 6x - 3 = 0$.

18. Find the foci of the hyperbola $5x^2 - 4y^2 = 20$.

19. Find the equation of the hyperbola that has its vertices at $(\pm 6, 0)$ and its foci at $(\pm 4\sqrt{3}, 0)$.

20. Find the equation of the hyperbola that has its vertices at $(0, 4)$ and $(-2, 4)$ and its foci at $(2, 4)$ and $(-4, 4)$.

21. How many real number solutions are there for the system $\begin{pmatrix} x^2 + y^2 = 16 \\ x^2 - 4y = 8 \end{pmatrix}$?

22. Solve the system $\begin{pmatrix} x^2 + 4y^2 = 25 \\ xy = 6 \end{pmatrix}$.

For Problems 23–25, graph each conic section.

23. $y^2 + 4y + 8x - 4 = 0$

24. $9x^2 - 36x + 4y^2 + 16y + 16 = 0$

25. $x^2 + 6x - 3y^2 = 0$

For Problems 1–10, evaluate each expression.

1. $\left(\dfrac{8}{27}\right)^{-2/3}$

2. $\sqrt{\dfrac{16}{9}}$

3. $\sqrt[6]{\dfrac{1}{64}}$

4. $(2^{-1} - 3^{-2})^{-1}$

5. $(2^{-3})^{-2}$

6. $9^{5/2}$

7. $\log_4 64$

8. $\ln e^2$

9. $\begin{vmatrix} 3 & -4 \\ 2 & -1 \end{vmatrix}$

10. $\begin{vmatrix} 2 & 1 & -3 \\ 4 & 2 & -1 \\ -1 & -2 & 6 \end{vmatrix}$

For Problems 11–18, perform the indicated operations and simplify when possible.

11. $3(2x - 1) - (3x + 1) + 2(-x - 1)$

12. $(-2 - x)(3 + 4x)$

13. $(x - 5)(3x^2 - 2x + 1)$

14. $(x^2 + x + 3)(2x^2 - x - 7)$

15. $(2x - 3)^3$

16. $\dfrac{-24x^3y^2 + 36x^2y^4}{4xy}$

17. $(3x^3 - 10x^2 + 10x + 12) \div (3x + 2)$

18. $(3x^4 - 2x^3 - 13x^2 + 7x + 6) \div (x - 2)$

For Problems 19–24, factor each polynomial completely. Indicate any that are not factorable using integers.

19. $-12 + 17x - 6x^2$

20. $x^2 - 3x + 2xy - 6y$

21. $2x^3 + 6x^2 - 36x$

22. $2x^2 + x + 5$

23. $8x^3 + 27$

24. $4x^4 - 5x^2 + 1$

For Problems 25–27, evaluate each algebraic expression for the given values of the variables.

25. $\dfrac{96x^3y^4}{48x^2y^3}$ for $x = -6$ and $y = 4$

26. $\dfrac{x - 2}{4} - \dfrac{x + 9}{7}$ for $x = 17$

27. $\dfrac{4a^2b}{9a^3b^2} \div \dfrac{8b}{3a}$ for $a = 5$ and $b = -2$

For Problems 28–30, find each of the indicated products or quotients, and express answers in the standard form of a complex number.

28. $(5 + 2i)(4 - 3i)$

29. $(-2 - i)(3 - 6i)$

30. $\dfrac{2 - i}{3 + 4i}$

For Problems 31–34, perform the indicated operations and express the answers in simplest radical form.

31. $\sqrt{8}(\sqrt{6} - 2\sqrt{3})$

32. $(2\sqrt{3} + 3\sqrt{5})(2\sqrt{3} - 3\sqrt{5})$

33. $\dfrac{5\sqrt{2}}{3\sqrt{6}}$

34. $\dfrac{\sqrt{5}}{\sqrt[3]{5}}$

35. Find the slope of the line $3x - 8y = -2$.

36. Find the coordinates of the point that is two-thirds of the distance from $(4, -3)$ to $(-2, 6)$.

37. Find the midpoint of the line segment whose endpoints are $(-3, -4)$ and $(5, 10)$.

38. A certain highway has a 4% grade. How many feet does it rise in a horizontal distance of 4000 feet?

39. Find the equation of the line determined by the two points $(-2, 4)$ and $(-4, 7)$.

40. Find the equation of the line that has a slope of $\dfrac{5}{6}$ and a y intercept of -3.

41. Find the equation of the line that is parallel to $-2x + 3y = -6$ and contains the point $(-1, 4)$.

42. Find the equation of the line that is perpendicular to the line $5x - 7y = 10$ and contains the point $(-3, -1)$.

43. What type of symmetry does the graph of $x = y^2 - 4$ possess?

44. What type of symmetry does the graph of $x^2y^2 = 6$ possess?

45. Find the center of the circle $x^2 + 12x + y^2 - 8y + 42 = 0$.

46. Find the length of a radius of the circle $x^2 - 10x + y^2 + 6y + 18 = 0$.

47. Write the equation of the circle that passes through the origin and has its center at $(-3, 4)$.

48. Find the equation of the circle for which the line segment with endpoints at $(4, -2)$ and $(6, 4)$ is a diameter of the circle.

49. Find the equations of the asymptotes for the hyperbola $9x^2 - y^2 = 14$.

50. If $f(x) = -x^2 + 7x - 4$, find $f(-3)$.

51. If $f(x) = 4x^2 - 2x - 7$, find $\dfrac{f(a + h) - f(a)}{h}$.

52. Express the domain of the function $f(x) = \dfrac{3}{x^2 - 4x + 4}$ using interval notation.

53. Express the domain of the function $f(x) = \sqrt{6 - x - x^2}$ using interval notation.

54. Find the vertex of the parabola $f(x) = 2x^2 + 8x + 15$.

55. Find the vertex of the parabola $f(x) = -x^2 + 16x - 70$.

56. Find the x intercepts of the parabola $f(x) = 2x^2 - 19x + 9$.

57. If $f(x) = 3x - 2$ and $g(x) = 2x^2 + 3x - 6$, find $f(g(-1))$ and $g(f(3))$.

58. If $f(x) = -3x^2 - x + 8$ and $g(x) = -2x + 5$, find $f(g(x))$ and $g(f(x))$.

59. If $f(x) = -\dfrac{2}{3}x + 1$, find the inverse of f.

60. Suppose that y varies jointly as x and z. If $y = 24$ when $x = 2$ and $z = -4$, find y when $x = -6$ and $z = -5$.

61. Suppose y is directly proportional to x and inversely proportional to the square root of w. If $y = \dfrac{2}{3}$ when $x = \dfrac{1}{2}$ and $w = 9$, find y when $x = 8$ and $w = \dfrac{1}{4}$.

62. Barium-140 has a half-life of 13 days. If there are 750 milligrams of barium initially, how many milligrams remain after 50 days? Express the answer to the nearest tenth of a milligram.

63. If a car depreciates 10% per year, how much will a $30,000 car be worth in 3 years?

64. The number of grams of a certain radioactive substance present after t seconds is given by the equation $Q(t) = 150e^{-0.2t}$. How many grams remain after 12 seconds? Express the answer to the nearest tenth of a gram.

65. If $f(x) = -3x + 4$ and $g(x) = \dfrac{4 - x}{3}$, are f and g inverse functions?

66. Is $x + 1$ a factor of $x^4 - 4x^3 - 22x^2 + 4x + 21$?

67. Is $x - 4$ a factor of $2x^3 - 3x^2 - 4x + 6$?

68. Find the x intercepts of the graph of the function $f(x) = x^3 - 2x^2 - 5x + 6$.

69. Find the equation of the oblique asymptote for the graph of the function $f(x) = \dfrac{x^2 + 2x + 1}{x - 1}$.

70. Find the partial fraction decomposition of $\dfrac{x + 10}{2x^2 - 9x + 4}$.

71. If $A = \begin{bmatrix} 2 & -1 \\ 3 & 4 \end{bmatrix}$ and $B = \begin{bmatrix} 6 & -2 \\ -1 & 5 \end{bmatrix}$, find $2A - 3B$.

72. If $A = \begin{bmatrix} 1 & -2 \\ 3 & -4 \end{bmatrix}$ and $B = \begin{bmatrix} -3 & -4 \\ 2 & -1 \end{bmatrix}$, find AB and BA.

73. Find A^{-1} if $A = \begin{bmatrix} 2 & 3 \\ 1 & 2 \end{bmatrix}$.

74. Identify the graph of each of the following equations as a circle, ellipse, parabola, or hyperbola.

a. $3x^2 + 9x + 2y^2 - 4y + 4 = 0$

b. $y^2 - 6y + 2x + 14 = 0$

c. $x^2 + 8x + 8y - 8 = 0$

d. $x^2 - 16x + y^2 + 12y + 99 = 0$

e. $7y^2 + 14y - 9x^2 - 72x - 200 = 0$

For Problems 75–79, graph each equation.

75. $4x^2 + 9y^2 - 16x - 18y - 11 = 0$

76. $x^2 - 6x + y^2 + 2y + 1 = 0$

77. $y^2 = x - 2$

78. $x^2 - y^2 - 2x + 4y - 12 = 0$

79. $y = -2x^2 + 12x - 15$

For Problems 80–90, graph each function.

80. $f(x) = \log_2(x + 2)$

81. $f(x) = -x(x + 1)(x - 3)$

82. $f(x) = 3|x - 1| - 2$ **83.** $f(x) = -\sqrt{x + 3} - 1$

84. $f(x) = e^x + 3$ **85.** $f(x) = -(x + 2)^3 + 1$

86. $f(x) = -3x + 4$ **87.** $f(x) = x^2 + 2x + 2$

88. $f(x) = (x - 2)^4 + 1$

89. $f(x) = \dfrac{2x}{x - 2}$ **90.** $f(x) = \dfrac{x^2 - 3}{x + 1}$

For Problems 91–114, solve each equation, system of equations, or inequality. Express solution sets of inequalities in interval notation.

91. $|3x - 1| = 8$ **92.** $(3x - 2)(2x + 1) = 0$

93. $\sqrt{2x} + x = 4$ **94.** $|2x + 6| \leq 12$

95. $(5x - 2)^2 = 18$ **96.** $2x^2 + x + 5 = 0$

97. $\begin{pmatrix} 3x - 2y = 4 \\ 2x + 5y = 7 \end{pmatrix}$ **98.** $(x + 2)(5x - 1) = 36$

99. $(x + 6)(x + 2)(x - 1) > 0$

100. $x^3 + 7x^2 + 4x - 12 = 0$

101. $\begin{pmatrix} x - 2y + z = 6 \\ -2x + y - z = -5 \\ 4x + 3y + 2z = 5 \end{pmatrix}$

102. $3^{3x+1} = 9^{x-2}$

103. $(4x + 3)^2 = -16$

104. $3x^2 - x - 3 = 0$ **105.** $x^2 + 2x < 48$

106. $\begin{pmatrix} y = 3x - 4 \\ 2x + 5y = -9 \end{pmatrix}$ **107.** $x^2 - 2x - 288 = 0$

108. $|7x - 2| > 12$

109. $(2x + 3)(x - 4) = (3x - 1)(x - 4)$

110. $4^{x-2} = 12$

111. $\log_2(x - 1) + \log_2(x + 4) = 1$

112. $(6x - 1)^2 = 45$

113. $\dfrac{x - 4}{x + 2} + 3 \geq 0$

114. $x^{2/3} - 3x^{1/3} - 4 = 0$

For Problems 115–125, use an equation, a system of equations, or an inequality to help solve each problem.

115. Six cans of soda and 3 bags of potato chips cost $8.37. At the same prices 8 cans of soda and 5 bags of potato chips cost $12.45. Find the price per can of soda and the price per bag of potato chips.

116. A recipe calls for $1\frac{1}{4}$ cups of sugar for every $2\frac{1}{2}$ cups of flour. How many cups of flour are needed if 3 cups of sugar are to be used?

117. Kaya has nickels, dimes, and quarters totaling $14. She has 5 more dimes than nickels, and the number of quarters is 2 more than twice the number of nickels. How many coins of each kind does she have?

118. Tina is 4 years older than Sherry. In 5 years the sum of their ages will be 48. Find their present ages.

119. One number is 72 larger than another number. If the larger number is divided by the smaller, the quotient is 6 and the remainder is 2. Find the numbers.

120. A retailer has some shirts that cost $24 each. She wants to sell them at a profit of 40% of the selling price. What price should she charge for the shirts?

121. How much water must be evaporated from 30 gallons of a 10% salt solution in order to obtain a 20% salt solution?

122. Josh rode his bicycle out into the country at a speed of 20 miles per hour and returned along the same route at 15 miles per hour. If the round trip took 5 hours and 50 minutes, how far out did he ride?

123. Suppose Mona invested $500 at 8% interest. At what rate must another $500 be invested so that the two investments yield more than $100 of yearly interest?

124. The time it takes to construct a highway varies directly with the length of the road and inversely with the number of workers. If it takes 100 workers 4 weeks to construct 2 miles of highway, how long will it take 80 workers to construct 10 miles of highway?

125. Suppose that the number of bacteria present in a certain culture after t minutes is given by the equation $Q(t) = Q_0 e^{0.04t}$, where Q_0 represents the initial number of bacteria. How long will it take for the number of bacteria to increase from 500 to 2000?

9

Sequences and Mathematical Induction

© Comstock Images / Jupiter Images

The concept of proof by mathematical induction can be depicted as an infinite series of dominoes being toppled.

Suppose that an auditorium has 35 seats in the first row, 40 seats in the second row, 45 seats in the third row, and so on for ten rows. The numbers 35, 40, 45, 50, . . . , 80 represent the number of seats per row from row 1 through row 10. This list of numbers has a constant difference of 5 between any two successive numbers; such a list is called an **arithmetic sequence.** (Used in this sense, the word "arithmetic" is pronounced with the accent on the syllable *met.*)

Suppose that a fungus culture growing under controlled conditions doubles in size each day. If today the size of the culture is 6 units, then the numbers 12, 24, 48, 96, 192 represent the size of the culture for the next 5 days. In this list of numbers, each number after the first is twice the previous number; such a list is called a **geometric sequence.** Arithmetic sequences and geometric sequences will be the center of our attention in this chapter.

9.1   Arithmetic Sequences

Objectives

- Write the terms of a sequence.
- Find the general term for an arithmetic sequence.
- Find a specific term for an arithmetic sequence.
- Find the sum of the terms of an arithmetic sequence.
- Determine the sum indicated by summation notation.

An **infinite sequence** is a function whose domain is the set of positive integers. For example, consider the function defined by the equation

$$f(n) = 5n + 1$$

where the domain is the set of positive integers. If we substitute the numbers of the domain in order, starting with 1, we can list the resulting ordered pairs:

$$(1, 6) \qquad (2, 11) \qquad (3, 16) \qquad (4, 21) \qquad (5, 26)$$

and so on. However, because we know we are using the domain of positive integers in order, starting with 1, there is no need to use ordered pairs. We can simply express the infinite sequence as

$$6, 11, 16, 21, 26, \ldots$$

Often the letter a is used to represent sequential functions, and the functional value of a at n is written a_n (this is read "a sub n") instead of $a(n)$. The sequence is then expressed

$$a_1, a_2, a_3, a_4, \ldots$$

where a_1 is the **first term,** a_2 is the **second term,** a_3 is the **third term,** and so on. The expression a_n, which defines the sequence, is called the **general term** of the sequence. Knowing the general term of a sequence enables us to find as many terms of the sequence as needed and also to find any specific terms. Consider the following example.

EXAMPLE 1   Find the first five terms of the sequence where $a_n = 2n^2 - 3$. Find the 20th term.

Solution

The first five terms are generated by replacing n with 1, 2, 3, 4, and 5.

$$a_1 = 2(1)^2 - 3 = -1 \qquad a_2 = 2(2)^2 - 3 = 5$$

$$a_3 = 2(3)^2 - 3 = 15 \qquad a_4 = 2(4)^2 - 3 = 29$$

$$a_5 = 2(5)^2 - 3 = 47$$

The first five terms are thus $-1, 5, 15, 29$, and 47. The 20th term is

$$a_{20} = 2(20)^2 - 3 = 797$$

▶ **Now work Problem 5.** ■

An **arithmetic sequence** (also called an arithmetic progression) is a sequence that has a common difference between successive terms. The following are examples of arithmetic sequences:

$1, 8, 15, 22, 29, \ldots$

$4, 7, 10, 13, 16, \ldots$

$4, 1, -2, -5, -8, \ldots$

$-1, -6, -11, -16, -21, \ldots$

The common difference in the first sequence is 7; that is, $8 - 1 = 7$, $15 - 8 = 7$, $22 - 15 = 7$, $29 - 22 = 7$, and so on. The common differences in the next three sequences are 3, -3, and -5, respectively.

In a more general setting, we say that the sequence

$$a_1, a_2, a_3, a_4, \ldots, a_n, \ldots$$

is an arithmetic sequence if and only if there is a real number d such that

$$a_{k+1} - a_k = d$$

for every positive integer k. The number d is called the **common difference.**

From the definition we see that $a_{k+1} = a_k + d$. In other words, we can generate an arithmetic sequence that has a common difference of d by starting with a first term a_1 and then simply adding d to each successive term.

First term: a_1

Second term: $a_1 + d$

Third term: $a_1 + 2d$ $\qquad (a_1 + d) + d = a_1 + 2d$

Fourth term: $a_1 + 3d$

$\vdots$

nth term: $a_1 + (n - 1)d$

Thus the **general term** of an arithmetic sequence is given by

$$a_n = a_1 + (n - 1)d$$

where a_1 is the first term, and d is the common difference. This formula for the general term can be used to solve a variety of problems involving arithmetic sequences.

E X A M P L E   2

Find the general-term expression for the arithmetic sequence $6, 2, -2, -6, \ldots$.

Solution

The common difference, d, is $2 - 6 = -4$, and the first term, a_1, is 6. We can substitute these values into $a_n = a_1 + (n - 1)d$ and simplify to obtain

$$a_n = a_1 + (n - 1)d$$

$$a_n = 6 + (n - 1)(-4)$$

$$a_n = 6 - 4n + 4$$

$$a_n = -4n + 10$$

▶ **Now work Problem 15.**   ■

E X A M P L E   3

Find the 40th term of the arithmetic sequence $1, 5, 9, 13, \ldots$.

Solution

Using $a_n = a_1 + (n - 1)d$, we obtain

$$a_{40} = 1 + (40 - 1)4$$

$$a_{40} = 1 + (39)(4)$$

$$a_{40} = 157$$

▶ **Now work Problem 25.**   ■

E X A M P L E   4

Find the first term of the arithmetic sequence where the fourth term is 26 and the ninth term is 61.

Solution

Using $a_n = a_1 + (n - 1)d$ with $a_4 = 26$ (the fourth term is 26) and $a_9 = 61$ (the ninth term is 61), we have

$$26 = a_1 + (4 - 1)d = a_1 + 3d$$

$$61 = a_1 + (9 - 1)d = a_1 + 8d$$

Solving the system of equations

$$\begin{pmatrix} a_1 + 3d = 26 \\ a_1 + 8d = 61 \end{pmatrix}$$

yields $a_1 = 5$ and $d = 7$. Thus the first term is 5.

▶ **Now work Problem 31.**   ■

■ Sums of Arithmetic Sequences

We often use sequences to solve problems, so we need to be able to find the sum of a certain number of terms of the sequence. Before we develop a general-sum formula for arithmetic sequences, let's consider an approach to a specific problem that we can then use in a general setting.

E X A M P L E 5

Find the sum of the first 100 positive integers.

Solution

We are being asked to find the sum of $1 + 2 + 3 + 4 + \cdots + 100$. Rather than adding in the usual way, let's find the sum in the following manner:

$$
\begin{array}{ccccccc}
1 + & 2 + & 3 + & 4 + & \cdots & + & 100 \\
100 + & 99 + & 98 + & 97 + & \cdots & + & 1 \\
\hline
101 + & 101 + & 101 + & 101 + & \cdots & + & 101
\end{array}
$$

$$\frac{\overset{50}{\cancel{100}}(101)}{2} = 5050$$

Note that we simply wrote the indicated sum forward and backward, and then we added the results. In so doing, we produced 100 sums of 101, but half of them are repeats. For example, $100 + 1$ and $1 + 100$ are both counted in this process. Thus we divide the product $(100)(101)$ by 2, which yields the final result of 5050. ■

The *forward–backward* approach we used in Example 5 can be used to develop a formula for finding the sum of the first n terms of any arithmetic sequence. Consider an arithmetic sequence $a_1, a_2, a_3, a_4, \ldots, a_n$ with a common difference of d. We use S_n to represent the sum of the first n terms and proceed as follows:

$$S_n = a_1 + (a_1 + d) + (a_1 + 2d) + \cdots + (a_n - 2d) + (a_n - d) + a_n$$

Now we write this sum in reverse:

$$S_n = a_n + (a_n - d) + (a_n - 2d) + \cdots + (a_1 + 2d) + (a_1 + d) + a_1$$

We add the two equations to produce

$$2S_n = (a_1 + a_n) + (a_1 + a_n) + (a_1 + a_n) + \cdots + (a_1 + a_n) + (a_1 + a_n) + (a_1 + a_n)$$

That is, we have n sums of $a_1 + a_n$, so

$$2S_n = n(a_1 + a_n)$$

from which we obtain a **sum formula:**

$$S_n = \frac{n(a_1 + a_n)}{2}$$

Using the *n*th-term formula, the sum formula, or both, we can solve a variety of problems involving arithmetic sequences.

E X A M P L E 6

Find the sum of the first 30 terms of the arithmetic sequence 3, 7, 11, 15, ...

Solution

Using $a_n = a_1 + (n - 1)d$, we can find the 30th term.

$$a_{30} = 3 + (30 - 1)4 = 3 + 29(4) = 119$$

Now we can use the sum formula.

$$S_{30} = \frac{30(3 + 119)}{2} = 1830$$

▶ **Now work Problem 35.**   ■

E X A M P L E 7

Find the sum $7 + 10 + 13 + \cdots + 157$.

Solution

To use the sum formula, we need to know the number of terms. The *n*th-term formula will do that for us.

$$a_n = a_1 + (n - 1)d$$
$$157 = 7 + (n - 1)3$$
$$157 = 7 + 3n - 3$$
$$157 = 3n + 4$$
$$153 = 3n$$
$$51 = n$$

Now we can use the sum formula.

$$S_{51} = \frac{51(7 + 157)}{2} = 4182$$

▶ **Now work Problem 43.**   ■

Keep in mind that we developed the sum formula for an arithmetic sequence by using the forward–backward technique, which we had previously used on a specific problem. Now that we have the sum formula, we have two choices when solving problems. We can either memorize the formula and use it, or simply use the forward–backward technique. If you choose to use the formula and some day you forget it, don't panic. Just use the forward–backward technique. In other words, understanding the development of a formula often enables you to do problems even when you forget the formula itself.

■ Summation Notation

Sometimes a special notation is used to indicate the sum of a certain number of terms of a sequence. The capital Greek letter *sigma*, Σ, is used as a **summation symbol.** For example,

$$\sum_{i=1}^{5} a_i$$

represents the sum $a_1 + a_2 + a_3 + a_4 + a_5$. The letter i is frequently used as the **index of summation;** the letter i takes on all integer values from the lower limit to the upper limit, inclusive. Thus

$$\sum_{i=1}^{4} b_i = b_1 + b_2 + b_3 + b_4$$

$$\sum_{i=3}^{7} a_i = a_3 + a_4 + a_5 + a_6 + a_7$$

$$\sum_{i=1}^{15} i^2 = 1^2 + 2^2 + 3^2 + \cdots + 15^2$$

$$\sum_{i=1}^{n} a_i = a_1 + a_2 + a_3 + \cdots + a_n$$

If $a_1, a_2, a_3, \ldots$ represents an arithmetic sequence, we can now write the **sum formula**

$$\sum_{i=1}^{n} a_i = \frac{n}{2}(a_1 + a_n)$$

E X A M P L E   8

Find the sum $\displaystyle\sum_{i=1}^{50} (3i + 4)$.

Solution

This indicated sum means

$$\sum_{i=1}^{50} (3i + 4) = [3(1) + 4] + [3(2) + 4] + [3(3) + 4] + \cdots + [3(50) + 4]$$

$$= 7 + 10 + 13 + \cdots + 154$$

Because this is an indicated sum of an arithmetic sequence, we can use our sum formula.

$$S_{50} = \frac{50}{2}(7 + 154) = 4025$$

▶   **Now work Problem 59.**   ■

EXAMPLE 9   Find the sum $\displaystyle\sum_{i=2}^{7} 2i^2$.

Solution

This indicated sum means

$$\sum_{i=2}^{7} 2i^2 = 2(2)^2 + 2(3)^2 + 2(4)^2 + 2(5)^2 + 2(6)^2 + 2(7)^2$$

$$= 8 + 18 + 32 + 50 + 72 + 98$$

This is not the indicated sum of an *arithmetic* sequence; therefore we simply add the numbers in the usual way. The sum is 278.

▶ **Now work Problem 67.**   ■

Example 9 suggests a note of caution. Be sure to analyze the sequence of numbers that is represented by the summation symbol. You may or may not be able to use a formula for adding the numbers.

CONCEPT QUIZ   For Problems 1–8, answer true or false.

1. An infinite sequence is a function whose domain is the set of all real numbers.

2. An arithmetic sequence is a sequence that has a common difference between successive terms.

3. The sequence 2, 4, 8, 16, . . . is an arithmetic sequence.

4. The odd whole numbers form an arithmetic sequence.

5. The terms of an arithmetic sequence are always positive.

6. The 6th term of an arithmetic sequence is equal to the first term plus 6 times the common difference.

7. The sum formula for n terms of an arithmetic sequence is n times the average of the first and last terms.

8. The indicated sum $\displaystyle\sum_{i=1}^{4} (2i - 7)^2$ is the sum of the first four terms of an arithmetic sequence.

Problem Set 9.1

For Problems 1–10, write the first five terms of the sequence that has the indicated general term.

1. $a_n = 3n - 7$

2. $a_n = 5n - 2$

3. $a_n = -2n + 4$

4. $a_n = -4n + 7$

5. $a_n = 3n^2 - 1$

6. $a_n = 2n^2 - 6$

7. $a_n = n(n - 1)$

8. $a_n = (n + 1)(n + 2)$

9. $a_n = 2^{n+1}$

10. $a_n = 3^{n-1}$

11. Find the 15th and 30th terms of the sequence where $a_n = -5n - 4$.

12. Find the 20th and 50th terms of the sequence where $a_n = -n - 3$.

13. Find the 25th and 50th terms of the sequence where $a_n = (-1)^{n+1}$.

14. Find the 10th and 15th terms of the sequence where $a_n = -n^2 - 10$.

For Problems 15–24, find the general term (the nth term) for each arithmetic sequence.

15. $11, 13, 15, 17, 19, \ldots$ **16.** $7, 10, 13, 16, 19, \ldots$

17. $2, -1, -4, -7, -10, \ldots$

18. $4, 2, 0, -2, -4, \ldots$

19. $\dfrac{3}{2}, 2, \dfrac{5}{2}, 3, \dfrac{7}{2}, \ldots$ **20.** $0, \dfrac{1}{2}, 1, \dfrac{3}{2}, 2, \ldots$

21. $2, 6, 10, 14, 18, \ldots$ **22.** $2, 7, 12, 17, 22, \ldots$

23. $-3, -6, -9, -12, -15, \ldots$

24. $-4, -8, -12, -16, -20, \ldots$

For Problems 25–30, find the required term for each arithmetic sequence.

25. The 15th term of $3, 8, 13, 18, \ldots$

26. The 20th term of $4, 11, 18, 25, \ldots$

27. The 30th term of $15, 26, 37, 48, \ldots$

28. The 35th term of $9, 17, 25, 33, \ldots$

29. The 52nd term of $1, \dfrac{5}{3}, \dfrac{7}{3}, 3, \ldots$

30. The 47th term of $\dfrac{1}{2}, \dfrac{5}{4}, 2, \dfrac{11}{4}, \ldots$

For Problems 31–42, solve each problem.

31. If the 6th term of an arithmetic sequence is 12 and the 10th term is 16, find the first term.

32. If the 5th term of an arithmetic sequence is 14 and the 12th term is 42, find the first term.

33. If the 3rd term of an arithmetic sequence is 20 and the 7th term is 32, find the 25th term.

34. If the 5th term of an arithmetic sequence is -5 and the 15th term is -25, find the 50th term.

35. Find the sum of the first 50 terms of the arithmetic sequence $5, 7, 9, 11, 13, \ldots$.

36. Find the sum of the first 30 terms of the arithmetic sequence $0, 2, 4, 6, 8, \ldots$.

37. Find the sum of the first 40 terms of the arithmetic sequence $2, 6, 10, 14, 18, \ldots$.

38. Find the sum of the first 60 terms of the arithmetic sequence $-2, 3, 8, 13, 18, \ldots$.

39. Find the sum of the first 75 terms of the arithmetic sequence $5, 2, -1, -4, -7, \ldots$.

40. Find the sum of the first 80 terms of the arithmetic sequence $7, 3, -1, -5, -9, \ldots$.

41. Find the sum of the first 50 terms of the arithmetic sequence $\dfrac{1}{2}, 1, \dfrac{3}{2}, 2, \dfrac{5}{2}, \ldots$.

42. Find the sum of the first 100 terms of the arithmetic sequence $-\dfrac{1}{3}, \dfrac{1}{3}, 1, \dfrac{5}{3}, \dfrac{7}{3}, \ldots$.

For Problems 43–50, find the indicated sum.

43. $1 + 5 + 9 + 13 + \cdots + 197$

44. $3 + 8 + 13 + 18 + \cdots + 398$

45. $2 + 8 + 14 + 20 + \cdots + 146$

46. $6 + 9 + 12 + 15 + \cdots + 93$

47. $(-7) + (-10) + (-13) + (-16) + \cdots + (-109)$

48. $(-5) + (-9) + (-13) + (-17) + \cdots + (-169)$

49. $(-5) + (-3) + (-1) + 1 + \cdots + 119$

50. $(-7) + (-4) + (-1) + 2 + \cdots + 131$

For Problems 51–58, solve each problem.

51. Find the sum of the first 200 odd whole numbers.

52. Find the sum of the first 175 positive even whole numbers.

53. Find the sum of all even numbers between 18 and 482, inclusive.

54. Find the sum of all odd numbers between 17 and 379, inclusive.

55. Find the sum of the first 30 terms of the arithmetic sequence with the general term $a_n = 5n - 4$.

56. Find the sum of the first 40 terms of the arithmetic sequence with the general term $a_n = 4n - 7$.

57. Find the sum of the first 25 terms of the arithmetic sequence with the general term $a_n = -4n - 1$.

58. Find the sum of the first 35 terms of the arithmetic sequence with the general term $a_n = -5n - 3$.

For Problems 59–70, find each sum.

59. $\displaystyle\sum_{i=1}^{45}(5i + 2)$

60. $\displaystyle\sum_{i=1}^{38}(3i + 6)$

61. $\displaystyle\sum_{i=1}^{30}(-2i + 4)$

62. $\displaystyle\sum_{i=1}^{40}(-3i + 3)$

63. $\displaystyle\sum_{i=4}^{32}(3i - 10)$

64. $\displaystyle\sum_{i=6}^{47}(4i - 9)$

65. $\displaystyle\sum_{i=10}^{20}4i$

66. $\displaystyle\sum_{i=15}^{30}(-5i)$

67. $\displaystyle\sum_{i=1}^{5}i^2$

68. $\displaystyle\sum_{i=1}^{6}(i^2 + 1)$

69. $\displaystyle\sum_{i=3}^{8}(2i^2 + i)$

70. $\displaystyle\sum_{i=4}^{7}(3i^2 - 2)$

■ ■ ■ THOUGHTS INTO WORDS

71. Before developing the formula $a_n = a_1 + (n - 1)d$, we stated the equation $a_{k+1} - a_k = d$. In your own words, explain what this equation says.

72. Explain how to find the sum $1 + 2 + 3 + 4 + \cdots + 175$ without using the sum formula.

73. Explain in words how to find the sum of the first n terms of an arithmetic sequence.

74. Explain how one can tell that a particular sequence is an arithmetic sequence.

■ ■ ■ FURTHER INVESTIGATIONS

The general term of a sequence can consist of one expression for certain values of n and another expression (or expressions) for other values of n. That is, a **multiple description** of the sequence can be given. For example,

$$a_n = \begin{cases} 2n + 3 & \text{for } n \text{ odd} \\ 3n - 2 & \text{for } n \text{ even} \end{cases}$$

means that we use $a_n = 2n + 3$ for $n = 1, 3, 5, 7, \ldots$ and we use $a_n = 3n - 2$ for $n = 2, 4, 6, 8, \ldots$. The first six terms of this sequence are 5, 4, 9, 10, 13, and 16.

For Problems 75–78, write the first six terms of each sequence.

75. $a_n = \begin{cases} 2n + 1 & \text{for } n \text{ odd} \\ 2n - 1 & \text{for } n \text{ even} \end{cases}$

76. $a_n = \begin{cases} \dfrac{1}{n} & \text{for } n \text{ odd} \\ n^2 & \text{for } n \text{ even} \end{cases}$

77. $a_n = \begin{cases} 3n + 1 & \text{for } n \leq 3 \\ 4n - 3 & \text{for } n > 3 \end{cases}$

78. $a_n = \begin{cases} 5n - 1 & \text{for } n \text{ a multiple of 3} \\ 2n & \text{otherwise} \end{cases}$

The multiple-description approach can also be used to give a **recursive description** for a sequence. A sequence is said to be **described recursively** if the first n terms are stated and then each succeeding term is defined as a function of one or more of the preceding terms. For example,

$$\begin{cases} a_1 = 2 \\ a_n = 2a_{n-1} \quad \text{for } n \geq 2 \end{cases}$$

means that the first term, a_1, is 2 and each succeeding term is 2 times the previous term. Thus the first six terms are 2, 4, 8, 16, 32, and 64.

For Problems 79–84, write the first six terms of each sequence.

79. $\begin{cases} a_1 = 4 \\ a_n = 3a_{n-1} \quad \text{for } n \geq 2 \end{cases}$

80. $\begin{cases} a_1 = 3 \\ a_n = a_{n-1} + 2 \quad \text{for } n \geq 2 \end{cases}$

81. $\begin{cases} a_1 = 1 \\ a_2 = 1 \\ a_n = a_{n-2} + a_{n-1} \quad \text{for } n \geq 3 \end{cases}$

82. $\begin{cases} a_1 = 2 \\ a_2 = 3 \\ a_n = 2a_{n-2} + 3a_{n-1} \quad \text{for } n \geq 3 \end{cases}$

83. $\begin{cases} a_1 = 3 \\ a_2 = 1 \\ a_n = (a_{n-1} - a_{n-2})^2 \quad \text{for } n \geq 3 \end{cases}$

84. $\begin{cases} a_1 = 1 \\ a_2 = 2 \\ a_3 = 3 \\ a_n = a_{n-1} + a_{n-2} + a_{n-3} \quad \text{for } n \geq 4 \end{cases}$

Answers to the Concept Quiz

1. False   **2.** True   **3.** False   **4.** True   **5.** False   **6.** False   **7.** True   **8.** False

9.2 Geometric Sequences

Objectives

■ Find the general term for a geometric sequence.

■ Determine the specified terms of a geometric sequence.

■ Find the common ratio of a geometric sequence.

■ Determine the sum of the terms of a geometric sequence.

■ Change a repeating decimal into $\dfrac{a}{b}$ form.

A **geometric sequence** or **geometric progression** is a sequence in which we obtain each term after the first by multiplying the preceding term by a common multiplier, called the **common ratio** of the sequence. We can find the common ratio of a geometric sequence by dividing any term (other than the first) by the preceding term.

The following geometric sequences have common ratios of 3, 2, $\dfrac{1}{2}$, and -4, respectively:

$$1, 3, 9, 27, 81, \ldots$$

$$3, 6, 12, 24, 48, \ldots$$

$$16, 8, 4, 2, 1, \ldots$$

$$-1, 4, -16, 64, -256, \ldots$$

In a more general setting, we say that the sequence $a_1, a_2, a_3, \ldots, a_n, \ldots$ is a geometric sequence if and only if there is a nonzero real number r such that

$$a_{k+1} = ra_k$$

for every positive integer k. The nonzero real number r is called the common ratio of the sequence.

The previous equation can be used to generate a general geometric sequence that has a_1 as a first term and r as a common ratio. We can proceed as follows:

First term: a_1

Second term: $a_1 r$

Third term: $a_1 r^2$ $(a_1 r)(r) = a_1 r^2$

Fourth term: $a_1 r^3$

$\vdots$

nth term: $a_1 r^{n-1}$

Thus the **general term** of a geometric sequence is given by

$$a_n = a_1 r^{n-1}$$

where a_1 is the first term, and r is the common ratio.

EXAMPLE 1

Find the general term for the geometric sequence $8, 16, 32, 64, \ldots$.

Solution

Using $a_n = a_1 r^{n-1}$, we obtain

$$a_n = 8(2)^{n-1} = (2^3)(2)^{n-1} = 2^{n+2}$$

▶ **Now work Problem 1.**   ■

EXAMPLE 2

Find the ninth term of the geometric sequence $27, 9, 3, 1, \ldots$.

Solution

Using $a_n = a_1 r^{n-1}$, we can find the ninth term as follows:

$$a_9 = 27\left(\frac{1}{3}\right)^{9-1} = 27\left(\frac{1}{3}\right)^8 = \frac{3^3}{3^8} = \frac{1}{3^5} = \frac{1}{243}$$

▶ **Now work Problem 13.**   ■

■ Sums of Geometric Sequences

As with arithmetic sequences, we often need to find the sum of a certain number of terms of a geometric sequence. Before we develop a general-sum formula for geometric sequences, let's consider an approach to a specific problem that we can then use in a general setting.

EXAMPLE 3   Find the sum $1 + 3 + 9 + 27 + \cdots + 6561$.

Solution

We let S represent the sum and proceed as follows:

$$S = 1 + 3 + 9 + 27 + \cdots + 6561 \tag{1}$$

$$3S = 3 + 9 + 27 + \cdots + 6561 + 19{,}683 \tag{2}$$

Equation (2) is the result of multiplying equation (1) by the common ratio, 3. Subtracting equation (1) from equation (2) produces

$$2S = 19{,}683 - 1 = 19{,}682$$

$$S = 9841$$

▶ **Now work Problem 33.**   ■

Now let's consider a general geometric sequence $a_1, a_1 r, a_1 r^2, \ldots, a_1 r^{n-1}$. By applying a procedure similar to the one we used in Example 3, we can develop a formula for finding the sum of the first n terms of any geometric sequence. We let S_n represent the sum of the first n terms.

$$S_n = a_1 + a_1 r + a_1 r^2 + \cdots + a_1 r^{n-1} \tag{3}$$

Next we multiply both sides of equation (3) by the common ratio, r.

$$r S_n = a_1 r + a_1 r^2 + a_1 r^3 + \cdots + a_1 r^n \tag{4}$$

We then subtract equation (3) from equation (4).

$$r S_n - S_n = a_1 r^n - a_1$$

When we apply the distributive property to the left side and then solve for S_n, we obtain

$$S_n(r - 1) = a_1 r^n - a_1$$

$$S_n = \frac{a_1 r^n - a_1}{r - 1}, \qquad r \neq 1$$

Therefore the sum of the first n terms of a geometric sequence with a first term a_1 and a common ratio r is given by

$$S_n = \frac{a_1 r^n - a_1}{r - 1}, \qquad r \neq 1$$

<table>
<tr><td>**E X A M P L E   4**</td><td>Find the sum of the first eight terms of the geometric sequence $1, 2, 4, 8, \ldots$.</td></tr>
</table>

Solution

Use the sum formula to obtain

$$S_8 = \frac{1(2)^8 - 1}{2 - 1} = \frac{2^8 - 1}{1} = 255$$

(▶) **Now work Problem 25.**   ∎

If the common ratio of a geometric sequence is less than 1, it may be more convenient to change the form of the sum formula. That is, the fraction

$$\frac{a_1 r^n - a_1}{r - 1}$$

can be changed to

$$\frac{a_1 - a_1 r^n}{1 - r}$$

by multiplying both the numerator and the denominator by -1. Thus by using

$$S_n = \frac{a_1 - a_1 r^n}{1 - r}$$

we can sometimes avoid unnecessary work with negative numbers when $r < 1$, as the next example illustrates.

<table>
<tr><td>**E X A M P L E   5**</td><td>Find the sum $1 + \dfrac{1}{2} + \dfrac{1}{4} + \cdots + \dfrac{1}{256}$.</td></tr>
</table>

Solution A

To use the sum formula, we need to know the number of terms, which can be found by counting them or by applying the nth-term formula, as follows:

$$a_n = a_1 r^{n-1}$$

$$\frac{1}{256} = 1\left(\frac{1}{2}\right)^{n-1}$$

$$\left(\frac{1}{2}\right)^8 = \left(\frac{1}{2}\right)^{n-1}$$

$$8 = n - 1 \qquad \text{If } b^n = b^m, \text{ then } n = m$$

$$9 = n$$

Now we use $n = 9$, $a_1 = 1$, and $r = \dfrac{1}{2}$ in the sum formula of the form

$$S_n = \frac{a_1 - a_1 r^n}{1 - r}$$

$$S_9 = \frac{1 - 1\left(\dfrac{1}{2}\right)^9}{1 - \dfrac{1}{2}} = \frac{1 - \dfrac{1}{512}}{\dfrac{1}{2}} = \frac{\dfrac{511}{512}}{\dfrac{1}{2}} = 1\frac{255}{256}$$

We can also do a problem like Example 5 without finding the number of terms; we use the general approach illustrated in Example 3. Solution B demonstrates this idea.

Solution B

We let S represent the desired sum.

$$S = 1 + \frac{1}{2} + \frac{1}{4} + \cdots + \frac{1}{256}$$

We can multiply both sides by the common ratio, $\dfrac{1}{2}$.

$$\frac{1}{2}S = \frac{1}{2} + \frac{1}{4} + \frac{1}{8} + \cdots + \frac{1}{256} + \frac{1}{512}$$

We then subtract the second equation from the first and solve for S.

$$\frac{1}{2}S = 1 - \frac{1}{512} = \frac{511}{512}$$

$$S = \frac{511}{256} = 1\frac{255}{256}$$

▶ **Now work Problem 35.**   ■

Summation notation can also be used to indicate the sum of a certain number of terms of a geometric sequence.

E X A M P L E   6   Find the sum $\displaystyle\sum_{i=1}^{10} 2^i$.

Solution

This indicated sum means

$$\sum_{i=1}^{10} 2^i = 2^1 + 2^2 + 2^3 + \cdots + 2^{10}$$

$$= 2 + 4 + 8 + \cdots + 1024$$

This is the indicated sum of a geometric sequence, so we can use the sum formula, with $a_1 = 2$, $r = 2$, and $n = 10$.

$$S_{10} = \frac{2(2)^{10} - 2}{2 - 1} = \frac{2(2^{10} - 1)}{1} = 2046$$

Now work Problem 39.

■ The Sum of an Infinite Geometric Sequence

Let's take the formula

$$S_n = \frac{a_1 - a_1 r^n}{1 - r}$$

and rewrite the right side by applying the property

$$\frac{a - b}{c} = \frac{a}{c} - \frac{b}{c}$$

Thus we obtain

$$S_n = \frac{a_1}{1 - r} - \frac{a_1 r^n}{1 - r} \tag{1}$$

Now let's examine the behavior of r^n for $|r| < 1$; that is, for $-1 < r < 1$. For example, suppose that $r = \frac{1}{2}$; then

$$r^2 = \left(\frac{1}{2}\right)^2 = \frac{1}{4} \qquad r^3 = \left(\frac{1}{2}\right)^3 = \frac{1}{8}$$

$$r^4 = \left(\frac{1}{2}\right)^4 = \frac{1}{16} \qquad r^5 = \left(\frac{1}{2}\right)^5 = \frac{1}{32}$$

and so on. We can make $\left(\frac{1}{2}\right)^n$ as close to zero as we please by choosing sufficiently large values for n. In general, for values of r such that $|r| < 1$, the expression r^n approaches zero as n gets larger and larger. Therefore the fraction $a_1 r^n/(1 - r)$ in equation (1) approaches zero as n increases. We say that **the sum of the infinite geometric sequence** is given by

$$S_\infty = \frac{a_1}{1 - r}, \qquad |r| < 1$$

EXAMPLE 7

Find the sum of the infinite geometric sequence

$$1, \frac{1}{2}, \frac{1}{4}, \frac{1}{8}, \ldots$$

Solution

Because $a_1 = 1$ and $r = \dfrac{1}{2}$, we obtain

$$S_\infty = \frac{1}{1 - \dfrac{1}{2}} = \frac{1}{\dfrac{1}{2}} = 2$$

▶ **Now work Problem 45.**   ∎

When we state that $S_\infty = 2$ in Example 7, we mean that as we add more and more terms, the sum approaches 2. Observe what happens when we calculate the sum up to five terms.

First term: 1

Sum of first two terms: $1 + \dfrac{1}{2} = 1\dfrac{1}{2}$

Sum of first three terms: $1 + \dfrac{1}{2} + \dfrac{1}{4} = 1\dfrac{3}{4}$

Sum of first four terms: $1 + \dfrac{1}{2} + \dfrac{1}{4} + \dfrac{1}{8} = 1\dfrac{7}{8}$

Sum of first five terms: $1 + \dfrac{1}{2} + \dfrac{1}{4} + \dfrac{1}{8} + \dfrac{1}{16} = 1\dfrac{15}{16}$

If $|r| > 1$, the absolute value of r^n increases without bound as n increases. Consider the following two examples and note the unbounded growth of the absolute value of r^n.

Let $r = 3$	Let $r = -2$	
$r^2 = 3^2 = 9$	$r^2 = (-2)^2 = 4$	
$r^3 = 3^3 = 27$	$r^3 = (-2)^3 = -8$	$|-8| = 8$
$r^4 = 3^4 = 81$	$r^4 = (-2)^4 = 16$	
$r^5 = 3^5 = 243$	$r^5 = (-2)^5 = -32$	$|-32| = 32$

If $r = 1$, then $S_n = na_1$, and as n increases without bound, $|S_n|$ also increases without bound. If $r = -1$, then S_n will be either a_1 or zero. Therefore we say that the sum of any infinite geometric sequence where $|r| \geq 1$ *does not exist.*

■ Repeating Decimals As Sums of Infinite Geometric Sequences

In Section 1.1 we defined rational numbers to be numbers that have either a terminating or a repeating decimal representation. For example,

$$2.23 \qquad 0.147 \qquad 0.\overline{3} \qquad 0.\overline{14} \qquad \text{and} \qquad 0.\overline{56}$$

are rational numbers. (Remember that $0.\overline{3}$ means 0.3333. . . .) Place value provides the basis for changing terminating decimals such as 2.23 and 0.147 to a/b form, where a and b are integers, and $b \neq 0$:

$$2.23 = \frac{223}{100} \quad \text{and} \quad 0.147 = \frac{147}{1000}$$

However, changing repeating decimals to a/b form requires a different technique, and our work with sums of infinite geometric sequences provides the basis for one such approach. Consider the following examples.

EXAMPLE 8

Change $0.\overline{14}$ to a/b form, where a and b are integers and $b \neq 0$.

Solution

The repeating decimal $0.\overline{14}$ can be written as the indicated sum of an infinite geometric sequence with first term 0.14 and common ratio 0.01:

$$0.14 + 0.0014 + 0.000014 + \cdots$$

Using $S_\infty = a_1/(1 - r)$, we obtain

$$S_\infty = \frac{0.14}{1 - 0.01} = \frac{0.14}{0.99} = \frac{14}{99}$$

Thus $0.\overline{14} = \dfrac{14}{99}$.

▶ **Now work Problem 59.** ∎

If the repeating block of digits does not begin immediately after the decimal point, as in $0.5\overline{6}$, we can make an adjustment in the technique we used in Example 8.

EXAMPLE 9

Change $0.5\overline{6}$ to a/b form, where a and b are integers, and $b \neq 0$.

Solution

The repeating decimal $0.5\overline{6}$ can be written

$$(0.5) + (0.06 + 0.006 + 0.0006 + \cdots)$$

where

$$0.06 + 0.006 + 0.0006 + \cdots$$

is the indicated sum of the infinite geometric sequence with $a_1 = 0.06$ and $r = 0.1$. Therefore

$$S_\infty = \frac{0.06}{1 - 0.1} = \frac{0.06}{0.9} = \frac{6}{90} = \frac{1}{15}$$

Now we can add 0.5 and $\frac{1}{15}$:

$$0.5\overline{6} = 0.5 + \frac{1}{15} = \frac{1}{2} + \frac{1}{15} = \frac{15}{30} + \frac{2}{30} = \frac{17}{30}$$

▶ **Now work Problem 63.** ∎

CONCEPT QUIZ For Problems 1–8, answer true or false.

1. The common ratio for a geometric sequence is found by dividing any term by the next successive term.

2. The 5th term of a geometric sequence is equal to the first term multiplied by the common ratio raised to the fourth power.

3. The common ratio of a geometric sequence could be zero.

4. The common ratio of a geometric sequence can be negative.

5. S_∞ denotes the sum of an infinite geometric sequence.

6. If the common ratio of an infinite geometric sequence is greater than 1, then the sum of the sequence does not exist.

7. For the sequence, $2, -2, 2, -2, 2, \ldots, S_9 = 2$.

8. Every repeating decimal can be changed into $\frac{a}{b}$ form where a and b are integers and $b \neq 0$.

Problem Set 9.2

For Problems 1–12, find the general term (the nth term) for each geometric sequence.

1. $3, 6, 12, 24, \ldots$

2. $2, 6, 18, 54, \ldots$

3. $3, 9, 27, 81, \ldots$

4. $2, 4, 8, 16, \ldots$

5. $\frac{1}{4}, \frac{1}{8}, \frac{1}{16}, \frac{1}{32}, \ldots$

6. $8, 4, 2, 1, \ldots$

7. $4, 16, 64, 256, \ldots$

8. $6, 2, \frac{2}{3}, \frac{2}{9}, \ldots$

9. $1, 0.3, 0.09, 0.027, \ldots$

10. $0.2, 0.04, 0.008, 0.0016, \ldots$

11. $1, -2, 4, -8, \ldots$

12. $-3, 9, -27, 81, \ldots$

For Problems 13–20, find the required term for each geometric sequence.

13. The 8th term of $\frac{1}{2}, 1, 2, 4, \ldots$

14. The 7th term of $2, 6, 18, 54, \ldots$

15. The 9th term of $729, 243, 81, 27, \ldots$

16. The 11th term of $768, 384, 192, 96, \ldots$

17. The 10th term of $1, -2, 4, -8, \ldots$

18. The 8th term of $-1, -\dfrac{3}{2}, -\dfrac{9}{4}, -\dfrac{27}{8}, \ldots$

19. The 8th term of $\dfrac{1}{2}, \dfrac{1}{6}, \dfrac{1}{18}, \dfrac{1}{54}, \ldots$

20. The 9th term of $\dfrac{16}{81}, \dfrac{8}{27}, \dfrac{4}{9}, \dfrac{2}{3}, \ldots$

For Problems 21–32, solve each problem.

21. Find the first term of the geometric sequence with 5th term $\dfrac{32}{3}$ and common ratio 2.

22. Find the first term of the geometric sequence with 4th term $\dfrac{27}{128}$ and common ratio $\dfrac{3}{4}$.

23. Find the common ratio of the geometric sequence with 3rd term 12 and 6th term 96.

24. Find the common ratio of the geometric sequence with 2nd term $\dfrac{8}{3}$ and 5th term $\dfrac{64}{81}$.

25. Find the sum of the first ten terms of the geometric sequence $1, 2, 4, 8, \ldots$.

26. Find the sum of the first seven terms of the geometric sequence $3, 9, 27, 81, \ldots$.

27. Find the sum of the first nine terms of the geometric sequence $2, 6, 18, 54, \ldots$.

28. Find the sum of the first ten terms of the geometric sequence $5, 10, 20, 40, \ldots$.

29. Find the sum of the first eight terms of the geometric sequence $8, 12, 18, 27, \ldots$.

30. Find the sum of the first eight terms of the geometric sequence $9, 12, 16, \dfrac{64}{3}, \ldots$.

31. Find the sum of the first ten terms of the geometric sequence $-4, 8, -16, 32, \ldots$.

32. Find the sum of the first nine terms of the geometric sequence $-2, 6, -18, 54, \ldots$.

For Problems 33–38, find each indicated sum.

33. $9 + 27 + 81 + \cdots + 729$

34. $2 + 8 + 32 + \cdots + 8192$

35. $4 + 2 + 1 + \cdots + \dfrac{1}{512}$

36. $1 + (-2) + 4 + \cdots + 256$

37. $(-1) + 3 + (-9) + \cdots + (-729)$

38. $16 + 8 + 4 + \cdots + \dfrac{1}{32}$

For Problems 39–44, find each indicated sum.

39. $\displaystyle\sum_{i=1}^{9} 2^{i-3}$

40. $\displaystyle\sum_{i=1}^{6} 3^{i}$

41. $\displaystyle\sum_{i=2}^{5} (-3)^{i+1}$

42. $\displaystyle\sum_{i=3}^{8} (-2)^{i-1}$

43. $\displaystyle\sum_{i=1}^{6} 3\left(\dfrac{1}{2}\right)^{i}$

44. $\displaystyle\sum_{i=1}^{5} 2\left(\dfrac{1}{3}\right)^{i}$

For Problems 45–56, find the sum of each infinite geometric sequence. If the sequence has no sum, so state.

45. $2, 1, \dfrac{1}{2}, \dfrac{1}{4}, \ldots$

46. $9, 3, 1, \dfrac{1}{3}, \ldots$

47. $1, \dfrac{2}{3}, \dfrac{4}{9}, \dfrac{8}{27}, \ldots$

48. $5, 3, \dfrac{9}{5}, \dfrac{27}{25}, \ldots$

49. $4, 8, 16, 32, \ldots$

50. $32, 16, 8, 4, \ldots$

51. $9, -3, 1, -\dfrac{1}{3}, \ldots$

52. $2, -6, 18, -54, \ldots$

53. $\dfrac{1}{2}, \dfrac{3}{8}, \dfrac{9}{32}, \dfrac{27}{128}, \ldots$

54. $4, -\dfrac{4}{3}, \dfrac{4}{9}, -\dfrac{4}{27}, \ldots$

55. $8, -4, 2, -1, \ldots$

56. $7, \dfrac{14}{5}, \dfrac{28}{25}, \dfrac{56}{125}, \ldots$

For Problems 57–68, change each repeating decimal to a/b form, where a and b are integers and $b \neq 0$. Express a/b in reduced form.

57. $0.\overline{3}$

58. $0.\overline{4}$

59. $0.\overline{26}$

60. $0.\overline{18}$

61. $0.\overline{123}$

62. $0.\overline{273}$

63. $0.2\overline{6}$

64. $0.4\overline{3}$

65. $0.2\overline{14}$

66. $0.3\overline{71}$

67. $2.\overline{3}$

68. $3.\overline{7}$

69. Explain the difference between an arithmetic sequence and a geometric sequence.

70. What does it mean to say that the sum of the infinite geometric sequence $1, \dfrac{1}{2}, \dfrac{1}{4}, \dfrac{1}{8}, \ldots$ is 2?

71. What do we mean when we say that the infinite geometric sequence $1, 2, 4, 8, \ldots$ has no sum?

72. Why don't we discuss the sum of an infinite arithmetic sequence?

Answers to the Concept Quiz

1. False   **2.** True   **3.** False   **4.** True   **5.** True   **6.** True   **7.** True   **8.** True

9.3   Another Look at Problem Solving

Objectives

■ Solve application problems involving arithmetic sequences.

■ Solve application problems involving geometric sequences.

In the previous two sections, many of the exercises fell into one of the following four categories.

1. Find the nth term of an arithmetic sequence:

$$a_n = a_1 + (n - 1)d$$

2. Find the sum of the first n terms of an arithmetic sequence:

$$S_n = \frac{n(a_1 + a_n)}{2}$$

3. Find the nth term of a geometric sequence:

$$a_n = a_1 r^{n-1}$$

4. Find the sum of the first n terms of a geometric sequence:

$$S_n = \frac{a_1 r^n - a_1}{r - 1}$$

In this section we want to use this knowledge of arithmetic sequences and geometric sequences to expand our problem-solving capabilities. Let's begin by restating some old problem-solving suggestions that continue to apply here; we will also consider some other suggestions that are directly related to problems that involve sequences of numbers. (We indicate the new suggestions with an asterisk.)

> **Suggestions for Solving Word Problems**
>
> **1.** Read the problem carefully and make certain that you understand the meanings of all the words. Be especially alert for any technical terms used in the statement of the problem.
>
> **2.** Read the problem a second time (perhaps even a third time) to get an overview of the situation being described and to determine the known facts, as well as what you are to find.
>
> **3.** Sketch a figure, diagram, or chart that might be helpful in analyzing the problem.
>
> *4. Write down the first few terms of the sequence to describe what is taking place in the problem. Be sure that you understand, term by term, what the sequence represents in the problem.
>
> *5. Determine whether the sequence is arithmetic or geometric.
>
> *6. Determine whether the problem is asking for a specific term of the sequence or for the sum of a certain number of terms.
>
> *7. Carry out the necessary calculations and check your answer for reasonableness.

As we solve some problems, these suggestions will become more meaningful.

P R O B L E M   1

Domenica started to work in 1998 at an annual salary of $22,500. She received a $1200 raise each year. What was her annual salary in 2007?

Solution

The following sequence represents her annual salary beginning in 1985:

  22,500,   23,700,   24,900,   26,100,   . . .

This is an arithmetic sequence, with $a_1 = 22,500$ and $d = 1200$. Because each term of the sequence represents her annual salary, we are looking for the tenth term.

$$a_{10} = 22,500 + (10 - 1)1200 = 22,500 + 9(1200) = 33,300$$

Her annual salary in 2007 was $33,000.

(▶) **Now work Problem 1.**   ■

P R O B L E M   2

An auditorium has 20 seats in the front row, 24 seats in the second row, 28 seats in the third row, and so on, for 15 rows. How many seats are there in the auditorium?

Solution

The following sequence represents the number of seats per row starting with the first row:

$$20, 24, 28, 32, \ldots$$

This is an arithmetic sequence, with $a_1 = 20$ and $d = 4$. Therefore the 15th term, which represents the number of seats in the 15th row, is given by

$$a_{15} = 20 + (15 - 1)4 = 20 + 14(4) = 76$$

The total number of seats in the auditorium is represented by

$$20 + 24 + 28 + \cdots + 76$$

We can use the sum formula for an arithmetic sequence to obtain

$$S_{15} = \frac{15}{2}(20 + 76) = 720$$

There are 720 seats in the auditorium.

▶ **Now work Problem 23.** ■

PROBLEM 3

Suppose that you save 25 cents the first day of a week, 50 cents the second day, and one dollar the third day and that you continue to double your savings each day. How much will you save on the seventh day? What will be your total savings for the week?

Solution

The following sequence represents your savings per day, expressed in cents:

$$25, 50, 100, \ldots$$

This is a geometric sequence, with $a_1 = 25$ and $r = 2$. Your savings on the seventh day is the seventh term of this sequence. Therefore, using $a_n = a_1 r^{n-1}$, we obtain

$$a_7 = 25(2)^6 = 1600$$

So you will save $16 on the seventh day. Your total savings for the 7 days is given by

$$25 + 50 + 100 + \cdots + 1600$$

We can use the sum formula for a geometric sequence to obtain

$$S_7 = \frac{25(2)^7 - 25}{2 - 1} = \frac{25(2^7 - 1)}{1} = 3175$$

Thus your savings for the entire week is $31.75.

▶ **Now work Problem 13.** ■

PROBLEM 4

A pump is attached to a container for the purpose of creating a vacuum. For each stroke of the pump, $\frac{1}{4}$ of the air that remains in the container is removed. To the nearest tenth of a percent, how much of the air remains in the container after six strokes?

Solution

Let's make a chart to help with the analysis of this problem:

First stroke:	$\frac{1}{4}$ of the air is removed	$1 - \frac{1}{4} = \frac{3}{4}$ of the air remains
Second stroke:	$\frac{1}{4}\left(\frac{3}{4}\right) = \frac{3}{16}$ of the air is removed	$\frac{3}{4} - \frac{3}{16} = \frac{9}{16}$ of the air remains
Third stroke:	$\frac{1}{4}\left(\frac{9}{16}\right) = \frac{9}{64}$ of the air is removed	$\frac{9}{16} - \frac{9}{64} = \frac{27}{64}$ of the air remains

The chart suggests two approaches to the problem.

Approach A   The sequence $\frac{1}{4}, \frac{3}{16}, \frac{9}{64}, \ldots$ represents, term by term, the fractional amount of air that is removed with each successive stroke. We can find the total amount removed and subtract it from 100%. The sequence is geometric with $a_1 = \frac{1}{4}$ and $r = \frac{3}{4}$.

$$S_6 = \frac{\frac{1}{4} - \frac{1}{4}\left(\frac{3}{4}\right)^6}{1 - \frac{3}{4}} = \frac{\frac{1}{4}\left[1 - \left(\frac{3}{4}\right)^6\right]}{\frac{1}{4}}$$

$$= 1 - \frac{729}{4096} = \frac{3367}{4096} = 82.2\%$$

Therefore $100\% - 82.2\% = 17.8\%$ of the air remains after six strokes.

Approach B   The sequence

$$\frac{3}{4}, \frac{9}{16}, \frac{27}{64}, \ldots$$

represents, term by term, the amount of air that remains in the container after each stroke. When we find the sixth term of this geometric sequence, we will have the answer to the problem. Because $a_1 = \frac{3}{4}$ and $r = \frac{3}{4}$, we obtain

$$a_6 = \frac{3}{4}\left(\frac{3}{4}\right)^5 = \left(\frac{3}{4}\right)^6 = \frac{729}{4096} = 17.8\%$$

Therefore 17.8% of the air remains after six strokes.

▶ **Now work Problem 25.**   ∎

It will be helpful for you to take another look at the two approaches we used to solve Problem 4. Note in Approach B that finding the sixth term of the sequence produced the answer to the problem without any further calculations. In Approach A we had to find the sum of six terms of the sequence and then subtract that amount from 100%. As we solve problems that involve sequences, we must understand what each particular sequence represents on a term-by-term basis.

Problem Set 9.3

Use your knowledge of arithmetic sequences and geometric sequences to help solve Problems 1–28.

1. A man started to work in 1980 at an annual salary of $12,000. He received a $900 raise each year. How much was his annual salary in 2001?

2. A woman started to work in 1990 at an annual salary of $16,500. She received a $1200 raise per year. How much was her annual salary in 2005?

3. State University had an enrollment of 9600 students in 1960. Each year the enrollment increased by 150 students. What was the enrollment in 2000?

4. Math University had an enrollment of 12,800 students in 1996. Each year the enrollment decreased by 75 students. What was the enrollment in 2003?

5. The enrollment at University X is predicted to increase at the rate of 10% per year. If the enrollment for 2008 was 5000 students, find the predicted enrollment for 2012. Express your answer to the nearest whole number.

6. If you pay $22,000 for a car and it depreciates 20% per year, how much will it be worth in 5 years? Express your answer to the nearest dollar.

7. A tank contains 16,000 liters of water. Each day one-half of the water in the tank is removed and not replaced. How much water remains in the tank at the end of 7 days?

8. If the price of a pound of coffee is $3.20, and the projected rate of inflation is 5% per year, how much per pound should we expect coffee to cost in 5 years? Express your answer to the nearest cent.

9. A tank contains 5832 gallons of water. Each day one-third of the water in the tank is removed and not replaced. How much water remains in the tank at the end of 6 days?

10. A fungus culture growing under controlled conditions doubles in size each day. How many units will the culture contain after 7 days if it originally contained 4 units?

11. Sue is saving quarters. She saves 1 quarter the first day, 2 quarters the second day, 3 quarters the third day, and so on for 30 days. How much money will she have saved in 30 days?

12. Suppose you save a penny the first day of a month, 2 cents the second day, 3 cents the third day, and so on for 31 days. What will be your total savings for the 31 days?

13. Suppose you save a penny the first day of a month, 2 cents the second day, 4 cents the third day, and continue to double your savings each day. How much will you save on the 15th day of the month? How much will your total savings be for the 15 days?

14. Eric saved a nickel the first day of a month, a dime the second day, and 20 cents the third day and then continued to double this daily savings each day for 14 days. What was his daily savings on the 14th day? What was his total savings for the 14 days?

15. Ms. Bryan invested $1500 at 12% simple interest at the beginning of each year for a period of 10 years. Find the total accumulated value of all the investments at the end of the 10-year period.

16. Mr. Woodley invested $1200 at 11% simple interest at the beginning of each year for a period of 8 years. Find the total accumulated value of all the investments at the end of the 8-year period.

17. An object falling from rest in a vacuum falls approximately 16 feet the first second, 48 feet the second second, 80 feet the third second, 112 feet the fourth second, and so on. How far will it fall in 11 seconds?

18. A raffle is organized so that the amount paid for each ticket is determined by the number on the ticket. The tickets are numbered with the consecutive odd whole numbers 1, 3, 5, 7, ... Each participant pays as many cents as the number on the ticket drawn. How much money will the raffle take in if 1000 tickets are sold?

19. Suppose an element has a half-life of 4 hours. This means that if n grams of it exist at a specific time, then only $\frac{1}{2}n$ grams remain 4 hours later. If at a particular moment we have 60 grams of the element, how many grams of it will remain 24 hours later?

20. Suppose an element has a half-life of 3 hours. (See Problem 19 for a definition of half-life.) If at a particular moment we have 768 grams of the element, how many grams of it will remain 24 hours later?

21. A rubber ball is dropped from a height of 1458 feet, and at each bounce it rebounds one-third of the height from which it last fell. How far has the ball traveled by the time it strikes the ground for the sixth time?

22. A rubber ball is dropped from a height of 100 feet, and at each bounce it rebounds one-half of the height from which it last fell. What distance has the ball traveled up to the instant it hits the ground for the eighth time?

23. A pile of logs has 25 logs in the bottom layer, 24 logs in the next layer, 23 logs in the next layer, and so on, until the top layer has 1 log. How many logs are in the pile?

24. A well driller charges $50.00 per foot for the first 10 feet, $55.00 per foot for the next 10 feet, $60.00 per foot for the next 10 feet, and so on, at a price increase of $5.00 per foot for succeeding intervals of 10 feet. How much does it cost to drill a well to a depth of 150 feet?

25. A pump is attached to a container for the purpose of creating a vacuum. For each stroke of the pump, $\frac{1}{3}$ of the air remaining in the container is removed. To the nearest tenth of a percent, how much of the air remains in the container after seven strokes?

26. Suppose that in Problem 25 each stroke of the pump removes $\frac{1}{2}$ of the air remaining in the container. What fractional part of the air has been removed after six strokes?

27. A tank contains 20 gallons of water. One-half of the water is removed and replaced with antifreeze. Then one-half of this mixture is removed and replaced with antifreeze. This process is continued eight times. How much water remains in the tank after the eighth replacement process?

28. The radiator of a truck contains 10 gallons of water. Suppose we remove 1 gallon of water and replace it with antifreeze. Then we remove 1 gallon of this mixture and replace it with antifreeze. This process is continued seven times. To the nearest tenth of a gallon, how much antifreeze is in the final mixture?

■ ■ ■ **THOUGHTS INTO WORDS**

29. Your friend solves Problem 6 as follows: If the car depreciates 20% per year, then at the end of 5 years it will have depreciated 100% and be worth zero dollars. How would you convince him that his reasoning is incorrect?

30. A contractor wants you to clear some land for a housing project. He anticipates that it will take 20 working days to do the job. He offers to pay you in one of two ways: (1) a fixed amount of $3000 or (2) a penny the first day, 2 cents the second day, 4 cents the third day, and so on, doubling your daily wages for the 20 days. Which offer should you take and why?

9.4 Mathematical Induction

Objectives

■ Know the necessary parts for a proof by mathematical induction.

■ Use mathematical induction to prove mathematical statements.

Is $2^n > n$ for all positive integer values of n? In an attempt to answer this question we might proceed as follows:

If $n = 1$, then $2^n > n$ becomes $2^1 > 1$, a true statement.

If $n = 2$, then $2^n > n$ becomes $2^2 > 2$, a true statement.

If $n = 3$, then $2^n > n$ becomes $2^3 > 3$, a true statement.

We can continue in this way as long as we want, but obviously we can never show in this manner that $2^n > n$ for *every* positive integer n. However, we do have a form of proof, called **proof by mathematical induction,** that can be used to verify the truth of many mathematical statements involving positive integers. This form of proof is based on the following principle.

Principle of Mathematical Induction

Let P_n be a statement in terms of n, where n is a positive integer. If

1. P_1 is true, and

2. the truth of P_k implies the truth of P_{k+1} for every positive integer k,

then P_n is true for every positive integer n.

The principle of mathematical induction — a proof that some statement is true for all positive integers — consists of two parts. First, we must show that the statement is true for the positive integer 1. Then we must show that if the statement is true for some positive integer, then it follows that it is also true for the next positive integer. Let's illustrate what this means.

E X A M P L E   1

Prove that $2^n > n$ for all positive integer values of n.

Proof

PART 1   If $n = 1$, then $2^n > n$ becomes $2^1 > 1$, which is a true statement.

PART 2   We must prove that if $2^k > k$, then $2^{k+1} > k + 1$ for all positive integer values of k. In other words, we should be able to start with $2^k > k$ and from that deduce $2^{k+1} > k + 1$. This can be done as follows:

$$2^k > k$$
$$2(2^k) > 2(k) \qquad \text{Multiply both sides by 2}$$
$$2^{k+1} > 2k$$

We know that $k \geq 1$ because we are working with positive integers. Therefore

$$k + k \geq k + 1 \qquad \text{Add } k \text{ to both sides}$$
$$2k \geq k + 1$$

Because $2^{k+1} > 2k$ and $2k \geq k + 1$, by the transitive property we conclude that

$$2^{k+1} > k + 1$$

Therefore using part 1 and part 2, we have proved that $2^n > n$ for *all* positive integers.

Now work Problem 12.

It will be helpful for you to look back over the proof in Example 1. Note that in part 1 we established that $2^n > n$ is true for $n = 1$. Then in part 2 we established that if $2^n > n$ is true for any positive integer, then it must be true for the next consecutive positive integer. Therefore, because $2^n > n$ is true for $n = 1$, it must be true for $n = 2$. Likewise, if $2^n > n$ is true for $n = 2$, then it must be true for $n = 3$, and so on, for *all* positive integers.

We can depict proof by mathematical induction with dominoes. Suppose that in Figure 9.1 we have infinitely many dominoes lined up. If we can push the first domino over (part 1 of a mathematical induction proof), and if the dominoes are spaced so that each time one falls over, it causes the next one to fall over (part 2 of a mathematical induction proof), then by pushing the first one over we will cause a chain reaction that will topple all of the dominoes (see Figure 9.2).

Figure 9.1

Figure 9.2

Recall that in the first three sections of this chapter we used a_n to represent the nth term of a sequence and S_n to represent the sum of the first n terms of a sequence. For example, if $a_n = 2n$, then the first three terms of the sequence are $a_1 = 2(1) = 2$, $a_2 = 2(2) = 4$, and $a_3 = 2(3) = 6$. Furthermore, the kth term is $a_k = 2(k) = 2k$ and the $(k + 1)$st term is $a_{k+1} = 2(k + 1) = 2k + 2$. Relative to this same sequence, we can state that $S_1 = 2$, $S_2 = 2 + 4 = 6$, and $S_3 = 2 + 4 + 6 = 12$.

There are numerous sum formulas for sequences that can be verified by mathematical induction. For such proofs, the following property of sequences is used:

$$S_{k+1} = S_k + a_{k+1}$$

This property states that **the sum of the first $k + 1$ terms is equal to the sum of the first k terms plus the $(k + 1)$st term.** Let's see how this can be used in a specific example.

E X A M P L E   2

Prove that $S_n = n(n + 1)$ for the sequence $a_n = 2n$, where n is any positive integer.

Proof

Part 1   If $n = 1$, then $1(1 + 1) = 2$, and 2 is the first term of the sequence $a_n = 2n$, so $S_1 = a_1 = 2$.

Part 2   Now we need to prove that if $S_k = k(k + 1)$, then $S_{k+1} = (k + 1)(k + 2)$. Using the property $S_{k+1} = S_k + a_{k+1}$, we can proceed as follows:

$$S_{k+1} = S_k + a_{k+1}$$
$$= k(k + 1) + 2(k + 1)$$
$$= (k + 1)(k + 2)$$

Therefore, using part 1 and part 2, we have proved that $S_n = n(n + 1)$ will yield the correct sum for any number of terms of the sequence $a_n = 2n$.

▶ **Now work Problem 1.**   ■

E X A M P L E   3

Prove that $S_n = 5n(n + 1)/2$ for the sequence $a_n = 5n$, where n is any positive integer.

Proof

Part 1   Because $5(1)(1 + 1)/2 = 5$, and 5 is the first term of the sequence $a_n = 5n$, we have $S_1 = a_1 = 5$.

Part 2   We need to prove that if $S_k = 5k(k + 1)/2$, then $S_{k+1} = \dfrac{5(k + 1)(k + 2)}{2}$.

$$S_{k+1} = S_k + a_{k+1}$$
$$= \frac{5k(k + 1)}{2} + 5(k + 1)$$
$$= \frac{5k(k + 1)}{2} + 5k + 5$$
$$= \frac{5k(k + 1) + 2(5k + 5)}{2}$$
$$= \frac{5k^2 + 5k + 10k + 10}{2}$$

$$= \frac{5k^2 + 15k + 10}{2}$$

$$= \frac{5(k^2 + 3k + 2)}{2}$$

$$= \frac{5(k + 1)(k + 2)}{2}$$

Therefore, using part 1 and part 2, we have proved that $S_n = 5n(n + 1)/2$ yields the correct sum for any number of terms of the sequence $a_n = 5n$.

▶ **Now work Problem 3.**   ■

EXAMPLE   4   Prove that $S_n = (4^n - 1)/3$ for the sequence $a_n = 4^{n-1}$, where n is any positive integer.

Proof

PART 1   Because $(4^1 - 1)/3 = 1$, and 1 is the first term of the sequence $a_n = 4^{n-1}$, we have $S_1 = a_1 = 1$.

PART 2   We need to prove that if $S_k = (4^k - 1)/3$, then $S_{k+1} = (4^{k+1} - 1)/3$.

$$S_{k+1} = S_k + a_{k+1}$$

$$= \frac{4^k - 1}{3} + 4^k$$

$$= \frac{4^k - 1 + 3(4^k)}{3} \ .$$

$$= \frac{4^k + 3(4^k) - 1}{3}$$

$$= \frac{4^k(1 + 3) - 1}{3}$$

$$= \frac{4^k(4) - 1}{3}$$

$$= \frac{4^{k+1} - 1}{3}$$

Therefore, using part 1 and part 2, we have proved that $S_n = (4^n - 1)/3$ yields the correct sum for any number of terms of the sequence $a_n = 4^{n-1}$.

▶ **Now work Problem 5.**   ■

As our final example of this section, let's consider a proof by mathematical induction involving the concept of divisibility.

Prove that for all positive integers n, the number $3^{2n} - 1$ is divisible by 8.

Proof

Part 1   If $n = 1$, then $3^{2n} - 1$ becomes $3^{2(1)} - 1 = 3^2 - 1 = 8$, and of course 8 is divisible by 8.

Part 2   We need to prove that if $3^{2k} - 1$ is divisible by 8, then $3^{2k+2} - 1$ is divisible by 8 for all integer values of k. This can be verified as follows. If $3^{2k} - 1$ is divisible by 8, then for some integer x, we have $3^{2k} - 1 = 8x$. Therefore

$$3^{2k} - 1 = 8x$$
$$3^{2k} = 1 + 8x$$
$$3^2(3^{2k}) = 3^2(1 + 8x) \qquad \text{Multiply both sides by } 3^2$$
$$3^{2k+2} = 9(1 + 8x)$$
$$3^{2k+2} = 9 + 9(8x)$$
$$3^{2k+2} = 1 + 8 + 9(8x) \qquad 9 = 1 + 8$$
$$3^{2k+2} = 1 + 8(1 + 9x) \qquad \text{Apply distributive}$$
$$3^{2k+2} - 1 = 8(1 + 9x) \qquad \text{property to } 8 + 9(8x)$$

Therefore $3^{2k+2} - 1$ is divisible by 8.

Using part 1 and part 2, we have proved that $3^{2n} - 1$ is divisible by 8 for all positive integers n.

▶ **Now work Problem 15.** ■

We conclude this section with a few final comments about proof by mathematical induction. Every mathematical induction proof is a two-part proof, and both parts are absolutely necessary. There can be mathematical statements that hold for one or the other of the two parts but not for both. For example, $(a + b)^n = a^n + b^n$ is true for $n = 1$, but it is false for every positive integer greater than 1. Therefore if we were to attempt a mathematical induction proof for $(a + b)^n = a^n + b^n$, we could establish part 1 but not part 2. Another example of this type is the statement that $n^2 - n + 41$ produces a prime number for all positive integer values of n. This statement is true for $n = 1, 2, 3, 4, \ldots, 40$, but it is false when $n = 41$ (because $41^2 - 41 + 41 = 41^2$, which is not a prime number).

It is also possible that part 2 of a mathematical induction proof can be established but not part 1. For example, consider the sequence $a_n = n$ and the sum formula $S_n = (n + 3)(n - 2)/2$. If $n = 1$, then $a_1 = 1$ but $S_1 = (4)(-1)/2 = -2$, so part 1 does not hold. However, it is possible to show that $S_k = (k + 3)(k - 2)/2$ implies $S_{k+1} = (k + 4)(k - 1)/2$. We will leave the details of this for you to do.

Finally, it is important to realize that some mathematical statements are true for all positive integers greater than some fixed positive integer other than 1. Back in Figure 9.1 perhaps we cannot knock down the first four dominoes, but we can knock down the fifth domino and every one thereafter. For example, we can prove by mathematical induction that $2^n > n^2$ for all positive integers $n > 4$. It requires a slight variation in the statement of the principle of mathematical induction. We will not concern ourselves with such problems in this text, but we want you to be aware of their existence.

CONCEPT QUIZ   For Problems 1–4, answer true or false.

1. Mathematical induction is used to prove mathematical statements involving positive integers.

2. A proof by mathematical induction consists of two parts.

3. Because $(a + b)^n = a^n + b^n$ is true for $n = 1$, it is true for all positive integer values of n.

4. To prove a mathematical statement involving positive integers by mathematical induction, the statement must be true for $n = 1$.

Problem Set 9.4

For Problems 1–10, use mathematical induction to prove each of the sum formulas for the indicated sequences. They are to hold for all positive integers n.

1. $S_n = \dfrac{n(n + 1)}{2}$ for $a_n = n$

2. $S_n = n^2$ for $a_n = 2n - 1$

3. $S_n = \dfrac{n(3n + 1)}{2}$ for $a_n = 3n - 1$

4. $S_n = \dfrac{n(5n + 9)}{2}$ for $a_n = 5n + 2$

5. $S_n = 2(2^n - 1)$ for $a_n = 2^n$

6. $S_n = \dfrac{3(3^n - 1)}{2}$ for $a_n = 3^n$

7. $S_n = \dfrac{n(n + 1)(2n + 1)}{6}$ for $a_n = n^2$

8. $S_n = \dfrac{n^2(n + 1)^2}{4}$ for $a_n = n^3$

9. $S_n = \dfrac{n}{n + 1}$ for $a_n = \dfrac{1}{n(n + 1)}$

10. $S_n = \dfrac{n(n + 1)(n + 2)}{3}$ for $a_n = n(n + 1)$

In Problems 11–20, use mathematical induction to prove that each statement is true for all positive integers n.

11. $3^n \geq 2n + 1$ **12.** $4^n \geq 4n$

13. $n^2 \geq n$ **14.** $2^n \geq n + 1$

15. $4^n - 1$ is divisible by 3

16. $5^n - 1$ is divisible by 4

17. $6^n - 1$ is divisible by 5

18. $9^n - 1$ is divisible by 4

19. $n^2 + n$ is divisible by 2

20. $n^2 - n$ is divisible by 2

■ ■ ■ **THOUGHTS INTO WORDS**

21. How would you describe proof by mathematical induction?

22. Compare inductive reasoning to proof by mathematical induction.

Answers to the Concept Quiz

1. True   **2.** True   **3.** False   **4.** False

There are four main topics in this chapter: arithmetic sequences, geometric sequences, problem solving, and mathematical induction.

■ Arithmetic Sequences

The sequence $a_1, a_2, a_3, a_4, \ldots$ is called **arithmetic** if and only if

$$a_{k+1} - a_k = d$$

for every positive integer k. In other words, there is a **common difference**, d, between successive terms.

The **general term** of an arithmetic sequence is given by the formula

$$a_n = a_1 + (n - 1)d$$

where a_1 is the first term, n is the number of terms, and d is the common difference.

The **sum** of the first n terms of an arithmetic sequence is given by the formula

$$S_n = \frac{n(a_1 + a_n)}{2}$$

Summation notation can be used to indicate the sum of a certain number of terms of a sequence. For example,

$$\sum_{i=1}^{5} 4i = 4(1) + 4(2) + 4(3) + 4(4) + 4(5)$$

■ Geometric Sequences

The sequence $a_1, a_2, a_3, a_4, \ldots$ is called **geometric** if and only if

$$a_{k+1} = ra_k$$

for every positive integer k. There is a **common ratio, r,** between successive terms.

The **general term** of a geometric sequence is given by the formula

$$a_n = a_1 r^{n-1}$$

where a_1 is the first term, n is the number of terms, and r is the common ratio.

The **sum** of the first n terms of a geometric sequence is given by the formula

$$S_n = \frac{a_1 r^n - a_1}{r - 1}, \qquad r \neq 1$$

The **sum of an infinite geometric sequence** is given by the formula

$$S_\infty = \frac{a_1}{1 - r} \qquad \text{for } |r| < 1$$

If $|r| \geq 1$, the sequence has no sum.

Repeating decimals (such as $0.\overline{4}$) can be changed to a/b form, where a and b are integers and $b \neq 0$, by treating them as the sum of an infinite geometric sequence. For example, the repeating decimal $0.\overline{4}$ can be written $0.4 + 0.04 + 0.004 + 0.0004 + \cdots$.

■ Problem Solving

Many of the problem-solving suggestions offered earlier in this text are still appropriate when we are solving problems that deal with sequences. However, there are also some special suggestions pertaining to sequence problems.

1. Write down the first few terms of the sequence to describe what is taking place in the problem. Drawing a picture or diagram may help with this step.

2. Be sure that you understand, term by term, what the sequence represents in the problem.

3. Determine whether the sequence is arithmetic or geometric. (Those are the only kinds of sequences we are working with in this text.)

4. Determine whether the problem is asking for a specific term or for the sum of a certain number of terms.

■ Mathematical Induction

Proof by mathematical induction relies on the following **principle of induction:**

Let P_n be a statement in terms of n, where n is a positive integer. If

1. P_1 is true, and

2. the truth of P_k implies the truth of P_{k+1}, for every positive integer k,

then P_n is true for every positive integer n.

Chapter 9   Review Problem Set

For Problems 1–10, find the general term (the nth term) for each sequence. These problems include both arithmetic sequences and geometric sequences.

1. $3, 9, 15, 21, \ldots$

2. $\dfrac{1}{3}, 1, 3, 9, \ldots$

3. $10, 20, 40, 80, \ldots$

4. $5, 2, -1, -4, \ldots$

5. $-5, -3, -1, 1, \ldots$

6. $9, 3, 1, \dfrac{1}{3}, \ldots$

7. $-1, 2, -4, 8, \ldots$

8. $12, 15, 18, 21, \ldots$

9. $\dfrac{2}{3}, 1, \dfrac{4}{3}, \dfrac{5}{3}, \ldots$

10. $1, 4, 16, 64, \ldots$

For Problems 11–16, find the required term of each of the sequences.

11. The 19th term of $1, 5, 9, 13, \ldots$

12. The 28th term of $-2, 2, 6, 10, \ldots$

13. The 9th term of $8, 4, 2, 1, \ldots$

14. The 8th term of $\dfrac{243}{32}, \dfrac{81}{16}, \dfrac{27}{8}, \dfrac{9}{4}, \ldots$

15. The 34th term of $7, 4, 1, -2, \ldots$

16. The 10th term of $-32, 16, -8, 4, \ldots$

For Problems 17–29, solve each problem.

17. If the 5th term of an arithmetic sequence is -19 and the 8th term is -34, find the common difference of the sequence.

18. If the 8th term of an arithmetic sequence is 37 and the 13th term is 57, find the 20th term.

19. Find the first term of a geometric sequence if the third term is 5 and the sixth term is 135.

20. Find the common ratio of a geometric sequence if the second term is $\dfrac{1}{2}$ and the sixth term is 8.

21. Find the sum of the first nine terms of the sequence $81, 27, 9, 3, \ldots$.

22. Find the sum of the first 70 terms of the sequence $-3, 0, 3, 6, \ldots$.

23. Find the sum of the first 75 terms of the sequence $5, 1, -3, -7, \ldots$.

24. Find the sum of the first ten terms of the sequence where $a_n = 2^{5-n}$.

25. Find the sum of the first 95 terms of the sequence where $a_n = 7n + 1$.

26. Find the sum $5 + 7 + 9 + \cdots + 137$.

27. Find the sum $64 + 16 + 4 + \cdots + \dfrac{1}{64}$.

28. Find the sum of all even numbers between 8 and 384, inclusive.

29. Find the sum of all multiples of 3 between 27 and 276, inclusive.

For Problems 30–33, find each indicated sum.

30. $\displaystyle\sum_{i=1}^{45} (-2i + 5)$

31. $\displaystyle\sum_{i=1}^{5} i^3$

32. $\displaystyle\sum_{i=1}^{8} 2^{8-i}$

33. $\displaystyle\sum_{i=4}^{75} (3i - 4)$

For Problems 34–36, solve each problem.

34. Find the sum of the infinite geometric sequence $64, 16, 4, 1, \ldots$.

35. Change $0.\overline{36}$ to reduced a/b form, where a and b are integers and $b \neq 0$.

36. Change $0.4\overline{5}$ to reduced a/b form, where a and b are integers and $b \neq 0$.

Solve each of Problems 37–40 by using your knowledge of arithmetic sequences and geometric sequences.

37. Suppose that your savings account contains \$3750 at the beginning of a year. If you withdrew \$250 per month from the account, how much will it contain at the end of the year?

38. Sonya decides to start saving dimes. She plans to save 1 dime the first day of April, 2 dimes the second day, 3 dimes the third day, 4 dimes the fourth day, and so on for the 30 days of April. How much money will she save in April?

39. Nancy decides to start saving dimes. She plans to save 1 dime the first day of April, 2 dimes the second day, 4 dimes the third day, 8 dimes the fourth day, and so on for the first 15 days of April. How much will she save in 15 days?

40. A tank contains 61,440 gallons of water. Each day one-fourth of the water is drained out. How much water remains in the tank at the end of 6 days?

For Problems 41–43, show a mathematical induction proof.

41. Prove that $5^n > 5n - 1$ for all positive integer values of n.

42. Prove that $n^3 - n + 3$ is divisible by 3 for all positive integer values of n.

43. Prove that

$$S_n = \frac{n(n + 3)}{4(n + 1)(n + 2)}$$

is the sum formula for the sequence

$$a_n = \frac{1}{n(n + 1)(n + 2)}$$

where n is any positive integer.

1. Find the 15th term of the sequence for which $a_n = -n^2 - 1$.

2. Find the 5th term of the sequence for which $a_n = 3(2)^{n-1}$.

3. Find the general term of the sequence $-3, -8, -13, -18, \ldots$.

4. Find the general term of the sequence $5, \dfrac{5}{2}, \dfrac{5}{4}, \dfrac{5}{8}, \ldots$.

5. Find the general term of the sequence $10, 16, 22, 28, \ldots$.

6. Find the 7th term of the sequence $8, 12, 18, 27, \ldots$.

7. Find the 75th term of the sequence $1, 4, 7, 10, \ldots$.

8. Find the number of terms in the sequence $7, 11, 15, \ldots, 243$.

9. Find the sum of the first 40 terms of the sequence $1, 4, 7, 10, \ldots$.

10. Find the sum of the first eight terms of the sequence $3, 6, 12, 24, \ldots$.

11. Find the sum of the first 45 terms of the sequence for which $a_n = 7n - 2$.

12. Find the sum of the first ten terms of the sequence for which $a_n = 3(2)^n$.

13. Find the sum of the first 150 positive even whole numbers.

14. Find the sum of the odd whole numbers between 11 and 193, inclusive.

15. Find the indicated sum $\displaystyle\sum_{i=1}^{50} (3i + 5)$.

16. Find the indicated sum $\displaystyle\sum_{i=1}^{10} (-2)^{i-1}$.

17. Find the sum of the infinite geometric sequence $3, \dfrac{3}{2}, \dfrac{3}{4}, \dfrac{3}{8}, \ldots$.

18. Find the sum of the infinite geometric sequence for which $a_n = 2\left(\dfrac{1}{3}\right)^{n+1}$.

19. Change $0.\overline{18}$ to reduced $\dfrac{a}{b}$ form, where a and b are integers and $b \neq 0$.

20. Change $0.2\overline{6}$ to reduced $\dfrac{a}{b}$ form, where a and b are integers and $b \neq 0$.

For Problems 21–23, solve each problem.

21. A tank contains 49,152 liters of gasoline. Each day three-fourths of the gasoline remaining in the tank is pumped out and not replaced. How much gasoline remains in the tank at the end of 7 days?

22. Suppose that you save a dime the first day of a month, $0.20 the second day, and $0.40 the third day and that you continue to double your savings per day for 15 days. Find the total amount that you will have saved at the end of 15 days.

23. A woman invests $350 at 12% simple interest at the beginning of each year for a period of 10 years. Find the total accumulated value of all the investments at the end of the 10-year period.

For Problems 24 and 25, show a mathematical induction proof.

24. $S_n = \dfrac{n(3n - 1)}{2}$ for $a_n = 3n - 2$

25. $9^n - 1$ is divisible by 8 for all positive integer values for n.

Appendix A: Binomial Theorem

In Chapter 0 we developed a pattern for expanding binomials, using Pascal's triangle to determine the coefficients of each term. Now we will be more precise and develop a general formula, called the *binomial formula*. In other words, we want to develop a formula that will allow us to expand $(a + b)^n$, where n is any positive integer.

Let's begin, as we did in Chapter 0, by looking at some specific expansions, which can be verified by direct multiplication.

$$(a + b)^1 = a + b$$

$$(a + b)^2 = a^2 + 2ab + b^2$$

$$(a + b)^3 = a^3 + 3a^2b + 3ab^2 + b^3$$

$$(a + b)^4 = a^4 + 4a^3b + 6a^2b^2 + 4ab^3 + b^4$$

$$(a + b)^5 = a^5 + 5a^4b + 10a^3b^2 + 10a^2b^3 + 5ab^4 + b^5$$

First, note the patterns of the exponents for a and b on a term-by-term basis. The exponents of a begin with the exponent of the binomial and *decrease* by 1, term by term, until the last term, which has $a^0 = 1$. The exponents of b begin with zero ($b^0 = 1$) and *increase* by 1, term by term, until the last term, which contains b to the power of the original binomial. In other words, the variables in the expansion of $(a + b)^n$ have the pattern

$$a^n, \quad a^{n-1}b, \quad a^{n-2}b^2, \quad \ldots, \quad ab^{n-1}, \quad b^n,$$

where for each term, the sum of the exponents of a and b is n.

Now let's consider the numerical coefficients of the terms of a binomial expansion. Before stating a formula that will yield these coefficients, we need to introduce the concept of **factorial notation.** If n is any positive integer, the symbol $n!$ (read, n *factorial*) is defined as follows.

$$n! = n(n - 1)(n - 2) \ldots 1$$

In other words, $n!$ is the product of the first n posititve integers. For example,

$$1! = 1, \quad 2! = 2 \cdot 1 = 2, \quad 3! = 3 \cdot 2 \cdot 1 = 6, \quad \text{and}$$

$$4! = 4 \cdot 3 \cdot 2 \cdot 1 = 24.$$

We also define $0! = 1$.

The numerical coefficients of the terms of a binomial expansion (often referred to as binomial coefficients) are denoted by $\binom{n}{k}$, where $k = 0, 1, 2, 3, \ldots, n$ and are given by the formula

$$\binom{n}{k} = \frac{n!}{k!(n-k)!}.$$

The symbol $\binom{n}{k}$ denotes the coefficient of the $(k + 1)$st term of the expansion of $(a + b)^n$. In other words, when $k = 0$, then $\binom{n}{0}$ denotes the coefficient of the first term; when $k = 1$, $\binom{n}{1}$ denotes the coefficient of the second term, and so on. The coefficients of the terms of the expansion of $(a + b)^5$ and $\binom{5}{0}, \binom{5}{1}, \binom{5}{2}, \binom{5}{3}, \binom{5}{4}$, and $\binom{5}{5}$, are calculated by the formula as follows.

$$\binom{5}{0} = \frac{5!}{0!(5-0)!} = \frac{5!}{0!5!} = 1$$

$$\binom{5}{1} = \frac{5!}{1!(5-1)!} = \frac{5!}{1!4!} = 5$$

$$\binom{5}{2} = \frac{5!}{2!(5-2)!} = \frac{5!}{2!3!} = 10$$

$$\binom{5}{3} = \frac{5!}{3!(5-3)!} = \frac{5!}{3!2!} = 10$$

$$\binom{5}{4} = \frac{5!}{4!(5-4)!} = \frac{5!}{4!1!} = 5$$

$$\binom{5}{5} = \frac{5!}{5!(5-5)!} = \frac{5!}{5!0!} = 1$$

Remark To evaluate an expression such as $\dfrac{9!}{2!7!}$, you may find the following format helpful.

$$\frac{9!}{2!7!} = \frac{9 \cdot 8 \cdot 7!}{2!7!} = \frac{9 \cdot \overset{4}{\cancel{8}}}{2 \cdot 1} = 36$$

A general expansion of $(a + b)^n$, often called the **binomial theorem,** can now be stated.

Binomial Theorem

For any binomial $(a + b)$ and any positive integer n,

$$(a + b)^n = \binom{n}{0} a^n + \binom{n}{1} a^{n-1}b + \binom{n}{2} a^{n-2}b^2 + \cdots + \binom{n}{n} b^n.$$

A formal proof of the binomial theorem requires mathematical induction, which is presented in Chapter 9. For now, let's simply illustrate how the binomial theorem can be used. Keep in mind that $\binom{n}{0} = 1$ and $\binom{n}{n} = 1$.

E X A M P L E   1

Expand $(a + b)^6$.

Solution

$$(a + b)^6 = a^6 + \binom{6}{1} a^5b + \binom{6}{2} a^4b^2 + \binom{6}{3} a^3b^3 + \binom{6}{4} a^2b^4 + \binom{6}{5} ab^5 + b^6$$

$$= a^6 + 6a^5b + 15a^4b^2 + 20a^3b^3 + 15a^2b^4 + 6ab^5 + b^6 \qquad ∎$$

E X A M P L E   2

Expand $(2x + 3y)^4$.

Solution

$$(2x + 3y)^4 = (2x)^4 + \binom{4}{1}(2x)^3(3y)^1 + \binom{4}{2}(2x)^2(3y)^2 + \binom{4}{3}(2x)(3y)^3 + (3y)^4$$

$$= 16x^4 + 96x^3y + 216x^2y^2 + 216xy^3 + 81y^4 \qquad ∎$$

E X A M P L E   3

Expand $(x - 2y^2)^5$.

Solution

We shall treat $(x - 2y^2)^5$ as $[x + (-2y^2)]^5$.

$$[x + (-2y^2)]^5 = x^5 + \binom{5}{1}x^4(-2y^2) + \binom{5}{2}x^3(-2y^2)^2 + \binom{5}{3}x^2(-2y^2)^3 + \binom{5}{4}x(-2y^2)^4 + (-2y^2)^5$$

$$= x^5 - 10x^4y^2 + 40x^3y^4 - 80x^2y^6 + 80xy^8 - 32y^{10} \qquad ∎$$

E X A M P L E   4

Expand $\left(a + \dfrac{1}{n} \right)^5$.

Solution

$$\left(a + \frac{1}{n}\right)^5 = a^5 + \binom{5}{1}a^4\left(\frac{1}{n}\right) + \binom{5}{2}a^3\left(\frac{1}{n}\right)^2 + \binom{5}{3}a^2\left(\frac{1}{n}\right)^3 + \binom{5}{4}a\left(\frac{1}{n}\right)^4 + \binom{5}{5}\left(\frac{1}{n}\right)^5$$

$$= a^5 + \frac{5a^4}{n} + \frac{10a^3}{n^2} + \frac{10a^2}{n^3} + \frac{5a}{n^4} + \frac{1}{n^5} \quad \blacksquare$$

EXAMPLE 5   Expand $(x^2 - 2y^3)^6$.

Solution

$$[x^2 + (-2y^3)]^6 = (x^2)^6 + \binom{6}{1}(x^2)^5(-2y^3) + \binom{6}{2}(x^2)^4(-2y^3)^2$$

$$+ \binom{6}{3}(x^2)^3(-2y^3)^3 + \binom{6}{4}(x^2)^2(-2y^3)^4$$

$$+ \binom{6}{5}(x^2)(-2y^3)^5 + \binom{6}{6}(-2y^3)^6$$

$$= x^{12} - 12x^{10}y^3 + 60x^8y^6 - 160x^6y^9 + 240x^4y^{12} - 192x^2y^{15} + 64y^{18} \quad \blacksquare$$

■ Finding Specific Terms

Sometimes it is convenient to be able to find a specific term of a binomial expansion without writing out the entire expansion. For example, suppose that we want the 6th term of the expansion $(a + b)^{12}$. We can reason as follows.

The 6th term will contain b^5. (Note in the binomial theorem that the *exponent of b is always one less than the number of the term*.) Since the sum of the exponents for a and b must be 12 (the exponent of the binomial), the 6th term will also contain a^7. The coefficient is $\binom{12}{5}$, where the 5 agrees with the exponent of b^5. Therefore, the 6th term is $\binom{12}{5}a^7b^5 = 792a^7b^5$.

EXAMPLE 6   Find the 4th term of $(3x + 2y)^7$.

Solution

The 4th term contains $(2y)^3$ and therefore also contains $(3x)^4$. The coefficient is $\binom{7}{3}$. Thus the 4th term is

$$\binom{7}{3}(3x)^4(2y)^3 = 35(81x^4)(8y^3)$$

$$= 22{,}680x^4y^3. \qquad \blacksquare$$

E X A M P L E 7 Find the 6th term of $(2x - y^2)^8$.

Solution

First, let's treat $(2x - y^2)^8$ as $[2x + (-y^2)]^8$. The 6th term contains $(-y^2)^5$ and $(2x)^3$.

The coefficient is $\binom{8}{5}$. Therefore, the 6th term is

$$\binom{8}{5}(2x)^3(-y^2)^5 = 56(8x^3)(-y^{10})$$

$$= -448x^3y^{10}. \qquad \blacksquare$$

Problem Set for Appendix A

For Problems 1–26, expand and simplify each binomial.

1. $(x + y)^8$

2. $(x + y)^9$

3. $(x - y)^6$

4. $(x - y)^4$

5. $(a + 2b)^4$

6. $(3a + b)^4$

7. $(x - 3y)^5$

8. $(2x - y)^6$

9. $(2a - 3b)^4$

10. $(3a - 2b)^5$

11. $(x^2 + y)^5$

12. $(x + y^3)^6$

13. $(2x^2 - y^2)^4$

14. $(3x^2 - 2y^2)^5$

15. $(x + 3)^6$

16. $(x + 2)^7$

17. $(x - 1)^9$

18. $(x - 3)^4$

19. $\left(1 + \dfrac{1}{n}\right)^4$

20. $\left(2 + \dfrac{1}{n}\right)^5$

21. $\left(a - \dfrac{1}{n}\right)^6$

22. $\left(2a - \dfrac{1}{n}\right)^5$

23. $(1 + \sqrt{2})^4$

24. $(2 + \sqrt{3})^3$

25. $(3 - \sqrt{2})^5$

26. $(1 - \sqrt{3})^4$

For Problems 27–36, write the first four terms of each expansion.

27. $(x - y)^{12}$

28. $(x + y)^{15}$

29. $(x - y)^{20}$

30. $(a - b)^{13}$

31. $(x^2 - 2y^3)^{14}$

32. $(x^3 - 3y^2)^{11}$

33. $\left(a + \dfrac{1}{n}\right)^9$

34. $\left(2 - \dfrac{1}{n}\right)^6$

35. $(-x + 2y)^{10}$

36. $(-a - b)^{14}$

For Problems 37–46, find the specified term for each binomial expansion.

37. The fourth term of $(x + y)^8$

38. The seventh term of $(x + y)^{11}$

39. The fifth term of $(x - y)^9$

40. The fourth term of $(x - 2y)^6$

41. The sixth term of $(3a + b)^7$

42. The third term of $(2x - 5y)^5$

43. The eighth term of $(x^2 - y^3)^{10}$

44. The ninth term of $(a + b^3)^{12}$

45. The seventh term of $\left(1 - \dfrac{1}{n}\right)^{15}$

46. The eighth term of $\left(1 - \dfrac{1}{n}\right)^{13}$

Answers to Odd-Numbered Problems and All Chapter Review, Chapter Test, and Cumulative Review Problems

CHAPTER 0

Problem Set 0.1 (page 16)

1. True   **3.** False   **5.** False   **7.** True
9. False   **11.** $\{46\}$ **13.** $\{0, -14, 46\}$
15. $\{\sqrt{5}, -\sqrt{2}, -\pi\}$ **17.** $\{0, -14\}$ **19.** $\subseteq$
21. $\subseteq$ **23.** $\not\subseteq$ **25.** $\subseteq$ **27.** $\not\subseteq$
29. $\subseteq$ **31.** $\not\subseteq$ **33.** $\{1\}$ **35.** $\{0, 1, 2, 3\}$
37. $\{\ldots, -2, -1, 0, 1\}$ **39.** $\varnothing$ **41.** $\{0, 1, 2\}$
43. a. 18   **c.** 39   **e.** 35
45. Commutative property of multiplication
47. Identity property of multiplication
49. Multiplication property of negative one
51. Distributive property
53. Commutative property of multiplication
55. Distributive property
57. Associative property of multiplication
59. -22 **61.** 100   **63.** -21 **65.** 8
67. 19   **69.** 66   **71.** -75 **73.** 34
75. 1   **77.** 11   **79.** 4

Problem Set 0.2 (page 28)

1. $\dfrac{1}{8}$ **3.** $-\dfrac{1}{1000}$ **5.** 27   **7.** 4

9. $-\dfrac{27}{8}$ **11.** 1   **13.** $\dfrac{16}{25}$ **15.** 4

17. $\dfrac{1}{100}$ or 0.01   **19.** $\dfrac{1}{100,000}$ or 0.00001   **21.** 81

23. $\dfrac{1}{16}$ **25.** $\dfrac{3}{4}$ **27.** $\dfrac{256}{25}$ **29.** $\dfrac{16}{25}$

31. $\dfrac{64}{81}$ **33.** 64   **35.** $\dfrac{1}{100,000}$ or 0.00001

37. $\dfrac{17}{72}$ **39.** $\dfrac{1}{6}$ **41.** $\dfrac{48}{19}$ **43.** $\dfrac{1}{x^4}$

45. $\dfrac{1}{a^2}$ **47.** $\dfrac{1}{a^6}$ **49.** $\dfrac{y^4}{x^3}$ **51.** $\dfrac{c^3}{a^3b^6}$

53. $\dfrac{y^2}{4x^4}$ **55.** $\dfrac{x^4}{y^6}$ **57.** $\dfrac{9a^2}{4b^4}$ **59.** $\dfrac{1}{x^3}$

61. $\dfrac{a^3}{b}$ **63.** $-20x^4y^5$ **65.** $-27x^3y^9$

67. $\dfrac{8x^6}{27y^9}$ **69.** $-8x^6$ **71.** $\dfrac{6}{x^3y}$ **73.** $\dfrac{6}{a^2y^3}$

75. $\dfrac{4x^3}{y^5}$ **77.** $-\dfrac{5}{a^2b}$ **79.** $\dfrac{1}{4x^2y^4}$ **81.** $\dfrac{x+1}{x^2}$

83. $\dfrac{y-x^2}{x^2y}$ **85.** $\dfrac{3b^3+2a^2}{a^2b^3}$ **87.** $\dfrac{y^2-x^2}{xy}$

89. $12x^{3a+1}$ **91.** 1   **93.** x^{2a} **95.** $-4y^{6b+2}$

97. x^b **99.** $(6.2)(10)^7$ **101.** $(4.12)(10)^{-4}$
103. 180,000   **105.** 0.0000023   **107.** 0.04
109. 30,000   **111.** 0.03
117. a. $(4.385)(10)^{14}$ **c.** $(2.322)(10)^{17}$ **e.** $(3.052)(10)^{12}$

Problem Set 0.3 (page 38)

1. $14x^2 + x - 6$ **3.** $-x^2 - 4x - 9$
5. $6x - 11$ **7.** $6x^2 - 5x - 7$ **9.** $-x - 34$
11. $12x^3y^2 + 15x^2y^3$ **13.** $30a^4b^3 - 24a^5b^3 + 18a^4b^4$
15. $x^2 + 20x + 96$ **17.** $n^2 - 16n + 48$
19. $sx + sy - tx - ty$ **21.** $6x^2 + 7x - 3$
23. $12x^2 - 37x + 21$ **25.** $x^2 + 8x + 16$
27. $4n^2 + 12n + 9$ **29.** $x^3 + x^2 - 14x - 24$
31. $6x^3 - x^2 - 11x + 6$ **33.** $x^3 + 2x^2 - 7x + 4$
35. $t^3 - 1$ **37.** $6x^3 + x^2 - 5x - 2$
39. $x^4 + 8x^3 + 15x^2 + 2x - 4$ **41.** $25x^2 - 4$
43. $x^4 - 10x^3 + 21x^2 + 20x + 4$ **45.** $4x^2 - 9y^2$
47. $x^3 + 15x^2 + 75x + 125$ **49.** $8x^3 + 12x^2 + 6x + 1$
51. $64x^3 - 144x^2 + 108x - 27$
53. $125x^3 - 150x^2y + 60xy^2 - 8y^3$
55. $a^7 + 7a^6b + 21a^5b^2 + 35a^4b^3 + 35a^3b^4 + 21a^2b^5 + 7ab^6 + b^7$
57. $x^5 - 5x^4y + 10x^3y^2 - 10x^2y^3 + 5xy^4 - y^5$

59. $x^4 + 8x^3y + 24x^2y^2 + 32xy^3 + 16y^4$

61. $64a^6 - 192a^5b + 240a^4b^2 - 160a^3b^3 + 60a^2b^4 - 12ab^5 + b^6$

63. $x^{14} + 7x^{12}y + 21x^{10}y^2 + 35x^8y^3 + 35x^6y^4 + 21x^4y^5 + 7x^2y^6 + y^7$

65. $32a^5 - 240a^4b + 720a^3b^2 - 1080a^2b^3 + 810ab^4 - 243b^5$ **67.** $3x^2 - 5x$ **69.** $-5a^4 + 4a^2 - 9a$

71. $5ab + 11a^2b^4$ **73.** $x^{2a} - y^{2b}$

75. $x^{2b} - 3x^b - 28$ **77.** $6x^{2b} + x^b - 2$

79. $x^{4a} - 2x^{2a} + 1$ **81.** $x^{3a} - 6x^{2a} + 12x^a - 8$

Problem Set 0.4 (page 49)

1. $2xy(3 - 4y)$ **3.** $(z + 3)(x + y)$

5. $(x + y)(3 + a)$ **7.** $(x - y)(a - b)$

9. $(3x + 5)(3x - 5)$ **11.** $(1 + 9n)(1 - 9n)$

13. $(x + 4 + y)(x + 4 - y)$

15. $(3s + 2t - 1)(3s - 2t + 1)$ **17.** $(x - 7)(x + 2)$

19. $(5 + x)(3 - x)$ **21.** Not factorable

23. $(3x - 5)(x - 2)$ **25.** $(10x + 7)(x + 1)$

27. $(x - 2)(x^2 + 2x + 4)$

29. $(4x + 3y)(16x^2 - 12xy + 9y^2)$ **31.** $4(x^2 + 4)$

33. $x(x + 3)(x - 3)$ **35.** $(3a - 7)^2$

37. $2n(n^2 + 3n + 5)$ **39.** $(5x - 3)(2x + 9)$

41. $(6a - 1)^2$ **43.** $(4x - y)(2x + y)$

45. Not factorable   **47.** $2n(n^2 + 7n - 10)$

49. $4(x + 2)(x^2 - 2x + 4)$

51. $(x + 3)(x - 3)(x^2 + 5)$

53. $2y(x + 4)(x - 4)(x^2 + 3)$

55. $(a + b + c + d)(a + b - c - d)$

57. $(x + 4 + y)(x + 4 - y)$

59. $(x + y + 5)(x - y - 5)$ **61.** $(10x + 3)(6x - 5)$

63. $3x(7x - 4)(4x + 5)$ **65.** $(x^a + 4)(x^a - 4)$

67. $(x^n - y^n)(x^{2n} + x^ny^n + y^{2n})$

69. $(x^a + 4)(x^a - 7)$ **71.** $(2x^n - 5)(x^n + 6)$

73. $(x^{2n} + y^{2n})(x^n + y^n)(x^n - y^n)$

75. a. $(x + 32)(x + 3)$ **c.** $(x - 21)(x - 24)$ **e.** $(x + 28)(x + 32)$

Problem Set 0.5 (page 61)

1. $\dfrac{2x}{3}$ **3.** $\dfrac{7y^3}{9x}$ **5.** $\dfrac{8x^4y^4}{9}$ **7.** $\dfrac{a + 4}{a - 9}$

9. $\dfrac{x(2x + 7)}{y(x + 9)}$ **11.** $\dfrac{x^2 + xy + y^2}{x + 2y}$ **13.** $-\dfrac{2}{x + 1}$

15. $\dfrac{2x + y}{x - y}$ **17.** $\dfrac{9x^2 - 6xy + 4y^2}{x - 5}$ **19.** $\dfrac{x}{2y^3}$

21. $-\dfrac{8x^3y^3}{15}$ **23.** $\dfrac{14}{27a}$ **25.** $5y$

27. $\dfrac{5(a + 3)}{a(a - 2)}$ **29.** $\dfrac{(x + 6y)^2(2x + 3y)}{y^3(x + 4y)}$

31. $\dfrac{3xy}{4(x + 6)}$ **33.** $\dfrac{x - 9}{42x^2}$ **35.** $\dfrac{8x + 5}{12}$

37. $\dfrac{7x}{24}$ **39.** $\dfrac{35b + 12a^3}{80a^2b^2}$ **41.** $\dfrac{12 + 9n - 10n^2}{12n^2}$

43. $\dfrac{9y + 8x - 12xy}{12xy}$ **45.** $\dfrac{13x + 14}{(2x + 1)(3x + 4)}$

47. $\dfrac{7x + 21}{x(x + 7)}$ **49.** $\dfrac{1}{a - 2}$ **51.** $\dfrac{5}{2(x - 1)}$

53. $\dfrac{2n + 10}{3(n + 1)(n - 1)}$ **55.** $\dfrac{1}{x + 1}$

57. $\dfrac{9x + 73}{(x + 3)(x + 7)(x + 9)}$

59. $\dfrac{3x^2 + 30x - 78}{(x + 1)(x - 1)(x + 8)(x - 2)}$ **61.** $\dfrac{x + 6}{(x - 3)^2}$

63. $\dfrac{-x^2 - x + 1}{(x + 1)(x - 1)}$ **65.** $\dfrac{-8}{(n^2 + 4)(n + 2)(n - 2)}$

67. $\dfrac{5x^2 + 16x + 5}{(x + 1)(x - 4)(x + 7)}$

69. a. $\dfrac{5}{x - 1}$ **c.** $\dfrac{5}{a - 3}$ **e.** $x + 3$ **71.** $\dfrac{5y^2 - 3xy^2}{x^2y + 2x^2}$

73. $\dfrac{x + 1}{x - 1}$ **75.** $\dfrac{n - 1}{n + 1}$ **77.** $\dfrac{-6x - 4}{3x + 9}$

79. $\dfrac{x^2 + x + 1}{x + 1}$ **81.** $\dfrac{a^2 + 4a + 1}{4a + 1}$ **83.** $-\dfrac{2x + h}{x^2(x + h)^2}$

85. $-\dfrac{1}{(x + 1)(x + h + 1)}$ **87.** $-\dfrac{4}{(2x - 1)(2x + 2h - 1)}$

89. $\dfrac{y + x}{x^2y - xy^2}$ **91.** $\dfrac{x^2y^2 + 2}{4y^2 - 3x}$

Problem Set 0.6 (page 73)

1. 9 **3.** 5 **5.** $\dfrac{6}{7}$ **7.** $-\dfrac{3}{2}$ **9.** $2\sqrt{6}$

11. $4\sqrt{7}$ **13.** $-6\sqrt{11}$ **15.** $\dfrac{3\sqrt{5}}{2}$

17. $2x\sqrt{3}$ **19.** $8x^2y^3\sqrt{y}$ **21.** $\dfrac{9y^3\sqrt{5x}}{7}$

23. $4\sqrt[3]{2}$ **25.** $2x\sqrt[3]{2x}$ **27.** $2x\sqrt[4]{3x}$

29. $\dfrac{2\sqrt{3}}{5}$ **31.** $\dfrac{\sqrt{14}}{4}$ **33.** $\dfrac{4\sqrt{15}}{5}$

35. $\dfrac{3\sqrt{2}}{7}$ **37.** $\dfrac{\sqrt{15}}{6x^2}$ **39.** $\dfrac{2\sqrt{15a}}{5ab}$

41. $\dfrac{3\sqrt[3]{2}}{2}$ **43.** $\dfrac{\sqrt[3]{18x^2 y}}{3x}$ **45.** $12\sqrt{3}$ **47.** $3\sqrt{7}$

49. $\dfrac{11\sqrt{3}}{6}$ **51.** $-\dfrac{89\sqrt{2}}{30}$ **53.** $48\sqrt{6}$

55. $10\sqrt{6} + 8\sqrt{30}$ **57.** $3x\sqrt{6y} - 6\sqrt{2xy}$

59. $13 + 7\sqrt{3}$ **61.** $30 + 11\sqrt{6}$ **63.** 16

65. $x + 2\sqrt{xy} + y$ **67.** $a - b$ **69.** $3\sqrt{5} - 6$

71. $\sqrt{7} + \sqrt{3}$ **73.** $\dfrac{-2\sqrt{10} + 3\sqrt{14}}{43}$

75. $\dfrac{x + \sqrt{x}}{x - 1}$ **77.** $\dfrac{x - \sqrt{xy}}{x - y}$ **79.** $\dfrac{6x + 7\sqrt{xy} + 2y}{9x - 4y}$

81. $\dfrac{2}{\sqrt{2x + 2h} + \sqrt{2x}}$ **83.** $\dfrac{1}{\sqrt{x + h - 3} + \sqrt{x - 3}}$

91. $4x^2$ **93.** $y^2\sqrt{3y}$ **95.** $2m^4\sqrt{7}$

97. $3d^3\sqrt{2d}$ **99.** $4n^{10}\sqrt{5}$

Problem Set 0.7 (page 80)

1. 7 **3.** 8 **5.** -4 **7.** 2 **9.** 64

11. 0.001 **13.** $\dfrac{1}{32}$ **15.** 2 **17.** $15x^{7/12}$

19. $y^{5/12}$ **21.** $64x^{3/4} y^{3/2}$ **23.** $4x^{4/15}$

25. $\dfrac{7}{a^{1/12}}$ **27.** $\dfrac{16x^{4/3}}{81y}$ **29.** $\dfrac{y^{3/2}}{x}$ **31.** $8a^{9/2} x^2$

33. $\sqrt[4]{8}$ **35.** $\sqrt[12]{x^7}$ **37.** $xy\sqrt[4]{xy^3}$ **39.** $a\sqrt[12]{a^5 b^{11}}$

41. $4\sqrt[6]{2}$ **43.** $\sqrt[6]{2}$ **45.** $\sqrt{2}$ **47.** $x\sqrt[12]{x^7}$

49. $\dfrac{5\sqrt[3]{x^2}}{x}$ **51.** $\dfrac{\sqrt[6]{x^3 y^4}}{y}$ **53.** $\dfrac{\sqrt[20]{x^{15} y^8}}{y}$

55. $\dfrac{5\sqrt[12]{x^9 y^8}}{4x}$ **57. a.** $\sqrt[6]{2}$ **c.** $\sqrt{x}$

61. $\dfrac{2x - 2}{(2x - 1)^{3/2}}$ **63.** $\dfrac{x}{(x^2 + 2x)^{3/2}}$ **65.** $\dfrac{4x}{(2x)^{4/3}}$

69. a. 13.391 **c.** 2.702 **e.** 4.304

Problem Set 0.8 (page 88)

1. $13 + 8i$ **3.** $3 + 4i$ **5.** $-11 + i$ **7.** $-1 - 2i$

9. $-\dfrac{3}{20} + \dfrac{5}{12}i$ **11.** $\dfrac{7}{10} - \dfrac{11}{12}i$ **13.** $4 + 0i$

15. $3i$ **17.** $i\sqrt{19}$ **19.** $\dfrac{2}{3}i$ **21.** $2i\sqrt{2}$

23. $3i\sqrt{3}$ **25.** $3i\sqrt{6}$ **27.** $18i$ **29.** $12i\sqrt{2}$

31. -8 **33.** $-\sqrt{6}$ **35.** $-2\sqrt{5}$ **37.** $-2\sqrt{15}$

39. $-2\sqrt{14}$ **41.** 3 **43.** $\sqrt{6}$ **45.** $-21 + 0i$

47. $8 + 12i$ **49.** $0 + 26i$ **51.** $53 - 26i$

53. $10 - 24i$ **55.** $-14 - 8i$ **57.** $-7 + 24i$

59. $-3 + 4i$ **61.** $113 + 0i$ **63.** $13 + 0i$

65. $-\dfrac{8}{13} + \dfrac{12}{13}i$ **67.** $1 - \dfrac{2}{3}i$ **69.** $0 - \dfrac{3}{2}i$

71. $\dfrac{22}{41} - \dfrac{7}{41}i$ **73.** $-1 + 2i$ **75.** $-\dfrac{17}{10} + \dfrac{1}{10}i$

77. $\dfrac{5}{13} - \dfrac{1}{13}i$ **83. a.** $2 + 11i$ **c.** $-11 + 2i$ **e.** $-7 - 24i$

Chapter 0 Review Problem Set (page 92)

1. $\dfrac{1}{125}$ **2.** $-\dfrac{1}{81}$ **3.** $\dfrac{16}{9}$ **4.** $\dfrac{1}{9}$ **5.** -8 **6.** $\dfrac{3}{2}$

7. $-\dfrac{1}{2}$ **8.** $\dfrac{1}{6}$ **9.** 4 **10.** -8 **11.** $12x^2 y$

12. $-30x^{7/6}$ **13.** $\dfrac{48}{a^{1/6}}$ **14.** $\dfrac{27y^{3/5}}{x^2}$ **15.** $\dfrac{4y^5}{x^5}$

16. $\dfrac{8y}{x^{7/12}}$ **17.** $\dfrac{16x^6}{y^6}$ **18.** $-\dfrac{a^3 b^1}{3}$ **19.** $4x - 1$

20. $-3x + 8$ **21.** $12a - 19$ **22.** $20x^2 - 11x - 42$

23. $-12x^2 + 17x - 6$ **24.** $-35x^2 + 22x - 3$

25. $x^3 + x^2 - 19x - 28$ **26.** $6x^3 - x^2 + 10x + 6$

27. $25x^2 - 30x + 9$ **28.** $9x^2 + 42x + 49$

29. $8x^3 - 12x^2 + 6x - 1$

30. $27x^3 + 135x^2 + 225x + 125$

31. $x^4 + 2x^3 - 6x^2 - 22x - 15$

32. $2x^4 + 11x^3 - 16x^2 - 8x + 8$ **33.** $-4x^2 y^3 + 8xy^2$

34. $-7y + 9xy^2$ **35.** $(3x + 2y)(3x - 2y)$

36. $3x(x + 5)(x - 8)$ **37.** $(2x + 5)^2$

38. $(x - y + 3)(x - y - 3)$ **39.** $(x - 2)(x - y)$

40. $(4x - 3y)(16x^2 + 12xy + 9y^2)$

41. $(3x - 4)(5x + 2)$ **42.** $3(x^3 + 12)$

43. Not factorable   **44.** $3(x + 2)(x^2 - 2x + 4)$

45. $(x + 3)(x - 3)(x + 2)(x - 2)$.

46. $(2x - 1 - y)(2x - 1 + y)$ **47.** $\dfrac{2}{3y}$ **48.** $\dfrac{-5a^2}{3}$

49. $\dfrac{3x + 5}{x}$ **50.** $\dfrac{2(3x - 1)}{x^2 + 4}$ **51.** $\dfrac{29x - 10}{12}$

52. $\dfrac{x - 38}{15}$ **53.** $\dfrac{-6n + 15}{5n^2}$ **54.** $\dfrac{-3x - 16}{x(x + 7)}$

55. $\dfrac{3x^2 - 8x - 40}{(x + 4)(x - 4)(x - 10)}$ **56.** $\dfrac{8x - 4}{x(x + 2)(x - 2)}$

57. $\dfrac{3xy - 2x^2}{5y + 7x^2}$ **58.** $\dfrac{3x - 2}{4x + 3}$ **59.** $-\dfrac{6x + 3h}{x^2(x + h)^2}$

60. $\dfrac{12}{(x^2 + 2)^{3/2}}$ **61.** $20\sqrt{3}$ **62.** $6x\sqrt{6x}$

63. $2xy\sqrt[3]{4xy^2}$ **64.** $\sqrt{3}$ **65.** $\dfrac{\sqrt{10x}}{2y}$

66. $\dfrac{15 - 3\sqrt{2}}{23}$ **67.** $\dfrac{24 - 4\sqrt{6}}{15}$ **68.** $\dfrac{3x + 6\sqrt{xy}}{x - 4y}$

69. $\sqrt[6]{5^5}$ **70.** $\sqrt[12]{x^{11}}$ **71.** $x^2\sqrt[6]{x^5}$ **72.** $x\sqrt[10]{xy^9}$

73. $\sqrt[6]{5}$ **74.** $\dfrac{\sqrt[12]{x^{11}}}{x}$ **75.** $-11 - 6i$

76. $-1 - 2i$ **77.** $1 - 2i$ **78.** $21 + 0i$
79. $26 - 7i$ **80.** $-25 + 15i$ **81.** $-14 - 12i$

82. $29 + 0i$ **83.** $0 - \dfrac{5}{3}i$ **84.** $-\dfrac{6}{25} + \dfrac{17}{25}i$

85. $0 + i$ **86.** $-\dfrac{12}{29} - \dfrac{30}{29}i$ **87.** $10i$ **88.** $2i\sqrt{10}$

89. $16i\sqrt{5}$ **90.** -12 **91.** $-4\sqrt{3}$ **92.** $2\sqrt{2}$
93. $600,000,000$ **94.** $800,000$

Chapter 0 Test (page 94)

1. a. $-\dfrac{1}{49}$ **b.** $\dfrac{8}{27}$ **c.** $\dfrac{8}{27}$ **d.** $\dfrac{3}{4}$ **2.** $-\dfrac{15}{x^4y^2}$
3. $-12x - 8$ **4.** $-30x^2 + 32x - 8$
5. $3x^3 + 4x^2 - 11x - 14$ **6.** $64x^3 - 48x^2 + 12x - 1$
7. $9x^3y + 12x^4y^2$ **8.** $3x(2x + 1)(3x - 4)$
9. $(5x + 2)(6x - 5)$ **10.** $8(x + 2)(x^2 - 2x + 4)$

11. $(x - 2)(x + y)$ **12.** $\dfrac{21x^5}{20}$ **13.** $\dfrac{x + 3}{x^2 + 2x + 4}$

14. $\dfrac{n - 8}{12}$ **15.** $\dfrac{23x + 6}{6x(x - 3)(x + 2)}$ **16.** $\dfrac{8 - 13n}{2n^2}$

17. $\dfrac{2y^2 - 5xy}{3y^2 + 4x}$ **18.** $12x^2\sqrt{7x}$ **19.** $\dfrac{5\sqrt{2}}{6}$

20. $\dfrac{4\sqrt{3} + 3\sqrt{2}}{5}$ **21.** $2xy\sqrt[3]{6xy^2}$ **22.** $-4 - 3i$

23. $34 - 18i$ **24.** $85 + 0i$ **25.** $\dfrac{1}{10} + \dfrac{7}{10}i$

CHAPTER 1

Problem Set 1.1 (page 104)

1. $\{-2\}$ **3.** $\left\{-\dfrac{1}{2}\right\}$ **5.** $\{7\}$ **7.** $\left\{-\dfrac{3}{2}\right\}$ **9.** $\left\{-\dfrac{10}{3}\right\}$

11. $\{-10\}$ **13.** $\{23\}$ **15.** $\left\{-\dfrac{21}{16}\right\}$ **17.** $\left\{\dfrac{3}{5}\right\}$

19. $\{-14\}$ **21.** $\{9\}$ **23.** $\left\{\dfrac{10}{7}\right\}$ **25.** $\{-10\}$

27. $\{1\}$ **29.** $\{-12\}$ **31.** $\{27\}$ **33.** $\left\{\dfrac{159}{5}\right\}$

35. $\{3\}$ **37.** $\{0\}$ **39.** $\left\{-\dfrac{2}{3}\right\}$ **41.** $\left\{\dfrac{1}{2}\right\}$

43. Chicken sandwich 30 grams, pasta salad 70 grams

45. 14, 16, and 18   **47.** 18 and 19
49. 10, 11, 12, and 13   **51.** 30°, 40°, and 110°
53. $9 per hour
55. 48 pennies, 21 nickels, and 11 dimes
57. 17 females and 26 males   **59.** 13 years old
61. Brad is 29 and Pedro is 23.

Problem Set 1.2 (page 117)

1. $\{1\}$ **3.** $\{9\}$ **5.** $\left\{\dfrac{10}{3}\right\}$ **7.** $\{4\}$ **9.** $\{14\}$

11. $\{9\}$ **13.** $\left\{\dfrac{1}{2}\right\}$ **15.** $\left\{\dfrac{1}{4}\right\}$ **17.** $\left\{\dfrac{2}{3}\right\}$

19. $\{-8\}$ **21.** $\varnothing$ **23.** $\{12\}$ **25.** $\{300\}$ **27.** $\{275\}$

29. $\left\{-\dfrac{66}{37}\right\}$ **31.** $\{6\}$ **33.** $w = \dfrac{P - 2l}{2}$

35. $\dfrac{A - 2\,lw}{2w + 2l}$ **37.** $h = \dfrac{A - 2\pi r^2}{2\pi r}$

39. $F = \dfrac{9C + 160}{5}$ or $F = \dfrac{9}{5}C + 32$

41. $T = \dfrac{NC - NV}{C}$ **43.** $T = \dfrac{I + klt}{kl}$

45. $R_n = \dfrac{R_1R_2}{R_1 + R_2}$ **47.** $1050   **49.** $900 and $1350
51. 168 two-wheel drive, 28 four-wheel drive   **53.** $65
55. $1540 per month   **57.** $32.20   **59.** $75
61. $30   **63.** 14 nickels and 29 dimes
65. 15 dimes, 45 quarters, and 10 half-dollars
67. $2000 at 5% and $3500 at 7%   **69.** $4000
71. 6 centimeters by 10 centimeters   **73.** 14 centimeters
83. a. $1.00   **b.** $11.33   **c.** $83.33   **d.** $400   **e.** $21,176.47

Problem Set 1.3 (page 131)

1. $\{-4, 7\}$ **3.** $\left\{-3, \dfrac{4}{3}\right\}$ **5.** $\left\{0, \dfrac{3}{2}\right\}$ **7.** $\left\{\pm\dfrac{2\sqrt{3}}{3}\right\}$

9. $\left\{\dfrac{-1 \pm 2\sqrt{5}}{2}\right\}$ **11.** $\left\{-\dfrac{5}{3}, \dfrac{2}{5}\right\}$ **13.** $\{2 \pm 2i\}$

15. $\left\{-\dfrac{7}{2}, \dfrac{1}{5}\right\}$ **17.** $\{4, 6\}$ **19.** $\{-5 \pm 3\sqrt{3}\}$

21. $\left\{\dfrac{3 \pm \sqrt{5}}{2}\right\}$ **23.** $\{-2 \pm i\sqrt{2}\}$

25. $\left\{\dfrac{-6 \pm \sqrt{46}}{2}\right\}$ **27.** $\{-16, 18\}$

29. $\left\{\dfrac{-5 \pm \sqrt{37}}{6}\right\}$ **31.** $\{-6, 9\}$ **33.** $\left\{-5, -\dfrac{1}{3}\right\}$

35. $\{1 \pm \sqrt{5}\}$ **37.** $\left\{\dfrac{3 \pm \sqrt{7}}{2}\right\}$ **39.** $\left\{\dfrac{3 \pm i\sqrt{19}}{2}\right\}$

41. $\{4 \pm 2\sqrt{3}\}$ **43.** $\left\{\dfrac{1}{2}\right\}$ **45.** $\left\{-\dfrac{3}{2}, \dfrac{1}{4}\right\}$

47. $\{-14, 12\}$ **49.** $\left\{\dfrac{3 \pm i\sqrt{47}}{4}\right\}$ **51.** $\left\{-1, \dfrac{5}{3}\right\}$

53. $\left\{\dfrac{-1 \pm \sqrt{2}}{2}\right\}$ **55.** $\{8 \pm 5\sqrt{2}\}$

57. $\{-10 \pm 5\sqrt{5}\}$ **59.** $\left\{\dfrac{1 \pm \sqrt{6}}{5}\right\}$

61. a. One real solution   **c.** One real solution
e. Two complex but nonreal solutions
g. Two unequal real solutions
63. 11 and 12   **65.** 12 feet   **67.** 10 meters and 24 meters
67. 10 meters and 24 meters   **69.** 8 inches by 14 inches
71. 7 meters wide and 18 meters long   **73.** 1 meter
75. 7 inches by 11 inches   **77.** 8 units

83. a. $r = \dfrac{\sqrt{A\pi}}{\pi}$ **c.** $t = \dfrac{\sqrt{2gs}}{g}$

e. $y = \dfrac{b\sqrt{x^2 - a^2}}{a}$

85. $k = \pm 4$
87. a. $\{-1.359, 7.359\}$ **c.** $\{-10.280, 4.280\}$
e. $\{-0.258, -7.742\}$ **g.** $\{0.191, 1.309\}$
i. $\{-0.422, 5.922\}$

Problem Set 1.4 (page 145)

1. $\{-12\}$ **3.** $\left\{\dfrac{37}{15}\right\}$ **5.** $\{-2\}$ **7.** $\{-8, 1\}$

9. $\left\{\dfrac{6}{29}\right\}$ **11.** $\left\{n \mid n \neq \dfrac{3}{2} \text{ and } n \neq 3\right\}$

13. $\left\{-4, \dfrac{4}{3}\right\}$ **15.** $\left\{-\dfrac{1}{4}\right\}$ **17.** $\varnothing$ **19.** $\{3\}$

21. 9 rows and 14 trees per row   **23.** $4\dfrac{1}{2}$ hours

25. 50 miles
27. 50 mph for the freight and 70 mph for the express
29. 3 liters
31. 3.5 liters of the 50% solution and 7 liters of the
80% solution

33. 5 quarts   **35.** $2\dfrac{2}{5}$ hours   **37.** 60 minutes

39. 9 hours   **41.** 8 hours   **43.** 7 golf balls
45. 60 hours

Problem Set 1.5 (page 154)

1. $\{-2, 1 \pm i\sqrt{3}\}$ **3.** $\left\{1, \dfrac{-1 \pm i\sqrt{3}}{2}\right\}$

5. $\{-2, -1, 2\}$ **7.** $\left\{\pm i, \dfrac{3}{2}\right\}$ **9.** $\left\{-\dfrac{5}{4}, 0, \pm\dfrac{\sqrt{2}}{2}\right\}$

11. $\{0, 16\}$ **13.** $\{1\}$ **15.** $\{6\}$ **17.** $\{3\}$ **19.** $\varnothing$

21. $\{-1\}$ **23.** $\left\{\dfrac{13}{2}\right\}$ **25.** $\{-15\}$ **27.** $\{9\}$

29. $\left\{\dfrac{2}{3}, 1\right\}$ **31.** $\{5\}$ **33.** $\{7\}$ **35.** $\{-2, -1\}$

37. $\{0\}$ **39.** $\{6\}$ **41.** $\{0, 4\}$ **43.** $\{\pm 1, \pm 2\}$

45. $\left\{\pm\dfrac{\sqrt{2}}{2}, \pm 2\right\}$ **47.** $\{\pm i\sqrt{5}, \pm\sqrt{7}\}$

49. $\{\pm\sqrt{2 + \sqrt{3}}, \pm\sqrt{2 - \sqrt{3}}\}$ **51.** $\{-125, 8\}$

53. $\left\{-\dfrac{8}{27}, \dfrac{27}{8}\right\}$ **55.** $\left\{-\dfrac{1}{6}, \dfrac{1}{2}\right\}$ **57.** $\{25, 36\}$

59. $\{4\}$ **61.** $\left\{\pm\dfrac{1}{2}, \pm 2\right\}$ **63.** $\left\{\pm\dfrac{1}{8}, \pm 1\right\}$

65. $\left\{-\dfrac{2}{3}, 5\right\}$ **67.** 12 inches   **69.** 320 feet

75. a. $\{\pm 1.62, \pm 0.62\}$ **c.** $\{\pm 1.78, \pm 0.56\}$
e. $\{\pm 8.00, \pm 6.00\}$

Problem Set 1.6 (page 167)

1. $(-\infty, -2]$

3. $(1, 4)$

5. $(0, 2)$

7. $[-2, -1]$

9. $(-\infty, 1) \cup (3, \infty)$

11. $(-2, \infty)$

13. $[-5, 4]$ **15.** $\left(-1, \dfrac{3}{2}\right)$ **17.** $(-11, 13)$ **19.** $(-1, 5)$

21. $(-\infty, -2)$ **23.** $\left[-\dfrac{5}{3}, \infty\right)$ **25.** $[7, \infty)$

27. $\left(-\infty, \dfrac{17}{5}\right]$ **29.** $\left(-\infty, \dfrac{7}{3}\right)$ **31.** $[-20, \infty)$

33. $(300, \infty)$ **35.** $\left(\dfrac{1}{5}, \dfrac{7}{5}\right)$ **37.** $[1, 5]$

39. $(-4, 1)$ **41.** $(-\infty, -3) \cup (5, \infty)$ **43.** $[-1, 2]$

45. $(-\infty, -4) \cup \left(\dfrac{1}{3}, \infty\right)$ **47.** $\left[\dfrac{2}{5}, \dfrac{4}{3}\right]$

49. $\left(-\infty, \dfrac{1}{2}\right) \cup \left(\dfrac{1}{2}, \infty\right)$ **51.** $(-\infty, -2) \cup (2, \infty)$

53. $(-6, 6)$ **55.** $(-\infty, \infty)$ **57.** $(-\infty, 0] \cup [2, \infty)$

59. $(-4, 0) \cup (0, \infty)$ **61.** $(-2, -1)$ **63.** $\left(-\infty, \dfrac{22}{3}\right]$

65. $(-4, 1) \cup (2, \infty)$ **67.** $(-\infty, -2] \cup \left[\dfrac{1}{2}, 5\right]$

69. $[-4, 0] \cup [6, \infty)$ **71.** $(-3, 2) \cup (2, \infty)$
73. Greater than 10% **75.** 98 or higher
77. Greater than or equal to 13.8 inches
79. Between $-4°F$ and $23°F$, inclusive
81. More than 250 miles

Problem Set 1.7 (page 177)

1. $(-\infty, -1) \cup (5, \infty)$ **3.** $\left(-2, \dfrac{1}{2}\right)$ **5.** $\left(\dfrac{1}{3}, 3\right]$

7. $[-3, -2)$ **9.** $(-\infty, -5) \cup (-2, \infty)$
11. $(-\infty, -5)$ **13.** $(-3, 2)$ **15.** $\{-4, 8\}$
17. $\left\{-\dfrac{13}{20}, \dfrac{3}{20}\right\}$ **19.** $\{-3, 4\}$ **21.** $\left\{-3, \dfrac{1}{3}\right\}$

23. $\varnothing$ **25.** $\left\{-\dfrac{10}{3}, 2\right\}$ **27.** $\{-3, 9\}$ **29.** $\left\{-\dfrac{1}{5}\right\}$

31. $\{-4, 1\}$ **33.** $\left\{\dfrac{1}{4}, \dfrac{7}{4}\right\}$ **35.** $\left\{-\dfrac{2}{5}, 4\right\}$ **37.** $\{-2, 0\}$

39. $\{-1\}$ **41.** $(-6, 6)$ **43.** $(-\infty, -8) \cup (8, \infty)$

45. $(-\infty, \infty)$ **47.** $(-\infty, -2) \cup (8, \infty)$ **49.** $[-3, 4]$

51. $\left(-\infty, -\dfrac{11}{3}\right) \cup \left(\dfrac{7}{3}, \infty\right)$ **53.** $\varnothing$ **55.** $\left(-\dfrac{1}{2}, \dfrac{7}{2}\right)$

57. $(-\infty, \infty)$ **59.** $(-\infty, -9] \cup [7, \infty)$ **61.** $(-6, 0)$

63. $(-\infty, -6) \cup (-2, \infty)$ **65.** $(-\infty, 0] \cup [4, \infty)$

67. $(-6, 4)$ **69.** $(-\infty, -1] \cup \left[\dfrac{5}{3}, \infty\right)$ **71.** $\varnothing$

73. $\left(-\infty, \dfrac{5}{4}\right) \cup \left(\dfrac{7}{2}, \infty\right)$ **75.** $(-\infty, -3) \cup (-3, -1)$

77. $\left[-\dfrac{2}{5}, 0\right) \cup \left(0, \dfrac{2}{3}\right]$ **79.** $\left(-\infty, \dfrac{2}{5}\right] \cup \left[\dfrac{2}{3}, \infty\right)$

Chapter 1 Review Problem Set (page 181)

1. $\{-14\}$ **2.** $\{-19\}$ **3.** $\left\{\dfrac{10}{7}\right\}$ **4.** $\{200\}$

5. $\left\{-1, \dfrac{5}{3}\right\}$ **6.** $\left\{\dfrac{5}{4}, 6\right\}$ **7.** $\{3 \pm i\}$

8. $\{-22, 18\}$ **9.** $\left\{-\dfrac{2}{5}, 0, \dfrac{1}{3}\right\}$ **10.** $\{-5\}$

11. $\left\{\dfrac{1}{2}, 6\right\}$ **12.** $\{\pm 3i, \pm\sqrt{5}\}$

13. $\left\{\pm\dfrac{\sqrt{5}}{5}, \pm\sqrt{2}\right\}$

14. $\left\{-1, 2, \dfrac{-5 \pm \sqrt{33}}{2}\right\}$ **15.** $\{2\}$

16. $\left\{-1, \dfrac{1}{2}\right\}$ **17.** $\{0\}$

18. $\left\{-\dfrac{6}{5}, \dfrac{8}{5}\right\}$ **19.** $\left\{\dfrac{2}{5}, 12\right\}$ **20.** $\left\{\dfrac{1}{4}, \dfrac{7}{4}\right\}$

21. $\{-\sqrt{2}, -1, \sqrt{2}\}$ **22.** $\left\{-64, \dfrac{27}{8}\right\}$

23. $(-8, \infty)$ **24.** $\left[-\dfrac{65}{4}, \infty\right)$ **25.** $\left(-\infty, -\dfrac{9}{2}\right)$

26. $(-\infty, 400]$ **27.** $[-2, 1]$ **28.** $\left(-\dfrac{2}{3}, 2\right)$

29. $(-3, 6)$ **30.** $(-\infty, -2] \cup [7, \infty)$

31. $(-\infty, -2) \cup (1, 4)$ **32.** $\left[-4, \dfrac{3}{2}\right)$

33. $\left(-\infty, \dfrac{1}{5}\right) \cup (2, \infty)$ **34.** $[-7, -3)$

35. $(-\infty, 4)$ **36.** $\left(-\infty, -\dfrac{1}{2}\right) \cup (2, \infty)$

37. $\left[-\dfrac{19}{3}, 3\right]$ **38.** $\left(-\dfrac{9}{2}, \dfrac{3}{2}\right)$

39. $(-1, 0) \cup \left(0, \dfrac{1}{3}\right)$ **40.** $\left(-\dfrac{3}{2}, \infty\right)$

41. 21, 23, and 25 **42.** 49 men
43. 7 centimeters by 12 centimeters
44. 13 nickels, 39 dimes, and 36 quarters
45. \$20 **46.** 20 gallons
47. Rosie is 14 years old and her mother is 33 years old.
48. \$3500 at 7% and \$4500 at 8.5% **49.** 95 or higher

50. Amy 4.5 hours, Angie 9 hours

51. $26\dfrac{2}{3}$ minutes **52.** 40 shares at \$15 per share

53. 54 mph for Mike and 52 mph for Larry
54. Cindy 4 hours and Bill 6 hours
55. 15 centimeters and 20 centimeters
56. 5 inches by 7 inches

Chapter 1 Test (page 183)

1. $\{2\}$ **2.** $\left\{-\dfrac{3}{2}, \dfrac{1}{5}\right\}$ **3.** $\left\{-\dfrac{7}{5}, \dfrac{3}{5}\right\}$

4. $\{-1\}$ **5.** $\left\{\dfrac{1 \pm i\sqrt{31}}{4}\right\}$ **6.** $\{-4, -1\}$

7. $\{600\}$ **8.** $\left\{-1, \dfrac{11}{3}\right\}$ **9.** $\left\{\dfrac{1 \pm \sqrt{7}}{3}\right\}$

10. $\{-9, 0, 2\}$ **11.** $\left\{-\dfrac{6}{7}, 3\right\}$ **12.** $\{8\}$

13. $\varnothing$ **14.** $\left\{-\dfrac{1}{4}, \dfrac{2}{3}\right\}$ **15.** $(-\infty, -35]$

16. $(3, \infty)$ **17.** $\left(-1, \dfrac{7}{3}\right)$

18. $\left(-\infty, -\dfrac{11}{4}\right] \cup \left[\dfrac{1}{4}, \infty\right)$ **19.** $\left[-\dfrac{1}{2}, 5\right]$

20. $(-\infty, -2) \cup \left(\dfrac{1}{3}, \infty\right)$ **21.** $[-10, -6]$

22. $\dfrac{2}{3}$ cup   **23.** 15 mph   **24.** 150 shares

25. 9 centimeters by 14 centimeters

CHAPTER 2

Problem Set 2.1 (page 193)

1. 10   **3.** -5 **5.** 5   **7.** 9   **9.** 7

11. $\dfrac{1}{3}$ **13.** -7 **15.** 5   **17.** $2\sqrt{13}$

19. 10   **21.** $\left(2, -\dfrac{5}{2}\right)$ **23.** $\left(\dfrac{15}{2}, -\dfrac{11}{2}\right)$

25. $\left(\dfrac{1}{12}, \dfrac{11}{12}\right)$ **27.** $(3, 5)$ **29.** $(2, 5)$

31. $\left(\dfrac{17}{8}, -7\right)$ **33.** $(-2, -3)$

35. $\left(3, \dfrac{13}{2}\right)$ **37.** $\left(4, \dfrac{25}{4}\right)$

43. $15 + 9\sqrt{5}$ **47.** 3 or -7 **49.** $(3, 8)$

51. Both midpoints are at $\left(\dfrac{7}{2}, \dfrac{5}{2}\right)$.

Problem Set 2.2 (page 204)

1.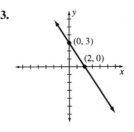

3.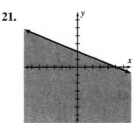

5.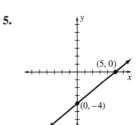

7.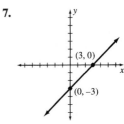

9.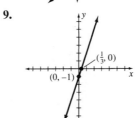

11.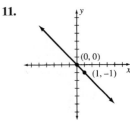

13.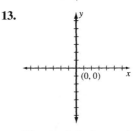

The graph is the y axis.

15.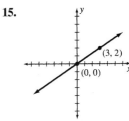

17.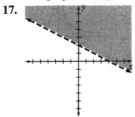

19.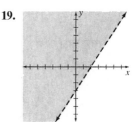

21.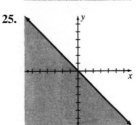

23.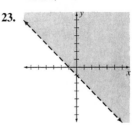

25.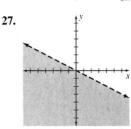

27.

29.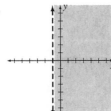

33.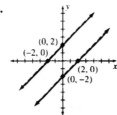

35.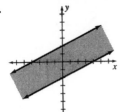

37.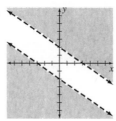

39.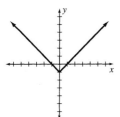

41.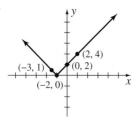

43.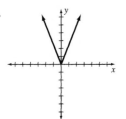

25. $2x + 3y = -1$ **27.** $0x + y = -2$ **29.** $x + 0y = 3$

31. $5x - 7y = -11$ **33.** $5x + 6y = 37$

35. $x + 5y = 14$ **37.** $0x + y = -3$ **39.** $x + 0y = -2$

41. $y = \frac{1}{2}x + 3$ **43.** $y = -\frac{3}{7}x + 2$ **45.** $y = 4x + \frac{3}{2}$

47. $y = -\frac{5}{6}x + \frac{1}{4}$ **49.** $5x - 4y = 20$ **51.** $x + 0y = -4$

53. $5x + 2y = 14$ **55.** $4x + y = -2$

57. $x + 0y = 1$ **59.** $x + 0y = -3$

61. Parallel   **63.** Perpendicular

65. Intersecting lines that are not perpendicular

67. Perpendicular   **69.** $m = \frac{2}{3}, b = -\frac{4}{3}$

71. $m = \frac{1}{2}, b = -\frac{7}{2}$ **73.** $m = -3, b = 0$

75. $m = \frac{7}{5}, b = -\frac{12}{5}$

77. a. 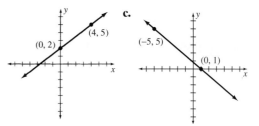   **c.**

e.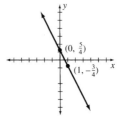

83. $x - y = -7, x + 5y = -19, 2x + y = 16$

85. $9x + 8y = -2, 6x - 7y = 11, 15x + y = 9$

87. 250 feet   **89.** 19 centimeters

95. a. $3x - y = 9$ **b.** $6x + 5y = 7$ **c.** $2x + 7y = 0$
  d. $3x - 8y = 23$

99. a. $2x - y = 4$ **b.** $3x + 7y = 19$
  c. $5x + 2y = 2$ **d.** $3x - 2y = 1$

Problem Set 2.3 (page 218)

1. $\frac{3}{4}$ **3.** -5 **5.** 0 **7.** $-\frac{b}{a}$ **9.** $x = \frac{23}{2}$

11. $x = -\frac{22}{9}$ **13–19.** Answers will vary.

21. $x - 3y = -10$ **23.** $2x - y = 0$

Problem Set 2.4 (page 231)

1. $(4, -3); (-4, 3); (-4, -3)$ **3.** $(-6, 1); (6, -1); (6, 1)$

5. $(0, -4); (0, 4); (0, -4)$ **7.** y axis   **9.** x axis

11. x axis, y axis, and origin   **13.** None   **15.** Origin

17. None   **19.** y axis   **21.** Origin   **23.** x axis

25. None   **27.** None   **29.** Origin

31.
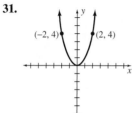

33.
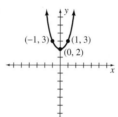

55.
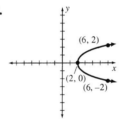

57.
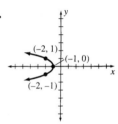

35.
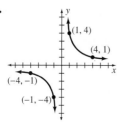

37.
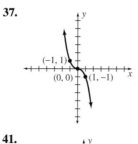

Problem Set 2.5 (page 243)

1. $x^2 + y^2 - 4x - 6y - 12 = 0$

3. $x^2 + y^2 + 2x + 10y + 17 = 0$

5. $x^2 + y^2 - 6x = 0$ **7.** $x^2 + y^2 - 49 = 0$

9. $(3, 5); r = 2$ **11.** $(-5, -7); r = 1$

13. $(5, 0); r = 5$ **15.** $(0, 0); r = 2\sqrt{2}$

17. $\left(\dfrac{1}{2}, 1\right); r = 2$ **19.** $(0, 2); r = \sqrt{6}$

21. $(1, -2); r = \dfrac{4\sqrt{3}}{3}$ **23.** $\left(-\dfrac{3}{2}, -\dfrac{7}{2}\right); r = \dfrac{5\sqrt{2}}{2}$

25. $x^2 + y^2 - 12x + 16y = 0$

27. $x^2 + y^2 - 6x - 6y - 67 = 0$

29. $x^2 + y^2 - 14x + 14y + 49 = 0$

31. $x^2 + y^2 + 6x - 10y + 9 = 0$ and $x^2 + y^2 + 6x + 10y + 9 = 0$

39.
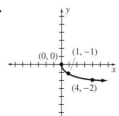

41.
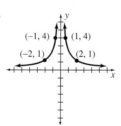

33.
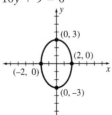

35.
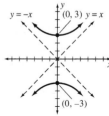

43.
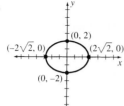

45.
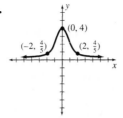

37.
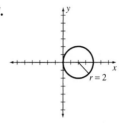

39.
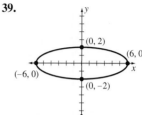

47.
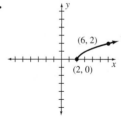

49.
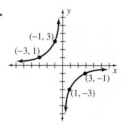

41.
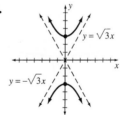

43.
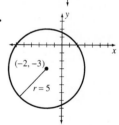

51.
(6, 2), (2, 0)

53.
(-1, 3), (-3, 1), (3, -1), (1, -3)

45.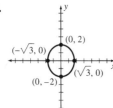

47.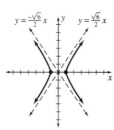

19.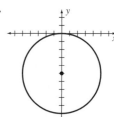

20.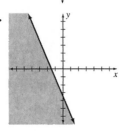

49.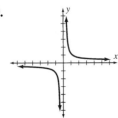

51.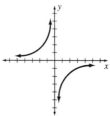

21.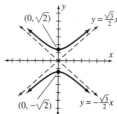

22.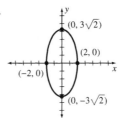

55. a. $(1, 4); r = 3$ **c.** $(-6, -4); r = 8$ **e.** $(0, 6); r = 9$

23. -5 **24.** $\dfrac{5}{7}$ **25.** $3x + 4y = 29$

Chapter 2 Review Problem Set (page 247)

1. 5   **2.** -5 **3.** $\left(9, \dfrac{1}{3}\right)$ **4.** $(-2, 6)$

7. x axis   **8.** None   **9.** x axis, y axis, and origin
10. y axis   **11.** Origin   **12.** y axis

26. $2x - y = -4$ **27.** $4x + 3y = -4$
28. $x + 2y = 3$ **29.** $x^2 + y^2 - 10x + 12y + 60 = 0$
30. $x^2 + y^2 - 4x - 6y - 4 = 0$
31. $x^2 + y^2 + 10x - 24y = 0$
32. $x^2 + y^2 + 8x + 8y + 16 = 0$

13.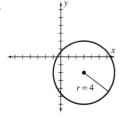

14.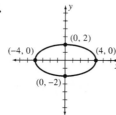

Chapter 2 Test (page 248)

1. 8   **2.** $(-4, 6)$ **3.** $(6, -4)$ **4.** $-\dfrac{4}{9}$

5. $\dfrac{2}{7}$ **6.** $3x + 4y = -12$ **7.** $11x - 3y = 23$

8. $x - 5y = -21$ **9.** $7x - 4y = 1$
10. $x + 0y = -2$ **11.** $x^2 + 6x + y^2 + 12y + 29 = 0$
12. $x^2 - 4x + y^2 - 8y + 10 = 0$
13. $x^2 - 8x + y^2 + 6y = 0$
14. Center at $(-8, 5)$ and radius of length three units
15. 5, $\sqrt{37}$, and $2\sqrt{5}$ units   **16.** 1 and 5

17. $\pm\sqrt{3}$ **18.** Six units   **19.** $y = \pm\dfrac{3}{4}x$

20. a. x axis **b.** Origin **c.** y axis
   d. x axis, y axis, and origin

15.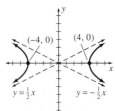

16.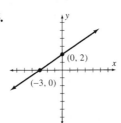

17.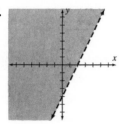

18.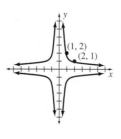

21.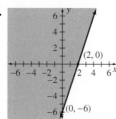

22.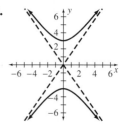

23.

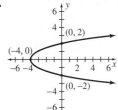

24.

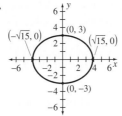

25.

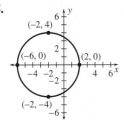

42. $(-\infty, -8) \cup (3, \infty)$ **43.** $\left(-\dfrac{3}{2}, \dfrac{1}{3}\right)$

44. $\left(-3, \dfrac{1}{2}\right) \cup (4, \infty)$ **45.** $\left(-1, \dfrac{2}{3}\right]$ **46.** $(1, 7]$

47. $\left(-\infty, -\dfrac{4}{3}\right) \cup (2, \infty)$ **48.** $\left(-\dfrac{9}{5}, 3\right)$

49.

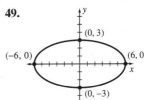

50.

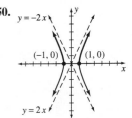

51.

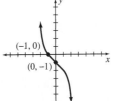

52.

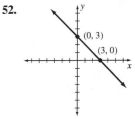

53.

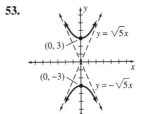

54.

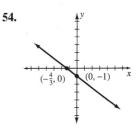

Cumulative Review Problem Set, Chapters 0–2 (page 249)

1. $\dfrac{1}{27}$ **2.** $-\dfrac{1}{16}$ **3.** $\dfrac{9}{4}$ **4.** $-\dfrac{2}{3}$ **5.** 9 **6.** $\dfrac{9}{16}$

7. $\dfrac{20}{x^2 y^3}$ **8.** $-\dfrac{56a}{b}$ **9.** $4x^4 y^2$ **10.** $\dfrac{5y^2}{x^4}$

11. $\dfrac{x^{1/3}}{17 y^{7/4}}$ **12.** $\dfrac{4a^8}{b^{14}}$ **13.** $-30\sqrt{2}$ **14.** $6xy\sqrt{3x}$

15. $2xy^2 \sqrt[3]{7xy}$ **16.** $\dfrac{3\sqrt{6}}{10}$ **17.** $\dfrac{\sqrt{21xy}}{7y}$

18. $-\dfrac{5(\sqrt{2} + 3)}{7}$ **19.** $\dfrac{6\sqrt{14} + 3\sqrt{42}}{2}$

20. $\dfrac{4x - 12\sqrt{xy}}{x - 9y}$ **21.** $\dfrac{3x^3 y^2}{8}$ **22.** $-\dfrac{3b^3}{8a^3}$

23. $\dfrac{5x + 1}{x}$ **24.** $\dfrac{21x + 5}{24}$ **25.** $\dfrac{10 - 3n}{6n^2}$

26. $\dfrac{5x^2 + 18x + 27}{(x + 9)(x - 3)(x + 3)}$ **27.** $\left\{-\dfrac{23}{4}\right\}$ **28.** $\{3\}$

29. $\left\{\dfrac{3}{7}\right\}$ **30.** $\left\{\pm\dfrac{2}{3}\right\}$ **31.** $\{-4, 0, 2\}$

32. $\left\{\dfrac{3}{7}, 4\right\}$ **33.** $\{\pm 4i, \pm 1\}$ **34.** $\left\{-\dfrac{1}{5}, 1\right\}$

35. $\left\{\dfrac{3 \pm \sqrt{17}}{4}\right\}$ **36.** $\left\{\dfrac{-13 \pm \sqrt{205}}{2}\right\}$

37. $\{1\}$ **38.** $\left\{\dfrac{1 \pm 2i}{2}\right\}$ **39.** $\left(-\infty, \dfrac{1}{8}\right)$

40. $\left[-\dfrac{5}{9}, \infty\right)$ **41.** $(-\infty, 250]$

55. $(-7, 4); r = 3$ **56.** $3x - 4y = -26$
57. $(-2, 6)$ **58.** $4x + 3y = 25$
59. $\$28.60; \$31.43$ **60.** $\$3000$ at 5% and $\$4500$ at 6%
61. Length of 8 inches and width of 4.5 inches
62. The side is 20 centimeters long, and the altitude is 8 centimeters long.
63. 16 milliliters   **64.** 30 miles   **65.** 3 hours

CHAPTER 3

Problem Set 3.1 (page 261)
1. $f(3) = -1; f(5) = -5; f(-2) = 9$
3. $g(3) = -20; g(-1) = -8; g(-4) = -41$

5. $h(3) = \dfrac{5}{4}; h(4) = \dfrac{23}{12}; h\left(-\dfrac{1}{2}\right) = -\dfrac{13}{12}$

7. $f(5) = 3; f\left(\dfrac{1}{2}\right) = 0; f(23) = 3\sqrt{5}$

9. $f(4) = 4; f(10) = 10; f(-3) = 9; f(-5) = 25$

11. $f(3) = 6; f(5) = 10; f(-3) = 6; f(-5) = 10$

13. $f(2) = 1; f(0) = 0; f\left(-\dfrac{1}{2}\right) = 0; f(-4) = -1$

15. -7 **17.** $\dfrac{1}{2}$ **19.** 0 **21.** $-2a - h + 4$

23. $6a + 3h - 1$ **25.** $3a^2 + 3ah + h^2 - 2a - h + 2$

27. $-\dfrac{2}{(a-1)(a+h-1)}$ **29.** $-\dfrac{2a+h}{a^2(a+h)^2}$

31. Yes   **33.** No   **35.** Yes   **37.** Yes

39. $D = \left\{x \mid x \geq \dfrac{4}{3}\right\}$
$R = \{f(x) \mid f(x) \geq 0\}$

41. $D = \{x \mid x \text{ is any real number}\}$
$R = \{f(x) \mid f(x) \geq -2\}$

43. $D = \{x \mid x \text{ is any real number}\}$
$R = \{f(x) \mid f(x) \text{ is any nonnegative real number}\}$

45. $D = \{x \mid x \text{ is any nonnegative real number}\}$
$R = \{f(x) \mid f(x) \text{ is any nonpositive real number}\}$

47. $D = \{x \mid x \geq 2\}$
$R = \{f(x) \mid f(x) \geq 3\}$

49. $D = \{x \mid x \text{ is any real number}\}$
$R = \{f(x) \mid f(x) \geq 5\}$

51. $D = \{x \mid x \text{ is any real number}\}$
$R = \{f(x) \mid f(x) \leq -6\}$

53. $D = \{x \mid x \neq -2\}$

55. $D = \left\{x \mid x \neq \dfrac{1}{2} \text{ and } x \neq -4\right\}$

57. $D = \{x \mid x \neq 2 \text{ and } x \neq -2\}$

59. $D = \{x \mid x \neq -3 \text{ and } x \neq 4\}$

61. $D = \left\{x \mid x \neq -\dfrac{5}{2} \text{ and } x \neq \dfrac{1}{3}\right\}$

63. $D = \left\{x \mid x \leq -\dfrac{1}{3}\right\}$

65. $D = \{x \mid x \text{ is any real number}\}$

67. $(-\infty, -4] \cup [4, \infty)$ **69.** $(-\infty, \infty)$

71. $(-\infty, -5] \cup [8, \infty)$

73. $\left(-\infty, -\dfrac{5}{2}\right] \cup \left[\dfrac{7}{4}, \infty\right)$

75. $[-4, 4]$ **77.** Odd   **79.** Neither
81. Neither   **83.** Even
85. Odd   **87.** $30.40
89. 12.57; 28.27; 452.39; 907.92   **91.** 48; 64; 48; 0
93. $55; $60; $67.50; $75   **95.** 125.66; 301.59; 804.25

Problem Set 3.2 (page 272)

1. 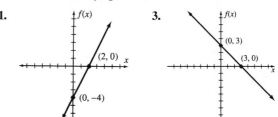   **3.**

5. 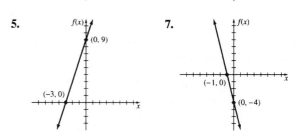   **7.**

9. 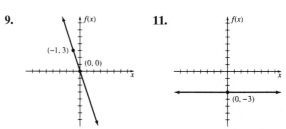   **11.**

13. 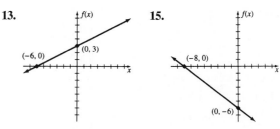   **15.**

17. $f(x) = \dfrac{2}{3}x + \dfrac{11}{3}$ **19.** $f(x) = -x - 4$

21. $f(x) = -\dfrac{1}{5}x + \dfrac{21}{5}$

23. a. $0.42   **c.** Answers may vary.
25. $f(x) = 0.25x + 30$ **27.** $26; $30.50; $50; $60.50
29. $f(p) = 0.8p$; $7.60; $12; $60; $10; $600

33.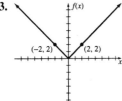

35.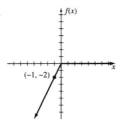

17.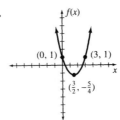

19.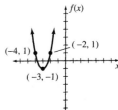

37.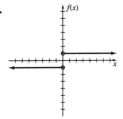

21.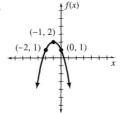

23.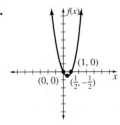

Problem Set 3.3 (page 284)

1.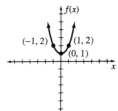

3.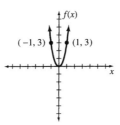

25.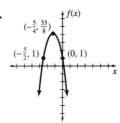

5.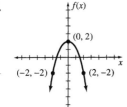

7.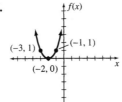

27.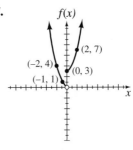

9.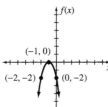

11.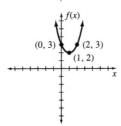

13.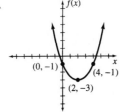

15.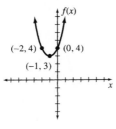

29.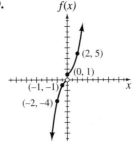

31.

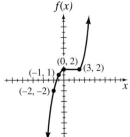

33.

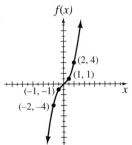

35.

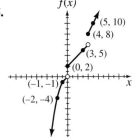

37.

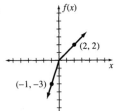

39.

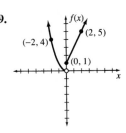

41.

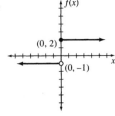

43.

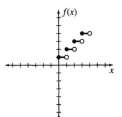

45.

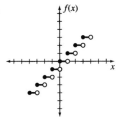

Problem Set 3.4 (page 295)

1.

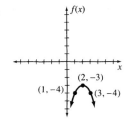

3.

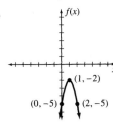

5.

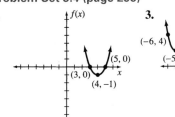

7.

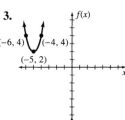

9.

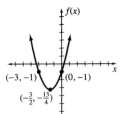

11.

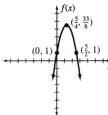

13.

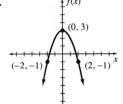

15.

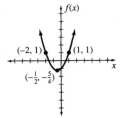

17.
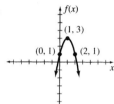

19.
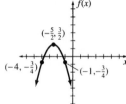

9.
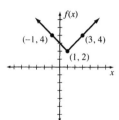

11.
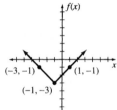

21.
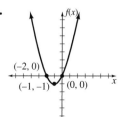

23.
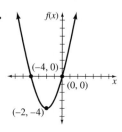

13.
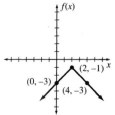

15.
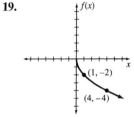

25. 3 and 5; $(4, -1)$ **27.** 6 and 8; $(7, -2)$

29. 4 and 6; $(5, 1)$ **31.** $-\dfrac{\sqrt{6}}{3}$ and $\dfrac{\sqrt{6}}{3}$; $(0, -2)$

33. -3 and 0; $\left(-\dfrac{3}{2}, -\dfrac{27}{4}\right)$

35. $7 + \sqrt{5}$ and $7 - \sqrt{5}$; $(7, -5)$

37. No x intercepts; $\left(\dfrac{9}{2}, -\dfrac{3}{4}\right)$

39. $\dfrac{1 + \sqrt{5}}{2}$ and $\dfrac{1 - \sqrt{5}}{2}$; $\left(\dfrac{1}{2}, 5\right)$ **41.** 70

43. 144 feet   **45.** 10 and 10   **47.** 25 and 25
49. 60 meters by 60 meters
51. 1100 subscribers at $13.75 per month
57. 75 feet

17.
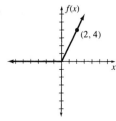

19.
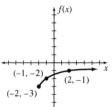

21.
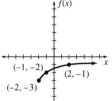

23.
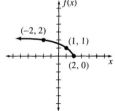

Problem Set 3.5 (page 309)

1.

3.

25.
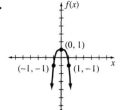

27.
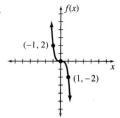

5.
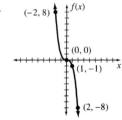

7.
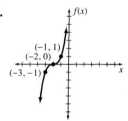

29.
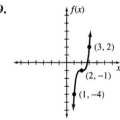

31. a.

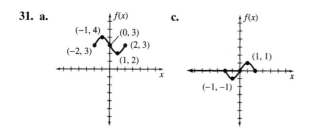

Problem Set 3.6 (page 316)

1. $8x - 2; -2x - 6; 15x^2 - 14x - 8; \dfrac{3x - 4}{5x + 2}$

3. $x^2 - 7x + 3; x^2 - 5x + 5; -x^3 + 5x^2 + 2x - 4;$
$\dfrac{x^2 - 6x + 4}{-x - 1}$

5. $2x^2 + 3x - 6; -5x + 4; x^4 + 3x^3 - 10x^2 + x + 5;$
$\dfrac{x^2 - x - 1}{x^2 + 4x - 5}$

7. $\sqrt{x - 1} + \sqrt{x}; \sqrt{x - 1} - \sqrt{x}; \sqrt{x^2 - x};$
$\dfrac{\sqrt{x(x - 1)}}{x}$

9. $(f \circ g)(x) = 6x - 2, D = \{\text{all reals}\}$
$(g \circ f)(x) = 6x - 1, D = \{\text{all reals}\}$

11. $(f \circ g)(x) = 10x + 2, D = \{\text{all reals}\}$
$(g \circ f)(x) = 10x - 5, D = \{\text{all reals}\}$

13. $(f \circ g)(x) = 3x^2 + 7, D = \{\text{all reals}\}$
$(g \circ f)(x) = 9x^2 + 24x + 17, D = \{\text{all reals}\}$

15. $(f \circ g)(x) = 3x^2 + 9x - 16, D = \{\text{all reals}\}$
$(g \circ f)(x) = 9x^2 - 15x, D = \{\text{all reals}\}$

17. $(f \circ g)(x) = \dfrac{1}{2x + 7}, D = \left\{x \mid x \neq -\dfrac{7}{2}\right\}$

$(g \circ f)(x) = \dfrac{7x + 2}{x}, D = \{x \mid x \neq 0\}$

19. $(f \circ g)(x) = \sqrt{3x - 3}, D = \{x \mid x \geq 1\}$
$(g \circ f)(x) = 3\sqrt{x - 2} - 1, D = \{x \mid x \geq 2\}$

21. $(f \circ g)(x) = \dfrac{x}{2 - x}, D = \{x \mid x \neq 0 \text{ and } x \neq 2\}$

$(g \circ f)(x) = 2x - 2, D = \{x \mid x \neq 1\}$

23. $(f \circ g)(x) = 2\sqrt{x - 1} + 1, D = \{x \mid x \geq 1\}$
$(g \circ f)(x) = \sqrt{2x}, D = \{x \mid x \geq 0\}$

25. $(f \circ g)(x) = x, D = \{x \mid x \neq 0\}$
$(g \circ f)(x) = x, D = \{x \mid x \neq 1\}$

27. $(f \circ g)(x) = \sqrt{x^2 - 9}, D = \{x \mid x \leq -3 \text{ or } x \geq 3\}$
$(g \circ f)(x) = x - 9, D = \{x \mid x \geq 9\}$

29. $4; 50$ **31.** $9; 0$ **33.** $\sqrt{11}; 5$

Problem Set 3.7 (page 328)

1. Yes   **3.** No   **5.** Yes   **7.** Yes
9. Yes   **11.** No   **13.** No
15. Domain of f: $\{1, 2, 5\}$
Range of f: $\{5, 9, 21\}$
$f^{-1} = \{(5, 1), (9, 2), (21, 5)\}$
Domain of f^{-1}: $\{5, 9, 21\}$
Range of f^{-1}: $\{1, 2, 5\}$
17. Domain of f: $\{0, 2, -1, -2\}$
Range of f: $\{0, 8, -1, -8\}$
f^{-1}: $\{(0, 0), (8, 2), (-1, -1), (-8, -2)\}$
Domain of f^{-1}: $\{0, 8, -1, -8\}$
Range of f^{-1}: $\{0, 2, -1, -2\}$
27. No   **29.** Yes   **31.** No
33. Yes   **35.** Yes
37. $f^{-1}(x) = x + 4$ **39.** $f^{-1}(x) = \dfrac{-x - 4}{3}$

41. $f^{-1}(x) = \dfrac{12x + 10}{9}$ **43.** $f^{-1}(x) = -\dfrac{3}{2}x$

45. $f^{-1}(x) = x^2$ for $x \geq 0$

47. $f^{-1}(x) = \sqrt{x - 4}$ for $x \geq 4$

49. $f^{-1}(x) = \dfrac{1}{x - 1}$ for $x > 1$

51. $f^{-1}(x) = \dfrac{1}{3}x$ **53.** $f^{-1}(x) = \dfrac{x - 1}{2}$

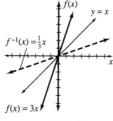

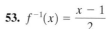

55. $f^{-1}(x) = \dfrac{x + 2}{x}$ **57.** $f^{-1}(x) = \sqrt{x + 4}$

for $x > 0$ for $x \geq -4$

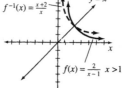

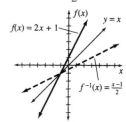
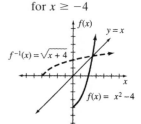

59. Increasing on $[0, \infty)$ and decreasing on $(-\infty, 0]$
61. Decreasing on $(-\infty, \infty)$
63. Increasing on $(-\infty, -2]$ and decreasing on $[-2, \infty)$
65. Increasing on $(-\infty, -4]$ and decreasing on $[-4, \infty)$

71. a. $f^{-1}(x) = \dfrac{x+9}{3}$ **c.** $f^{-1}(x) = -x + 1$

e. $f^{-1}(x) = -\dfrac{1}{5}x$

Chapter 3 Review Problem Set (page 334)

1. 7; 4; 32

2. a. -5 **b.** $4a + 2h - 1$ **c.** $-6a - 3h + 2$

3. The domain is the set of all real numbers, and the range is the set of all real numbers greater than or equal to 5.

4. The domain is the set of all real numbers except $\dfrac{1}{2}$ and -4.

5. $(-\infty, 2] \cup [5, \infty)$

6.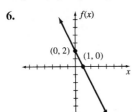

7.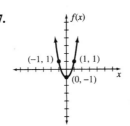

8.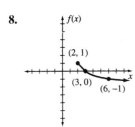

9.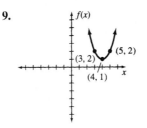

10.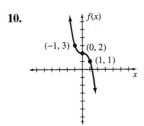

11.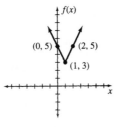

12.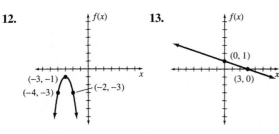

13.

14.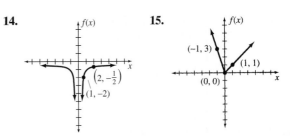

15.

16. $x^2 - 2x$; $-x^2 + 6x + 6$; $2x^3 - 5x^2 - 18x - 9$; $\dfrac{2x+3}{x^2 - 4x - 3}$

17. $(f \circ g)(x) = -6x + 12$; $D = \{\text{all reals}\}$
$(g \circ f)(x) = -6x + 25$; $D = \{\text{all reals}\}$

18. $(f \circ g)(x) = 25x^2 - 40x + 11$; $D = \{\text{all reals}\}$
$(g \circ f)(x) = 5x^2 - 29$; $D = \{\text{all reals}\}$

19. $(f \circ g)(x) = \sqrt{x - 3}$; $D = \{x | x \geq 3\}$
$(g \circ f)(x) = \sqrt{x - 5} + 2$; $D = \{x | x \geq 5\}$

20. $(f \circ g)(x) = \dfrac{x+2}{-3x - 5}$; $D = \left\{x \middle| x \neq -2 \text{ and } x \neq -\dfrac{5}{3}\right\}$

$(g \circ f)(x) = \dfrac{x-3}{2x - 5}$; $D = \left\{x \middle| x \neq 3 \text{ and } x \neq \dfrac{5}{2}\right\}$

21. a. Neither   **b.** Odd   **c.** Even   **d.** Neither

22. $f(5) = 23$; $f(0) = -2$; $f(-3) = 13$

23. $f(g(6)) = -2$; $g(f(-2)) = 0$

24. $(f \circ g)(1) = 1$; $(g \circ f)(-3) = 5$

25. $f(x) = \dfrac{2}{3}x - \dfrac{16}{3}$ **26.** $f(x) = 2x + 15$

27. \$0.72   **28.** $f(x) = 0.7x$; \$45.50; \$33.60; \$10.85

29. 2 and 8   **30.** 112 students   **31.** Yes

32. No   **33.** Yes   **34.** Yes

35. $f^{-1}(x) = \dfrac{x - 5}{4}$ **36.** $f^{-1}(x) = \dfrac{-x - 7}{3}$

37. $f^{-1}(x) = \dfrac{6x + 2}{5}$ **38.** $f^{-1}(x) = \sqrt{-2 - x}$

39. Increasing on $(-\infty, 4]$ and decreasing on $[4, \infty)$

40. Increasing on $[3, \infty)$

Chapter 3 Test (page 336)

1. $\dfrac{11}{6}$ **2.** 11   **3.** $6a + 3h + 2$

4. $\left\{x \middle| x \neq -4 \text{ and } x \neq \dfrac{1}{2}\right\}$ **5.** $\left\{x \middle| x \leq \dfrac{5}{3}\right\}$

6. $2x^2 + 2x - 6$; $-2x^2 + 4x + 4$; $6x^3 - 5x^2 - 14x + 5$

7. $-21x - 2$ **8.** $8x^2 + 38x + 48$

9. $\dfrac{3x}{2 - 2x}$ **10.** 12; 7

11. a. Even **b.** Odd **c.** Neither **d.** Even

12. $\{x|x \neq 0 \text{ and } x \neq 1\}$ **13.** 18; 10; 0

14. $x^3 + 4x^2 - 11x + 6; x + 6$

15. $f(x) = -\dfrac{5}{6}x - \dfrac{14}{3}$ **16.** $f^{-1}(x) = -\dfrac{1}{3}x - 2$

17. $f^{-1}(x) = \dfrac{3}{2}x + \dfrac{9}{10}$

18. $s(c) = 1.35c; \$17.55$ **19.** 6 and 54

20. The graph of $g(x) = (x - 6)^3 - 4$ is the graph of $f(x) = x^3$ translated six units to the right and four units downward.

21. The graph of $g(x) = -\sqrt{x + 5} + 7$ is the graph of $f(x) = \sqrt{x}$ reflected across the x axis and then translated five units to the left and seven units upward.

22.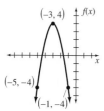

23.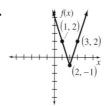

24.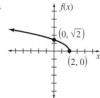

25.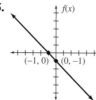

Cumulative Review Problem Set, Chapters 0–3 (page 337)

1. 9 **2.** $\dfrac{9}{7}$ **3.** $\dfrac{1}{8}$ **4.** $\dfrac{1}{4}$ **5.** $\dfrac{72}{17}$

6. -0.4 **7.** $\dfrac{3}{2}$ **8.** 27 **9.** 4 **10.** 81

11. -8 **12.** -1 **13.** 2 **14.** 75

15. -31 **16.** $\dfrac{4x^2y}{9}$ **17.** $\dfrac{3x + 7}{4x - 9}$

18. $\dfrac{2(x^2 - 2x + 4)}{x - 2}$ **19.** $\dfrac{y - 2}{x}$ **20.** $\dfrac{27}{8a^2}$

21. $\dfrac{x + 4}{x(x + 5)}$ **22.** $\dfrac{16x + 43}{90}$ **23.** $\dfrac{35a - 44b}{60a^2b}$

24. $\dfrac{2}{x - 4}$ **25.** $\dfrac{2y - 3xy}{3x + 4xy}$ **26.** $\dfrac{5y^2 - 3xy^2}{x^2y + 2x^2}$

27. $\dfrac{3a^2 - 2a + 1}{2a - 1}$ **28.** $-\dfrac{12}{x^3y}$ **29.** $\dfrac{8y}{x^5}$

30. $-\dfrac{a^3}{9b}$ **31.** $\dfrac{2\sqrt{2}}{5}$ **32.** $\dfrac{2\sqrt{2}}{7}$

33. $4xy^3\sqrt{3xy}$ **34.** $2(\sqrt{5} + \sqrt{3})$

35. $2xy\sqrt[3]{6xy^2}$ **36.** $\sqrt[3]{2}$ **37.** $40 + 13i$

38. $2 + 14i$ **39.** $0 - \dfrac{5}{4}i$ **40.** $\dfrac{2}{5} + \dfrac{6}{5}i$

41. $\left\{-\dfrac{21}{16}\right\}$ **42.** $\left\{\dfrac{40}{3}\right\}$ **43.** $\{6\}$

44. $\left\{-\dfrac{5}{2}, 3\right\}$ **45.** $\left\{-3, \dfrac{5}{3}\right\}$ **46.** $\{-6, 0, 6\}$

47. $\left\{\dfrac{1 \pm 3\sqrt{5}}{3}\right\}$ **48.** $\left\{\dfrac{-5 \pm 4i\sqrt{2}}{2}\right\}$

49. $\left\{\dfrac{3 \pm i\sqrt{23}}{4}\right\}$ **50.** $\{-5, 7\}$ **51.** $\left\{-6, \dfrac{1}{2}\right\}$

52. $\{-5, 8\}$ **53.** $\{-17\}$ **54.** $\left\{\dfrac{-7 \pm \sqrt{41}}{2}\right\}$

55. $\{12\}$ **56.** $\{-3\}$ **57.** $\left\{\pm\dfrac{\sqrt{5}}{2}, \pm\dfrac{\sqrt{3}}{3}\right\}$

58. $\{\pm\sqrt{3}, 4\}$ **59.** $\left(-\infty, -\dfrac{11}{5}\right) \cup (3, \infty)$

60. $[-4, 2]$ **61.** $\left[-\dfrac{5}{3}, 1\right]$ **62.** $(-8, 3)$

63. $(-\infty, 3)$ **64.** $\left[-\dfrac{9}{11}, \infty\right)$

65. $\left(-5, -\dfrac{1}{2}\right) \cup (2, \infty)$ **66.** $(-\infty, 3] \cup (7, \infty)$

67. $(-6, -3)$ **68.** $(-\infty, -3) \cup \left(0, \dfrac{1}{2}\right)$

69. $(-2, 4); r = 3$ **70.** 6 **71.** $(5, 8)$

72. $\dfrac{2}{5}$ **73.** $3x + 5y = -11$

74. $x^2 - 4x + y^2 - 14y + 28 = 0$ **75.** 43; 24

76. $2\sqrt{x + 2} - 1; \sqrt{2x + 1}$

77. $(-\infty, -10] \cup [3, \infty)$ **78.** $-2a + 6 - h$

79.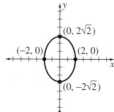

80.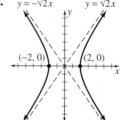

81.

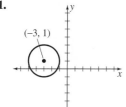

82.

83.

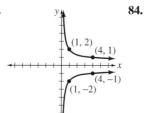

84.

85.

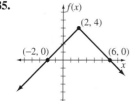

86.

87.

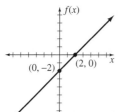

88.

89.

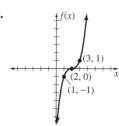

90.

91. 17, 19, and 21
92. 14 nickels, 20 dimes, and 29 quarters
93. $48°$ and $132°$ **94.** $600
95. $1700 at 8% and $2000 at 9%
96. 66 mph and 76 mph   **97.** 4 quarts
98. 69 or less   **99.** $-3, 0,$ or 3   **100.** 1 inch
101. $1050 and $1400   **102.** 3 hours
103. 30 shares at $10 per share   **104.** 37
105. $10°, 60°,$ and $110°$

CHAPTER 4

Problem Set 4.1 (page 347)

1. $Q: 4x + 5$
$R: 0$

3. $Q: t^2 + 2t - 4$
$R: 0$

5. $Q: 2x + 5$
$R: 1$

7. $Q: 3x - 4$
$R: 3x - 1$

9. $Q: 5y - 1$
$R: -8y - 2$

11. $Q: 4a + 6$
$R: 7a - 19$

13. $Q: 3x + 4y$
$R: 0$

15. $Q: 3x + 4$
$R: 0$

17. $Q: x + 6$
$R: 14$

19. $Q: 4x - 3$
$R: 2$

21. $Q: x^2 - 1$
$R: 0$

23. $Q: 3x^3 - 4x^2 + 6x - 13$
$R: 12$

25. $Q: x^2 - 2x - 3$
$R: 0$

27. $Q: x^3 + 7x^2 + 21x + 56$
$R: 167$

29. $Q: x^2 + 3x + 2$
$R: 0$

31. $Q: x^4 + x^3 + x^2 + x + 1$
$R: 0$

33. $Q: x^4 + x^3 + x^2 + x + 1$
$R: 2$

35. $Q: 2x^2 + 2x - 3$
$R: \dfrac{9}{2}$

37. $Q: 4x^3 + 2x^2 - 4x - 2$
$R: 0$

Problem Set 4.2 (page 352)

1. $f(2) = -2$ **3.** $f(-4) = -105$ **5.** $f(-2) = 9$
7. $f(6) = 74$ **9.** $f(3) = 200$ **11.** $f(-1) = 5$
13. $f(7) = -5$ **15.** $f(-2) = -27$
17. $f\left(\dfrac{1}{2}\right) = -2$ **19.** Yes   **21.** Yes

23. No   **25.** Yes   **27.** Yes
29. $(x + 2)(x + 6)(x - 1)$
31. $(x - 3)(2x - 1)(3x + 2)$ **33.** $(x + 1)^2(x - 4)$
35. $k = 6$ **37.** $k = -30$
39. Let $f(x) = x^{12} - 4096$; then $f(-2) = 0$; therefore $x + 2$ is a factor of $f(x)$.
41. Let $f(x) = x^n - 1$. Since $1^n = 1$ for all positive integral values of n, $f(1) = 0$ and $x - 1$ is a factor.
43. a. Let $f(x) = x^n - y^n$. Therefore $f(y) = y^n - y^n = 0$ and $x - y$ is a factor of $f(x)$.
c. Let $f(x) = x^n + y^n$. Therefore $f(-y) = (-y)^n + y^n = -y^n + y^n = 0$ when n is odd, and $x - (-y) = x + y$ is a factor of $f(x)$.
47. $f(1 + i) = 2 + 6i$
51. a. $f(4) = 137; f(-5) = 11; f(7) = 575$
c. $f(4) = -79; f(5) = -162; f(-3) = 110$

Problem Set 4.3 (page 365)

1. $\{-2, -1, 2\}$ **3.** $\left\{-\dfrac{3}{2}, \dfrac{1}{3}, 1\right\}$ **5.** $\left\{-7, \dfrac{2}{3}, 2\right\}$

7. $\{-1, 4\}$ **9.** $\{-3, 1, 2, 4\}$ **11.** $\{-2, 1 \pm \sqrt{7}\}$

13. $\{-1, 2, 1 \pm \sqrt{3}\}$ **15.** $\left\{-\dfrac{4}{3}, 0, \dfrac{1}{2}, 3\right\}$

17. $\{-1, 2, 1 \pm i\}$ **19.** $\left\{-1, \dfrac{3}{2}, 2, \pm i\right\}$

27. a. $\{-4, -2, 1\}$ **c.** $\left\{-4, -2, \dfrac{3}{2}\right\}$

29. 2 positives *or* 2 nonreal complex solutions
31. 1 negative and 2 nonreal complex solutions
33. 1 positive and 2 negative *or* 1 positive and 2 nonreal complex solutions
35. 1 negative and 2 positive and 2 nonreal complex solutions *or* 1 negative and 4 nonreal complex solutions
37. 1 positive and 1 negative and 4 nonreal complex solutions
39. $x^4 + 2x^3 - 9x^2 - 2x + 8 = 0$
41. $12x^3 - 37x^2 - 3x + 18 = 0$
43. $x^4 + 12x^3 + 54x^2 + 108x + 81 = 0$
45. $x^3 + 13x + 34 = 0$
47. $x^4 + 4x^3 + 14x^2 + 4x + 13 = 0$

Problem Set 4.4 (page 377)

1.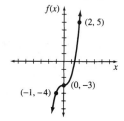

3.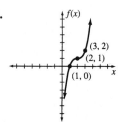

5.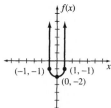

7.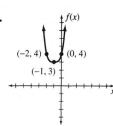

9.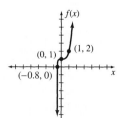

11.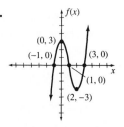

13.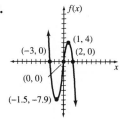

15.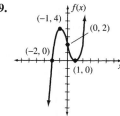

17.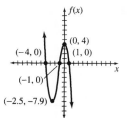

19.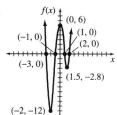

21.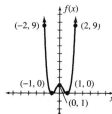

23.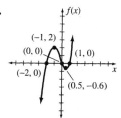

25.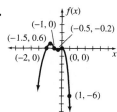

27.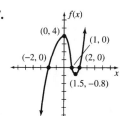

29.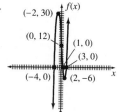

31.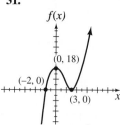

33.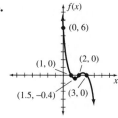

35. a. 60
 c. $f(x) > 0$ for $(-4, 3) \cup (5, \infty)$
 $f(x) < 0$ for $(-\infty, -4) \cup (3, 5)$
37. a. 432
 c. $f(x) > 0$ for $(-3, 4) \cup (4, \infty)$
 $f(x) < 0$ for $(-\infty, -3)$
39. a. 8
 c. $f(x) > 0$ for $(-\infty, -2) \cup (-2, 1) \cup (2, \infty)$
 $f(x) < 0$ for $(1, 2)$
41. a. 512
 c. $f(x) > 0$ for $(-2, 4) \cup (4, \infty)$
 $f(x) < 0$ for $(-\infty, -2)$
45. a. 1.6　**c.** $-5.5, 0.7, 4.8$　**e.** $-1.5, 5.5$
51. a. $-2, 1,$ and 4; $f(x) > 0$ for $(-2, 1) \cup (4, \infty)$ and
 $f(x) < 0$ for $(-\infty, -2) \cup (1, 4)$
 c. 2 and 3; $f(x) > 0$ for $(3, \infty)$ and
 $f(x) < 0$ for $(-\infty, 2) \cup (2, 3)$
 e. $-3, -1,$ and 2; $f(x) > 0$ for $(-\infty, -3) \cup (2, \infty)$
 and $f(x) < 0$ for $(-3, -1) \cup (-1, 2)$
53. a. -3.3; $(0.5, 3.1)$ and $(-1.8, 10.1)$
 c. $-2.2, 2.2$;　$(-1.4, -8.0)$ and $(0.0, -4.0)$ and
 $(1.4, -8.0)$
55. 32 units

Problem Set 4.5 (page 389)

1.
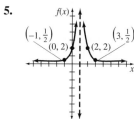

3.
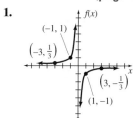

5.
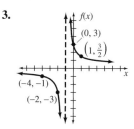

7.
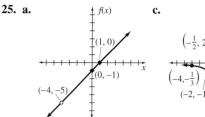

9.
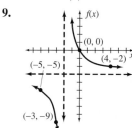

11.
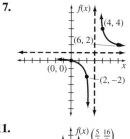

13.
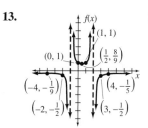

15.
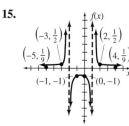

17.
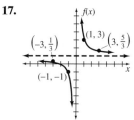

19.
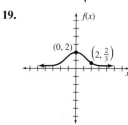

21.
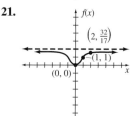

25. a.
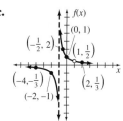

c.
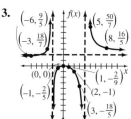

Problem Set 4.6 (page 398)

1.
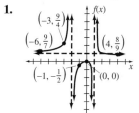

3.
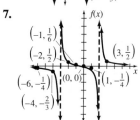

5.
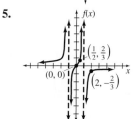

7.
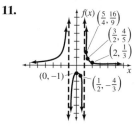

9.

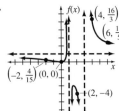

11.

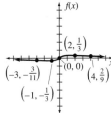

13.

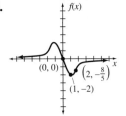

15.

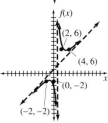

17.

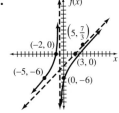

19.

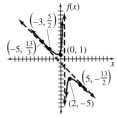

19. 2 positive and 2 negative solutions *or* 2 positive and 2 nonreal complex solutions *or* 2 negative and 2 nonreal complex solutions *or* 4 nonreal complex solutions

20. 1 negative and 4 nonreal complex solutions

21.

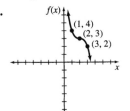

22.

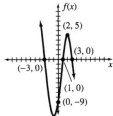

23.

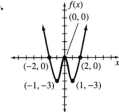

24.

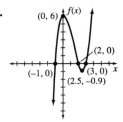

25.

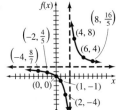

26.

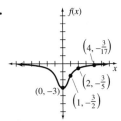

27.

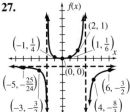

28.

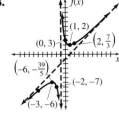

29. 9   **30.** 441
31. 128 pounds   **32.** 15 hours

Problem Set 4.7 (page 406)

1. $y = kx^3$ **3.** $A = klw$ **5.** $V = \dfrac{k}{P}$

7. $V = khr^2$ **9.** 24   **11.** $\dfrac{22}{7}$ **13.** $\dfrac{1}{2}$

15. 7   **17.** 6   **19.** 8   **21.** 96
23. 5 hours   **25.** 2 seconds   **27.** 24 days
29. 28   **31.** \$2400   **37.** 2.8 seconds   **39.** 1.4

Chapter 4 Review Problem Set (page 410)

1. $Q: 3x^2 - 5x + 4$ **2.** $Q: 2a - 1$
  $R: 4$ $R: 5$
3. $Q: 3x^2 - x + 5$ **4.** $Q: 5x^2 - 3x - 3$
  $R: 3$ $R: 16$
5. $Q: -2x^3 + 9x^2 - 38x + 151$
  $R: -605$
6. $Q: -3x^3 + 9x^2 - 32x + 96$
  $R: -279$
7. $f(1) = 1$ **8.** $f(-3) = -197$ **9.** $f(-2) = 20$
10. $f(8) = 0$ **11.** Yes   **12.** No   **13.** Yes

14. Yes   **15.** $\{-3, 1, 5\}$ **16.** $\left\{-\dfrac{7}{2}, -1, \dfrac{5}{4}\right\}$

17. $\{1, 2, 1 \pm 5i\}$ **18.** $\{-2, 3 \pm \sqrt{7}\}$

Chapter 4 Test (page 412)

1. $Q: 2x^2 - 3x - 4;$ **2.** $Q: 3x^3 - x^2 - 2x - 6;$
  $R: 0$ $R: 3$
3. $Q: 4x^3 + 8x^2 + 9x + 18;$
  $R: 40$
4. -24 **5.** 5   **6.** 39   **7.** No   **8.** No
9. Yes   **10.** No
11. 1 positive, 1 negative, and 2 nonreal complex solutions

12. $-7, 0,$ and $\dfrac{2}{3}$ **13.** $x = -3$ **14.** $f(x) = 5$

15. y axis   **16.** Origin   **17.** $f(x) = 4x - 3$

18. -4 **19.** 1694   **20.** $96

21.

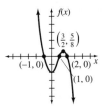

22.

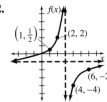

23.

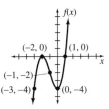

24.

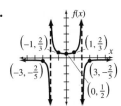

25.

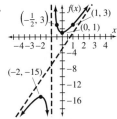

Cumulative Review Problem Set,
Chapters 0 – 4 (page 413)

1. $\dfrac{64}{27}$ **2.** $-\dfrac{2}{3}$ **3.** $-\dfrac{1}{25}$ **4.** 16   **5.** $\dfrac{1}{27}$

6. $\dfrac{8}{27}$ **7.** $\dfrac{9}{4}$ **8.** $-\dfrac{1}{36}$ **9.** 16

10. $-\dfrac{4}{3}$ **11.** $(-\infty, -6] \cup \left[\dfrac{1}{2}, \infty\right)$

12. $(f \circ g)(-2) = 26$ and $(g \circ f)(3) = 59$

13. $(f \circ g)(x) = -2x + 8$ and $D = \{x \mid x \neq 4\}$

$(g \circ f)(x) = -\dfrac{x}{4x + 2}$ and

$D = \left\{x \mid x \neq 0 \text{ and } x \neq -\dfrac{1}{2}\right\}$

14. $f^{-1}(x) = \dfrac{-x + 7}{2}$ **15.** $2a + h + 7$

16. $f(9) = 33$ **17.** $3x^4 + 9x^3 + 2x^2 - x - 2$

18. No   **19.** 30   **20.** $(-3, 2);\ r = 3$

21. $x + 3y = 2$ **22.** $4x + 3y = 5$ **23.** 16 units

24. $y = \pm\dfrac{1}{3}x$ **25.** 12   **26.** $\dfrac{2}{7}$

27. $6\frac{2}{3}$ days   **28.** $f(x) = x + 3$

29. 10 nickels, 15 dimes, and 32 quarters

30. $125   **31.** $1\frac{1}{3}$ quarts   **32.** 45 miles

33. 4 hours   **34.** $\left\{\dfrac{3}{5}\right\}$ **35.** $\left\{\dfrac{-13 \pm \sqrt{193}}{2}\right\}$

36. $\{-7, 0, 2\}$ **37.** $\left\{-\dfrac{5}{2}, -1, \dfrac{2}{3}\right\}$ **38.** $\left\{-1, \dfrac{5}{2}\right\}$

39. $\{0\}$ **40.** $\left\{-\dfrac{4}{3}, \dfrac{2}{3}\right\}$ **41.** $\{1\}$ **42.** $\{11\}$

43. $\{\pm 3i, \pm\sqrt{6}\}$ **44.** $\left\{-\dfrac{13}{2}, 4\right\}$

45. $\left\{1, 2, \dfrac{-1 \pm i\sqrt{11}}{2}\right\}$ **46.** $(-\infty, -5)$

47. $\left[-\dfrac{11}{17}, \infty\right)$ **48.** $(-3, 6)$

49. $[-3, 1] \cup [2, \infty)$ **50.** $\left(-\infty, -\dfrac{5}{2}\right) \cup \left(\dfrac{7}{2}, \infty\right)$

51. $\left[-\dfrac{10}{3}, 2\right]$ **52.** $\left(-\infty, \dfrac{3}{4}\right] \cup (2, \infty)$

53. $(-\infty, 4) \cup \left(\dfrac{15}{2}, \infty\right)$

54.

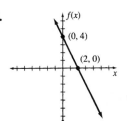

55.

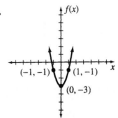

56.

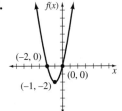

57.

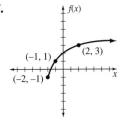

58.

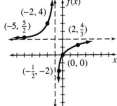

59.

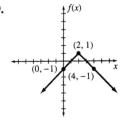

60.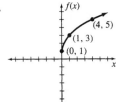

61.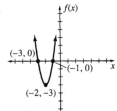

62.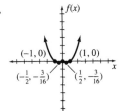

63.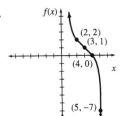

64.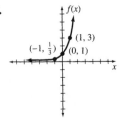

CHAPTER 5

Problem Set 5.1 (page 424)

1. {3}   **3.** {3}   **5.** {4}   **7.** {2}

9. {−2}   **11.** {−3}   **13.** {−2}

15. $\left\{\dfrac{5}{3}\right\}$ **17.** $\left\{\dfrac{3}{2}\right\}$ **19.** {−9}

21. $\left\{\dfrac{4}{9}\right\}$ **23.** $\left\{\dfrac{4}{3}\right\}$ **25.** $\left\{\dfrac{2}{3}\right\}$

27.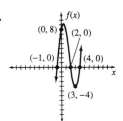

29.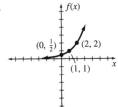

31.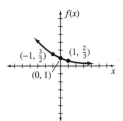

33.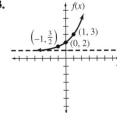

35.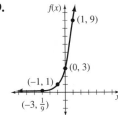

37.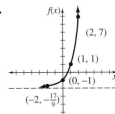

39.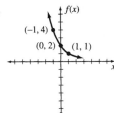

41.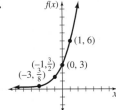

43.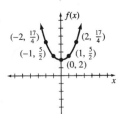

45.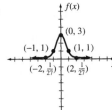

47.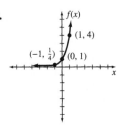

49.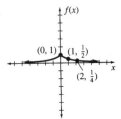

51.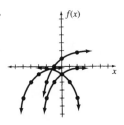

53.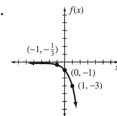

Problem Set 5.2 (page 435)

1. a. $1.45 **b.** $5.22 **c.** $2.68 **d.** $2.36 **e.** $26,766
f. $287,400 **g.** $77.64
3. $334.56 **5.** $480.31 **7.** $1379.76 **9.** $1221.00
11. $1356.59 **13.** $12,515.01 **15.** $567.63
17. $1422.36 **19.** $3644.24 **21.** $12,999.40
23. $6688.37 **25.** $3624.66 **27.** $674.93 **29.** 5.9%
31. 8.06% **33.** 8.25% compounded quarterly
35. 50 grams; 37 grams **37.** 2226; 3320; 7389
39. 2000
41. a. 6.5 pounds per square inch
   c. 13.6 pounds per square inch
43. a. Approximately 100 times brighter
   c. Approximately 10 billion times brighter

45.
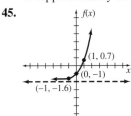

47.
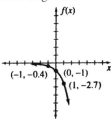

49.
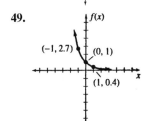

53.

	4%	6%	8%	10%
5 years	$1221	1350	1492	1649
10 years	1492	1822	2226	2718
15 years	1822	2460	3320	4482
20 years	2226	3320	4953	7389
25 years	2718	4482	7389	12,182

55.

	4%	6%	8%	10%
Compounded annually	$1480	1791	2159	2594
Compounded semiannually	1486	1806	2191	2653
Compounded quarterly	1489	1814	2208	2685
Compounded monthly	1491	1819	2220	2707
Compounded continuously	1492	1822	2226	2718

57.
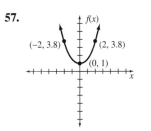

59.
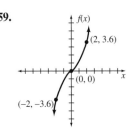

Problem Set 5.3 (page 448)

1. $\log_3 9 = 2$ **3.** $\log_5 125 = 3$
5. $\log_2\left(\dfrac{1}{16}\right) = -4$ **7.** $\log_{10} 0.01 = -2$

9. $2^6 = 64$ **11.** $10^{-1} = 0.1$ **13.** $2^{-4} = \dfrac{1}{16}$

15. 2 **17.** -1 **19.** 1 **21.** $\dfrac{1}{2}$ **23.** $\dfrac{1}{2}$

25. $-\dfrac{1}{8}$ **27.** $\dfrac{3}{2}$ **29.** $-\dfrac{3}{2}$ **31.** 7 **33.** 0

35. $\{25\}$ **37.** $\{32\}$ **39.** $\{9\}$ **41.** $\{1\}$

43. $\left\{\dfrac{1}{8}\right\}$ **45.** $\left\{\dfrac{1}{9}\right\}$ **47.** 5.1293 **49.** 6.9657

51. 1.4037 **53.** 7.4512 **55.** 6.3219
57. -0.3791 **59.** 0.5766 **61.** 2.1531
63. 0.3949 **65.** $\log_b x + \log_b y + \log_b z$
67. $2\log_b x + 3\log_b y$
69. $\dfrac{1}{2}\log_b x + \dfrac{1}{2}\log_b y$ **71.** $\dfrac{1}{2}\log_b x - \dfrac{1}{2}\log_b y$
73. $2\log_b x + \log_b y - \log_b z$
75. $\dfrac{2}{3}\log_b x - \dfrac{1}{3}\log_b y - \dfrac{1}{3}\log_b z$ **77.** $\log_b\left(\dfrac{xy}{z}\right)$

79. $\log_b\left(\dfrac{x}{yz}\right)$ **81.** $\log_b(x\sqrt{y})$

83. $\log_b\left(\dfrac{x^2\sqrt{x-1}}{(2x+5)^4}\right)$ **85.** $\left\{\dfrac{9}{4}\right\}$ **87.** $\{25\}$

89. $\{4\}$ **91.** $\left\{-\dfrac{22}{5}\right\}$ **93.** $\varnothing$ **95.** $\{-1\}$

97. $\left\{\dfrac{19}{8}\right\}$ **99.** $\{9\}$ **101.** $\{1\}$

Problem Set 5.4 (page 457)

1. 0.8597 **3.** 1.7179 **5.** 3.5071 **7.** -0.1373
9. -3.4685 **11.** 411.43 **13.** 90, 095
15. 79.543 **17.** 0.048440 **19.** 0.0064150
21. 1.6094 **23.** 3.4843 **25.** 6.0638
27. -0.7765 **29.** -3.4609 **31.** 1.6034
33. 3.1346 **35.** 108.56 **37.** 0.48268
39. 0.035994

41.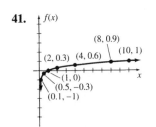

43.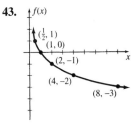

45.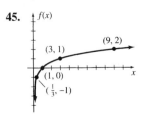

47.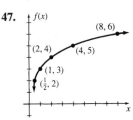

49.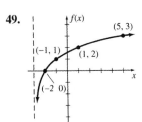

51.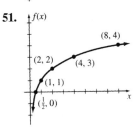

53.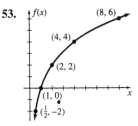

55.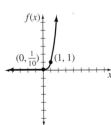

57.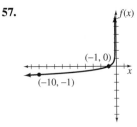

59.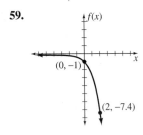

61.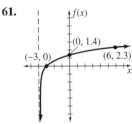

63. 1.77   **65.** 1.06   **67.** 27.47   **69.** 7.00

Problem Set 5.5 (page 469)

1. {3.17}   **3.** {2.99}   **5.** {1.81}   **7.** {1.41}
9. {1.41}   **11.** {3.10}   **13.** {1.82}

15. {7.84}   **17.** {10.32}   **19.** {2}   **21.** $\left\{\dfrac{29}{8}\right\}$

23. $\left\{\dfrac{-1+\sqrt{65}}{2}\right\}$ **25.** $\{2\sqrt{2}\}$ **27.** $\{-1+\sqrt{5}\}$

29. $\{\sqrt{2}\}$ **31.** {6}   **33.** {1, 100}   **35.** 2.402
37. 0.461   **39.** 2.657   **41.** 1.211   **43.** 11.7 years
45. 27.5 years   **47.** 11.8%   **49.** 6.6 years
51. 1.5 hours   **53.** 20.3 years
55. Approximately 200 million times the reference intensity
57. Approximately 8 times   **59.** 5.8   **61.** 10
69. $x = \log(y \pm \sqrt{y^2 - 1})$

Chapter 5 Review Problem Set (page 474)

1. 32   **2.** −125   **3.** 81   **4.** 3   **5.** −2   **6.** $\dfrac{1}{3}$ **7.** $\dfrac{1}{4}$

8. −5   **9.** 1   **10.** 12   **11.** {5}   **12.** $\left\{\dfrac{1}{9}\right\}$ **13.** $\left\{\dfrac{7}{2}\right\}$

14. {3.40}   **15.** {8}   **16.** $\left\{\dfrac{1}{11}\right\}$ **17.** {1.95}   **18.** {1.41}

19. {1.56}   **20.** {20}   **21.** $\{10^{100}\}$ **22.** {2}   **23.** $\left\{\dfrac{11}{2}\right\}$

24. {0}   **25.** 0.3680   **26.** 1.3222   **27.** 1.4313   **28.** 0.5634

29. a. $\log_b x - 2\log_b y$ **b.** $\dfrac{1}{4}\log_b x + \dfrac{1}{2}\log_b y$

  c. $\dfrac{1}{2}\log_b x - 3\log_b y$

30. a. $\log_b x^3 y^2$ **b.** $\log_b\left(\dfrac{\sqrt{y}}{x^4}\right)$ **c.** $\log_b\left(\dfrac{\sqrt{xy}}{z^2}\right)$

31. 1.58   **32.** 0.63   **33.** 3.79   **34.** −2.12

35. 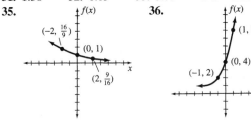   **36.**

37. 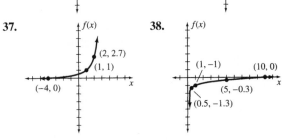   **38.**

39.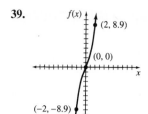

40.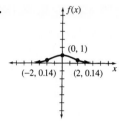

6. -2 **7.** $\dfrac{1}{5}$ **8.** $\dfrac{4}{3}$ **9.** $\dfrac{1}{x-3}$

10. $\dfrac{7x+25}{x(x+3)}$ **11.** $\dfrac{x^2+4}{x(x+7)}$ **12.** $\dfrac{18n+11}{42}$

13. $11\sqrt{2}$ **14.** $\dfrac{3x-2y}{4x^2+7xy}$ **15.** $5x+6y=2$

16. $\dfrac{2}{9}$ **17.** $f(x)=\dfrac{9}{5}x-\dfrac{2}{5}$ **18.** $3x+5y=46$

19. $f(x)=-2x+1$

20. $f(g(-1))=35$ and $g(f(2))=49$

21. $f(g(-3))=2$ and $g(f(8))=3$

22. $f(g(x))=14x^2-21x-31$

23. $g(f(x))=3x^4-2x^2-3$ **24.** $\dfrac{5\sqrt{3}}{21}$

41.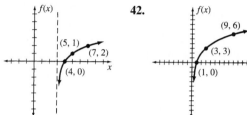

42.

25. $5xy\sqrt{2x}$ **26.** $2xy\sqrt[3]{7x^2y}$ **27.** $\dfrac{\sqrt[3]{6}}{2}$

28. $\dfrac{\sqrt{6}+\sqrt{3}}{3}$ **29.** $8-27i$

30. $-\dfrac{1}{5}-\dfrac{7}{5}i$ **31.** $117+0i$ **32.** $0+\dfrac{3}{2}i$

33. 4.14 **34.** $f^{-1}(x)=2x+12$

35. $3x^2-5x-4$ **36.** x^2+7x-2

43. $1656.03 **44.** $3561.18 **45.** $8985.50

46. $5563.85 **47.** $12,298.02 **48.** $20,387.11

49. Approximately 7.3 years

50. Approximately 16.9 years

51. Approximately 8.7%

52. 61,070; 67,493; 74,591

53. Approximately 4.8 hours **54.** 133 grams

55. 8.1

37. $\left\{-\dfrac{7}{3},2\right\}$ **38.** $\{-10,7\}$ **39.** $\left\{-\dfrac{1}{4}\right\}$

40. $\left\{\dfrac{66}{19}\right\}$ **41.** $\left\{\dfrac{1}{2},4\right\}$ **42.** $\left\{\dfrac{-7\pm\sqrt{61}}{2}\right\}$

Chapter 5 Test (page 476)

1. $\dfrac{1}{2}$ **2.** 1 **3.** 1 **4.** -1 **5.** $\{-3\}$

43. $\left\{\dfrac{2}{45}\right\}$ **44.** $\{400\}$ **45.** $\{310.5\}$ **46.** $\varnothing$

6. $\left\{-\dfrac{3}{2}\right\}$ **7.** $\left\{\dfrac{8}{3}\right\}$ **8.** $\{243\}$ **9.** $\{2\}$

47. $\left\{-\dfrac{5}{3},\dfrac{3}{2}\right\}$ **48.** $\{-2\pm i\}$ **49.** $\{-12,18\}$

10. $\left\{\dfrac{2}{5}\right\}$ **11.** 4.1919 **12.** 0.2031

50. $\left\{-\dfrac{7}{3},7\right\}$ **51.** $\left\{-3,-\dfrac{2}{3}\right\}$ **52.** $\left\{-9,0,\dfrac{3}{2}\right\}$

13. 0.7325 **14.** $\{5.17\}$ **15.** $\{10.29\}$

16. 4.0069 **17.** $8199.53 **18.** 50 times

19. 9.3 years **20.** $6342.08

21. 13.5 years **22.** 7.8 hours **23.** 4813 grams

53. $\{-2,1\}$ **54.** $\left\{-\dfrac{7}{3}\right\}$ **55.** $\left\{\dfrac{-3+\sqrt{37}}{2}\right\}$

24. 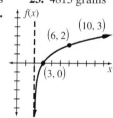 **25.**

56. $\{9\}$ **57.** $\{-125,64\}$ **58.** $\left\{\pm 3i,\pm\dfrac{\sqrt{14}}{2}\right\}$

59. $\{-4,2,5\}$ **60.** $\{-3,-2,1,6\}$ **61.** $\left\{\dfrac{1}{2},4,5\right\}$

62. $\left(-\infty,-\dfrac{3}{2}\right)\cup(4,\infty)$ **63.** $(-\infty,-7]\cup[0,5]$

64. $(-\infty,-6]\cup(4,\infty)$ **65.** $(-3,-1)$

66. $\left(-\infty,-\dfrac{3}{8}\right]\cup\left[\dfrac{9}{8},\infty\right)$ **67.** $\left(-\dfrac{1}{2},\dfrac{3}{10}\right)$

Cumulative Review Problem Set, Chapters 0–5 (page 477)

1. 0.2 **2.** 6 **3.** 4 **4.** $\dfrac{36}{13}$ **5.** $\dfrac{4}{729}$

68. $\left[-\dfrac{3}{7},\infty\right)$ **69.** $(-\infty,\infty)$ **70.** $(-\infty,-19]$

71. $\left[-\frac{5}{4}, \frac{11}{4}\right]$ **72.** $\left(-\infty, -\frac{7}{2}\right) \cup (3, \infty)$

73. $\left(-\frac{7}{2}, -2\right) \cup \left(-2, -\frac{5}{4}\right)$ **74.** 8 units

75. $y = \pm 2x$ **76.** $5\sqrt{2}$ units   **77.** $\left(-1, \frac{3}{2}\right)$

78. x axis   **79.** Origin   **80.** $(-8, 7)$

81. 30   **82.** $\frac{21}{4}$ **83.** $-2, 3, 7$ **84.** $-2, 0, 4$

85. $x = 3, f(x) = 2$ **86.** $x = 3, f(x) = x + 5$

87.   **88.**

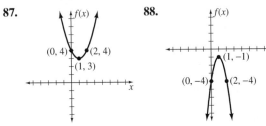

89.   **90.**

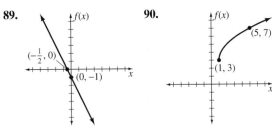

91.   **92.**

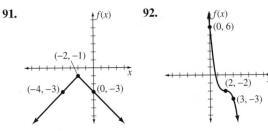

93.   **94.**

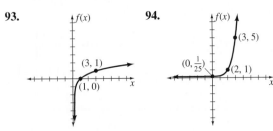

95. Any four consecutive whole numbers will work.
96. 7, 8, 9, and 10   **97.** 60 males and 140 females
98. 89 or higher   **99.** $8.50 per hour
100. $1300   **101.** $1092   **102.** $79
103. 1.6 liters   **104.** 40 hours

CHAPTER 6

Problem Set 6.1 (page 490)

1. $\{(7, 9)\}$ **3.** $\{(-4, 7)\}$ **5.** $\{(6, 3)\}$
7. $a = -3$ and $b = -4$
9. $\left\{\left(k, \frac{2}{3}k - \frac{4}{3}\right) \middle| k \text{ is any real number}\right\}$, a dependent system
11. $u = 5$ and $t = 7$ **13.** $\{(2, -5)\}$
15. $\varnothing$, an inconsistent system   **17.** $\left\{\left(-\frac{3}{4}, -\frac{6}{5}\right)\right\}$
19. $\{(3, -4)\}$ **21.** $\{(2, 8)\}$ **23.** $\{(-1, -5)\}$
25. $\varnothing$, an inconsistent system
27. $a = \frac{5}{27}$ and $b = -\frac{26}{27}$ **29.** $s = -6$ and $t = 12$
31. $\left\{\left(-\frac{1}{2}, \frac{1}{3}\right)\right\}$ **33.** $\left\{\left(\frac{13}{22}, \frac{2}{11}\right)\right\}$
35. $\{(-4, 2)\}$ **37.** $\{(5, 5)\}$
39. $\varnothing$, an inconsistent system   **41.** $\{(12, -24)\}$
43. $t = 8$ and $u = 3$ **45.** $\{(200, 800)\}$
47. $\{(400, 800)\}$ **49.** $\{(3.5, 7)\}$ **51.** 17 and 36
53. $15°$ and $75°$ **55.** 72   **57.** 34   **59.** 12
61. 2500 student and 500 nonstudent tickets
63. $750 at 6% and $1200 at 8%.   **65.** 3 mph
67. $22.00
69. 30 five-dollar bills and 18 ten-dollar bills
75. $\{(4, 6)\}$ **77.** $\{(2, -3)\}$ **79.** $\left\{\left(\frac{1}{4}, -\frac{2}{3}\right)\right\}$

Problem Set 6.2 (page 501)

1. $\{(-4, -2, 3)\}$ **3.** $\{(-2, 5, 2)\}$
5. $\{(4, -1, -2)\}$ **7.** $\{(3, 1, 2)\}$ **9.** $\{(-1, 3, 5)\}$
11. $\{(-2, -1, 3)\}$ **13.** $\{(0, 2, 4)\}$
15. $\{(4, -1, -2)\}$ **17.** $\{(-4, 0, -1)\}$
19. $\{(2, 2, -3)\}$
21. 4 pounds of pecans, 4 pounds of almonds, and 12 pounds of peanuts
23. 7 nickels, 13 dimes, and 22 quarters
25. $40°, 60°,$ and $80°$
27. $5000 at 6%, $1000 at 7%, and $15,000 at 8%
29. 50 of type A, 75 of type B, and 150 of type C
31. $x^2 + y^2 - 4x - 20y - 2 = 0$

Problem Set 6.3 (page 513)

1. Yes   **3.** Yes   **5.** No   **7.** No
9. Yes   **11.** $\{(-1, -5)\}$ **13.** $\{(3, -6)\}$
15. $\varnothing$ **17.** $\{(-2, -9)\}$ **19.** $\{(-1, -2, 3)\}$
21. $\{(3, -1, 4)\}$ **23.** $\{(0, -2, 4)\}$
25. $\{(-7k + 8, -5k + 7, k) \mid k \text{ is any real number}\}$
27. $\{(-4, -3, -2)\}$ **29.** $\{(4, -1, -2)\}$

31. $\{(1, -1, 2, -3)\}$ **33.** $\{(2, 1, 3, -2)\}$
35. $\{(-2, 4, -3, 0)\}$ **37.** $\varnothing$
39. $\{(-3k + 5, -1, -4k + 2, k)\,|\,k$ is any real number$\}$
41. $\{(-3k + 9, k, 2, -3)\,|\,k$ is any real number$\}$
45. $\{(17k - 6, 10k - 5, k)\,|\,k$ is any real number$\}$
47. $\left\{\left(-\dfrac{1}{2}k + \dfrac{34}{11}, \dfrac{1}{2}k - \dfrac{5}{11}, k\right)\,\middle|\,k$ is any real number$\right\}$
49. $\varnothing$

Problem Set 6.4 (page 524)

1. 22   **3.** -29 **5.** 20   **7.** 5   **9.** -2

11. $-\dfrac{2}{3}$ **13.** -25 **15.** 58   **17.** 39   **19.** -12

21. -41 **23.** -8 **25.** 1088   **27.** -140
29. 81   **31.** 146   **33.** Property 6.3
35. Property 6.2   **37.** Property 6.4
39. Property 6.3   **41.** Property 6.5

Problem Set 6.5 (page 533)

1. $\{(1, 4)\}$ **3.** $\{(3, -5)\}$ **5.** $\{(2, -1)\}$

7. $\varnothing$ **9.** $\left\{\left(-\dfrac{1}{4}, \dfrac{2}{3}\right)\right\}$ **11.** $\left\{\left(\dfrac{2}{17}, \dfrac{52}{17}\right)\right\}$

13. $\{(9, -2)\}$ **15.** $\left\{\left(2, -\dfrac{5}{7}\right)\right\}$

17. $\{(0, 2, -3)\}$ **19.** $\{(2, 6, 7)\}$ **21.** $\{(4, -4, 5)\}$
23. $\{(-1, 3, -4)\}$ **25.** Infinitely many solutions
27. $\left\{\left(-2, \dfrac{1}{2}, -\dfrac{2}{3}\right)\right\}$ **29.** $\left\{\left(3, \dfrac{1}{2}, -\dfrac{1}{3}\right)\right\}$
31. $\{(-4, 6, 0)\}$ **37.** $\{(0, 0, 0)\}$
39. Infinitely many solutions

Problem Set 6.6 (page 541)

1. $\dfrac{4}{x - 2} + \dfrac{7}{x + 1}$ **3.** $\dfrac{3}{x + 1} - \dfrac{5}{x - 1}$

5. $\dfrac{1}{3x - 1} + \dfrac{6}{2x + 3}$ **7.** $\dfrac{2}{x - 1} + \dfrac{3}{x + 2} - \dfrac{4}{x - 3}$

9. $\dfrac{-1}{x} + \dfrac{2}{2x - 1} - \dfrac{3}{4x + 1}$ **11.** $\dfrac{2}{x - 2} + \dfrac{5}{(x - 2)^2}$

13. $\dfrac{4}{x} + \dfrac{7}{x^2} - \dfrac{10}{x + 3}$ **15.** $\dfrac{-3}{x^2 + 1} - \dfrac{2}{x - 4}$

17. $\dfrac{3}{x + 2} - \dfrac{2}{(x + 2)^2} + \dfrac{1}{(x + 2)^3}$

19. $\dfrac{2}{x} + \dfrac{3x + 5}{x^2 - x + 3}$ **21.** $\dfrac{2x}{x^2 + 1} + \dfrac{3 - x}{(x^2 + 1)^2}$

Chapter 6 Review Problem Set (page 543)

1. $\{(3, -7)\}$ **2.** $\{(-1, -3)\}$ **3.** $\{(0, -4)\}$
4. $\left\{\left(\dfrac{23}{3}, -\dfrac{14}{3}\right)\right\}$ **5.** $\{(4, -6)\}$

6. $\left\{\left(-\dfrac{6}{7}, -\dfrac{15}{7}\right)\right\}$ **7.** $\{(-1, 2, -5)\}$

8. $\{(2, -3, -1)\}$ **9.** $\{(5, -4)\}$ **10.** $\{(2, 7)\}$
11. $\{(-2, 2, -1)\}$ **12.** $\{(0, -1, 2)\}$
13. $\{(-3, -1)\}$ **14.** $\{(4, 6)\}$ **15.** $\{(2, -3, -4)\}$
16. $\{(-1, 2, -5)\}$ **17.** $\{(5, -5)\}$

18. $\{(-12, 12)\}$ **19.** $\left\{\left(\dfrac{5}{7}, \dfrac{4}{7}\right)\right\}$

20. $\{(-10, -7)\}$ **21.** $\{(1, 1, -4)\}$
22. $\{(-4, 0, 1)\}$ **23.** $\varnothing$ **24.** $\{(-2, -4, 6)\}$
25. -34 **26.** 13   **27.** -40 **28.** 16

29. 51   **30.** 125   **31.** $\dfrac{1}{x} - \dfrac{2}{x^2} + \dfrac{4}{x + 2}$

32. $\dfrac{3x + 1}{x^2 + 4} - \dfrac{5}{2x - 1}$ **33.** 72

34. \$900 at 10% and \$1600 at 12%
35. 20 nickels, 32 dimes, and 54 quarters
36. $25°, 45°,$ and $110°$

Chapter 6 Test (page 545)

1. III   **2.** I   **3.** III   **4.** II   **5.** 8

6. $-\dfrac{7}{12}$ **7.** -18 **8.** 112

9. Infinitely many   **10.** $\{(-2, 4)\}$

11. $\{(3, -1)\}$ **12.** $x = -12$ **13.** $y = -\dfrac{13}{11}$
14. No   **15.** $x = 14$ **16.** $y = 13$
17. Infinitely many   **18.** None

19. $\left\{\left(\dfrac{11}{5}, 6, -3\right)\right\}$ **20.** $\{(-2, -1, 0)\}$

21. $x = 1$ **22.** $\dfrac{3}{2x - 1} + \dfrac{4}{x - 6}$
23. 30   **24.** 2 liters   **25.** 22 quarters

Cumulative Review Problem Set, Chapters 0–6 (page 547)

1. $\dfrac{324}{121}$ **2.** 2   **3.** 5   **4.** -3 **5.** $\dfrac{3}{2}$

6. 81   **7.** 6   **8.** 3   **9.** $\dfrac{8}{27}$ **10.** -0.2

11. -52 **12.** $-\dfrac{7}{12}$ **13.** $\dfrac{5}{18}$ **14.** 21

15. $\dfrac{2}{3}$ **16.** $\dfrac{x + 4}{x(x + 2)(x - 2)}$ **17.** $\dfrac{x^2 - x + 1}{x + 2}$

18. $\dfrac{17x - 17}{24}$ **19.** $\dfrac{3}{2x^3 y^2}$ **20.** $\dfrac{b^7}{4a^3}$

21. $-\dfrac{30}{x^3 y^3}$ **22.** $48\sqrt{2}$ **23.** $3xy\sqrt[3]{2y^2}$

24. $\dfrac{\sqrt{2}}{4}$ **25.** $x + 2y = 10$ **26.** $f(x) = \dfrac{11}{6}x + \dfrac{2}{3}$

27. $f(g(-2)) = -199$ and $g(f(3)) = 81$

28. $f(g(x)) = 2x^2 + x + 4$ and
$g(f(x)) = 2x^2 + 29x + 102$

29. $10 + 30i$ **30.** $f^{-1}(x) = -\dfrac{1}{3}x + \dfrac{10}{3}$

31. 6 units   **32.** 14   **33.** 0

34. $Q: x^2 + 7x - 1;\ R: 0$

35. $Q: 2x^3 - 5x^2 + 7x - 8;\ R: 14$ **36.** Yes

37. No   **38.** Yes   **39.** $y = \pm 3x$ **40.** -6

41. -1 **42.** 6.66   **43.** $x = 1$ and $f(x) = 2$

44. y axis   **45.** $f(x) = 2x - 2$ **46.** $\left\{\dfrac{11}{8}\right\}$

47. $\left\{-\dfrac{53}{11}\right\}$ **48.** $\left\{-\dfrac{11}{2}, \dfrac{1}{2}\right\}$ **49.** $\{(-1, 4)\}$

50. $\left\{-\dfrac{7}{2}, 0, \dfrac{2}{7}\right\}$ **51.** $\left\{-2, \dfrac{3 \pm i\sqrt{15}}{4}\right\}$

52. $\left\{-\dfrac{1}{2}, \dfrac{5}{3}\right\}$ **53.** $\{-7, -1\}$ **54.** $\{(1, 2, -3)\}$

55. $\{-13\}$ **56.** $\{1\}$ **57.** $\{3.21\}$

58. $\{-1 + \sqrt{14}\}$ **59.** $\left[\dfrac{17}{2}, \infty\right)$ **60.** $\left(-\dfrac{3}{2}, \dfrac{5}{2}\right)$

61. $\left(-\infty, -\dfrac{8}{3}\right) \cup \left(\dfrac{4}{3}, \infty\right)$ **62.** $[3, \infty)$

63. $[-4, 2] \cup [7, \infty)$ **64.** $(-\infty, 1) \cup (2, \infty)$

65. $(4, 9)$ **66.** $[-8, 3]$

67.

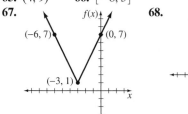

68.

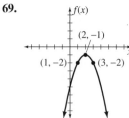

69.

70.

71.

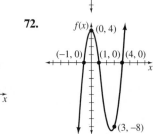

72.

73.

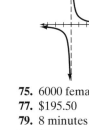

74.

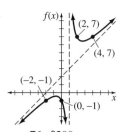

75. 6000 females and 7200 males   **76.** $200

77. $195.50   **78.** 7 centimeters by 12 centimeters

79. 8 minutes   **80.** 5 liters   **81.** More than 200 miles

82. $18,000   **83.** Approximately 11.7 years

CHAPTER 7

Problem Set 7.1 (page 556)

1. $\begin{bmatrix} 3 & -5 \\ 8 & 3 \end{bmatrix}$ **3.** $\begin{bmatrix} -2 & 21 \\ -7 & 2 \end{bmatrix}$ **5.** $\begin{bmatrix} -2 & 1 \\ -3 & 19 \end{bmatrix}$

7. $\begin{bmatrix} -1 & -5 \\ 2 & 3 \end{bmatrix}$ **9.** $\begin{bmatrix} -12 & -14 \\ -18 & -20 \end{bmatrix}$ **11.** $\begin{bmatrix} 2 & -11 \\ -7 & 0 \end{bmatrix}$

13. $AB = \begin{bmatrix} 4 & -6 \\ 8 & -12 \end{bmatrix}$, $BA = \begin{bmatrix} -5 & 5 \\ 3 & -3 \end{bmatrix}$

15. $AB = \begin{bmatrix} -5 & -18 \\ -4 & 42 \end{bmatrix}$, $BA = \begin{bmatrix} 19 & -39 \\ -16 & 18 \end{bmatrix}$

17. $AB = \begin{bmatrix} 14 & -28 \\ 7 & -14 \end{bmatrix}$, $BA = \begin{bmatrix} 0 & 0 \\ 0 & 0 \end{bmatrix}$

19. $AB = \begin{bmatrix} -14 & -7 \\ -12 & -1 \end{bmatrix}$, $BA = \begin{bmatrix} -2 & -3 \\ -32 & -13 \end{bmatrix}$

21. $AB = \begin{bmatrix} 1 & 0 \\ 0 & 1 \end{bmatrix}$, $BA = \begin{bmatrix} 1 & 0 \\ 0 & 1 \end{bmatrix}$

23. $AB = \begin{bmatrix} 0 & -\dfrac{5}{3} \\ \dfrac{17}{6} & -3 \end{bmatrix}$, $BA = \begin{bmatrix} 0 & -\dfrac{17}{6} \\ \dfrac{5}{3} & -3 \end{bmatrix}$

25. $AB = \begin{bmatrix} 1 & 0 \\ 0 & 1 \end{bmatrix}$, $BA = \begin{bmatrix} 1 & 0 \\ 0 & 1 \end{bmatrix}$

27. $AB = \begin{bmatrix} 3 & -2 \\ 4 & 5 \end{bmatrix}$, $BA = \begin{bmatrix} 5 & 4 \\ -2 & 3 \end{bmatrix}$

29. $AD = \begin{bmatrix} 1 & 1 \\ 9 & 9 \end{bmatrix}$, $DA = \begin{bmatrix} 3 & 7 \\ 3 & 7 \end{bmatrix}$

49. $A^2 = \begin{bmatrix} -1 & -4 \\ 8 & 7 \end{bmatrix}$, $A^3 = \begin{bmatrix} -9 & -11 \\ 22 & 13 \end{bmatrix}$

Problem Set 7.2 (page 564)

1. $\begin{bmatrix} 3 & -7 \\ -2 & 5 \end{bmatrix}$ **3.** $\begin{bmatrix} -5 & 8 \\ 2 & -3 \end{bmatrix}$ **5.** $\begin{bmatrix} -\dfrac{2}{5} & \dfrac{1}{5} \\ \dfrac{3}{10} & \dfrac{1}{10} \end{bmatrix}$

7. Does not exist

9. $\begin{bmatrix} -\dfrac{5}{7} & \dfrac{2}{7} \\ -\dfrac{4}{7} & \dfrac{3}{7} \end{bmatrix}$

11. $\begin{bmatrix} -\dfrac{3}{5} & \dfrac{1}{5} \\ 1 & 0 \end{bmatrix}$

17. AB does not exist; $BA = \begin{bmatrix} 20 \\ 2 \\ -30 \end{bmatrix}$

13. $\begin{bmatrix} -\dfrac{4}{5} & \dfrac{3}{5} \\ \dfrac{1}{5} & -\dfrac{2}{5} \end{bmatrix}$

15. $\begin{bmatrix} 2 & -\dfrac{5}{3} \\ 1 & -\dfrac{2}{3} \end{bmatrix}$

17. $\begin{bmatrix} \dfrac{1}{2} & \dfrac{1}{2} \\ \dfrac{1}{2} & -\dfrac{1}{2} \end{bmatrix}$

19. $AB = \begin{bmatrix} 9 & -12 \\ -12 & 16 \\ 6 & -8 \end{bmatrix}$; BA does not exist.

19. $\begin{bmatrix} 30 \\ 36 \end{bmatrix}$ **21.** $\begin{bmatrix} 0 \\ 5 \end{bmatrix}$ **23.** $\begin{bmatrix} -4 \\ 13 \end{bmatrix}$ **25.** $\begin{bmatrix} -4 \\ -13 \end{bmatrix}$

21. $\begin{bmatrix} -\dfrac{1}{5} & \dfrac{3}{10} \\ \dfrac{2}{5} & -\dfrac{1}{10} \end{bmatrix}$ **23.** $\begin{bmatrix} 4 & -1 \\ -7 & 2 \end{bmatrix}$

27. $\{(2, 3)\}$ **29.** $\{(-2, 5)\}$ **31.** $\{(0, -1)\}$

33. $\{(-1, -1)\}$ **35.** $\{(4, 7)\}$ **37.** $\left\{\left(-\dfrac{1}{3}, \dfrac{1}{2}\right)\right\}$

39. $\{(-9, 20)\}$

25. $\begin{bmatrix} -\dfrac{4}{5} & -\dfrac{1}{5} \\ \dfrac{3}{5} & \dfrac{2}{5} \end{bmatrix}$ **27.** $\begin{bmatrix} \dfrac{7}{2} & -3 & \dfrac{1}{2} \\ -\dfrac{1}{2} & 0 & \dfrac{1}{2} \\ -\dfrac{1}{2} & 1 & -\dfrac{1}{2} \end{bmatrix}$

Problem Set 7.3 (page 573)

1. $\begin{bmatrix} 1 & 3 & -3 \\ 3 & -6 & 7 \end{bmatrix}$; $\begin{bmatrix} 3 & -5 & 11 \\ -7 & 6 & 3 \end{bmatrix}$; $\begin{bmatrix} 1 & 10 & -13 \\ 11 & -18 & 16 \end{bmatrix}$;
$\begin{bmatrix} 10 & -12 & 30 \\ -18 & 12 & 16 \end{bmatrix}$

3. $[-1 \;\; -7 \;\; 13 \;\; 7]$; $[5 \;\; 5 \;\; -5 \;\; 17]$;
$[-5 \;\; -20 \;\; 35 \;\; 9]$; $[14 \;\; 8 \;\; -2 \;\; 58]$

5. $\begin{bmatrix} 8 & -3 & -2 \\ 9 & 2 & -3 \\ 7 & 5 & 21 \end{bmatrix}$; $\begin{bmatrix} -2 & -1 & 4 \\ -11 & 6 & -11 \\ -7 & 5 & -3 \end{bmatrix}$; $\begin{bmatrix} 21 & -7 & -7 \\ 28 & 2 & -2 \\ 21 & 10 & 54 \end{bmatrix}$;
$\begin{bmatrix} 2 & -6 & 10 \\ -24 & 20 & -36 \\ -14 & 20 & 12 \end{bmatrix}$

29. $\begin{bmatrix} -50 & -9 & 11 \\ -23 & -4 & 5 \\ 5 & 1 & -1 \end{bmatrix}$ **31.** Does not exist

7. $\begin{bmatrix} 0 & 2 \\ -1 & 10 \\ 1 & -9 \\ 2 & 9 \end{bmatrix}$; $\begin{bmatrix} -2 & -2 \\ 5 & -4 \\ -11 & 1 \\ -16 & 13 \end{bmatrix}$; $\begin{bmatrix} 1 & 6 \\ -5 & 27 \\ 8 & -23 \\ 13 & 16 \end{bmatrix}$; $\begin{bmatrix} -6 & -4 \\ 14 & -2 \\ -32 & -6 \\ -46 & 48 \end{bmatrix}$

33. $\begin{bmatrix} \dfrac{4}{7} & -1 & -\dfrac{9}{7} \\ -\dfrac{3}{14} & \dfrac{1}{2} & \dfrac{6}{7} \\ \dfrac{2}{7} & 0 & -\dfrac{1}{7} \end{bmatrix}$ **35.** $\begin{bmatrix} \dfrac{1}{2} & 0 & 0 \\ 0 & \dfrac{1}{4} & 0 \\ 0 & 0 & \dfrac{1}{10} \end{bmatrix}$

37. $\{(-3, 2)\}$ **39.** $\{(2, 5)\}$ **41.** $\{(-1, -2, 1)\}$

43. $\{(-2, 3, 5)\}$ **45.** $\{(-4, 3, 0)\}$

47. a. $\{(-1, 2, 3)\}$ **c.** $\{(-5, 0, -2)\}$ **e.** $\{(1, -1, -1)\}$

51. a. Double Jeopardy   **b.** Drop The Course
c. The Price Is Right   **d.** Martini Shaken Not Stirred

9. $AB = \begin{bmatrix} 11 & -8 & 14 \\ 4 & -16 & 8 \\ -28 & 22 & -36 \end{bmatrix}$; $BA = \begin{bmatrix} -20 & 21 \\ 8 & -21 \end{bmatrix}$

11. $AB = \begin{bmatrix} 22 & -8 & 1 & 3 \\ -42 & 36 & -26 & -20 \end{bmatrix}$; BA does not exist.

13. $AB = \begin{bmatrix} -12 & 5 & -5 \\ 14 & -2 & 4 \\ -10 & 13 & -5 \end{bmatrix}$; $BA = \begin{bmatrix} -1 & 0 & -6 \\ 10 & -2 & 16 \\ -8 & 5 & -16 \end{bmatrix}$

15. $AB = [-9]$; $BA = \begin{bmatrix} -2 & 1 & -3 & -4 \\ -6 & 3 & -9 & -12 \\ 4 & -2 & 6 & 8 \\ -8 & 4 & -12 & -16 \end{bmatrix}$

Problem Set 7.4 (page 584)

1.

3.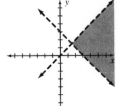

5.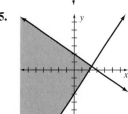

7.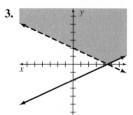

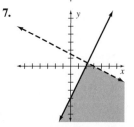

9.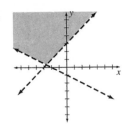

11.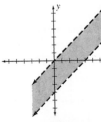

8. $\begin{bmatrix} 26 & -36 \\ -15 & 32 \end{bmatrix}$ **9.** $\begin{bmatrix} -27 \\ 26 \end{bmatrix}$

10. *EF* does not exist.   **14.** $\begin{bmatrix} 4 & -5 \\ -7 & 9 \end{bmatrix}$

15. $\begin{bmatrix} -3 & 4 \\ 7 & -9 \end{bmatrix}$ **16.** $\begin{bmatrix} -\dfrac{3}{8} & \dfrac{1}{8} \\ \dfrac{1}{4} & \dfrac{1}{4} \end{bmatrix}$

13.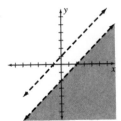

15.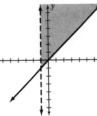

17. Inverse does not exist.   **18.** $\begin{bmatrix} \dfrac{5}{7} & -\dfrac{3}{7} \\ -\dfrac{4}{7} & \dfrac{1}{7} \end{bmatrix}$

17. $\varnothing$

19.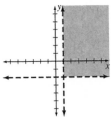

19. $\begin{bmatrix} \dfrac{2}{7} & \dfrac{1}{7} \\ -\dfrac{1}{3} & 0 \end{bmatrix}$ **20.** $\begin{bmatrix} \dfrac{39}{8} & -\dfrac{17}{8} & -\dfrac{1}{8} \\ 2 & -1 & 0 \\ \dfrac{1}{8} & \dfrac{1}{8} & \dfrac{1}{8} \end{bmatrix}$

21. $\begin{bmatrix} 8 & -8 & 5 \\ -3 & 2 & -1 \\ -1 & -1 & 1 \end{bmatrix}$ **22.** Inverse does not exist.

21.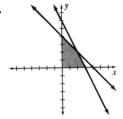

23.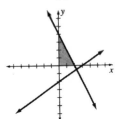

23. $\begin{bmatrix} -\dfrac{20}{3} & -\dfrac{7}{3} & \dfrac{1}{3} \\ \dfrac{1}{3} & -\dfrac{2}{3} & -\dfrac{1}{3} \\ -\dfrac{5}{3} & -\dfrac{1}{3} & \dfrac{1}{3} \end{bmatrix}$ **24.** $\{(-2, 6)\}$

25. $\{(4, -1)\}$ **26.** $\{(2, -3, -1)\}$
27. $\{(-3, 2, 5)\}$ **28.** $\{(-4, 3, 4)\}$

25. Minimum of 8 and maximum of 52
27. Minimum of 0 and maximum of 28   **29.** 63
31. 340   **33.** 2   **35.** 98
37. $5000 at 9% and $5000 at 12%
39. 300 of type A and 200 of type B
41. 12 units of A and 16 units of B

29.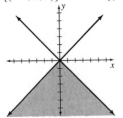

30.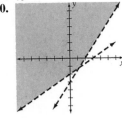

Chapter 7 Review Problem Set (page 589)

1. $\begin{bmatrix} 7 & -5 \\ -3 & 10 \end{bmatrix}$ **2.** $\begin{bmatrix} 3 & 3 \\ 3 & -6 \end{bmatrix}$ **3.** $\begin{bmatrix} 2 & 1 \\ -6 & 8 \\ -2 & 2 \end{bmatrix}$

31.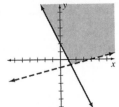

32.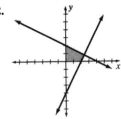

4. $\begin{bmatrix} 19 & -11 \\ -6 & 22 \end{bmatrix}$ **5.** $\begin{bmatrix} 7 & 1 \\ -14 & 20 \\ 1 & -2 \end{bmatrix}$

6. $\begin{bmatrix} -11 & -3 & 15 \\ 24 & 2 & -20 \\ -40 & -5 & 38 \end{bmatrix}$ **7.** $\begin{bmatrix} 16 & -26 \\ 0 & 13 \end{bmatrix}$

33. 37   **34.** 56   **35.** 57   **36.** 1700
37. 75 one-gallon and 175 two-gallon freezers

Chapter 7 Test (page 591)

1. $\begin{bmatrix} 9 & -1 \\ 4 & -6 \end{bmatrix}$ **2.** $\begin{bmatrix} -11 & 13 \\ -8 & 14 \end{bmatrix}$

3. $\begin{bmatrix} -1 & -3 & 11 \\ -4 & -5 & 18 \\ 37 & -1 & 9 \end{bmatrix}$

4. BC does not exist.   **5.** $\begin{bmatrix} -35 \\ 8 \end{bmatrix}$

6. $\begin{bmatrix} -5 & 8 \\ 4 & -3 \end{bmatrix}$ **7.** $\begin{bmatrix} 4 & 9 \\ 13 & -16 \\ 24 & 23 \end{bmatrix}$ **8.** $\begin{bmatrix} -3 & -5 \\ -20 & 8 \end{bmatrix}$

9. $\begin{bmatrix} 8 & 33 \\ -12 & 13 \end{bmatrix}$ **10.** $\begin{bmatrix} 1 & -34 \\ 16 & -19 \end{bmatrix}$ **11.** $\begin{bmatrix} -3 & 2 \\ -5 & 3 \end{bmatrix}$

12. $\begin{bmatrix} 7 & 5 \\ 3 & 2 \end{bmatrix}$ **13.** $\begin{bmatrix} 4 & \frac{3}{2} \\ 1 & \frac{1}{2} \end{bmatrix}$ **14.** $\begin{bmatrix} \frac{4}{7} & -\frac{5}{7} \\ -\frac{1}{7} & \frac{3}{7} \end{bmatrix}$

15. $\begin{bmatrix} -\frac{4}{3} & -\frac{5}{3} & 1 \\ -\frac{4}{3} & -\frac{8}{3} & 1 \\ \frac{1}{3} & \frac{2}{3} & 0 \end{bmatrix}$ **16.** $\begin{bmatrix} 1 & 2 & -10 \\ 0 & 1 & -3 \\ 0 & 0 & 1 \end{bmatrix}$

17. $\{(8, -12)\}$ **18.** $\{(-6, -14)\}$ **19.** $\{(9, 13)\}$

20. $\left\{ \left(\frac{7}{3}, -\frac{1}{3}, \frac{13}{3} \right) \right\}$ **21.** $\{(-1, 2, 1)\}$

22.
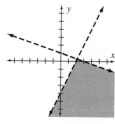

23.
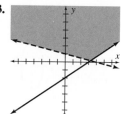

24.

25. 4050

CHAPTER 8

Problem Set 8.1 (page 602)

1. $V(0, 0)$, $F(2, 0)$, $x = -2$
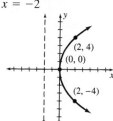

3. $V(0, 0)$, $F(0, -3)$, $y = 3$
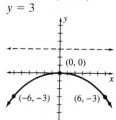

5. $V(0, 0)$, $F\left(-\frac{1}{2}, 0\right)$, $x = \frac{1}{2}$

7. $V(0, 0)$, $F\left(0, \frac{3}{2}\right)$, $y = -\frac{3}{2}$
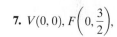

9. $V(0, 2)$, $F(0, 3)$, $y = 1$
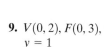

11. $V(0, -2)$, $F(0, -4)$, $y = 0$
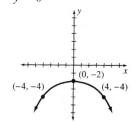

13. $V(2, 0)$, $F(5, 0)$, $x = -1$
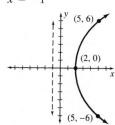

15. $V(1, 2)$, $F(1, 3)$, $y = 1$
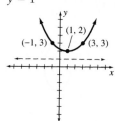

17. $V(-3, 1), F(-3, -1),$
$y = 3$

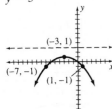

19. $V(3, 1), F(0, 1),$
$x = 6$

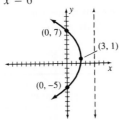

9. $F(0, \sqrt{33}),$
$F'(0, -\sqrt{33})$

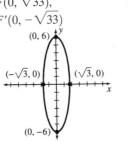

11. $F(2, 0),$
$F'(-2, 0)$

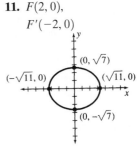

21. $V(-2, -3), F(-1, -3), x = -3$

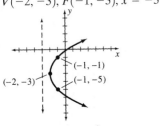

13. $F(1 + \sqrt{5}, 2),$
$F'(1 - \sqrt{5}, 2)$

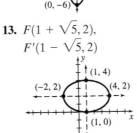

15. $F(-2, -1 + 2\sqrt{3})$
$F'(-2, -1 - 2\sqrt{3})$

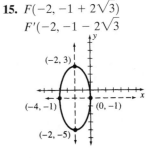

23. $x^2 = 12y$ **25.** $y^2 = -4x$
27. $x^2 + 12y - 48 = 0$ **29.** $x^2 - 6x - 12y + 21 = 0$
31. $y^2 - 10y + 8x + 41 = 0$ **33.** $x^2 = -y$
35. $y^2 = 10x$ **37.** $x^2 - 14x - 8y + 73 = 0$
39. $y^2 + 6y - 12x + 105 = 0$
41. $x^2 + 18x + y + 80 = 0$ **43.** $x^2 = 750(y - 10)$
45. $10\sqrt{2}$ feet   **47.** 62.5 feet

17. $F(3 + \sqrt{3}, 0),$
$F'(3 - \sqrt{3}, 0)$

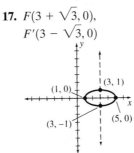

Problem Set 8.2 (page 613)

For Problems 1–21, the foci are indicated above the graph and the vertices and endpoints of the minor axes are indicated on the graph.

1. $F(\sqrt{3}, 0), F'(-\sqrt{3}, 0)$

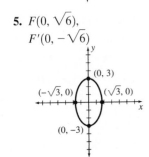

3. $F(0, \sqrt{5}), F'(0, -\sqrt{5})$

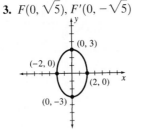

19. $F(4, -1 + \sqrt{7}),$
$F'(4, -1 - \sqrt{7})$

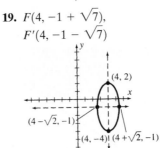

5. $F(0, \sqrt{6}),$
$F'(0, -\sqrt{6})$

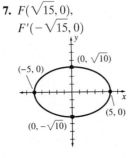

7. $F(\sqrt{15}, 0),$
$F'(-\sqrt{15}, 0)$

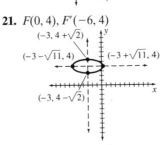

21. $F(0, 4), F'(-6, 4)$

23. $16x^2 + 25y^2 = 400$ **25.** $36x^2 + 11y^2 = 396$
27. $x^2 + 9y^2 = 9$ **29.** $100x^2 + 36y^2 = 225$
31. $7x^2 + 3y^2 = 75$
33. $3x^2 - 6x + 4y^2 - 8y - 41 = 0$
35. $9x^2 + 25y^2 - 50y - 200 = 0$ **37.** $3x^2 + 4y^2 = 48$

39. $\dfrac{10\sqrt{5}}{3}$ feet

Problem Set 8.3 (page 624)

For Problems 1–21, the foci and equations of the asymptotes are indicated above the graphs. The vertices are given on the graphs.

1. $F(\sqrt{13}, 0)$,
$F'(-\sqrt{13}, 0)$,
$y = \pm\dfrac{2}{3}x$

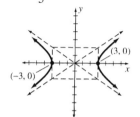

3. $F(0, \sqrt{13})$,
$F'(0, -\sqrt{13})$,
$y = \pm\dfrac{2}{3}x$

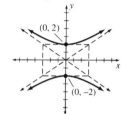

5. $F(0, 5)$,
$F'(0, -5)$,
$y = \pm\dfrac{4}{3}x$

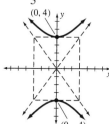

7. $F(3\sqrt{2}, 0)$,
$F'(-3\sqrt{2}, 0)$,
$y = \pm x$

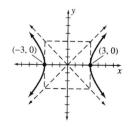

9. $F(0, \sqrt{30})$,
$F'(0, -\sqrt{30})$,
$y = \pm\dfrac{\sqrt{5}}{5}x$

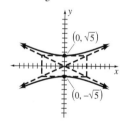

11. $F(\sqrt{10}, 0)$,
$F'(-\sqrt{10}, 0)$,
$y = \pm 3x$

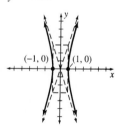

13. $F(3 + \sqrt{13}, -1)$,
$F'(3 - \sqrt{13}, -1)$,
$2x - 3y = 9$ and
$2x + 3y = 3$

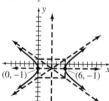

15. $F(-3, 2 + \sqrt{5})$,
$F'(-3, 2 - \sqrt{5})$,
$2x - y = -8$ and
$2x + y = -4$

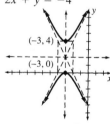

17. $F(2 + \sqrt{6}, 0)$,
$F'(2 - \sqrt{6}, 0)$,
$\sqrt{2}x - y = 2\sqrt{2}$ and
$\sqrt{2}x + y = 2\sqrt{2}$

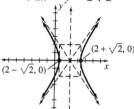

19. $F(0, -5 + \sqrt{10})$,
$F'(0, -5 - \sqrt{10})$,
$3x - y = 5$ and
$3x + y = -5$

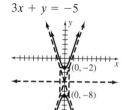

21. $F(-2 + \sqrt{2}, -2)$,
$F'(-2 - \sqrt{2}, -2)$,
$x - y = 0$ and
$x + y = -4$

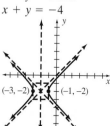

23. $5x^2 - 4y^2 = 20$ **25.** $16y^2 - 9x^2 = 144$
27. $3x^2 - y^2 = 3$ **29.** $4y^2 - 3x^2 = 12$
31. $7x^2 - 16y^2 = 112$
33. $5x^2 - 40x - 4y^2 - 24y + 24 = 0$
35. $3y^2 - 30y - x^2 - 6x + 54 = 0$
37. $5x^2 - 20x - 4y^2 = 0$ **39.** Circle
41. Straight line **43.** Ellipse **45.** Hyperbola
47. Parabola

Problem Set 8.4 (page 633)

1. $\{(1, 2)\}$ **3.** $\{(1, -5), (-5, 1)\}$
5. $\{(2 + i\sqrt{3}, -2 + i\sqrt{3}), (2 - i\sqrt{3}, -2 - i\sqrt{3})\}$
7. $\{(-6, 7), (-2, -1)\}$ **9.** $\{(-3, 4)\}$

11. $\left\{\left(\dfrac{-1 + i\sqrt{3}}{2}, \dfrac{-7 - i\sqrt{3}}{2}\right),\right.$
$\left.\left(\dfrac{-1 - i\sqrt{3}}{2}, \dfrac{-7 + i\sqrt{3}}{2}\right)\right\}$

13. $\{(-1, 2)\}$ **15.** $\{(-6, 3), (-2, -1)\}$

17. $\{(5, 3)\}$ **19.** $\{(1, 2), (-1, 2)\}$

21. $\{(-3, 2)\}$ **23.** $\{(2, 0), (-2, 0)\}$

25. $\{(\sqrt{2}, \sqrt{3}), (\sqrt{2}, -\sqrt{3}), (-\sqrt{2}, \sqrt{3}), (-\sqrt{2}, -\sqrt{3})\}$

27. $\{(1, 1), (1, -1), (-1, 1), (-1, -1)\}$

29. $\left\{\left(2, \dfrac{3}{2}\right), \left(\dfrac{3}{2}, 2\right)\right\}$

31. $\{(9, -2)\}$ **33.** $\{(\ln 2, 1)\}$

35. $\left\{\left(\dfrac{1}{2}, \dfrac{1}{8}\right), (-3, -27)\right\}$ **43.** $\{(-2.3, 7.4)\}$

45. $\{(6.7, 1.7), (9.5, 2.1)\}$ **47.** None

Chapter 8 Review Problem Set (page 636)

1. $F(4, 0), F'(-4, 0)$

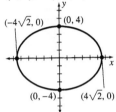

2. $F(-3, 0)$

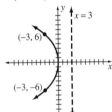

3. $F(0, 2\sqrt{3}),$
$F'(0, -2\sqrt{3}),$
$y = \pm\dfrac{\sqrt{3}}{3}x$

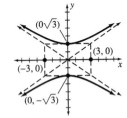

4. $F(\sqrt{15}, 0),$
$F'(-\sqrt{15}, 0),$
$y = \pm\dfrac{\sqrt{6}}{3}x$

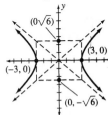

5. $F(0, \sqrt{6}),$
$F'(0, -\sqrt{6})$

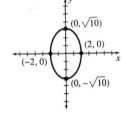

6. $F\left(0, \dfrac{1}{2}\right)$

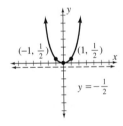

7. $F(4 + \sqrt{6}, 1), F'(4 - \sqrt{6}, 1),$
$\sqrt{2}x - 2y = 4\sqrt{2} - 2$ and $\sqrt{2}x + 2y = 4\sqrt{2} + 2$

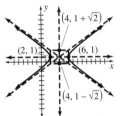

8. $F(3, -2 + \sqrt{7}), F'(3, -2 - \sqrt{7})$

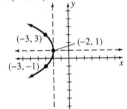

9. $F(-3, 1), x = -1$

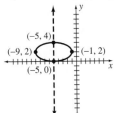

10. $F(-1, -5), y = -1$

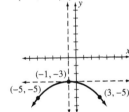

11. $F(-5 + 2\sqrt{3}, 2), F'(-5 - 2\sqrt{3}, 2)$

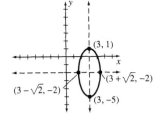

12. $F(-2, -2 + \sqrt{10}), F'(-2, -2 - \sqrt{10}),$
$\sqrt{6}x - 3y = 6 - 2\sqrt{6}$ and $\sqrt{6}x + 3y = -6 - 2\sqrt{6}$

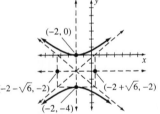

13. $y^2 = -20x$ **14.** $y^2 + 16x^2 = 16$

15. $25x^2 - 2y^2 = 50$ **16.** $4x^2 + 3y^2 = 16$

17. $3x^2 = 2y$ **18.** $9y^2 - x^2 = 9$

19. $9x^2 - 108x + y^2 - 8y + 331 = 0$

20. $y^2 + 4y - 8x + 36 = 0$

21. $3y^2 + 24y - x^2 - 10x + 20 = 0$

22. $x^2 + 12x - y + 33 = 0$

23. $4x^2 + 40x + 25y^2 = 0$

24. $4x^2 - 32x - y^2 + 48 = 0$ **25.** $\{(-1, 4)\}$

26. $\{(3, 1)\}$ **27.** $\{(-1, -2), (-2, -3)\}$

28. $\left\{\left(\dfrac{4\sqrt{2}}{3}, \dfrac{4}{3}i\right), \left(\dfrac{4\sqrt{2}}{3}, -\dfrac{4}{3}i\right), \left(-\dfrac{4\sqrt{2}}{3}, \dfrac{4}{3}i\right),\right.$
$\left. \left(-\dfrac{4\sqrt{2}}{3}, -\dfrac{4}{3}i\right)\right\}$

29. $\{(0, 2), (0, -2)\}$

30. $\left\{\left(\dfrac{\sqrt{15}}{5}, \dfrac{2\sqrt{10}}{5}\right), \left(\dfrac{\sqrt{15}}{5}, -\dfrac{2\sqrt{10}}{5}\right),\right.$
$\left. \left(-\dfrac{\sqrt{15}}{5}, \dfrac{2\sqrt{10}}{5}\right), \left(-\dfrac{\sqrt{15}}{5}, -\dfrac{2\sqrt{10}}{5}\right)\right\}$

Chapter 8 Test (page 637)

1. $(0, -5)$ **2.** $(-3, 2)$ **3.** $x = -3$

4. $(6, 0)$ **5.** $(-2, -1)$ **6.** $y = 4$

7. $y^2 + 8x = 0$ **8.** $x^2 - 6x + 12y - 39 = 0$

9. $(0, 6)$ and $(0, -6)$ **10.** Six units

11. $(-7, 1)$ and $(-3, 1)$ **12.** $(-2\sqrt{3}, 0)$ and $(2\sqrt{3}, 0)$

13. $(-5, 8)$ **14.** $25x^2 + 9y^2 = 900$

15. $x^2 - 12x + 4y^2 + 16y + 36 = 0$ **16.** $y = \pm\dfrac{3}{2}x$

17. $(-1, 6)$ and $(-1, 0)$ **18.** $(\pm 3, 0)$

19. $x^2 - 3y^2 = 36$

20. $8x^2 + 16x - y^2 + 8y - 16 = 0$ **21.** 2

22. $\left\{(3, 2), (-3, -2), \left(4, \dfrac{3}{2}\right), \left(-4, -\dfrac{3}{2}\right)\right\}$

23.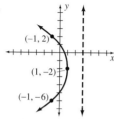

24.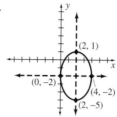

25.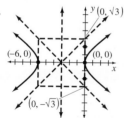

Cumulative Review Problem Set, Chapters 0–8 (page 638)

1. $\dfrac{9}{4}$ **2.** $\dfrac{4}{3}$ **3.** $\dfrac{1}{2}$ **4.** $\dfrac{18}{7}$ **5.** 64

6. 243   **7.** 3   **8.** 2   **9.** 5   **10.** 15

11. $x - 6$ **12.** $-6 - 11x - 4x^2$

13. $3x^3 - 17x^2 + 11x - 5$

14. $2x^4 + x^3 - 2x^2 - 10x - 21$

15. $8x^3 - 36x^2 + 54x - 27$ **16.** $-6x^2y + 9xy^3$

17. $x^2 - 4x + 6$ **18.** $3x^3 + 4x^2 - 5x - 3$

19. $(-3 + 2x)(4 - 3x)$ **20.** $(x + 2y)(x - 3)$

21. $2x(x + 6)(x - 3)$ **22.** Not factorable

23. $(2x + 3)(4x^2 - 6x + 9)$

24. $(x + 1)(x - 1)(2x + 1)(2x - 1)$ **25.** -48

26. $\dfrac{1}{28}$ **27.** $\dfrac{1}{24}$ **28.** $26 - 7i$

29. $-12 + 9i$ **30.** $\dfrac{2}{25} - \dfrac{11}{25}i$ **31.** $4\sqrt{3} - 4\sqrt{6}$

32. -33 **33.** $\dfrac{5\sqrt{3}}{9}$ **34.** $\sqrt[6]{5}$ **35.** $\dfrac{3}{8}$

36. $(0, 3)$ **37.** $(1, 3)$ **38.** 160 feet

39. $3x + 2y = 2$ **40.** $5x - 6y = 18$

41. $2x - 3y = -14$ **42.** $7x + 5y = -26$

43. x axis   **44.** x axis, y axis, and origin

45. $(-6, 4)$ **46.** $r = 4$

47. $x^2 + 6x + y^2 - 8y = 0$

48. $x^2 - 10x + y^2 - 2y + 16 = 0$ **49.** $y = \pm 3x$

50. -34 **51.** $8a + 4h - 2$

52. $(-\infty, 2) \cup (2, \infty)$ **53.** $[-3, 2]$

54. $(-2, 7)$ **55.** $(8, -6)$ **56.** $\dfrac{1}{2}$ and 9

57. -23 and 113

58. $f(g(x)) = -12x^2 + 62x - 72$ and $g(f(x)) = 6x^2 + 2x - 11$

59. $f^{-1}(x) = -\dfrac{3}{2}x + \dfrac{3}{2}$ **60.** -90 **61.** 64

62. 52.1 milligrams   **63.** \$21,870

64. 13.6 grams   **65.** Yes   **66.** Yes   **67.** No

68. $-2, 1,$ and 3   **69.** $y = x + 3$

70. $\dfrac{-3}{2x - 1} + \dfrac{2}{x - 4}$ **71.** $\begin{bmatrix} -14 & 4 \\ 9 & -7 \end{bmatrix}$

72. $AB = \begin{bmatrix} -7 & -2 \\ -17 & -8 \end{bmatrix}$; $BA = \begin{bmatrix} -15 & 22 \\ -1 & 0 \end{bmatrix}$

73. $\begin{bmatrix} 2 & -3 \\ -1 & 2 \end{bmatrix}$

74. a. Ellipse   **b.** Parabola   **c.** Parabola   **d.** Circle   **e.** Hyperbola

75.

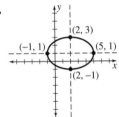

76.

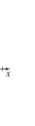

87.

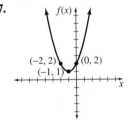

88.

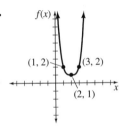

77.

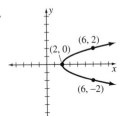

78.

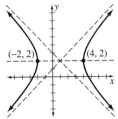

89.

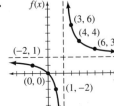

90.

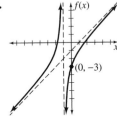

79.

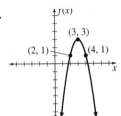

80.

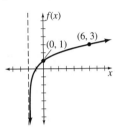

91. $\left\{-\dfrac{7}{3}, 3\right\}$ **92.** $\left\{-\dfrac{1}{2}, \dfrac{2}{3}\right\}$ **93.** $\{2\}$

94. $[-9, 3]$ **95.** $\left\{\dfrac{2 \pm 3\sqrt{2}}{5}\right\}$

96. $\left\{\dfrac{-1 \pm i\sqrt{39}}{4}\right\}$ **97.** $\left\{\left(\dfrac{34}{19}, \dfrac{13}{19}\right)\right\}$

98. $\left\{-\dfrac{19}{5}, 2\right\}$ **99.** $(-6, -2) \cup (1, \infty)$

81.

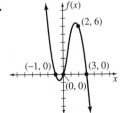

82.

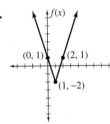

100. $\{-6, -2, 1\}$ **101.** $\{(0, -1, 4)\}$ **102.** $\{-5\}$

103. $\left\{\dfrac{-3 \pm 4i}{4}\right\}$ **104.** $\left\{\dfrac{1 \pm \sqrt{37}}{6}\right\}$

105. $(-8, 6)$ **106.** $\left\{\left(\dfrac{11}{17}, -\dfrac{35}{17}\right)\right\}$

107. $\{-16, 18\}$ **108.** $\left(-\infty, -\dfrac{10}{7}\right) \cup (2, \infty)$

83.

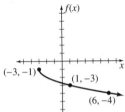

84.

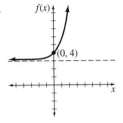

109. $\{4\}$ **110.** $\{3.79\}$ **111.** $\left\{\dfrac{-3 + \sqrt{33}}{2}\right\}$

112. $\left\{\dfrac{1 \pm 3\sqrt{5}}{6}\right\}$ **113.** $(-\infty, -2) \cup \left[-\dfrac{1}{2}, \infty\right)$

114. $\{-1, 64\}$
115. $0.75 per can of soda and $1.29 per bag of
     potato chips
116. 6 cups   **117.** 20 nickels, 25 dimes, and 42 quarters
118. Tina is 21 and Sherry is 17.   **119.** 14 and 86
120. $40   **121.** 15 gallons   **122.** 50 miles
123. Higher than 12%   **124.** 25 weeks
125. Approximately 34.7 minutes

85.

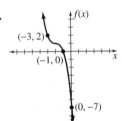

86.

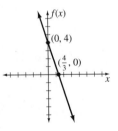

CHAPTER 9

Problem Set 9.1 (page 649)
1. $-4, -1, 2, 5, 8$ **3.** $2, 0, -2, -4, -6$

5. 2, 11, 26, 47, 74   **7.** 0, 2, 6, 12, 20
9. 4, 8, 16, 32, 64   **11.** $a_{15} = -79$; $a_{30} = -154$
13. $a_{25} = 1$; $a_{50} = -1$ **15.** $a_n = 2n + 9$
17. $a_n = -3n + 5$ **19.** $a_n = \dfrac{n+2}{2}$
21. $a_n = 4n - 2$ **23.** $a_n = -3n$ **25.** 73
27. 334   **29.** 35   **31.** 7   **33.** 86
35. 2700   **37.** 3200   **39.** -7950 **41.** 637.5
43. 4950   **45.** 1850   **47.** -2030 **49.** 3591
51. 40,000   **53.** 58,250   **55.** 2205
57. -1325 **59.** 5265   **61.** -810 **63.** 1276
65. 660   **67.** 55   **69.** 431
75. 3, 3, 7, 7, 11, 11   **77.** 4, 7, 10, 13, 17, 21
79. 4, 12, 36, 108, 324, 972   **81.** 1, 1, 2, 3, 5, 8
83. 3, 1, 4, 9, 25, 256

Problem Set 9.2 (page 660)

1. $a_n = 3(2)^{n-1}$ **3.** $a_n = 3^n$ **5.** $a_n = \left(\dfrac{1}{2}\right)^{n+1}$
7. $a_n = 4^n$ **9.** $a_n = (0.3)^{n-1}$
11. $a_n = (-2)^{n-1}$ **13.** 64   **15.** $\dfrac{1}{9}$
17. -512 **19.** $\dfrac{1}{4374}$ **21.** $\dfrac{2}{3}$ **23.** 2
25. 1023   **27.** 19,682   **29.** $394\dfrac{1}{16}$
31. 1364   **33.** 1089   **35.** $7\dfrac{511}{512}$ **37.** -547
39. $127\dfrac{3}{4}$ **41.** 540   **43.** $2\dfrac{61}{64}$ **45.** 4
47. 3   **49.** No sum   **51.** $\dfrac{27}{4}$ **53.** 2
55. $\dfrac{16}{3}$ **57.** $\dfrac{1}{3}$ **59.** $\dfrac{26}{99}$ **61.** $\dfrac{41}{333}$
63. $\dfrac{4}{15}$ **65.** $\dfrac{106}{495}$ **67.** $\dfrac{7}{3}$

Problem Set 9.3 (page 666)

1. $28,900   **3.** 15,600   **5.** 7321
7. 125 liters   **9.** 512 gallons   **11.** $116.25
13. $163.84; $327.67   **15.** $28,900
17. 1936 feet   **19.** $\dfrac{15}{16}$ of a gram   **21.** 2910 feet
23. 325 logs   **25.** 5.9%   **27.** $\dfrac{5}{64}$ of a gallon

Problem Set 9.4 (page 673)

These problems are proofs by mathematical induction and require class discussion.

Chapter 9 Review Problem Set (page 676)

1. $a_n = 6n - 3$ **2.** $a_n = 3^{n-2}$ **3.** $a_n = 5 \cdot 2^n$
4. $a_n = -3n + 8$ **5.** $a_n = 2n - 7$
6. $a_n = 3^{3-n}$ **7.** $a_n = -(-2)^{n-1}$
8. $a_n = 3n + 9$ **9.** $a_n = \dfrac{n+1}{3}$ **10.** $a_n = 4^{n-1}$
11. 73   **12.** 106   **13.** $\dfrac{1}{32}$ **14.** $\dfrac{4}{9}$
15. -92 **16.** $\dfrac{1}{16}$ **17.** -5 **18.** 85
19. $\dfrac{5}{9}$ **20.** 2 or -2 **21.** $121\dfrac{40}{81}$ **22.** 7035
23. $-10,725$ **24.** $31\dfrac{31}{32}$ **25.** 32,015
26. 4757   **27.** $85\dfrac{21}{64}$ **28.** 37,044
29. 12,726   **30.** -1845 **31.** 225   **32.** 255
33. 8244   **34.** $85\dfrac{1}{3}$ **35.** $\dfrac{4}{11}$ **36.** $\dfrac{41}{90}$
37. $750   **38.** $46.50   **39.** $3276.70
40. 10,935 gallons

Chapter 9 Test (page 678)

1. -226 **2.** 48   **3.** $a_n = -5n + 2$
4. $a_n = 5(2)^{1-n}$ **5.** $a_n = 6n + 4$
6. $\dfrac{729}{8}$ or $91\dfrac{1}{8}$ **7.** 223   **8.** 60 terms
9. 2380   **10.** 765   **11.** 7155   **12.** 6138
13. 22,650   **14.** 9384   **15.** 4075   **16.** -341
17. 6   **18.** $\dfrac{1}{3}$ **19.** $\dfrac{2}{11}$ **20.** $\dfrac{4}{15}$
21. 3 liters   **22.** $3276.70   **23.** $5810
24. and **25.** Instructor supplies proof.

Appendix A (page 683)

1. $x^8 + 8x^7y + 28x^6y^2 + 56x^5y^3 + 70x^4y^4 + 56x^3y^5 + 28x^2y^6 + 8xy^7 + y^8$
3. $x^6 - 6x^5y + 15x^4y^2 - 20x^3y^3 + 15x^2y^4 - 6xy^5 + y^6$
5. $a^4 + 8a^3b + 24a^2b^2 + 32ab^3 + 16b^4$
7. $x^5 - 15x^4y + 90x^3y^2 - 270x^2y^3 + 405xy^4 - 243y^5$
9. $16a^4 - 96a^3b + 216a^2b^2 - 216ab^3 + 81b^4$
11. $x^{10} + 5x^8y + 10x^6y^2 + 10x^4y^3 + 5x^2y^4 + y^5$
13. $16x^8 - 32x^6y^2 + 24x^4y^4 - 8x^2y^6 + y^8$
15. $x^6 + 18x^5 + 135x^4 + 540x^3 + 1215x^2 + 1458x + 729$
17. $x^9 - 9x^8 + 36x^7 - 84x^6 + 126x^5 - 126x^4 + 84x^3 - 36x^2 + 9x - 1$
19. $1 + \dfrac{4}{n} + \dfrac{6}{n^2} + \dfrac{4}{n^3} + \dfrac{1}{n^4}$

21. $a^6 - \dfrac{6a^5}{n} + \dfrac{15a^4}{n^2} - \dfrac{20a^3}{n^3} + \dfrac{15a^2}{n^4} - \dfrac{6a}{n^5} + \dfrac{1}{n^6}$

23. $17 + 12\sqrt{2}$ **25.** $843 - 589\sqrt{2}$

27. $x^{12} + 12x^{11}y + 66x^{10}y^2 + 220x^9y^3$

29. $x^{20} - 20x^{19}y + 190x^{18}y^2 - 1140x^{17}y^3$

31. $x^{28} - 28x^{26}y^3 + 364x^{24}y^6 - 2912x^{22}y^9$

33. $a^9 + \dfrac{9a^8}{n} + \dfrac{36a^7}{n^2} + \dfrac{84a^6}{n^3}$

35. $x^{10} - 20x^9y + 180x^8y^2 - 960x^7y^3$ **37.** $56x^5y^3$

39. $126x^5y^4$ **41.** $189a^2b^5$ **43.** $120x^6y^{21}$

Index

Solution set
  of an equation, 96
  of an inequality, 157
  of a quadratic equation, 120
  of a system, 481
Square matrix, 516, 569
Square root, 64
Standard form
  of complex numbers, 82
  of equation of a circle, 233
  of equation of a straight line, 197, 218
  of a quadratic equation, 120
Subset, 5
Substitution of functions, 312
Substitution method, 482
Substitution property of equality, 97
Subtraction
  of complex numbers, 83
  of functions, 311
  of matrices, 552
  of polynomials, 31
  of radical expressions, 67
  of rational expressions, 54
Suggestions for solving word problems, 101, 111, 136, 663
Sum
  of an arithmetic sequence, 646
  of cubes, 48

  of a geometric sequence, 654
  of infinite geometric sequence, 657
Summation notation, 648
Symmetric property of equality, 97
Symmetry, 224–226
Synthetic division, 343
System(s)
  of linear equations in two variables, 481
  of linear equations in three variables, 494
  of linear inequalities, 576
  of nonlinear equations, 626

Term(s)
  addition of like, 13
  of an algebraic expression, 13
  like, 13
  similar, 13
Test number, 164
Transformations, 298
Transitive property of equality, 97
Transverse axis of a hyperbola, 616
Triangular form of a system, 510
Trinomial, 30
Turning points, 372

Union of sets, 162
Upper bound, 366

Variable, 2
Variation
  constant of, 400
  direct, 400
  inverse, 402
  joint, 404
Vertex of a parabola, 274, 286, 594
Vertical asymptote, 382
Vertical line test, 254
Vertical stretching and shrinking, 305
Vertical translation, 276, 302

Whole numbers, 3

x axis reflection, 303
x axis symmetry, 225
x intercept, 100, 197

y axis reflection, 304
y axis symmetry, 224
y intercept, 197

Zero
  addition property of, 8
  as an exponent, 20
  multiplication property of, 9
Zeros of a function, 368